Mem-elements for Neuromorphic Circuits with Artificial Intelligence Applications

A volume in the *Advances in Nonlinear Dynamics and Chaos* series

Series editors

Ahmad Taher Azar
Sundarapandian Vaidyanathan

Visit the Series webpage at https://www.elsevier.com/books/book-series/advances-in-nonlinear-dynamics-and-chaos-andc

Mem-elements for Neuromorphic Circuits with Artificial Intelligence Applications

Edited by

Christos Volos

Viet-Thanh Pham

ELSEVIER

ACADEMIC PRESS
An imprint of Elsevier

Academic Press is an imprint of Elsevier
125 London Wall, London EC2Y 5AS, United Kingdom
525 B Street, Suite 1650, San Diego, CA 92101, United States
50 Hampshire Street, 5th Floor, Cambridge, MA 02139, United States
The Boulevard, Langford Lane, Kidlington, Oxford OX5 1GB, United Kingdom

Library of Congress Cataloging-in-Publication Data
A catalog record for this book is available from the Library of Congress

British Library Cataloguing-in-Publication Data
A catalogue record for this book is available from the British Library

ISBN: 978-0-12-821184-7

For information on all Academic Press publications
visit our website at https://www.elsevier.com/books-and-journals

Publisher: Mara Conner
Acquisitions Editor: Sonnini R. Yura
Editorial Project Manager: Rafael G. Trombaco
Production Project Manager: Prem Kumar Kaliamoorthi
Designer: Mark Rogers

Typeset by VTeX

Contents

3. On the FPGA implementation of chaotic oscillators based on memristive circuits

Omar Guillén-Fernández, Esteban Tlelo-Cuautle, Gustavo Rodríguez-Gómez, Luis Gerardo de la Fraga, and Rui Li

4. Microwave memristive components for smart RF front-end modules

Milka M. Potrebić and Dejan V. Tošić

5. The modeling of memcapacitor oscillator motion with ANN and its nonlinear control application

Akif Akgul, Murat Erhan Cimen, Irene M. Moroz, and Ali Fuat Boz

Part II
Applications of mem-elements

13. Memristor, mem-systems and neuromorphic applications: a review

Viet-Thanh Pham, Christos Volos, and Tomasz Kapitaniak

14. Guidelines for benchmarking non-ideal analog memristive crossbars for neural networks

Alex James

List of contributors

Cyrille Ainamon, Institute of Mathematics and Physical Sciences, University of Abomey-Calavi, Porto Novo, Benin

Akif Akgul, Department of Computer Engineering, Faculty of Engineering, Hitit University, Corum, Turkey

S. Aldana, Electronics and Computer Technology Department, Science Faculty, Granada University, Granada, Spain

Bo-cheng Bao, School of Microelectronics and Control Engineering, Changzhou University, Changzhou, China

Stelios Bekiros, European University Institute, Department of Economics, Florence, Italy
Rimini Centre for Economic Analysis (RCEA), Wilfrid Laurier University, Waterloo, ON, Canada

Ciro Fabian Bermúdez-Márquez, Faculty of Electronics Sciences, Meritorious Autonomous University of Puebla, Puebla, Mexico

Ali Fuat Boz, Department of Electrical and Electronic Engineering, Sakarya University of Applied Sciences, Sakarya, Turkey

F. Campabadal, Institute of Microelectronics of Barcelona, IMB-CNM (CSIC), Bellaterra, Spain

André Chéagé Chamgoué, School of Geology and Mining Engineering, University of Ngaoundere, Meiganga, Cameroon

Cheng-jie Chen, School of Microelectronics and Control Engineering, Changzhou University, Changzhou, China

Mo Chen, School of Microelectronics and Control Engineering, Changzhou University, Changzhou, China

Long Cheng, Wuhan National Laboratory for Optoelectronics, School of Optical and Electronic Information, Huazhong University of Science and Technology, Wuhan, China

Murat Erhan Cimen, Department of Electrical and Electronic Engineering, Sakarya University of Applied Sciences, Sakarya, Turkey

Fernando Corinto, Department of Electronics and Telecommunications, Politecnico di Torino, Turin, Italy

Toan Dao Thanh, Faculty of Electrical-Electronic Engineering, University of Transport and Communications, Hanoi, Vietnam

Carola de Benito, Physics Department, University of Balearic Islands, Palma, Spain

Luis Gerardo de la Fraga, CINVESTAV, San Pedro Zacatenco, Mexico

S. Dinesh Vijay, Department of Nonlinear Dynamics, School of Physics, Bharathidasan University, Tiruchirappalli, India

Tukaram D. Dongale, School of Nanoscience and Biotechnology, Shivaji University, Kolhapur, India
School of Electrical Engineering, Korea University, Seoul, Republic of Korea

Dongming Gan, School of Engineering Technology, Purdue University, West Lafayette, IN, United States

M.B. González, Institute of Microelectronics of Barcelona, IMB-CNM (CSIC), Bellaterra, Spain

G. González-Cordero, Electronics and Computer Technology Department, Science Faculty, Granada University, Granada, Spain

Omar Guillén-Fernández, INAOE, Puebla, Mexico

Hadi Jahanshahi, Department of Mechanical Engineering, University of Manitoba, Winnipeg, MB, Canada

Alex James, Indian Institute of Information Technology and Management, Kerala, India

F. Jiménez-Molinos, Electronics and Computer Technology Department, Science Faculty, Granada University, Granada, Spain

Victor Kamdoum Tamba, Department of Telecommunication and Network Engineering, IUT - Fotso Victor of Bandjoun, University of Dschang, Bandjoun, Cameroon
Research Unit of Automation and Applied Computer, Department of Electrical Engineering, IUT - FV Bandjoun, University of Dschang, Bandjoun, Cameroon

Francois Kapche Tagne, Department of Telecommunication and Network Engineering, IUT - Fotso Victor of Bandjoun, University of Dschang, Bandjoun, Cameroon

Tomasz Kapitaniak, Division of Dynamics, Lodz University of Technology, Lodz, Poland

Tae Geun Kim, School of Electrical Engineering, Korea University, Seoul, Republic of Korea

S. Leo Kingston, Division of Dynamics, Lodz University of Technology, Lodz, Poland

Guillaume Honore Kom, Research Unit of Automation and Applied Computer (RU-AIA), Electrical Engineering Department of IUT-FV, University of Dschang, Bandjoun, Cameroon

Ingrid Ornela Lowe Kombou, Research Unit of Automation and Applied Computer (RU-AIA), Electrical Engineering Department of IUT-FV, University of Dschang, Bandjoun, Cameroon
Research Unit of Condensed Matter, Electronics and Signal Processing (RU-LAMACETS), Department of Physics, Faculty of Science, University of Dschang, Dschang, Cameroon

Olga Krestinskaya, King Abdullah University of Science and Technology, Thuwal, Saudi Arabia

Rui Li, UESTC, Chengdu, China

Yi Li, Wuhan National Laboratory for Optoelectronics, School of Optical and Electronic Information, Huazhong University of Science and Technology, Wuhan, China

Akshay Kumar Maan, Griffith University, Brisbane, QLD, Australia

M. Maestro-Izquierdo, Institute of Microelectronics of Barcelona, IMB-CNM (CSIC), Bellaterra, Spain

D. Maldonado, Electronics and Computer Technology Department, Science Faculty, Granada University, Granada, Spain

Arsene Loic Mbanda Biamou, Research Unit of Condensed Matter, Electronics and Signal Processing, Department of Physics, Faculty of Science, University of Dschang, Dschang, Cameroon

Justin Roger Mboupda Pone, Research Unit of Automation and Applied Computer (RU-AIA), Electrical Engineering Department of IUT-FV, University of Dschang, Bandjoun, Cameroon

Xiangshui Miao, Wuhan National Laboratory for Optoelectronics, School of Optical and Electronic Information, Huazhong University of Science and Technology, Wuhan, China

Foutse Momo, Biomedical Engineering, Energy and Modeling Laboratory (BEEMo.Lab.), Higher Institute of Science and Technology (HIST), University of Mountains, Bangangte, Cameroon

M. Moner Al Chawa, Chair of Fundamentals of Electrical Engineering, Technical University of Dresden, Dresden, Germany

Gayatri M. More, School of Nanoscience and Biotechnology, Shivaji University, Kolhapur, India

Irene M. Moroz, Mathematical Institute, Oxford University, Oxford, United Kingdom

Jesus Manuel Muñoz-Pacheco, Faculty of Electronics Sciences, Meritorious Autonomous University of Puebla, Puebla, Mexico

Hector Nistazakis, Faculty of Physics, Department of Electronics, Computers, Telecommunications and Control, National and Kapodistrian University, Athens, Greece

Manuela Corazon Nkeing, Research Unit of Condensed Matter, Electronics and Signal Processing, Department of Physics, Faculty of Science, University of Dschang, Dschang, Cameroon

Armand Nzeukou Takougang, Laboratory of Industrial Systems and Environmental Engineering, Fotso Victor University Institute of Technology, University of Dschang, Bandjoun, Cameroon

Jean Bio Chabi Orou, Institute of Mathematics and Physical Sciences, University of Abomey-Calavi, Porto Novo, Benin

Kiran D. Pawar, School of Nanoscience and Biotechnology, Shivaji University, Kolhapur, India

Viet-Thanh Pham, Faculty of Electrical and Electronic Engineering, Phenikaa Institute for Advanced Study (PIAS), Phenikaa University, Hanoi, Vietnam
Phenikaa Research and Technology Institute (PRATI), A&A Green Phoenix Group, Hanoi, Vietnam
Division of Dynamics, Lodz University of Technology, Lodz, Poland

Rodrigo Picos, Industrial Engineering and Construction Department, University of Balearic Islands, Palma, Spain

S. Poblador, Institute of Microelectronics of Barcelona, IMB-CNM (CSIC), Bellaterra, Spain

Milka M. Potrebić, University of Belgrade, School of Electrical Engineering, Belgrade, Serbia

Gustavo Rodríguez-Gómez, INAOE, Puebla, Mexico

J.B. Roldán, Electronics and Computer Technology Department, Science Faculty, Granada University, Granada, Spain

Binoy Krishna Roy, Department of Electrical Engineering, NIT Silchar, Cachar, Assam, India

Piyush Pratap Singh, Department of Electrical Engineering, NIT Meghalaya, Shillong, Meghalaya, India

Samaneh Soradi-Zeid, Faculty of Industry and Mining (Khash), University of Sistan and Baluchestan, Zahedan, Iran

Stavros G. Stavrinides, School of Science and Technology, International Hellenic University, Thessaloniki, Greece

Sifeu Takougang Kingni, Department of Mechanical, Petroleum and Gas Engineering, Faculty of Mines and Petroleum Industries, University of Maroua, Maroua, Cameroon

K. Thamilmaran, Department of Nonlinear Dynamics, School of Physics, Bharathidasan University, Tiruchirappalli, India

Alain Tiedeu, Signal, Image and Systems Laboratory, Higher Technical Teachers' Training College (HTTTC) of Ebolowa, University of Yaoundé 1, Yaoundé, Cameroon

Arpita P. Tiwari, Centre of Interdisciplinary Research, D.Y. Patil University, Kolhapur, India

Esteban Tlelo-Cuautle, INAOE, Puebla, Mexico

Dejan V. Tošić, University of Belgrade, School of Electrical Engineering, Belgrade, Serbia

Christos Volos, Laboratory of Nonlinear Systems, Circuits and Complexity (LaNSCom), Department of Physics, Aristotle University of Thessaloniki, Thessaloniki, Greece

Paul Woafo, Laboratory of Modelling and Simulation in Engineering, Biomimetics and Prototypes (LaMSEBP), Department of Physics, Faculty of Science, University of Yaoundé I, Yaoundé, Cameroon

Quan Xu, School of Microelectronics and Control Engineering, Changzhou University, Changzhou, China

René Yamapi, Fundamental Physics Laboratory, Department of Physics, Faculty of Science, University of Douala, Douala, Cameroon

Ling Yang, Wuhan National Laboratory for Optoelectronics, School of Optical and Electronic Information, Huazhong University of Science and Technology, Wuhan, China

Amin Yousefpour, School of Mechanical Engineering, College of Engineering, University of Tehran, Tehran, Iran

M. Zabala, Institute of Microelectronics of Barcelona, IMB-CNM (CSIC), Bellaterra, Spain

Preface

In 1971, Leon Chua, in a seminal paper, extrapolated a conceptual symmetry between the nonlinear resistor (voltage vs. current), nonlinear capacitor (voltage vs. charge), and nonlinear inductor (magnetic flux linkage vs. current). He then inferred the possibility of a new element as another fundamental nonlinear circuit element linking magnetic flux and charge. In contrast to a linear (or nonlinear) resistor the new element has a dynamic relationship between current and voltage including a memory of past voltages or currents.

This new element's resistance depends on the integral of the input applied to the terminals. Since the element "remembers" the amount of current that last passed through, it was tagged by Chua with the name "memristor". Another way of describing a memristor is as any passive two-terminal circuit element that maintains a functional relationship between the time integral of current (called charge) and the time integral of voltage (often called flux, as it is related to magnetic flux). The slope of this function is called the memristance and is similar to variable resistance.

The memristor definition is based solely on the fundamental circuit variables of current and voltage and their time-integrals, just like the resistor, capacitor, and inductor. Unlike those three elements, however, which are allowed in linear time-invariant system theory, memristors of interest have a dynamic function with memory and may be described as some function of net charge. There is no such thing as a standard memristor. Instead, each device implements a particular function, wherein the integral of voltage determines the integral of current, and vice versa. A linear time-invariant memristor, with a constant value for memristance, is simply a conventional resistor. Manufactured devices are never purely memristors (ideal memristor), but also exhibit some capacitance and resistance. In 2009, Di Ventra, Pershin, and Chua extended the notion of memristive systems to capacitive and inductive elements in the form of "memcapacitors" and "meminductors", whose properties depend on the state and history of the system, further extended in 2013 by Di Ventra and Pershin.

In 2008, a team at HP Labs claimed to have found Chua's missing memristor based on an analysis of a thin film of titanium dioxide, thus connecting the operation of ReRAM devices to the memristor concept. The HP result was published in the scientific journal Nature. Following this claim, Chua has argued that the memristor definition could be generalized to cover all forms of two-terminal nonvolatile memory devices based on resistance switching effects. Chua also argued that the memristor is the oldest known circuit element, with its effects predating the resistor, capacitor, and inductor. There are, however, some serious doubts as to whether a genuine memristor can actually exist in physical reality.

Memristor patents include applications in programmable logic, signal processing, neural networks, control systems, reconfigurable computing, brain–computer interfaces, and RFID. Also, memristive devices are potentially used for stateful logic implication, allowing a replacement for CMOS-based logic computation. Several early works have been reported in this direction. Furthermore, Chua published a tutorial underlining the broad span of complex phenomena and applications that memristors span and how they can be used as non-volatile analog memories and can mimic classic habituation and learning phenomena.

This book tries to illustrate recent advances and achievements in the field of mem-elements (memristor, memcapacitor, meminductor) and their applications in nonlinear dynamical systems, computer science, analog and digital systems, and especially in neuromorphic circuits and artificial intelligence.

The proposed book is divided into two sections; the editors have invited top researchers in each category based on Scopus and Engineering Village database to contribute with good surveys or new contributions. The first section of the book, which includes 12 chapters, covers the recent advances in the field of mem-elements, the implementation of their emulators and their use in dynamical systems as nonlinear elements, while the second section, which has 13 chapters, includes interesting applications of mem-elements in neuromorphic circuits, computer science and artificial intelligence.

Stated simply, this book aims to bridge the gap between different interdisciplinary applications of mem-elements, starting from mathematical concepts, modeling, analysis, and up to the realization and experimental work on mem-elements for Neuromorphic Circuits and Artificial Intelligence Applications. Therefore, it is our hope that this book can serve as an updated and handy reference for researchers from academia and industry, who are working in the research areas—dynamical systems, artificial intelligence, electrical engineering, computer science, information technology and neuromorphic systems. The book can also be used at the graduate or advanced undergraduate level as a textbook or major reference for courses such as electrical circuits, nonlinear dynamical systems, mathematical modeling, computational science, and numerical simulation. Both the novice and the expert reader should find this book a useful reference in the aforementioned fields.

Christos Volos
Viet-Thanh Pham

Part I

Mem-elements and their emulators

Chapter 1

The fourth circuit element was found: a brief history

Christos Volos[a], **Viet-Thanh Pham**[b,c], **and Hector Nistazakis**[d]
[a]*Laboratory of Nonlinear Systems, Circuits and Complexity (LaNSCom), Department of Physics, Aristotle University of Thessaloniki, Thessaloniki, Greece,* [b]*Faculty of Electrical and Electronic Engineering, Phenikaa Institute for Advanced Study (PIAS), Phenikaa University, Hanoi, Vietnam,* [c]*Phenikaa Research and Technology Institute (PRATI), A&A Green Phoenix Group, Hanoi, Vietnam,* [d]*Faculty of Physics, Department of Electronics, Computers, Telecommunications and Control, National and Kapodistrian University, Athens, Greece*

1.1 Memristor – the first step

The history of electrical circuits' theory has been written based on the three fundamental passive elements, *capacitor*, *resistor*, and *inductor*. The capacitor, was the first circuit element that was discovered in 1745 by the German Ewald Georg von Kleist, who found that charge could be stored by connecting a high-voltage electrostatic generator by a wire to a volume of water in a hand-held glass jar [1]. Next year, the Dutch physicist Pieter van Musschenbroek invented a similar capacitor, which was named the Leyden jar, after the University of Leyden, where he worked [2]. Daniel Gralath was the first to combine several jars in parallel to increase the charge storage capacity [3], while the famous Benjamin Franklin investigated further the Leyden jar and he adopted the term "battery" [4].

In 1826, Georg Simon Ohm gave a mathematical description of conduction in circuits modeled of Fourier's study of heat conduction. This work continued Ohm's deduction of results from experimental evidence. He was able to propose laws, which went a long way to explaining results of others working on galvanic electricity. This work certainly was a first step in a comprehensive theory, which Ohm was able to give in his famous book, published in the following year, called "Die Galvanische Kette, mathematisch bearbeitet" [5] which means "Galvanic Chain, Mathematically Worked" and contained what is known today as the "*Ohm laws*". In more detail, he investigated that the flow of electric current through a conductor experiences a certain amount of resistance. The magnitude of resistance is dictated by electric properties of the material and by the material geometry. Henceforth, the conductors that exhibit the property of resisting current flow are called resistors.

Mem-elements for Neuromorphic Circuits with Artificial Intelligence Applications
https://doi.org/10.1016/B978-0-12-821184-7.00009-8

Inductance was first found by Michael Faraday in 1831, in a simple yet strange way [6]. He wrapped a paper cylinder with wire, attached the ends of the wire to a galvanometer and moved a magnet in and out of the cylinder. The galvanometer reacted to this, revealing the production of a small current. Shortly after this discovery, Reverend Nicholas Calland invented the inductor coil [7]. The earliest version of the inductor consisted of a coil with two terminals at the ends which stored energy inside a magnetic field when a current passed through the coil.

Therefore, from this last invention and for almost 140 years the electrical circuits' theory was spinning around these three basic passive elements. Then in 1971, Leon Chua, a Professor of electrical engineering at the University of California, Berkley, predicted the existence of a fourth fundamental element. Chua became famous for his work on nonlinear circuit theory, as in the decade of 1960 he had established its mathematical foundation [8]. It was an exciting period for electronics engineers, who were working on exotic nonlinear devices, such as the Esaki diode, Josephson junction, varactor and thyristor. For Chua's work on nonlinear circuits and elements, he is acknowledged as the father of nonlinear circuit theory and a large number of awards have been awarded to him, including the IEEE Gustav Robert Kirchhoff Award, a number of other major awards, nine honorary doctorates at major universities around the world and numerous visiting professorships [8].

Unfortunately, during that period circuit theory was concerned only with the prediction of the voltage $v(t)$ and current $i(t)$ associated with the external terminals of the linear and nonlinear devices interconnected in a circuit and not with the internal physical variables associated with the individual devices in the circuit. Thus, basic nonlinear circuit elements were defined from a black box perspective, independent of their internal composition, material, geometry and architecture.

The basic principles that would predict the fourth fundamental circuit element, the memristor, were first reported by Chua in 1969 in a book [9], and it took him a year to derive and prove mathematically the unique theoretic properties and memory attributed to the memristor. Next year, Chua submitted a seminal paper, which presented the results of his work [10]. As he claimed from the circuit-theoretic point of view, the three fundamental two-terminal circuit elements are defined in terms of a relationship between two of the four fundamental circuit variables, namely the current i, the voltage v, the charge q, and the flux-linkage φ. Out of the six possible combinations of these four variables, five have led to well-known equations. In more detail, two of these relationships are already given by Eqs. (1.1)–(1.2), while three other equations are given, respectively, by the axiomatic definition of the three classical circuit elements, namely, the resistor (R), the capacitor (C) and the inductor (L) (see Eqs. (1.3)–(1.5)). Only one relationship remained undefined, which was the relationship between

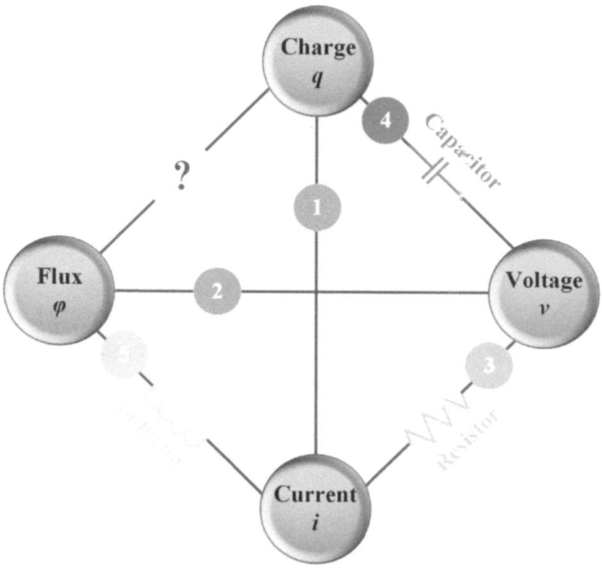

FIGURE 1.1 The three fundamental two-terminal circuit elements and the fourth unknown element till the discovery by Prof. Leon Chua. Circles represent Eqs. (1.1)–(1.5), which combine the four fundamental circuit variables.

φ and q (Fig. 1.1). We have

$$dq = i\,dt \tag{1.1}$$

$$d\varphi = v\,dt \tag{1.2}$$

$$dv = R\,di \tag{1.3}$$

$$dq = C\,dv \tag{1.4}$$

$$d\phi = L\,di \tag{1.5}$$

Therefore, Chua concluded that "*From the logical as well as axiomatic points of view, it is necessary for the sake of completeness to postulate the existence of a fourth basic two-terminal circuit element which is characterized by a φ–q curve*" [10].

This element was named, by Chua, *memristor* because it behaves somewhat like a nonlinear resistor with memory. The proposed symbol of the memristor is shown in Fig. 1.2.

1.2 Properties of memristor

Also, Chua in his seminal paper [10] claimed, according to the aforementioned analysis, that by definition a memristor is characterized by a relation of the type:

$$g(\varphi, q) = 0 \tag{1.6}$$

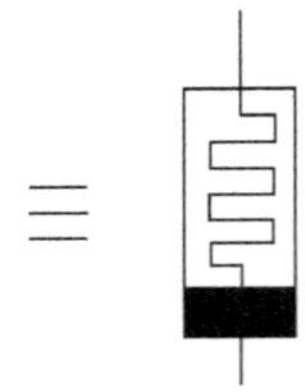

FIGURE 1.2 The symbol of the memristor.

So, a memristor is called charge-controlled or flux-controlled, if this relation can be expressed as a single-valued function of the charge q or flux φ, respectively. The voltage across a charge-controlled memristor is given by

$$v(t) = M(q(t))\, i(t) \tag{1.7}$$

where

$$M(q) = \frac{d\varphi(q)}{dq} \tag{1.8}$$

which is called the incremental memristance, since it has the unit of resistance.

In the same way the current through a flux-controlled memristor is given by

$$i(t) = W(\varphi(t))\, v(t) \tag{1.9}$$

where

$$W(\varphi) = \frac{dq(\varphi)}{d\varphi} \tag{1.10}$$

which is called incremental memductance, since it has the unit of conductance.

From Eqs. (1.8) and (1.10) the reader can observe that the value of the incremental memristance or memductance at any time t_o depends upon the time integral of the memristor's current or voltage, respectively, from $t = -\infty$ to $t = t_o$. Therefore, while the memristor behaves like an ordinary resistor at a given instant of time t_o, its resistance or conductance depends on the complete past history of the memristor current or voltage, respectively. This observation motivated Chua to give the name *memory resistor*, or *memristor* to this new circuit element. Also, Chua has proved theoretically that a memristor is a nonlinear element because its i–v characteristic is similar to that of a Lissajous pattern. So, as the memristor's voltage $v(t)$ or current $i(t)$ is specified, the memristor behaves like a linear time-varying resistor. Furthermore, when the memristor's φ–q characteristic curve is a straight line, then $M(q) = R$, or $W(p) = G$, and the memristor reduces to a linear time-invariant resistor.

1.2.1 Passivity criterion

In this section the class of memristors, which might be discovered as a physical element without power supply, is investigated through the following passivity criterion according to Chua's analysis in his seminal paper [10].

Theorem. *A memristor is characterized by a differentiable charge-controlled φ–q curve is passive if, and only if, its incremental memristance $M(q)$ is non-negative; i.e., $M(q) > 0$.*

Proof. The power dissipated by a memristor is given by

$$p(t) = v(t)i(t) \tag{1.11}$$

By using Eq. (1.7) in Eq. (1.11), the power is

$$p(t) = M(q(t))[i(t)]^2 \tag{1.12}$$

Therefore, if $M(q(t)) \geq 0$, then $p(t) \geq 0$ and the memristor is passive. For proving the converse the existence of a value q_o in order to suggest that $M(q_o) < 0$ is considered. In this case the differentiability of the φ–q curve implies that there exists an $\varepsilon > 0$ such that $M(q_o + \Delta q) < 0$, with $|\Delta q| < \varepsilon$. By driving the memristor with a current $i(t)$, which is zero for $t < \hat{t}$, and such that $q(t) = q_o + \Delta q(t)$ for $t > t_o > \hat{t}$, then $\int_{-\infty}^{t} p(\tau)d\tau < 0$ for sufficiently large t, and hence the memristor is active. □

The passivity criterion shows that only memristors characterized by a monotonically increasing φ–q curve can exist in a device form without internal power supplies.

1.2.2 Closure theorem

Another interesting theorem related with memristor, which has been formulated by Chua in his paper [10], is the closure theorem.

Theorem. *A one-port containing only memristors is equivalent to a memristor.*

Proof. If we let i_j, v_j, q_j, and φ_j denote the current, voltage, charge, and flux of the jth charge-controlled memristor, where $j = 1, 2, \ldots, b$, and if we let i and v denote the port current and port voltage of the one-port, then $(n-1)$ independent Kirchhoff current law equations can be written (assuming the network is connected). So,

$$a_{j0}i + \sum_{k=1}^{b} a_{jk}i_k = 0, \quad j = 1, 2, \ldots, n-1 \tag{1.13}$$

where a_{jk} is either 1, -1, or 0 and n is the total number of nodes.

In the same way a system of $(b - n + 2)$ independent Kirchhoff voltage law equations can be written:

$$\beta_{j0}v + \sum_{k=1}^{b} \beta_{jk} v_k = 0, \quad j = 1, 2, \ldots, b - n + 2 \tag{1.14}$$

where β_{jk} is either 1, -1, or 0.

By integrating Eqs. (1.13) and (1.14) with respect to time and substituting $\varphi_k = \varphi_k(q_k)$ for φ_k in the resulting expressions, we obtain

$$\sum_{k=1}^{b} a_{jk} q_k = Q_i - a_{j0} q, \quad j = 1, 2, \ldots, n - 1 \tag{1.15}$$

$$\beta_{j0}\varphi + \sum_{k=1}^{b} \beta_{jk} \varphi_k(q_k) = \Phi_j, \quad j = 1, 2, \ldots, b - n + 2 \tag{1.16}$$

where Q_j and Φ_j are arbitrary constants of integration. Eqs. (1.15) and (1.16) together constitute a system of $(b + 1)$ independent nonlinear functional equations with $(b + 1)$ unknowns. Hence, solving for φ, we obtain a relation $f(q, \varphi) = 0$. □

1.2.3 Existence and uniqueness theorem

Finally, Chua has proved the theorem of existence and uniqueness in the case of memristors [10]. According to this, we have the following result.

Theorem. *Any network containing only memristors with positive incremental memristances has one, and only one, solution.*

Proof. For the proof of this theorem the reader can see the corresponding proof in Ref. [11] for a network containing only nonlinear resistors. □

Therefore, Chua has proved theoretically that a memristor is a nonlinear element because its i–v characteristic is similar to that of a Lissajous pattern. So, a memristor with a non-constant M describes a resistor with a memory, more precisely a resistor which resistance depends on the amount of charge that has passed through the device. Typical responses of a memristor to sinusoidal inputs are depicted in Fig. 1.3. The "*pinched hysteresis loop current–voltage characteristic*" is an important fingerprint of a memristor. If any device has an i–v hysteresis curve like this, then it is either a memristor or a memristive device. Another signature of the memristor is that the "pinched hysteresis loop" shrinks with the increase of the excitation frequency. The fundamentality of the memristor can also be deduced from Fig. 1.3, as it is impossible to make a network of capacitors, inductors and resistors with an i–v behavior forming a pinched hysteresis curve.

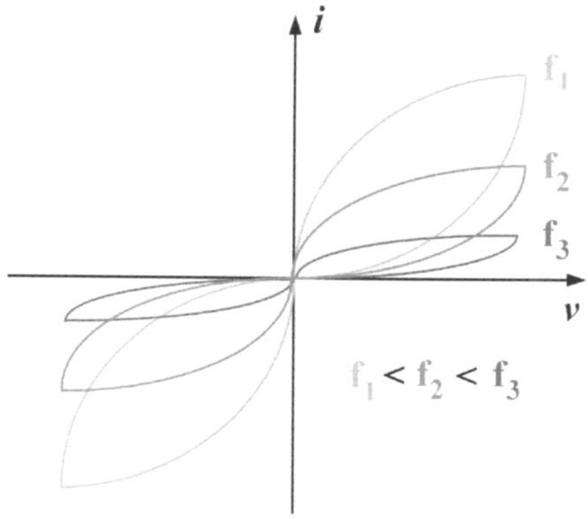

FIGURE 1.3 Typical i–v characteristic curves of a memristor driven by a simusoidal voltage input of various frequencies.

In conclusion, some of the more interesting properties of memristor are:

- Non-linear relationship between current (i) and voltage (v).
- Similar to classical circuit elements, a system of memristors can also be described as a single memristor.
- Reduces to resistor for large frequencies as evident in the i–v characteristic curve.
- Does not store energy.
- Memory capacities based on different resistances produced by the memristor.
- Non-volatile memory is possible if the magnetic flux and charge through the memristor have a positive relationship ($M > 0$).

1.3 Memristive systems

In 1976 Chua and Kang introduced a more generalized class of systems, in regard to the original definition of a memristor, called memristive systems [12]. These systems have a state variable indicated by w that describes the physical properties of the device at any time. A memristive system is characterized by two equations,

- the quasi-static conduction equations that relate the voltage across the device to the current through it at any particular time and
- the dynamical equation, which explicitly asserts that the state variable w is a time varying function f of itself and possibly the current (or voltage) through the device.

An nth-order current-controlled memristive one-port is represented by

$$\begin{cases} v = R(w, i, t)i \\ \dfrac{dw}{dt} = f(w, i, t) \end{cases} \tag{1.17}$$

where $w \in R^n$ is the n-dimensional state variable of the system.

Also, the nth-order voltage-controlled memristive one-port is defined as

$$\begin{cases} i = G(w, v, t)v \\ \dfrac{dw}{dt} = f(w, v, t) \end{cases} \tag{1.18}$$

These systems are unconventional in the sense that, while they behave like resistive devices, they can be endowed with a rather exotic variety of dynamic characteristics.

Therefore, similarly to a memristor, a memristive system has the following properties:

- The memristive system should have a dc curve passing through the origin.
- For any periodic excitation the i–v characteristic curve should pass through the origin.
- As the excitation frequency increases toward infinity the memristive system has a linear behavior.
- The small signal impedance of a memristive system can be resistive, capacitive, or inductive depending on the operating bias point.

1.4 The first physical model of the memristor

Unfortunately, research on memristor was given a lower priority not only by Chua, who had already been awarded the W.R.G. Baker Award by the Institute of Electrical and Electronics Engineers for the potential of the memristor in 1973, but also by the whole research community. This happened due to the fact that the memristor was an exotic new circuit element, as research in circuit theory had been dominated mainly by linear networks. Also, it was not surprising that this element was not even discovered in a device form because it was unnatural to associate charge with flux. As a proof of principle, only three memristor models using operational amplifiers and off-the-shelf electronic components had been presented by Chua in his seminal paper [10]. However, the challenge of fabricating a passive monolithic memristor remained unfulfilled till 2008.

That year, Hewlett–Packard scientists, working at their laboratories in Palo Alto California, announced in Nature [13] that a physical model of memristor had been realized. This research team began reading Chua's papers and trying to understand what was the cause of the pinched hysteresis loops in the devices that they had already made. The big breakthrough and their most significant contribution came in 2006 when they realized that the time derivative of the state variable in Chua's dynamical state equation was comparable to the drift velocity of oxygen vacancies in a titanium dioxide resistive switch.

In more detail, in their scheme, a memory effect is achieved in solid-state thin film two terminal passive device. The memristor, which was realized by HP researchers, is made of a titanium dioxide layer which is located between

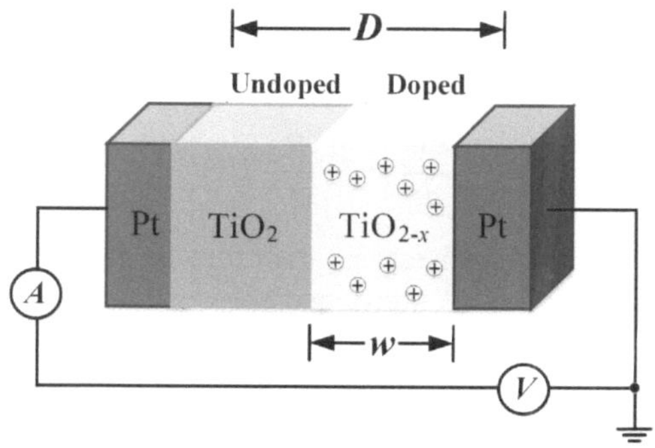

FIGURE 1.4 Cross-section of the first HP memristor consisting of a high conductive (doped) and a low conductive (undoped) part placed in between two platinum electrodes. The boundary between the two parts is dynamic and is moved back and forth by the passing charge carriers.

two platinum electrodes (Fig. 1.4). This layer is of the dimension of several nanometers and if an oxygen dis-bonding occurs, its conductance will rise instantaneously. However, without doping, the layer behaves as an isolator. The area of oxygen dis-bonding is referred to as space-charge region and changes its dimension if an electrical field is applied. This is done by a drift of the charge carriers. The smaller the insulating layer, the higher the conductance of the memristor. Also, the tunnel effect plays a crucial role. Without an external influence the extension of the space-charge region does not change.

The internal state x is the extent of the space-charge region, which is restricted in the interval [0, 1] and can be described by the equation

$$x = \frac{w}{D}, \quad 0 \leq x \leq 1, \quad x \in R \tag{1.19}$$

where w is the absolute extent of the space-charge region and D is the absolute extent of the titanium dioxide layer. The memristance can be described by the following equation:

$$M(x) = R_{\text{on}}x + R_{\text{off}}(1 - x) \tag{1.20}$$

where R_{on} is the resistance of the maximum conducting state and R_{off} represents the opposite case. Therefore, when $x = 0$, $R = R_{\text{off}}$, and when $x = 1$, $R = R_{\text{on}}$. The vector containing the internal states of the memristor is one dimensional. For this reason scalar notation is used.

The state equation is

$$\frac{dx}{dt} = \frac{\mu_v R_{\text{on}}}{D^2} i(t) \tag{1.21}$$

where μ_v is the oxygen vacancy mobility and $i(t)$ is the current through the device. By using Eq. (1.19) the previous equation can be rewritten as

$$\frac{dw}{dt} = \frac{\mu_v R_{\text{on}}}{D} i(t) \tag{1.22}$$

The dynamics of the memristor can therefore be modeled through the time dependence of the width w of the doped region. Integrating Eq. (1.22) with respect to time,

$$w = w_0 + \frac{\mu_v R_{\text{on}}}{D} q(t) \tag{1.23}$$

where w_0 is the initial width of the doped region at $t = 0$ and q is the amount of charges that have passed through the device. Substituting (1.19), (1.22) into Eq. (1.20) gives

$$M(q) = R_0 - \frac{\mu_v R_{on} \Delta R}{D^2} q(t) \tag{1.24}$$

where

$$R_0 = R_{\text{on}} \frac{w_0}{D} + R_{\text{off}} \left(1 - \frac{w_0}{D}\right) \tag{1.25}$$

and $\Delta R = R_{\text{off}} - R_{\text{on}}$. The term R_0 refers to the net resistance at $t = 0$ that serves as the device's memory. This term is associated with the memristive state, which is essentially established through a collective contribution, i.e. it depends directly on the amount of all charges that have flown through the device. As a result, we can say that the memristor has the feature to "remember" whether it was on or off after its power is turned on or off, respectively.

1.5 Memristor's applications

The announcement of the fabrication of the first physical memristor brought a revolution in various scientific fields, as many phenomena in systems, such as in thermistors, of which the internal state depends on temperature [14], spintronic devices, which resistance varies according to their spin polarization [15], and molecules whose resistance changes according to their atomic configuration [16], could be explained now with the use of the memristor. Also, electronic circuits with memory circuit elements could simulate processes typical of biological systems, such as learning and associative memory [17] and the adaptive behavior of unicellular organisms [18].

Therefore, due to the fact that memristor is a two terminal and variable resistance element, it could be used for the following applications.

- *Non-volatile memory applications* (*NVRAMs*) [19]. Memristors can retain memory states, and data, in power-off modes. Non-volatile random access memory is pretty much the first to-market memristor application that will be seen.

- *In remote sensing and low-power applications* [20,21]. Coupled with memcapacitors and meminductors, the complementary circuits to the memristor, which allow the storage of charge, memristors can possibly allow for nanoscale low-power memory and distributed state storage, as a further extension of NVRAM capabilities.
- *Crossbar latches as transistor replacements or augmentors* [22]. Solid-state memristors can be combined into devices called crossbar latches, which could replace transistors in future computers, taking up a much smaller area.
- *Analog computation and circuit applications* [23]. Analog computations embodied a whole area of research which, unfortunately, were not as scalable, reproducible, or dependable as digital solutions. Memristor applications will now allow us to revisit a lot of the analog science that was abandoned in the mid-1960s.
- *Programmable logic and signal processing* [24,25]. The memristive applications in these areas will remain relatively the same, because it will only be a change in the underlying physical architecture, allowing their capabilities to expand.
- *Analog filter applications* [26]. The memristors can be programmed with low frequency pulses to set its resistance using trick to separate high frequency signal path from the programming pulses, alike to those tricks used with varicap to separate a high frequency signal path from bias voltage. In this way the circuit could be reconfigured while running.
- *Circuits which mimic neuromorphic and biological systems* [27,28]. This is a very large area of research, because a large part of the analog science, given in detail above, has to do with advances in cognitive psychology, Artificial Intelligence (AI) modeling, machine learning and recent neurology advances. The ability to map peoples' brain activities under MRI, CAT, and EEG scans is leading to a treasure trove of information about how our brains work. Simple electronic circuits based on an LC network and memristors have been built and used recently to model experiments on adaptive behavior of unicellular organisms. Some types of learning circuits find applications anywhere from pattern recognition to Artificial Neural Networks (ANNs).

1.6 Conclusion

A brief history of the invention in 1971 of the fourth fundamental circuit element, which was called memristor, as well as its fabrication 37 years later, has been presented in this chapter. Also, the basic properties of this new element and especially its characteristic fingerprint, the pinched hysteresis i–v characteristic curve, have been discussed. Furthermore, the more generalized class of systems, in regard to the original definition of a memristor, called memristive systems, were mathematically described. Finally, the applications of this new element in various fields have been highlighted.

References

[1] W.H. Smith, A History of Science Volume II, Part VI: The Leyden Jar Discovered, Gutenberg Ebook History of Science, 2013.

[2] J.F. Keithley, The Story of Electrical and Magnetic Measurements: From 500 BC to the 1940s, John Wiley & Sons, 1999.

[3] P. Benjamin, A History of Electricity: (The Intellectual Rise in Electricity) From Antiquity to the Days of Benjamin Franklin, J. Wiley & Sons, 1895.

[4] B. Franklin, Experiments & Observations on Electricity: Letter IV to Peter Collinson, 2009.

[5] G.S. Ohm, Die galvanische Kette, mathematisch bearbeitet, TH Riemann, 1827.

[6] S.P. Thompson, Michael Faraday: His Life and Work, The Macmillan, 1898.

[7] D.P. Currier, A Biographical History of Induction Coils, 1857, Essay by Dean P. Currier.

[8] https://people.eecs.berkeley.edu/~chua/.

[9] L.O. Chua, Introduction to Nonlinear Network Theory, McGraw–Hill, 1969.

[10] L.O. Chua, Memristor – the missing circuit element, IEEE Transactions on Circuit Theory 18 (5) (1971) 507–519.

[11] C.A. Desoer, J. Katzenelson, Nonlinear RLC networks, Bell System Technical Journal 44 (1) (1965) 161–198.

[12] L.O. Chua, S. Kang, Memristive devices and systems, Proceedings of the IEEE 64 (1976) 209–223.

[13] D. Strukov, G. Snider, G. Stewart, R. Williams, The missing memristor found, Nature 453 (2008) 80–83.

[14] M. Sapoff, R. Oppenheim, Theory and application of self-heated thermistors, Proceedings of the IEEE 51 (1963) 1292–1305.

[15] Y.V. Pershin, M.D. Ventra, Spin memristive systems: spin memory effects in semiconductor spintronics, Physical Review B 78 (11) (2008) 1–4.

[16] Y. Chen, G. Jung, D. Ohlberg, X. Li, D. Stewart, J. Jeppesen, K. Nielsen, J. Stoddart, R. Williams, Nanoscale molecular-switch crossbar circuits, Nanotechology 14 (4) (2003) 462–468.

[17] Y.V. Pershin, M.D. Ventra, Experimental demonstration of associative memory with memristive neural networks, Neural Networks 23 (7) (2010) 881–886.

[18] Y.V. Pershin, S.L. Fontaine, M.D. Ventra, Memristive model of amoeba learning, Physical Review E 80 (2) (2000) 1–6.

[19] H. Abbas, Y. Abbas, S.N. Truong, K.S. Min, M.R. Park, J. Cho, T.S. Yoon, C.J. Kang, A memristor crossbar array of titanium oxide for non-volatile memory and neuromorphic applications, Semiconductor Science and Technology 32 (6) (2017) 065014.

[20] C. Wen, J. Hong, S. Yao, T. Niu, Y. Ju, A novel exposure sensor based on reverse series memristor, Sensors and Actuators A, Physical 278 (2018) 25–32.

[21] Y. Halawani, B. Mohammad, D. Homouz, M. Al-Qutayri, H. Saleh, Modeling and optimization of memristor and STT-RAM-based memory for low-power applications, IEEE Transactions on Very Large Scale Integration (VLSI) Systems 24 (3) (2015) 1003–1014.

[22] Q. Xia, W. Robinett, M.W. Cumbie, N. Banerjee, T.J. Cardinali, J.J. Yang, W. Wu, X. Li, W.M. Tong, D.B. Strukov, G.S. Snider, Memristor – CMOS hybrid integrated circuits for reconfigurable logic, Nano Letters 9 (10) (2009) 3640–3645.

[23] M. Hu, C.E. Graves, C. Li, Y. Li, N. Ge, E. Montgomery, N. Davila, H. Jiang, R.S. Williams, J.J. Yang, Q. Xia, Memristor-based analog computation and neural network classification with a dot product engine, Advanced Materials 30 (9) (2018) 1705914.

[24] C. Li, M. Hu, Y. Li, H. Jiang, N. Ge, E. Montgomery, J. Zhang, W. Song, N. Dávila, C.E. Graves, Z. Li, Analogue signal and image processing with large memristor crossbars, Nature Electronics 1 (1) (2018) 52.

[25] J. Rajendran, H. Manem, R. Karri, G.S. Rose, Memristor based programmable threshold logic array, in: 2010 IEEE/ACM International Symposium on Nanoscale Architectures, 2010, June, pp. 5–10.

[26] A. Ascoli, Memristor-based filtering applications, in: Proc. of 14th Latin American Test Workshop (LATW), 2013, pp. 1–6.
[27] Z. Wang, S. Joshi, S.E. Savel'ev, H. Jiang, R. Midya, P. Lin, M. Hu, N. Ge, J.P. Strachan, Z. Li, Q. Wu, Memristors with diffusive dynamics as synaptic emulators for neuromorphic computing, Nature Materials 16 (1) (2017) 101–108.
[28] S.H. Jo, T. Chang, I. Ebong, B.B. Bhadviya, P. Mazumder, W. Lu, Nanoscale memristor device as synapse in neuromorphic systems, Nano Letters 10 (4) (2010) 1297–1301.

Chapter 2

Implementing memristor emulators in hardware

Stavros G. Stavrinides[a], Rodrigo Picos[b], Fernando Corinto[c], M. Moner Al Chawa[d], and Carola de Benito[e]
[a]*School of Science and Technology, International Hellenic University, Thessaloniki, Greece,* [b]*Industrial Engineering and Construction Department, University of Balearic Islands, Palma, Spain,* [c]*Department of Electronics and Telecommunications, Politecnico di Torino, Turin, Italy,* [d]*Chair of Fundamentals of Electrical Engineering, Technical University of Dresden, Dresden, Germany,* [e]*Physics Department, University of Balearic Islands, Palma, Spain*

2.1 Introduction

The apparent symmetry between the relations of the four fundamental electrical magnitudes, namely the current i, the voltage v, the charge q and the flux φ, was a pattern that passed unnoticed for many years in circuit theory. It was this idea that led Leon Chua, in the early 1970s, to present the axiomatic introduction and the related description of a fourth (missing at that moment) electrical element, named the memristor [1]. Its name originated from the fact that such an element should behave as a resistor endowed with memory, having these two properties, resistance and the feature of memory, being unified in one element. In fact, memristors had been described many years ago [2], though they had never been in the mainstream of electrical or circuit theory. Besides, Chua's work led to the generalization of a class of devices as well as systems that are inherently nonlinear and governed by a state-dependent, algebraic relation accompanied by a set of differential equations, which are called memristive systems or devices [3].

As a result of the inherent memory feature embodied in memristors, these novel devices are expected to be one of the key technology enablers of a technological breakthrough in integrated circuit (IC) performance growth, beyond and more than Moore [4]. Among others, they are expected to provide us with a solution to the classical problem of the bottleneck in data transmission between memories and processors. The IoT and other edge computing applications are expected to be areas where the introduction of memristors and memristive devices would be beneficial, if not changing radically the related technological landscape. Thus, an increasing number of memristor-based applications has already been proposed; indicatively, new kind of memories (ReRAMs, MRAM, etc.) [5–7], innovative new sensor devices [8,9], or fundamental elements in bio-inspired systems (ANNs) [10].

Mem-elements for Neuromorphic Circuits with Artificial Intelligence Applications
https://doi.org/10.1016/B978-0-12-821184-7.00010-4

On the other hand memristive devices can be nowadays implemented in a wide range of technologies, from spintronics [11] to organic materials [12,13] and many different oxides [14–17]. However, and up to the best of our knowledge, very few foundries are including memristors in their design repertoire, as it customarily happens with other passive elements, like resistors or capacitors. This is due to the inadequate level of maturity of all the up-to-day proposed memristive technologies and it is something that would change in the near future.

Simulating a new design incorporating memristors is not something straightforward; many good models have been proposed, both using the classical approach that utilizes current and voltage [18–22], or the later introduced charge and flux approach [23–26], which historically had been also used in oxide breakdown. However, most of these models appear to demonstrate drawbacks that make simulation of large circuits rather difficult or even impractical [27]. Besides, some of the main problems of current memristor device-technologies are: the variability they exhibit from cycle to cycle [28,29], and the short number of cycles they can withstand (between 10^6 and 10^8 cycles).

Therefore, a good option to be considered is using actual circuits that have the ability to reproduce memristive behavior. Some researchers have opted to implement memristor models into FPGAs or ASICs [30–32] to improve simulation time, but this approach is bulky and requires complex implementations, since modeling of analog behavior using digital circuitry is an ungainly approach, especially in the case of designs including large number of elements.

Another option to be considered is using circuits mimicking memristive behavior. There are many approaches proposed, in the literature [5,33–37], as well as research, showing that some well-known electrical elements can be considered as memristors [38]. Most of these proposals [5,36,37] use active elements as operational amplifiers, OTAs, or even more advanced elements as current conveyors, that make implementation complex and power consuming. Others use passive elements [33–35], but then they are usually limited to short-term memory (volatility), because they rely on RC networks to store the state.

These emulators provide with an option for having properly operating memristors that could be fully utilized in a trustworthy way, in circuit and system design. In Fig. 2.1 appears an indicative compilation of various analog emulators that have been presented, recently. In this figure, these memristor emulators have been categorized according to:

- their volatility, i.e. exhibiting long (LTM) or short (STM) term memory (mainly because they store their current state on simple RC networks);
- their floating or not-floating voltage reference (the latter, in the case of having one terminal grounded);
- the requirement of external biasing, leading to an active (mainly due to op-amps utilization) or passive circuit design.

It has to be noticed that all the references marked *floating* can be also used as *grounded*, since in the latter case, one needs only to connect a terminal to

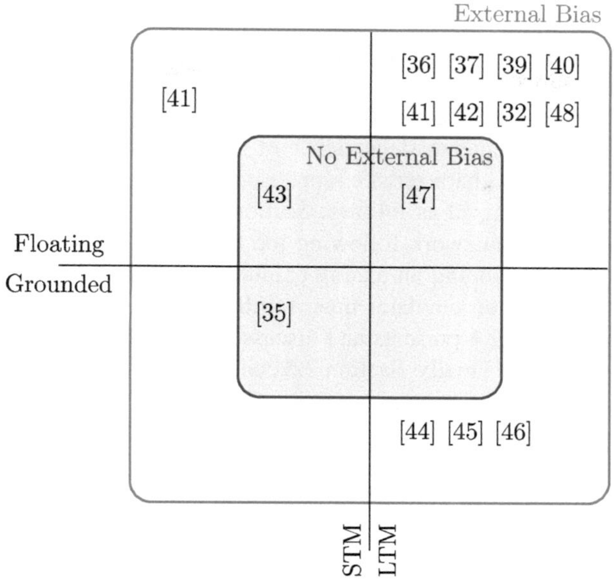

FIGURE 2.1 Indicative survey of the memristor emulator landscape ([32,35–37,39–48]). Notice that all the emulators marked as *Floating* can also be used as *Grounded*. It is also possible to make all the *LTM* emulators to operate as *STM* ones, according to the discussion in the text.

ground. Additionally, the references marked *long-term memory* (LTM) can be transformed into *short-term memory* (STM) by using a discharge mechanism for the capacitors storing the memristor's charge (for instance, adding a resistor R in parallel to the capacitor C would mean a decay time constant of value RC).

On the other hand, it seems that passive memristor-emulators, next to their design simplicity, stability and much higher frequency-performance, are closer to real-world devices. This is because they can mimic both symmetric and non-symmetric hysteretic memristive behaviors and include extra electrical state variables, leading to the implementation of extended memristors, that most of the memristive devices are. However, they bear a significant disadvantage; they are usually limited to short-term memory, thus demonstrating an inherent volatility. It is then worth highlighting the correlation between long-term memory and external bias requirement, which applies for most of the cases, but for [47]. It is also interesting to point out that some emulators, claiming to be "floating", are in fact grounded, because of the circuit configuration; thus introducing a third terminal in the emulated device. In most of these, the problem is that they use a transistor as the variable resistance. The gate of those transistors is connected to the reference of the element, implementing the control variable (usually the voltage at a capacitor), which is grounded in almost all cases. Therefore, an apparition of a common-mode effect, in our opinion, makes them non-strictly floating.

In this chapter, we review a methodology [47] that allows one to design circuits emulating a memristive system. Beginning from the basic equations that describe memristive behavior [49], we follow the construction of a framework, which can be used for the generation of analog circuits, mimicking memristive behavior [50], i.e. frequency dependence of the resistance and a pinched hysteresis loop in the i–v characteristic representation.

The work is organized as follows: Section 2.2 deals with the introduction of the flux-charge framework following the formalism proposed in [49], and its utilization in developing an analog emulator. Section 2.3 discusses the requirements that such an emulator must fulfill to reproduce the fingerprints of a memristor. Section 2.4 presents and discusses the results of a specific design using this framework. Finally, Section 2.5 concludes this work.

2.2 Memristor modeling framework

A fundamental theoretical framework for studying memristors and circuits presenting memristive behavior in the flux–charge (φ–q) domain, was developed by Corinto et al. in [49]. In that paper, the authors explain the reason why the flux–charge (φ–q) domain demonstrates advantages in studying memristor elements, compared to the current–voltage (i–v) one.

On the other hand, utilizing the classification proposed in [51], memristors are classified according to their proximity to the original definition of memristor. Thus, three main categories of memristor devices emerge, namely the ideal, the generic, and the extended memristor. In the same work [49], the essential mathematical framework describing their behavior was also developed. The introduction of this extending categorization emerged as a necessity, in order to include theoretically the description of pinched, hysteretic behaviors demonstrated by various elements, not only in circuit theory and electronics but also in nature.

Among the three different categories presented above, the class of extended memristors is the most general one and it refers to memristors that have extra state variables (next to φ and q). For the specific case of flux-controlled memristors, they are described by Eqs. (2.1) to (2.3):

$$i = G(\varphi, v, \mathbf{x}) \cdot v \tag{2.1}$$

$$\dot{\mathbf{x}} = \mathbf{g}_{\varphi}(\varphi, v, \mathbf{x}) \tag{2.2}$$

$$\dot{\varphi} = v \tag{2.3}$$

The nonlinear conductance G, appearing in Eq. (2.1), represents the inverse memristance M of an extended memristor, while in the above set of equations, v is the voltage applied to memristor, φ is the flux or voltage first momentum, and $\mathbf{x}$ stands for a *set* of extra state variables, which includes physical magnitudes according to the memristive system; indicatively they could be the internal temperature, the radius of a conducting filament, or any other non-electrical variable

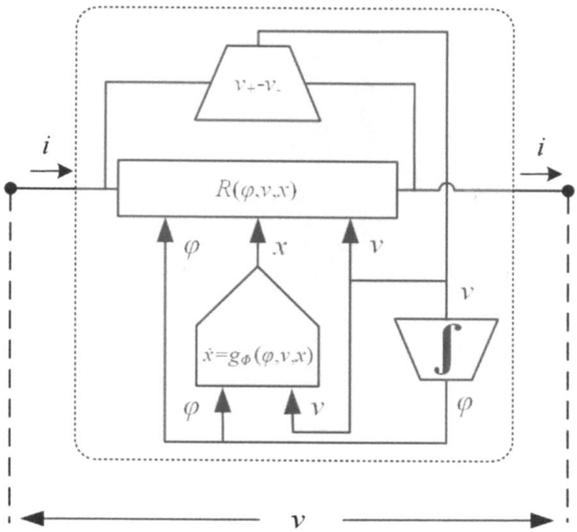

FIGURE 2.2 Graphical representation in the form of block-diagram, of the implementation of Eqs. (2.1)–(2.3), explicitly showing the relation between the state variables, in the case of a flux-controlled extended memristor.

describing the state of the memristor. Those state variables are the ones defining the difference for extended memristors. In addition, the dynamics of the state variables **x** are described by Eq. (2.2). It is noted that all the real-world memristor devices that have appeared until now, are indeed extended memristors.

Taking into account the Lagrangian L and the Jacobian **J**, these are defined in Eqs. (2.4) and (2.5), respectively:

$$L(\varphi, v, \mathbf{x}) = \frac{\partial f(\varphi, v, \mathbf{x})}{\partial v} \quad (2.4)$$

$$\mathbf{J}(\varphi, v, \mathbf{x}) = \left(\frac{\partial f(\varphi, v, \mathbf{x})}{\partial x_1}, \ldots, \frac{\partial f(\varphi, v, \mathbf{x})}{\partial x_n} \right) \quad (2.5)$$

A proper memristor emulator circuit requires the extra state variables (vector **x**) to be mapped somehow onto internal (to the circuit) electrical variables, stored within the circuit, which should not be directly accessible from the outside. These variables are then used to modify the value of the exhibited resistance R of the emulating circuit. Based on the method presented in [47], a block diagram illustrating the implementation of Eqs. (2.1)–(2.3) is presented in Fig. 2.2. This figure clearly illustrates the relations between all the state variables of an extended memristor, in the case of a flux-controlled memristor. The dual configuration (a charge-controlled memristor emulator) is shown in Fig. 2.3. If no parasitic effects are present, then extended memristors can be simplified to generic memristors (or, simply, memristors). That is because function f is only

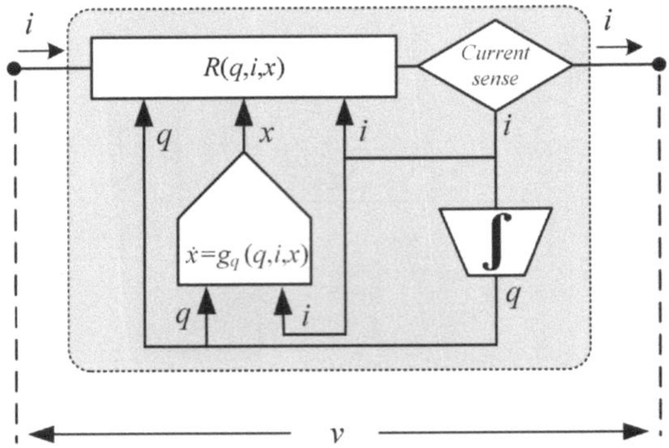

FIGURE 2.3 The dual graph in the form of block-diagram, in the case of a charge-controlled memristor.

dependent on flux φ and the state variables $\mathbf{x}$, thus $L = 0$. Finally, ideal memristors (those corresponding to the original definition described by [1]) are generic memristors that demonstrate no other state-variable dependence, thus, $J = 0$.

A special case of Eq. (2.2) is often referred to as the power–off plot (POP) equation and determines the memory capability of the system under no excitation; in this case for $v = 0$ or $\varphi =$ constant. It is apparent that if the POP equation is zero, the system presents a long-term memory since the state variable will not change with time, while if it is different from zero, the system is capable of exhibiting only short-term memory.

It is noted that the above framework has been already used in successfully modeling different kinds of memristive systems, further improving the generalized framework for compact modeling in the flux-charge space [52]. Other relevant works using this approach could also be found, like in the case of [53] where a charge-dependent mobility model was used to describe a memristor, [24] or [26] which presented semi-empirical models for ReRAMs as memristors, [54] that described a Monte Carlo model for ReRAMs, [55] which derives a delay model for memristor-based memories utilizing a flux-charge description, or [23] where a model for phase change memories is presented. Finally, two examples of experimental characterization of a memristive system utilizing the flux-charge notation, are presented in [38], where a light bulb is determined to be a generic memristor and in [56] the influence of waveform frequency and shape are discussed.

2.3 The fingerprints of a memristor

At the beginning of this century, there was a stir in the electron device compact modeling community to define how good a model was. This led to the definition of some criteria by the Compact Model Council (CMC) [57,58]. These criteria addressed not only how the models should be implemented and documented—to be easily shared between all the CMC members—but also what kind of electrical behavior they should reproduce. Thus, a good question to ask anyone claiming that his device is a memristor, is whether it fulfils the requirements that the mathematical framework imposes on any memristor or memristive device.

There are two different sets of conditions coming from the equations discussed in the previous section: the explicit conditions are those arising as a consequence of the equations; and the implicit ones are those that are assumed in order to derive the equations. In the next sections, we will discuss both sets.

2.3.1 The classical requirements: explicit conditions

As has been already discussed in the literature [49,50], the equations defining a memristor designate two characteristic fingerprints:

1. The i–v curve must be pinched: it has to cross the origin $(0, 0)$. That is, zero voltage means zero current.
2. The area of the loop tends to zero at high frequency.

Referring to the first condition, it is the most usual way to experimentally determine if a device has a memristive behavior or not. In Leon Chua's words [59]: "If it's NOT pinched, it's NOT a memristor". This condition can be adapted in a straightforward manner to the charge-flux space. In these variables, the condition reads "A constant flux implies a constant charge".

Remark 1. It is important to observe that a *pinched hysteresis loop is not a memristor model*, but it is just the response of the device to a specific zero-mean input. Thus, the zero-coincidence property of memristor voltage and current, implying pinched i–v curves is just an important test from the experimental point of view, but pinched hysteresis loops do not have any predictability property [60].

The second condition is less often considered, but it is as important as the first one, from the experimental point of view. Thus, it has to be shown that different frequencies cause different loop areas. As has been shown in [49] for high frequencies, memristor's behavior tends to that of a linear resistor. At low frequencies, Eq. (2.2) defines its behavior. In the latter case and for a system with long-term memory, the area of the loop tends to be maximized. On the other hand, if the system exhibits only short-term memory, the device tends to behave as a nonlinear resistor.

A very characteristic example of a short-term memory element is the resistive filament of a light bulb [38,62]; at low frequency it quickly reaches the thermal equilibrium with the ambient, while at higher frequencies it does not have the essential time to heat, thus keeping its resistance constant. Some experimental i–v waveforms from a resistive light bulb are plotted in Fig. 2.4, illustrating this behavior. Equivalently, an example of a long-term memory device would be a ReRAM device [60], as seen in Fig. 2.5, where experimental data from a device are plotted along with the model behavior [61].

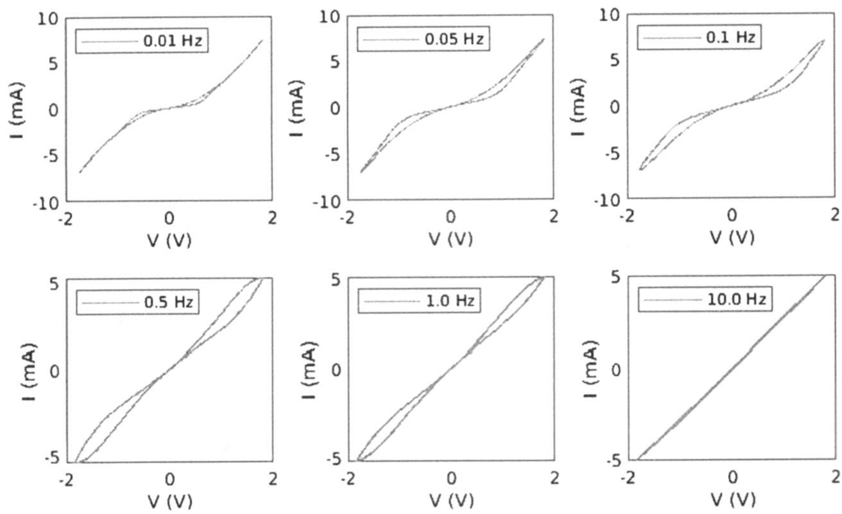

FIGURE 2.4 Hysteresis of a light bulb at different frequencies. At low frequencies, the bulb acts as a non-linear resistance, while at higher frequencies it acts as a linear resistor.

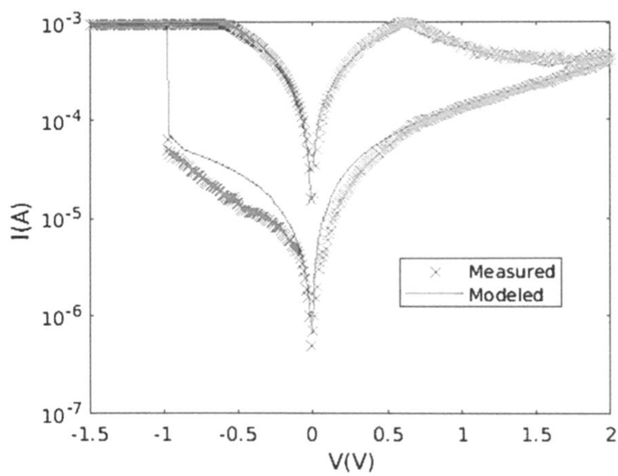

FIGURE 2.5 Hysteresis of memristive device, operating as a ReRAM. The measured characteristics are the symbols, while the line corresponds to the model in [61].

2.3.2 Additional requirements: implicit conditions

In the case of memristors or memristive devices, there are some implicit assumptions arising from Eqs. (2.1)–(2.5) that are interesting to explicitly mention, due to the imposition of specific restrictions in circuit design:

1. The memristor device has to be a two-terminal device. That means that there cannot be any internal ground reference, so the only electrical variable allowed (externally) is the voltage difference between the device terminals.

Notice that this implication may exclude many of the so-called memristors proposed till now, since they are in fact 3-terminal devices (because of the ground reference incorporated), as can be most of the organic memristors.

2. The memristor device has no other external input or output ports than its two electrical terminals. This condition is also implied by the equation governing the state variables, entailing that, for instance, the external temperature cannot play a role as a variable. Obviously, this is not true, but it is commonly accepted that temperature is actually an external constant and not a variable.
3. The memristor device must not generate net energy. This condition is not stated into the equations, but it is a physical requirement, since the memristor is one of the four fundamental elements, thus, it is a passive device.

2.4 Designing memristor emulators

In order to design a memristor emulator, the previous requirements must be taken into account. The implementation of the defining equations can be achieved by mapping or transposing the function-describing sub-blocks in Fig. 2.3 or Fig. 2.2 onto specific circuit blocks. As has already been pointed out, the implicit conditions apposed in Section 2.3.2 require a more careful consideration, since they introduce the demand for a fully differential implementation, with no ground reference inside the circuit.

2.4.1 The diode bridge

The common mode in the memristor emulator circuits presented so far, arises from different points in design methodologies:

- Some of the designs use a capacitor or an equivalent, as the element to make the integration. This capacitor is usually connected to ground, thus creating an additional path for the current. In fact, this approach adds an effective third terminal for the device.
- Other designs use a voltage-controlled element (a.k.a. a MOS transistor) as the variable resistance, with the gate connected to a control variable. The problem in this case has multiple dimensions:
  - Firstly, the value of the control variable must take into account the common-mode implication, or else the current through the transistor will depend on the value of the control variable and the value of the source terminal.
  - Secondly, the bulk voltage may modify the threshold voltage if connected to a fixed value, instead to the source terminal.

In order to remove the common mode from the circuit, the designer has to solve the above described problems. An efficient way to do so is by using a diode bridge [43] or alternatively an improved bridge-version, using MOS transistors controlled by complementary signals, as shown in Fig. 2.6. These circuits allow

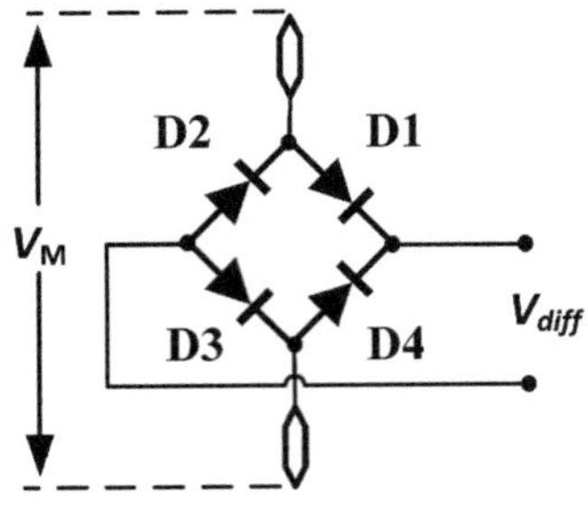

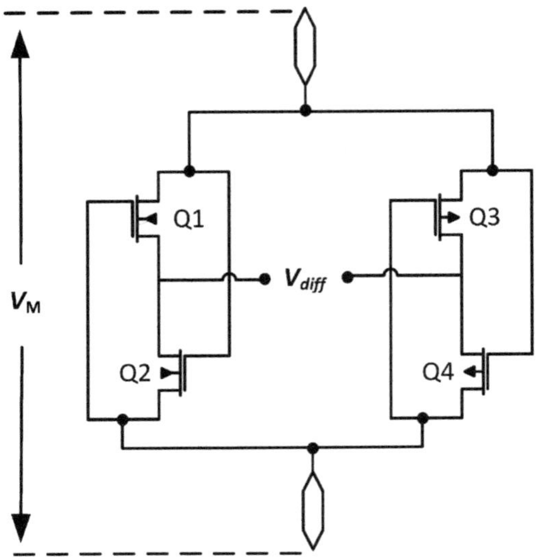

FIGURE 2.6 Two different options for a full-wave rectifier: (a) iode bridge based configuration [43]; and (b) MOS based configuration controlled by complementary signals.

the designer to obtain the maximum and the minimum values of the input signal. This way, the internal circuit can always be referenced to them, thus removing the common-mode voltage. The main problem with this solution is that it forces an additional element in the current path, with a threshold voltage. We believe, however, that this is a price worth paying to get the common mode removed in an easy way. In addition, as we will see in the next section, the diode bridge is also a very useful element for synthesizing passive memristor elements.

2.4.2 A short-term memory emulator

The realization of memristors in the form of active circuits had been presented in Chua's seminal paper [1]. Such an implementation was based on a linear

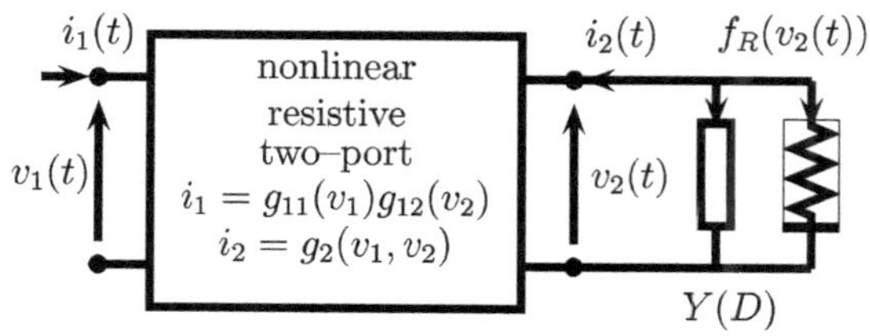

FIGURE 2.7 Novel implementation of a memristor based on a special class of (passive) nonlinear resistive two-port connected to a nonlinear dynamic one-port.

dynamic two-port (mutator) connected to a nonlinear resistive (or dynamic) one-port. As reported in the previous section, an original realization of memristor having only passive circuits has been proposed in [43]. This novel implementation of memristor relies on a special class of (passive) nonlinear resistive two-port connected to a nonlinear dynamic one-port (see Fig. 2.7). In particular, we consider:

- voltage-controlled nonlinear resistive two-ports described as follows[1]:

$$i_1(t) = g_{11}(v_1(t))g_{12}(v_2(t)) \tag{2.6}$$
$$i_2(t) = g_2(v_1(t), v_2(t)) \tag{2.7}$$

  where $g_{11}(v_1(t))$ is a function such that $g_{11}(0) = 0$. It follows that

$$g_{11}(v_1(t)) = \tilde{g}_{11}(v_1(t))v_1(t) \tag{2.8}$$

- nonlinear dynamic one-ports obtained by the parallel connection of a linear dynamic bipole and a voltage-controlled nonlinear resistor, that is (note that $i_2(t)$ is opposite to the associated reference direction),

$$-i_2(t) = Y(D)\, v_2(t) + f_R(v_2(t)) \tag{2.9}$$

  where $D = d(\cdot)/dt$ the first-order time-derivative operator, $Y(D)$ is the differential operator associated to the admittance of the linear dynamic bipole and $f_R(v_2(t))$ is the nonlinear constitutive equation of the nonlinear resistor.

It turns out that (2.6)–(2.9) can be rewritten as follows:

$$0 = Y(D)\, v_2 + f_R(v_2) + g_2(v_1, v_2) \tag{2.10}$$
$$i_1(t) = [\tilde{g}_{11}(v_1)g_{12}(v_2)]\, v_1(t) = g(v_1, v_2)\, v_1(t) \tag{2.11}$$

where Eq. (2.10) (Eq. (2.11)) corresponds to Eq. (2.2) (Eq. (2.1)) and the state $\mathbf{x}(t)$ is represented by $v_2(t)$. Thereby, the design of nonlinear resistive two-ports such that i_1 can be factored into the product of $g_{12}(v_2)$ and $g_{11}(v_1(t)) =$

[1] Current-controlled time-invariant memristor can be obtained, *mutatis mutandis*, considering current-controlled nonlinear resistive two-port connected to a nonlinear dynamic one-port.

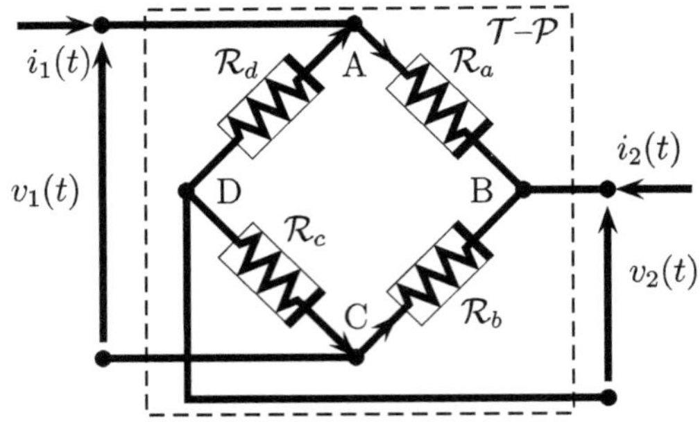

FIGURE 2.8 Nonlinear resistive two-port connected so as to satisfy Eqs. (2.6)–(2.8). The current in each bipole has the direction specified by the arrow and voltage is defined by the associated reference direction.

$[\tilde{g}_{11}(v_1)]\, v_1(t)$ is crucial to implement the memristor *state/voltage-dependent Ohm's law*. The next section presents a simple class of nonlinear resistive two-ports that fulfills these properties.

Remark 2. The purely-passive electronic circuit with nonlinearity provided by a simple network of 4 diodes arranged in a Graetz bridge configuration and dynamics supplied by inductor and capacitor of a RLC series filter (see details in [43]) is a special case of (2.10)–(2.11), where the nonlinear resistive one-port, described by $f_R(v_2)$, is replaced by an open-circuit.

2.4.2.1 Nonlinear resistive two-port for memristor implementation

Let us consider a nonlinear resistive two-port composed of four identical voltage-controlled nonlinear resistor ($\mathcal{R}_k$, with $k = \{a, b, c, d\}$) connected as shown in Fig. 2.8. The current i_k in each bipole has the direction specified by the arrow. The direction of voltage v_k across each bipole is defined by the associated reference direction. Let us assume that the characteristic equation of $\mathcal{R}_k$ is defined as follows:

$$i_k = I\left[\alpha \exp(\gamma\, v_k) + \beta \exp(-\gamma v_k) - (\alpha + \beta)\right] \tag{2.12}$$

where $I \in \mathbf{R}$ is a current, $\gamma^{-1} \in \mathbf{R}$ is a voltage, α and β are real constants. It is worth noting that (2.12) defines the ideal diode characteristics for $(\alpha, \beta, \gamma) = (1, 0, 1/V_T)$ (with V_T denoting the thermal voltage).

The four-terminal multipole with nodes $\{A, B, C, D\}$ can be seen as a nonlinear resistive two-port by considering the pairs (A, C) and (B, D) as the input and the output port, respectively. Current and voltage at the input (output) port are denoted by i_1 and v_1 (i_2 and v_2), respectively.

Following the procedure presented in [43], it can be readily derived that, under the condition[2] $i_1(t) \neq 0$, Kirchhoff's voltage law equation $v_a + v_d = v_b + v_c$,

[2] It turns out that $i_1(t)$ given in (2.6) has to be different from zero.

Kirchhoff's current law equation $i_a + i_b = i_c + i_d$ and (2.12) imply the constrains $v_a = v_c$ and $v_b = v_d$ from which it follows that $i_a = i_c$ and $i_b = i_d$.

These constraints allow us to write i_1 and i_2 in terms of v_1 and v_2, i.e. the voltage-controlled characteristic of the nonlinear resistive two-port $\mathcal{T}$–$\mathcal{P}$. By observing that $v_1 = v_a - v_b$ and $v_2 = -v_a - v_b$, it is readily derived that

$$v_a = 0.5\,(v_1 - v_2) \tag{2.13}$$

$$v_b = -0.5\,(v_1 + v_2) \tag{2.14}$$

Using (2.13) and (2.14), some algebraic manipulations yield

$$\begin{aligned} i_1 = i_a - i_b &= I\,\{\alpha[\exp(\gamma\, v_a) - \exp(\gamma\, v_b)] + \\ &\qquad \beta[\exp(-\gamma\, v_a) - \exp(-\gamma\, v_b)]\} \\ &= 2I\,\sinh\left(\frac{\gamma\, v_1}{2}\right)\left[\alpha\,\exp\left(-\frac{\gamma v_2}{2}\right) - \beta\,\exp\left(\frac{\gamma v_2}{2}\right)\right] \end{aligned} \tag{2.15}$$

$$\begin{aligned} i_2 = -i_a - i_b &= -I\,\{\alpha[\exp(\gamma\, v_a) + \exp(\gamma\, v_b)] + \\ &\qquad \beta[\exp(-\gamma\, v_a) + \exp(-\gamma\, v_b)] - 2(\alpha + \beta)\} \\ &= -2I\,\cosh\left(\frac{\gamma\, v_1}{2}\right)\left[\alpha\,\exp\left(-\frac{\gamma v_2}{2}\right) + \beta\,\exp\left(\frac{\gamma v_2}{2}\right)\right] \\ &\quad + 2I\,(\alpha + \beta) \end{aligned} \tag{2.16}$$

It turns out that Eq. (2.16) corresponds to Eq. (2.7) and Eq. (2.15) is in the form given in (2.11). In particular, we can identify (such that γI is a conductance)

$$\tilde{g}_{11}(v_1) = \gamma I \sum_{q=0}^{+\infty}\left(\frac{1}{2}\right)^{2q}\frac{(\gamma v_1)^{2q}}{(2q+1)!} \tag{2.17}$$

$$g_{12}(v_2) = \left[\alpha\,\exp\left(-\frac{\gamma v_2}{2}\right) - \beta\,\exp\left(\frac{\gamma v_2}{2}\right)\right] \tag{2.18}$$

$$\begin{aligned} g_2(v_1, v_2) &= 2I\,(\alpha + \beta) - 2I\,\cosh\left(\frac{\gamma\, v_1}{2}\right)\cdot \\ &\quad \left[\alpha\,\exp\left(-\frac{\gamma v_2}{2}\right) + \beta\,\exp\left(\frac{\gamma v_2}{2}\right)\right] \end{aligned} \tag{2.19}$$

Thereby, if the output port of the nonlinear resistive two-port shown in Fig. 2.8, is terminated with a nonlinear dynamic bipole defined by (2.9) then the circuit at the input port is equivalent to a memristor described by (2.10)–(2.11) with $\tilde{g}_{11}(v_1)$, $g_{12}(v_2)$ and $g_2(v_1, v_2)$ given in (2.17)–(2.19). The special case $(\alpha, \beta, \gamma) = (1, 0, 1/V_T)$ is reported in [43] where it is shown that the resulting circuit manifests the typical pinched hysteretic current–voltage loop characterizing a memristor.

2.4.3 A long-term memory emulator

As has already been discussed above, the goal is to create an analog memristor emulator, as close as possible to memristor devices, as these are defined in Section 2.2, being at the same time fully compatible to the current technology. As a consequence, the designed circuit should exhibit the following merits:

- It has to be compatible with the current technological state of the art, thus it has to be implemented only by CMOS transistors.
- It cannot have any external power supply, other than that demanded for the body biasing.
- It must be a truly floating emulator.

A combination of the method presented in the previous section for passive emulators and the method presented in [47], describing the design of a charge controlled memristor emulator, allows for designing circuits that fulfill all the above requirements. In the present case, we will focus on the design of a flux-controlled analog memristor emulator based on the block diagram presented in Fig. 2.2, (in contrast to the charge-controlled implementation, described in Fig. 2.3 and appearing in [47]).

2.4.3.1 Circuit description and operation

The proposed circuit design appears in Fig. 2.9. This circuit presents a compact topology, consisting of only 9 CMOS transistors in total, plus a capacitor. This capacitor C was implemented using a Poly-Metal1-Metal2 structure. It is noted that the voltage V_C on capacitor C acts as the internal state variable, essential for emulating extended memristors [49]. It is noted that, as another design option, this capacitor could have been implemented as the gate capacitance of a MOS transistor. In the former case, the capacitor would exhibit a linear behavior, while in the latter its behavior would decline from linearity.

The proposed circuit (Fig. 2.9) can be described as a two-pole cell, exactly as a memristor device, operated by an external excitation voltage signal. Thanks to the use of a diode bridge-like structure, no grounding is demanded, thus, the proposed topology is a floating one. In addition, no external dc-bias is required for operating the emulator circuit, therefore the proposed circuit is a two-pole memristor emulator indeed. The state variable that controls its state (memristance/memconductance), through the flux (integral of the voltage–voltage momentum), is the charge integrated at the capacitor, which is apparently a function of the integral of the driving voltage V_M. Therefore, it is a flux-controlled memristor emulator (Figs. 2.2 and 2.9).

The circuit can be decomposed into three different fundamental entities: a voltage rectifier, a variable resistance, and an integrator that drives the dynamics of this resistance; providing a clear and direct correspondence to Fig. 2.2. The non-linearity of the implemented circuit provides with the desirable nonlinear behavior for the state variable $\mathbf{x}$ and function $\mathbf{g}$, further ensuring correspondence to the operation illustrated by the scheme appearing in Fig. 2.2.

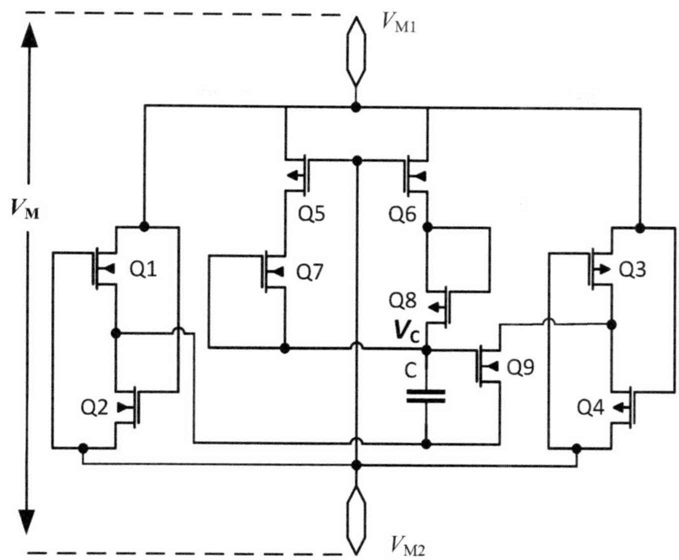

FIGURE 2.9 The circuit schematic of an analog, passive, extended memristor emulator. Notice that the drains of Q1 and Q2 are connected to the state variable, voltage V_C. The circuit has been implemented by using only CMOS transistors and a capacitor. PMOS transistor bulk was connected to the maximum possible voltage, while NMOS bulks were connected to the lowest possible voltage, as usual in microelectronic circuit design.

The voltage rectifier is used to remove the problem of the common mode discussed in the introduction, since the voltage at the capacitor is always referenced to one of the voltages at the terminals and not to any external point, like the ground. Transistors Q1, Q2, Q3, and Q4 form this voltage rectifier, which is an alternative and improved version of the diode bridge, requiring a voltage drop of only one threshold voltage, instead of the usual two threshold voltages (note that the four transistors are not connected in a diode mode). During the positive semi-period (V_{M1} is higher than V_{M2}) and when V_M is higher by at least a threshold voltage, then Q3 (PMOS) and Q2 (NMOS) will be active, while Q1 and Q4 will be in cut off region. During the negative semi-period (V_{M1} is lower than V_{M2}), the current path goes through Q4 (PMOS) and Q1 (NMOS), while Q2 and Q3 will be in cut off region.

The integrator is formed by transistors Q5, Q6, Q7, Q8 and the capacitor C. Its structure and operation is as follows: during the positive semi-period (V_{M1} is higher than V_{M2}) and Q6 (PMOS) will be active, while Q5 (NMOS) will be inactive. This way a current proportional to the voltage between the terminals would flow into the capacitor, making its voltage V_C to be proportional to the integral of V_M. Again, during the negative semi-period (V_{M1} is lower than V_{M2}), the NMOS Q5 will be active, while the PMOS Q6 will be inactive. This way, the memristor's current (proportional to the voltage between the emulator's terminals) will flow and get integrated out of the capacitor.

To ensure a long-term memory behavior, transistors Q7 and Q8 are used in a diode configuration to prevent charge losses out of the implemented capacitor. As a consequence, the POP equation (Eq. (2.2) for $v = 0$) is ensured to be equal to zero, since there is no current into (or out of) the capacitor C for $V_M = 0$; thus ensuring that the presented system has long-term memory effects, as opposite to most of the other CMOS-only presented emulators. As a side comment, it is also worth noticing that this emulator can be modified in order to present a short-term memory (STM) by adding a discharge element between the terminals of the capacitor C. Notice that this discharge would not mean any charge flowing outside the emulator, but only a controlled internal current path.

The varying resistance defining memristor dynamics is connected between the terminals of this voltage rectifier, and it is implemented using the Q9 (NMOS). The gate of this transistor is controlled by the output of the integrator, while its source is connected to the same node as the one of the terminals of the capacitor. This way, the behavior of Q9 depends on difference between V_C and the minimum of (V_{M1}, V_{M2}), which is different from the reference of simply V_{M1} or V_{M2} used in most of the other proposed topologies.

2.4.3.2 Physical layout

The circuit schematic in Fig. 2.9 has been implemented in a mixed-signal 3.3V, 0.35 μm technology, utilizing the Cadence Design Suite. A possible layout implementation is presented in Fig. 2.10. The dimensions of the transistors implemented in this version of the emulator, are provided in Table 2.1.

The proposed layout version of the emulator cell has a size of 39 μm × 37 μm, which is fairly low. It has to be noted that the body of all NMOS transistors has been connected to the minimum voltage possible in the system (−1.65 V), and the PMOS body is also connected to the maximum possible voltage (+1.65 V). This connection, however, does not mean an external power source to the system, but it is simply a technological requirement.

The effects introduced by the parasitic components accompanying the transistors used, are not important in this design. This is due to the fact that there exists no need for the transistors to be exactly tuned to a given set of parameters neither is there any set of transistors that need to be paired. The only parasitic effect worth noticing is a small charge injection into the control capacitor C when Q_5 or Q_6 switch off.

2.4.3.3 Results

The proposed analog memristor-cell has been tested to check if it could emulate the well-known signatures of memristor [50] presented in section 2.3. As has been previously mentioned, the main fingerprints are the *pinched* (zero voltage for zero current and vice verso) hysteresis i–v loop, the area of which tends

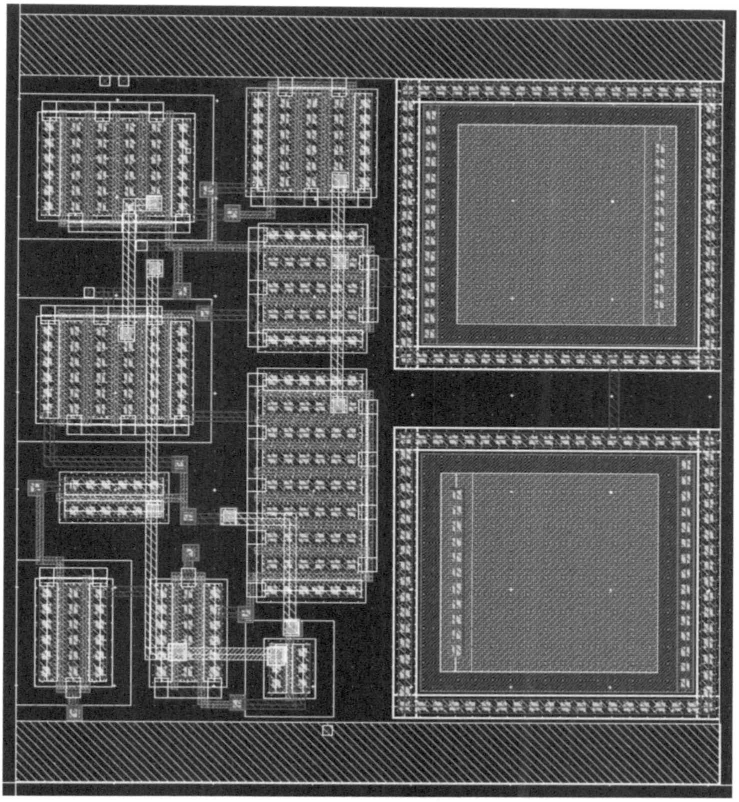

FIGURE 2.10 The cell-layout implementing the analog memristor emulator of Fig. 2.9, in a mixed-signal 0.35 μm technology. The area occupied by this specific implementation is 37 μm x 39 μm.

to zero (it becomes a simple line, without any hysteresis) at higher driving frequencies, further tending to an ohmic behavior. These tests were performed by post-layout simulations using the cadence design suite, with the PDK corresponding to a mixed-signal 3.3V, 0.35 μm technology.

Initially, the characteristic *i–v* curves, at four different driving frequencies, were simulated. These curves are shown in Fig. 2.11. From this figure becomes apparent that all four characteristic curves are passing through the origin (0 V, 0 A), thus they are being pinched. Also, the lobes become narrower with higher driving signal frequency, as expected for memristors [49,50].

The presented emulator exhibits a frequency-span up to some hundreds of kHz. This span is higher than the ones of just a few Hz at which real memristor devices are currently operating. Furthermore, the frequency-span of this emulator can be decided by the designer, since it depends upon the size and geometry of the cell's transistors, as well as the utilized technology. Regarding the form of the hysteresis loops, these appear to be non-symmetric due to the differences in

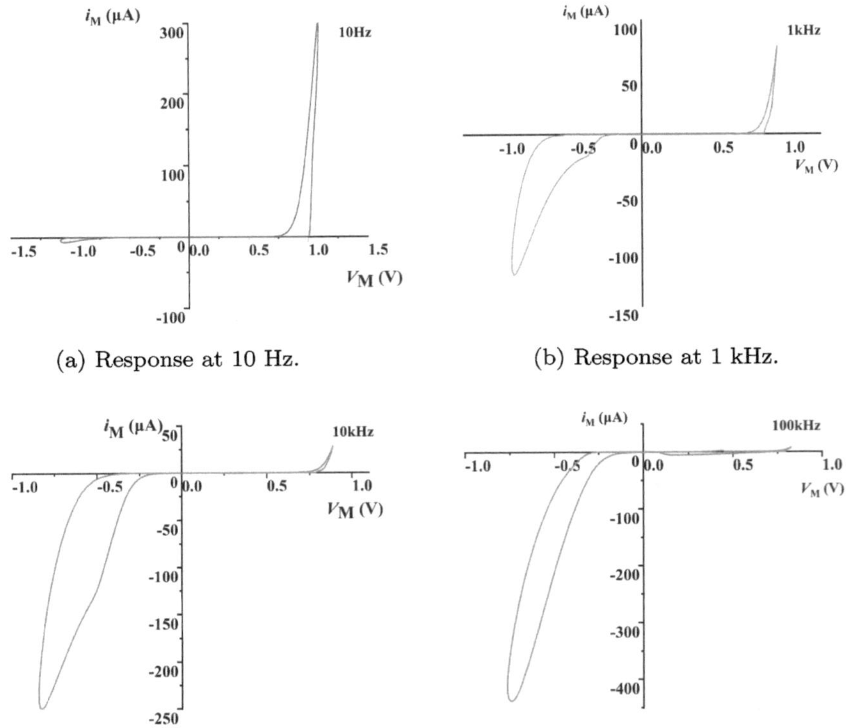

(a) Response at 10 Hz.

(b) Response at 1 kHz.

(c) Response at 10 kHz.

(d) Response at 100 kHz.

FIGURE 2.11 Characteristic curves for four different sinusoidal driving signals at driving frequencies of 100 kHz, 10 kHz 1 kHz, and 100 Hz. The horizontal axis shows the driving voltage in Volts, while the vertical one the current through the cell, in Amperes.

the frequency behavior between the P and N transistors. This asymmetric behavior is desirable when real memristor devices emulation is demanded, as most of real devices are asymmetric. In any case, the differences in frequency behavior between the P and N transistors can be set against by their relative geometry.

Another important test for the emulator is its behavior in terms of the demonstrated volatility. To check this feature, we have used two different tests. In the first, four driving pulse-series, of different duty cycles (DC), in specific 0.05, 0.25, and 0.45, all over a pulse frequency of 10 kHz, were applied. The amplitude was set to +1.6 V and −1.6 V. In Fig. 2.12 the behavior of the memristor-resistance is presented in all four cases. In the second test, a number of pulses at a fixed duty cycle of 0.45, were applied for four different frequencies (10 Hz, 1 kHz, 10 kHz, 1000 kHz). The corresponding results are plotted and appear in Fig. 2.13; a behavior typical for non-volatile memristors is revealed, i.e. the current through the device (in this case the cell) gets a specific value after it is pulsed. As expected, the shorter the effective pulse, the smaller the

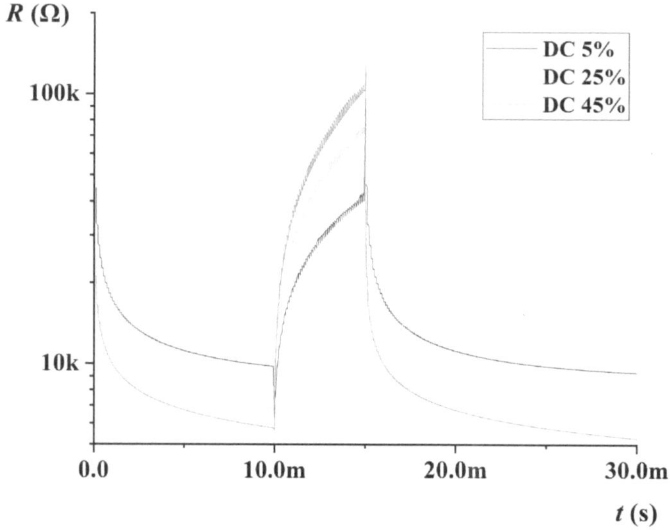

FIGURE 2.12 Time evolution of resistance for three pulse timeseries with duty cycles of 0.05 (gray, dark gray in print version), 0.25 (red, light gray in print version) and 0.45 (blue, mid gray in print version). From $t = 0.01$ to 0.015 s the pulses are positive, thus leading to an increase of memristance, while the pulses are negative elsewhere, decreasing memristance.

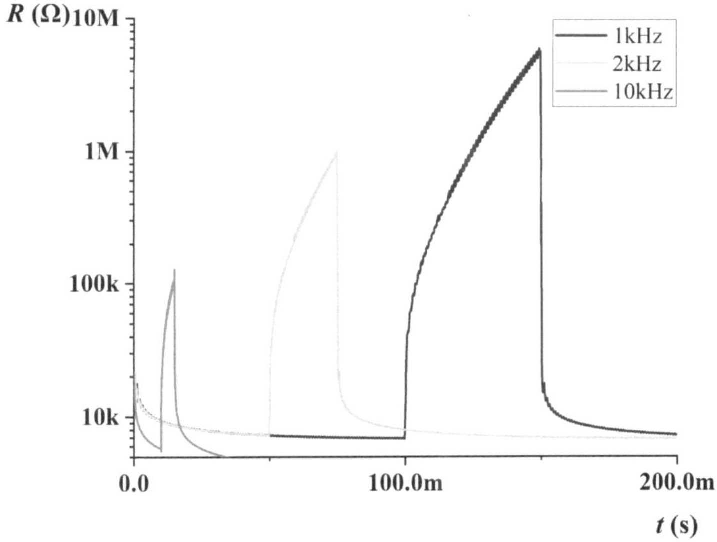

FIGURE 2.13 Evolution of resistance as a function of the number of pulses for three pulse timeseries with duty cycles 0.45, and three different frequencies of 1 kHz (gray, dark gray in print version), 2 kHz (red, light gray in print version) and 10 kHz (blue, mid gray in print version). The pulses between 2000 and 3000 are positive, and cause an increase of memristance, while the pulses are negative elsewhere, decreasing memristance.

TABLE 2.1 Transistor width W for the circuit in Fig. 2.9. All the lengths are $L = 0.35\ \mu m$.

Transistor	Type	W
Q1	PMOS	20 μm
Q2	NMOS	20 μm
Q3	NMOS	25 μm
Q4	NMOS	25 μm
Q5	NMOS	2.5 μm
Q6	NMOS	5 μm
Q7	PMOS	10 μm
Q8	PMOS	12.5 μm
Q9	NMOS	40 μm

change in resistance, showing that the device has the ability to be programmed to a specific resistive state by using a series of pulses, as in real memristors [5,6].

2.5 Conclusion

Beginning from a discussion on the general framework for memristor modeling, the general set of equations that serve for this purpose was recalled. Based on these equations, two block-diagrams for the cases of flux- and charge-controlled memristor emulators were drawn. A way to develop memristor emulators with only passive elements has been also discussed, based on those equations. In order to improve these emulators in terms of long-term memory capabilities, we have then further developed the idea, and a modular approach has been utilized so that the memristor emulators can be decomposed into basic analog building blocks. Then these blocks have been translated into circuital elements, aiming into designing a memristor emulator (cell) with specific, interesting characteristics: to be a floating and passive cell, and to demonstrate the feature of long-term memory. In this context, the floating property means that the common mode between the two terminals is minimized, while the passive property means that it does not need any external energy supply, other than bulk biasing. Finally, the long-term memory is due to the fact that the capacitor has no path to discharge when the external voltage is zero. Simulation results with both periodic and pulsed signals show the effectiveness of proposed method and its implementation as a circuit, which can be qualified as a memristor.

It is apparent from its performance profile that this memristor emulator can be used to implement many different bio-inspired topologies, as well as other non-Neumann approaches. For instance, this circuit would allow easy implementation of more than 600 cells per mm^2 for neural network implementation or crossbar arrays within the studied technology. More advanced nodes would improve this density.

References

[1] L.O. Chua, Memristor – the missing circuit element, IEEE Transactions on Circuit Theory 18 (1971) 507–519.

[2] T. Prodromakis, C. Toumazou, L. Chua, Two centuries of memristors, Nature Materials 11 (2012) 478–481.

[3] L. Chua, S.M. Kang, Memristive devices and systems, Proceedings of the IEEE 64 (1976) 209–223, https://doi.org/10.1109/PROC.1976.10092.

[4] R. Tetzlaff, Memristors and Memristive Systems, Springer, 2013.

[5] H. Kim, M.P. Sah, C. Yang, L.O. Chua, Memristor-based multilevel memory, in: 2010 12th International Workshop on Cellular Nanoscale Networks and Their Applications (CNNA), IEEE, 2010, pp. 1–6.

[6] S. Stathopoulos, A. Khiat, M. Trapatseli, S. Cortese, A. Serb, I. Valov, T. Prodromakis, Multibit memory operation of metal-oxide bi-layer memristors, Scientific Reports 7 (2017) 1–7.

[7] K. Eshraghian, K.-R. Cho, O. Kavehei, S.-K. Kang, D. Abbott, S.-M.S. Kang, Memristor mos content addressable memory (mcam): hybrid architecture for future high performance search engines, IEEE Transactions on Very Large Scale Integration (VLSI) Systems 19 (2011) 1407–1417.

[8] S. Carrara, D. Sacchetto, M.-A. Doucey, C. Baj-Rossi, G. De Micheli, Y. Leblebici, Memristive-biosensors: a new detection method by using nanofabricated memristors, Sensors and Actuators B, Chemical 171 (2012) 449–457.

[9] I. Gupta, A. Serb, A. Khiat, R. Zeitler, S. Vassanelli, T. Prodromakis, Memristive integrative sensors for neuronal activity, arXiv preprint, arXiv:1507.06832, 2015.

[10] Y.V. Pershin, M. Di Ventra, Experimental demonstration of associative memory with memristive neural networks, Neural Networks 23 (2010) 881–886.

[11] J. Grollier, D. Querlioz, M.D. Stiles, Spintronic nanodevices for bioinspired computing, Proceedings of the IEEE 104 (2016) 2024–2039.

[12] B. Sun, X. Zhang, G. Zhou, P. Li, Y. Zhang, H. Wang, Y. Xia, Y. Zhao, An organic nonvolatile resistive switching memory device fabricated with natural pectin from fruit peel, Organic Electronics 42 (2017) 181–186.

[13] S. Battistoni, A. Dimonte, V. Erokhin, Organic memristor based elements for bio-inspired computing, in: Advances in Unconventional Computing, Springer, 2017, pp. 469–496.

[14] D.B. Strukov, G.S. Snider, D.R. Stewart, R.S. Williams, The missing memristor found, Nature 453 (2008) 80–83.

[15] C. Dias, H. Lv, R. Picos, P. Aguiar, S. Cardoso, P. Freitas, J. Ventura, Bipolar resistive switching in Si/Ag nanostructures, Applied Surface Science 424 (2017) 122–126.

[16] S. Brivio, G. Tallarida, E. Cianci, S. Spiga, Formation and disruption of conductive filaments in a hfo2/tin structure, Nanotechnology 25 (2014) 385705.

[17] B. Mohammad, M.A. Jaoude, V. Kumar, D.M. Al Homouz, H.A. Nahla, M. Al-Qutayri, N. Christoforou, State of the art of metal oxide memristor devices, Nanotechnology Reviews 5 (2016) 311–329.

[18] I. Messaris, A. Serb, A. Khiat, S. Nikolaidis, T. Prodromakis, A compact verilog – a reram switching model, arXiv preprint, arXiv:1703.01167, 2017.

[19] P.S. Georgiou, S.N. Yaliraki, E.M. Drakakis, M. Barahona, Window functions and sigmoidal behaviour of memristive systems, International Journal of Circuit Theory and Applications 44 (2016) 1685–1696.

[20] A. Ascoli, R. Tetzlaff, L. Chua, Continuous and differentiable approximation of a tao memristor model for robust numerical simulations, in: Emergent Complexity from Nonlinearity, in Physics, Engineering and the Life Sciences, Springer, 2017, pp. 61–69.

[21] F. Jimenez-Molinos, M. Villena, J. Roldan, A. Roldan, A spice compact model for unipolar rram reset process analysis, IEEE Transactions on Electron Devices 62 (2015) 955–962.

[22] Q. Li, A. Serb, T. Prodromakis, H. Xu, A memristor spice model accounting for synaptic activity dependence, PLoS ONE 10 (2015) e0120506.

[23] J. Secco, F. Corinto, A. Sebastian, Flux-charge memristor model for phase change memory, IEEE Transactions on Circuits and Systems II, Express Briefs 65 (2017) 111–114.

[24] R. Picos, J.B. Roldan, M.M. Al Chawa, P. Garcia-Fernandez, F. Jimenez-Molinos, E. Garcia-Moreno, Semiempirical modeling of reset transitions in unipolar resistive-switching based memristors, Radioengineering 24 (2015) 421.

[25] E. Garcia-Moreno, R. Picos, M.M. Al-Chawa, Spice model for unipolar rram based on a flux-controlled memristor, in: 2015 IEEE International Autumn Meeting on Power, Electronics and Computing (ROPEC), IEEE, 2015, pp. 1–4.

[26] M.M. Al Chawa, R. Picos, J.B. Roldan, F. Jimenez-Molinos, M.A. Villena, C. de Benito, Exploring resistive switching-based memristors in the charge–flux domain: a modeling approach, International Journal of Circuit Theory and Applications 46 (2018) 29–38, https://doi.org/10.1002/cta.2397.

[27] Z. Kolka, D. Biolek, V. Biolkova, Z. Biolek, Evaluation of memristor models for large crossbar structures, in: Radioelektronika (RADIOELEKTRONIKA), 2016 26th International Conference, IEEE, 2016, pp. 91–94.

[28] R. Picos, J. Roldan, M. Al Chawa, F. Jimenez-Molinos, E. Garcia-Moreno, A physically based circuit model to account for variability in memristors with resistive switching operation, in: Conference on Design of Circuits and Integrated Systems (DCIS), IEEE, 2016, pp. 1–6.

[29] R. Naous, M. Al-Shedivat, K.N. Salama, Stochasticity modeling in memristors, IEEE Transactions on Nanotechnology 15 (2016) 15–28.

[30] I. Vourkas, A. Abusleme, V. Ntinas, G.C. Sirakoulis, A. Rubio, A digital memristor emulator for fpga-based artificial neural networks, in: IEEE International Verification and Security Workshop (IVSW), IEEE, 2016, pp. 1–4.

[31] R. Ranjan, P.M. Ponce, A. Kankuppe, B. John, L.A. Saleh, D. Schroeder, W.H. Krautschneider, Programmable memristor emulator asic for biologically inspired memristive learning, in: 2016 39th International Conference on Telecommunications and Signal Processing (TSP), IEEE, 2016, pp. 261–264.

[32] Z. Kolka, J. Vavra, V. Biolkova, A. Ascoli, R. Tetzlaff, D. Biolek, Programmable emulator of genuinely floating memristive switching devices, in: 2019 26th IEEE International Conference on Electronics, Circuits and Systems (ICECS), IEEE, 2019, pp. 217–220.

[33] A. Ascoli, F. Corinto, R. Tetzlaff, A class of versatile circuits, made up of standard electrical components, are memristors, International Journal of Circuit Theory and Applications 44 (2016) 127–146.

[34] A. Ascoli, R. Tetzlaff, L. Chua, W. Yi, R. Williams, Memristor emulators: a note on modeling, in: Advances in Memristors, Memristive Devices and Systems, Springer, 2017, pp. 1–17.

[35] J. Kalomiros, S.G. Stavrinides, F. Corinto, A two-transistor non-ideal memristor emulator, in: 5th International Conference on Modern Circuits and Systems Technologies (MOCAST), IEEE, 2016, pp. 1–4.

[36] C. Sánchez-López, J. Mendoza-Lopez, M. Carrasco-Aguilar, C. Muñiz-Montero, A floating analog memristor emulator circuit, IEEE Transactions on Circuits and Systems II, Express Briefs 61 (2014) 309–313.

[37] M.T. Abuelma'atti, Z.J. Khalifa, A continuous-level memristor emulator and its application in a multivibrator circuit, AEU International Journal of Electronics and Communications 69 (2015) 771–775.

[38] A. Theodorakakos, S.G. Stavrinides, E. Hatzikraniotis, R. Picos, A non-ideal memristor device, in: 2015 International Conference on Memristive Systems (MEMRISYS), IEEE, 2015, pp. 1–2.

[39] H. Kim, M.P. Sah, C. Yang, S. Cho, L.O. Chua, Memristor emulator for memristor circuit applications, IEEE Transactions on Circuits and Systems I, Regular Papers 59 (2012) 2422–2431.

[40] A. Yeşil, Y. Babacan, F. Kaçar, A new ddcc based memristor emulator circuit and its applications, Microelectronics Journal 45 (2014) 282–287.

[41] H. Sözen, U. Çam, Electronically tunable memristor emulator circuit, Analog Integrated Circuits and Signal Processing 89 (2016) 655–663.

[42] Z.G. Cam, H. Sedef, A new floating memristance simulator circuit based on second generation current conveyor, Journal of Circuits, Systems, and Computers 26 (2017) 1750029.

[43] F. Corinto, A. Ascoli, Memristive diode bridge with lcr filter, Electronics Letters 48 (2012) 824–825.

[44] V. Saxena, A compact cmos memristor emulator circuit and its applications, in: 2018 IEEE 61st International Midwest Symposium on Circuits and Systems (MWSCAS), IEEE, 2018, pp. 190–193.

[45] Y. Babacan, F. Kaçar, Floating memristor emulator with subthreshold region, Analog Integrated Circuits and Signal Processing 90 (2017) 471–475.

[46] Y. Babacan, A. Yesil, F. Kacar, Memristor emulator with tunable characteristic and its experimental results, AEU International Journal of Electronics and Communications 81 (2017) 99–104.

[47] M. Al Chawa, C. de Benito, M. Roca, R. Picos, S. Stavrinides, Design and implementation of passive memristor emulators using a charge-flux approach, in: 2018 IEEE International Symposium on Circuits and Systems (ISCAS), IEEE, 2018, pp. 1–5.

[48] G. Svetoslavov, O. Camps, S.G. Stavrinides, R. Picos, A switched capacitor memristive emulator, IEEE Transactions on Circuits and Systems I (2020), https://doi.org/10.1109/TCSII.2020.3032632.

[49] F. Corinto, P.P. Civalleri, L.O. Chua, A theoretical approach to memristor devices, IEEE Journal on Emerging and Selected Topics in Circuits and Systems 5 (2015) 123–132.

[50] D. Biolek, Z. Biolek, V. Biolková, Z. Kolka, Some fingerprints of ideal memristors, in: 2013 IEEE International Symposium on Circuits and Systems (ISCAS), IEEE, 2013, pp. 201–204.

[51] L.O. Chua, Everything you wish to know about memristors but are afraid to ask, Radioengineering 24 (2015) 319.

[52] S. Shin, K. Kim, S.-M. Kang, Compact models for memristors based on charge-flux constitutive relationships, IEEE Transactions on Computer-Aided Design of Integrated Circuits and Systems 29 (2010) 590–598.

[53] R. Picos, M.M. Al Chawa, M. Roca, E. Garcia-Moreno, A charge-dependent mobility memristor model, in: Proceedings of the 10th Spanish Conference on Electron Devices, CDE'2015, IEEE, 2015.

[54] R. Picos, J.B. Roldan, M.M. Al Chawa, F. Jimenez-Molinos, M. Villena, E. Garcia-Moreno, Exploring ReRAM-based memristors in the charge-flux domain, a modeling approach, in: Proceedings of International Conference on Memristive Systems, MEMRISYS'2015, 2015.

[55] C. de Benito, M.M. Al Chawa, R. Picos, E. Garcia-Moreno, A procedure to calculate a delay model for memristive switches, in: Workshop on Memristor Technology, Design, Automation and Computing, 2017.

[56] M.M. Al Chawa, A. Rodriguez-Fernandez, M. Bargallo, F. Campabadal, C. de Benito, S. Stavrinides, E. Garcia-Moreno, R. Picos, Waveform and frequency effects on reset transition in bipolar reram in flux-charge space, in: 2017 International Conference on Memristive Systems (MEMRISYS), IEEE, 2017.

[57] B. Brooks, K. Green, J. Krick, T. Vrotsos, D. Weiser, Standardization and validation of compact models, in: Technical Proceedings of the International Conference on Modeling and Simulation of Microsystems, vol. 1, 2002, pp. 653–656.

[58] O. Camps, R. Picos, M. Roca, B. Iniguez, E. Garcia-Moreno, A web application for mosfet compact model validation using cmc tests, in: 2008 7th International Caribbean Conference on Devices, Circuits and Systems, IEEE, 2008, pp. 1–6.

[59] L. Chua, If it's pinched it's a memristor, Semiconductor Science and Technology 29 (2014) 104001.

[60] L.O. Chua, Resistance switching memories are memristors, Applied Physics A 102 (2011) 765–783.

[61] M.M. Al Chawa, R. Picos, A simple quasi-static compact model of bipolar reram memristive devices, IEEE Transactions on Circuits and Systems II, Express Briefs (2019).

[62] R. Picos, S.G. Stavrinides, A. Theodorakakos, E. Hatzikraniotis, The experimental model of a non-ideal memristor, in: 2016 5th International Conference on Modern Circuits and Systems Technologies (MOCAST), IEEE, 2016, pp. 1–4.

Chapter 3

On the FPGA implementation of chaotic oscillators based on memristive circuits

Omar Guillén-Fernández[a], Esteban Tlelo-Cuautle[a], Gustavo Rodríguez-Gómez[a], Luis Gerardo de la Fraga[b], and Rui Li[c]
[a]*INAOE, Puebla, Mexico,* [b]*CINVESTAV, San Pedro Zacatenco, Mexico,* [c]*UESTC, Chengdu, China*

3.1 Introduction

The circuit element-called the memristor was introduced in 1971 by Leon O. Chua [1], completing a theoretical quartet of fundamental electrical components which comprises also the resistor (R), capacitor (C), and inductor (L), but exhibits peculiar behavior that is characterized by a relationship between the charge q(t) and the flux-linkage φ(t). This characteristic was not exploited until the new millennium. One example was the investigation of the non-linear dynamics of a Bernoulli cell presented in 2000 [2], encountered in log-domain structures. Basically, the cell comprises an NPN bipolar-junction transistor and an emitter-connected grounded capacitor, which dynamic behavior was modeled by a differential equation of the Bernoulli form, and it has memristive type dynamic behavior. On this direction, the neuron modeled by Hodgkin–Huxley nerve axon membrane dynamics was also successfully simulated as a characteristic example of memristive behavior.

The investigations on memristive systems were passive until 2008, when G. Chen summarized recent developments on Chua's memristor [3]. In that year, Hewlett Packard Labs presented the first physical memristor [4], and Chua published a paper on memristor oscillators [5], highlighting that the memristor has attracted phenomenal worldwide attention since its debut on May 2008 in the journal Nature [6], where the author summarized many potential applications including super-dense non-volatile computer memory and neural synapses.

From 2008, the fourth or missing memristor circuit element [6,7] has been exploited to take advantage of its wide range of hysteretic current–voltage behavior. The authors in [8] emphasized the promising applications in the fields of nanoelectronics, computer logic and neuromorphic computers, and proposed its use in memristor-based chaotic circuits. For the first time, the authors derived memristor-based chaotic circuits from the canonical Chua circuit, and

Mem-elements for Neuromorphic Circuits with Artificial Intelligence Applications
https://doi.org/10.1016/B978-0-12-821184-7.00011-6

concluded that they provide a memristor-based framework for the development of applications in secure communications with chaos. One year later, the community introduced the concepts of circuit elements with memory: memristors, memcapacitors, and meminductors [9]. Basically, the authors extended the notion of memristive systems to capacitive and inductive elements, namely, capacitors and inductors whose properties depend on the state and history of the system. The important point was the observation that all these elements typically show pinched hysteretic loops in the two constitutive variables that define them: current–voltage for the memristor, charge–voltage for the memcapacitor, and current–flux for the meminductor. These elements and their combination in circuits opened new applications in neuromorphic devices, and their electrical characteristics were embedded in simulation programs with emphasis on integrated circuit design [10].

Nowadays, researchers take advantage of the programmable/configurable circuits like field-programmable gate arrays (FPGAs), which allow fast prototyping and realization of digital/analog systems at relatively low development cost while providing good performance. For instance, a compact memristor-CMOS hybrid look-up-table (LUT) design and potential application in FPGA, was introduced in [11]. The advantage was emphasized as: Due to the conventional LUT using the static random access memory (SRAM) cell, FPGAs almost reach the limitation in terms of the density, speed, and configuration overhead, and therefore they are quite useful to propose an improved memristor-based LUT (MLUT). A memristor-based chaotic system and its FPGA implementation are given in [12], which were developed using verilog hardware description language (HDL). The FPGA implementation of infinitely many coexisting chaotic attractors of a dual memristive Shinriki oscillator is given in [13]. Also, the authors in [14], introduced the FPGA implementation of a simple chaotic oscillator with a fractional-order-memristor component. The chaotic oscillators based on memristive circuits can be used in the generation of random sequences, as shown in [15], which introduced a true random bit generator (TRBG) based on a memristive chaotic circuit implemented on an FPGA board. The TRBG was modeled and co-simulated on the Xilinx System Generator (XSG) platform and implemented on the Xilinx Kintex-7 KC705 FPGA Evaluation Board. Other FPGA-based implementations of memristive systems can be found in recent literature [16,17], however, almost all of them do not detail the implementation taking into account the characteristics of the numerical methods, as done in this chapter.

The memristive systems found applications in programmable analog circuits [18], chaotic systems for image encryption [19], and so on. Their physical properties have been analyzed [20] and used to introduce circuit models as in [21–24], this has been also done for the meminductor [25–27], of which basic fingerprints are the flux–current pinched hysteresis loops. Advanced modeling was introduced in [28], as equivalent statistical circuits for passive memory arrays of memristive devices. Another direction is the introduction of fractional-order memristive circuits [29]. All those models and FPGA implementations are suitable for the design of neural networks, as recently shown in [30], dealing with fractional-order memristor-based neural networks, or recurrent neural networks as in [31]. Some neural networks have been synchronized in recent papers [32,33], and it can be extended to memristor-based echo state networks [34]. In this case, the challenge is the learning technique for the different kinds of neural networks, which has also been improved as shown in [35].

Section 3.2 shows the mathematical models of memristive systems in 3D, 4D, and 5D. Section 3.3 describes numerical methods of one-step and multi-step type and shows the analyses of their stability regions. Section 3.4 shows the complete dynamical analysis of one memristive system and its FPGA implementation is detailed in Section 3.5. This chaotic oscillator based on the memristor is used in Section 3.6 to implement a chaotic secure communication system to transmit an image. Finally, the conclusions are given in Section 3.7.

3.2 3D, 4D, and 5D memristive systems

Historically, the mathematical models associated to continuous-time dynamical systems have originated from the fields of physics, chemistry and other engineering areas. It is well known that in the real world the majority of those mathematical models have non-linear nature and therefore their behavior can be captured by the formulation of non-linear ordinary differential equations (ODEs) whose analytical solution is complex or sometimes impossible to find. Besides, the solution of the mathematical models of dynamical systems that generate non-linear phenomena, like memristor-based chaotic oscillators, can be obtained applying numerical methods.

Chaotic oscillators have the main property of high sensitivity to the initial conditions at which the mathematical model is solved by numerical methods. For example, Fig. 3.1 shows the evolution in time of the solution of a mathematical model using two initial conditions that have a difference of 10^{-6}, i.e. $y(0) = 0.5$ and $y(0) = 0.500001$. In this case, the solution evolves differently for each initial condition and one can see that the trajectories that are infinitesimally near at the beginning, they diverge in an exponential behavior as the time increases.

Continuous-time chaotic oscillators can be modeled by mathematical equations consisting of at least three ODEs, and its dynamics can be evaluated computing Lyapunov exponents. In such a case, if the system has one positive Lyapunov exponent, it is said that chaotic behavior is guaranteed. A seminal paper to determine Lyapunov exponents from a time series was published by Wolf and colleagues [36], and it can be applied to any dynamical system of order equal or higher than three. If the chaotic system has more than three ODEs, they generate hyper-chaotic behavior if and only if they have at least two positive Lyapunov exponents. In this section, four mathematical models are given to simulate chaotic oscillators based on memristor and having three, four and five ODEs, so that they have three dimensions (3D), 4D, and 5D, respectively.

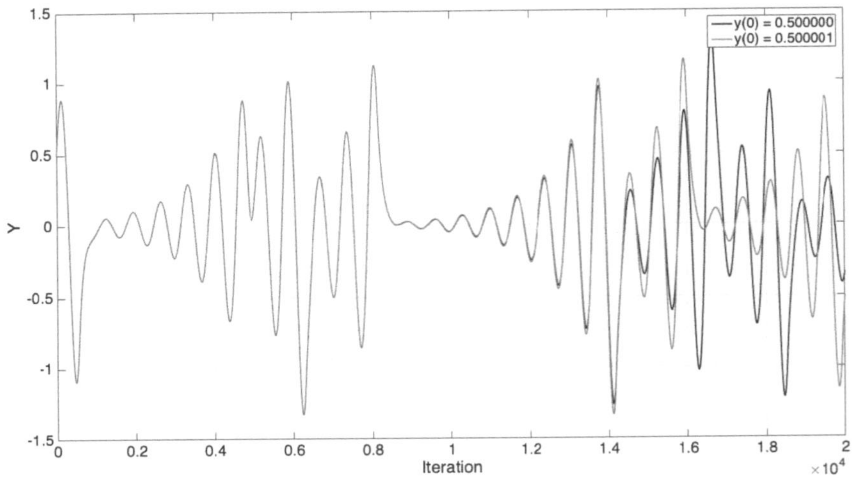

FIGURE 3.1 Effect of the solution of a chaotic dynamical system solved by two initial conditions with a difference of 10^{-6}.

There are recent papers that perform the implementation of memristive-based chaotic oscillators, and which have been applied to data encryption. For example, the mathematical model consisting of three ODEs given in (3.1) was introduced in [37], and the authors concluded that the dynamical system generates chaotic behavior if the values of the coefficients are set to $a = 0.2$, $b = \frac{1}{3}$, and $c = 0.4$:

$$\begin{aligned} \dot{x} &= ax + y \\ \dot{y} &= -bx - byz^2 \\ \dot{z} &= -cy - cz + yz \end{aligned} \tag{3.1}$$

A four-dimensional (4D) memristor-based chaotic oscillator is described by the ODEs given in (3.2). This mathematical model was introduced in [38], and the dynamical system generates chaotic behavior if the coefficient values are set to $a = 10.0$, $b = 9.0$, $c = 8.7$, and the non-linear function $w(x) = a_1 + b_1 x^2$, for which $a_1 = 4.0$ and $b_1 = 0.01$. We have

$$\begin{aligned} \dot{x} &= az + xy \\ \dot{y} &= 1 - x^2 - z^2 - w^2 \\ \dot{z} &= -bx + yz \\ \dot{w} &= cz - w - zw(x) \end{aligned} \tag{3.2}$$

Another 4D memristive-based chaotic oscillator is given in (3.3), which was introduced in [39]. The chaotic behavior arises by setting the following values: $\alpha = 4.0$, $\beta = 1.0$, $\gamma = 0.65$, $c = 0.2$, $d = 10.0$ and $W(w(n)) = d$ if $|w(n)| > 1$ or $W(w(n)) = c$ if $|w(n)| < 1$. Both (3.2) and (3.3) can generate hyper-chaotic behavior because they can have two positive Lyapunov exponents. We have

$$\begin{aligned}\dot{x} &= \alpha(y - xW(w))\\ \dot{y} &= z - x\\ \dot{z} &= -\beta y + \gamma z\\ \dot{w} &= x\end{aligned} \tag{3.3}$$

A 5D memristive-based chaotic system is given in (3.4). This mathematical model was introduced by the authors in [40]. It can have up to three positive Lyapunov exponents and therefore can generate hyper-chaotic behavior. The non-linear function $W(\phi) = e + 3n\phi^2$, and to guarantee chaotic behavior the coefficients are set to the following values: $a = 1.0$, $b = 1.0$, $c = 0.7$, $m = 1.0$, $d = 0.2$, $e = 0.1$, and $n = 0.01$. We have

$$\begin{aligned}\dot{x} &= ax + dW(\phi)y + yz - cw\\ \dot{y} &= yz - xz\\ \dot{z} &= -z - mxy + b\\ \dot{w} &= x\\ \dot{\phi} &= y\end{aligned} \tag{3.4}$$

The solution of the ODEs described above can be obtained by applying numerical methods. As they are initial value problems of the form $\dot{x} = f(x)$, the initial conditions should be near the equilibrium points. However, in the classification of chaotic oscillators one can find two groups: self-excited and hidden attractors. The former ones have well-defined equilibrium points, but the hidden attractors do not have explicit equilibrium points or they are difficult to compute. This is another challenge in the area of chaos [41].

3.3 Numerical methods

The solution of the initial value problems, as the mathematical models given above describing 3D, 4D, and 5D memristive-based chaotic oscillators, can be obtained through applying numerical integration methods. The challenge is the selection of an appropriate method to better approximate the solution of the non-linear ODEs. One can decide on the application of one-step or multi-step methods, and in each case, the step-size is the first problem and it is related to the stability of the method. In linear dynamical systems, the step-size can be determined from the evaluation of the eigenvalues, which are related to the natural frequencies of the system. In non-linear dynamical systems, the problem is complex and one needs to analyze the problem of the form $\dot{x} = f(x, t)$, which requires of both, an initial condition given by $x(0) = \eta$, and the estimation of the step-size $h = t_{n+1} - t_n$. In some problems, h can vary and be adapted to reduce numerical errors. An initial value problem must accomplish the continuity of the non-linear functions in the interval $a \leq x \leq b$, and it must guarantee the principle of existence and uniqueness, as detailed in [42]. The number of ODEs determine the dimension of the mathematical model and the memristive-based chaotic oscillators can be modeled in higher orders performing transformations as described in [43].

The main goal of a numerical method is to provide a good approximation of the behavior of a continuous-time dynamical system through a computational method in a digital machine. The errors with respect to the ideal or correct solution of an initial value problem can be associated to the truncation of the computer arithmetic, the truncation of higher-order terms in the numerical method, and the type of method, e.g. one-step and multi-step. In both cases, the solution is found at discrete points related to the step-size h and therefore the fitting of these points provides a different numerical error when comparing to the theoretical solution in the continuous interval $a \leq x \leq b$, where the bounds a and b are finite [42]. For instance, the general linear multi-step method is given in (3.5), where the coefficients α_j and β_j are constants. Assuming that $\alpha_k = 1$ and not both α_0 and β_0 are zero, then if $\beta_k = 0$ the method is explicit, and if $\beta_k \neq 0$ the method is said to be of implicit type. Otherwise, the method is explicit and in this case, the evaluation of the next iteration depends on past discrete values. We have

$$\sum_{j=0}^{k} \alpha_j y_{n+j} = h \sum_{j=0}^{k} \beta_j f_{n+j} \tag{3.5}$$

In other words, when the explicit numerical methods compute the next value y_{n+k}, one must use just the past values y_{n+j}, f_{n+j}, with $j = 0, 1, ..., k-1$. The past values have already been computed previously. By contrast, the implicit methods compute y_{n+k} updating values at the same iteration index (using an explicit method) and past values to approach the solution by $y_{n+k} = h\beta_k f(x_{n+k}, y_{n+k}) + g$, where g is a function of values previously computed at y_{n+j}, f_{n+j}. The explicit and implicit methods have different accuracy and convergence regions that basically depend on the step-size.

3.3.1 One-step methods

The simplest explicit and one-step method that can be derived from (3.5), is when $k = 1$, and this numerical method is well-known as the Forward-Euler or the single-step method, and its iterative formulae is given in (3.6). Its approximation is based in the evaluation of the slope that estimated in the function being extrapolated from an actual value (at iteration n) to a next value (at iteration $n + 1$). This explicit method is derived from Taylor series in which the higher-order terms, from the second power, are truncated. For this reason, this is the method with the lowest exactness due to the fact that the error when comparing the approximated solution with the analytical one can be very high if the step-size h is not chosen appropriately, e.g. sufficiently small.

The simplest implicit and one-step method is known as Backward-Euler, it requires and additional computation to approximate the solution of the problem at iteration $n+1$, and its iterative formulae is given in (3.7). As one sees, in this kind of implicit methods, the non-linear function denoted by $f(x_{n+1}, y_{n+1})$, must be approximated applying an explicit method as the Forward-Euler given in (3.6), and afterwards y_{n+1} in (3.7) can be obtained:

$$y_{n+1} = y_n + h \cdot f(x_n, y_n) \tag{3.6}$$

$$y_{n+1} = y_n + h \cdot f(x_{n+1}, y_{n+1}) \tag{3.7}$$

Among the different types of implicit and explicit numerical methods, the solution of the majority of problems in engineering, which are formulated as initial value problems, can be obtained applying Runge–Kutta algorithms, which are derived from the Taylor series expansion. The most used go from order one to order four, and they require different number of evaluations at the same time-step. In another side one can find the methods proposed by Runge and developed by Kutta and Heun, which are called long pass methods but are of one-step type [44,45]. The fourth-order Runge–Kutta method is the one most used due to its high exactness in approximating the solution of ODEs. It has the disadvantages of losing linearity, error analysis and the extra evaluations of the function to approximate mid-points. For these reasons it is considered more difficult than a lineal multi-step method. There exist implicit Runge–Kutta methods whose main advantage is the improvement of the stability characteristics. Table 3.1 shows that the most used Runge–Kutta methods having orders from one to four.

TABLE 3.1 Runge–Kutta algorithms.

Order	Algorithm
1st	$y_{j+1} = y_j + hf(y_j, t_j)$
2nd	$y_{j+1} = y_j + hf(y_j + \frac{h}{2} f(y_j, t_j), t_k + \frac{h}{2})$
3rd	$y_{j+1} = y_j + \frac{h}{4}(k_1 + 3k_3)$ $k_1 = f(y_j, t_j)$ $k_2 = f(y_j + \frac{hk_1}{3}, t_j + \frac{h}{3})$ $k_3 = f(y_j + \frac{2hk_2}{3}, t_j + \frac{2h}{3})$
4th	$y_{j+1} = y_j + \frac{h}{6}(k_1 + 2k_2 + 2k_3 + k_4)$ $k_1 = f(y_j, t_j)$ $k_2 = f(y_j + \frac{hk_1}{2}, t_j + \frac{h}{2})$ $k_3 = f(y_j + \frac{hk_2}{2}, t_j + \frac{h}{2})$ $k_4 = f(y_j + hk_3, t_j + h)$

3.3.2 Multi-step methods

This kind of numerical methods re-use past information and then more than one discrete points are averaged to estimate the next step. Compared to the fourth-order Runge–Kutta method, which requires the evaluation of four nested functions at each step to estimate the value of y_{k+1}, the general linear multi-step method evaluates different number of functions depending on the order of the method, but the computed functions are re-used to estimate the next iteration. In this manner, from (3.5) one can derive explicit and implicit multi-step methods. In the side of explicit-type methods one can find the ones called Nth-order Adams–Bashforth, which are derived by setting $\alpha_k = 1$ and $\beta_k = 0$ in (3.5). The resulting iterative equations are given in Table 3.2, listing from order one to order six. It is worth mentioning that the first-order Adams–Bashforth method is equivalent to the Forward-Euler one.

TABLE 3.2 Adams Bashforth algorithms.

Order	Algorithm
1st	$y_{j+1} = y_j + hf(y_j, t_j)$
2nd	$y_{j+1} = y_j + \frac{h}{2}\{3f(y_j, t_j) - f(y_{j-1}, t_{j-1})\}$
3rd	$y_{j+1} = y_j + \frac{h}{12}\{23f(y_j, t_j) - 16f(y_{j-1}, t_{j-1}) + 5f(y_{j-2}, t_{j-2})\}$
4th	$y_{j+1} = y_j + \frac{h}{24}\{55f(y_j, t_j) - 59f(y_{j-1}, t_{j-1}) + 37f(y_{j-2}, t_{j-2})$ $-9f(y_{j-3}, t_{j-3})\}$
5th	$y_{j+1} = y_j + \frac{h}{720}\{1901f(y_j, t_j) - 2774f(y_{j-1}, t_{j-1}) + 2616f(y_{j-2}, t_{j-2})$ $-1274f(y_{j-3}, t_{j-3}) + 251f(y_{j-4}, t_{j-4})\}$
6th	$y_{j+1} = y_j + \frac{h}{1440}\{4277f(y_j, t_j) - 7923f(y_{j-1}, t_{j-1}) + 9982f(y_{j-2}, t_{j-2})$ $-7298f(y_{j-3}, t_{j-3}) + 2877f(y_{j-4}, t_{j-4}) - 475f(y_{j-5}, t_{j-5})\}$

On the other hand, in the side of implicit-type methods one can find the ones called Nth-order Adams–Moulton, which are generated from (3.5) by setting $\alpha_k = 1$ and $\beta_k \neq 0$. Their solution requires the estimation of $f(y_{j+1}, t_{j+1})$ that can be done applying an explicit method. Depending on the order of the implicit method, one can evaluate the use of the explicit ones. For example: If one uses the sixth-order Adams–Moulton method given in Table 3.3, the prediction of $f(y_{j+1}, t_{j+1})$ can be done using any of the Adams–Bashforth methods listed in Table 3.2, but for sure the less expensive computationally speaking is using the first-order or Forward-Euler method. The selection of the order of the implicit and explicit methods depends on the problem at hand, and also one can find methods that allow adaptation of the order and the step-size.

3.3.3 Stability of the numerical methods

The one-step and multi-step numerical integration methods described above performs the approximation of a solution through discrete points, so that the

TABLE 3.3 Adams Moulton algorithms.

Order	Algorithm
1st	$y_{j+1} = y_j + hf(y_{j+1}, t_{j+1})$
2nd	$y_{j+1} = y_j + \frac{h}{2}\{f(y_{j+1}, t_{j+1}) + f(y_j, t_j)\}$
3rd	$y_{j+1} = y_j + \frac{h}{12}\{5f(y_{j+1}, t_{j+1}) + 8f(y_j, t_j)) - f(y_{j-1}, t_{j-1})\}$
4th	$y_{j+1} = y_j + \frac{h}{24}\{9f(y_{j+1}, t_{j+1}) + 19f(y_j, t_j)) - 5f(y_{j-1}, t_{j-1})$ $+f(y_{j-2}, t_{j-2})\}$
5th	$y_{j+1} = y_j + \frac{h}{720}\{251f(y_{j+1}, t_{j+1}) + 646f(y_j, t_j) - 264f(y_{j-1}, t_{j-1})$ $+106f(y_{j-2}, t_{j-2}) - 19f(y_{j-3}, t_{j-3})\}$
6th	$y_{j+1} = y_j + \frac{h}{1440}\{475f(y_{j+1}, t_{j+1}) + 1427f(y_j, t_j) - 798f(y_{j-1}, t_{j-1})$ $+482f(y_{j-2}, t_{j-2}) - 173f(y_{j-3}, t_{j-3}) + 27f(y_{j-4}, t_{j-4})\}$

question is: What precision can be reached? Or in other words, one must answer: How many errors does the method generate? In general, two types of error can be evaluated: the local error that is introduced by a single step during the execution of the integration process and the global error, which is the overall error caused by repeated application of the integration formulae [43]. Both the local and the global errors are divided into rounded-off and truncation errors. The round-off error is inevitable and results from performing real arithmetic on a digital computer. The truncation error is the local error that would result if the numerical algorithm is implemented on an infinite-precision computer. In general any numerical method can avoid these kind of errors, but some methods possess more errors than the others. However, even if the error is low, this will not serve at all if the method is numerically unstable. In this case, when the method is unstable the sum of all the small errors can become unlimited and the trajectory of the discrete points can have an inevitable divergency.

The computation of the stability region of a numerical method can be obtained as follows: One should apply the numerical method to a first-order lineal equation of the type $y' = f(x, y)$, with $y(0) = y_0$, and where $f(x, y) = \lambda y$, and for which λ is an eigenvalue that can be a complex number. For example, when applying the Forward-Euler method given in (3.6) to the linear equation, then one defines $f(x) = \lambda x$, and the discretization gives the iterative formula $y_{n+1} = y_n + h\lambda y_n = (1 + h\lambda)y_n$. In this case, the Forward-Euler method is stable if it accomplishes the inequality: $|1+h\lambda| < 1$, which is called stability region and it can be illustrated as shown in Fig. 3.2 (a).

The evaluation of the stability region of the Backward-Euler method given in (3.7), is performed in the same manner as for the Forward-Euler method. Therefore, using the same linear equation, one gets the iterative equation: $y_{n+1} = y_n + h\lambda y_{n+1} = (\frac{1}{1-h\lambda})y_n$, so that the stability region is defined by $|1 - h\lambda| > 1$, and shown in Fig. 3.2 (b).

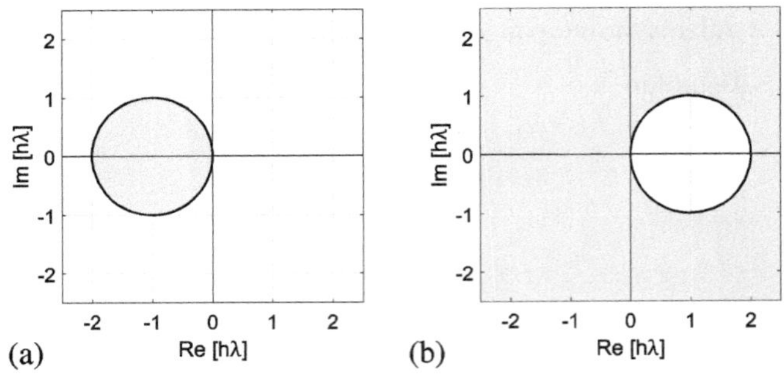

FIGURE 3.2 Stability region of the: (a) Forward-Euler and (b) Backward-Euler methods.

As one sees, the stability region of the numerical methods is plotted in the complex plane because the eigenvalue λ can be real or complex. From Fig. 3.2 one can appreciate that the stability region relates inversely the eigenvalue with the step-size h value. In this case, if the eigenvalues of a function augment, this means that the step-size must decrease and vice versa. This means that before the application of a numerical method, one must analyze the eigenvalues of the dynamical system to estimate an appropriate step-size that accomplishes the stability of the numerical method. However, this stability criterion can be taken as guarantee when solving linear problems. But, fortunately, this stability criterion is also accomplished by the majority of non-linear problems. Fig. 3.3 shows the stability regions of the Runge–Kutta method from its first to its fourth order.

In Fig. 3.3, it can be appreciated that, as the order augments, the stability region also augments, so that this property makes the fourth-order Runge–Kutta method the one most used.

The stability regions of the Adams–Bashforth methods are shown in Fig. 3.4, and the ones for the Adams Moulton methods are given in Fig. 3.5. The stability regions are computed in a similar way to the Euler methods. In fact, as one can infer, the equivalents methods of the one-step type for these families are the Forward- and Backward-Euler. For this reason, Figs. 3.4 and 3.5 show the stability regions of the multi-step methods from the second to the fifth order.

The solution of initial value problems applying numerical methods gives rise to a trade-off between the expected result and the payment to get the most exact one. For this reason this section shows the most classical numerical methods, which have the lowest consumption of computer resources to provide the best approximation of the solution. The analysis of the stability regions of the one-step and multi-step methods helps to choose an adequate step-size h, to guarantee stability of the method. In [46], one can find the comparison of different numerical methods (Forward-Euler, fourth-order Runge–Kutta, third-order Adams–Bashforth and second-order Adams–Moulton), to approximate the initial value problem: $\frac{dy}{dt} = -y^2$, with $y(0) = 1$, and of which the analytical solution is $y(t) = 1/t$, and its evaluation at $t = 4$ shows $y(4) = 0.2$. In that reference

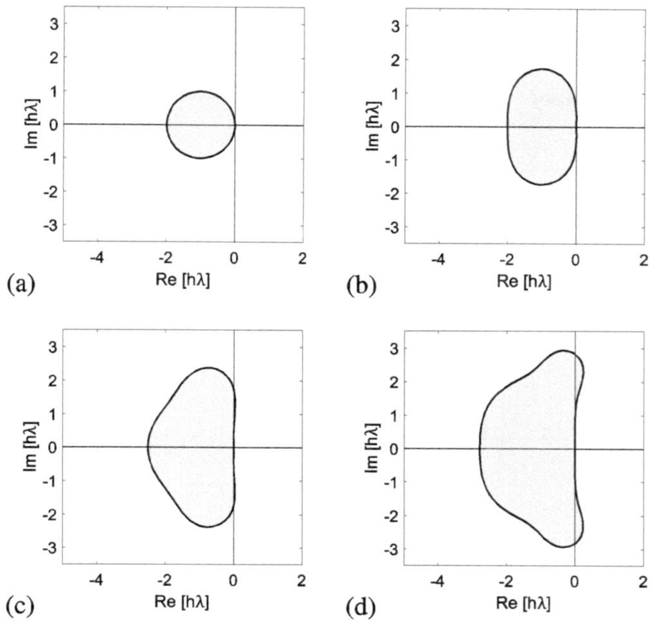

FIGURE 3.3 Stability regions of Runge–Kutta methods: (a) 1st, (b) 2nd, (c) 3rd, and (d) 4th-order.

one can appreciate that the fourth-order Runge–Kutta method generates the lowest numerical error, generating errors around 10^{-6} from $h = 2^{-1}$. This is a very high time-step value compared to the one using Forward-Euler, which reaches that error value only with $h = 2^{-14}$. This means that it will take longer time to go from the discrete time 0 to 4 in steps of 2^{-14}. This is the main reason why researchers prefer using the fourth-order Runge–Kutta method, which allows the use of higher step-sizes without abandoning the stability region, and which provides better exactness of the results.

The solution of non-linear dynamical systems as the memristor-based ones requires a previous analysis of the mathematical equations to choose an adequate step-size that is related to the eigenvalues. One must also take into account practical issues when programming the numerical methods and solving chaotic systems, which are high sensitive to the initial conditions, as mentioned in Section 3.2. In this case, one must perform a rigorous analysis, good programming and selection of the step-size to avoid practical problems as computational chaos and superstability. The computational chaos is a kind of false chaotic behavior that arises when choosing a bad step-size so that chaotic behavior arises even if the problem is not of chaotic nature. The superstability arises when the bad selection of the step-size makes that a chaotic system loses its behavior and converges toward one point [47,48].

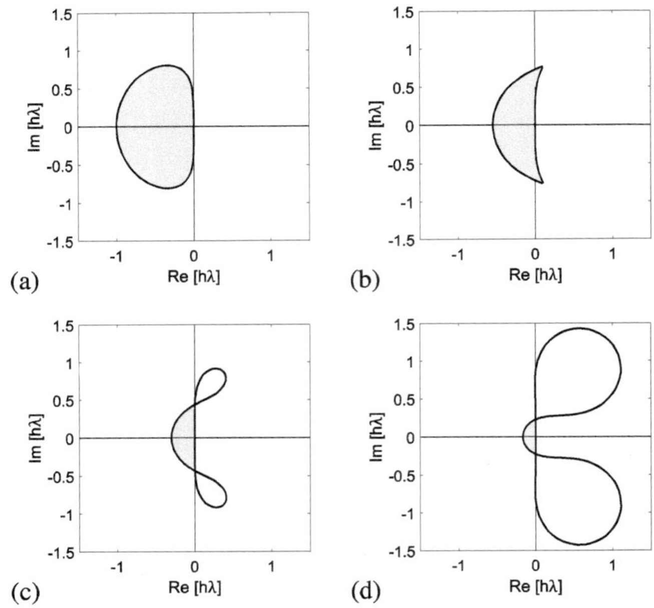

FIGURE 3.4 Stability regions of Adams–Basforth methods: (a) second, (b) third, (c) fourth, and (d) fifth order.

3.4 Analysis of memristive-based chaotic oscillators

The first analysis that is performed for a dynamical system as the memristive-based chaotic oscillators described in Section 3.2 is the computation of the equilibrium points, which are required to find the eigenvalues and then estimate the step-size. Let us consider the 3D memristive system given in (3.1), which is rewritten in (3.8). The equilibrium points are computed equating to zero all the derivatives, as follows:

$$\begin{aligned} \dot{x} &= 0.2x + y = 0 \\ \dot{y} &= -\tfrac{1}{3}x - \tfrac{1}{3}yz^2 = 0 \\ \dot{z} &= -0.4y - 0.4z + yz = 0 \end{aligned} \tag{3.8}$$

The solution of this system, finding the static values of the state variables x, y, and z, leads us to find three equilibrium points. The first equilibrium point is located at the origin: $x = 0$, $y = 0$, and $z = 0$. The second is located at $x = -2.4357$, $y = 0.4871$, and $z = 2.2360$; and the third equilibrium point is given by $x = -1.6965$, $y = 0.3393$, and $z = -2.2360$. These three equilibrium points are used to evaluate three Jacobians to find the corresponding eigenvalues that help to evaluate the stability of the equilibrium points and estimate the step-size to perform the numerical solution of the initial value problem. The Jacobian matrix is given in (3.9), from which one must replace the state variables x, y,

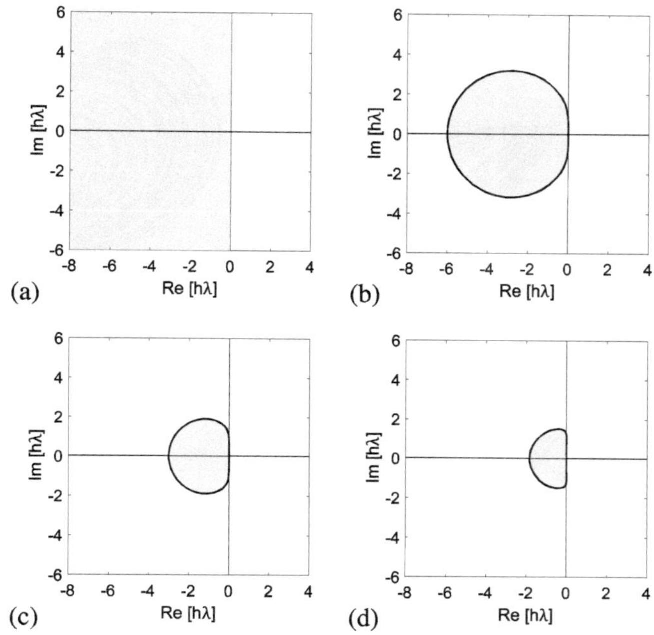

FIGURE 3.5 Stability regions of Adams–Moulton Methods: (a) second, (b) third, (c) fourth, and (d) fifth order.

and z with the sets of equilibrium points to derive the characteristic equation evaluation by the relationship $|\lambda I - J| = 0$. Using the equilibrium point at the origin one gets (3.10), which can be solved to compute the three eigenvalues associated to one equilibrium point. One can do the same using the other two equilibrium points to find the nine eigenvalues:

$$J = \begin{bmatrix} 0.2 & 1 & 0 \\ -\frac{1}{3} & -\frac{1}{3}z^2 & -\frac{2}{3}yz \\ 0 & -0.4+z & -0.4+y \end{bmatrix} \tag{3.9}$$

$$\lambda^3 + 0.2\lambda^2 + 0.2533\lambda + 0.1333 = 0 \tag{3.10}$$

The evaluation of the eigenvalues using the three equilibrium points leads us to the first set of eigenvalues given by $\lambda_{1,2} = 0.1000 \pm 0.5686j$ and $\lambda_3 = -0.4$. The second set of eigenvalues becomes $\lambda_{1,2} = -0.7795 \pm 0.9368j$ and $\lambda_3 = 0.1795$. The third set of eigenvalues is given as $\lambda_{1,2} = -0.8426 \pm 0.9891j$ and $\lambda_3 = 0.1579$. In order that the dynamical system given in (3.8) can be determined as a chaotic system, each set of eigenvalues must be in equilibrium, it means that, if the eigenvalue is purely real and positive, the real part of the complex eigenvalues must be negative, and vice versa. This is one of the criteria

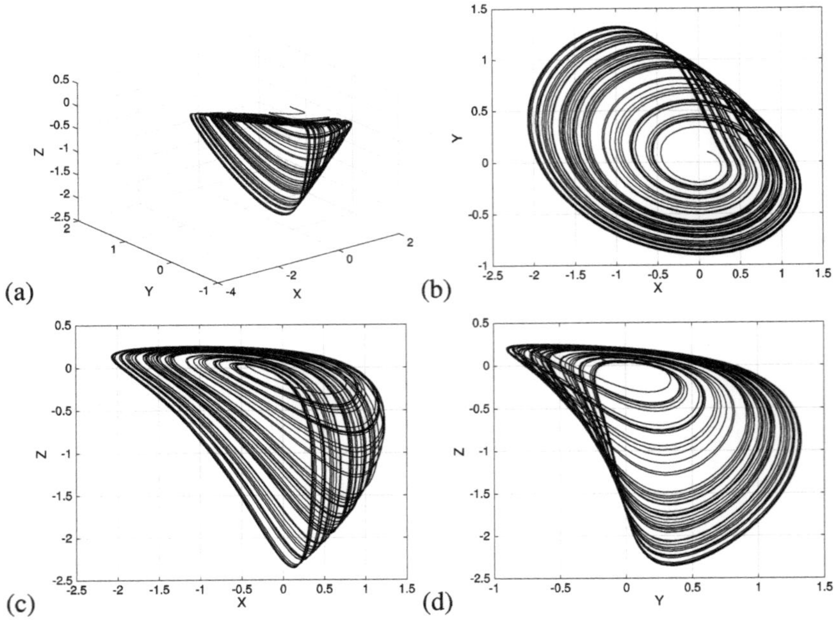

FIGURE 3.6 Simulation of the memristive-based chaotic oscillator given in (3.1) applying Forward-Euler with initial conditions $x_0 = y_0 = z_0 = 0.1$, and $h = 0.01$. Phase-space portrait of the state variables: (a) $x - y - z$, (b) $x - y$, (c) $x - z$, and (d) $y - z$.

to evaluate the chaotic behavior of a dynamical system [43]. Afterwards, one must analyze the stability of the numerical method. For example, choosing the Forward-Euler method, the stability is defined by ($|1 - h\lambda| > 1$), and substituting the eigenvalues of the system, one finds that the step-size that accomplishes those eigenvalues is $0 < h < 0.9982$. As one can infer, if the step-size accomplishes the inequality, it is an indication that the numerical method is stable and can be applied to solve the initial value problem. Fig. 3.6 shows the simulation of (3.8) applying Forward-Euler with $h = 0.01$. In a similar way, one can perform the analysis of the other memristive-based chaotic oscillators given in Section 3.2, and their respective simulations are shown in Figs. 3.7, 3.8, and 3.9, for the dynamical systems given in (3.2), (3.3), and (3.4), respectively.

The chaotic behavior is guaranteed if the memristive system has one positive Lyapunov exponent. This is done evaluating the Lyapunov exponents spectrum of (3.1) applying Wolf's method [36]. The simulation is performed until a final time $t = 1000$ with steps of $h = 0.01$. The results show that the associated Lyapunov exponents are equal to $LE_1 = 0.0250$, $LE_2 = 0.0040$, and $LE_3 = -0.3968$, and since the maximum Lyapunov exponent is positive, then the dynamical system is chaotic [49]. The whole Lyapunov spectrum is shown in Fig. 3.10. These values can be used to evaluate the Kaplan–Yorke dimension associated to a chaotic system and the equation is given by (3.11). In

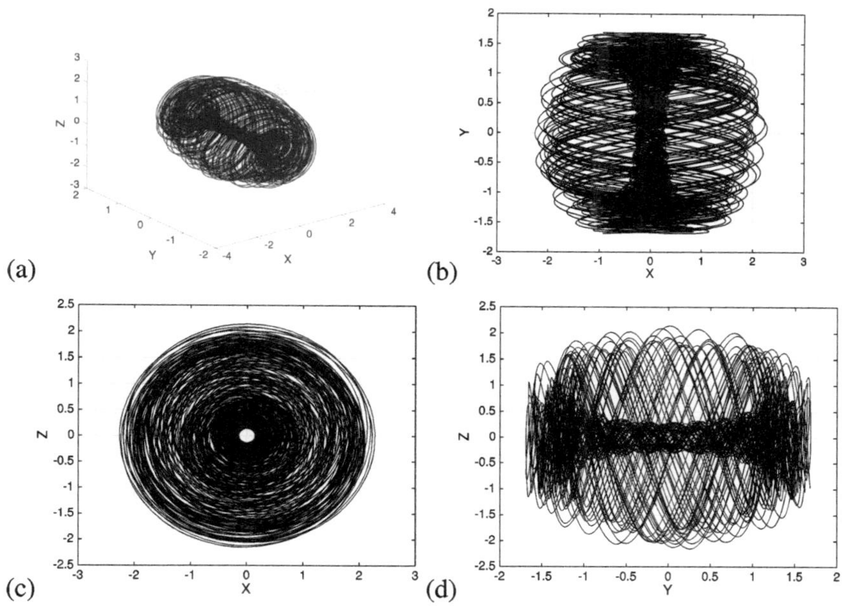

FIGURE 3.7 Simulation of the memristive-based chaotic oscillator given in (3.2) applying Forward-Euler with initial conditions $x_0 = y_0 = z_0 = w_0 = 0.1$, and $h = 0.01$. Phase-space portrait of the state variables: (a) $x - y - z$, (b) $x - y$, (c) $x - z$, and (d) $y - z$.

this case, the Lyapunov exponents are ordered from the maximum to the minimum, i.e. $\lambda_1, \lambda_1, ..., \lambda_n$, and j is the index to accomplish that $\sum_{i=1}^{j} \lambda_i \geq 0$ and $\sum_{i=1}^{j+1} \lambda_i < 0$. We have

$$D_{KY} = j + \frac{\sum_{i=1}^{j} \lambda_i}{|\lambda_{j+1}|} = 2 + \frac{0.0250 + 0.0040}{0.3968} = 2.07 \quad (3.11)$$

3.5 FPGA implementation of memristive systems

The digital implementation of the memristive-based chaotic oscillators given in Section 3.2 using FPGAs has advantages, such as fast verification of the chaotic behavior, fast prototyping, exploitation of the processing speed, high computational power and programming flexibility [50]. As mentioned in Section 3.1, the FPGA implementation of dynamical systems requires choosing the appropriate numerical method to discretize the equations and then describe the building blocks for synthesis purposes. The FPGA-based implementation of memristive systems is then suitable to develop applications in data encryption and secure communications, as shown in [51]. In this manner, applying the Forward-Euler to discretize (3.1), the iterative equations are given in (3.12). Applying another

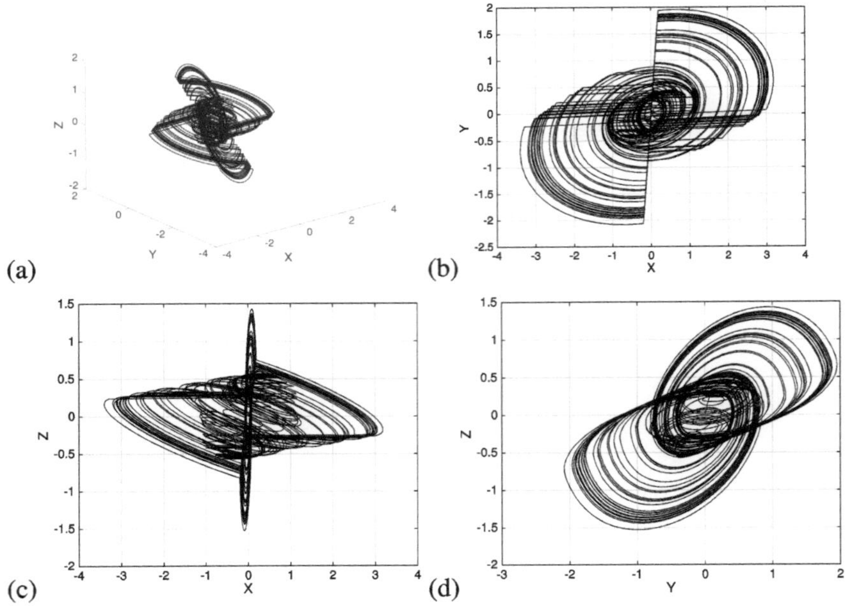

FIGURE 3.8 Simulation of the memristive-based chaotic oscillator given in (3.3) applying Forward-Euler with initial conditions $x_0 = y_0 = z_0 = w_0 = 0.1$, and $h = 0.01$. Phase-space portrait of the state variables: (a) $x - y - z$, (b) $x - y$, (c) $x - z$, and (d) $y - z$.

method like the fourth-order Runge–Kutta leads to a more complex iterative equation because to solve the $n + 1$ iteration one uses four function evaluations and this causes the time and resources to increase. For instance, the block diagram shown in Fig. 3.11 shows the hardware synthesis of (3.12), where one can see adders, multipliers, and single-constant multipliers (SCM). These SCM blocks reduce hardware resources as shown in [46], because the multipliers are implemented by adders and shift-registers:

$$\begin{aligned} x_{n+1} &= x_n + h[ax + y] \\ y_{n+1} &= y_n + h[-bx - byz^2] \\ z_{n+1} &= z_n + h[-cy - cz + yz] \end{aligned} \tag{3.12}$$

To increase the processing capabilities, the blocks in Fig. 3.11 must be described using machine-level software. In this chapter, the VHDL programming language is used as already shown in [46]. The step-size $h = 0.01$ is used and a numeric representation as fixed-point of 3.29 to process words of 32 bits. In this format one has one sign bit, 2 bits to represent the integer part, and 29 bits to represent the fractional part. The VHDL codes are similar to the ones already given in [46], and the time series of the state variables are shown in Fig. 3.12. The phase-space portraits are shown in Figs. 3.13 and 3.14, applying

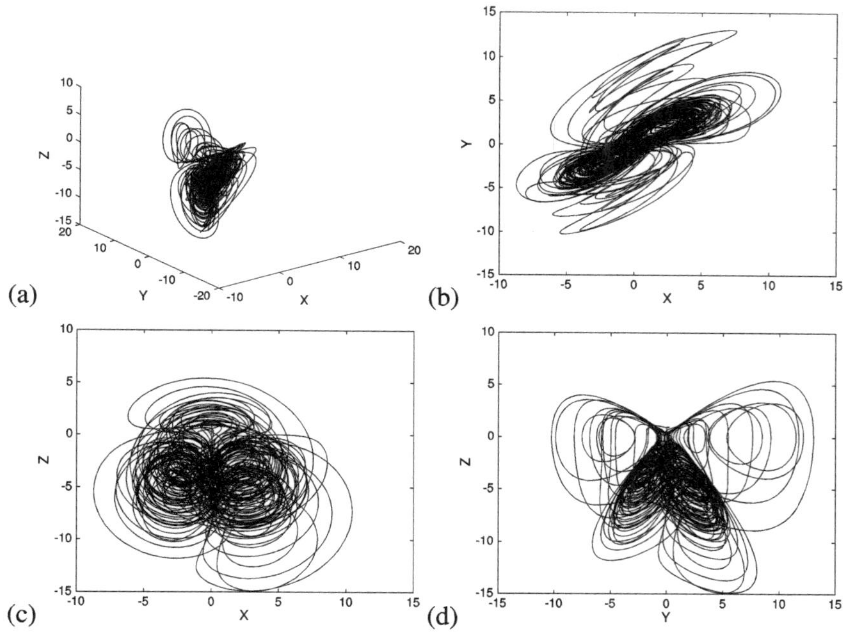

FIGURE 3.9 Simulation of the memristive-based chaotic oscillator given in (3.1) applying Forward-Euler with initial conditions $x_0 = y_0 = z_0 = w_0 = \phi_0 = 0.1$, and $h = 0.01$. Phase-space portrait of the state variables: (a) $x - y - z$, (b) $x - y$, (c) $x - z$, and (d) $y - z$.

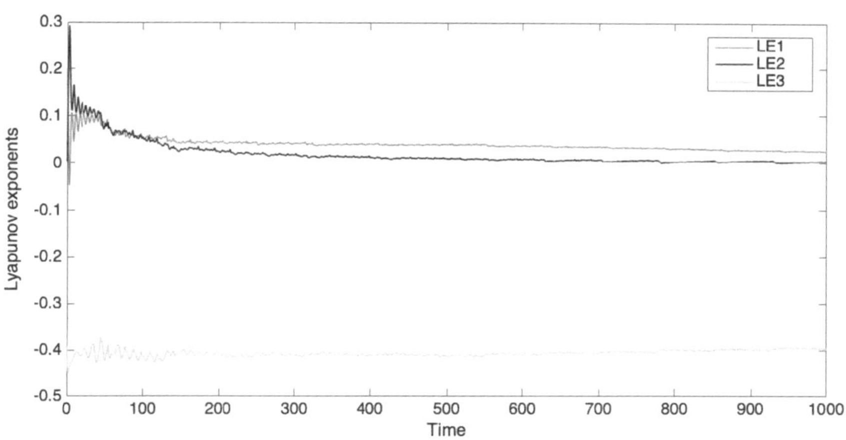

FIGURE 3.10 Lyapunov exponents spectrum of the memristive system given in (3.1).

the Forward-Euler and fourth-order Runge–Kutta methods, respectively. These are experimental results using the FPGA Cyclone IV EP4CGX150DF31C7 and observed in a Teledyne oscilloscope. Table 3.4 shows the hardware resources related to the implementation using the Forward-Euler and fourth-order Runge–

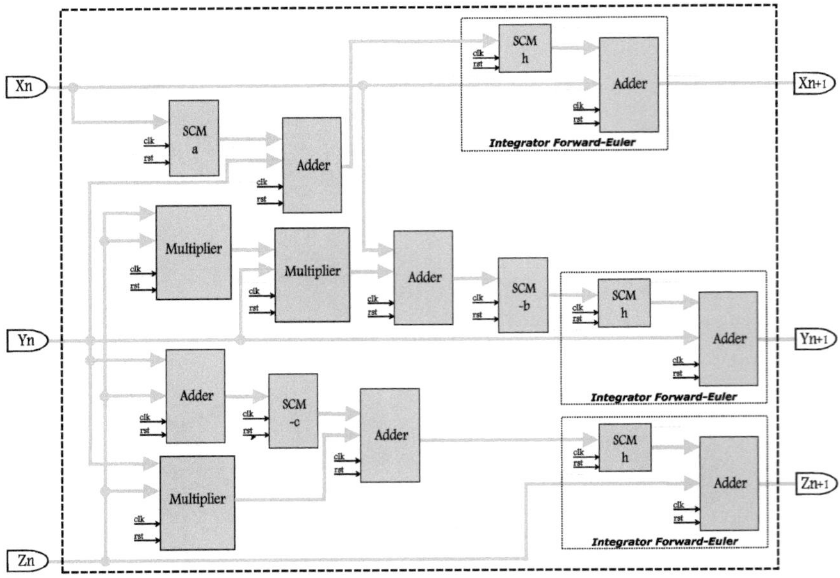

FIGURE 3.11 Block diagram of discretizing (3.12) using Forward-Euler.

Kutta methods. The synthesizer is Altera Quartus II 13.0. As one sees, the fourth-order Runge–Kutta method uses four times more resources (logic elements, registers and 9 × 9 bit multipliers) than Forward-Euler, but recall that it is more exact. The maximum frequency is multiplied by the number of clock cycles that are required to process the data from the input to the output, so that the processing speed or latency is listed in the last column in nanoseconds (ns). While the fourth-order Runge–Kutta use 35 clock cycles equivalent to 700 ns, the Forward-Euler use 8 clock cycles equivalent to 160 ns; in other words, the first one is slower than the second one for generating one iteration but the approximation to the analytic solution is better.

TABLE 3.4 FPGA resources of the memristive-based chaotic oscillator given in (3.1), by applying two numerical methods and using the FPGA Cyclone IV GX EP4CGX150DF31C7.

Numerical method	Logic elements	Registers	9*9 bits multiplier	Maximum frequency (MHz)	Cycles	Iteration latency with 50 MHz (ns)
Forward-Euler	2065	769	24	73.6	8	160
Runge Kutta 4th-order	8466	3433	96	69.2	35	700

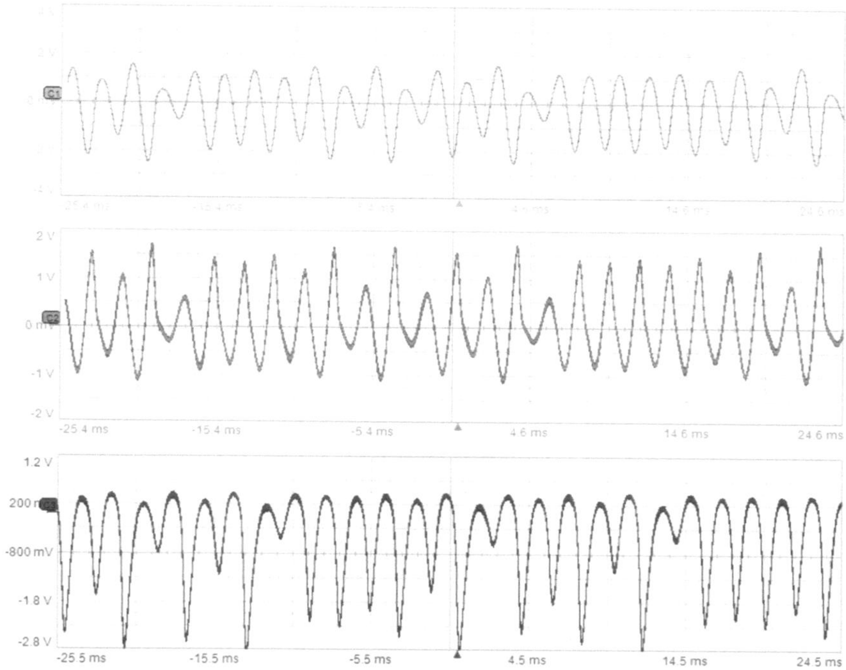

FIGURE 3.12 Time series of the state variables x, y, and z from the experimental data provided by the FPGA implementation of (3.1).

3.6 Memristive-based secure communication system

Two chaotic oscillators can be synchronized in a master–slave topology to implement a secure communication system, as described in the seminal paper [52]. In this section, the synchronization of two memristive-based chaotic oscillators is performed applying Hamiltonian forms and observer approach [53]. The case of this study is the memristive system given in (3.1), so that the master block is denoted by the state variables x_m, y_m and z_m, and the Hamiltonian form with a destabilizing vector is given by (3.13), where H is proposed to be $H(x) = 1/2[x^2 + y^2 + z^2]$, and $J(x)$ and $S(x)$ are matrices representing the conservative and nonconservative parts of the system, respectively, and must satisfy $J(x) + J^T(x) = 0$ and $S(x) = S^T(x)$. It leads to (3.14)

$$\dot{x} = J(x)\frac{\partial H}{\partial x} + S(x)\frac{\partial H}{\partial x} + F(x), \quad x \in R^n \tag{3.13}$$

$$\begin{bmatrix} x_m \\ y_m \\ z_m \end{bmatrix} = \begin{bmatrix} 0 & 2/3 & 0 \\ -2/3 & 0 & 1/5 \\ 0 & -1/5 & 0 \end{bmatrix} \frac{\partial H}{\partial x} + \begin{bmatrix} 1/5 & 1/3 & 0 \\ 1/3 & 0 & -1/5 \\ 0 & -1/5 & -2/5 \end{bmatrix} \frac{\partial H}{\partial x}$$

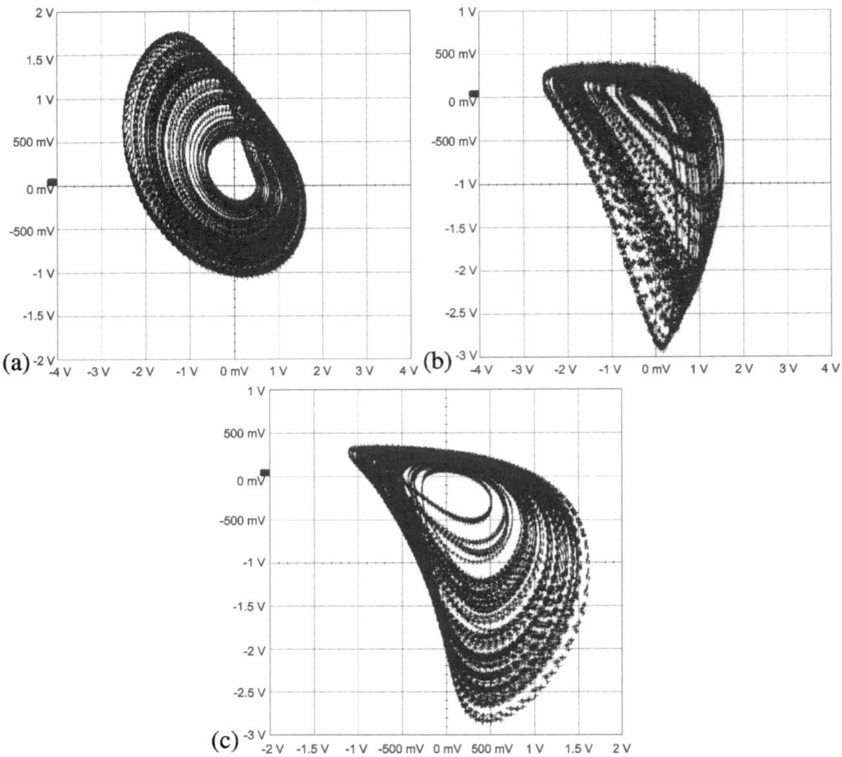

FIGURE 3.13 Experimental chaotic attractors of (3.1), implemented on an FPGA using Forward-Euler with $h = 0.01$. Portrait (a) $x - y$ with 0.5V/Div-1V/Div, (b) $x - z$ with 0.5V/Div-1V/Div, and (c) $y - z$ with 0.5V/Div-0.5V/Div.

$$+\begin{bmatrix} 0 \\ -1/3 * y_m * z_m^2 \\ y_m * z_m \end{bmatrix} \tag{3.14}$$

If one considers the system with a linear output, one gets (3.15), where y is a vector denoting the output of the system. In addition, if ξ is the estimated state vector of x and η the estimated output in terms of ξ, then an observer to (3.13) can be given by (3.16), where K is a vector of constant gains. We have

$$\begin{aligned} \dot{x} &= J(y)\frac{\partial H}{\partial x} + S(y)\frac{\partial H}{\partial x} + F(y), \quad x \in R^n \\ y &= C\frac{\partial H}{\partial x}, \quad y \in R^m \end{aligned} \tag{3.15}$$

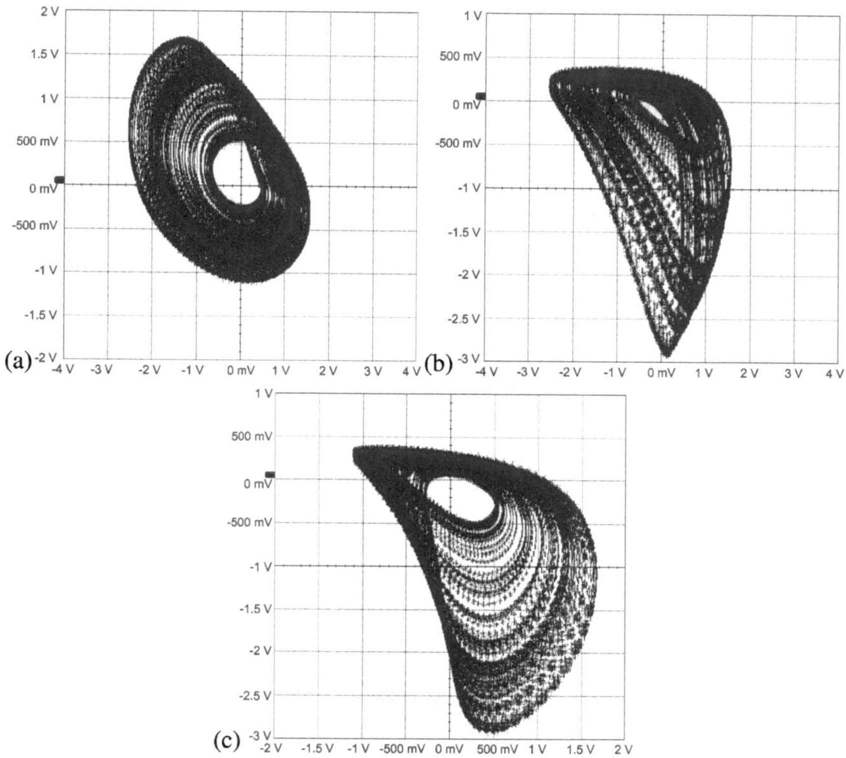

FIGURE 3.14 Experimental chaotic attractors of (3.1), implemented on an FPGA using the fourth-order Runge–Kutta method with $h = 0.01$. Portrait (a) $x - y$ with 0.5V/Div-1V/Div, (b) $x - z$ with 0.5V/Div-1V/Div, and (c) $y - z$ with 0.5V/Div-0.5V/Div.

$$\begin{aligned} \dot{\xi} &= J(y)\frac{\partial H}{\partial \xi} + S(y)\frac{\partial H}{\partial \xi} + F(y) + K(y - \eta) \\ \eta &= C\frac{\partial H}{\partial \xi} \end{aligned} \tag{3.16}$$

The synchronization by Hamiltonian forms is achieved after accomplishing the theorems given in the seminal paper [53]. Transforming the slave system to its original form, one gets (3.17), where $a = 0.2$, $b = \frac{1}{3}$, and $c = 0.4$. We have

$$\begin{aligned} \dot{x_s} &= ax_s + y_s + 5[x_m - x_s] \\ \dot{y_s} &= -bx_s - by_s z_s^2 + 5[y_m - y_s] \\ \dot{z_s} &= -cy_s - c_s z_s + y_s z_s + 5[z_m - z_s] \end{aligned} \tag{3.17}$$

The synchronization of systems in master–slave topology occurs when the trajectories of the state variables of the master and slave systems meet in the same time with a minimum synchronization error. The simulation of the master–

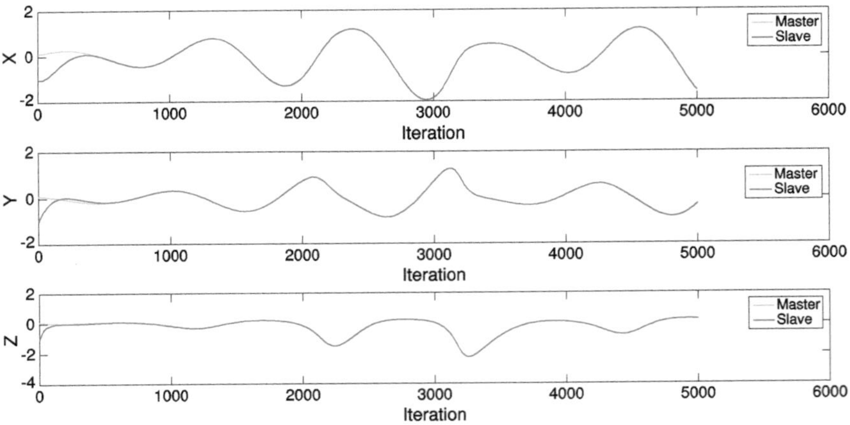

FIGURE 3.15 Time series of the three state variables of the master and slave memristive-based chaotic oscillators showing minimum synchronization error at iteration 400.

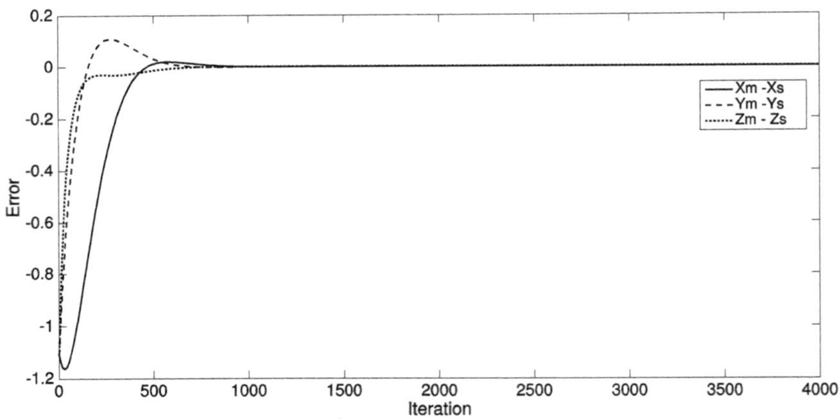

FIGURE 3.16 Synchronization error between the state variables of the master and slave memristive-based chaotic oscillators.

slave system applying Forward-Euler with $h = 0.01$ is shown in Fig. 3.15, where it can be appreciated that the time series of both the master and the slave systems are in good agreement, and the error is minimum after iteration 400. Considering that a good synchronization occurs while the minimum error exists in the shortest possible time, the synchronization in 400 iterations with a clock of 50 MHz require in the order of microseconds to have master and slave synchronized. The experimental synchronization error is shown in Fig. 3.16.

The complete FPGA implementation of the memristive-based secure communication system is sketched in Fig. 3.17, which is detailed in [46]. The FPGA for transmission is the Cyclone IV EP4CGX150DF31C7, and the image transmission results are shown in Figs. 3.18 and 3.19, where the first column

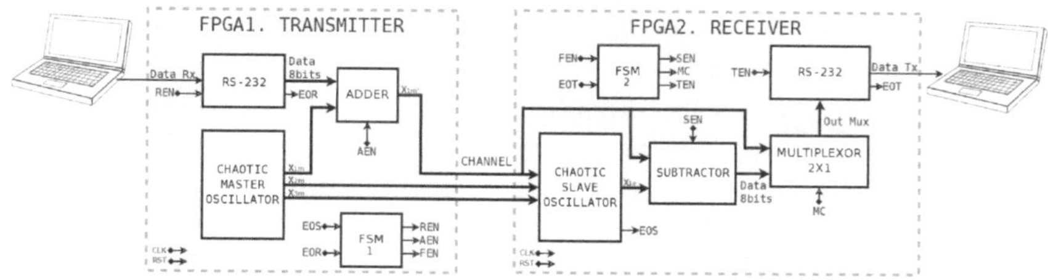

FIGURE 3.17 Hardware description of the memristive-based secure communication system to transmit color images.

FIGURE 3.18 Original Lena image (left), masked image (center), and recovered image (right).

FIGURE 3.19 Original Baboon image (left), masked image (center), and recovered image (right).

represents the original image, the second column represents the chaotic channel (the original image plus master chaotic signal) and the last column represents the recovery image with a slave chaotic signal. A correlation analysis between the original images and the chaotic channel gives 0.0035, and the correlation between the original and recovered images gives 1. This means that the master chaotic signal realizes a good encryption to the images and the original image was completely recovered with a slave system perfectly synchronized.

3.7 Conclusions

This chapter reviewed the history on the introduction of the memristor as the four missing circuit element to complete the relationships among the voltage, current, charge and flux of the RLC elements. The memristor was used to simulate 3D, 4D, and 5D memristive-based chaotic oscillators, for which their dynamical analyses were done and the Lyapunov spectrum was computed for the 3D dynamical system to verify that it has chaotic behavior. The FPGA-based implementation of the discretized ODEs of the 3D dynamical system were showed in a block diagram consisting of adders, multipliers and SCM blocks. The chaotic secure communication system shown in Fig. 3.17 has been implemented by using FPGA-based memristive systems. As a result, one can conclude that the hardware resources, frequency of operation and latency of the FPGA implementation of memristive-based systems depend on the selection of the correct numerical method and its associated step-size.

References

[1] L. Chua, Memristor – the missing circuit element, IEEE Transactions on Circuit Theory 18 (5) (Sep. 1971) 507–519.

[2] E.M. Drakakis, A.J. Payne, A Bernoulli cell-based investigation of the non-linear dynamics in log-domain structures, Analog Integrated Circuits and Signal Processing 22 (2–3) (Mar 2000) 127–146.

[3] G. Chen, Leon Chua's memristor, IEEE Circuits and Systems Magazine 8 (2) (2008) 55.

[4] Michael Banks, Hewlett–Packard brings 'memristor' to market, Physics World 23 (10) (Oct 2010) 11.

[5] Makoto Itoh, Leon O. Chua, Memristor oscillators, International Journal of Bifurcation and Chaos 18 (11) (Nov 2008) 3183–3206.

[6] Neil D. Mathur, The fourth circuit element, Nature 455 (7217) (Oct 30 2008) E13.

[7] Dmitri B. Strukov, Gregory S. Snider, Duncan R. Stewart, R. Stanley Williams, The missing memristor found, Nature 453 (7191) (May 1 2008) 80–83.

[8] Bharathwaj Muthuswamy, Pracheta P. Kokate, Memristor-based chaotic circuits, IETE Technical Review 26 (6) (Nov-Dec 2009) 417–429.

[9] Massimiliano Di Ventra, Yuriy V. Pershin, Leon O. Chua, Circuit elements with memory: memristors, memcapacitors, and meminductors, Proceedings of the IEEE 97 (10) (Oct 2009) 1717–1724.

[10] Zdenek Biolek, Dalibor Biolek, Viera Biolkova, SPICE model of memristor with nonlinear dopant drift, Radioengineering 18 (2, 2) (Jun 2009) 210–214.

[11] Yanwen Guo, Xiaoping Wang, Zhigang Zeng, A compact memristor-CMOS hybrid look-up-table design and potential application in FPGA, IEEE Transactions on Computer-Aided Design of Integrated Circuits and Systems 36 (12) (Dec 2017) 2144–2148.

[12] Xu Ya-Ming, Wang Li-Dan, Duan Shu-Kai, A memristor-based chaotic system and its field programmable gate array implementation, Acta Physica Sinica 65 (12) (Jun 20 2016).

[13] Qiusen Jin, Fuhong Min, Chunbiao Li, Infinitely many coexisting attractors of a dual memristive Shinriki oscillator and its FPGA digital implementation, Chinese Journal of Physics 62 (Dec 2019) 342–357.

[14] Karthikeyan Rajagopal, Sifeu Takougang Kingni, Abdul Jalil M. Khalaf, Yasser Shekofteh, Fahimeh Nazarimehra, Coexistence of attractors in a simple chaotic oscillator with fractional-order-memristor component: analysis, FPGA implementation, chaos control and synchronization, The European Physical Journal Special Topics 228 (10, SI) (Oct 2019) 2035–2051.

[15] Baris Karakaya, Arif Gulten, Mattia Frasca, A true random bit generator based on a memristive chaotic circuit: analysis, design and FPGA implementation, Chaos, Solitons and Fractals 119 (Feb 2019) 143–149.

[16] Umer Farooq, M. Hassan Aslam, Muhammad Usman, On the comparison of memristor-transistor hybrid and transistor-only heterogeneous FPGAs, Journal of King Saud University: Computer and Information Sciences 31 (4) (Oct 2019) 514–527.

[17] Mohammed F. Tolba, Mohammed E. Fouda, Haneen G. Hezayyin, Ahmed H. Madian, Ahmed G. Radwan, Memristor FPGA IP core implementation for analog and digital applications, IEEE Transactions on Circuits and Systems II, Express Briefs 66 (8) (Aug 2019) 1381–1385.
[18] Yuriy V. Pershin, Massimiliano Di Ventra, Practical approach to programmable analog circuits with memristors, IEEE Transactions On Circuits and Systems I, Regular Papers 57 (8) (Aug 2010) 1857–1864.
[19] Zhaohui Lin, Hongxia Wang, Efficient image encryption using a chaos-based PWL memristor, IETE Technical Review 27 (4) (Jul-Aug 2010) 318–325.
[20] Ricardo Riaza, Nondegeneracy conditions for active memristive circuits, IEEE Transactions on Circuits and Systems II, Express Briefs 57 (3) (Mar 2010) 223–227.
[21] C. Sanchez-Lopez, J. Mendoza-Lopez, M.A. Carrasco-Aguilar, C. Muniz-Montero, A floating analog memristor emulator circuit, IEEE Transactions on Circuits and Systems II, Express Briefs 61 (5) (May 2014) 309–313.
[22] Xuliang Zhang, Zhangcai Huang, Juebang Yu, Memristor model for SPICE, IEICE Transactions on Electronics E93C (3) (Mar 2010) 355–360.
[23] C. Sanchez-Lopez, L.E. Aguila-Cuapio, A 860 kHz grounded memristor emulator circuit, AEU International Journal of Electronics and Communications 73 (2017) 23–33.
[24] C. Sanchez-Lopez, M.A. Carrasco-Aguilar, C. Muniz-Montero, A 16 Hz–160 kHz memristor emulator circuit, AEU International Journal of Electronics and Communications 69 (9) (2015) 1208–1219.
[25] Dalibor Biolek, Zdenek Biolek, Viera Biolkova, PSPICE modeling of meminductor, Analog Integrated Circuits and Signal Processing 66 (1) (Jan 2011) 129–137.
[26] C. Sanchez-Lopez, A 500 kHz frequency-shift keying modulator based on memductor, Analog Integrated Circuits and Signal Processing 100 (3) (Sep 2019) 527–536.
[27] C. Sanchez-Lopez, V.H. Carbajal-Gomez, M.A. Carrasco-Aguilar, F.E. Morales-Lopez, PID controller design based on memductor, AEU International Journal of Electronics and Communications 101 (2019) 9–14.
[28] Sangho Shin, Kyungmin Kim, Sung-Mo Kang, Data-dependent statistical memory model for passive array of memristive devices, IEEE Transactions on Circuits and Systems II, Express Briefs 57 (12) (Dec 2010) 986–990.
[29] Ivo Petras, Fractional-order memristor-based Chua's circuit, IEEE Transactions on Circuits and Systems II, Express Briefs 57 (12) (Dec 2010) 975–979.
[30] M. Syed Ali, Govindasamy Narayanan, Vineet Shekher, Hamed Alsulami, Tareq Saeed, Dynamic stability analysis of stochastic fractional-order memristor fuzzy BAM neural networks with delay and leakage terms, Applied Mathematics and Computation 369 (Mar 15 2020).
[31] Yuxiao Wang, Yuting Cao, Zhenyuan Guo, Shiping Wen, Passivity and passification of memristive recurrent neural networks with multi-proportional delays and impulse, Applied Mathematics and Computation 369 (Mar 15 2020).
[32] Xiaona Song, Jingtao Man, Shuai Song, Junwei Lu, Integral sliding mode synchronization control for Markovian jump inertial memristive neural networks with reaction–diffusion terms, Neurocomputing 378 (Feb 22 2020) 324–334.
[33] Yanchao Shi, Jinde Cao, Finite-time synchronization of memristive Cohen–Grossberg neural networks with time delays, Neurocomputing 377 (Feb 15 2020) 159–167.
[34] Shiping Wen, Rui Hu, Yin Yang, Tingwen Huang, Zhigang Zeng, Yong-Duan Song, Memristor-based echo state network with online least mean square, IEEE Transactions on Systems, Man and Cybernetics, Systems 49 (9) (Sep 2019) 1787–1796.
[35] Shiping Wen, Shuixin Xiao, Yin Yang, Zheng Yan, Zhigang Zeng, Tingwen Huang, Adjusting learning rate of memristor-based multilayer neural networks via fuzzy method, IEEE Transactions on Computer-Aided Design of Integrated Circuits and Systems 38 (6) (Jun 2019) 1084–1094.
[36] Alan Wolf, Jack B. Swift, Harry L. Swinney, John A. Vastano, Determining Lyapunov exponents from a time series, Physica D: Nonlinear Phenomena 16 (3) (1985) 285–317.

[37] Renping Wu, Chunhua Wang, A new simple chaotic circuit based on memristor, International Journal of Bifurcation and Chaos 26 (09) (2016) 1650145.

[38] Karthikeyan Rajagopal, Laarem Guessas, Anitha Karthikeyan, Ashokkumar Srinivasan, Girma Adam, Fractional order memristor no equilibrium chaotic system with its adaptive sliding mode synchronization and genetically optimized fractional order pid synchronization, Complexity (2017) 2017.

[39] Barics Karakaya, Arif Gülten, Mattia Frasca, A true random bit generator based on a memristive chaotic circuit: analysis, design and fpga implementation, Chaos, Solitons and Fractals 119 (2019) 143–149.

[40] Fei Yu, Li Liu, Binyong He, Yuanyuan Huang, Changqiong Shi, Shuo Cai, Yun Song, Sichun Du, Qiuzhen Wan, Analysis and fpga realization of a novel 5d hyperchaotic four-wing memristive system, active control synchronization, and secure communication application, Complexity (2019) 2019.

[41] Christos Volos, V-T. Pham, E. Zambrano-Serrano, J.M. Munoz-Pacheco, Sundarapandian Vaidyanathan, E. Tlelo-Cuautle, Analysis of a 4-d hyperchaotic fractional-order memristive system with hidden attractors, in: Advances in Memristors, Memristive Devices and Systems, Springer, 2017, pp. 207–235.

[42] John D. Lambert, Computational Methods in Ordinary Differential Equations, Wiley, 1973.

[43] Thomas S. Parker, Leon Chua, Practical Numerical Algorithms for Chaotic Systems, Springer Science & Business Media, 2012.

[44] Carl Runge, Über die numerische auflösung von differentialgleichungen, Mathematische Annalen 46 (2) (1895) 167–178.

[45] Wilhelm Kutta, Beitrag zur naherungsweisen integration totaler differentialgleichungen, Zeitschrift für Mathematik und Physik 46 (1901) 435–453.

[46] Esteban Tlelo-Cuautle, Ana Dalia Pano-Azucena, Omar Guillén-Fernández, Alejandro Silva-Juárez, Analog/Digital Implementation of Fractional Order Chaotic Circuits and Applications, Springer, 2020.

[47] Robert M. Corless, What good are numerical simulations of chaotic dynamical systems?, Computers & Mathematics with Applications 28 (10–12) (1994) 107–121.

[48] C. Varsakelis, P. Anagnostidis, On the susceptibility of numerical methods to computational chaos and superstability, Communications in Nonlinear Science and Numerical Simulation 33 (2016) 118–132.

[49] Sandra Rugonyi, Klaus-Jürgen Bathe, An evaluation of the Lyapunov characteristic exponent of chaotic continuous systems, International Journal for Numerical Methods in Engineering 56 (1) (2003) 145–163.

[50] Aceng Sambas, Sundarapandian Vaidyanathan, Esteban Tlelo-Cuautle, Sen Zhang, Omar Guillen-Fernandez, Yuyun Hidayat, Gugun Gundara, et al., A novel chaotic system with two circles of equilibrium points: multistability, electronic circuit and fpga realization, Electronics 8 (11) (2019) 1211.

[51] Omar Guillén-Fernández, Ashley Meléndez-Cano, Esteban Tlelo-Cuautle, Jose Cruz Núñez-Pérez, Jose de Jesus Rangel-Magdaleno, On the synchronization techniques of chaotic oscillators and their fpga-based implementation for secure image transmission, PLoS ONE 14 (2) (2019).

[52] Louis M. Pecora, Thomas L. Carroll, Synchronization in chaotic systems, Physical Review Letters 64 (8) (1990) 821.

[53] Hebertt Sira-Ramirez, César Cruz-Hernández, Synchronization of chaotic systems: a generalized Hamiltonian systems approach, International Journal of Bifurcation and Chaos 11 (05) (2001) 1381–1395.

Chapter 4

Microwave memristive components for smart RF front-end modules

Milka M. Potrebić and Dejan V. Tošić
University of Belgrade, School of Electrical Engineering, Belgrade, Serbia

Nowadays, a smart system cannot be designed without using the control based on artificial intelligence. The smart-city concept of life is expected in the near future [1]. This concept is based on intelligent technologies and cloud (centralized) radio access networks, which affect the development of novel mobile networks such as 5G. The novel communication networks have to be able to share capacity across a wide range of indoor and outdoor smart-city and IoT applications [2].

The applicable 5G frequency range is extremely wide from ultra-high frequencies to millimeter waves. 5G New Radio covers novel operating frequencies beyond the LTE maximum frequency of 6 GHz and especially millimeter wave frequencies. The three main spectrum bands are the low-band below 1 GHz, the mid-band between 2.5 and 5 GHz, and millimeter wave above 20 GHz [3].

5G communication promises novel advantages, such as faster speed, lower latency, and wider coverage. These features enable realization of IoT, virtual reality applications, factory automation, and autonomous vehicles, which require reliability, security, efficiency, and quality of service. In order to enable increased wireless data transfer needed for the new 5G wireless standard, a spectrum in the millimeter wave range is required.

Next radio applications require wider channel bandwidths, higher operating frequencies, multi-antenna access, etc. 5G RF front-ends have complex design featuring a small footprint, low power consumption, affordable price, and advanced filter specifications. All the RF front-end components will be impacted by the arrival of the new standard. RF engineers have to satisfy tighter requirements for designing components such as filters, antennas, phase shifters, power amplifiers, oscillators, and switches.

5G microwave circuits such as dielectric resonators and oscillators are fabricated of superconducting and ceramic materials, while filter implementations are based on dielectric materials such as liquid crystal and ceramic materials.

Mem-elements for Neuromorphic Circuits with Artificial Intelligence Applications
https://doi.org/10.1016/B978-0-12-821184-7.00012-8

These materials are used to prevent the system from signal losses, safeguarding signal integrity and improving performance [4].

For each RF band, antenna module with phase shifters and various filter realizations are required which increase the number of components in a limited footprint space. Consequently, the next challenge for RF/microwave circuit design is miniaturization. In order to optimize the energy consumption in the front-end module, nonvolatile components could be used as RF memristive switches. Memristors are potential candidates to enable reconfigurability and tunability of RF/microwave devices [5]. In comparison with traditional microwave switches, memristors are expected to be used as highly adaptable nanoscale switches. Memristors can change their state (ON or OFF) promptly and have low insertion loss. One of the main features that memristors demonstrate is a possibility of hot switching, which means that memristors might be reprogrammed on-the-fly [6]. The first RF memristor implementation was reported by Pi [7,8]. Recent research efforts, inspired by memristor technology, have been reported in [9], [10] as non-volatile RF and millimeter wave switches based on MoS_2 atomristors and monolayer hBN.

Comparative analysis of memristor models and memristor state transitions is presented in [11] and [12]. In [13], a new memristive two-port element is introduced for an efficient construction of RF single-pole double-throw switches. A scalable lumped model, that predicts the steady-state high-frequency behavior of nanoscale radio-frequency memristive switches, was reported in [14].

The potential design of microwave passive circuits with memristors was presented in [15–17]. Microstrip antenna design with memristor [18] is one of numerous realizations at RF/microwave frequencies. Quasi-lumped microwave inductor realization was reported in [19] and it uses radio-frequency memristive switches to implement tunable inductors.

In this chapter, we explore possible memristor applications in the design of RF/microwave components such as reconfigurable multi-band filters and digital phase shifters. These components are the infrastructure for smart RF front-ends and designed for the sub-6 GHz bands, i.e. mid-band 5G.

A potential memristor-based application is design of digital phase shifter. Main-line memristor mounted type loaded-line phase shifter was analyzed as shown in [20]. The main idea is to replace PIN diodes, acting as traditional RF/microwave switches, with memristors in order to reduce power consumption.

For microwave filter design, memristors could be used to realize reconfigurability of the multi-band bandpass planar filters. A dual-band bandpass filter was realized with the memristor-based dual-mode resonators [21]. A switchable planar filter for multi-band receiver application is an example of how memristors can be utilized for reconfigurability [22]. This filter consists of two subnetworks realized by interdigital microstrip bandpass filters. The corresponding memristor setup circuitry is optimized in order to minimize the circuitry influence on the desired filter frequency response.

Advantages of the proposed memristive designs are that they could be used at millimeter waves above 20 GHz. The physical dimensions of the analyzed digital phase shifters and filters could be rescaled to millimeter wave frequencies. The analyzed components could be fabricated using materials for designing 5G systems. These materials offer improved frequency, stability, supply current and voltage, operating temperature, etc. Memristive switches could be good device candidates for millimeter wave frequencies, because memristors are non-volatile components with nanosecond switching speed, high power handling (20 dBm), high figure-of-merit cutoff frequency (129 THz), and heater-less ambient integration [10].

4.1 RF/microwave model of memristive switch and PIN diode

The memristor as a two-terminal electric circuit element is defined by the relation between magnetic flux and electric charge. A main memristor parameter is memristance, which is determined by the history of applied current and/or voltage, the implementation technology, and the conditions during implementation. Memristor switching mechanism is typically based on formatting or rupturing of one or more conductive filaments between two electrodes. This process is related to electrochemical reactions or the migration of mobile ions that modulate the interfacial properties. The memristor demonstrates many useful features (see Table 4.1). It is not a volatile component. The memristor switching energy is around picojoules and the switching speed is around nanoseconds. It is high endurance component (about 1000 cycles) which has excellent scalability and CMOS compatibility. Such characteristics enable memristors to become promising candidates for application in different fields of electrical engineering.

Fabrication process of an RF memristor is presented in [7,8] with its characterization at high frequencies. The RF memristor is a nanoscale device, but for measurements the switch is placed on the coplanar waveguide feed lines with the pads of the footprint area of approximately 320 μm × 300 μm [7]. For these dimensions, the optimal surface-mounted packages are 0603 (0.6 mm × 0.3 mm) and 1005 (1 mm × 0.5 mm). For example, if the surface-mounted package 0603 is used, then the gap width between the memristor terminal pads is set to 0.5 mm.

The memristor exhibits nonvolatile bipolar resistance switching that crosses the origin in the I–V curve (Fig. 4.1). The memristor was programmed with a DC voltage sweep. The memristor was programmed in the ON state by applying a maximum voltage of 3 V. The device was in the OFF state by applying a minimal voltage of −0.4 V.

The memristance programming circuitry can be functionally presented by a serial connection of a resistor ($R = 10\ \mathrm{k\Omega}$), an inductor (RF choke, $L_{\mathrm{choke}} = 30\ \mathrm{nH}$), and a programmable voltage source. This research does not consider the realization details of the programming circuitry. It is taken into account that the circuitry elements might affect the device characteristics at the operating frequency band.

TABLE 4.1 Memristor parameters and features [21], [7,8].

Parameter/Feature		Value
Frequency range		up to 110 GHz
Nonvolatile		+
Hot switching		+
Switching	Energy	~pJ
	Speed	<ns
Endurance		$\sim 10^{12}$ cycles
Compatible with CMOS technology		+
Programming	ON state – Maximum voltage	3 V
	OFF state – Minimum voltage	−0.4 V
ON/OFF conductance ratio		$\leq 10^{12}$
Measured @ 40 GHz	OFF state – Isolation	30 dB
	ON state – Insertion loss	0.3 dB
Equivalent model	ON state	linear resistor – R_{ON}
	OFF state	linear capacitor – C_{OFF}
Min/ Mean/ Max	R_{ON} [Ω]	2.1/ 3.6/ 6.9
	C_{OFF} [fF]	1/ 1.37/ 1.6
Maximum dissipation		34 dBm-20log$_{10}$ R_{ON}

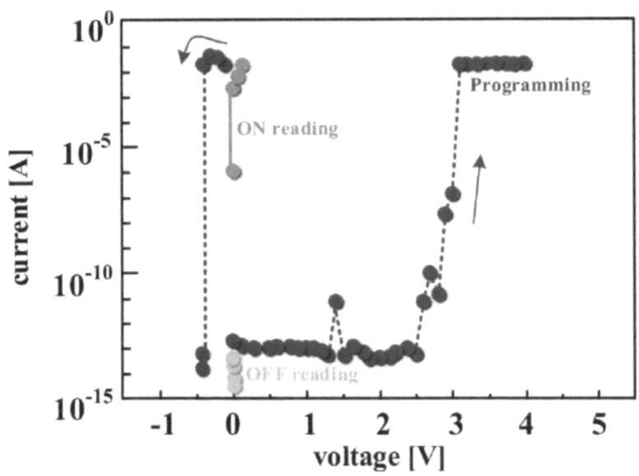

FIGURE 4.1 Programming of the memristive RF switch with a DC voltage sweep [7,8].

A memristor model is presented at RF/microwave frequencies (Pi's model). Pi's model of a memristive switch is shown in Fig. 4.2, where are presented the ON state and the OFF state. The PIN diode switch model is also shown in Fig. 4.1 for the ON state and the OFF state.

At RF/microwave frequencies, memristor is equivalent to a resistor in the ON state and to a capacitor in the OFF state. Microwave circuits require a switch with low ON state resistance (R_{ON}) and low OFF state capacitance (C_{OFF}), in

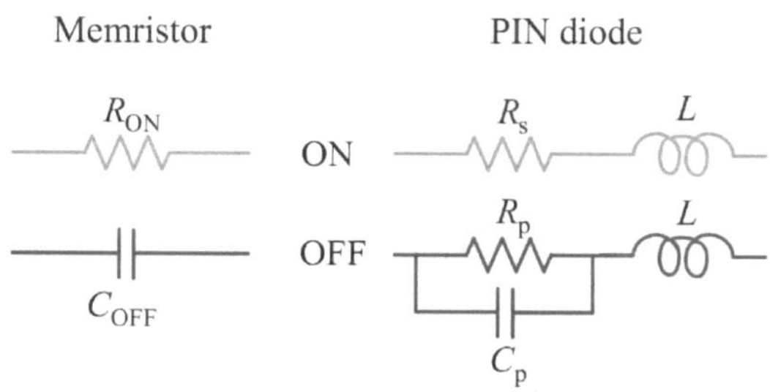

FIGURE 4.2 Equivalent circuit for memristors and PIN diodes at RF/microwave frequencies for the ON and the OFF state.

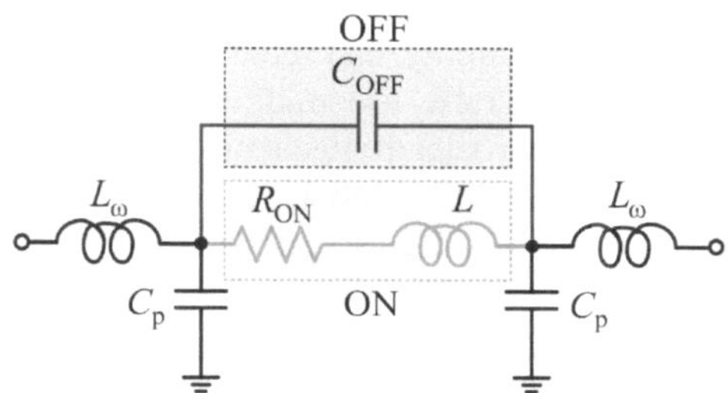

FIGURE 4.3 Memristor model as presented in [23].

order to achieve a high cutoff frequency (f_c):

$$f_c = \frac{1}{2\pi R_{ON} C_{OFF}}. \quad (4.1)$$

Realized memristors mean values are $R_{ON} = 3.6\ \Omega$, $C_{OFF} = 1.37$ fF, thereby enabling $f_c = 35.2$ THz. The model was analyzed in the frequency range from 10 MHz to 110 GHz. The memristor was characterized at 40 GHz. The measured isolation was 30 dB (in the OFF state), and insertion loss was 0.3 dB (in the ON state). The critical memristor state is the ON state, because the device is limited with maximum dissipation. For the ON state, measured memristance values are in the range of 3–14.5 Ω, and maximum dissipation can be calculated such as

$$34\ \text{dBm} - 20\log_{10} R_{ON}. \quad (4.2)$$

A novel memristor model is presented in [23], which takes into account more effects such as: 1) S_iO_2 parasitic capacitance, 2) S_i substrate capacitance, 3) the fringe capacitance between ground planes and the signal line, and 4) the filament and the electrodes effects. Fig. 4.3 shows the equivalent scheme of the proposed memristor model with the next element values: $R_{ON} = 2.56\ \Omega$, $C_{OFF} = 1.168$ fF, $L = 52$ fH, $L_\omega = 3.1$ pH, $C_p = 1.15$ fF.

4.2 Memristive phase-shifter realization

A possible application of memristive switches is design of phase shifters. RF phase shifters are indispensable parts of multi-element antenna arrays which steer the beam. The main idea is to replace traditional RF/microwave switches, as PIN diodes, with memristors in order to reduce power consumption [20]. Different phase-shifter realizations, as presented in [24], could be designed using different configurations and technologies. Analyzed phase shifters are realized as main-line mounted loaded-line. Phase shifters should be reciprocal networks and they need to change the transmission phase angle between defined two ports. Their design should include features such as minimal insertion loss, equal amplitude for all phase states, accurate phase angle, low switching time, specified power handling, temperature stability, and reliability.

Digital phase shifters generally use traditional switches realized as PIN diodes, RF transistors, and microelectromechanical systems – MEMS. Some of traditional switches have limitations such as long time of switching states, low working frequencies and so on. Some side effects of PIN diodes are large power consumption because they need constant current supply in order to sustain stable working requirements.

4.2.1 Planar main-line memristor mounted type loaded-line phase shifter

Phase shifters as two-port networks control the phase difference between the input and the output ports. Phase shifters are defined as digital as the differential phase shift can be changed only a few predestined discrete values. On the other hand, analog phase shifters are defined when the differential phase shift can change continuous by a corresponding continuous variation of the control signal. Digital phase shifters are used for the computer control of beam scanning in phased-array antenna systems [24].

At microwave frequencies, digital phase shifters can be designed using: 1) ferromagnetic materials for obtaining switchable phase shift, 2) semiconductor, or 3) MEMS devices. In comparison with ferrite realizations, phase shifters based on semiconductor or MEMS devices are more compact, have lower switching times and require lower drive power.

Phase shifters based on semiconductor devices can be of the reflection or the transmission type. For reflection type design, the basic realization cell is a one-port network and the phase shift of the reflected signal is changed by the control signal. Nevertheless, transmission type realization uses a transmitted signal and can be divided in the next groups: switched-line type, loaded-line, switched-network type, and amplifier type.

The analyzed phase-shifter realization with memristor is based on the loaded line presented in Fig. 4.4a. The main transmission line is loaded by two identical variable susceptances, which distance is approximately a quarter wavelength at the center frequency. This realization is used for phase-shift values

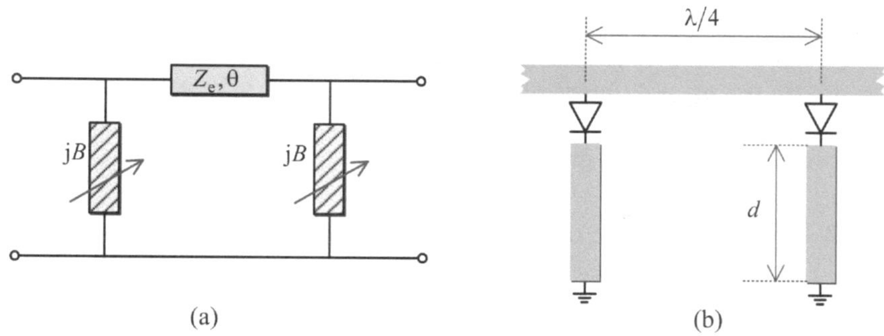

FIGURE 4.4 (a) Basic loaded-line phase shifter, and (b) main-line PIN diode mounted type loaded-line phase-shifter design [20].

up to 45°. Otherwise, the impedance of the transmission line becomes impractical to implement [25] phase-shift values beyond 45°. Based on implementation, the loaded-line phase shifter can be implemented as: main-line mounted, stub-mounted and switchable-stub mounted realizations [26]. The analyzed implementation is the main-line mounted type phase shifter (see Fig. 4.4b). The susceptance value can be tuned by adjusting the stub length d. First, the phase shifter with PIN diodes as traditional switches is analyzed.

This phase shifter can be analyzed using wave-amplitude transmission matrices, and it is found that [27]

$$T_{\mathrm{IO}} = \left(\left(1 + \mathrm{j}\frac{\bar{B}}{2} \right)^2 \mathrm{e}^{\mathrm{j}\theta} + \frac{\bar{B}^2}{4} \mathrm{e}^{-\mathrm{j}\theta} \right)^{-1}, \tag{4.3}$$

where $\bar{B}$ represents the normalized shunt susceptance, T_{IO} represents the transmission coefficient from input to output port, and θ is the electrical length of the main transmission line. For $|T_{\mathrm{IO}}| = 1$ it can be found that

$$\tan\theta = \frac{2}{\bar{B}'}. \tag{4.4}$$

From Eqs. (4.3) and (4.4), it can be found that

$$T_{\mathrm{IO}} = -\mathrm{e}^{\mathrm{j}\theta} = \mathrm{e}^{-\mathrm{j}(\pi-\theta)} \tag{4.5}$$

When the diodes are switched ON, the change in phase shift is

$$\Delta\phi = (\pi - \theta) - \theta = \pi - 2\theta. \tag{4.6}$$

In the ON state, the diode impedance is a resistance in series with inductance. This could be added as part of the susceptance that is switched into the line. In the OFF state, the diode impedance is equivalent to the shunt capacitance. In such a way, the shunt capacitance must be small in order to make the phase shifter work regularly. The memristive switch operates in the same manner as the diode. The main difference is that the memristor in the ON state does not have a series inductance as in [7] or has a several orders of magnitude lower inductance than the stub's inductance [23].

4.2.2 Implementation of main-line memristor mounted type loaded-line phase shifter and results

The analyzed phase shifter (Fig. 4.4) is implemented in microstrip technology with memristive switches [20]. A three-dimensional (3D) model and the equivalent electrical circuit are shown in Fig. 4.5. Frequency domain analysis of the phase shifter has been performed in Cadence AWR Microwave Office [28] for both switches: memristive and PIN diode switches. The main difference between these realizations is that PIN diodes need bias circuitry.

Significant parasitic effects were introduced into the phase-shifter model. The circuit for programming memristor state is modeled by a serial connection of a resistor, an RF choke, and a voltage source. The short circuits are realized with via-holes which are modeled as a cylindrical via with a diameter of 0.3 mm. Gaps in the transmission line are modeled as capacitive π-networks. The distance between the memristor terminal pads is set to be 0.5 mm, while the port impedances (Z_0) are 50 Ω (see explanation in Section 4.1).

A phase-shifter device can be realized as a cascade of single phase-shift cells. Every cell is designed for single phase-shift value. Fig. 4.6 presents the potential connection between the phase shifter and the other system parts. Memristor programming can be controlled with signals C_{ntr1} and C_{ntr2} (from CPU). Programming could be achieved using PWM (Pulse Width Modulation) or some amplitude variable signal by using DAC (Digital-to-Analog Converter). If the memristor changes state to OFF, the corresponding phase-shifter cell is immobilized. If the memristor changes state to ON, the corresponding phase-shifter cell is "demobilized". In order to achieve the specified phase shift between input and output ports of the phase shifter, the user can combine phase shifts using target cells.

The device implemented is realized using microstrip technology with relevant parameters given in Table 4.2: relative permittivity (ε_r), substrate thickness (h), loss tangent (tanδ), conductor thickness (t).

TABLE 4.2 Substrate parameters.

ε_r	$\tan\delta$	h [mm]	t [μm]	σ [MS/m]
10.8	0.0018	1.27	18	14

The losses due to the skin effect and surface roughness are taken into consideration by setting the conductivity to $\sigma = 14$ MS/m. In Cadence AWR Microwave Office this conductivity is equivalent to the parameter Rho value of 3.

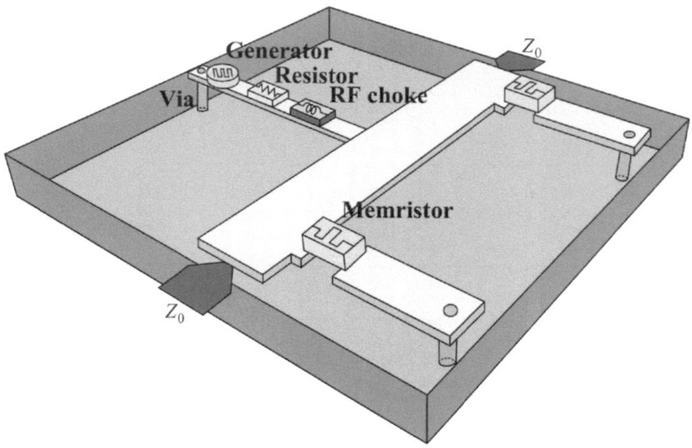

(a)

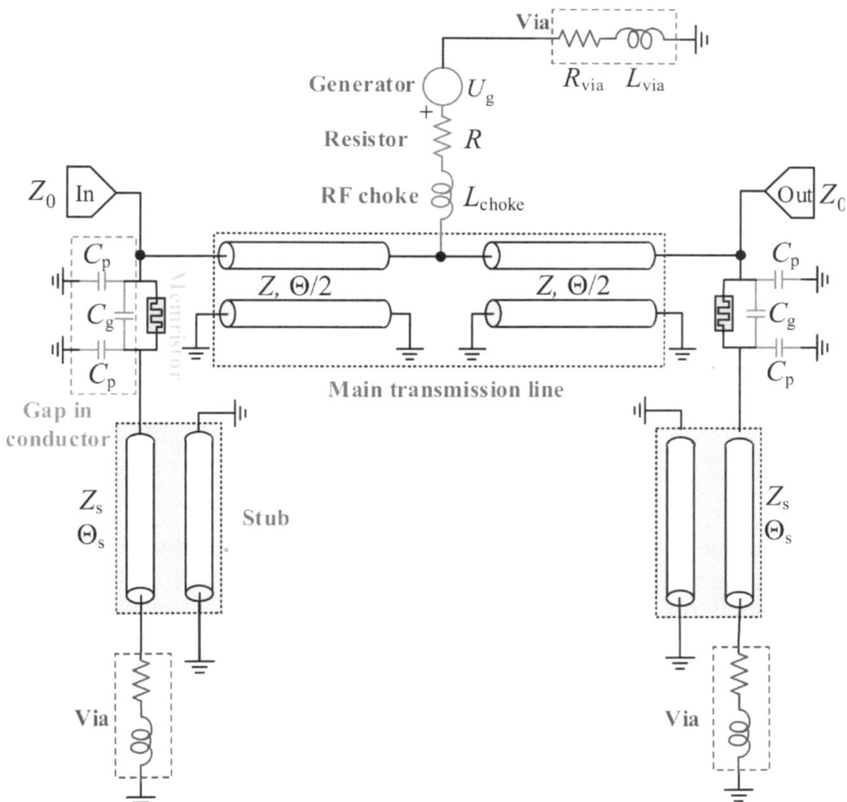

(b)

FIGURE 4.5 Phase shifter with a single cell - single phase-shift value: (a) 3D model and (b) equivalent electrical circuit [20].

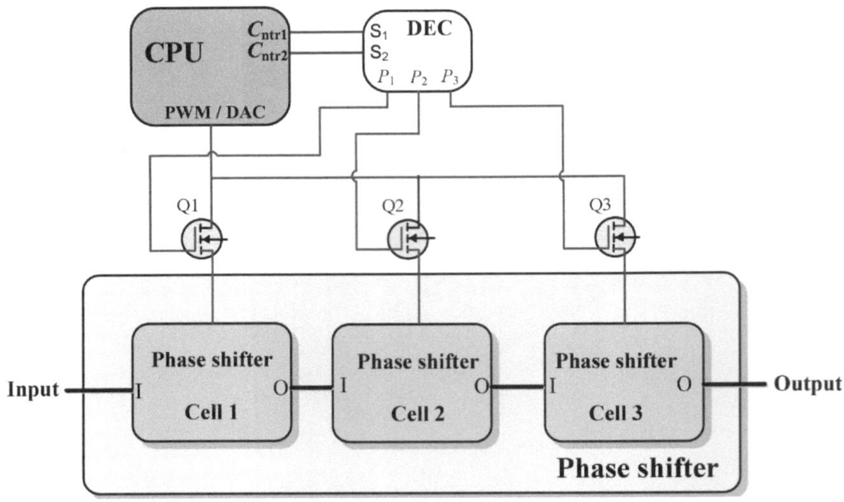

FIGURE 4.6 System block diagram of phase shifter with unit cells: connection between phase shifter and other system parts [20].

TABLE 4.3 Transmission line parameters.

Phase shift $\Delta\phi$		45°		22.5°		11.25°	
Switch		Mem.	PIN	Mem.	PIN	Mem.	PIN
Main TL	Θ [°]	132	146	90.5	77	80.3	70
	L [mm]	7.6	8.4	5.2	4.4	4.6	4
	Z [Ω]	43.5	47.7	33	42	33	31
	w [mm]	1.5	1.25	2.5	1.6	2.5	2.8
Stub	Θ_S [°]	19	26	25.4	37.4	41.3	26
	L_S [mm]	1.1	1.5	1.5	2.2	2.4	2.5
	Z_S [Ω]	52.6	50.8	51	79	60.2	73
	w_S [mm]	1	1.1	1.1	0.3	0.7	0.43

Rho is the metal bulk resistivity normalized to gold. The memristor model has the parameters $R_{ON} = 2.56\ \Omega$, $C_{OFF} = 1.168$ fF, $L = 52$ fH, $L_{\omega} = 3.1$ pH, $C_P = 1.15$ fF (see [23]). The PIN diode model (MA4PBL027 PIN diode [29]) has the parameters $L = 0.15$ nH, $R_S = 0.05\ \Omega$, $R_P = 100\ k\Omega$ and $C_P = 0.04$ pF. Transmission line (TL) parameters are shown in Table 4.3.

According to [7,8], the RF memristor switch can be modeled at RF and microwave frequencies by a lumped linear time-invariant circuit element (see Fig. 4.2). Therefore, a linear microwave circuit simulator can be used to verify the phase-shifter performance.

Fig. 4.7 presents the frequency response of the phase-shifter implementation with memristors and PIN diodes. The analyzed frequency range is from 5 to 5.6 GHz. The discrete values of the phase shifts are 45°, 22.5°, and 11.25°.

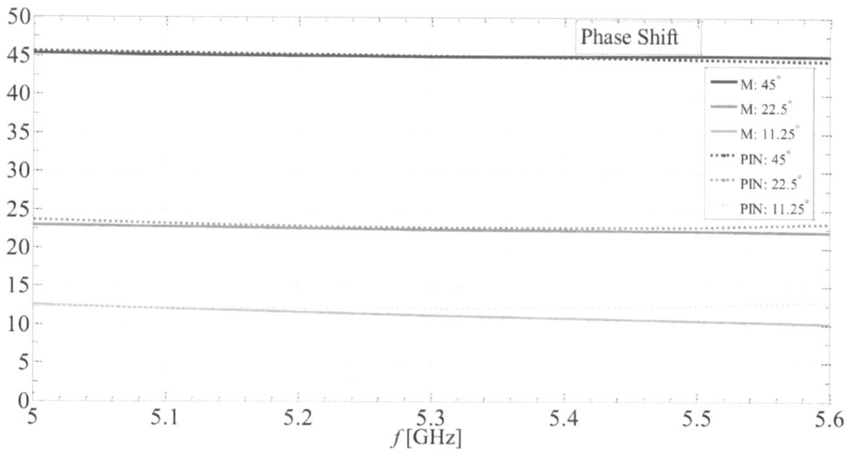

FIGURE 4.7 Phase response (in degrees) over the frequency range 5–5.6 GHz. Solid lines – memristor as a switch, dotted lines – PIN diode as a switch.

TABLE 4.4 Phase-shift deviations for maximum and minimum memristor values.

Phase shift	45°	22.5°	11.25°
$R_{ON} = 7\ \Omega$, $C_{OFF} = 1.6$ fF	6% (2.68°)	2%	2%
$R_{ON} = 2\ \Omega$, $C_{OFF} = 1$ fF	1% (0.31°)	negligible	negligible

For both switches, each discrete state has almost the constant phase shift over the specified frequency band. The maximum deviation is visible for the 11.25° phase shift, and it is around ±1°.

Based on results presented in [7], the memristance value is not easily reproducible. The measured resistance values in the ON state are between 2 Ω and 7 Ω, and capacitance values in the OFF states are between 1 fF and 1.6 fF. Sensitivity analysis was conducted, using maximum ($R_{ON} = 7\ \Omega$, $C_{OFF} = 1.6$ fF) and minimum ($R_{ON} = 2\ \Omega$, $C_{OFF} = 1$ fF) values, in order to observe how phase shift is stable. Fig. 4.8 presents the results obtained for maximum values.

The phase-shift deviations for maximum and minimum memristor values are given in Table 4.4.

The circuit for programming memristance does not impact the phase-shift deviation, because it has an RF choke.

For the given memristor maximum dissipation (about 23 dBm (0.2 W) for $R_{ON} = 3.6\ \Omega$), we compute the power handling capability of the phase shifter. Every phase-shifter cell was analyzed independently. It has been found that the critical cell is one for a phase shift of 45° degrees. Therefore, the maximum input voltage of the RF/microwave signal is about 34 V.

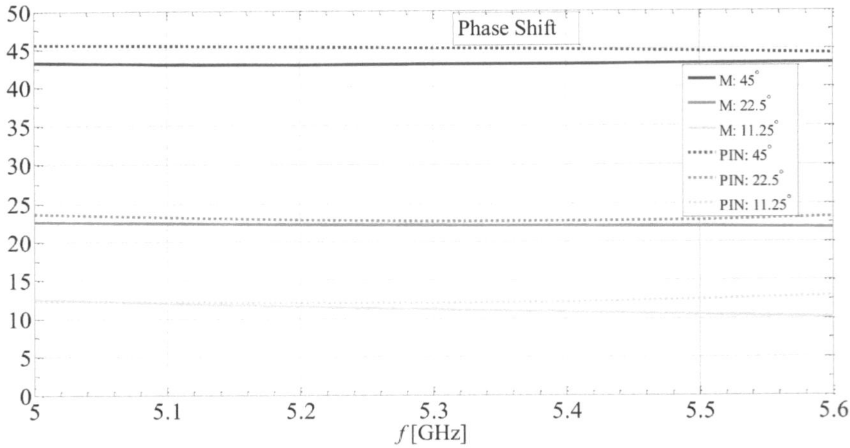

FIGURE 4.8 Phase response (in degrees) over the frequency range 5 – 5.6 GHz. Solid lines – memristor as a switch, with maximum values in ON and OFF states, as reported in [7]. Dotted lines – PIN diode as a switch.

4.3 Reconfigurable dual-band bandpass microwave filter

One of the main components of multi-band receiver is a dual-band bandpass filter (2B-BPF) which uses memristors in order to achieve reconfigurability [22]. This filter has two subnetworks which consist of the low-band bandpass filter (L-BPF) and the high-band bandpass filter (H-BPF). The first filter L-BPF has the center frequency of 2.44 GHz and the 3 dB fractional bandwidth of 10%. The second filter H-BPF has the center frequency of 5.3 GHz and the 3 dB fractional bandwidth of 5%. 2B-BPF has two operating modes: dual-band mode (L-BPF – H-BPF) and single low-band mode (L-BPF). L-BPF and H-BPF can be independently designed. The block diagram of 2B-BPF is presented in Fig. 4.9. The filter is realized with two single-band bandpass filters which are connected in parallel. The operation mode is selected with a memristive RF switch. The proposed filter is in L-BPF mode if the memristor is in the ON state. In that mode the filter has single-band bandpass characteristic. If the memristor is in the OFF state, the filter has dual-band bandpass characteristics (L-BPF – H-BPF mode).

L-BPF and H-BPF filters are realized as the fifth-order Chebyshev interdigital bandpass filters [30]. Interdigital bandpass filters consist of five coupled resonators. In this design, each bandpass filter is realized using Dishal's method [31], which is based on three fundamental parameters: the synchronous tuning frequency of each resonator, the couplings coefficient between adjacent resonators, and the external quality factor of the first and last resonators. The filter structure was tuned or optimized using knowledge of these fundamental parameters. When the required coupling coefficients and the external quality factors are known, Dishal's method uses a simulated model to generate a design curve for each parameter.

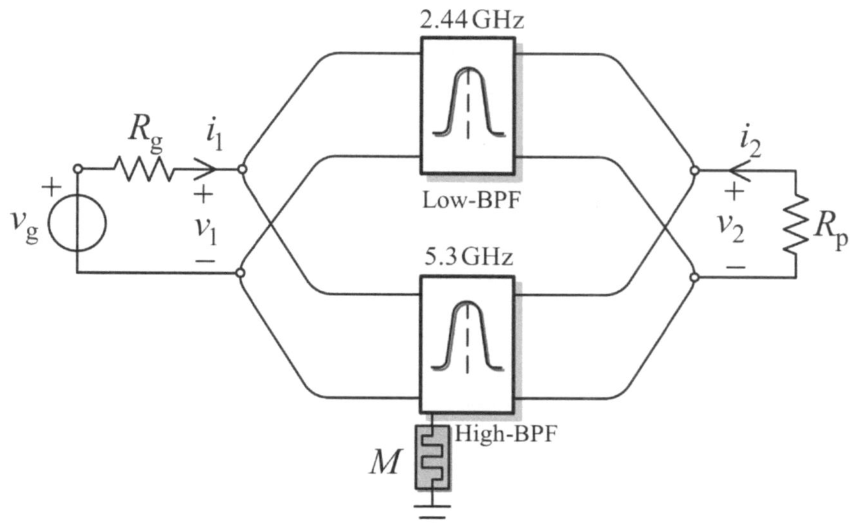

FIGURE 4.9 Block diagram of a reconfigurable dual-band bandpass filter (2B-BPF) with memristor.

The coupling scheme is presented in Fig. 4.10. The coupling between non-adjacent resonators is negligible. The equivalent filter parameters such as the resonator external quality factors at the filter input (m_{01}) and output (m_{56}) and the coupling coefficients between the adjacent resonators (m_{12}, m_{23}, m_{34}, m_{45}) are found to be as follows [30,31]:

$$m_{01} = q_1 = g_0 g_1 / FBW, \qquad m_{56} = q_5 = g_5 g_6 / FBW, \tag{4.7}$$

$$m_{i,j} = FBW / \sqrt{g_i g_{i+1}}, \quad i = 1, 2, 3, 4. \tag{4.8}$$

Parameters g_i ($i = 0$ to 6) are lowpass filter prototype coefficients given for a normalized lowpass cutoff frequency of 1 rad/s. The filter fractional bandwidth (FBW) is defined as $FBW = (f_{\text{UPPER}} - f_{\text{LOWER}}) / f_0$. Frequencies f_{UPPER} and f_{LOWER} are the passband-edge frequencies, while f_0 is the center passband frequency.

The resonators are quarter-wavelength grounded transmission lines at the filter center frequency.

The memristor is connected at the open end of the second H-BPF resonator. When the memristor is in the ON state, the resonator becomes an eighth-wavelength resonator. For the OFF state, the resonator is a quarter-wavelength resonator (see Fig. 4.11). The resonator is designed to resonate at the center frequency of H-BPF when the memristor is in the OFF state, which allows for a through path for the filter. When the memristor is in the ON state, signals from the H-BPF band are suppressed.

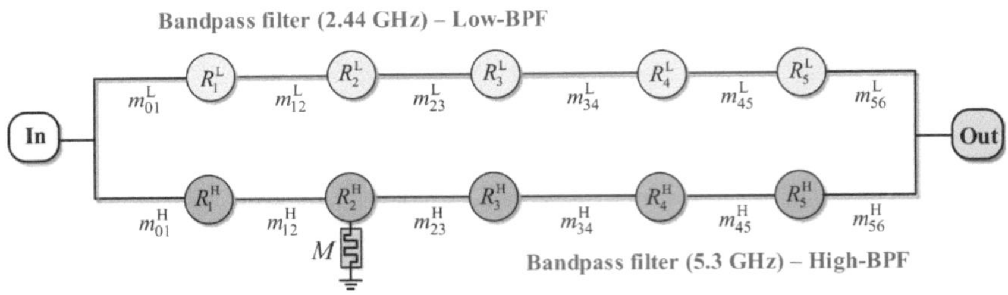

FIGURE 4.10 Coupling scheme of the reconfigurable dual-band bandpass filter (2B-BPF).

As described in Section 4.1, the RF memristor is modeled by a resistor ($R_{ON} = 3.6\ \Omega$) in the ON state, and a capacitor ($C_{OFF} = 1.37$ fF) in the OFF state [7,8]. The memristance programming circuit is presented by a serial connection of a resistor, an inductor (RF choke), and a voltage source. The effect of the gap in the planar conductor of 0.5 mm (between the memristor terminal pads) is also introduced in the model, as shown in Fig. 4.11.

According to Fig. 4.11, all the parasitics have been taken into account during modeling. The resonator dimensions, such as length and width, should be adjusted in order to satisfy the specification. The input admittance of the equivalent resonant circuit with RF memristor is

$$\underline{Y} = \underline{Y}_1 + \underline{Y}_2 \tag{4.9a}$$

$$\underline{Y}_1 = \frac{\underline{Y}_C \underline{Z}_{via} \sinh\left(\underline{\gamma} D\right) + \cosh\left(\underline{\gamma} D\right)}{\underline{Z}_{via} \cosh\left(\underline{\gamma} D\right) + \underline{Z}_C \sinh\left(\underline{\gamma} D\right)}, \tag{4.9b}$$

$$\underline{Z}_{via} = R_{via} + j\omega L_{via} \tag{4.9c}$$

$$\underline{Y}_2 = \underline{Y}_{2via} + \underline{Y}_{2M}, \tag{4.9d}$$

$$\underline{Y}_{2via} = j\omega C_p + \frac{1}{\underline{Z}_{via} + R + j\omega L_{choke}}, \tag{4.9e}$$

$$\underline{Y}_{2M} = \frac{1}{\underline{Z}_M + \underline{Z}_{via}/\left(1 + j\omega C_p \underline{Z}_{via}\right)} \tag{4.9f}$$

$$\underline{Z}_M = \frac{\underline{Z}_{Memristor}}{1 + j\omega C_g \underline{Z}_{Memristor}} \tag{4.9g}$$

$$\underline{Z}_{Memristor} = \begin{cases} R_{ON} & \text{ON state} \\ 1/(j\omega C_{OFF}) & \text{OFF state.} \end{cases} \tag{4.9h}$$

Eqs. (4.9) are used for balancing the resonator with the memristor, i.e. satisfying the condition of resonance $\underline{Y} = \underline{Y}_1 + \underline{Y}_2 = 0$ [22].

The filter layout is presented in Fig. 4.12. The physical dimensions are found by the parameter extraction technique that is explained in [31] (see Table 4.5).

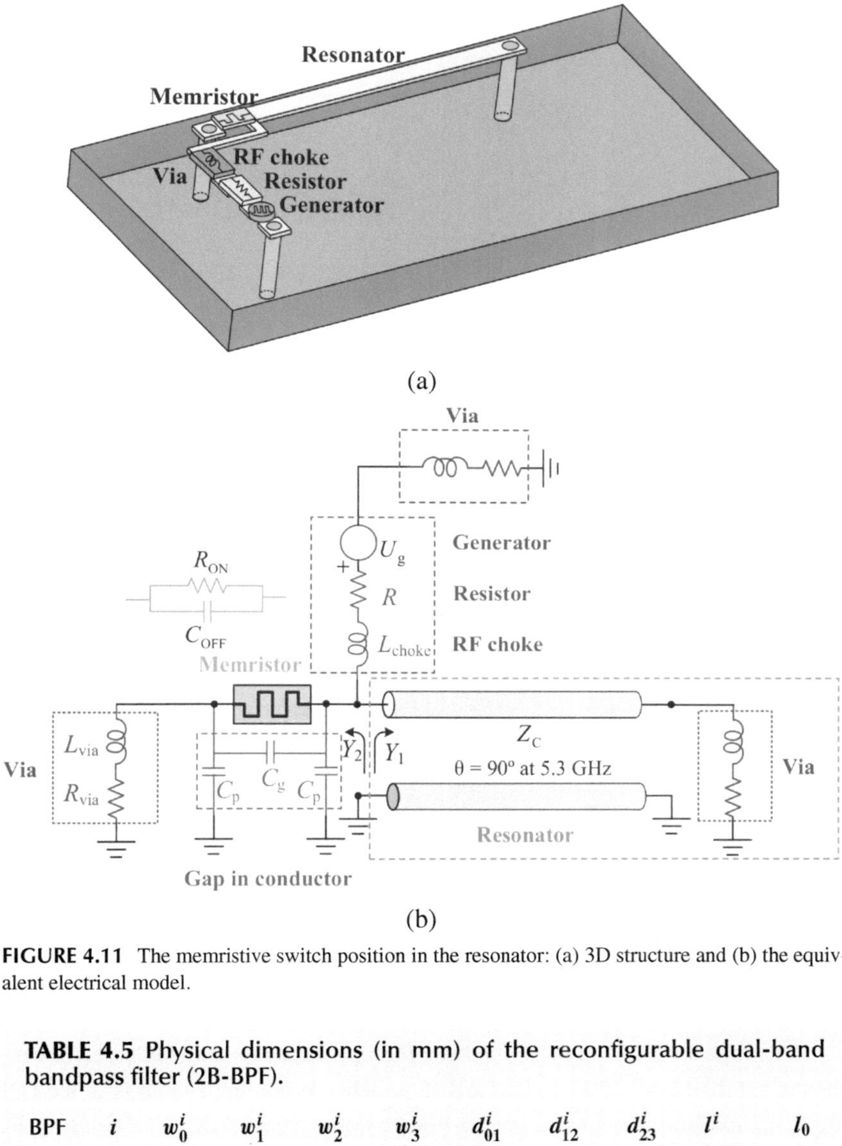

FIGURE 4.11 The memristive switch position in the resonator: (a) 3D structure and (b) the equivalent electrical model.

TABLE 4.5 Physical dimensions (in mm) of the reconfigurable dual-band bandpass filter (2B-BPF).

BPF	i	w_0^i	w_1^i	w_2^i	w_3^i	d_{01}^i	d_{12}^i	d_{23}^i	l^i	l_0
Low	L	1.65	1.16	2.31	2.36	0.39	1.55	1.73	17.37	1.5
High	H	1.94	2.38	2.75	2.76	0.24	1.82	2.19	6.18	

The first resonator and the last (fifth) resonator are capacitively coupled to the feed lines implemented as 50 Ω (Z_0) transmission lines.

The filter is realized using the stripline technology, characterized by the parameters presented in Table 4.6.

TABLE 4.6 Stripline substrate parameters [32].

ε_r	$\tan\delta$	H_1 [mm]	H_2 [mm]	$H_1 + H_2$ [mm]	t [μm]	σ [MS/m]
2.33	0.001	1.575	1.575	3.15	17	20

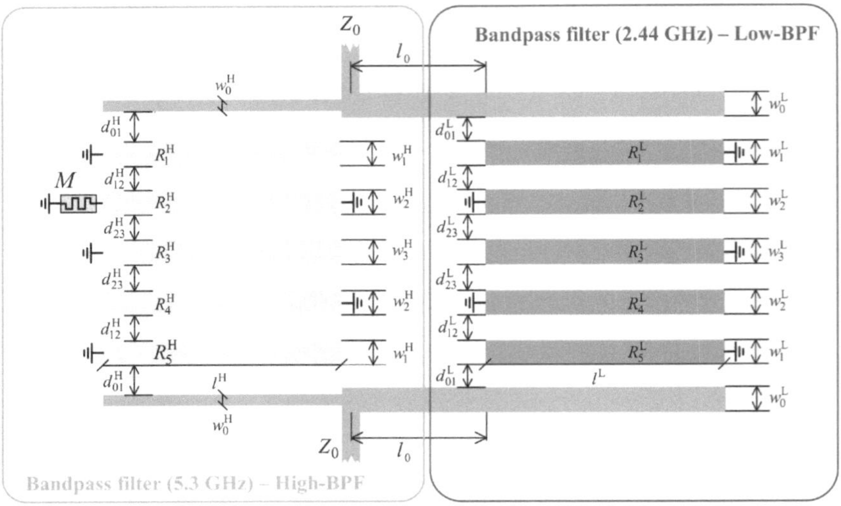

FIGURE 4.12 Layout of the reconfigurable dual-band bandpass filter (2B-BPF).

The conductor conductivity is set to 20 MS/m to account for the losses due to the surface roughness and the skin effect. Via-hole grounding is realized using a cylindrical via with a stripline pad and diameter of 0.3 mm.

Cadence AWR Microwave Office linear circuit simulator [28] is used to find the frequency response of the reconfigurable filter (2B-BPF); see Figs. 4.13 and 4.14.

When the switch is in the OFF state, the insertion loss at the center frequencies is 1.2 dB (2.44 GHz) and 2.5 dB (5.3 GHz). When the switch is in the ON state, the minimum stopband attenuation is higher than 35 dB. The bandwidths satisfy the filter specification in both operational modes.

For the proposed filter, the maximum input voltage of the RF/microwave signal is 85 V at 5.3 GHz. Up to this voltage value the input signal should not change the memristor state. This means that memristor dissipation would not exceed the maximum value of +23 dBm (R_{ON} = 3.6 Ω) as presented in [7,8].

The realized analysis of the reconfigurable filter showed that the memristor might be used as an RF/microwave switch.

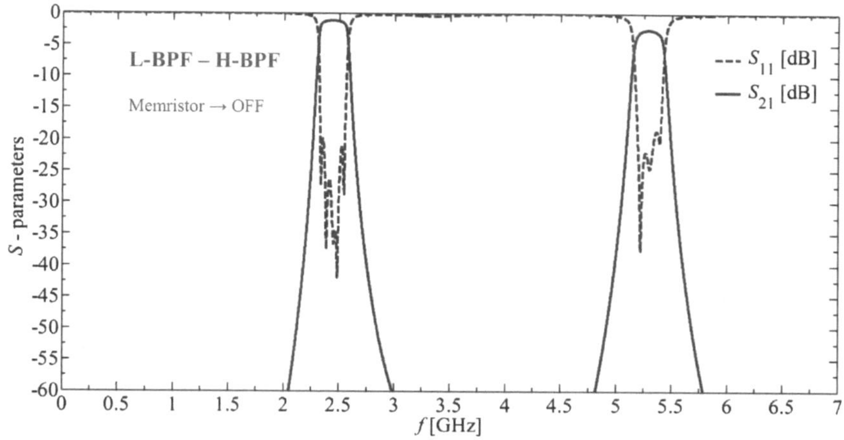

FIGURE 4.13 Frequency response of 2B-BPF in L-BPF – H-BPF mode. The memristor switch is in the OFF state and it is modeled by a capacitor with C_{OFF} = 1.37 fF.

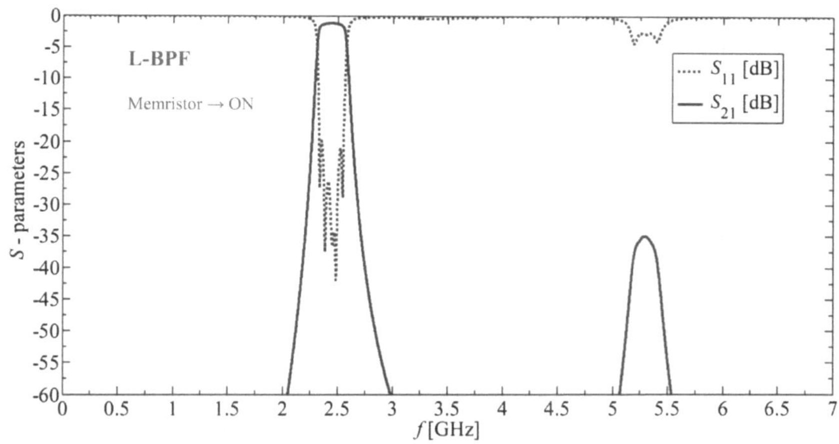

FIGURE 4.14 Frequency response of 2B-BPF in L-BPF mode. The memristor switch is in the ON state and it is modeled by a resistor with R_{ON} = 3.6 Ω.

4.4 Dual-band bandpass filter with multilayer dual-mode resonator enhanced with RF memristor

A memristive multilayer dual-mode resonator is a unit cell for designing reconfigurable multi-band filters [21]. The memristor is used as an RF/microwave switch. A dual-band bandpass filter is designed with the proposed memristive dual-mode resonators. The memristor setup circuit is optimized in order to minimize the circuit influence on the specified filter response.

4.4.1 Dual-mode resonator with memristor

A dual-mode resonator (2MR) is used for filter miniaturization. This resonator has two resonating frequencies which correspond to the even-mode (f_{even}) and odd-mode (f_{odd}) frequencies. These modes are not coupled to each other. Filters designed with one dual-mode resonator are the second order filters. In this research, a resonator with the central short-circuited stub i.e. perturbation element (see Fig. 4.15) [33], [30], [34] is used. This design is convenient for adjustment of the desired filter specification by optimizing resonator dimensions. The resonator design starts with a search for the optimal modal frequencies. The 2MR structure is symmetrical (Fig. 4.15), which is convenient for using the even- and odd-mode analysis and equivalent circuits are presented in Fig. 4.15. The resonant frequencies can be found from the following equations [21]:

$$\mathrm{Im}(Y_{\text{in\_even}}) = 0 \quad \text{and}\,\mathrm{Im}(Y_{\text{in\_odd}}) = 0. \tag{4.10}$$

The resonant frequency f_{odd} can be calculated as

$$Z_2/Z_1 = \tan(\Theta_1^{\text{odd}})\tan(\Theta_2^{\text{odd}}), \tag{4.11}$$

where the electrical lengths at f_{odd} are $\Theta_1^{\text{odd}} = \beta_1^{\text{odd}} D_1 = 2\pi f_{\text{odd}}\sqrt{\varepsilon_{\text{reff1}}} D_1/c_0$ and $\Theta_2^{\text{odd}} = \beta_2^{\text{odd}} D_2 = 2\pi f_{\text{odd}}\sqrt{\varepsilon_{\text{reff2}}} D_2/c_0$. When we have $Z_2 < Z_1$ ($Z_2 = \alpha Z_1$, $\alpha < 1$) then $\alpha = \tan(\Theta_1^{\text{odd}})\tan(\Theta_2^{\text{odd}})$ and f_{odd} can be calculated from the relation

$$\alpha = \tan(f_{\text{odd}} \times 2\pi\sqrt{\varepsilon_{\text{reff1}}} D_1/c_0)\tan(f_{\text{odd}} \times 2\pi\sqrt{\varepsilon_{\text{reff2}}} D_2/c_0) < 1, \tag{4.12}$$

where $c_0 = 2.99792458 \cdot 10^8$ m/s and $\varepsilon_{\text{reff}}$ is the effective relative permittivity of the substrate.

When the electrical length Θ_3^{even} is small ($\Theta_3^{\text{even}} << 1$) and the characteristic impedance Z_3 is approximately 20 Ω, it can be assumed that the third short-circuited stub (see Fig. 4.15) can be presented with a series inductor of an impedance $Z_3^{\text{even}} \approx j2Z_3\Theta_3^{\text{even}}$ ($\tan(\Theta_3^{\text{even}}) \approx \Theta_3^{\text{even}}$). The resonant frequency f_{even} can be found as

$$\tan(\Theta_1^{\text{even}})\tan(\Theta_2^{\text{even}}) - \alpha + 2\beta\Theta_3^{\text{even}}\tan(\Theta_2^{\text{even}}) + 2\alpha\beta\Theta_3^{\text{even}}\tan(\Theta_1^{\text{even}}) = 0 \tag{4.13}$$

assuming the next relations: $Z_2 < Z_3 < Z_1$, $Z_2 = \alpha Z_1$, $Z_3 = \beta Z_1$, and $\alpha < \beta < 1$. The electrical lengths at f_{even} are equal to $\Theta_i^{\text{even}} = 2\pi f_{\text{even}}\sqrt{\varepsilon_{\text{reff}i}} D_i/c_0$, where $i = 1, 2, 3$. Eq. (4.13) can be rearranged as

$$\begin{aligned}
&\tan(f_{\text{even}} \times 2\pi\sqrt{\varepsilon_{\text{reff1}}} D_1/c_0)\tan(f_{\text{even}} \times 2\pi\sqrt{\varepsilon_{\text{reff2}}} D_2/c_0) - \alpha + \\
&\quad + f_{\text{even}} \times 4\beta\pi\sqrt{\varepsilon_{\text{reff3}}} D_3/c_0 \tan(f_{\text{even}} \times 2\pi\sqrt{\varepsilon_{\text{reff2}}} D_2/c_0) + \\
&\quad + f_{\text{even}} \times 4\alpha\beta\pi\sqrt{\varepsilon_{\text{reff3}}} D_3/c_0 \tan(f_{\text{even}} \times 2\pi\sqrt{\varepsilon_{\text{reff1}}} D_1/c_0) = 0
\end{aligned} \tag{4.14}$$

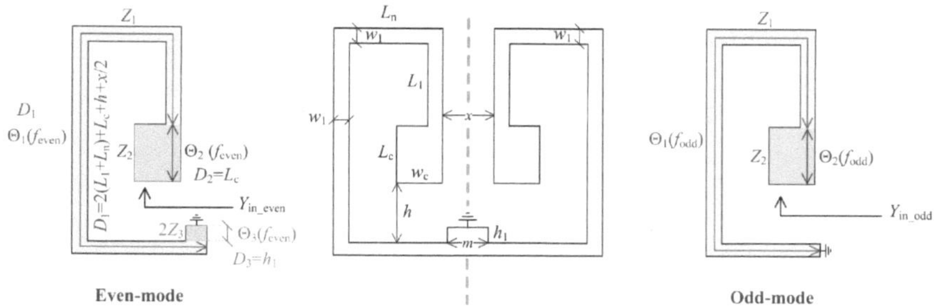

FIGURE 4.15 Dual-mode resonator (2MR).

Eqs. (4.12) and (4.14) are used for design of the dual-mode resonator and calculation of even- and odd-mode frequencies.

In order to present the design steps, 2MR is realized on microstrip technology using a RT/duroid 5880 substrate with the parameters presented in Table 4.7 [32]. In order to introduce the losses due to the surface roughness and the skin effect, the conductor conductivity is set to 20 MS/m. Via-holes are modeled with cylindrical holes of diameter D with microstrip pads.

TABLE 4.7 RT/duroid 5880 substrate parameters.

ε_r	$\tan\delta$	h [mm]	t [μm]	σ [MS/m]
2.2	0.0009	0.508	18	20

In order to find the specified f_{odd}, design procedure begins with optimization of the resonator dimensions D_1, D_2, w_1 (i.e. Z_1) and w_c (i.e. Z_2). The perturbation element has no effect on the odd-mode frequency. The second step is to analyze the even-mode frequency as a function of the perturbation element dimensions D_3, m (i.e. Z_3). The found resonator dimensions are given in Table 4.8.

TABLE 4.8 Dimensions of dual-mode resonator.

D_1 [mm]	D_2 [mm]	D_3 [mm]	w_1 [mm]	w_c [mm]	m [mm]
26	5.1	1	1.1	2.5	5.2

Fig. 4.16 shows the amplitude response of the parameter S_{21} obtained by simulation with the two peaks which correspond to the resonant frequencies of 2MR (f_{even}, f_{odd}). The frequency f_{even} has been analyzed as a function of the parameter D_3 (see Fig. 4.16). For these analyses, the resonator is weakly coupled with the two defined ports, i.e. the resonator is almost unloaded.

The next design step is a miniaturization process of a dual-mode resonator (shown in Fig. 4.15). The resonator might be folded in two layers along the axis of symmetry [33]. The structure consists of two dielectric layers. The common

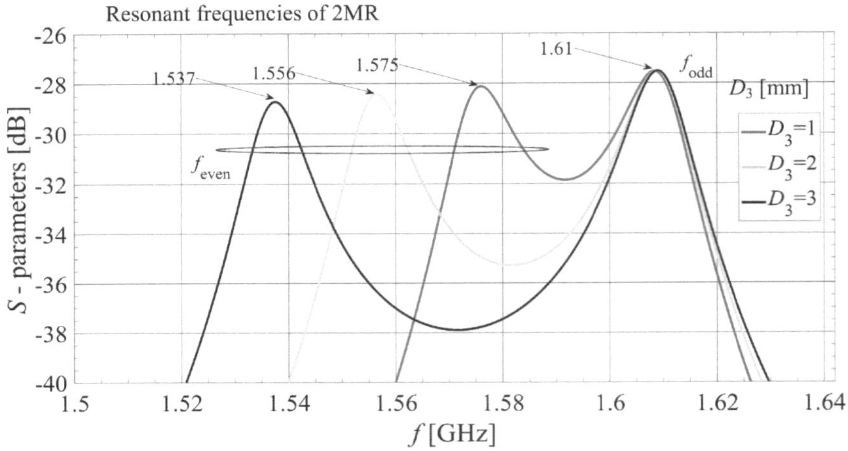

FIGURE 4.16 Amplitude response of the parameter S_{21} as a function of D_3 [mm].

ground plane (GND) is in the middle of the structure, i.e. between the two dielectric layers. The halves of the resonator are printed on different sides of the dielectric layers.

The multilayer 2MR realization accomplishes a footprint reduction around 50%, while the characteristics of the original resonator do not change. None of the resonator dimensions are changed and the desired frequencies f_{even} and f_{odd} have the same value determined by Eqs. (4.12) and (4.14). The multilayer realization has additional vias for the connection between the resonator parts, which are printed on different sides of the dielectric layers.

Next design step is modification of multilayer dual-mode resonator with the memristor (m2MR) in order to realize a reconfigurable resonator (see Fig. 4.17).

The conceptualization of the memristor programming circuit is shown in Fig. 4.17. When the memristor is in the OFF state the even- and odd-mode frequencies are the same as for the original dual-mode resonator. When the memristor is in the ON state, the filter realized with one m2MR has suppressed passband.

The memristor is modeled by a capacitor in the OFF state (C_{OFF} = 1.37 fF), and a resistor in the ON state (R_{ON} = 3.6 Ω) (see Table 4.1). The memristance programming circuitry is realized by a series connection of a resistor and a programmable voltage source. The RF choke is modeled as a linear inductor (see details in Section 4.1).

The memristor-based multilayer dual-mode resonator is shown in Fig. 4.17. An open-circuited stub with an electrical length of around 90°, at the center frequency of the desired passband, has been added.

The parts of the resonator are electrically connected by via holes, as presented in Fig. 4.17. Via^{con} connects parts of the resonator without electrical

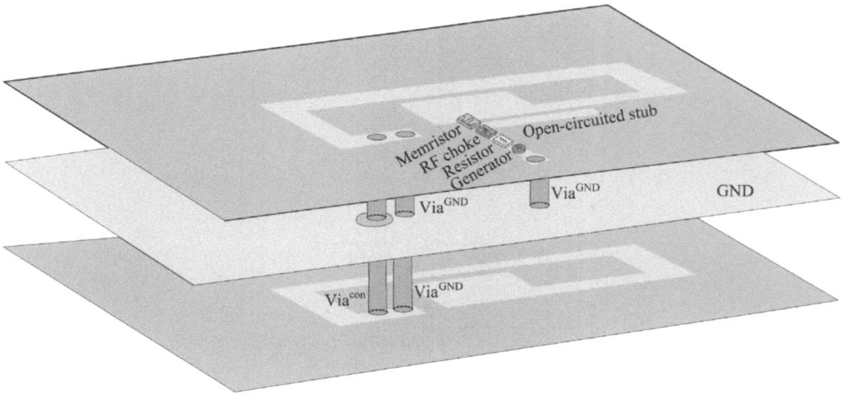

FIGURE 4.17 Memristor-based multilayer dual-mode resonator (m2MR) [21].

contact to the ground plane. An anti-pad has been made in ground plane around Via^{con}. Via^{GND} is used for grounding the top and bottom parts of the resonator.

There are two working modes of the filter. The first working mode is the memristor programming mode and the second is the RF/microwave operating mode, i.e. the cold switching mode of operation has been proposed.

If the memristor is being programmed, the open-circuited stub is almost equal to an open circuit, thus having no influence on the memristance setup process.

In the RF/microwave operating mode, the input impedance of the open-circuited stub is almost equal to zero. This means that one memristor terminal is connected to the virtual ground plane and the other terminal is connected to the resonator. The memristor programming circuit is virtually short-circuited and it does not affect the resonator.

If the memristor is in the ON state, the resonator is loaded with a resistor R_{ON} (see Fig. 4.18). The resonant frequencies are approximately calculated assuming $R_{\text{ON}} \to 0$

$$\begin{aligned}
&\alpha \sin(2\Theta_1)\left(1/2 - \cos^2(\Theta_2)\right) + (\sin(2\Theta_2)/2)\left(\sin^2(\Theta_1) - \alpha^2 \cos^2(\Theta_1)\right) + \\
&+ \Theta_3\left(\sin(2\Theta_1)\sin(2\Theta_2)\left(\alpha^2\beta/2 + \beta/2\right) - \alpha\beta\cos(2\Theta_1)\cos(2\Theta_2)\right) = 0
\end{aligned} \tag{4.15}$$

where electrical lengths of the resonator segments are denoted with Θ_1, Θ_2, Θ_3. If it is assumed that $Z_2 = \alpha Z_1$, $Z_3 = \beta Z_1$ with condition $\alpha < \beta < 1$, the new solution can be found as another resonant frequency. However, the amplitude response peak is significantly suppressed, which is the specified feature of this memristor-based resonator. The optimum position of the connection between the memristor and the resonator is presented in Fig. 4.18.

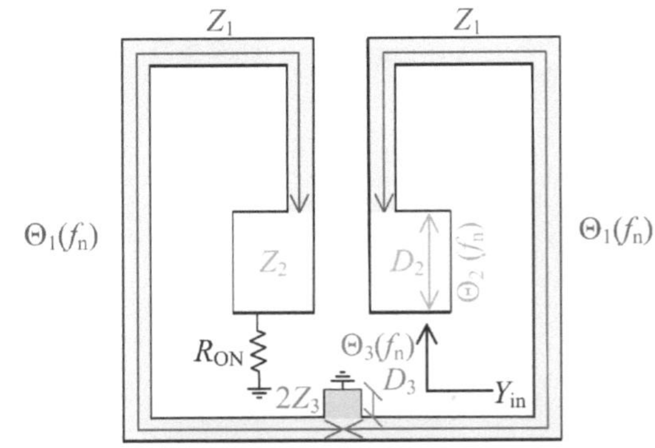

FIGURE 4.18 Memristor-based dual-mode microstrip resonator [21].

As presented in Table 4.1, memristive switches demonstrate the possibility of hot switching, which means that they can be reprogrammed on-the-fly.

Sensitivity analysis of resonator frequencies is investigated as a function of R_{ON} and C_{OFF}. For all analyzed simulations, the resonator is weakly coupled with the two ports. For the OFF state, a variation of the capacitor value (C_{OFF} = 1, 1.37, 1.6 fF; see Table 4.1), does not change the resonator frequencies. For the ON state, the amplitude response peaks are significantly suppressed for different resistor values (R_{ON} = 2.1, 3.6, 6.9 Ω; see Table 4.1).

4.4.2 Dual-band bandpass filter with multilayer dual-mode resonator enhanced with RF memristor

The multilayer reconfigurable filter with two pass bands is presented. The filter center frequencies are 1.6 GHz (L-BandPass) and 3.5 GHz (H-BandPass). The fractional bandwidths are approximately 7%. The specified frequency bands are used for communication services such as GPS navigation and WiMAX access.

The filter is designed as a parallel connection of two single-band bandpass filters. Each single-band filter is designed with a memristor-based multilayer dual-mode resonator (see Fig. 4.17). RF memristors are used to achieve filter reconfigurability.

This filter has two passbands and two RF memristors, and it operates with four states: 1) single bandpass filter at 1.6 GHz – L-BandPass state, 2) single bandpass filter at 3.5 GHz – H-BandPass state, 3) dual-band bandpass filter – Dual-Band state, and 4) bandstop filter with rejection greater than 30 dB up to 7.5 GHz – BandStop state. The filter layout is presented in Fig. 4.19. The equivalent circuit is presented in Fig. 4.20. The operating state is selected by memristors M_{Low} and M_{High}; see Table 4.9 and Fig. 4.19.

A resonator can be independently realized for each passband, because the resonator positions on the substrate are determined to eliminate the unwanted

TABLE 4.9 Operating state selection of memristor-based reconfigurable multilayer dual-mode bandpass filter.

STATE	M_{Low}	M_{High}
L-BandPass	OFF	ON
H-BandPass	ON	OFF
Dual-Band	OFF	OFF
BandStop	ON	ON

coupling. Multilayer technology is used to reduce the filter footprint. The footprint reduction of each resonator is about 50% of the microstrip design.

This filter is realized in microstrip technology. The technology parameters are shown in Table 4.7. The reconfigurable filter is realized in several steps as presented in Fig. 4.21:

1. Design of a dual-mode resonator in microstrip technology (Fig. 4.15).
2. Design of multilayer dual-mode resonator in order to minimize the resonator footprint [33].
3. Design of memristor-based multilayer dual-mode resonator in order to achieve reconfigurability (Fig. 4.17).
4. Design of two single-band bandpass filters: L-BandPass and H-BandPass. The coupling matrix for single-band bandpass filter ($[m]$) is shown in Fig. 4.21 [30]. The input and output filter ports (i.e. source and load) are coupled to the even-resonator mode (marked 1) and odd-resonator mode (marked 2). These coupling coefficients in matrix $[m]$ are m_{S_1}, m_{S_2}, m_{L_1} and m_{L_2}. The resonator modes are not coupled to each other. In this case, for an asynchronously tuned filter, diagonal matrix entries (m_{11} and m_{22}) are nonzero. The relations for finding external quality factors and resonant frequencies for both modes are shown in Fig. 4.21, step #4. The assumption is that coupling between the source and the load is negligible.
5. Design of dual-band bandpass filter as a parallel connection of two single-band bandpass filters. The coupling matrix for the dual-band bandpass filter is

$$[m]^{\text{dual-band}} = \begin{bmatrix} 0 & m_{S_1} & m_{S_2} & m_{S_3} & m_{S_4} & 0 \\ m_{S_1} & m_{11} & 0 & 0 & 0 & m_{L_1} \\ m_{S_2} & 0 & m_{22} & 0 & 0 & m_{L_2} \\ m_{S_3} & 0 & 0 & m_{33} & 0 & m_{L_3} \\ m_{S_4} & 0 & 0 & 0 & m_{44} & m_{L_4} \\ 0 & m_{L_1} & m_{L_2} & m_{L_3} & m_{L_4} & 0 \end{bmatrix}. \tag{4.16}$$

There is coupling between source/load and resonator modes, but coupling between the different resonators is negligible (see Fig. 4.21). In order to sat-

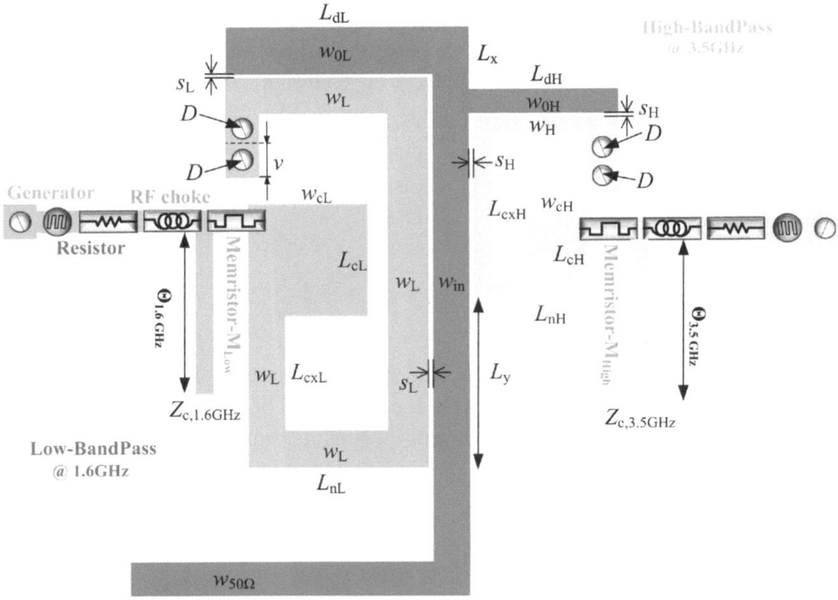

FIGURE 4.19 Reconfigurable memristor-based multilayer dual-mode bandpass filter [21].

isfy the filter specification (i.e. the coupling matrix), the filter dimensions are optimized, and they are presented in Table 4.10.

A memristance programming circuit and an open-circuited stub are connected to each dual-mode resonator. The open-circuited stub is used to realize virtual ground for RF/microwave operating mode and to eliminate influence of the programming circuit. For programming mode, the open-circuit stub is almost an open circuit without any impact on the memristance setup process. Parameters of both open-circuited stubs are tuned in order to eliminate undesired pass bands and to widen the filter rejection band. The optimized parameters of the open-circuited stubs are $Z_{c,1.6\ GHz} = 26\ \Omega$, $\Theta_{1.6\ GHz} = 60°$, $Z_{c,3.5\ GHz} = 63\ \Omega$, $\Theta_{3.5\ GHz} = 68°$. In order to minimize the length of the open-circuited stubs, a cascade connection of stepped transmission-lines can be used. Meandering is another way for miniaturization of the open-circuited stubs. When the characteristic impedance of an open-circuited stub is too low, it would be suitable to print two parallel transmission lines with twice the characteristic impedance. A multilayer realization allows freely optimizing the position of the transmission lines, which is important for a layout miniaturization.

An RF memristor is modeled by a lumped linear time-invariant element (see Table 4.1). The frequency response of the reconfigurable filter is found by using a linear microwave circuit simulator [28]. The circuit simulations are performed at the microwave frequencies with the following assumptions:

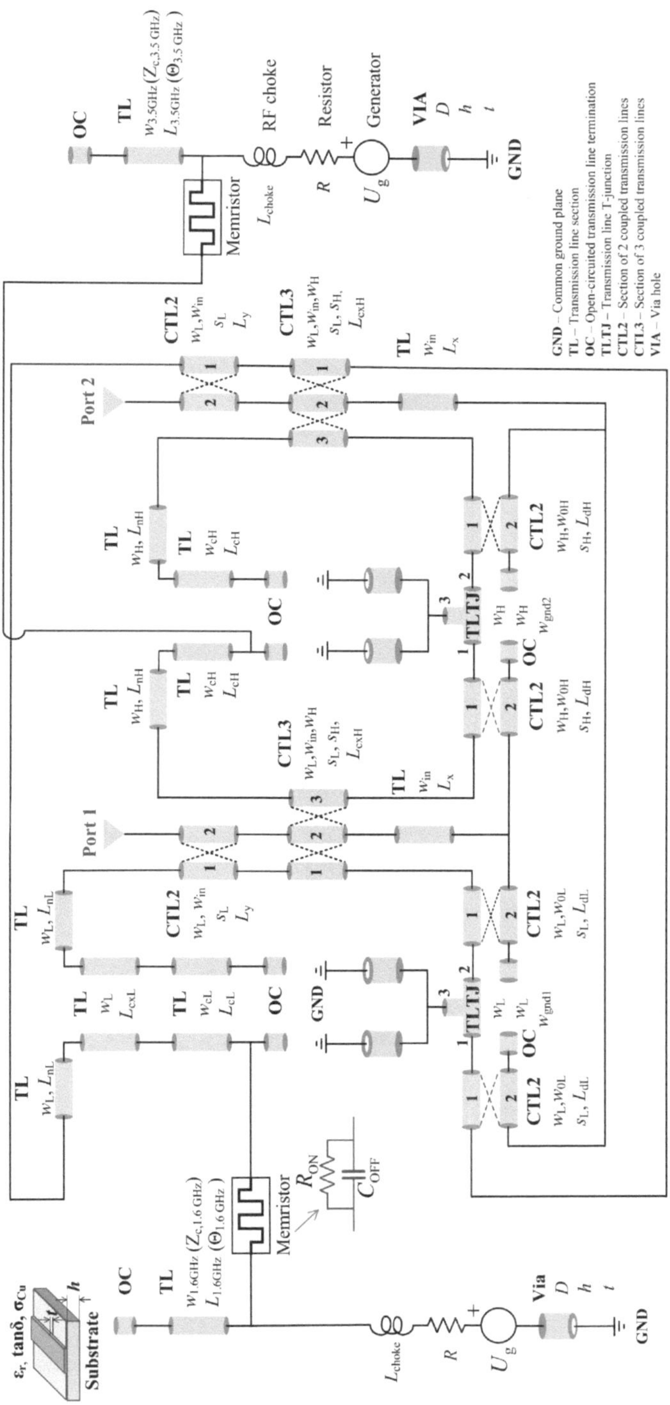

FIGURE 4.20 Equivalent circuit of reconfigurable memristor-based multilayer dual-mode bandpass filter. Substrate – RT/duroid 5880 (see Table 4.7).

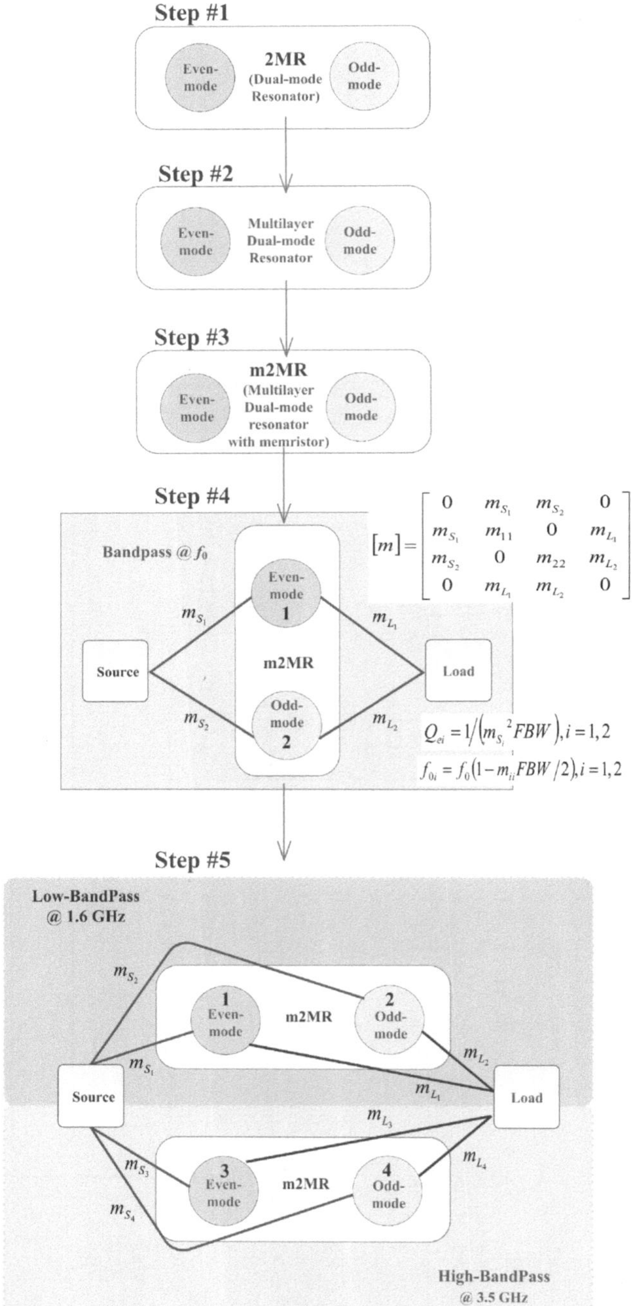

FIGURE 4.21 Design steps of memristor-based reconfigurable multilayer dual-mode bandpass filter.

TABLE 4.10 Reconfigurable filter parameters (see Fig.s 5.19 and 5.20). All dimensions are in mm.

	L-BandPass / H-BandPass								Other dimensions							
i	w_i	s_i	L_{di}	L_{ni}	w_{ci}	L_{ci}	L_{cxi}	w_{0i}	L_x	L_y	D	v	w_{in}	w_{gnd1}	w_{gnd2}
L	1.1	0.1	7.0	4.6	2.52	5.09	3.8	1.4	1.42	5.5	0.6	1.0	0.9	7.5	4
H	0.4	0.05	4.4	2.8	1.43	2.76	3.25	0.7							

TABLE 4.11 Reconfigurable filter characteristics for different operating states.

STATE	Insertion loss	Stopband attenuation
L-BandPass	2.35 dB (1.6 GHz)	
H-BandPass	2.25 dB (3.5 GHz)	
Dual-Band	2.25 dB (1.6 GHz) 1.70 dB (3.5 GHz)	
BandStop		>30 dB (up to 7.5 GHz)

1. The memristor is modeled as in Table 4.1 ($R_{ON} = 3.6\ \Omega$ and $C_{OFF} = 1.37$ fF),
2. The filter equivalent circuit is presented in Fig. 4.20 (see Table 4.10 for dimensions),
3. The memristance programming circuit is realized as presented in Section 4.1 (a serial connection of a resistor of 10 kΩ, a programmable voltage source, an RF choke of 30 nH).

Fig. 4.22 presents the amplitude response in dB for S_{11} and S_{21} of the reconfigurable filter. The filter characteristics are given in Table 4.11 for all operating states. The filter bandwidths satisfy the specification in operating states: L-BandPass, H-BandPass, and Dual-Band BandPass.

One of the main filter characteristics is the power-handling capability. The ON state is analyzed as being critical for power handling. The maximum voltage of the input RF/microwave signal should not change the memristor state from ON to OFF. Three filter operating states are critical for this analysis: L-BandPass, H-BandPass, and BandStop. The dual-band state is not analyzed because both memristors are in the OFF state. For H-BandPass filter state, the critical case is at 3.5 GHz, when M_{Low} is in the ON state (M_{High} is in the OFF state). The maximum voltage of M_{Low} (L-BandPass memristor) is about 20 V, which corresponds to the maximum memristor dissipation of +23 dBm (for ON memristor state when $R_{ON} = 3.6\ \Omega$).

The presented results of the filter reconfigurability inspired us to consider the memristor as a candidate for a novel microwave switch.

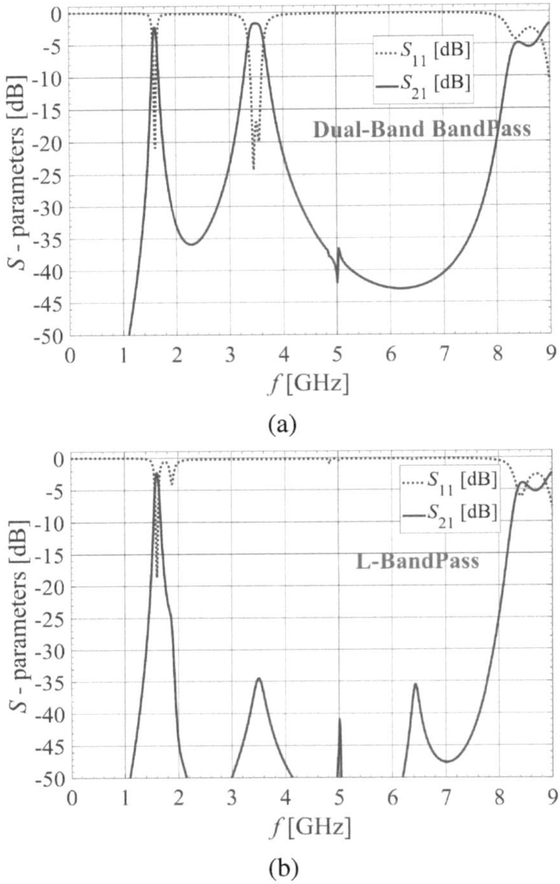

FIGURE 4.22 Frequency responses of the reconfigurable dual-band bandpass filter: (a) Dual-Band BandPass state, (b) L-BandPass state, (c) H-BandPass state, (d) BandStop state.

4.4.3 Conclusion

The memristor is the fourth fundamental electric circuit element, which has attracted significant attention because it can offer a lot of potential applications in various areas of electrical engineering. Possible memristor applications have been shown which are relevant to the RF and microwave component design. Memristors are mostly used as programmable RF/microwave switches with two distinctive states: ON state and OFF state.

RF/microwave filters are parts of RF front-end modules which process high-frequency sinusoidal signals, so it can be expected that the initially set memristance can be considered almost unchanged. Therefore, it can be expected that traditional microwave switches might be replaced by memristors in the RF/microwave filter design. Additionally, the memristor nonlinear modeling is not

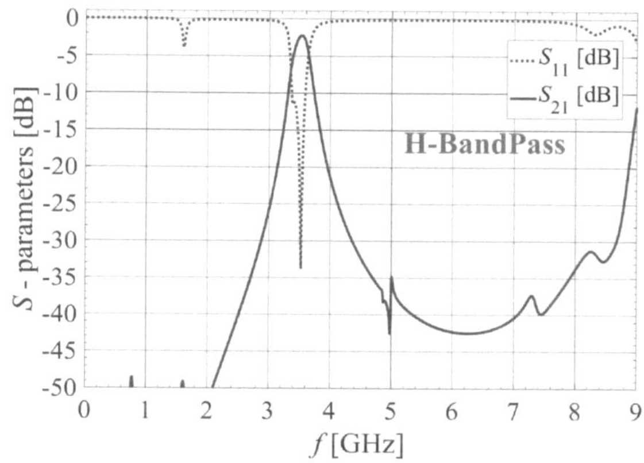

(c)

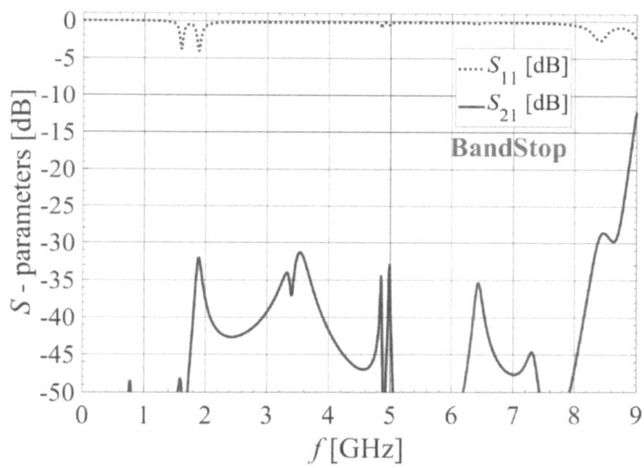

(d)

FIGURE 4.22 *(continued)*

critical, because the memristive switch behaves as a time-invariant linear circuit element at the RF and microwave frequencies.

Reconfigurable dual-band microwave filters have been proposed for wireless communication systems, using memristive switches. These filters are realized in planar technologies such as microstrip and multilayer. The filters are realized as a parallel connection of two single-band bandpass filters. Resonators are enhanced with memristors in order to achieve the filter reconfigurability. A single bandpass filter is designed with single-mode or dual-mode resonators enhanced with memristors. The memristor setup circuit is optimized in order to minimize the circuit influence on the specified filter frequency response. The reconfigurable filters have operated in a few modes which have been selected by memristors. When the reconfigurable filter has been enhanced with one memristor, there were two operating modes: a single-band bandpass mode and a dual-band bandpass mode. The reconfigurable filter with two memristors has had four modes: two single-band bandpass modes for low and high band, a dual-band bandpass mode, and a stopband mode. The filter passband is suppressed, when the memristor is in the ON state. Multilayer filter realization has been used to miniaturize filter footprint area.

The maximum input signal voltage at the memristor has been analyzed in order to determine the limit at which the memristor changes its state. The critical case is when the memristor is in the ON state at the center frequency of passband. The maximum input voltage at memristor is found in order not to exceed the maximum dissipation for the ON memristor state.

Extensive simulations of the proposed filter solutions with memristors have verified the expected behavior and reconfigurability. Accordingly, it can be concluded that the memristor is a promising circuit element for RF/microwave filter design.

Another possible application of the memristor is in the realization of main-line mounted loaded-line phase shifters. The most common application of phase shifters is in the computer control of beam scanning, for example in the phased-array antenna systems. Digital phase shifters allow us to set discrete values of phase shift, using switches. The main idea has been to replace PIN diodes, acting as RF/microwave switches, with memristors in order to reduce power consumption. As a proof-of-concept, circuit-level simulations have been performed to validate the expected functionality. Parasitic effects caused by the memristor programming circuitry, which might be relevant at RF/microwave frequencies, have been taken into consideration.

Memristors could be used instead of traditional RF and microwave switches, because of their specific characteristics: nanoscale dimensions, low power consumption, low insertion loss in the ON state and high isolation in the OFF state at RF and microwave frequencies.

Acknowledgments

A part of this research was supported by the Ministry of Education, Science and Technological Development of the Republic of Serbia.

References

[1] Cobham Delivers First Smart City Coverage Solution in Berlin, Microwave Journal, https://www.microwavejournal.com/articles/29854-cobham-delivers-first-smart-city-coverage-solution-in-berlin. (Accessed 21 July 2020).

[2] Antenna Coupler for Smart City Street Lights: 5G NB IoT Ready, RF Savvy Technologies LLC, Santa Rosa, Calif, https://www.microwavejournal.com/articles/32953-article-headline. (Accessed 21 July 2020).

[3] H. Holma, A. Toskala, T. Nakamura (Eds.), 5G Technology: 3GPP New Radio, John Wiley & Sons, Hoboken, NJ, USA, 2020.

[4] Selection of PCB materials for 5G, Microwave Journal, E-book, February 2018, https://rogerscorp.com/-/media/project/rogerscorp/documents/articles/english/advanced-connectivity-solutions/selection-of-pcb-materials-for-5g.pdf. (Accessed 21 July 2020).

[5] S. Vaidyanathan, C. Volos (Eds.), Advances in Memristors, Memristive Devices and Systems, Studies in Computational Intelligence, vol. 701, Springer, Cham, Switzerland, 2017.
[6] New nanoscale programmable switches promise faster, more versatile chip-scale devices, DARPA, https://www.darpa.mil/news-events/2015-08-06. (Accessed 21 July 2020).
[7] S. Pi, M. Ghadiri-Sadrabadi, J.C. Bardin, Q. Xia, Nanoscale memristive radiofrequency switches, Nature Communications 6 (2015) 1–9.
[8] S. Pi, M. Ghadiri-Sadrabadi, J.C. Bardin, Q. Xia, Memristors as radiofrequency switches, in: Proceedings of the 2016 IEEE International Symposium on Circuits and Systems (ISCAS), Montreal, Canada, May 22 – May 25, 2016, pp. 377–380.
[9] M. Kim, R. Ge, X. Wu, X. Lan, J. Tice, J.C. Lee, D. Akinwande, Zero-static power radio-frequency switches based on MoS 2 atomristors, Nature Communications 9 (2018) 1–7.
[10] M. Kim, E. Pallecchi, R. Ge, X. Wu, G. Ducournau, J.C. Lee, H. Happy, D. Akinwande, Analogue switches made from boron nitride monolayers for application in 5G and terahertz communication systems, Nature Electronics (2020) 1–7.
[11] I. Marković, M. Potrebić, D. Tošić, Z. Cvetković, Comparison of memristor models for microwave circuit simulations, in: Proceedings of the 13th International Conference on Applied Electromagnetics (ПEC), Niš, Serbia, August 30 – September 1, 2017, pp. 1–4.
[12] I. Marković, M. Potrebić, D. Tošić, Memristor state transition in reconfigurable microwave filter, in: Proceedings of the 2017 IEEE 30th International Conference on Microelectronics (MIEL), Niš, Serbia, October 9 – October 11, 2017, pp. 71–74.
[13] J. Vavra, J. Bajer, D. Biolek, RF single-pole double-throw switch based on two-port memistor, in: Proceedings of the 2018 IEEE Radio and Antenna Days of the Indian Ocean (IEEE RADIO), Mauritius, October 15 – October 18, 2018, pp. 1–4.
[14] N. Wainstein, S. Kvatinsky, A lumped RF model for nanoscale memristive devices and non-volatile single-pole double-throw switches, IEEE Transactions on Nanotechnology 17 (5) (2018) 873–883.
[15] M. Potrebić, D. Tošić, Application of memristors in microwave passive circuits, Radioengineering 24 (2) (2015) 408–419.
[16] M. Potrebić, D. Tošić, D. Biolek, RF/microwave applications of memristors, in: S. Vaidyanathan, C. Volos (Eds.), Advances in Memristors, Memristive Devices and Systems, Springer, Cham, Switzerland, 2017, pp. 159–185 (Chapter 7).
[17] M. Potrebić, D. Tošić, Potential applications of memristors in microwave circuits, in: Proceedings of the 13th International Conference on Applied Electromagnetics (ПEC), Niš, Serbia, August 30 – September 1, 2017, I2 1-4 (invited lecture).
[18] K.D. Xu, Y.H. Zhang, L. Wang, M.Q. Yuan, Y. Fan, W.T. Joines, et al., Two memristor SPICE models and their applications in microwave devices, IEEE Transactions on Nanotechnology 13 (3) (2014) 607–616.
[19] N. Wainstein, S. Kvatinsky, TIME—Tunable Inductors Using MEmristors, IEEE Transactions on Circuits and Systems I 65 (5) (2018) 1505–1515.
[20] I. Marković, M. Potrebić, D. Tošić, Main-line memristor mounted type loaded-line phase shifter realization, Microelectronic Engineering 185–186 (2018) 48–54.
[21] M. Potrebić, D. Tošić, A. Plazinić, Reconfigurable multilayer dual-mode bandpass filter based on memristive switch, AEU International Journal of Electronics and Communications 97 (2018) 290–298.
[22] M. Potrebić, D. Tošić, D. Biolek, Reconfigurable microwave filters using memristors, International Journal of Circuit Theory and Applications 46 (1) (2018) 113–121.
[23] N. Wainstein, S. Kvantinsky, An RF memristor model and memristive single-pole double-throw switches, in: Proceedings of the 2017 IEEE International Symposium on Circuits and Systems (ISCAS), Baltimore, USA, May 28 – May 31, 2017, pp. 1–4.
[24] I. Bahl, P. Bhartia, Microwave Solid Slate Circuit Design, 2nd ed., John Wiley & Sons. Inc., New Jersey, 2003.
[25] S. Lucyszyn, Advanced RF MEMS, Cambridge University Press, Cambridge (UK), 2010, pp. 307–343.

[26] I.J. Bahl, K.C. Gupta, Design of loaded-line p-i-n diode phase shifter circuits, IEEE Transactions on Microwave Theory and Techniques 28 (3) (1980) 219–224.
[27] R. Collin, Foundations for Microwave Engineering, 2nd ed., John Wiley & Sons Inc., New York, 2001.
[28] Cadence AWR design environment, Cadence Design Systems, Inc., 2655 Seely Avenue, San Jose, CA, 95134, USA, https://www.awr.com/awr-software/products/awr-design-environment. (Accessed 21 July 2020).
[29] Silicon beam lead PIN diode, https://cdn.macom.com/datasheets/MA4PBL027.pdf. (Accessed 21 July 2020).
[30] J.-S. Hong, Microstrip Filters for RF/Microwave Applications, 2nd ed., John Wiley & Sons, Hoboken, 2011.
[31] D.G. Swanson, Narrow-band microwave filter design, IEEE Microwave Magazine 8 (5) (2007) 105–114.
[32] Rogers Corporation, Rogers, Connecticut, USA, http://www.rogerscorp.com. (Accessed 21 July 2020).
[33] A. Plazinić, M. Potrebić, D. Tošić, Compact microwave multilayer dual-band bandpass filter with folded dual-mode resonators, Journal of Optoelectronics and Advanced Materials 19 (5–6) (2017) 352–358.
[34] K.C. Lee, H.T. Su, M.K. Haldar, A review of centrally loaded multimode microstrip resonators for bandpass filter design, AEU International Journal of Electronics and Communications 69 (10) (2015) 1533–1540.

Chapter 5

The modeling of memcapacitor oscillator motion with ANN and its nonlinear control application

Akif Akgul[a], Murat Erhan Cimen[b], Irene M. Moroz[c], and Ali Fuat Boz[b]
[a]*Department of Computer Engineering, Faculty of Engineering, Hitit University, Corum, Turkey,* [b]*Department of Electrical and Electronic Engineering, Sakarya University of Applied Sciences, Sakarya, Turkey,* [c]*Mathematical Institute, Oxford University, Oxford, United Kingdom*

5.1 Introduction

Design of novel chaotic systems with different features has attracted lots of interest in recent years. These systems can be categorized as without equilibrium points [1,2], only one stable equilibria [3], two stable equilibrium [4], multi-scroll attractors [5,6], different kinds of symmetry [7] etc. Discussions have also been presented of using memristors, memcapacitors, cubic nonlinear resistors, and piecewise linear functions used as a nonlinear factor to form chaos and hyperchaos [8–12]. Chaotic systems have been utilized in fields such as oscillator [13,14], random number generator [15,16], security [17], data hiding [18–20], FPGA [21], control [22], optimization [23], modeling [24], and Artificial Neural Networks (ANNs) [25].

Today, modeling is used in many different fields for specific purposes. ANN methodology has many important features such as learning from data, generalization and working with unlimited number of variables. ANN can provide linear and non-linear modeling without any prior knowledge of input and output variables. In particular, ANN is also used for modeling of nonlinear elements, for examination of complex or multivariable systems and examination of parameters that are difficult or impossible to measure, such as industry, medicine and control areas [26–29]. While creating mathematical models, the expression of physical systems with differential equations and the analysis of the obtained equations with the help of computer sometimes are encountered problems due to simplifying systems, approximations and assumptions. Furthermore, it is more difficult to model nonlinear systems directly from time series information. This can be modeled by combining multiple independent linear equations [30]. In this regard, based on the studies on artificial intelligence, the concept of intelligent modeling has been developed and found the application field.

Mem-elements for Neuromorphic Circuits with Artificial Intelligence Applications
https://doi.org/10.1016/B978-0-12-821184-7.00013-X

Modeling with intelligent structures is based on input–output relationships and that can be used even if there is limited information about the system. The models which were obtained have the ability to adapt to changing conditions and produce results owing to their ability to generalize when an unknown input is applied. In this context, the chaotic memcapacitor oscillator was investigated [31]. A video was created in which an object follows the states of the chaotic memcapacitor oscillator depending on time. After that, object's images which move along the states are tracked using image processing methods. The position data of this system states were taken as time series and an artificial neural network were trained by inputs and outputs. Thus, the chaotic memcapacitor oscillator (CMO) can be modeled with an ANN model which contains multiple inputs–outputs.

5.2 Chaotic memcapacitor oscillator and its dynamical analysis

In this section, a memcapacitor-controlled 3D chaotic oscillator with two unstable equilibria is introduced [31]. To demonstrate the presence of chaotic oscillations, various dynamic properties of the system are investigated such as equilibrium points and bifurcation. We have

$$\begin{aligned} R\frac{dq_{C_m}}{dt} &= V_C + \left(G - \frac{1}{R}\right)(\alpha - \beta\sigma)\, q_{C_m}, \\ C\frac{dV_C}{dt} &= \frac{1}{R}\left((\alpha - \beta\sigma)\, q_{C_m} - V_C\right) - i_L, \\ L\frac{di_L}{dt} &= V_C. \end{aligned} \tag{5.2.1}$$

The CMO is given in Fig. 5.1 and Kirchoff's law is applied to the circuit. In the circuit, R is the resistance, G is the conductance, L is the inductance and C is the capacitance. C_m is the memcapacitor. The current flowing through the circuit are i_g, i_R, i_{Cm}, i_L. Thus Eq. (5.2.1) is obtained. For the sake of

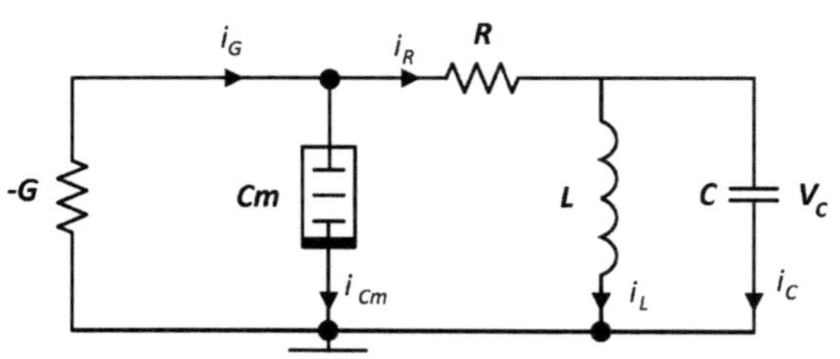

FIGURE 5.1 Memcapacitor-based chaotic oscillator.

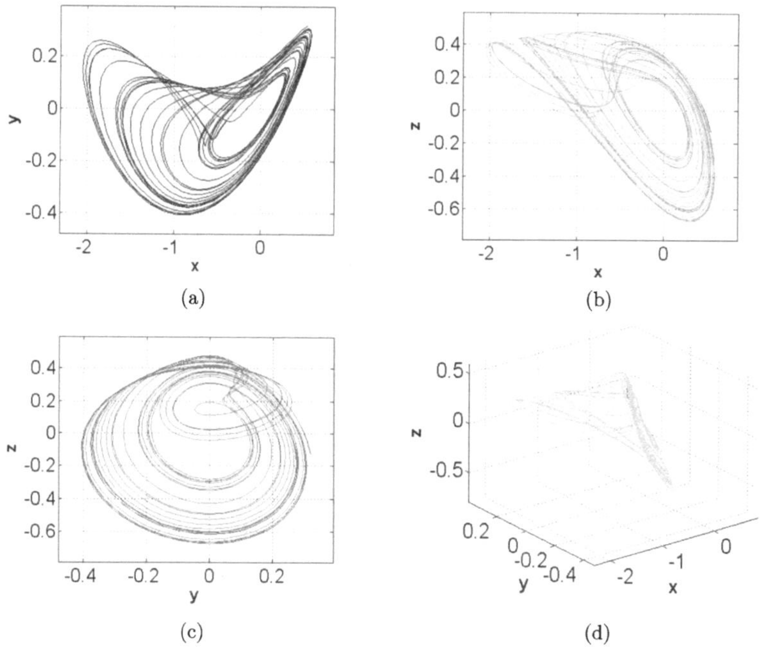

FIGURE 5.2 2D and 3D phase planes of the CMO: (a) x–y; (b) x–z; (c) y–z; (d) x–y–z.

simplicity, Eq. (5.2.1) is transformed to Eq. (5.2.2):

$$\begin{aligned} \dot{x} &= a_1 x + a_2 x^2 + a_3 y, \\ \dot{y} &= a_4 x + a_5 x^2 - y + z, \\ \dot{z} &= a_6 y, \end{aligned} \tag{5.2.2}$$

where the parameters are defined by $a_1 = -1.638$, $a_2 = -0.936$, $a_3 = 4.5$, $a_4 = 0.7$, $a_5 = -0.4$, and $a_6 = -1.75$ for chaotic oscillations. The initial conditions are chosen as $[0.1, 0.1, 0.1]$. Fig. 5.2 shows the 2D and 3D phase planes of system (5.2.2).

5.2.1 Equilibrium points

The system is

$$\dot{x} = a_1 x + a_2 x^2 + a_3 y = F(x, y, z), \tag{5.2.3a}$$

$$\dot{y} = a_4 x + a_5 x^2 - y + z = G(x, y, z), \tag{5.2.3b}$$

$$\dot{z} = a_6 y = H(x, y, z). \tag{5.2.3c}$$

The equilibrium states are found by setting the LHS of (5.2.3) to zero. Eq. (5.2.3c) gives $y_0 = 0$, while (5.2.3a) gives $x_0 = 0$ or $x_e = -a_1/a_2$. Then, (5.2.3b) gives $z_0 = 0$ or $z_e = -x_e(a_4 + a_5 x_e)$.

The linear stability of the transition states can be found by calculating the Jacobian matrix of the third order:

$$J = \begin{pmatrix} a_1 + 2a_2 x & a_3 & 0 \\ a_4 + 2a_5 x & -1 & 1 \\ 0 & a_6 & 0 \end{pmatrix}, \tag{5.2.4}$$

where $x = x_0$ for the trivial equilibrium state, or $x = x_e$ for the nontrivial state. Linear stability is found by computing the determinant of $J - \lambda I_3$ for each equilibrium state. Here λ give the eigenvalues and I_3 is the $(3, 3)$ identity matrix.

(i) For $x_0 = 0$, the characteristic equation is obtained:

$$\lambda^3 - (a_1 - 1)\lambda^2 - (a_1 + a_6 + a_3 a_4)\lambda + a_1 a_6 = 0. \tag{5.2.5}$$

For the given set of parameter values which produce chaotic dynamics:

$$(a_1, a_2, a_3, a_4, a_5, a_6) = (-1.638, -0.936, 4.5, 0.7, 0.4, -1.75), \tag{5.2.6}$$

the roots of the characteristic equation (5.2.5) are $\lambda_1 = -2.8973$, $\lambda_{2,3} = 0.1297 \pm 0.9862i$, so that we have a saddle-focus.

(ii) For x_e, a cubic characteristic equation is obtained:

$$\lambda^3 + (1 + a_1)\lambda^2 + (a_1 - a_6 - a_3 a_4 + 2a_3 a_4 x_e)\lambda - a_1 a_6 = 0. \tag{5.2.7}$$

For the given set of parameter values, the roots of Eq. (5.2.6) become $\lambda_1 = 4.1728$, $\lambda_2 = -3.3284$, $\lambda_3 = -0.2064$, namely a saddle with two unstable directions and one stable direction.

5.2.2 Bifurcation analysis

In this section, the possibility codimension one steady and Hopf bifurcations for each of these equilibria will be investigated. Steady state bifurcations occur when either $a_1 = 0$ or $a_6 = 0$ for each of these equilibria. For the trivial equilibrium, a Hopf bifurcation occurs when $\lambda = i\omega$. If we substitute this value into Eq. (5.2.5) and equate real and imaginary parts, we find that ω^2 satisfies

$$\omega^2{}_R = \frac{a_1 a_6}{1 - a_1} > 0, \quad \omega^2{}_I = -(a_1 + a_6 + a_3 a_4) > 0. \tag{5.2.8}$$

Since both of these expressions for ω^2 must be equal, we obtain the curve

$$a_1{}^2 + a_1(a_3 a_4 - 1) - (a_6 + a_3 a_4) = 0. \tag{5.2.9}$$

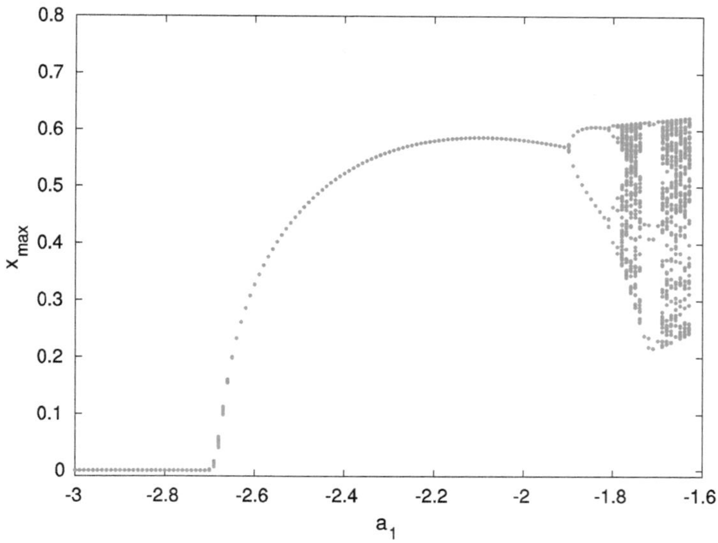

FIGURE 5.3 A bifurcation transition diagram of x_{max} as a_1 decreases.

Fixing all parameters except for a_1 (so that we take a_1 as the bifurcation parameter), Eq. (5.2.9) has the solution $a_1 = -2.6736$ and $a_1 = 0.5236$. Substituting $a_1 = -2.6736$ into Eq. (5.2.8) we find that $\omega^2 = 1.2736$ for both quantities in Eq. (5.2.8). We therefore have a Hopf bifurcation for this value of a_1. If, however, we select the second root $a_1 = 0.5236$, we now have $\omega^2 = -1.9234$. Since this value of ω^2 violates the condition of positivity, there is no Hopf bifurcation for this value of a_1.

Moreover, substituting $a_1 = -2.6736$ into the characteristic Eq. (5.2.5) gives the roots $\lambda_1 = -3.6736$, $\lambda_{2,3} = \pm 1.1286i$, so that the Hopf bifurcation is supercritical. (For the second value $a_1 = 0.5236$, (5.2.5) gives $\lambda_{1,2} = \pm 1.3869$, $\lambda_3 = -0.4763$, i.e. a saddle point.)

Repeating the same procedure for the nontrivial equilibrium solution, we find the equivalent to Eq. (5.2.8) is

$$\Omega^2{}_R = -\frac{a_1 a_6}{1+a_1} > 0, \quad \Omega^2{}_I = a_1 - a_6 - a_3 a_4 + 2a_3 a_4 x_e > 0. \tag{5.2.10}$$

Elimination of Ω^2 results in a quadratic equation

$$a_1{}^2(1 - \frac{2a_3 a_4}{a_2}) + a_1(1 - a_3 a_4 - \frac{2a_3 a_4}{a_2}) - (a_6 + a_3 a_4) = 0. \tag{5.2.11}$$

Again fixing all parameters, except for a_1 gives $a_1 = -0.8148$ (not possible since $\Omega^2{}_R < 0$) or $a_1 = 0.2223$ (which gives $\Omega^2 = 0.3185$).

We now show a bifurcation transition plot as a_1 varies (Fig. 5.3) by plotting successive maxima of x over each cycle. There is evidence of a Hopf bifurcation from the trivial equilibrium state x_0 as well as period doubling cascades for $a_1 > -1.9$. Moreover, we can identify this supercritical Hopf bifurcation as that associated with the trivial equilibrium state, described above.

5.3 Nonlinear feedback control

We now illustrate how to use nonlinear feedback control to drive the chaotic dynamics of the system to either the trivial or the nontrivial equilibrium state. Adding control signals u_1, u_2, and u_3 to the right hand sides of Eqs. (5.2.3)a–c gives

$$\dot{x} = a_1 x + a_2 x^2 + a_3 y + u_1, \tag{5.3.1a}$$

$$\dot{y} = a_4 x + a_5 x^2 - y + z + u_2, \tag{5.3.1b}$$

$$\dot{z} = a_6 y + u_3. \tag{5.3.1c}$$

Writing the equilibrium state generally as $\mathbf{x_e} = (x_e, y_e, z_e)$, we introduce error perturbations $\mathbf{x_e}$: $e_1 = x - x_e$, $e_2 = y - y_e$, $e_3 = z - z_e$ about this state. Substituting for x, y, and z into Eqs. (5.3.1) results in a set of equations for e_j:

$$\dot{e_1} = a_1 e_1 + a_2 (e_1{}^2 + 2 e_1 x_e) + a_3 e_2 + u_1, \tag{5.3.2a}$$

$$\dot{e_2} = a_4 e_1 + a_5 (e_1{}^2 + 2 e_1 x_e) - e_2 + e_3 + u_2, \tag{5.3.2b}$$

$$\dot{e_3} = a_6 e_2 + u_3, \tag{5.3.2c}$$

where we have used the equilibrium conditions to simplify Eqs. (5.3.2).

We now introduce the Lyapunov function

$$V(e_1, e_2, e_3) = \frac{1}{2}(e_1{}^2 + e_2{}^2 + e_3{}^2), \tag{5.3.3}$$

a positive definite quantity. The objective is to choose controls u_1, u_2, and u_3 to make dV/dt negative definite, and so ensure that the error system (5.3.2) decays asymptotically to zero, so that system (5.3.1) converges to the chosen equilibrium state $\mathbf{x_e}$.

Differentiating V w.r.t. t and substituting for $\dot{e_1}$, $\dot{e_2}$, and $\dot{e_3}$ gives, after some simplification,

$$\dot{V} = (a_1 e_1{}^2 - e_2{}^2 - e_3{}^2) + e_1 (a_2 e_1{}^2 + 2 a_2 x_e e_1 + (a_3 + a_4) e_2 + u_1) \tag{5.3.4a}$$

$$+ e_2 (2 a_5 e_1 x_e + a_5 e_1{}^2 + u_2) + e_3 (e_2 (1 + a_6) + e_3 + u_3). \tag{5.3.4b}$$

Choosing

$$u_1 = -(a_2 e_1{}^2 + 2 a_2 x_e e_1 + (a_3 + a_4) e_2), \tag{5.3.5a}$$

$$u_2 = -(2 a_5 e_1 x_e + a_5 e_1{}^2), \tag{5.3.5b}$$

$$u_3 = -(e_2 (1 + a_6) + e_3), \tag{5.3.5c}$$

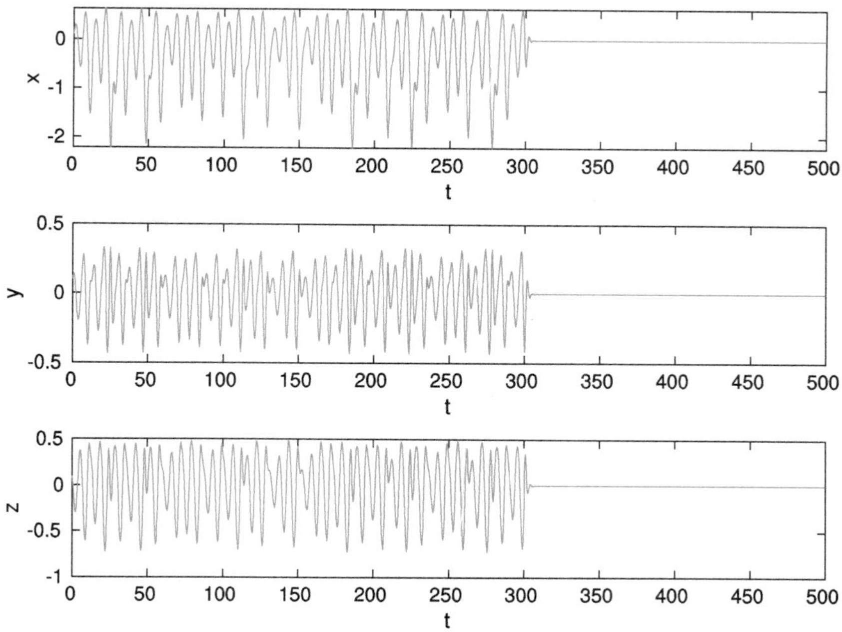

FIGURE 5.4 The time series of x, y, and z for the given set of parameter values for the trivial equilibrium state $\mathbf{x_e} = (0, 0, 0)$. The uncontrolled system (1.1) is integrated for 300 s, before applying the control and switching to system (2.1) for the next 200 s.

reduces Eq. (5.3.6) to

$$\dot{V} = a_1 {e_1}^2 - {e_2}^2 - {e_3}^2. \tag{5.3.6}$$

Moreover, since $a_1 < 0$ for the chaotic example, $\dot{V}$ is negative definite, and system (2.1) evolves to the chosen equilibrium state.

We illustrate the nonlinear controller for each of the two equilibria.

In each case, the errors e_j decay to zero, and the system evolves either to $(0, 0, 0)$ or to $(-1.75, 0, 0)$, respectively. With nonlinear feedback control that is constructed using the Lyapunov function, the system can be controlled, moreover, equilibrium points might be changed during system operation (Figs. 5.4, 5.5).

5.4 Chaotic motion extraction from video and ANN

Object tracking is generally the tracking of an object in a video or sequence of images. Videos are a combination of many images. Therefore, tracking of any object on the video is provided by determining the position of the object in the image. Object tracking methods are basically divided into three categories [32,33]. These are point-based, core-based and silhouette-based methods,

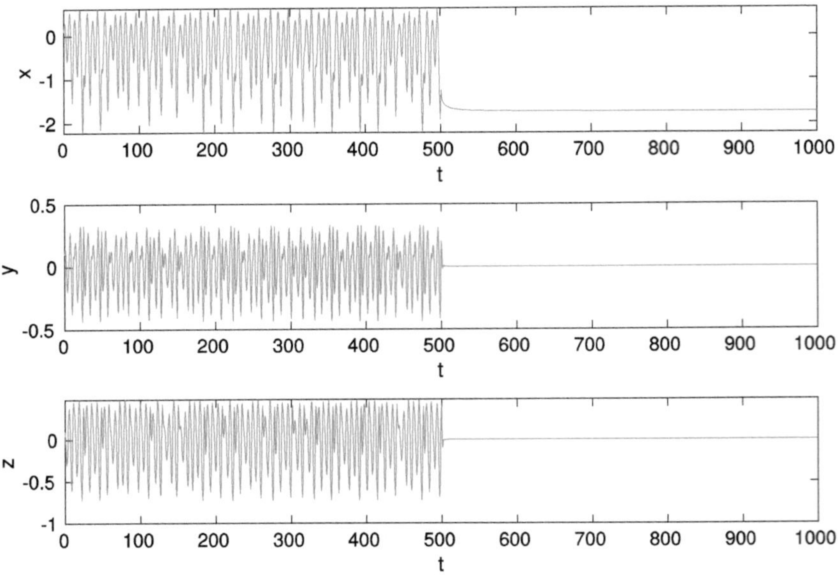

FIGURE 5.5 The time series of x, y, and z for the given set of parameter values for the nontrivial equilibrium state $\mathbf{x_e}$. The uncontrolled system (1.1) is again integrated for 300 s, before applying the control and switching to system (2.1) for the next 200 s.

respectively. In the point-based tracking method, the object to be tracked is expressed with points.

The data such as the positions of these points in the next image and their distance from each other are expected to be parallel to each other in the subsequent video frame. Object tracking is provided based on this information. In core-based methods, the object to be traced with the help of geometric shape is framed. Meaningful information of the object part in this frame is calculated and the object is followed with the help of the initial shape. In this method, it is sufficient to extract the object information contained within the object's geometric shape.

Silhouette-based methods are generally used in cases where the object which is to be traced cannot be expressed in a certain geometric shape, such as human or animal. The main purpose of this method is to search for this information in subsequent images by extracting edge information or shape information to identify the object. This method is very sensitive to shape. Hence, these categories have advantages and disadvantages. Nevertheless, the purposes of these methods are still the same; obtaining the position information of the desired image within the video. This is done with image processing techniques. Many techniques are available, such as image matching, template matching [34], directional gradient histogram [35,36], HAAR cascade classifier [37,38], principal component analysis [39], independent component analysis [40], support vector

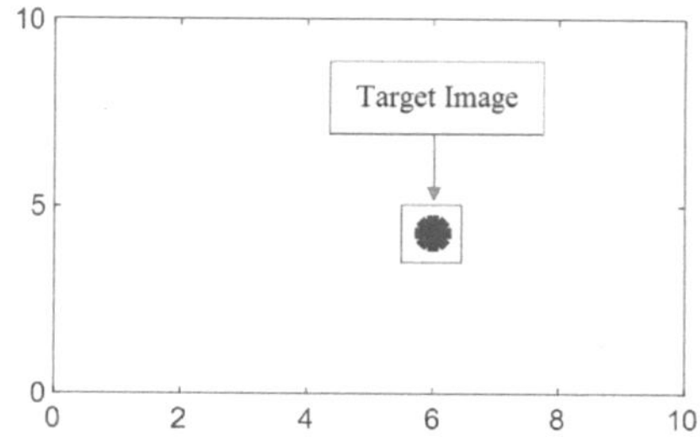

FIGURE 5.6 Sample picture.

machines [41,42], SIRF [43,44], and SURF [32,45]. For the purpose of exemplary illustration, we determine the moving target image in Fig. 5.6 to obtain location information.

Artificial neural networks are generally an information processing technology based on the human brain's ability to work and think. ANN, in other words, are computer programs that imitate biological neural networks. Examples such as macroeconomic forecasts, evaluation of bank loans, exchange rate estimates, and risk analyzes also find application in financial matters. It is used in medical science, in the analysis of medical signals and cancerous cells, in defense industry applications such as target tracking, recognizing objects or images, and determining flight trajectories of military aircraft. ANN applications are generally used in prediction, classification, data association, data interpretation, and data filtering processes [46,28,29].

5.4.1 Perceptron architecture

The basic structure of ANN is called a perceptron. This structure is shown in Fig. 5.7. It consists of inputs, weights, operator, and activation function. The input data passes through the weights and is processed by an operator on the neuron to generate a network value. This network value is transferred to the ANN output via a function called activation function. So, due to it having a single neuron, it is also called a single-layer neural network.

Units of ANN are inputs, weights operator, activation function, and output. Inputs are numerical values that are entered to ANN. Weights are certain value and these values are multiplied with inputs. After multiplying the inputs and weights, these values combine in the neuron. At this point, the operator comes into action. The operator can use different functions such as minimum, maximum, summation, multiplication. An operator can use different functions such as the minimum given in Eq. (5.4.1), the maximum given in Eq. (5.4.2), the summation given in Eq. (5.4.3), and multiplication given in Eq. (5.4.4). The operator generates a NET value based on the selected operator process. This generated value is transferred to the system output via the activation function. The activation function is generally one of selected functions such as linear, step, sigmoid,

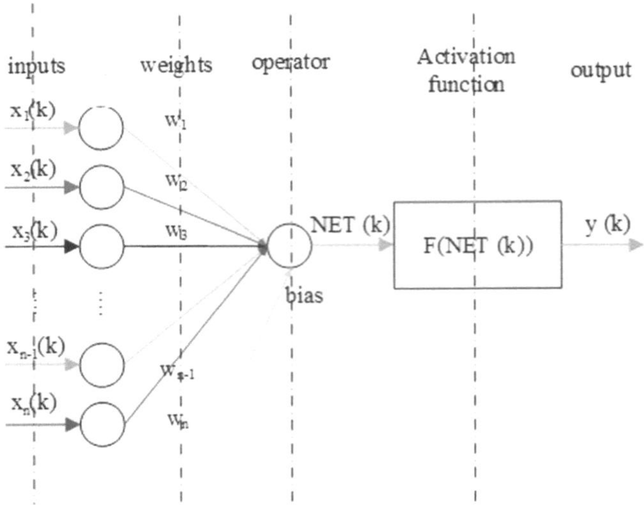

FIGURE 5.7 Perceptron neural network.

tangent hyperbolic. We have

$$NET = \min(x_1(k), x_2(k), \ldots, x_n(k)) \tag{5.4.1}$$

$$NET = \max(x_1(k), x_2(k), \ldots, x_n(k)) \tag{5.4.2}$$

$$NET = \left(\prod_{i=1}^{n} w_i x_i(k)\right) bias \tag{5.4.3}$$

$$NET = \sum_{i=1}^{n} w_i x_i(k) + bias \tag{5.4.4}$$

The activation function can be expressed with Eq. (5.4.5), which takes the linear sum of the values coming into the network as in Eq. (5.4.3) or multiplication of the values coming into the network as in Eq. (5.4.4). The activation function in the form of a step is given in Eq. (5.4.6). The logarithmic activation function is given in Eq. (5.4.7). Besides, sinus, sigmoid, tangent hyperbolic functions are given in Eqs. (5.4.8), (5.4.9), and (5.4.10); these are differentiable and continuous functions, respectively. In general, in this study, the sigmoid activation function is used as in Eq. (5.4.9) and the summation operator is used as in Eq. (5.4.4). We have

$$f(NET) = NET \tag{5.4.5}$$

$$f(NET) = \begin{cases} 0 & NET < 0 \\ 1 & NET \geq 0 \end{cases} \tag{5.4.6}$$

$$f(NET) = \begin{cases} \log(1 - NET) & NET < 0 \\ \log(1 + NET) & NET \geq 0 \end{cases} \tag{5.4.7}$$

$$f(NET) = \sin(NET) \tag{5.4.8}$$

$$f(NET) = \frac{1}{1 + e^{-NET}} \tag{5.4.9}$$

$$f(NET) = \frac{e^{NET} - e^{-NET}}{e^{NET} + e^{-NET}} = \frac{e^{2NET} - 1}{e^{2NET} + 1} \tag{5.4.10}$$

Firstly, forward calculation is made in the network structure. This is done by multiplying the inputs and the weights. Input variables are $x_1, x_2, \ldots, x_n$ and the number of variables is n. This structure has $w_1, w_2, \ldots, w_n$ weights that enable the inputs to be transferred to the neuron. The calculation of the network value using the values that reach the neuron through these variables is done as in Eq. (5.4.11). Linearly, it is the summation of multiplications of inputs and weights. In addition, a bias value has been added to the network as an add-on. This network value produced in the neuron is activated by the activation function and transferred to the neural network output as in Eq. (5.4.12). We have

$$\begin{aligned} NET = {} & x_1(k)w_1 + x_2(k)w_2 + x_3(k)w_3 + \ldots \\ & x_{n-1}(k)w_{n-1} + x_n(k)w_n + bias \\ NET = {} & \textstyle\sum_{i=1}^{n} x_i(k) + bias \end{aligned} \tag{5.4.11}$$

$$y(k) = \varphi(NET) = \frac{1}{1 + e^{-NET}} \tag{5.4.12}$$

5.4.2 Multilayer artificial neural networks

The structure in which multiple neurons are created in ANN is called multi layer artificial neural networks (MANN), such as given in Fig. 5.8. This structure contains more neurons and layers, unlike the single-layer perceptron structure. In this structure, since the inputs are not transferred directly to the output, this MANN contains a hidden layer. The output value of the network is calculated on each layer and transferred to the output.

5.4.3 Delayed artificial neural networks

The unit delay is expressed as the delay of the signal as an example in discrete time signals. That is, the previous value is preserved in memory and transferred

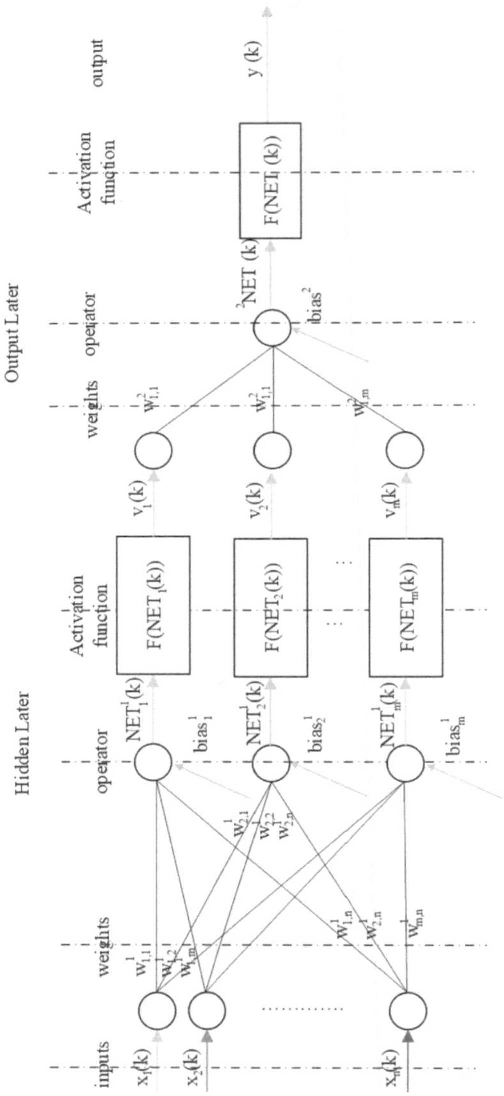

FIGURE 5.8 Multilayer artificial neural network.

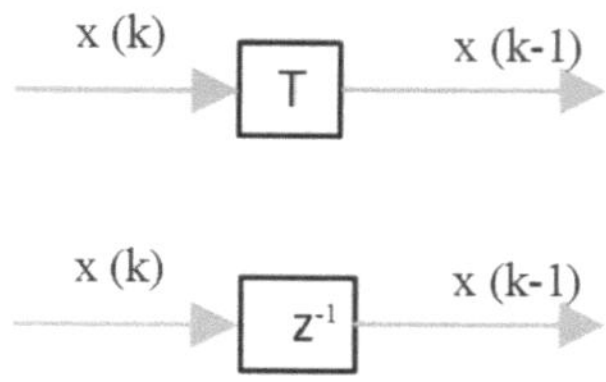

FIGURE 5.9 Unit delay.

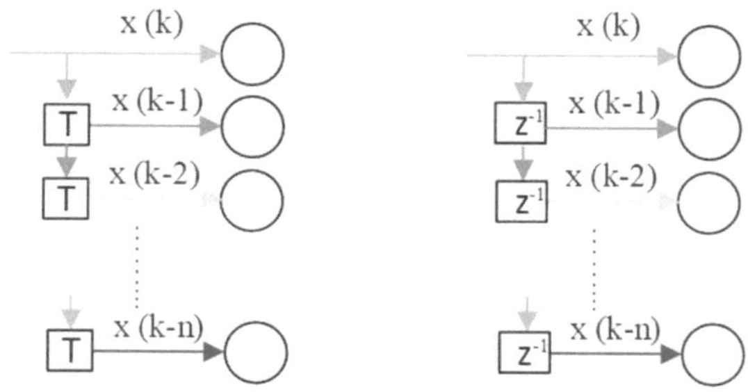

FIGURE 5.10 Unit delay at neuron.

to the output after a sampling. This transform can be given by T or z^{-1} as in Fig. 5.9. When multiple delays are desired, states are preserved in memory by adding more unit delay elements. Then, from the end of the delay, the input signal is transferred to the output of the system. More than one input delay in the input signal is shown in Fig. 5.10. These delays could be at inputs, output, and feedbacks inputs.

5.4.4 Training artificial neural networks

In ANN, there are outputs that are required to produce in response to the given inputs. These outputs are certain and are expected to generate values when any input to the ANN arrives. It is also expected to make estimates for different input data. There are many methods in training ANN. In general, methods such as the backpropagation method (BP) [46], the Levenberg–Marquardt method (LM) [47], and the fast propagation method [48,49], which are gradient based methods, are used. In addition, artificial neural networks can be trained with recently developed heuristic algorithms [50].

5.5 Identification of memcapacitor system with ANN

The dynamic structure of CMO obtained from time series was modeled with ANN. For this, time series of CMO is obtained with Runge–Kutta 4 (RK4) method. The states of this system depending on time are given in Fig. 5.11.

An error function to be minimized has been determined between the output of the ANN and the chaotic system outputs in order to train the ANN with the

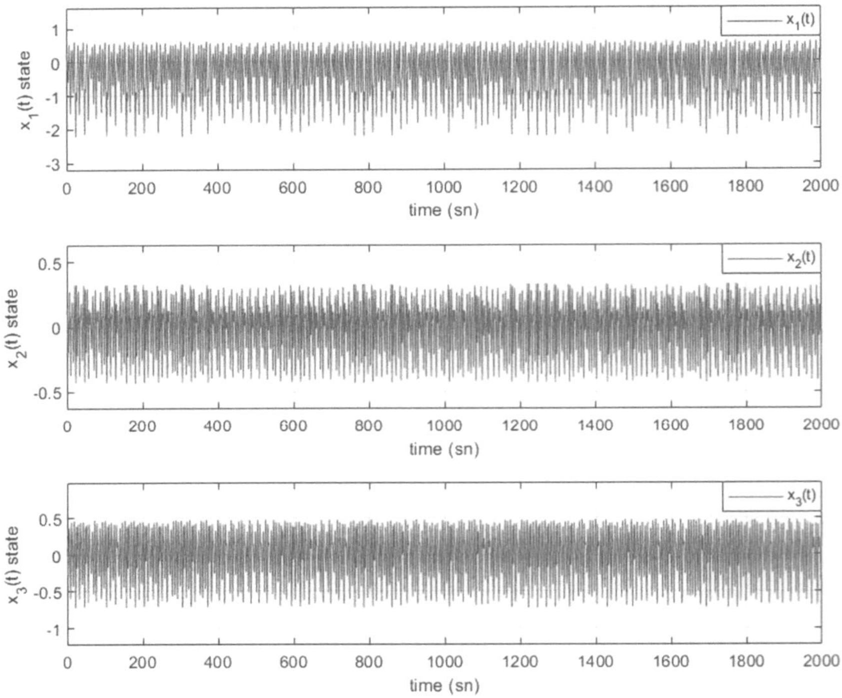

FIGURE 5.11 Chaotic memcapacitor oscillator states.

time series produced by the CMO. When the value of this error function is zero, the CMO is parametrically designed as an ANN model. In this study, this error function was chosen as mean square error (MSE). The convergence of the value of this error function to zero is only possible by determining the correct weights. For this, a backpropagation algorithm and Levenberg Marquardt algorithms are used in this study. The training structure is given in Fig. 5.12 in general.

The ANN model created is given in Fig. 5.13. The model has inputs that are the states of the system, 2 hidden layers, and outputs. In the model, the number of inputs is determined as 3, the number of feedback delays is determined as 10, the number of neurons in the first layer is determined as 10, 10 in the hidden layer and 3 outputs in the output layer. For the training of ANN, nntraintool in nntool given in Fig. 5.14 and Fig. 5.15 is used. Besides, a similar and brief ANN design in Fig. 5.13 is available in Fig. 5.14 and Fig. 5.15. The ANN model of the CMO was trained with the backpropagation method in Fig. 5.14 and the Levenberg–Marquardt method in Fig. 5.15. Parameters are given in Table 5.1 for these algorithms to be trained by the given data. Training results are given in Table 5.2. According to these results, although the iteration number of the backpropagation method is 1000 times higher than the Levenberg–Marquardt method, the Levenberg–Marquardt method gave better results in terms of per-

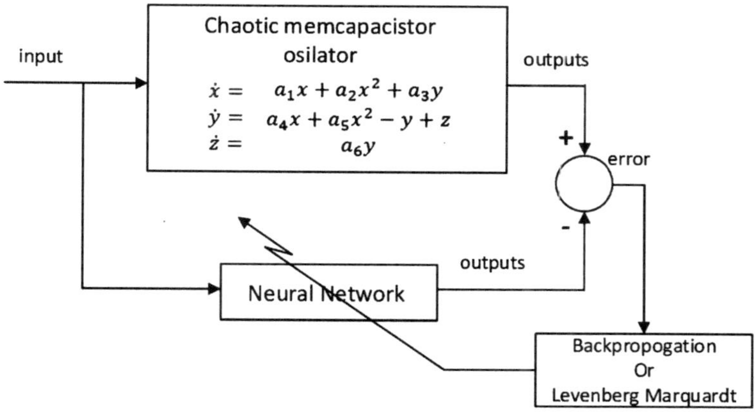

FIGURE 5.12 Artificial neural network training.

TABLE 5.1 Training parameters of ANN.

	Backpropagation	Levenberg–Marquardt
Iteration number	1000000	1000
Training results	–	0.001

TABLE 5.2 Training results of ANN model.

	Backpropagation	Levenberg–Marquardt
Time	1165 s	565 s
Performance	0.000422	0.000279
Gradient	0.000525	0.000279

formance and gradient. Also, in terms of training performance, the Levenberg Marquardt runs more slowly but gives better results in fewer iterations. If the results obtained with this modeled ANN structure did not give good results, then the number of layers, feedback delays, neurons or the number of iterations used in training could be changed in the ANN model.

With the trained MANN, it is desired to obtain the states of the CMO again. The x and y states of ANN and real CMO trained by backpropagation method are given in Fig. 5.16. The results of the ANN trained with the Levenberg–Marquardt method on the X–Y axes are given in Fig. 5.17. As can be seen, the MANN trained with the Levenberg–Marquardt method follows the trajectory created by CMO more accurately.

With the trained MANN, one desires to obtain the states of the CMO again. The z and x states of ANN and real CMO trained by the backpropagation method are given in Fig. 5.18. The results of the ANN trained with the Levenberg–Marquardt method on the Z–X axes are given in Fig. 5.19. As can be seen, the

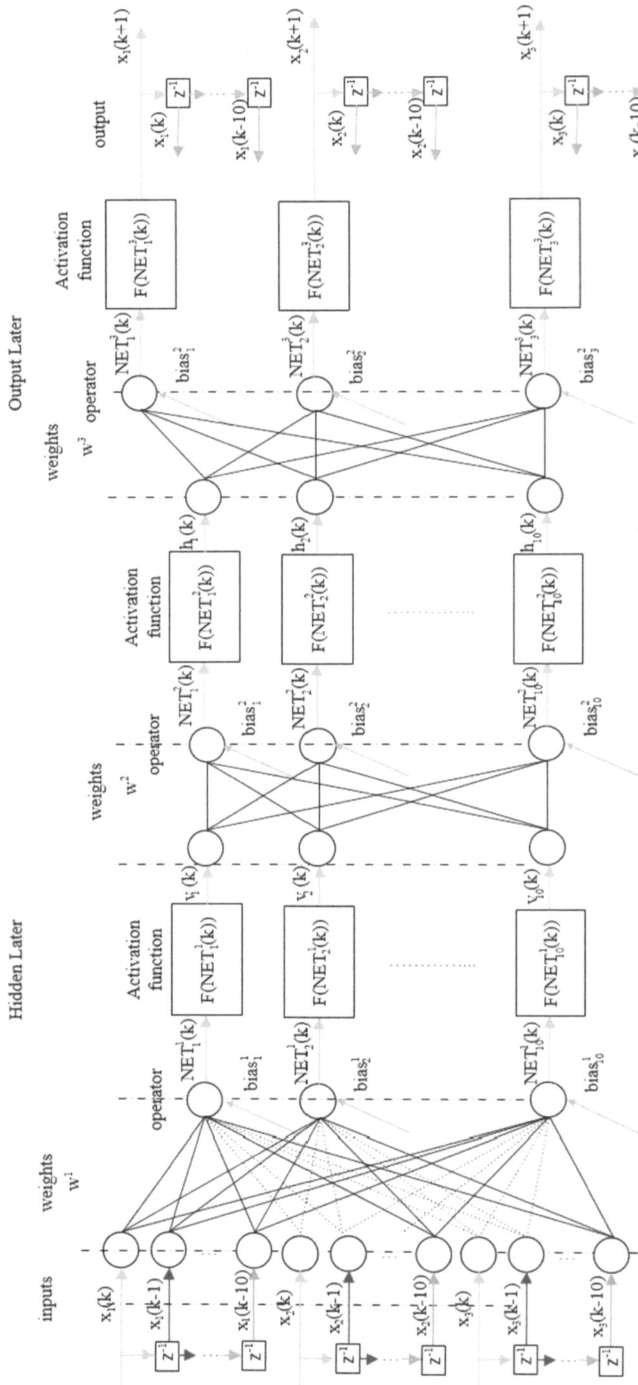

FIGURE 5.13 Artificial neural network model of chaotic memcapacitor oscillator system.

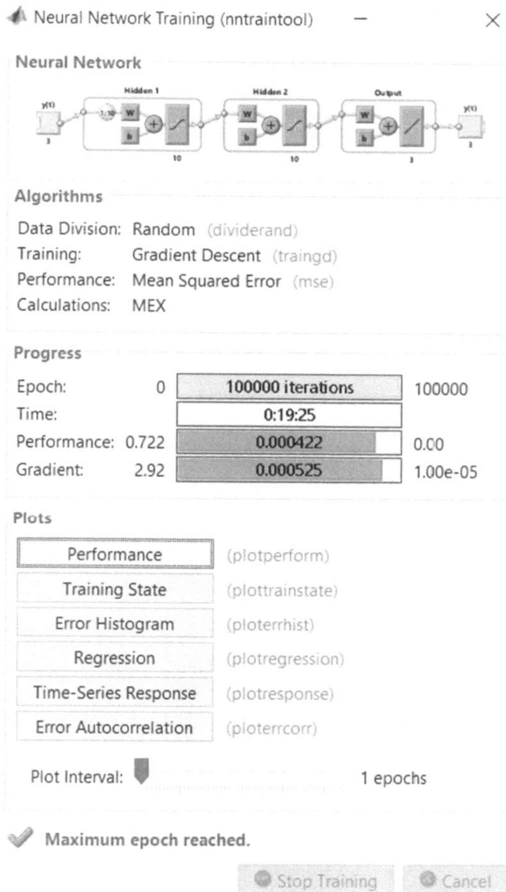

FIGURE 5.14 Neural network training with backpropagation.

MANN trained with the Levenberg–Marquardt method follows the trajectory created by CMO more accurately.

With the trained MANN, one desires to obtain the states of the CMO again. The z and y states of ANN and real CMO trained by the backpropagation method are given in Fig. 5.20. The results of the ANN trained with the Levenberg–Marquardt method on the Z–Y axes are given in Fig. 5.21. As can be seen, the MANN trained with the Levenberg–Marquardt method follows the trajectory created by CMO more accurately.

With the trained MANN, one desires to obtain the states of the CMO again. The x, y, and z states of ANN and real CMO trained by the backpropagation method are given in Fig. 5.22. The results of the ANN trained with the Levenberg–Marquardt method on the X–Y–Z axes are given in Fig. 5.23. As

FIGURE 5.15 Neural network training with Levenberg–Marquardt.

can be seen, the MANN trained with the Levenberg–Marquardt method follows the trajectory created by CMO more accurately.

5.6 Conclusions

In this study, we have investigated a CMO with its engineering applications. The dynamical properties of the CMO are realized with phase portraits, equilibrium points and bifurcation. After that, we have implemented nonlinear feedback control and the modeling of the CMO with ANN as engineering applications. In the modeling with ANN, an object movement, which follows states of CMO, is produced as a video that is composed of sequences of images in time. The object in the images which might be anything, is recognized by image processing techniques. After that position information of the object in the image is ob-

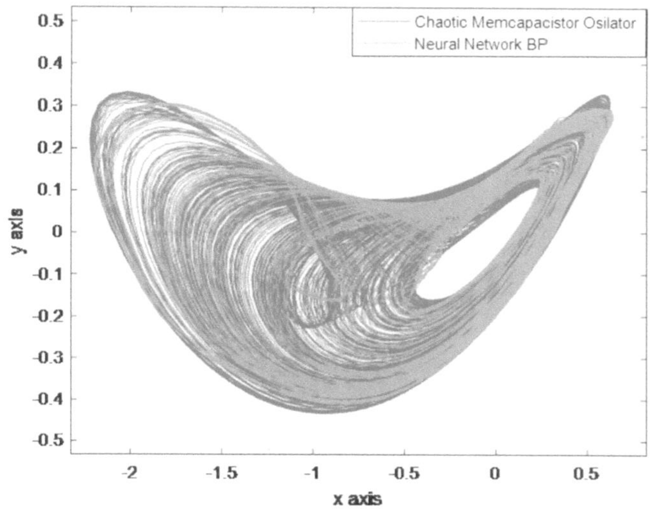

FIGURE 5.16 Real CMO response and ANN with backpropagation training responses X–Y axis.

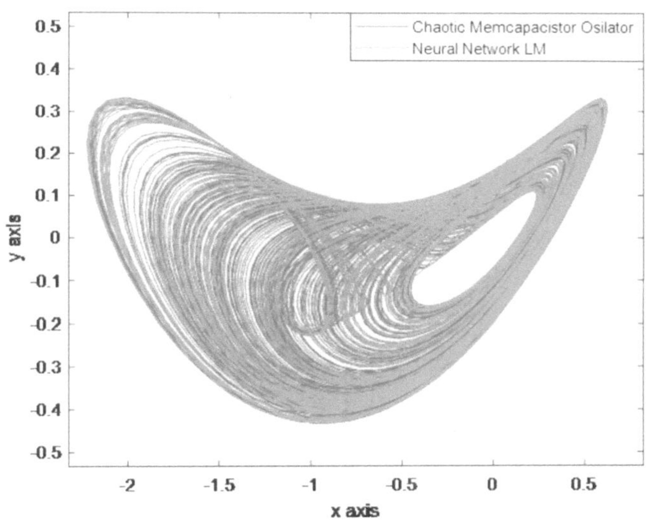

FIGURE 5.17 Real CMO response and ANN with Levenberg–Marquardt training responses X–Y axis.

tained. To generate the movement of the CMO, MANN is modeled. Training of the MANN, backpropagation, and Levenberg–Marquardt methods are used. The Levenberg–Marquardt method gave much better results and performance than backpropagation.

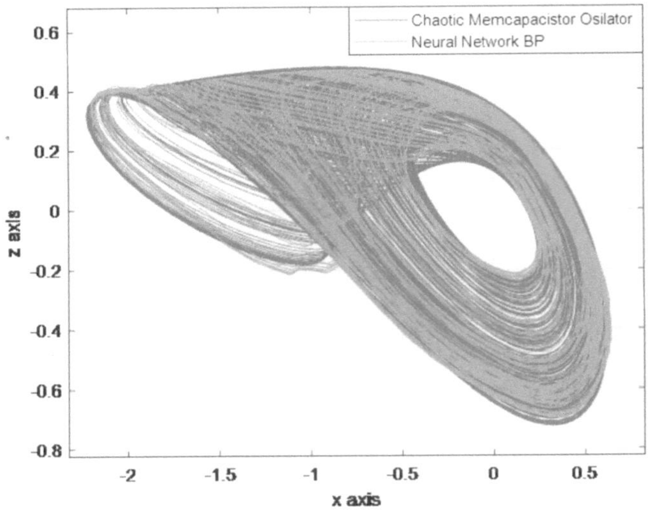

FIGURE 5.18 Real CMO response and ANN with backpropagation training responses Z–X axis.

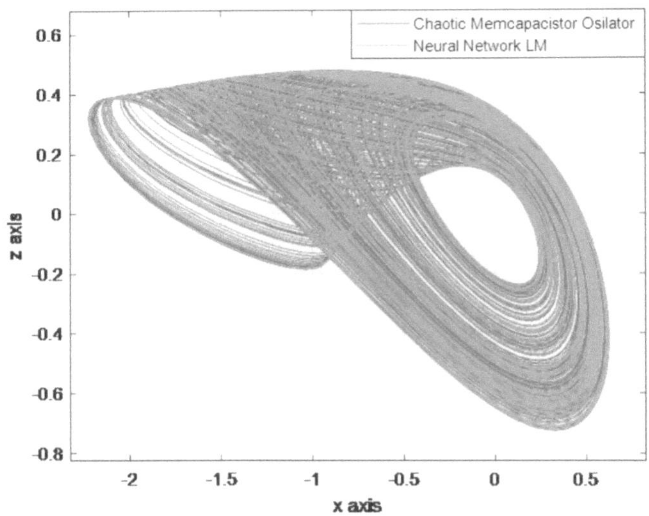

FIGURE 5.19 Real CMO response and ANN with Levenberg–Marquardt training responses Z–X axis.

References

[1] Shuili Ren, Shirin Panahi, Karthikeyan Rajagopal, Akif Akgul, Viet-Thanh Pham, Sajad Jafari, A new chaotic flow with hidden attractor: the first hyperjerk system with no equilibrium, Zeitschrift für Naturforschung A 73 (3) (2018) 239–249.

[2] Xiong Wang, Ünal Çavuşoğlu, Sezgin Kacar, Akif Akgul, Viet-Thanh Pham, Sajad Jafari, Fawaz E. Alsaadi, Xuan Quynh Nguyen, S-box based image encryption application using a

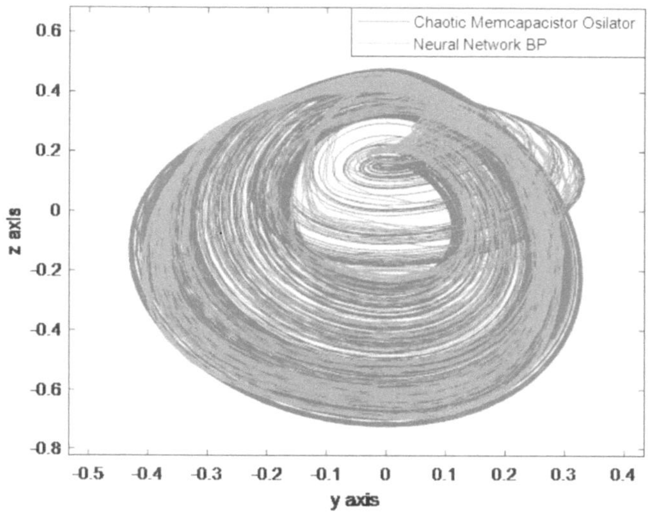

FIGURE 5.20 Real CMO response and ANN with backpropagation training responses Z–Y axis.

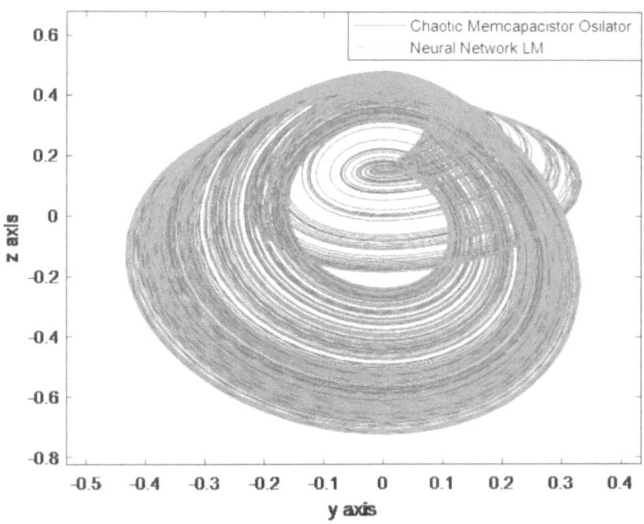

FIGURE 5.21 Real CMO response and ANN with Levenberg–Marquardt training responses Z–Y axis.

chaotic system without equilibrium, Applied Sciences 9 (4) (2019) 781.

[3] Xiong Wang, Guanrong Chen, A chaotic system with only one stable equilibrium, Communications in Nonlinear Science and Numerical Simulation 17 (3) (2012) 1264–1272.

[4] Xiong Wang, Akif Akgul, Serdar Cicek, Viet-Thanh Pham, Duy Vo Hoang, A chaotic system with two stable equilibrium points: dynamics, circuit realization and communication application, International Journal of Bifurcation and Chaos 27 (08) (2017) 1750130.

[5] Karthikeyan Rajagopal, Ali Durdu, Sajad Jafari, Yilmaz Uyaroglu, Anitha Karthikeyan, Akif

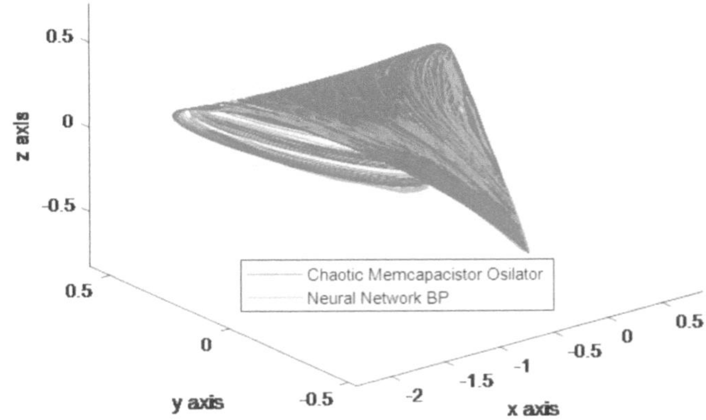

FIGURE 5.22 Real CMO response and ANN with backpropagation training responses Z–Y–X axis.

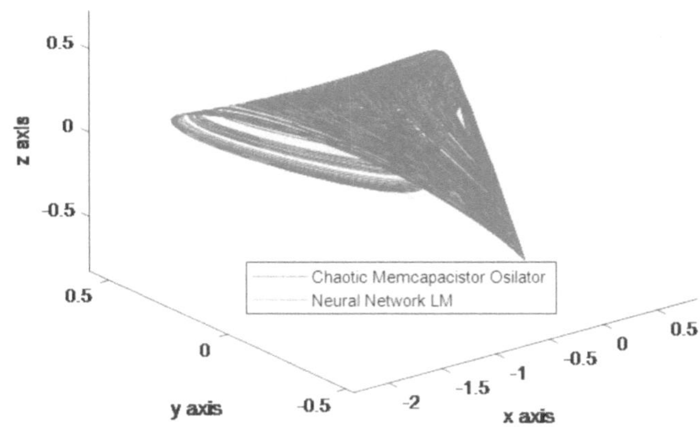

FIGURE 5.23 Real CMO response and ANN with Levenberg–Marquardt training responses Z–Y–X axis.

Akgul, Multiscroll chaotic system with sigmoid nonlinearity and its fractional order form with synchronization application, International Journal of Non-Linear Mechanics 116 (2019) 262–272.

[6] İhsan Pehlivan, Ersin Kurt, Qiang Lai, Aziz Basaran, Mustafa Cagri Kutlu, A multiscroll chaotic attractor and its electronic circuit implementation, Chaos Theory and Applications 1 (1) (2019) 29–37.

[7] Michael Field, Martin Golubitsky, Symmetry in Chaos: A Search for Pattern in Mathematics, Art, and Nature, Vol. 111, SIAM, 2009.

[8] Zhouchao Wei, Karthikeyan Rajagopal, Wei Zhang, Sifeu Takougang Kingni, Akif Akgül,

Synchronisation, electronic circuit implementation, and fractional-order analysis of 5d ordinary differential equations with hidden hyperchaotic attractors, Pramana 90 (4) (2018) 50.
[9] Metin Varan, Akif Akgul, Control and synchronisation of a novel seven-dimensional hyperchaotic system with active control, Pramana 90 (4) (2018) 54.
[10] Karthikeyan Rajagopal, Murat Tuna, Anitha Karthikeyan, İsmail Koyuncu, Prakash Duraisamy, Akif Akgul, Dynamical analysis, sliding mode synchronization of a fractional-order memristor Hopfield neural network with parameter uncertainties and its non-fractional-order fpga implementation, The European Physical Journal Special Topics 228 (10) (2019) 2065–2080.
[11] Akif Akgul, Chaotic oscillator based on fractional order memcapacitor, Journal of Circuits, Systems, and Computers (2019) 1950239.
[12] Karthikeyan Rajagopal, Akif Akgul, Irene M. Moroz, Zhouchao Wei, Sajad Jafari, Iqtadar Hussain, A simple chaotic system with topologically different attractors, IEEE Access 7 (2019) 89936–89947.
[13] Karthikeyan Rajagopal, Sajad Jafari, Akif Akgul, Anitha Karthikeyan, Serdar Çiçek, Yasser Shekofteh, A simple snap oscillator with coexisting attractors, its time-delayed form, physical realization, and communication designs, Zeitschrift für Naturforschung A 73 (5) (2018) 385–398.
[14] Jay Prakash Singh, Jit Koley, Akif Akgul, Bilal Gurevin, Binoy Krishna Roy, A new chaotic oscillator containing generalised memristor, single op-amp and rlc with chaos suppression and an application for the random number generation, The European Physical Journal Special Topics 228 (10) (2019) 2233–2245.
[15] Akif Akgül, Coskun Arslan, Burak Arıcıoğlu, Design of an interface for random number generators based on integer and fractional order chaotic systems, Chaos Theory and Applications 1 (1) (2019) 1–18.
[16] Akif Akgul, Chunbiao Li, Ihsan Pehlivan, Amplitude control analysis of a four-wing chaotic attractor, its electronic circuit designs and microcontroller-based random number generator, Journal of Circuits, Systems, and Computers 26 (12) (2017) 1750190.
[17] Shahtaj Shaukat, Ali Arshid, Amna Eleyan, Syed Aziz Shah, Jawad Ahmad, et al., Chaos theory and its application: an essential framework for image encryption, Chaos Theory and Applications 2 (1) (2020) 17–22.
[18] Sundarapandian Vaidyanathan, Akif Akgul, Sezgin Kaçar, U. Çavuşoğlu, A new 4-d chaotic hyperjerk system, its synchronization, circuit design and applications in rng, image encryption and chaos-based steganography, The European Physical Journal Plus 133 (2) (2018) 46.
[19] Akif Akgul, Sezgin Kacar, Burak Aricioglu, A new two-level data hiding algorithm for high security based on a nonlinear system, Nonlinear Dynamics 90 (2) (2017) 1123–1140.
[20] Akif Akgul, Irene M. Moroz, Ali Durdu, A novel data hiding method by using a chaotic system without equilibrium points, Modern Physics Letters B 33 (29) (2019) 1950357.
[21] Emre Gungor, Enver Çavuş, Ihsan Pehlivan, A logistic map Runge Kutta-4 solution for FPGA using fixed point representation, Chaos Theory and Applications 1 (1) (2019) 19–28.
[22] Yusuf Adiyaman, Selcuk Emiroglu, Muhammed Kürşad Uçar, Muhammed Yildiz, Dynamical analysis, electronic circuit design and control application of a different chaotic system, Chaos Theory and Applications 2 (1) (2020) 10–16.
[23] Zeynep Garip Batık, Murat Erhan Cimen, Durmus Karayel, Ali Fuat Boz, The chaos-based whale optimization algorithms global optimization, Chaos Theory and Applications 1 (1) (2019) 51–63.
[24] Selcuk Coskun, Sinan Tuncel, Ihsan Pehlivan, Akif Akgul, Microcontroller-controlled electronic circuit for fast modelling of chaotic equations, Electronics World 121 (1947) (2015) 24–25.
[25] Murat Alçın, İhsan Pehlivan, İsmail Koyuncu, Hardware design and implementation of a novel ann-based chaotic generator in fpga, Optik 127 (13) (2016) 5500–5505.
[26] Murat Erhan Cimen, Zeynep Batık Garip, Muhammed Ali Pala, Ali Fuat Boz, Akif Akgul, Modelling of a chaotic system motion in video with artificial neural networks, Chaos Theory and Applications 1 (1) (2019) 38–50.

[27] Murat Erhan Cimen, Sezgin Kacar, Emre Guleryuz, Bilal Gürevin, Akif Akgul, Modeling of chaotic motion video with artificial neural networks, Co-chair, page 193.
[28] Pilar Gómez-Gil, Juan Manuel Ramírez-Cortes, Saúl E Pomares Hernández, Vicente Alarcón-Aquino, A neural network scheme for long-term forecasting of chaotic time series, Neural Processing Letters 33 (3) (2011) 215–233.
[29] Ercan Öztemel, Yapay sinir ağlari, PapatyaYayincilik, Istanbul, 2003.
[30] Oliver Nelles, Nonlinear System Identification: From Classical Approaches to Neural Networks and Fuzzy Models, Springer Science & Business Media, 2013.
[31] Karthikeyan Rajagopal, Akif Akgul, Sajad Jafari, Burak Aricioglu, A chaotic memcapacitor oscillator with two unstable equilibriums and its fractional form with engineering applications, Nonlinear Dynamics 91 (2) (2018) 957–974.
[32] Yiğithan Dedeoğlu, B Uğur Töreyin, Uğur Güdükbay, A Enis Çetin, Videoda nesne sınıflandırması için siluet tabanlı yöntem, Signal Processing and Communications Applications (2006) 1–4.
[33] Rafael C. Gonzales, Richard E. Woods, Digital Image Processing, 2002.
[34] Roberto Brunelli, Template Matching Techniques in Computer Vision: Theory and Practice, John Wiley & Sons, 2009.
[35] Navneet Dalal, Bill Triggs, Cordelia Schmid, Human detection using oriented histograms of flow and appearance, in: European Conference on Computer Vision, Springer, 2006, pp. 428–441.
[36] Zhen Lei, Tao Fang, Deren Li, Histogram of oriented gradient detector with color-invariant gradients in Gaussian color space, Optical Engineering 49 (10) (2010) 109701.
[37] Li Cuimei, Qi Zhiliang, Jia Nan, Wu Jianhua, Human face detection algorithm via Haar cascade classifier combined with three additional classifiers, in: 2017 13th IEEE International Conference on Electronic Measurement & Instruments (ICEMI), IEEE, 2017, pp. 483–487.
[38] Sander Soo, Object Detection Using Haar-Cascade Classifier, Institute of Computer Science, University of Tartu, 2014, pp. 1–12.
[39] Fernando De la Torre, Michael J. Black, Robust principal component analysis for computer vision, in: Proceedings Eighth IEEE International Conference on Computer Vision, ICCV 2001, Vol. 1, IEEE, 2001, pp. 362–369.
[40] Marian Stewart Bartlett, Javier R. Movellan, Terrence J. Sejnowski, Face recognition by independent component analysis, IEEE Transactions on Neural Networks 13 (6) (2002) 1450–1464.
[41] Christos Kyrkou, Theocharis Theocharides, A parallel hardware architecture for real-time object detection with support vector machines, IEEE Transactions on Computers 61 (6) (2011) 831–842.
[42] Bernd Heisele, Thomas Serre, Sam Prentice, Tomaso Poggio, Hierarchical classification and feature reduction for fast face detection with support vector machines, Pattern Recognition 36 (9) (2003) 2007–2017.
[43] Adhiyaman Manickam, Ezhilmaran Devarasan, Gunasekaran Manogaran, Malarvizhi Kumar Priyan, R. Varatharajan, Ching-Hsien Hsu, Raja Krishnamoorthi, Score level based latent fingerprint enhancement and matching using sift feature, Multimedia Tools and Applications 78 (3) (2019) 3065–3085.
[44] Masoud Z. Lifkooee, Ömer M. Soysal, Kazim Sekeroglu, Video mining for facial action unit classification using statistical spatial–temporal feature image and log deep convolutional neural network, Machine Vision and Applications 30 (1) (2019) 41–57.
[45] Resul Daş, Berna Polat, Gürkan Tuna, Derin öğrenme ile resim ve videolarda nesnelerin tanınması ve takibi, Fırat Üniversitesi Mühendislik Bilimleri Dergisi 31 (2) (2019) 571–581.
[46] Jacek M. Zurada, Introduction to Artificial Neural Systems, Vol. 8, West St. Paul, 1992.
[47] Hao Yu, Bogdan M. Wilamowski, Levenberg–Marquardt training, Industrial Electronics Handbook 5 (12) (2011) 1.
[48] T-H. Cho, Richard W. Conners, Philip A. Araman, Fast backpropagation learning using steep activation functions and automatic weight reinitialization, in: Conference Proceed-

ings 1991 IEEE International Conference on Systems, Man, and Cybernetics, IEEE, 1991, pp. 1587–1592.

[49] Omaima N. Ahmad Al-Allaf, Fast backpropagation neural network algorithm for reducing convergence time of bpnn image compression, in: ICIMU 2011: Proceedings of the 5th International Conference on Information Technology & Multimedia, IEEE, 2011, pp. 1–6.

[50] Nilanjan Dey, Amira Ashour, Siddhartha Bhattacharyya, Applied Nature-Inspired Computing: Algorithms and Case Studies, Springer, 2020.

Chapter 6

Rich dynamics of memristor based Liénard systems

S. Leo Kingston and Tomasz Kapitaniak
Division of Dynamics, Lodz University of Technology, Lodz, Poland

6.1 Introduction

In the past few decades, studies on memristor based oscillators received notable attention in the different disciplines. The fundamental theoretical relationship of the memristor element has been introduced by the circuit theorist Leon O. Chua in 1971 [1]. However, the physical realization of the memristor was achieved by Stanley Williams and his coworkers at the HP laboratory in the year 2008 [2]. After the physical implementation of this memristor, in the same year, Itoh and Chua [3] proposed that memristor also a suitable nonlinear element for designing different nonlinear oscillators. This pioneering work triggered the research in both scientific and technological communities to focus on memristors and their applications [4–6]. The local activities of the memristor are classified into two different groups namely, elements and emulators. Recently a nanoscale based Nbo_2 Mott memristor was physically implemented, which is more suitable for analog computing applications [7]. Owing to the technical difficulties in manufacturing this nanoscale device the commercial memristor element has not been available till today. To overcome these issues, locally active memristor emulators have been designed to explore memristor theory and successfully applied in different scenarios such as neural networks, memory storage, chaotic circuit designing, and secure communications [8–11]. Memristors are used to study different electrical activities and memory characteristics of neurons [12]. These memristors are behaving like synopsis, which are used to connect two different neurons. These memristor synopses promote the synchronization stability of neuron and the neuron firing states continuously changing by varying the memristence [13]. Some advanced neuron models once considered memristors as autapses which show various firing pattern transitions [14].

In recent years, the memristor emulator circuit was used to investigate the significant features of a real memristor [15]. This memristor emulators circuit is made up of the off-the-shelf components such as resistors, capacitors, operational amplifiers, and analog device multipliers which is successfully implemented in well known nonlinear systems, for example, Duffing oscillators

Mem-elements for Neuromorphic Circuits with Artificial Intelligence Applications
https://doi.org/10.1016/B978-0-12-821184-7.00014-1

[16], Liénard systems [17,18], Chua's circuits [19,20], and networks [21]. These models manifest the diversity of novel phenomena, namely strange attractors, bursting oscillation, mixed-mode oscillations, coexisting attractors, extreme events, and hidden attractors [16–18,22–25].

The coexistence of multiple periodic attractors has been identified in the memristor based Liénard system [18] for the different choices of initial conditions. These attractors are located away from the fixed points of the system and these types of attractors are known as *hidden attractors*. The emergence of hidden attractors in different memristor based oscillators was reported in the literature [18,26–28]. Further the mixed-mode oscillations (MMOs) are the waveform that consist of an alternate sequence of large and small amplitude oscillations. These MMOs are denoted as L^s, here L and s are large and small amplitude oscillations, respectively. The appearance of MMOs in different model systems [29–32] follows three predominant sequences: the alternate appearance of periodic MMOs and chaotic states, successive period-adding sequence of MMOs and Farey sequences [17,31,33]. The quasiperiodic motions are characterized by many incommensurate basic frequencies, which can be of two-, three-, four-frequency quasiperiodicity, etc. [34]. The applications of quasiperiodic motion are involved in optoelectronics, lasers, magnetic films, klystron generators, and others (see [34] and the references therein). The emergence of the higher dimensional torus in a single system is rarely reported in the literature [35]. Besides, the large expansion of chaotic attractors from bounded motion has been studied in different models [36–38]. It would help to understand some catastrophic transitions and their possible earlier predictions [39].

Based on the above understanding and motivation, in this chapter, we studied some intriguing dynamics of memristor emulator based nonlinear systems, such as hidden attractors, mixed-mode oscillations, higher dimensional tori, and large expanded chaotic attractors. It is worthwhile to explore some inherent properties of different memristor emulator based nonlinear systems to attain further advancement in memristor based neurocomputing applications.

6.2 Model system

The memristor element has two different functional relationships between electric charge (q) and magnetic flux (ϕ). The terminal voltage (v) and current (i) of the memristor is governed by the relations

$$v = M(q)i \tag{6.1}$$

and

$$i = W(\phi)v, \tag{6.2}$$

where $v = \frac{d\phi}{dt}$, and $i = \frac{dq}{dt}$. Here, $M(q)$, and $W(\phi)$ are the different nonlinear functions which represent *memristance* and *memductance*, respectively, and these functions are defined as $M(q) = \frac{d\phi(q)}{dq}$ and $W(\phi) = \frac{dq(\phi)}{\phi}$. It is well established in the literature that the memristor emulator circuits mimic the real characteristics behavior of the memristor element. Hence, the memristor emulator circuits reveals the pinched hysteresis loop in $(v\text{–}i)$ plane, whereas Itoh and Chua reported that the existence of some other characteristics curves of the memristor in the $(q\text{–}\varphi)$ plane [3]. There are distinct types of nonlinearities used to explain the charge and flux relationship of the memristor emulator, namely nonsmooth piecewise linearity, smooth cubic nonlinearity, and smooth piecewise quadratic nonlinearity (see [16] and the references therein). In this study, we have taken the charge control memristor $M(q)$ and using the smooth cubic nonlinearity which is defined as follows:

$$\phi(q) = \omega_0^2 q + \beta q^3. \tag{6.3}$$

In order to realize the smooth cubic nonlinearity function in our considered model system, we have used the same circuit model as reported in [16,18]. Moreover, to explore the rich dynamical behaviors in memristor emulator based nonlinear systems, we have taken two different classes of the Liénard system. The French mathematical physicist studied the different motions in nonlinear dynamical systems. He particularly explained the criteria for ensuring the existence, uniqueness, and stability of periodic solutions of a general class of equations as follows:

$$\ddot{x} + f(x)\dot{x} + x = 0. \tag{6.4}$$

Eq. (6.4) is referred to as the Liénard equation or Liénard type oscillator, and the familiar van der Pol oscillator is a special case of the Liénard system [40]. Furthermore, in 1942 Levinson and Smith introduced the existence of periodic solutions in the more general form of the Liénard system as follows:

$$\ddot{x} + f(x)\dot{x} + g(x) = 0. \tag{6.5}$$

In Eq. (6.5), $f(x)$ and $g(x)$ are the two different functions depending on the particular system model. This Liénard system is a paradigmatic model for different nonlinear systems, and recently its different intriguing dynamics have been reported in the literature (see Refs. [17,18,31,32,36]). From the viewpoint of application, different advantages have been explained in chemical [41] and optoelectronic oscillator [42] type Liénard systems. In the present study, first we consider the following class of Liénard systems [32]:

$$\ddot{x} + \alpha x^2 \dot{x} + \omega_0^2 x + \beta x^3 = A\sin(\omega t); \tag{6.6}$$

Eq. (6.6) satisfies the general form of Liénard system equation (6.5), where $f(x) = \alpha x^2$ and $g(x) = \omega_0^2 x + \beta x^3$. In order to explore the memristor effect in the system equation (6.6), we have replaced the memristor function $M(q)x$ for

$g(x)$. Hence the memristor emulator based Liénard system is represented by the following equation:

$$\ddot{x} + \alpha x^2 \dot{x} + M(q)x = A\sin(\omega t) \tag{6.7}$$

here, $M(q) = \frac{d\phi(q)}{dq} = \omega_0^2 + 3\beta q^2$ represents the memristance function. Further for numerical investigation, we used the normalized variables as $q = x_1$, $x = x_2$, and $\dot{x} = x_3$, and Eq. (6.7) written in the following form of first order equations:

$$\begin{aligned} \dot{x_1} &= x_2, \\ \dot{x_2} &= x_3, \\ \dot{x_3} &= -\alpha x_2^2 x_3 - (\omega_0^2 + 3\beta x_1^2)x_2 + A\sin(\omega t). \end{aligned} \tag{6.8}$$

We solved Eq. (6.8) numerically using a fourth order Runge–Kutta algorithm with adaptive step size and its relevant dynamics will be explained in the following sections.

6.2.1 Stability analysis

In order to find the stability of the system equation (6.8), we set the external forcing term $A\sin(\omega t)$ as zero. The fixed points of the system are calculated by choosing the initial conditions as $\dot{x_1} = \dot{x_2} = \dot{x_3} = 0$, and the obtained three equilibrium points are x_1^*, 0, and 0. Here the fixed point value of x_1^* can be either zero or nonzero. Furthermore, for calculating the eigenvalues of the system, the Jacobian matrix of Eq. (6.8), in the absence of external forcing term, is written as

$$\det(M - \lambda I) = \begin{pmatrix} 0-\lambda & 1 & 0 \\ 0 & 0-\lambda & 1 \\ 0 & -2\alpha x_2 x_3 - \omega_0^2 - 3\beta x_1^{*2} & -\alpha x_2^2 - \lambda \end{pmatrix} = 0. \tag{6.9}$$

The characteristic equation for the above matrix is

$$\lambda^3 + (\alpha x_2^2)\lambda^2 + (2\alpha x_2 x_3 + \omega_0^2 + 3\beta x_1^2)\lambda = 0 \tag{6.10}$$

and the eigenvalues for Eq. (6.10) are

$$\begin{aligned} \lambda_1 &= 0 \\ \lambda_{2,3} &= \frac{1}{2}[-\alpha x_2^2 \pm \sqrt{\alpha^2 x_2^4 - 8\alpha x_2 x_3 - 4\omega_0^2 - 12\beta x_1^{*2}}] \end{aligned} \tag{6.11}$$

First, we started with finding the stability of the system equation (6.8) at $(\dot{x_1}, \dot{x_2}, \dot{x_3}) = (0, 0, 0)$. For locating the different stability of the system, we have to change the three parameters $(\alpha, \beta, \omega_0^2)$ in the wider range. The obtained stability phase diagram of the system in the three parameter phase diagram is depicted

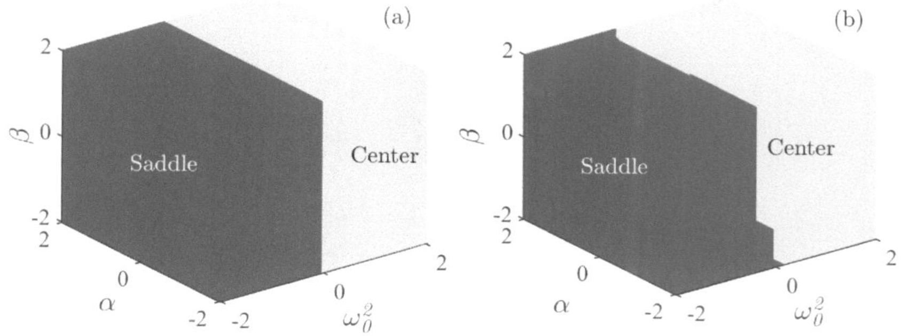

FIGURE 6.1 Stability phase diagram in three parameter space to distinguish different stability region of the system equation (6.8) in the absence of external forcing. (a) Stability regions for x^* values as zero, whereas (b) represents stability regions for nonzero x^* values.

in Fig. 6.1(a). Here the system exhibits two different types of stability, *Saddle* and *Center type*, respectively. Similarly, we have calculated the stability region for $(\dot{x_1}, \dot{x_2}, \dot{x_3}) = (x^*, 0, 0)$ (i.e.) for nonzero x^* value. In this case the system reveals similar stability like zero x^* values. However, the existing stability regions differ with respect to the three different parameters of the system. The three parameter phase diagram for nonzero x^* value is shown in Fig. 6.1(b). The stability phase diagrams of Fig. 6.1(a), and Fig. 6.1(b) are used to understand the appearance of different stability of the system in the wider parameter regions.

6.2.2 Hidden attractors

The memristor based autonomous Liénard system equation (6.8) exhibits coexistence multiple hidden attractors for the particular choice of system parameters. We obtained a saddle state $(0, \pm 0.0043)$ at $(\dot{x_1}, \dot{x_2}, \dot{x_3}) = (0.4772, 0, 0)$ for the fixed system parameters value as $\alpha = 0.3$, $\beta = 0.5$, and $\omega_0^2 = 0.3$. However, the system signifies multiple periodic attractors for the different choice of initial conditions, which is shown in Fig. 6.2. It can be noticed from Fig. 6.2 that all the attractors are not approaching the fixed point of the system and are located away from the fixed point, hence these attractors are known as *hidden attractors*. Besides, the system has exhibited infinite numbers of hidden attractors, since the model satisfied that the periodic line invariant type behavior [43]. Also, we include the external forcing term in the system equation (6.8), it reveals different complex dynamics, which we will discuss in the next sections.

6.3 Mixed-mode oscillations

In this section, we explain the emergence of mixed-mode oscillation in the memristor based Liénard system equation (6.8). The system exhibits numerous types

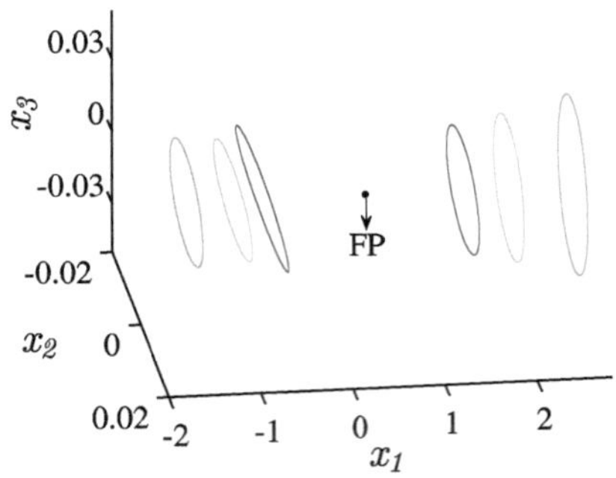

FIGURE 6.2 Coexistence of multiple hidden attractors in memristor based autonomous Liénard system for parameter values as α = 0.3, β = 0.5, and ω_0^2 = 0.3 and FP refers the fixed point of the system.

of primary and secondary MMOs for the two different set of system parameters. Especially, we are focusing on the appearance of distinct sequences of MMOs for the fixed system parameters and varying the control parameter namely frequency, and amplitude of the external forcing term.

6.3.1 Frequency scanning

For observing the different types of MMOs, the system parameters of Eq. (6.8) are fixed as $\alpha = 0.1$, $\beta = 0.7$, $\omega_0^2 = 0.1$, $A = 0.15$ and we have varying ω in the range $\omega \in (0.15, 0.485)$. The system exhibits numerous types of an alternate sequence of large L and small s amplitude oscillations. In most of the system, the occurrence of MMOs follows two predominant sequences such as an alternate sequence of periodic MMOs and chaotic states as well as the successive period sequence of MMOs [17,31]. In our considered model system, by varying the specific range of forcing frequency, we noticed that the system exhibits an alternate sequence of periodic MMOs and chaotic states. In order to confirm this sequence, we have plotted the inter spike interval (ISI) bifurcation diagram in the range of control parameter, $0.15 \leq \omega \leq 0.485$, and the obtained ISI bifurcation diagram as shown in Fig. 6.3(a). For the larger value of control parameter, say $\omega = 0.485$, the system exhibits chaotic states. When we gradually decrease the ω value, the system turns into 1^1 periodic MMO via saddle-node bifurcation. Further decreasing the control parameter 1^1 periodic MMOs become 2^2 periodic MMOs and transit to chaos via the period-doubling route [31]. Here all the periodic MMOs are also known as primary MMOs. Similarly different primary MMOs (1^2, 1^3, 1^4, 1^5, 1^6, 1^7, and 1^8 MMOs) arises in between the chaotic states as shown in Fig. 6.3(a). Furthermore, to confirm the existence of an alternate sequence of periodic MMOs and chaotic states, we have drawn a one parameter bifurcation diagram in the (ω–x_1) plane as shown in Fig. 6.3(b), which exactly matches with the ISI bifurcation diagram of Fig. 6.3(a). The tem-

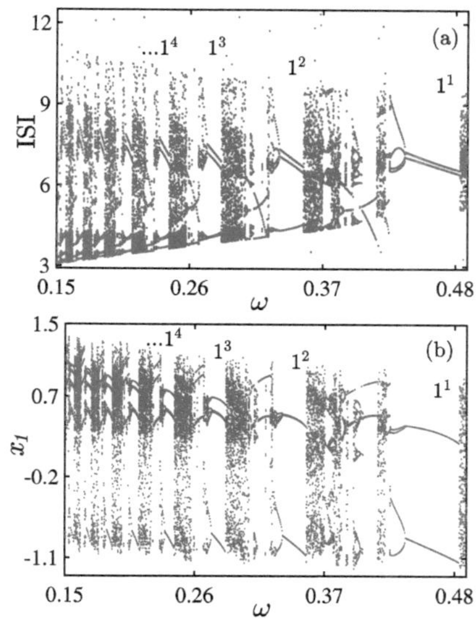

FIGURE 6.3 (a) Inter spike interval, and (b) one parameter bifurcation diagram for the specific range of control parameter ω, which confirms the appearance of alternate sequence of periodic MMOs and chaotic states.

poral dynamics of different primary MMOs depicted in Fig. 6.4(a)–(h) and its corresponding ω values are given in the figure caption.

Interestingly, in between these primary MMOs there exist secondary MMOs in the system for a narrow range of the control parameter values. In general these secondary MMOs manifest more than one large amplitude oscillations in their dynamics. The time series of different secondary MMOs are shown in Fig. 6.5(a)–(f) and the values of the control parameter for each secondary MMOs are $\omega = 0.413$ (2^1), $\omega = 0.3225$ (2^2), $\omega = 0.267$ (2^3), $\omega = 0.23$ (2^4), $\omega = 0.203$ (2^5), $\omega = 0.18$ (2^6), and $\omega = 0.1545$ (2^7). We display an inverted time series for both primary (Fig. 6.4) and secondary (Fig. 6.5) MMOs for the purpose of clear visualization. The various combinations of primary and secondary MMOs for the different ω values are listed in Table 6.1. Furthermore, for characterizing these periodic MMOs and chaotic states, we have drawn the return map of an inter spike interval for both periodic MMOs and chaotic states. The regular structure of the ISI return map represents periodic MMOs, i.e. for $\omega = 0.1575$ (cf. Fig. 6.6(a)). However, the randomly distributed ISI return map reveals a chaotic state for $\omega = 0.158$, which is shown in Fig. 6.6(b). Besides, the existence of primary and secondary MMOs in between the chaotic states interpreted as the Devil's staircase structure based on the firing numbers has been reported in [31,44]. The firing number is defined as $F = \frac{L}{L+s}$. We have calculated the firing number for both primary and secondary MMOs and obtained the Devil's staircase structure as shown in Fig. 6.7(a), and Fig. 6.7(b). The existing parameter region of MMOs is wider for the larger values of ω, whereas this

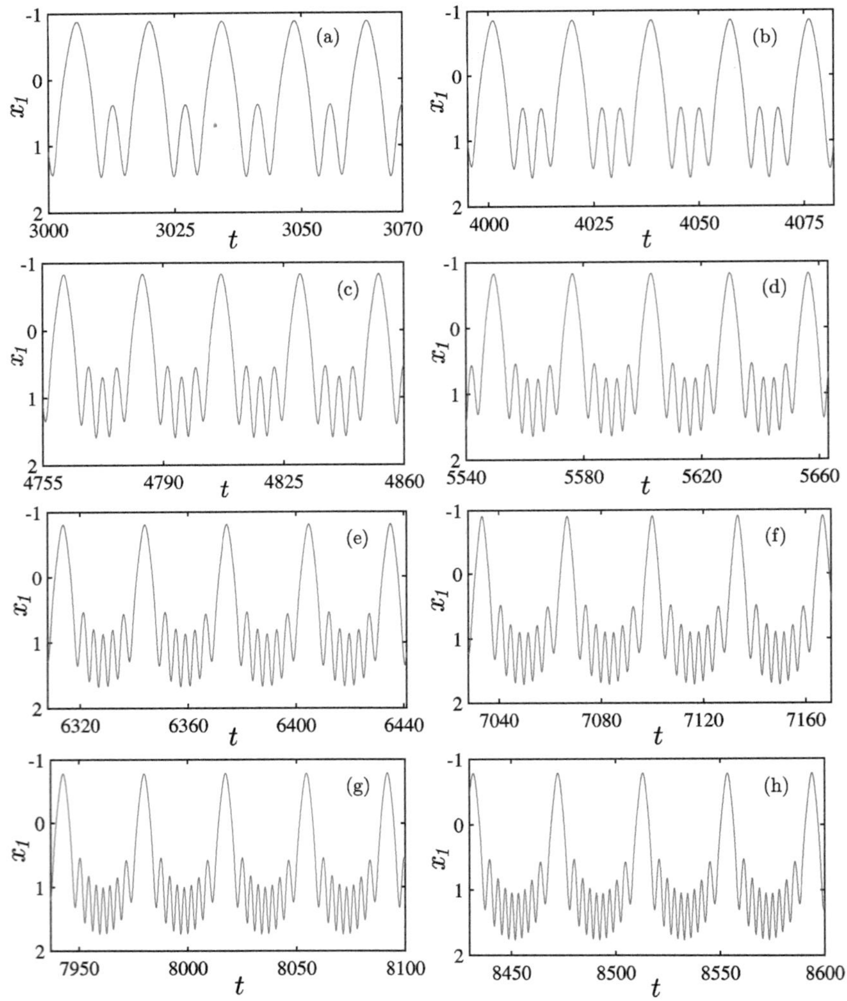

FIGURE 6.4 Time series of primary MMOs for the different ω values, (a) 1^1 ($\omega = 0.439$), (b) 1^2 ($\omega = 0.3345$), (c) 1^3 ($\omega = 0.2745$), (d) 1^4 ($\omega = 0.2355$), (e) 1^5 ($\omega = 0.207$), (f) 1^6 ($\omega = 0.1855$), (g) 1^7 ($\omega = 0.1685$), and (h) 1^8 ($\omega = 0.1550$) MMOs.

region gradually decreases by reducing the control parameter ω to the smaller value, which is depicted in Fig. 6.7(a) for primary MMOs, and Fig. 6.7(b) for secondary MMOs.

Furthermore, we have derived the scaling law following [31] for the periodic MMOs, for a specific range of the control parameter ω. The obtained scaling law behavior is shown in Fig. 6.8. In this calculation, ω_{ref} denotes the critical parameter value where the system exhibits maximum numbers of small amplitude oscillations, and ω_n signifies the control parameter value at which the

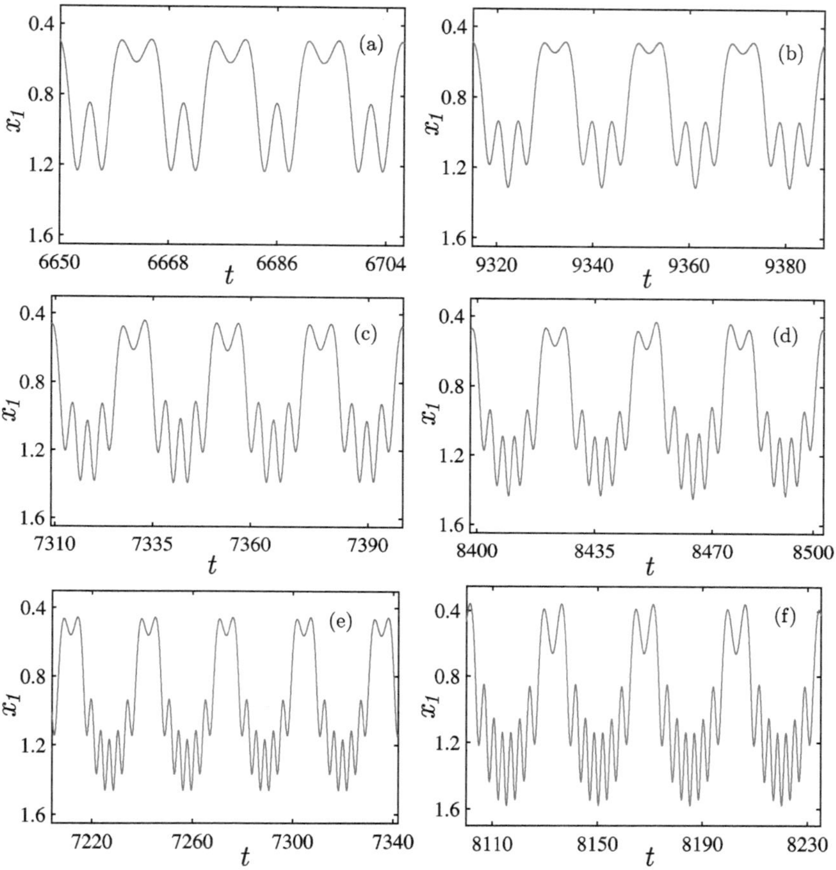

FIGURE 6.5 Time series of secondary MMOs for the different ω values, (a) 2^1 ($\omega = 0.413$), (b) 2^2 ($\omega = 0.3225$), (c) 2^3 ($\omega = 0.267$), (d) 2^4 ($\omega = 0.23$), (e) 2^5 ($\omega = 0.203$), $\omega = 0.18$ (2^6), and (f) 2^7 ($\omega = 0.1545$) MMOs.

system transits from periodic MMOs to chaotic state. The existing regions of periodic MMOs gradually decreased by turning the control parameter range ω from higher to smaller values, which are presented in Fig. 6.8(a). A similar scaling law behavior for periodic MMOs illustrated in the semilog scale is shown in Fig. 6.8(b). The obtained scaling law values are fitted with a linear least square fitting and its scaling exponent value is -1.7589 ± 0.1643.

6.3.2 Amplitude scanning

Moreover, we obtained a similar sequence of mixed-mode oscillations in the system equation, Eq. (6.8), for another set of fixed system parameters and varying the control parameter of the forcing amplitude A. In order to explore this

TABLE 6.1 Different combination of primary and secondary MMOs for the control parameter values, ω.

ω	L	s	MMO sequence (L^s)
0.439	1	1	1^1
0.426	2	2	2^2
0.413	2	1	2^1
0.403	2	1	2^1
0.3345	1	2	1^2
0.334	2	4	2^4
0.3325	2	2	2^2
0.318	2	2	2^2
0.2855	1	3	1^3
0.274	2	6	2^6
0.267	2	3	2^3
0.2425	1	4	1^4
0.235	2	8	2^8
0.23	2	4	2^4
0.212	1	5	1^5
0.2065	2	10	2^{10}
0.203	2	5	2^5
0.1895	1	6	1^6
0.1845	2	12	2^{12}
0.18	2	6	2^6
0.1715	1	7	1^7
0.168	2	14	2^{14}
0.165	2	6	2^6
0.1575	1	8	1^8
0.1545	2	16	2^{16}
0.1515	2	7	2^7

dynamics, we fixed the system parameters as $\alpha = 0.1$, $\beta = 0.5$, $\omega_0^2 = 0.1$, $\omega = 0.1$, varying A in the range $0.02 \leq A \leq 0.1$. In this case, the secondary MMOs are dominantly represented over the primary MMOs. For locating the existing regions of both primary and secondary MMOs, once again we have drawn the inter spike interval bifurcation diagram in the $(A–ISI)$ plane for the specific range of A values as shown in Fig. 6.9(a). The ISI bifurcation diagram of Fig. 6.9(a) clearly shows the appearance of an alternate sequence of periodic MMOs and chaotic states. Furthermore, the emergence of this sequence has been verified by plotting one parameter bifurcation diagram in the $(A–x_1)$

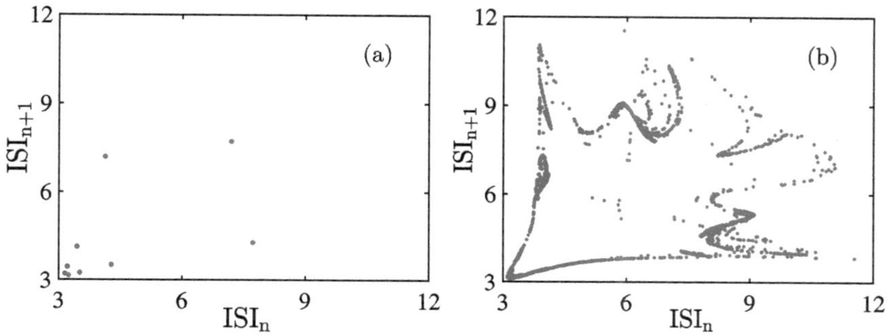

FIGURE 6.6 Inter spike interval return map of (a) periodic MMOs for $\omega = 0.1575$, and (b) chaotic state for $\omega = 0.158$.

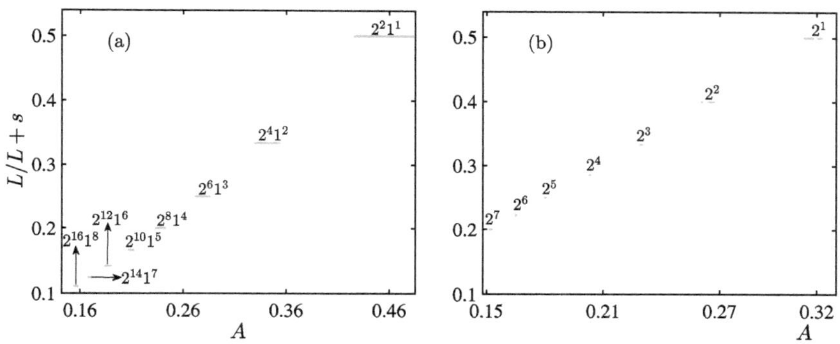

FIGURE 6.7 Devil's staircase structure based on the firing numbers as a function of ω, for (a) primary MMOs, and (b) secondary MMOs.

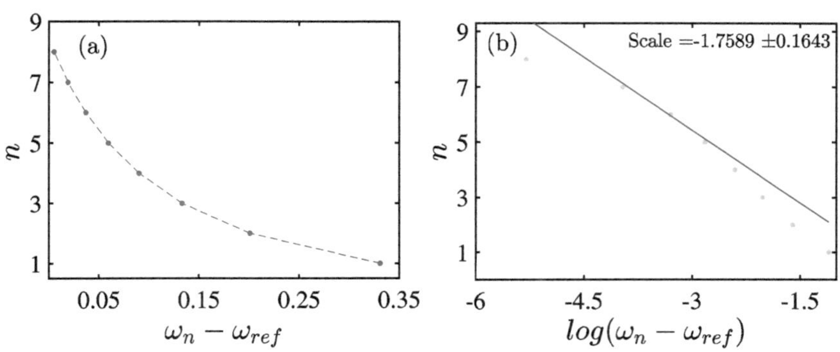

FIGURE 6.8 Scaling behavior for periodic MMOs (a) normal scale and (b) semilog scale. The critical parameter value $\omega_{ref} \approx 0.1545$, the values are fitted in semilog scale and the scaling exponent value is -1.7589 ± 0.1643.

plane which is depicted in Fig. 6.9(b). It can be noticed from Fig. 6.9(a) and Fig. 6.9(b) that the system exhibits only one primary MMO (1^7) and different

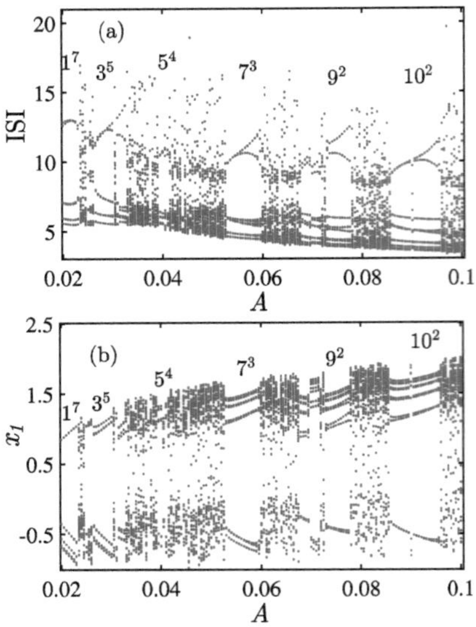

FIGURE 6.9 (a) Inter spike interval, and (b) one parameter bifurcation diagram for the specific range of A values, which confirms the appearance of alternate sequence of periodic MMOs and chaotic states.

types of secondary MMOs such as $3^5, 5^4, 7^3, 9^2$, and 10^2 for the specific range of control parameter values. The time series of 1^7 MMOs for $A = 0.02$, 2^6 MMOs for $A = 0.025$, and 3^5 MMOs for $A = 0.027$ are shown in Fig. 6.10(a)–(c). In this case also the primary as well as secondary MMOs appears in between the chaotic states. Besides, to confirm the appearance of an alternate sequence of periodic MMOs and chaotic states the Devil's staircase structure is plotted for the range of $A \in (0.02, 0.1)$, which is shown in Fig. 6.11. This Devil's staircase structure manifests the existing regions of different periodic MMOs and chaotic states. Also, the disparate types of primary and secondary MMOs for the specific values of A are listed in Table 6.2.

6.3.3 Successive period-adding sequence of MMOs

In addition to the alternate sequence of periodic MMOs and chaotic states, the system exhibits a successive period-adding sequence of MMOs. For identifying this sequence, we fixed the system parameters as $\alpha = 0.5$, $\beta = 0.5$, $\omega_0^2 = 0.5$, $A = 1.0$ and varying the control parameter ω in the range $\omega \in (1, 0.12)$. Here we noticed that the system proves the successive period-adding sequence of periodic MMOs from 1^1 to 1^{21}. The time series of 1^2, 1^4, 1^8, 1^{10}, 1^{12}, and 1^{16} MMOs are shown in Fig. 6.12(a)–(f). The specific control parameter ω values for the different periodic MMOs are illustrated in Table 6.3. Furthermore, for identifying the existing regions of different periodic MMOs, we have plotted the inter spike interval bifurcation diagram as well as one parameter bifurcation

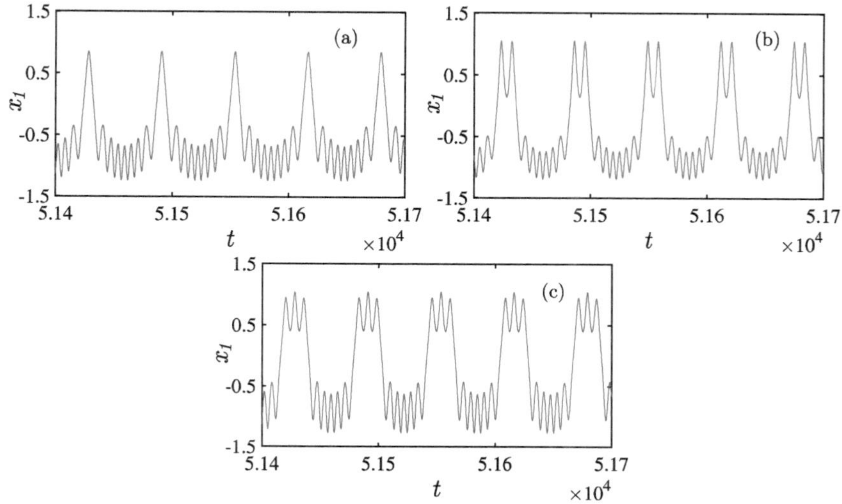

FIGURE 6.10 Time series of (a) 1^7 primary MMOs, (b) and (c) are the 2^6 and 3^5 secondary MMOs.

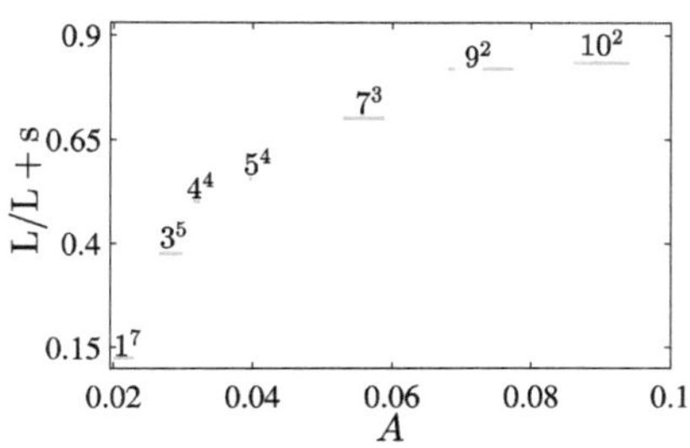

FIGURE 6.11 Devil's staircase structure for different primary and secondary MMOs.

diagram for the range of control parameter $1.0 \leq \omega \leq 0.12$, which is presented in Fig. 6.13(a), and Fig. 6.13(b). These ISI and one parameter bifurcation diagrams depicted the appearance of the period-adding sequence of MMOs. Moreover, we have calculated the scaling law for the period-adding sequences of MMOs and the resultant scaling law is shown in Fig. 6.14(a). In Fig. 6.14, ω_n represents the values of the control parameter at which the system transits from one periodic state to another periodic state, whereas ω_{ref} denotes the control parameter value when the system exhibits a large number of small amplitude MMOs. In this case, we consider the ω_{ref} value as 0.12. The similar scaling law plot presented in the semilog scale is shown in Fig. 6.14(b). The acquired scaling law values are fitted with a linear least square fit and the obtained scaling exponent value is 4.7328 ± 1.6192.

TABLE 6.2 Different types of primary and secondary MMOs for the control parameter values, *A*.

A	L	s	MMO sequence (L^s)
0.02	1	7	1^7
0.025	2	6	2^6
0.027	3	5	3^5
0.0315	2	1	4^4
0.0375	5	3	5^3
0.04	5	4	5^4
0.044	6	3	6^3
0.053	7	3	7^3
0.069	9	2	9^2
0.071	8	3	8^3
0.077	9	2	9^2
0.0.095	10	2	10^2

6.4 Higher dimensional torus and large expanded chaotic attractor

In this section, we explain the existence of higher dimensional torus and large expanded chaotic attractor in another class of memristor based Liénard system as follows:

$$\begin{aligned} \dot{x_1} &= x_2 \\ \dot{x_2} &= x_3 \\ \dot{x_3} &= -\alpha x_2 x_3 - (\omega_0^2 + 3\beta x_1^2)x_2 + A\sin(\omega t). \end{aligned} \tag{6.12}$$

The origin and detailed stability analysis of system equation (6.12) is reported in [18]. In order to observe the higher dimensional torus in system equation (6.12), we set the parameter as $\alpha = 0.5$, $\beta = 0.4975$, $\omega_0^2 = 0.5896$, $A = 0.5$, and $\omega = 0.14$. The phase portraits of the higher dimensional torus and its corresponding Poincaré surface of section are shown in Fig. 6.15(a) and 6.15(b), which signifies the existence of more complicated dynamics in the system. Moreover, to ensure the presence of a higher dimensional torus, we have plotted the power spectra for the system state x_3, which is demonstrated in Fig. 6.15(c). This power spectra display the existence of multiple incommensurate frequencies in the system. This is an intrinsic nature of the higher dimensional torus. For further confirmation of this phenomenon, we have calculated the Lyapunov exponent for the fixed parameter values of the system. The obtained Lyapunov exponent values are $\lambda_1 = 0.00002$, $\lambda_2 = 0.0$, and $\lambda_3 = -0.00002$, which proves the existence of a higher dimensional torus in the system. For the appearance of

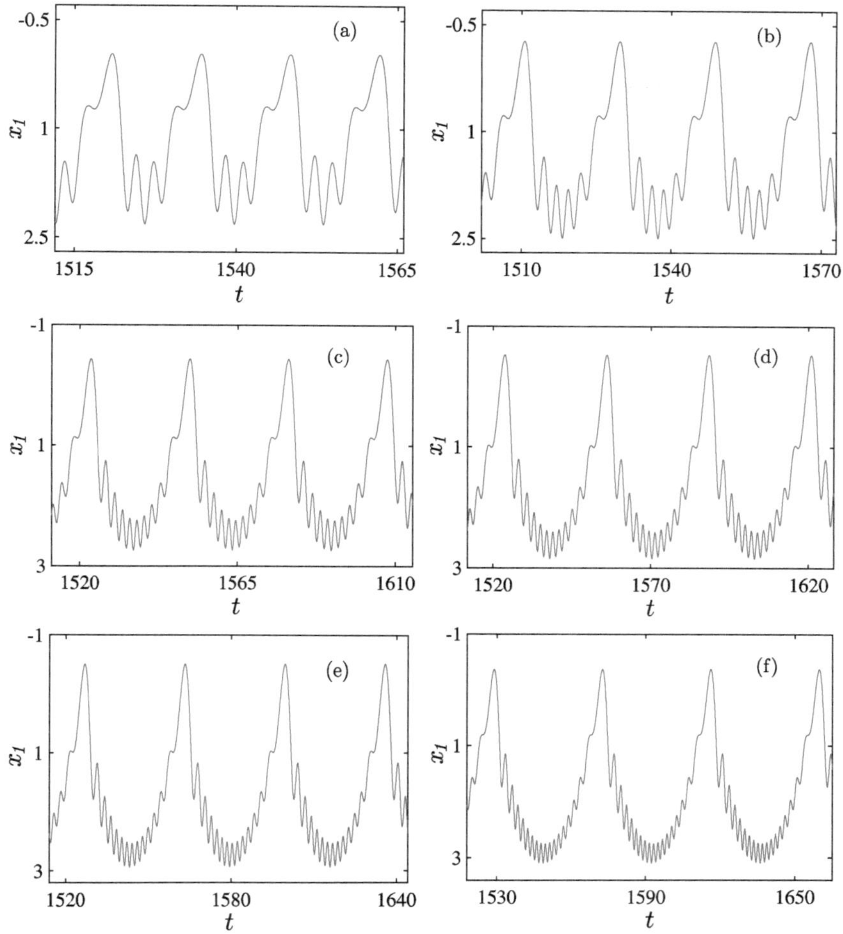

FIGURE 6.12 Time series of successive period-adding sequence of MMOs for the control parameter (a) $\omega = 0.581$ (1^2), (b) $\omega = 0.329$ (1^4), (c) $\omega = 0.223$ (1^8), (d) $\omega = 0.194$ (1^{10}), (e) $\omega = 0.173$ (1^{12}), and (f) $\omega = 0.144$ (1^{16}) [17].

the higher dimensional torus at least two Lyapunov exponents must be zero in the considered model system.

Finally, we obtained the large expanded chaotic attractor in the system equation (6.12). To observe this dynamics, the system parameters are fixed as $\alpha = 0.9$, $\beta = 0.6$, $\omega_0^2 = 0.05$, $A = 0.5$, and $\omega = 0.35$. For the specific choice of the system parameters, we noticed that the system exhibits rare but recurrent large expanded events from the bounded chaotic motion and its peaks of the time series (P_n) illustrated in Fig. 6.16(a). In this dynamics, most of the time system oscillates in the maximum amplitude ≈ 6 and rarely expanded up to the maximum amplitude of 245. The return map of the system state variable x_3 is pre-

TABLE 6.3 Disparate periodic MMOs for the control parameter values, ω.

ω	L	s	MMO sequence (L^s)
0.581	1	1	1^1
0.455	1	2	1^2
0.38	1	3	1^1
0.329	1	4	1^4
0.292	1	5	1^5
0.264	1	6	1^6
0.241	1	7	1^7
0.223	1	8	1^8
0.207	1	9	1^9
0.194	1	10	1^{10}
0.183	1	11	1^{11}
0.173	1	12	1^{12}
0.165	1	13	1^{13}
0.157	1	14	1^{14}
0.150	1	15	1^{15}
0.144	1	16	1^{16}
0.138	1	17	1^{17}
0.131	1	18	1^{18}
0.128	1	19	1^{19}
0.124	1	20	1^{20}
0.12	1	21	1^{21}

sented in Fig. 6.16(b). This return map represented the large expansion of events from bounded chaotic motion, and the magnified region of bounded chaotic motion as shown in the insert of Fig. 6.16(b). We have calculated the Lyapunov exponent for the above set of system parameters and obtained one positive Lyapunov exponent value $((\lambda_1, \lambda_2, \lambda_3) = (0.0507, 0.0, -0.0560))$, which confirms the existence of chaotic dynamics in the system.

The occurrence of large amplitude oscillation from the bounded motion in different models is confirmed by using the statistical measurement of the significant height (H_s) threshold criteria [36,38]. If any events cross this significant height (H_s) of mean plus four to eight times of standard deviations, which are known as extreme or large expanded events. Here, we have calculated the H_s value as mean plus eight times of standard deviation of the peaks of time series ($\mathrm{H_s} = \langle P_n \rangle + 8\sigma$) and plotted as horizontal red line (dark gray dashed line in print version) Fig. 6.16(a). The return map of Fig. 6.16(b) clearly shows that large expanded chaotic attractor from the bounded motion. The insert of

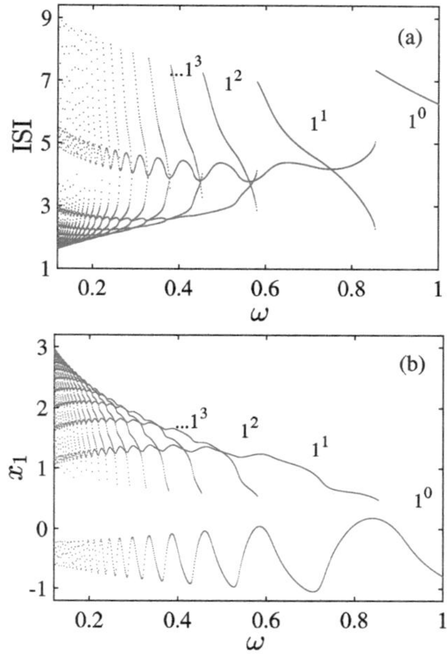

FIGURE 6.13 (a) Inter spike interval, and (b) one parameter bifurcation diagram for the specific range of control parameter ω, which confirms the existence of a successive period-adding sequence of MMOs [17].

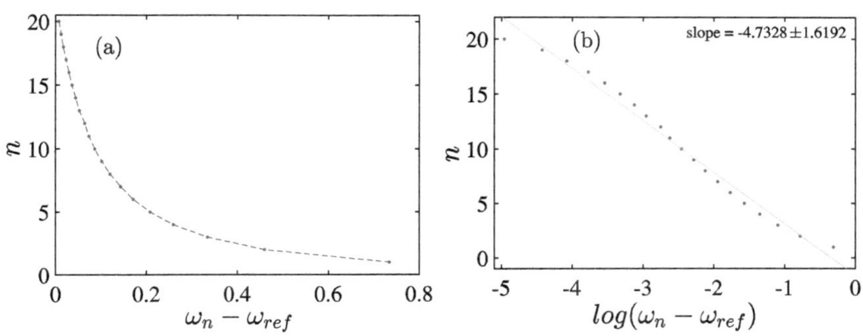

FIGURE 6.14 Scaling behavior for successive period-adding sequence of MMOs in (a) normal scale, and (b) semilog scale. The critical parameter value $\omega_{ref} \approx 0.12$, the values are fitted in semilog scale and the scaling exponent value is 4.7328 ± 1.6192 [17].

Fig. 6.16(b) represents the bounded chaotic motion. Besides, we determined the probability distribution function for the peaks of the system state variable x_3, for this calculation, we have taken a very long time series as shown in Fig. 6.16(c). The probability distribution function of Fig. 6.16(c) manifests a long-tail distribution, which confirms the occurrence of a large expanded chaotic attractor in the considered model system.

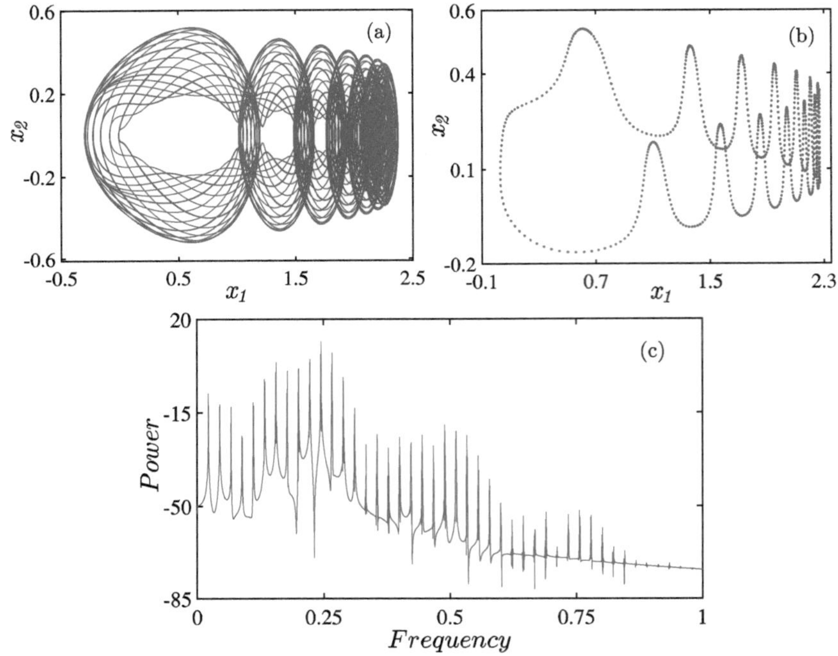

FIGURE 6.15 Higher dimensional torus: (a) phase portraits, (b) Poincaré surface of section in the $(x_1–x_2)$ planes, and (c) represents its corresponding power spectra.

6.5 Conclusion

In this chapter, we have studied rich varieties of complex dynamics in two different classes of the memristor emulator based Liénard system. The autonomous Liénard system reveals the coexistence of multiple hidden attractors for a precise choice of system parameters. After including the external sinusoidal force, the system exhibits disparate types of primary and secondary mixed-mode oscillations. Especially, we investigated two different transitions of MMOs, namely alternate sequence of periodic MMOs and chaotic states as well as the successive period-adding sequence of MMOs. The existence of different periodic MMOs and chaotic states is confirmed by bifurcation diagrams, inter spike interval return maps, the Devil's staircase structures, and scaling behaviors. The higher dimensional torus and large expanded chaotic attractor have been identified in another class of memristor based Liénard systems. The appearance of multiple incommensurate frequencies in higher dimensional torus is corroborated by using power spectra and Lyapunov exponent method. The large expanded events from the bounded chaotic motion manifest the long-tail probability distribution function. It is believed that our investigation would give some intuition to explore various neuronal behaviors in the memristor emulator based nonlinear models for possible neurocomputing applications.

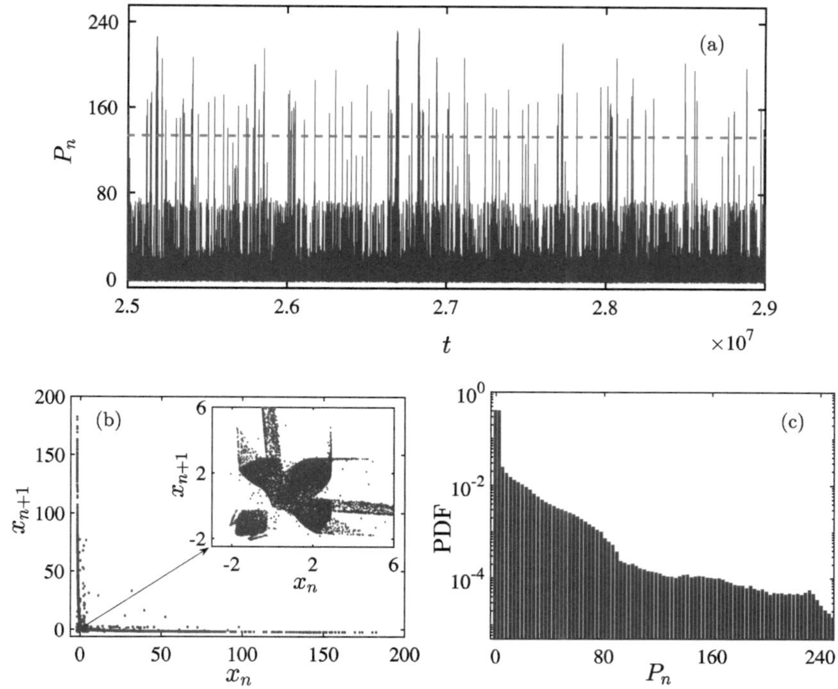

FIGURE 6.16 (a) Time series of large expanded chaotic attractor, (b) return map, and (c) probability distribution function for peaks of the time series (P_n) signifies long-tail distribution.

Acknowledgment

T.K. and S.L.K. have been supported by the National Science Centre, Poland, MAESTRO Programme (Project No. 2013/08/A/ST8/00780) and the OPUS Programme (No. 2018/29/B/ST8/00457). S.L.K acknowledges PLGrid Infrastructure (Poland) for the computation facility.

References

[1] L. Chua, Memristor – the missing circuit element, IEEE Trans. Circuit Theory 18 (1971) 507–519.

[2] D.B. Strukov, G.S. Snider, D.R. Stewart, R.S. Williams, The missing memristor found, Nature 453 (2008) 80–83.

[3] M. Itoh, L.O. Chua, Memristor oscillators, Int. J. Bifurc. Chaos 18 (2008) 3183–3206.

[4] A. Adamatzky, L. Chua, Memristor Networks, Springer Science & Business Media, 2013.

[5] R. Kozma, R.E. Pino, G.E. Pazienza, Advances in Neuromorphic Memristor Science and Applications, Springer Science & Business Media, 2012.

[6] J.H. Yoon, Z. Wang, K.M. Kim, H. Wu, V. Ravichandran, Q. Xia, C.S. Hwang, J.J. Yang, An artificial nociceptor based on a diffusive memristor, Nat. Commun. 9 (2018) 417.

[7] S. Kumar, J.P. Strachan, R.S. Williams, Nature 548 (2017) 318.

[8] A. Thomas, Memristor-based neural networks, J. Phys. D, Appl. Phys. 46 (2013) 093001.

[9] D.B. Strukov, Endurance-write-speed tradeoffs in nonvolatile memories, Appl. Phys. A 122 (2016) 302.
[10] A. Buscarino, L. Fortuna, M. Frasca, L.V. Gambuzza, A chaotic circuit based on Hewlett–Packard memristor, Chaos 22 (2012) 023136.
[11] J.W. Sun, Y. Shen, Q. Yin, C.J. Xu, Compound synchronization of four memristor chaotic oscillator systems and secure communication, Chaos 23 (2013) 013140.
[12] Y. Tan, C. Wang, A simple locally active memristor and its application in HR neurons, Chaos 30 (2020) 053118.
[13] G. Ren, Y. Xu, C. Wang, Synchronization behavior of coupled neuron circuits composed of memristors, Nonlinear Dyn. 88 (2017) 893–901.
[14] H. Bao, A. Hu, W. Liu, B. Bao, Hidden bursting firings and bifurcation mechanisms in memristive neuron model with threshold electromagnetic induction, IEEE Trans. Neural Netw. Learn. Syst. 31 (2019) 502–511.
[15] C. Yang, H. Choi, S. Park, M.P. Sah, H. Kim, L.O. Chua, A memristor emulator as a replacement of a real memristor, Semicond. Sci. Technol. 30 (2015) 015007.
[16] S. Sabarathinam, C.K. Volos, K. Thamilmaran, Implementation and study of the nonlinear dynamics of a memristor-based Duffing oscillator, Nonlinear Dyn. 87 (2017) 37–49.
[17] S.L. Kingston, K. Suresh, K. Thamilmaran, Mixed-mode oscillations in memristor emulator based Liénard system, AIP Conf. Proc. 1942 (2018) 060008.
[18] S.L. Kingston, K. Suresh, K. Thamilmaran, T. Kapitaniak, Extreme and critical transition events in the memristor based Liénard system, Eur. Phys. J. Spec. Top. 229 (2020) 1033–1044.
[19] B. Muthuswamy, Implementing memristor based chaotic circuits, Int. J. Bifurc. Chaos 20 (2010) 1335–1350.
[20] B. Muthuswamy, L.O. Chua, Simplest chaotic circuit, Int. J. Bifurc. Chaos 20 (2010) 1567–1580.
[21] E. Gale, B.L. Costello, A. Adamatzky, Emergent spiking in non-ideal memristor networks, Microelectron. J. 45 (2014) 1401–1415.
[22] N. Wang, G.S. Zhang, H. Bao, Bursting oscillations and coexisting attractors in a simple memristor-capacitor-based chaotic circuit, Nonlinear Dyn. 97 (2019) 1477–1494.
[23] B. Bao, P. Wu, H. Bao, Q. Xu, M. Chen, Numerical and experimental confirmations of quasi-periodic behavior and chaotic bursting in third-order autonomous memristive oscillator, Chaos Solitons Fractals 106 (2018) 161–170.
[24] C.K. Volos, A. Akgul, V.T. Pham, M.S. Baptista, Antimonotonicity, crisis and multiple attractors in a simple memristive circuit, J. Circuits Syst. Comput. 27 (2017) 1850026.
[25] S. Zhang, Y.C. Zeng, A simple jerk-like system without equilibrium: asymmetric coexisting hidden attractors, bursting oscillation and double full Feigenbaum remerging trees, Chaos Solitons Fractals 120 (2019) 25–40.
[26] F. Yuan, G. Wang, X. Wang, Extreme multistability in a memristor-based multi-scroll hyperchaotic system, Chaos 26 (2016) 073107.
[27] D.A. Prousalis, C.K. Volos, I.N. Stouboulos, I.M. Kyprianidis, Hyperchaotic memristive system with hidden attractors and its adaptive control scheme, Nonlinear Dyn. 90 (2017) 1681–1694.
[28] P. Saha, D. Saha, A. Ray, A. Chowdhury, Memristive nonlinear system and hidden attractor, Eur. Phys. J. Spec. Top. 224 (2015) 1563.
[29] R.K. Upadhyay, A. Mondal, W.W. Teka, Mixed mode oscillations and synchronous activity in noise induced modified Morris–Lecar neural system, Int. J. Bifurc. Chaos 27 (2017) 1730019.
[30] M. Desroches, B. Krauskopf, H.M. Osinga, Mixed-mode oscillations and slow manifolds in the self-coupled FitzHugh–Nagumo system, Chaos 18 (2008) 015107.
[31] S.L. Kingston, K. Thamilmaran, Bursting oscillations and mixed-mode oscillations in driven Liénard system, Int. J. Bifurc. Chaos 27 (2017) 1730025.
[32] S.D. Vijay, S.L. Kingston, K. Thamilmaran, Different transitions of bursting and mixed-mode oscillations in Liénard system, Int. J. Electron. Commun. 111 (2019) 152898.

[33] M.T. Koper, P. Gaspard, The modeling of mixed-mode and chaotic oscillations in electrochemical systems, J. Chem. Phys. 96 (1992) 7797–7813.
[34] A. Kuznetsov, S. Kuznetsov, N. Shchegoleva, N. Stankevich, Dynamics of coupled generators of quasiperiodic oscillations: different types of synchronization and other phenomena, Physica D 398 (2019) 1–12.
[35] V. Anishchenko, S. Nikolaev, J. Kurths, Winding number locking on a two-dimensional torus: synchronization of quasiperiodic motions, Phys. Rev. E 73 (2006) 056202.
[36] S.L. Kingston, K. Thamilmaran, P. Pal, U. Feudel, S.K. Dana, Extreme events in the forced Liénard system, Phys. Rev. E 96 (2017) 052204.
[37] K. Suresh, A.N. Pisarchik, Extreme events in systems with discontinuous boundaries, Phys. Rev. E 98 (2018) 032203.
[38] A. Mishra, S.L. Kingston, C. Hens, T. Kapitaniak, U. Feudel, S.K. Dana, Routes to extreme events in dynamical systems: dynamical and statistical characteristics, Chaos 30 (2020) 063114.
[39] N.M. Alvarez, S. Borkar, C. Masoller, Predictability of extreme intensity pulses in optically injected semiconductor lasers, Eur. Phys. J. Spec. Top. 226 (2017) 1971.
[40] L. Zhang, Y. Wang, A note on periodic solutions of a forced Liénard type equation, ANZIAM J. 51 (2010) 350–368.
[41] S. Ghosh, D.S. Ray, Liénard-type chemical oscillator, Eur. Phys. J. B 87 (2014) 65.
[42] B. Romeira, J. Figueiredo, C.N. Ironside, J. Javaloyes, Dynamics of Liénard Optoelectronic Oscillators, Selected Topics in Nonlinear Dynamics and Theoretical Electrical Engineering, Springer-Verlag, Berlin, 2013, pp. 117–136.
[43] V. Varshney, S. Sabarathinam, A. Prasad, K. Thamilmaran, Infinite number of hidden attractors in memristor-based autonomous Duffing oscillator, Int. J. Bifurc. Chaos 28 (2018) 1850013.
[44] V. Petrov, S.K. Scott, K. Showalter, Mixed-mode oscillations in chemical systems, J. Chem. Phys. 97 (1992) 6191–6198.

Chapter 7

Hidden extreme multistability generated from a novel memristive two-scroll chaotic system

Victor Kamdoum Tamba[a,b], Francois Kapche Tagne[a], Arsene Loic Mbanda Biamou[c], Manuela Corazon Nkeing[c], and Armand Nzeukou Takougang[d]

[a]*Department of Telecommunication and Network Engineering, IUT - Fotso Victor of Bandjoun, University of Dschang, Bandjoun, Cameroon,* [b]*Research Unit of Automation and Applied Computer, Department of Electrical Engineering, IUT - FV Bandjoun, University of Dschang, Bandjoun, Cameroon,* [c]*Research Unit of Condensed Matter, Electronics and Signal Processing, Department of Physics, Faculty of Science, University of Dschang, Dschang, Cameroon,* [d]*Laboratory of Industrial Systems and Environmental Engineering, Fotso Victor University Institute of Technology, University of Dschang, Bandjoun, Cameroon*

7.1 Introduction

Dynamical systems are usually investigated in order to point out their complex behaviors such as chaos, hyperchaos, transient chaos, bursting oscillations, mixed mode oscillations, multistability and extreme multistability. Among these complex and intriguing behaviors, extreme multistability, meaning the coexistence of infinite kinds of stable attractors, is a striking physical phenomenon extremely depending on the initial conditions, which has been reported in many nonlinear dynamical systems [1–13]. The construction of chaotic or hyperchaotic systems with extreme multistability has become a great research topic and received much attention very recently. The extreme multistability is obtained in a system by keeping constant all its parameters and varying only its initial conditions to produce diverse stable states including fixed point, limit-cycles with different periods, chaos, hyperchaos, and so on [8]. Extreme multistable systems have been exploited for some engineering applications such as image processing and decryption [9,10]. Some important chaotic and hyperchaotic systems have been constructed with extreme hidden attractors by introducing the flux or charge-controlled memristor into several existing systems. For example, Bao and coworkers introduced an interesting dynamical system with one saddle and two stable node-foci by using the flux-controlled memristor [7]. The authors

Mem-elements for Neuromorphic Circuits with Artificial Intelligence Applications
https://doi.org/10.1016/B978-0-12-821184-7.00015-3

demonstrated that the proposed system experiences various and striking behaviors including transient transition behavior and transient chaos. Experimental measurements as well as PSIM-based circuit simulations are provided in order to validate the theoretical and numerical analyses. The same strategy is used in [11] to introduce and investigate a 6D hyperchaotic system. The authors reported some interesting dynamical behaviors such as transient chaos, bursting and offset boosting phenomenon. Some Pspice-based circuit simulations are provided to support the theoretical and numerical results. Very recently, Zhang and Li [13] examined the hidden extreme multistability in a novel 4D fractional-order chaotic system using theoretical, numerical and experimental approaches. Kamdem and coworkers [14] suggested and investigated the dynamics of a new 4D memristive hyperchaotic system without equilibrium. They demonstrated that the proposed hyperchaotic system exhibits very long transient chaotic regimes, rare bursting oscillations schemes and coexistence of four attractors. A low-cost microcontroller-based implementation for digital engineering applications has been performed to confirm the theoretical and numerical investigations. All this recent work proves the great interests in the study of dynamical systems with memristors. The memristor is modeled as a resistor with memory, which has potentiation and depression characteristics similar to those of biological synapses. The study of neuromorphic circuits implementing neural networks with memristors for artificial intelligence (AI) applications is currently growing at an exponential pace. This exponential evolution is mainly motivated by the applications of AI in several fields including business management and optimization, stock markets, image recognition, speech processing, medical diagnosis, global climate forecast, autonomous cars, just to name a few [15]. In the field of neuromorphic engineering, the memristors have used to build improved neuromorphic circuits (i.e. circuits with low computing power, physical size and energy consumption). The choice of memristor is justified by its particular plasticity behavior, which emulates the dynamics of biological synapses. Interesting work on memristors for neuromorphic circuits and artificial intelligence applications is reported in [15].

Inspired by the above discussions, in this chapter, we introduce and examine the dynamics of a novel memristive chaotic system with hidden extreme multistability. The system under consideration is derived by introducing the well-known smooth flux-controlled memristor into a new two-scroll chaotic system with two nonlinearities reported very recently by Vaidyanathan and colleagues [16]. The rich and complex behaviors of the proposed system are analyzed with the help of some nonlinear techniques such as bifurcation diagrams, Lyapunov exponents, and phase portraits. It is demonstrated that the system experiences rich and striking dynamics including hidden extreme multistability, offset boosting dynamics and remerging period-doubling bifurcation.

The rest of the chapter is organized as follows. In Section 7.2, we present the memristive two-scroll chaotic system and its basic properties. Section 7.3 is devoted to the dynamical analysis of the proposed memristive two-scroll chaotic system. Hidden bifurcation dynamics, hidden extreme multistability, offset boosting dynamics and remerging period-doubling bifurcation are pointed out. To validate the theoretical and numerical analyses, an electronic hardware circuit capable to describe the complete dynamics of the memristive two-scroll chaotic system is designed and implemented in Section 7.4. The chapter is concluded in Section 7.5.

7.2 Memristive two-scroll chaotic system and its basic properties

In this chapter, a novel four-dimensional memristive two-scroll chaotic system with only seven terms is introduced. The system is derived from the one proposed in [16] defined as

$$\begin{aligned} \dot{x} &= a(y - x) \\ \dot{y} &= xz \\ \dot{z} &= c - by^4 \end{aligned} \tag{7.1}$$

In system (7.1) x, y, and z represent the state variables and a, b, and c are the system parameters. System (7.1) exhibits two-scroll chaotic attractors for the parameter values fixed as: $a = 6$, $b = 1$, and $c = 50$.

In order to generate extreme multistability in system (7.1), a flux-controlled memristor is added in its first equation. The resulting system is defined as

$$\begin{aligned} \dot{x} &= a\,(y - W(w)x) \\ \dot{y} &= xz \\ \dot{z} &= c - by^4 \\ \dot{w} &= dx \end{aligned} \tag{7.2}$$

In system (7.2), w and d are the new state variable and system parameter, respectively. System (7.2) is a seven terms system with four constant parameters and three nonlinearities. It is important to mention that system (7.2) has the minimum number of terms required to produce chaotic dynamics in a four-dimensional autonomous system. Such 4D system with only seven terms is rarely described in the literature. This represents one of the advantages of the proposed system (7.2).

The first equation of system (7.2) contains a flux-controlled memristor characterized by the following relation:

$$W(w) = \alpha + \beta w^2 \tag{7.3}$$

where α and β are positive coefficients characterizing the dynamics of the memristor. The parameter α determines the polarity of the memristor. When $\alpha > 0$, the memristor is passive, while for $\alpha < 0$, the memristor is active. A detailed analysis of the flux-controlled memristor is given in [6].

The volume contraction rate of system (7.2) is defined by the following Lie derivative:

$$\nabla V = \frac{\partial \dot{x}}{\partial x} + \frac{\partial \dot{y}}{\partial y} + \frac{\partial \dot{z}}{\partial z} + \frac{\partial \dot{w}}{\partial w} = -(\alpha + \beta w^2) \tag{7.4}$$

System (7.2) is dissipative for $\alpha + \beta w^2 > 0$. It is easy to prove that system (7.2) is invariant under the transformation $(x, y, z, w) \to (-x, -y, z, -w)$. Thus, it is symmetric to the coordinate axis z. When we set the right-hand side of system (7.2) to zero, we found that it has no real solutions. This means that all the attractors generated from system (7.2) are hidden [17–22] and thus, standard methods such as Shil'nikov theorem cannot be used to explain its chaotic dynamics [23].

7.3 Dynamical analysis of memristive two-scroll chaotic system

7.3.1 Two-parameter Lyapunov exponents analysis

In order to have a global idea about the dynamical behavior of system (7.2), we provide in this subsection a two-parameter Lyapunov analysis. This analysis helps to simplify the designing and implementation of the system. A two-parameter analysis is carried out when two different parameters of the system are varied simultaneously as presented in Fig. 7.1a and 7.1b, for $d = 0.0001$, $\alpha = 1$, and $\beta = 0.02$. The initial conditions are chosen as $(x(0) = y(0) = z(0) = w(0) = 0.1)$.

From Fig. 7.1, we observe that the complexity of system (7.2) increases progressively in terms of two-parameters. Each color characterizes a specific behavior of the system. Indeed, the blue region represents stable periodic states while the rest of regions are associated to chaotic states. In the engineering point of view, such two-parameter Lyapunov exponents diagrams can help to design and implement the system with a desired dynamical behavior.

7.3.2 One parameter bifurcation analysis

In order to investigate the effect of parameter a on the dynamical behavior of system (7.2), we provide as shown in Fig. 7.2 the bifurcation diagram with the three first Lyapunov exponents versus parameter a varying from 5 to 20 for $b = 1$, $c = 50$, $d = 0.0001$, $\alpha = 1$, and $\beta = 0.02$. The initial conditions are selected as $(x(0) = y(0) = z(0) = w(0) = 0.1)$.

When the parameter a is decreased from 20 to 5, the behavior of system (7.2) starts from period-1 limit cycle and breaks into chaos via common period-doubling routes to chaos. The chaotic windows are intersected with some narrow windows of periodic states. It can be seen that the bifurcation diagram well coincides with the three first Lyapunov exponents. It is important to note that system (7.2) displays only one positive Lyapunov exponent for this range of values of

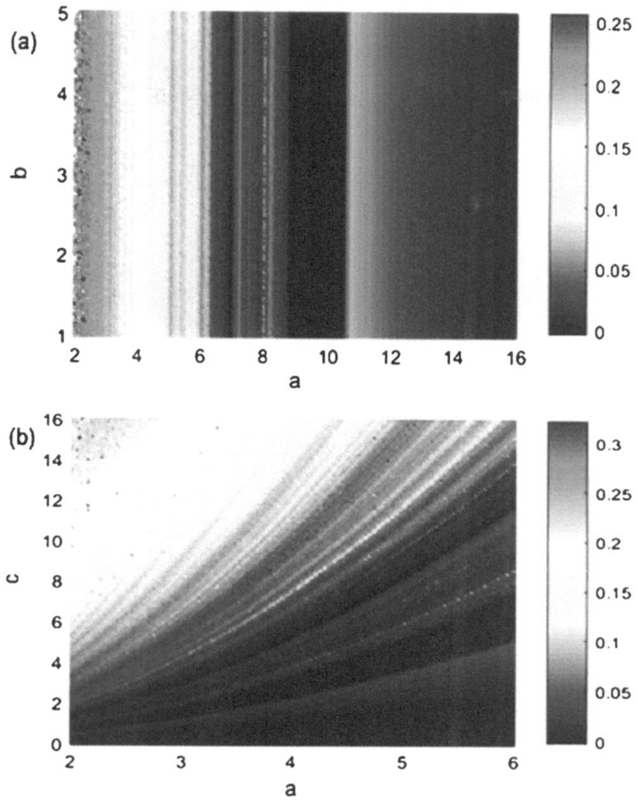

FIGURE 7.1 Two-parameter Lyapunov exponents diagram of system (7.2) obtained by varying simultaneously two parameters in different plans. (a) In the a–b plane for $c = 50$ and (b) in the a–c plane for $b = 1$. The values of other parameters are fixed as $d = 0.0001$, $\alpha = 1$, and $\beta = 0.02$. The initial conditions are chosen as $(x(0) = y(0) = z(0) = w(0) = 0.1)$. (For interpretation of the colors in the figure, the reader is referred to the web version of this chapter.)

parameters as we can see in Fig. 7.2b. This means that the possibility to have hyperchaotic behavior for these values of the system parameters is excluded although it is a four-dimensional system.

To confirm the period-doubling route to chaos revealed by the bifurcation diagram, some phase portraits of system (7.2) are computed in the $(x$–$z)$ plane as shown in Fig. 7.3.

Fig. 7.3 confirms very well the period-doubling route to chaos predicted by the bifurcation diagram.

7.3.3 Emergence of hidden extreme multistability

In this subsection, we investigate the dynamics of system (7.2) by focalizing our attention on extreme multistability. The fourth initial condition $w(0)$ affects

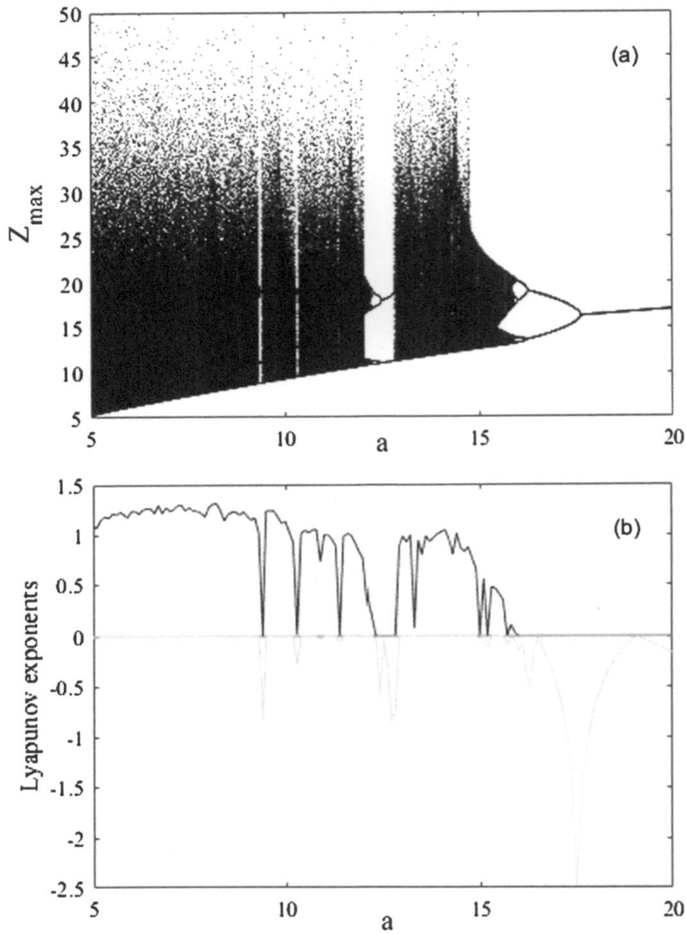

FIGURE 7.2 One parameter bifurcation diagram (a) and three first Lyapunov exponents (b) versus the parameter a describing the route to chaos in system (7.2) for $b = 1$, $c = 50$, $d = 0.0001$, $\alpha = 1$, and $\beta = 0.02$. The initial conditions are chosen as $(x(0) = y(0) = z(0) = w(0) = 0.1)$.

considerably the behavior of system (7.2). Thus, it is chosen here as the main bifurcation parameter. The system parameters are set as $a = 6$, $b = 1$, $c = 50$, $d = 0.0001$, $\alpha = 1$, and $\beta = 0.02$. The bifurcation diagram with the corresponding largest Lyapunov exponent showing the hidden extreme multistability in system (7.2) versus the fourth initial condition $w(0)$ is shown in Fig. 7.4 with other initial conditions chosen as $(x(0), y(0), z(0)) = (0.1)$.

From Fig. 7.4, we notice that when the fourth initial condition $w(0)$ is increasing from -10 to 10, chaos occurs in system (7.2) via period-doubling bifurcation and disappears via remerging period-doubling bifurcation. Fig. 7.4 shows that there are infinite number of coexisting hidden attractors in system (7.2)

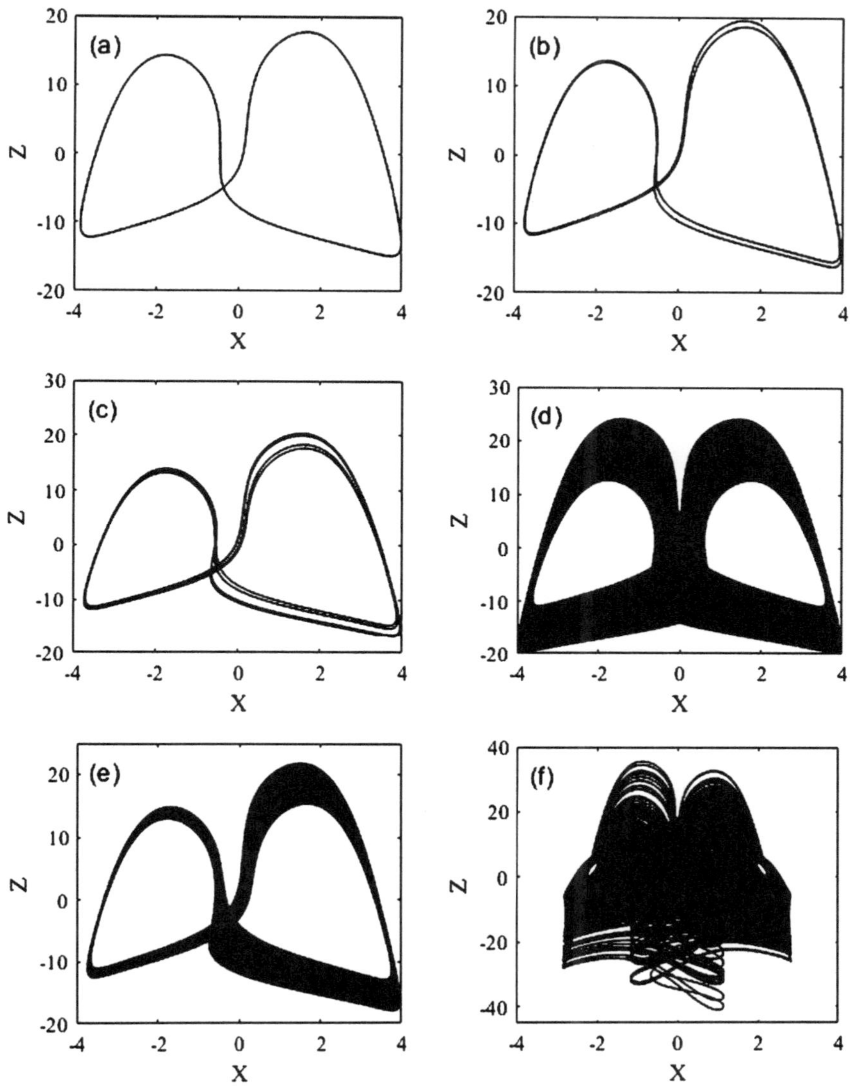

FIGURE 7.3 Phase portraits computed in the (x–z) plane confirming the period-doubling route to chaos in system (7.2). (a) Period-1 for $a = 18$, (b) period-2 for $a = 16.4$, (c) period-4 for $a = 16$, (d) two-scroll chaotic attractor for $a = 15.5$, (e) one-scroll chaotic attractor for $a = 15$, and (f) another two-scroll chaotic attractor for $a = 5.5$. The other system parameters are: $b = 1$, $c = 50$, $d = 0.0001$, $\alpha = 1$, and $\beta = 0.02$. The initial conditions are chosen as ($x(0) = y(0) = z(0) = w(0) = 0.1$).

for these values of parameters setting. In order words, the dynamics of system (7.2) is extremely sensitive to tiny variation of the fourth initial condition $w(0)$. Fig. 7.4 represents a great method to characterize extreme multistability in a dynamical system. It helps to choose the values of the initial conditions depending

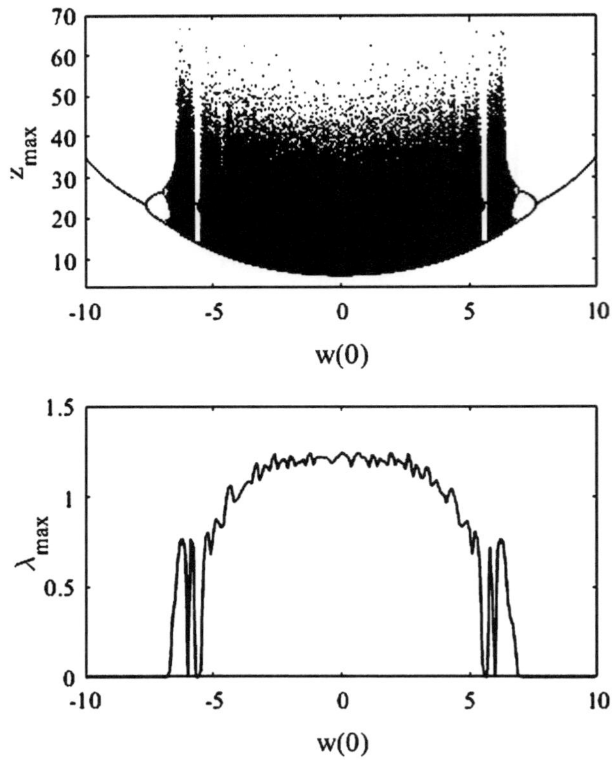

FIGURE 7.4 Bifurcation digram (a) associated with the largest Lyapunov exponent (b) showing the hidden extreme multistability in system (7.2) with respect to the fourth initial condition $w(0)$ varying from -10 to 10 for $a = 6, b = 1, c = 50, d = 0.0001, \alpha = 1$, and $\beta = 0.02$. The other initial conditions are chosen as $(x(0), y(0), z(0)) = (0.1)$.

to the desired state. By selecting some specific values of initial conditions $w(0)$ in the interval -10 to 10, we compute several typical hidden attractors as shown in Fig. 7.5 with the following values of the system parameters: $a = 6$, $b = 1$, $c = 50$, $d = 0.0001$, $\alpha = 1$ and $\beta = 0.02$. The rest of the initial conditions are selected as $(x(0), y(0), z(0)) = (0.1)$.

Fig. 7.5 confirms the occurrence of hidden extreme multistability in system (7.2) as predicted by the bifurcation diagram of Fig. 7.4. We confirm also from Fig. 7.5 that the hidden chaos occurs in system (7.2) via period-doubling bifurcation and disappears via remerging period-doubling bifurcation. It is important to mention that although hidden extreme multistability is commonly reported in many nonlinear dynamical systems, no investigation has been presented about this proposed memristive two-scroll chaotic system with hidden extreme multistability to the author's knowledge.

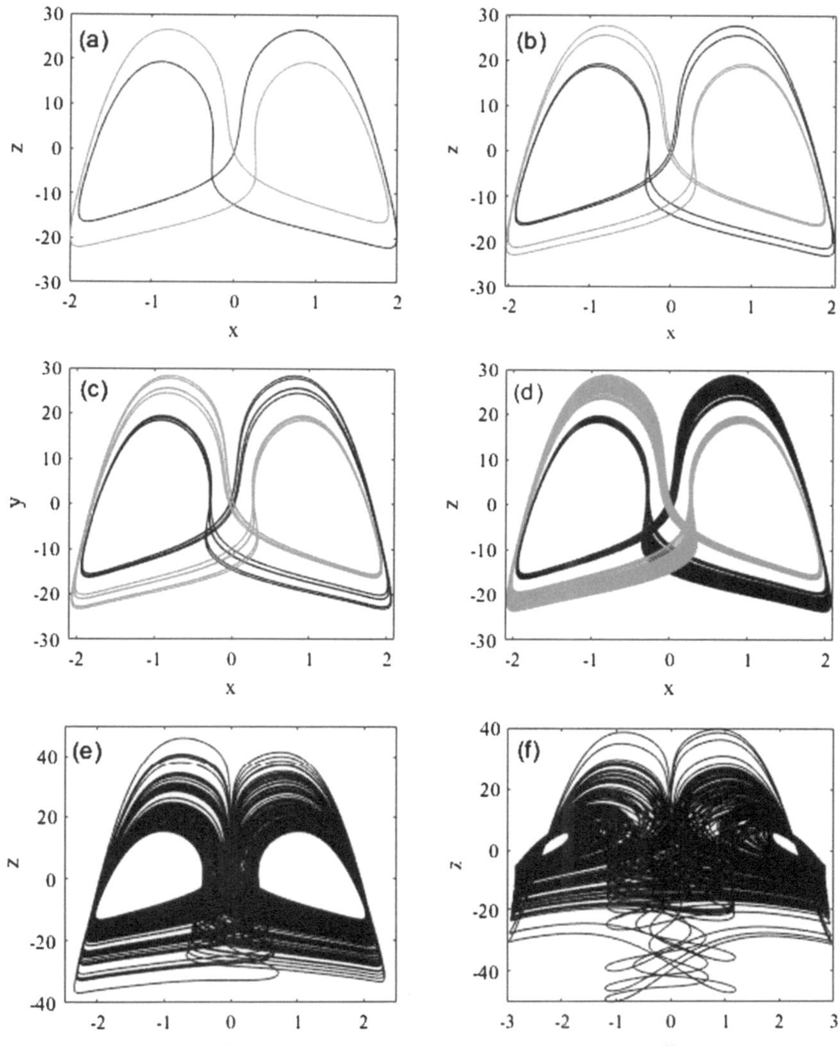

FIGURE 7.5 Typical phase portraits of system (7.2) computed with some discrete values of the fourth initial condition $w(0)$ selected in the interval -10 to 10, showing hidden extreme multistability. (a) A pair of period-1 limit cycle for $w(0) = \pm 6.975$, (b) a pair of period-2 limit cycle for $w(0) = \pm 6.84$, (c) a pair of period-4 limit cycle for $w(0) = \pm 6.75$, (d) a pair of one scroll chaotic attractors for $w(0) = \pm 6.70$, (e) two-scroll chaotic attractors for $w(0) = 6.005$, and (f) another two-scroll chaotic attractors for $w(0) = 0.20$. The system parameters are $a = 6$, $b = 1$, $c = 50$, $d = 0.0001$, $\alpha = 1$ and $\beta = 0.02$. The rest of the initial conditions are chosen as $(x(0), y(0), z(0)) = (0.1)$.

7.3.4 Remerging period-doubling bifurcation

Remerging period-doubling bifurcation (i.e. antimonotonicity) is reported in the literature as one of the most basic bifurcation complements in the theory and application of nonlinear dynamical systems [24]. This phenomenon is manifested by the creation and annihilation of periodic orbits via reversals period-doubling cascades when the bifurcation parameter is adjusted [25]. Usually the remerging period-doubling bifurcation is obtained by varying one bifurcation parameter of the system in a certain range for some discrete values of another parameter of the same system. Here, the remerging period-doubling bifurcation diagram is pointed out by varying the fourth initial condition $w(0)$ for some discrete values of parameter a. The bifurcation diagrams with respect to the fourth initial condition $w(0)$ for remerging Feigenbaum tree of the maxima of the state variable z are provided in Fig. 7.6 for $b = 1$, $c = 50$, $d = 0.0001$, $\alpha = 1$, and $\beta = 0.02$. The rest of initial conditions are chosen as $(x(0), y(0), z(0)) = (0.1)$.

The phenomenon of forward period-doubling bifurcation sequences followed by reverse period-doubling sequences is clearly observed in Figs. 7.6a–h. For $a = 16.10$, $a = 16.00$, and $a = 15.97$, primary bubble, period-4 bubble and period-8 bubble are created, respectively. When a is increased discretely from 15.97 to 15.50, the bifurcation diagram becomes more complex and an infinite trees appear (see Fig. 7.6d to h).

7.3.5 Offset boosting control

The hidden memristive chaotic system (7.2) has the special feature of offset boosting because the state variable z appears only in its second equation. This means that the amplitude of the state variable z can be adjusted by adding a control parameter n into the second equation of system (7.2) as follows:

$$\begin{aligned} \dot{x} &= a\,(y - W(w)x) \\ \dot{y} &= x(z + n) \\ \dot{z} &= c - by^4 \\ \dot{w} &= dx \end{aligned} \tag{7.5}$$

It is important to mention that this modification has no influence on the dynamics of system (7.2). It just provides a controllable ability of the amplitudes of the system. When $b = 1$, $c = 50$, $d = 0.0001$, $\alpha = 1$, $\beta = 0.02$ and initial conditions chosen as $(x(0) = y(0) = z(0) = w(0) = 0.1)$, the influence of the offset boosting parameter n on the dynamics of system (7.5) is investigated by plotting the bifurcation diagram versus the parameter a for some discrete values of n as shown in Fig. 7.7.

From Fig. 7.7, it is visible that when the control offset boosting parameter n is equal to zero (see the black curve), system (7.5) and system (7.2) have the same dynamics. While the positive and negative values of the control offset boosting parameter n translate the state variable z, respectively in the negative

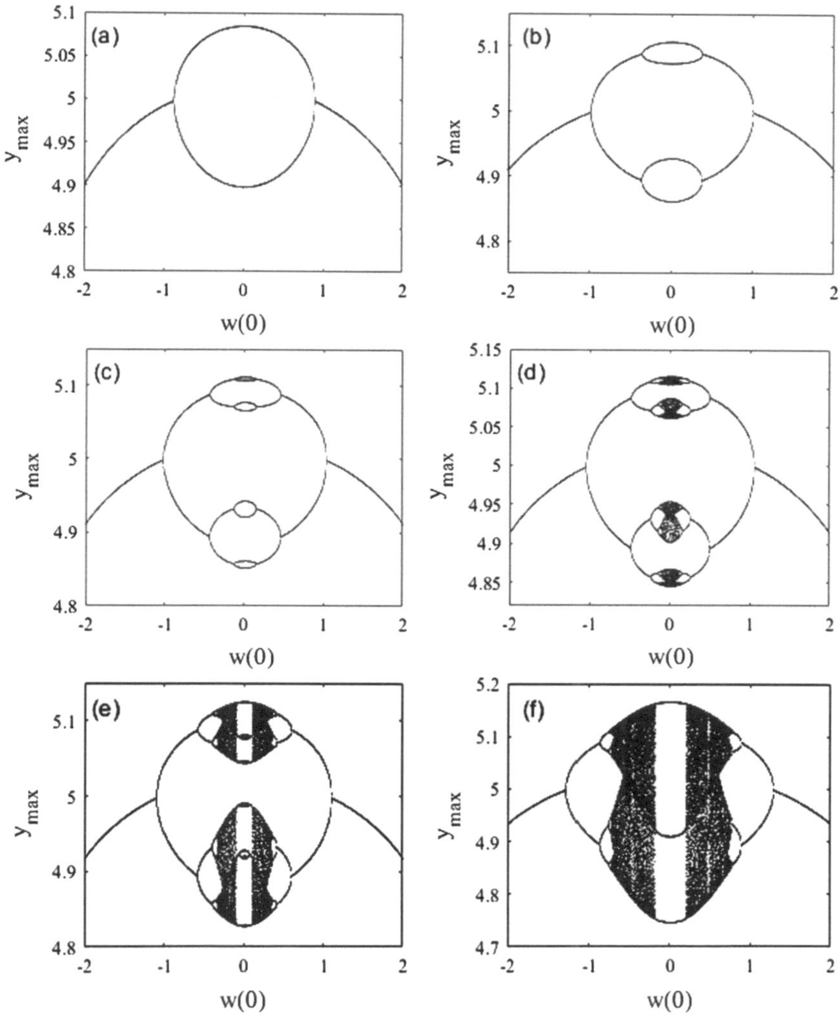

FIGURE 7.6 Bifurcation diagrams (a)–(h) with respect to the fourth initial conditions $w(0)$ for remerging Feigenbaum tree obtained with some discrete values of parameter a. (a) Primary bubble for $a = 16.10$; (b) period-4 bubble for $a = 16.00$; (c) period-8 bubble for $a = 15.97$; (d), (e), (f), (g), and (h) full Feigenbaum remerging tree for $a = 15.95$, $a = 15.90$, $a = 15.80$, $a = 15.70$, and $a = 15.50$, respectively.

and positive directions (see the curves in red and blue (light gray and mid gray in print version)). In order to more illustrate the offset boosting dynamics, we provide in Fig. 7.8 some phase portraits for $a = 6$, $b = 1$, $c = 50$, $d = 0.0001$, $\alpha = 1$, and $\beta = 0.02$. The initial conditions are those chosen in Fig. 7.7.

From Fig. 7.8, we observe that the state variable z is really boosted when the offset boosting parameter n takes some specific values. We confirm that a

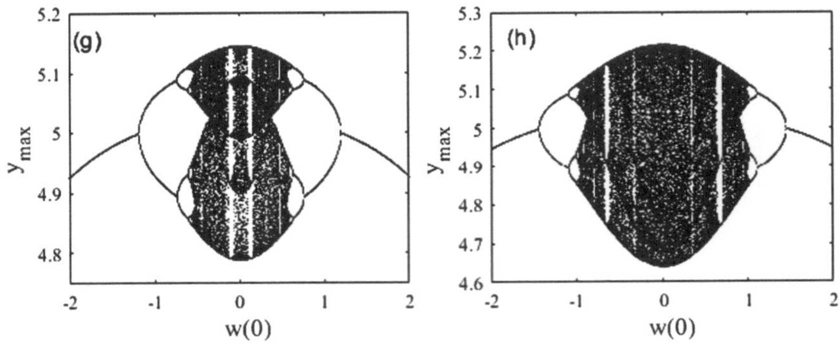

FIGURE 7.6 *continued*

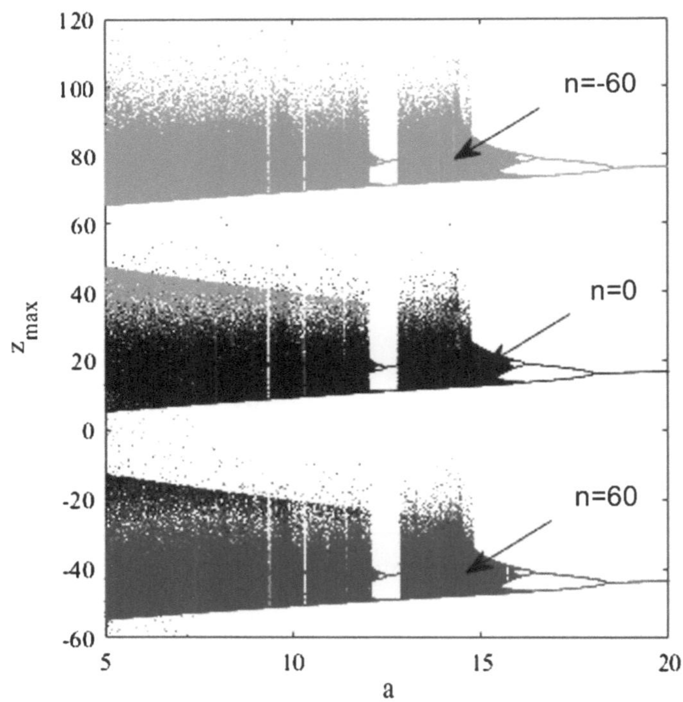

FIGURE 7.7 Bifurcation diagrams versus parameter a illustrating the offset boosting dynamics in system (7.5). The values of the system parameters are set as $b = 1$, $c = 50$, $d = 0.0001$, $\alpha = 1$, $\beta = 0.02$ and initial conditions ($x(0) = y(0) = z(0) = w(0) = 0.1$). The curves in blue, black, and red (mid gray, dark gray, light gray in print version) are, respectively, obtained for $n = -60$, $n = 0$, and $n = 60$.

positive value of the control boosting parameter n boosts the state variable z in the negative direction, while negative value of the controller n boosts the variable z in the positive direction as predicted by the bifurcation diagrams in

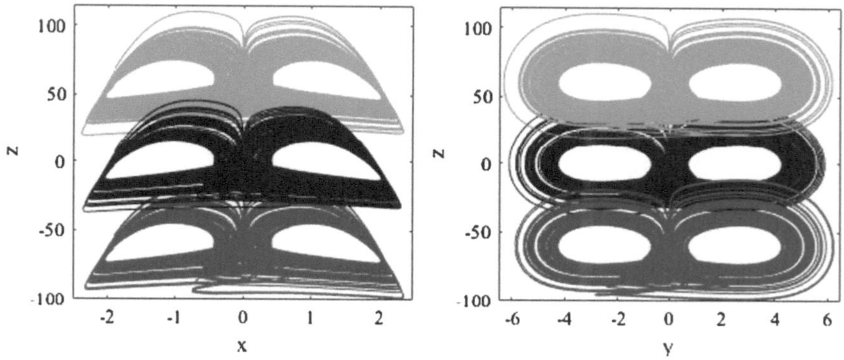

FIGURE 7.8 Phase portraits in the (x–z) and (y–z) planes illustrating the offset boosting dynamics in system (7.5). The different phase portraits are obtained with $n = 0$ (black curve), $n = -60$ (blue curve (mid gray in print version)) and $n = 60$ (red curve (light gray in print version)). The system parameters are $a = 6$, $b = 1$, $c = 50$, $d = 0.0001$, $\alpha = 1$, and $\beta = 0.02$. The initial conditions are those chosen in Fig. 7.7.

Fig. 7.7. The offset boosting has been reported in some chaotic systems [26–29]. It can be used in some relevant engineering applications where a single constant can be tuned to achieve a desired boosting level.

7.4 Circuit design and experimental measurements

The hardware implementation of theoretical chaotic models [30,31] plays an important role in engineering applications such as image encryption, chaos-based communications, random bits generation, just to name a few. In this section, we design and implement an electronic circuit in order to reproduce the dynamical behaviors of system (7.2) obtained with theoretical and numerical methods. The proposed circuit designed with electronic elements capable to mimic the dynamics of system (7.2) is provided in Fig. 7.9.

The circuit includes 2 quadruple operational amplifiers (TL084), 4 capacitors, 11 resistors, a DC source used to realize the constant term and 5 analog multiplier chips (AD 633JN) exploited to construct the nonlinear terms. To define the state equations describing the circuit given in Fig. 7.9, we assume that capacitors and operational amplifiers operate in their linear working domain. Considering this assumption and by applying Kirchhoff's laws to the circuit, the following circuit equations are obtained:

$$\begin{cases} \frac{dV_x}{d\tau} = -\frac{V_x}{R_1C} - \frac{0.01V_w^2V_x}{R_2C} + \frac{V_y}{R_3C} \\ \frac{dV_y}{d\tau} = \frac{0.1V_zV_x}{R_4C} \\ \frac{dV_z}{d\tau} = -\frac{0.001V_y^4}{R_5C} + \frac{V_0}{R_6C} \\ \frac{dV_w}{d\tau} = \frac{V_x}{R_7C} \end{cases} \tag{7.6}$$

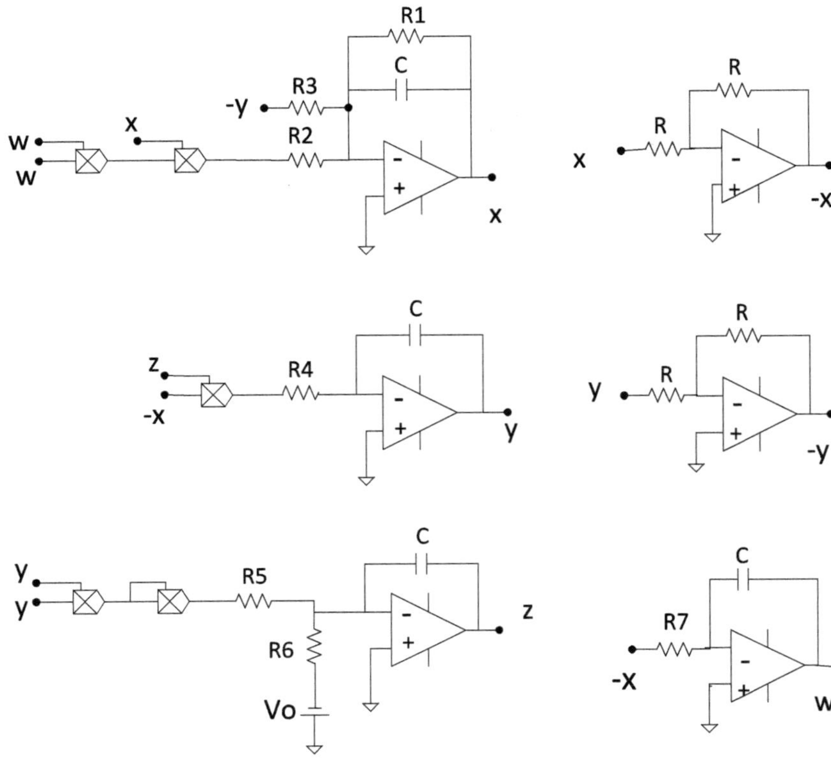

FIGURE 7.9 Proposed electronic circuit design for hidden memristive chaotic system (7.2).

where V_x, V_y, V_z, and V_w are the output voltages of the ideal operational amplifiers. Obviously, the circuit equations (7.6) correspond to system (7.2). With the values of system (7.2) chosen as $b = 1$, $c = 50$, $d = 0.0001$, $\alpha = 1$, and $\beta = 0.02$, the corresponding values of circuit elements are calculated as $C = 10$ nF, $R = 10$ kΩ, $R_2 = 83.33$ Ω, $R_3 = 1.66$ KΩ, $R_4 = 1$ KΩ, $R_5 = 10$ Ω, $R_6 = 200$ Ω, $R_7 = 100$ MΩ, and R_1 is a control resistor which will be adjusted to obtain the different behaviors developed by the circuit.

By fixing the circuit component values as calculated above, the experimental phase portraits confirming the period-doubling route to chaos in the circuit are shown in Fig. 7.10 for some particular values of the control resistor R_1.

In Fig. 7.10, we can confirm the transition to chaos via period-doubling bifurcation as predicted by numerical results. Fig. 7.10 serves to prove that the designed electronic circuit describes very well the complex dynamical behaviors of the proposed memristive two-scroll chaotic system with hidden extreme multistability. It is important to mention that the tiny deviation between numerical and experimental results may be attributed to the precision on the values of electronic components used for the real implementation of the circuit.

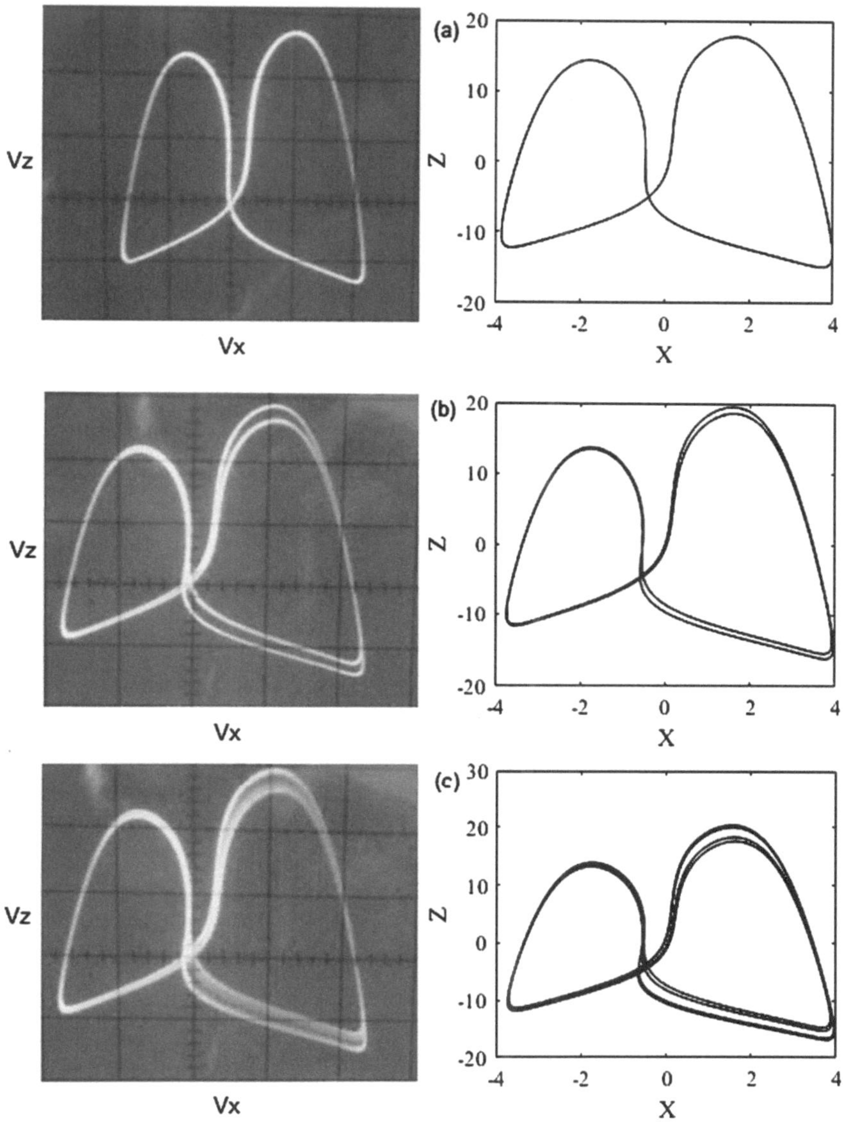

FIGURE 7.10 Experimental phase portraits confirming period-doubling route to chaos (left) captured from the circuit using a dual trace oscilloscope in the XY mode. The corresponding numerical phase portraits are provided in the right for comparison. (a) Period-1 for $R_1 = 555.55\ \Omega$, (b) period-2 for $R_1 = 609.75\ \Omega$, (c) period-4 for $R_1 = 625\ \Omega$, (d) two-scroll chaotic attractors for $R_1 = 645.16\ \Omega$, (e) one-scroll chaotic attractors for $R_1 = 666.66\ \Omega$, and (f) another two-scroll chaotic attractors for $R_1 = 2\ \text{k}\Omega$. The values of the other electronic components are $C = 10$ nF, $R = 10\ \text{k}\Omega$, $R_2 = 83.33\ \Omega$, $R_3 = 1.66\ \text{K}\Omega$, $R_4 = 1\ \text{K}\Omega$, $R_5 = 10\ \Omega$, $R_6 = 200\ \Omega$, $R_7 = 100\ \text{M}\Omega$.

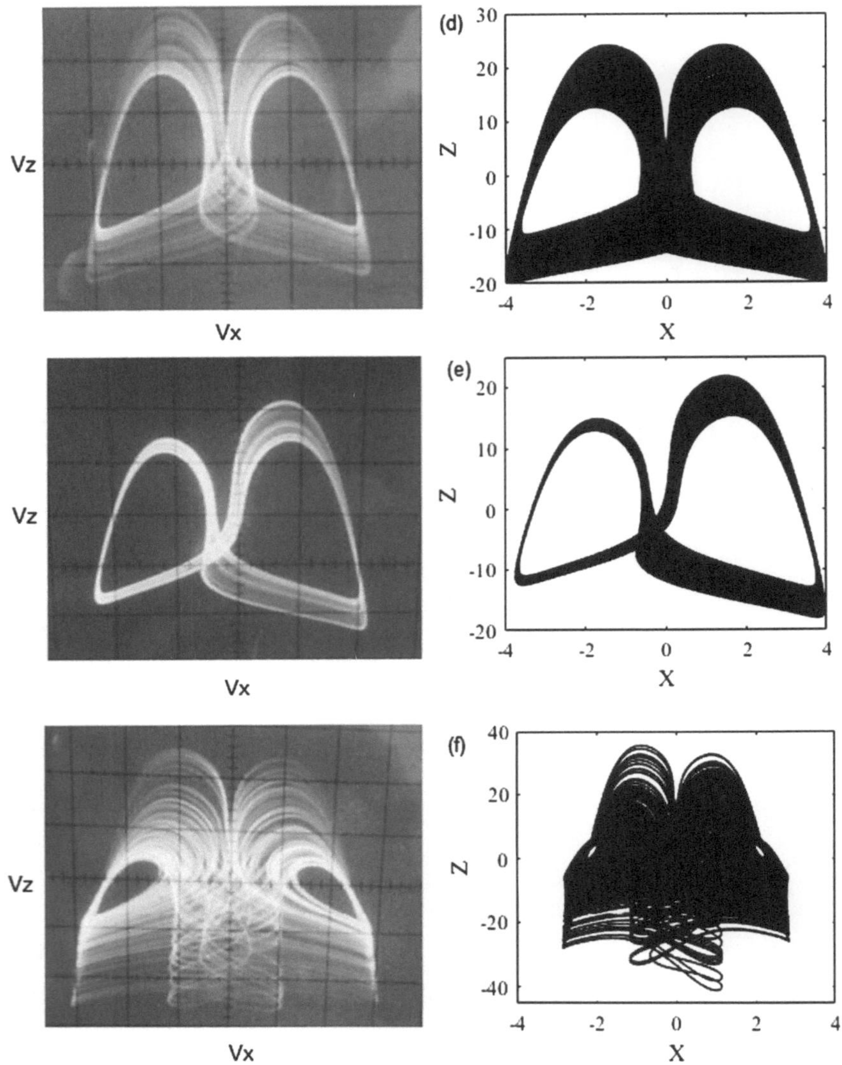

FIGURE 7.10 *continued*

7.5 Conclusion

In this chapter, we considered and investigated the dynamics of a 4D memristive two-scroll chaotic system with hidden extreme multistability. The system in consideration has been found very elegant with only seven terms. The basic characteristics, theoretical and numerical investigations of the system have been pointed out. It has been demonstrated that the system under scrutiny experiences rich and intriguing dynamical behaviors such as hidden extreme multistability, offset boosting dynamics and remerging period-doubling bifurcation. These rich behaviors have been illustrated using two-parameter Lyapunov exponent diagram, phase portraits and bifurcation diagrams associated with Lyapunov exponents. To support theoretical/numerical investigations, we designed and implemented an electronic circuit capable to describe the complete dynamics of the proposed memristive two-scroll chaotic system with hidden extreme multistability. Experimental results have been in accordance with those obtained theoretically and numerically. Such simple systems with only seven terms and hidden extreme multistability can be exploited for many real time applications such as chaos-based communication systems, image and voice encryption and random number generators. Since the dynamical behavior of developed system is already mastered, this system will be exploited to develop a practical neuromorphic application for artificial intelligence in our future work.

References

[1] C. Hens, S.K. Dana, U. Feudel, Extreme multistability: attractors manipulation and robustness, Chaos 25 (5) (2015) 053112.
[2] H. Sun, S.K. Scott, K. Showalter, Uncertain destination dynamics, Physical Review E 60 (4) (1999) 3876–3880.
[3] M.S. Patel, U. Patel, A. Sen, G.C. Sethia, C. Hens, S.K. Dana, Experimental observation of extreme multistability in an electronic system of two coupled Rössler oscillators, Physical Review E 89 (2) (2014) 022918.
[4] B. Bao, O. Xu, H. Bao, M. Chen, Extreme multistability in a memristive circuit, Electronics Letters 52 (12) (2016) 1008–1010.
[5] B. Bao, T. Jiang, Q. Xu, M. Chen, H. Wu, Y. Hu, Coexisting infinitely many attractors in active band-pass filter-based memristive circuit, Nonlinear Dynamics 86 (3) (2016) 1711–1723.
[6] F. Yuan, G. Wang, X. Wang, Extreme multistability in a memristor-based multi-scroll hyperchaotic system, Chaos 26 (7) (2016) 073107.
[7] B. Bao, H. Bao, N. Wang, M. Chen, Q. Xu, Hidden extreme multistability in memristive hyperchaotic system, Chaos, Solitons and Fractals 94 (2017) 102–111.
[8] H. Bao, N. Wang, B. Bao, M. Chen, P. Jin, G. Wang, Initial condition-dependent dynamics and transient period in memristor-based hypogenetic jerk system with four line equilibria, Communications in Nonlinear Science and Numerical Simulation 57 (2018) 264–275.
[9] S. Morfu, B. Nofiele, P. Marquié, On the use of multistability for image processing, Physics Letters A 367 (3) (2007) 192–198.
[10] Chuang Li, Fuhong Min, Qiusen Jin, Hanyuan Ma, Extreme multistability analysis of memristor-based chaotic system and its application in image decryption, AIP Advances 7 (2017) 12504.
[11] B.A. Mezatio, M. Tingue Motchongom, B.R. Wafo Tekam, R. Kengne, R. Tchitnga, A. Fomethe, A novel memristive 6D hyperchaotic autonomous system with hidden extreme multistability, Chaos, Solitons and Fractals 120 (2019) 100–115.
[12] P. Louodop, R. Tchitnga, F.F. Fagundes, M. Kountchou, V. Kamdoum Tamba, C.L. Pando, H.A. Cerdeira, Extreme multistability in a Josephson-junction-based circuit, Physical Review E 99 (2019) 042208.
[13] X. Zhang, Z. Li, Hidden extreme multistability in a novel 4D fractional-order chaotic system, International Journal of Non-Linear Mechanics (2019).
[14] P.D. Kamdem Kuate, Q. Lai, H. Fotsin, Complex behaviors in a new 4D memristive hyperchaotic system without equilibrium and its microcontroller-based implementation, The European Physical Journal Special Topics 228 (2019) 2171–2184.

[15] J. Suñé, Memristors for Neuromorphic Circuits and Artificial Intelligence Applications, Materials, 2020, https://doi.org/10.3390/books978-3-03928-577-8.
[16] S. Vaidyanathan, A. Sambas, S. Zhang, Y. Zeng, M.A. Mohamed, M. Mamat, A new two-scroll chaotic system with two nonlinearities: dynamical analysis and circuit simulation, Telkomnika 17 (5) (2019) 2465–2474.
[17] S. Jafari, J.C. Sprott, Simple chaotic flows with a line equilibrium, Chaos, Solitons and Fractals 57 (2013) 79–84.
[18] V.T. Pham, S. Vaidyanathan, C.K. Volos, S. Jafari, Hidden attractors in a chaotic system with an exponential nonlinear term, The European Physical Journal Special Topics 224 (2015) 1507–1517.
[19] V.T. Pham, S. Jafari, C.K. Volos, T. Kapitaniak, A gallery of chaotic systems with an infinite number of equilibrium points, Chaos, Solitons and Fractals 93 (2016) 58–63.
[20] V.T. Pham, S. Vaidyanathan, C.K. Volos, S. Jafari, S. Takougang Kingni, A no-equilibrium hyperchaotic system with a cubic nonlinear term, International Journal for Light and Electron Optics 127 (2016) 3259–3265.
[21] V.V. Pham, S. Jafari, X. Wang, J. Ma, A chaotic system with different shapes of equilibria, International Journal of Bifurcation and Chaos 26 (2016) 1650069.
[22] S. Bouali, Hidden structure and complex dynamics of hyperchaotic attractors, Annual Review of Chaos Theory, Bifurcations and Dynamical Systems 6 (2016) 48–58.
[23] C.P. Silva, Shil'nikov's theorem – a tutorial, IEEE Transactions on Circuits and Systems I 40 (1993) 675–682.
[24] S.P. Dawson, C. Grebogi, J.A. Yorke, I. Kan, H. Koçak, Antimonotonicity: inevitable reversals of period-doubling cascades, Physics Letters A 162 (1992) 249–254.
[25] S.P. Dawson, Geometric mechanism for antimonotonicity in scalar maps with two critical points, Physical Review E 48 (1993) 1676–1680.
[26] C. Li, J.C. Sprott, Amplitude control approach for chaotic signals, Nonlinear Dynamics 73 (2013) 1335–1341.
[27] C. Li, J.C. Sprott, Variable-boostable chaotic flows, Optik International Journal for Light and Electron Optics 27 (2016) 10389–10398.
[28] V.T. Pham, A. Akgul, C.K. Volos, S. Jafari, T. Kapitaniak, Dynamics and circuit realization of a no-equilibrium chaotic system with a boostable variable, International Journal of Electronics and Telecommunications 78 (2017) 134–140.
[29] V. Kamdoum Tamba, R. Karthikeyan, V.T. Pham, D.V. Hoang, Chaos in a system with an absolute nonlinearity and chaos synchronization, Advances in Mathematical Physics 5985489 (2018) 12.
[30] V. Kamdoum Tamba, V.T. Pham, D.V. Hoang, S. Jafari, F.E. Alsaadi, F.E. Alsaadi, Dynamic system with no equilibrium and its chaos anti-synchronization, Journal for Control, Measurement, Electronics, Computing and Communications 59 (1) (2018) 35–42.
[31] S.T. Kingni, L. Keuninckx, P. Woafo, G. Van der Sande, J. Danckaert, Dissipative chaos, Shilnikov chaos and bursting oscillations in a three-dimensional autonomous system: theory and electronic implementation, Nonlinear Dynamics 73 (2013) 1111–1123.

Chapter 8

Extreme multistability, hidden chaotic attractors and amplitude controls in an absolute memristor Van der Pol–Duffing circuit: dynamical analysis and electronic implementation

Justin Roger Mboupda Pone[a], Foutse Momo[b], Guillaume Honore Kom[a], Ingrid Ornela Lowe Kombou[a,c], Sifeu Takougang Kingni[d], and Alain Tiedeu[e]

[a]*Research Unit of Automation and Applied Computer (RU-AIA), Electrical Engineering Department of IUT-FV, University of Dschang, Bandjoun, Cameroon* , [b]*Biomedical Engineering, Energy and Modeling Laboratory (BEEMo.Lab.), Higher Institute of Science and Technology (HIST), University of Mountains, Bangangte, Cameroon* , [c]*Research Unit of Condensed Matter, Electronics and Signal Processing (RU-LAMACETS), Department of Physics, Faculty of Science, University of Dschang, Dschang, Cameroon* , [d]*Department of Mechanical, Petroleum and Gas Engineering, Faculty of Mines and Petroleum Industries, University of Maroua, Maroua, Cameroon* , [e]*Signal, Image and Systems Laboratory, Higher Technical Teachers' Training College (HTTTC) of Ebolowa, University of Yaoundé 1, Yaoundé, Cameroon*

8.1 Introduction

The Van der Pol–Duffing (VdPD) circuit [1] is a well-known chaotic circuit which is equivalent to Chua's circuit [2]. It is made of five elements connected in parallel namely, two capacitors, an inductor, a resistor and an active nonlinear resistor constructed using a set of diodes and an operational amplifier [3]. The VdPD circuit is governed by a common nonlinear differential equation which can be used to describe several physically interesting systems in engineering, biology, physics, neurology and so on [4–12]. In order to obtain interesting dynamics in the VdPD circuit, its implementation has been modified in at least two physically interesting situations: (i) By adding a serial resistor to its induc-

Mem-elements for Neuromorphic Circuits with Artificial Intelligence Applications
https://doi.org/10.1016/B978-0-12-821184-7.00016-5

tor and the offset currents to its operational amplifier Ref. [1] and (ii) by adding a resistor in parallel with its inductor [13].

Due to its potential applications in secure communications [14], mathematical modeling and analysis [15–20], chemical and biological systems [21,22], the memristor has attracted wide attention in the scientific community since its physical realization in 2008 [23–25]. The authors of Ref. [14] replaced the nonlinear resistor of the VdPD circuit with a resistor in parallel Ref. [14] by a flux-controlled memristor in order to have a circuit more suitable for applications related to high frequency chaotic oscillators and secure communications. Recently, some authors of this contribution proposed a memristor-based chaotic VdPD circuit with series and parallel resistors [26] built by replacing the nonlinear resistor of a VdPD circuit by an ideal and active flux-controlled memristor with absolute value nonlinearity [27]. The authors of Ref. [26] demonstrated that the absolute memristor VdPD circuit exhibits bistable one-scroll self-excited chaotic attractors, double-scroll self-excited chaotic attractors, bistable periodic attractors and the antimonotonicity phenomenon. The circuit of an ideal and active flux-controlled memristor with an absolute memristance function used in Ref. [26] has been taken in Ref. [28] and was made of a buffer, an integrator connected to two resistors as well as a capacitor, an absolute function unit circuit, one multiplier and a current inverter connected to three resistors. By using an absolute memristor made of a buffer, an integrator connected to one resistor as well as a capacitor, an absolute function unit circuit, one multiplier and a current inverter connected to three resistors, an absolute memristor VdPD circuit is proposed and analyzed in this chapter.

It is shown in this chapter that the proposed absolute memristor VdPD circuit can exhibit the extreme multistability phenomenon, hidden chaotic attractors and the amplitude control feature. We recalled that the proposed VdPD is built with an active flux-controlled memristor with an absolute memristance function. It is well known that memristor system technologies have been developed, notably ReRAM. In addition, artificial intelligence applications are implemented in a circuit using RAM. It could be used as for neuromorphic circuits with artificial intelligence applications owing its mem-element [31].

The rest of this chapter is structured as follows. In Section 8.2, an absolute memristor VdPD circuit is presented and its rate-equations are derived and investigated. Section 8.3 deals with electronic implementation of the absolute memristor VdPD circuit and some concluding remarks end this paper in Section 8.4.

8.2 Theoretical analysis of an absolute memristor autonomous Van der Pol–Duffing circuit

The absolute memristor autonomous Van der Pol–Duffing circuit is shown in Fig. 8.1.

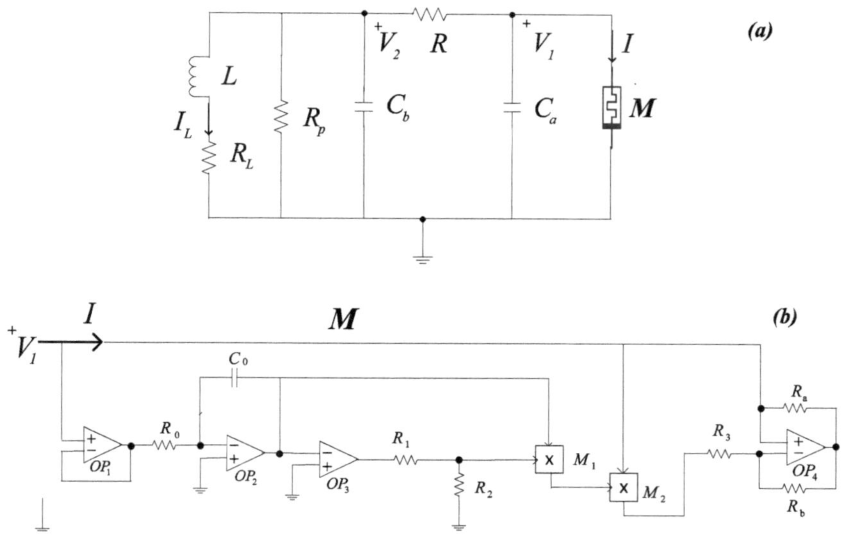

FIGURE 8.1 The schematic diagrams of: (a) the proposed absolute memristor-based unified autonomous Van der Pol–Duffing circuit and (b) the ideal and active flux-controlled memristor M used in (a).

A unified autonomous Van der Pol–Duffing circuit consists of two capacitors C_a, C_b, an inductor L with internal resistor R_L, two linear resistors R, R_P and a nonlinear resistor [3,13,29,30]. The circuit of Fig. 8.1(a) is obtained by replacing the nonlinear resistor of unified autonomous Van der Pol–Duffing circuit by an ideal and active flux-controlled memristor M with an absolute memristance function [28]. An ideal and active flux-controlled memristor with an absolute memristance function is shown in Fig. 8.1(b). The circuit of the ideal and active flux-controlled memristor consists a buffer OP_1 with a resistor R_0, an integrator OP_2 connected to a resistor R_0, an absolute function unit circuit (made of a comparator OP_3 with two resistors R_1 and R_2), two multipliers M_1, M_2, and two resistors R_a and R_b surrounding an operational amplifier OP_4. Applying Kirchhoff's law to the circuit of Fig. 8.1(b), the following equations are obtained:

$$I = M\left(V_0\right) V_1 = \left(\frac{1}{R_3} - \frac{g_1 g_2}{R_3}\left|V_0\right|\right) V_1, \tag{8.1a}$$

$$\frac{dV_0}{dt'} = -\frac{1}{R_0 C_0} V, \tag{8.1b}$$

where V_0 is the voltage across the capacitor C_0 in the memristor circuit, $M(V_0)V_1 = \left(\frac{1}{R_3} - \frac{g_1 g_2}{R_3}|V_0|\right) V_1$, is the memristance and g_1, g_2 are the gains of multipliers M_1 and M_2, respectively. Applying Kirchhoff's law on the proposed absolute memristor autonomous Van der Pol–Duffing circuit in Fig. 8.1(a), the following state equations can be derived:

$$\frac{dV_1}{dt'} = \frac{1}{RC_a}(V_2 - V_1) + \frac{1}{C_a}M(V_0)V_1 \tag{8.2a}$$

$$\frac{dV_1}{dt'} = \frac{1}{RC_a}(V_1 - V_2) - \frac{1}{RC_b}\frac{R}{R_p}V_2 - \frac{I_L}{C_b}, \tag{8.2b}$$

$$\frac{dI_L}{dt'} = \frac{V_2}{L} - \frac{R_L I_L}{L} \tag{8.2c}$$

$$\frac{dV_0}{dt'} = \frac{1}{R_0 C_0}V_1. \tag{8.2d}$$

The mathematical model (8.2) is simplified using the normalized variables:

$$x = \frac{V_1}{V_{\text{ref}}},\ y = \frac{V_2}{V_{\text{ref}}},\ z = \frac{RI_L}{V_{\text{ref}}},\ w = \frac{V_0}{V_{\text{ref}}},\ t = \frac{t'}{RC_b},\ V_{\text{ref}} = 1V \tag{8.3}$$

And the simplified version of the set of Eqs. (8.2) can be derived as

$$\frac{dx}{dt} = a(y - x + xM(w)), \tag{8.4a}$$

$$\frac{dy}{dt} = x - ey - z, \tag{8.4b}$$

$$\frac{dz}{dt} = b(y - dz), \tag{8.4c}$$

$$\frac{dw}{dt} = -cx, \tag{8.4d}$$

where $a = \frac{Cb}{Ca}$, $b = \frac{RC_b}{L}$, $c = RC_b R_0 C_0$, $d = R_L R$, $e = 1 + \frac{R}{R_p}$, $\alpha = \frac{R}{R_3}$, $\beta = \frac{g_1 g_2 R}{R_3}$ and $M(w) = \alpha - \beta|w|$ are the system parameters. System (8.4) has a natural symmetry under the transformation $S(x, y, z, w) \rightarrow (-x, -y, -z, -w)$. System (8.4) has a line equilibrium given by

$$E = \{(x, y, z, w) \in R^4 \ / \ x = y = z = 0,\ w = w^*\}. \tag{8.5}$$

Therefore, system (8.3) belongs to the family of chaotic systems with hidden attractors. The Jacobian matrix of system (8.5) at equilibrium point $E = (0, 0, 0, w^*)$ is given by

$$J = \begin{pmatrix} a[-1 + M(w^*)] & a & 0 & -a\beta x^* \text{sign}(w^*) \\ 1 & -e & -1 & 0 \\ 0 & b & -bd & 0 \\ -c & 0 & 0 & 0 \end{pmatrix} \tag{8.6}$$

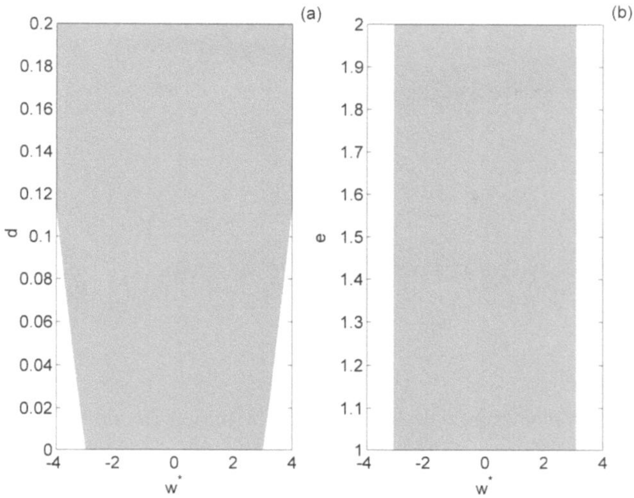

FIGURE 8.2 Stability region distributions of the three non-zero eigenvalues in the space: (a) (w^*, d) for $e = 1.2$ and (b) (w^*, e) $d = 0.01$. The white area is stable, while the gray one is unstable.

The characteristic equation of the Jacobian matrix (8.6) on this line equilibrium is

$$\lambda(\lambda^3 + \delta_1\lambda^2 + \delta_2\lambda + \delta_3) = 0, \tag{8.7}$$

where the expressions for δ_i $(i = 1, 2, 3)$ are given by

$$\begin{aligned} \delta_1 &= a + e + bd - a(\alpha - \beta|w^*|), \\ \delta_2 &= b + a(e - 1) + bd(a + e) - a(e + bd)(\alpha - \beta|w^*|), \\ \delta_3 &= ab[-d + (1 + de)(1 - \alpha + \beta|w^*|)]. \end{aligned} \tag{8.8}$$

One of the four eigenvalues always equals zero and the other three eigenvalues can be found by solving

$$\lambda^3 + \delta_1\lambda^2 + \delta_2\lambda + \delta_3 = 0, \tag{8.9}$$

and applying the Routh–Hurwitz criterion, Eq. (8.9) has all roots with negative real parts if and only if

$$\delta_1 > 0,\ \delta_3 > 0,\ \delta_1\delta_2 - \delta_3 > 0. \tag{8.10}$$

Let $a = 17.8571$, $b = 28$, $c = 37$, $\alpha = 1.3$, $\beta = 0.1$ and by varying the parameters d, e, w^*, in Fig. 8.2 the stability region distributions of the three non-zero eigenvalues are plotted.

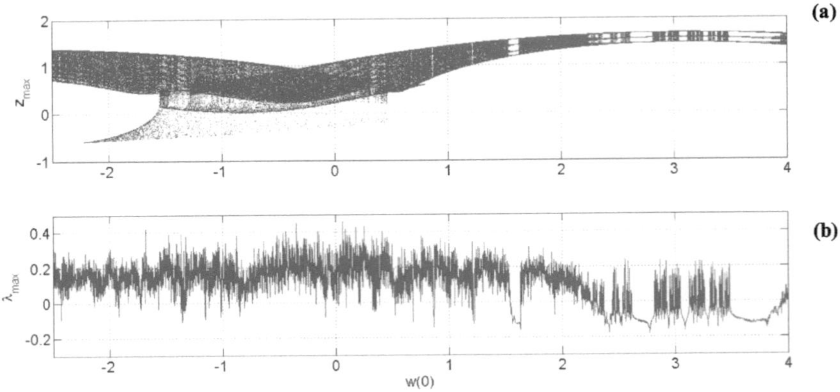

FIGURE 8.3 Bifurcation diagram depicting the local maxima of $z(t)$ (a) and (b) the maximum Lyapunov exponent plot λ_{max} versus the initial state $w(0)$. The bifurcation diagram is obtained by scanning the initial state $w(0)$. The system (8.2) parameters are $a = 17.8571$, $b = 28$, $c = 37$, $d = 0.01$, $e = 1.2$, $\alpha = 1.3$, $\beta = 0.1$ and the other initial states are $x(0) = y(0) = z(0) = 0.1$.

In Fig. 8.2, the white regions are stable regions, while the gray regions are unstable regions. In Fig. 8.3, the bifurcation diagrams depicting the local maxima of $x(t)$ and the three largest Lyapunov exponents (LLE) versus the initial state $w(0)$.

In Fig. 8.3, by varying the initial state $w(0)$ it is noticed that the system (8.4) exhibits various dynamical behaviors illustrating the striking and rare feature namely extreme multistability as illustrated by the phase portraits of the system (8.4) in Fig. 8.4.

Fig. 8.4 shows that system (8.4) has sensitivity towards the initial state $w(0)$ and its displays the extreme multistability feature. The basin of attraction for different attractors competing in system (8.4) in the ($x = 0$, $y = 0$) and ($z = 0$, $y = 0$) planes are presented in Fig. 8.5.

In Fig. 8.5, the initial conditions in the dark blue region lead to the line equilibrium, those in the light blue, green, yellow and red regions lead to the strange attractor and those in the blue region lead to the periodic orbits. The two parameters bifurcation diagram depicting the LLE (λ_{max}) is plotted in Fig. 8.6 to analyze the sensitivity towards some couple of parameters of the system (8.4).

System (8.4) can display a line equilibrium point, periodic and chaotic behaviors as indicated in Fig. 8.6. The bifurcation diagram and the related LLE with respect to d (for $e = 1.2$) and e (for $d = 0.01$) are plotted in Figs. 8.7 and 8.8, respectively.

The bifurcation diagrams of the output $z(t)$ in Figs. 8.7 and 8.8 show a reverse period-doubling route to chaos interspersed with periodic windows followed by chaotic oscillations. The chaotic behavior is illustrated in Fig. 8.9 for a specific value of parameters $d = 0.01$ and $e = 1.2$.

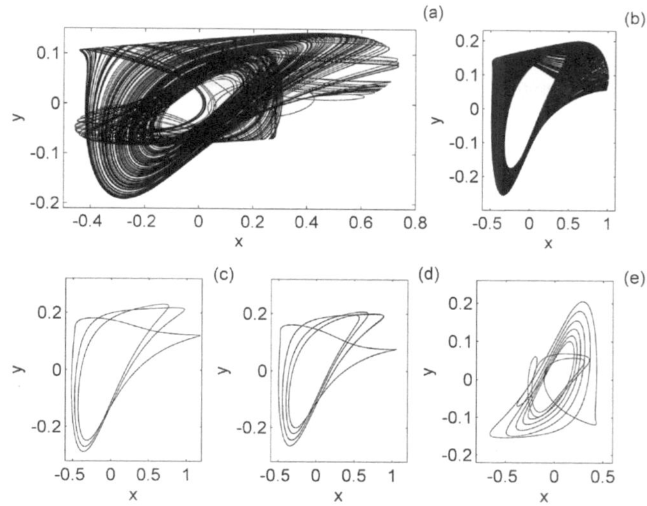

FIGURE 8.4 Phase portraits illustrating various dynamical behaviors of system (8.4) in the (x, y) plane for different values of the initial state $w(0)$: (a) $w(0) = 0$, (b) $w(0) = 1.3$, (c) $w(0) = 3$, (d) $w(0) = 1.59$ and (e) $w(0) = -1.33$. The other initial conditions are $x(0) = y(0) = z(0) = 0.1$.

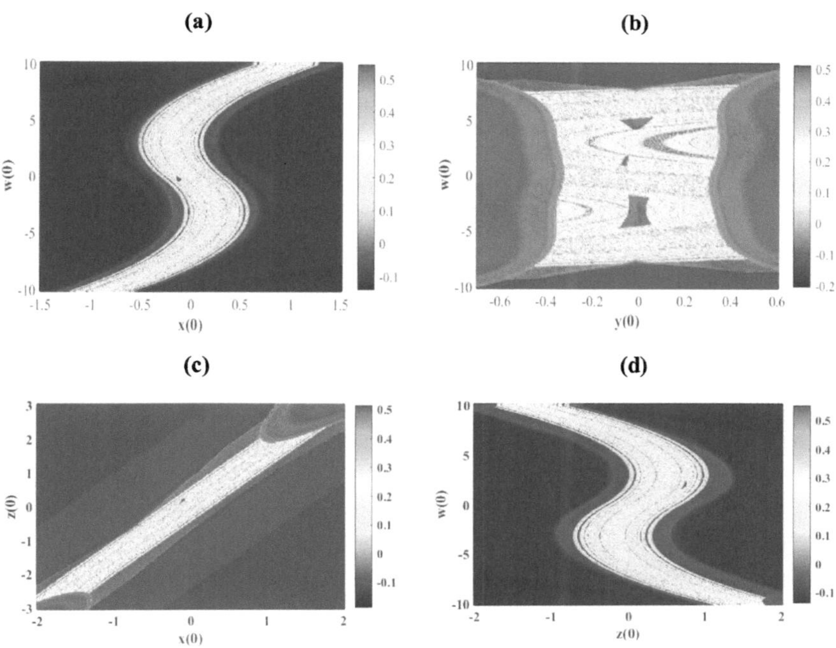

FIGURE 8.5 Structure of the cross sections of the basin of attraction in different planes of the system: (a) $(w(0), x(0))$, (b) $(w(0), y(0))$, (c) $(z(0), x(0))$, and (d) $(w(0), z(0))$. The right columns represent the largest Lyapunov exponent λ_{max}. The parameters of system (8.2) are for a = 17.5871, b = 28, c = 37, $\alpha = 1.3$, $\beta = 0.1$, d = 0.01, e = 1.2. For the interpretation of the references to color in this figure, the reader is referred to the web version of this book chapter.

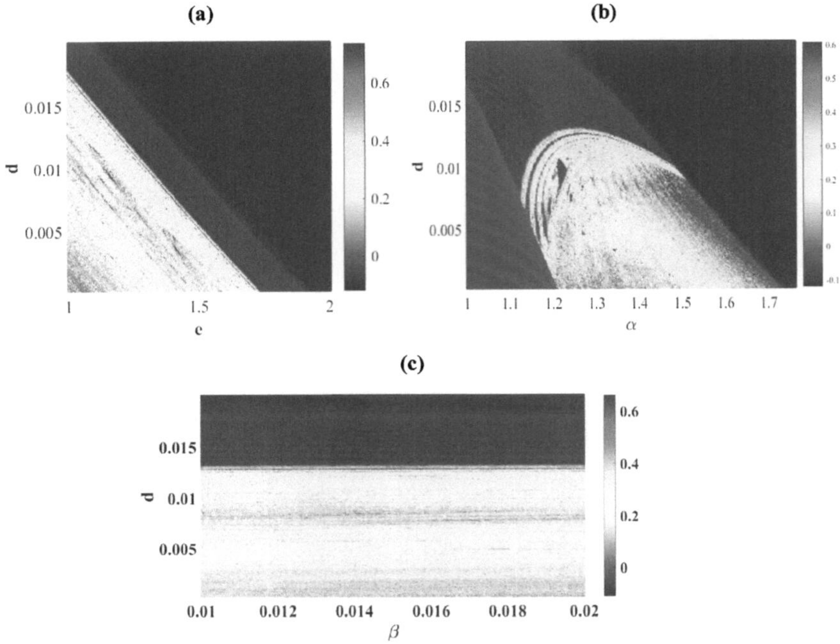

FIGURE 8.6 The 2-parameters bifurcation diagrams in the (a) (e, d), (b) (d, α), (c) (d, β) planes depicting various dynamical regions of the system (8.4). The right columns represent the largest Lyapunov exponent λ_{max}. The line equilibrium point is represented by dark blue color, periodic behavior is represented by light blue color, while chaotic behavior is represented by red, yellow, green and blue colors. The system parameters are a = 17.5871, b = 28, c = 37. For the interpretation of the references to color in this figure, the reader is referred to the web version of this book chapter.

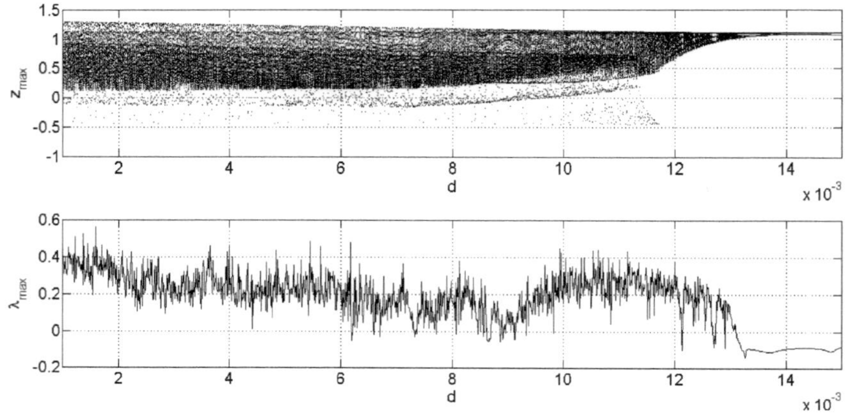

FIGURE 8.7 Bifurcation diagram depicting the local maxima of $z(t)$ and corresponding graph of LLE (λ_{max}) versus the parameter d for $a = 17.8571$; $b = 28$; $c = 37$; $e = 1.2$; $\alpha = 1.3$, and $\beta = 0.1$.

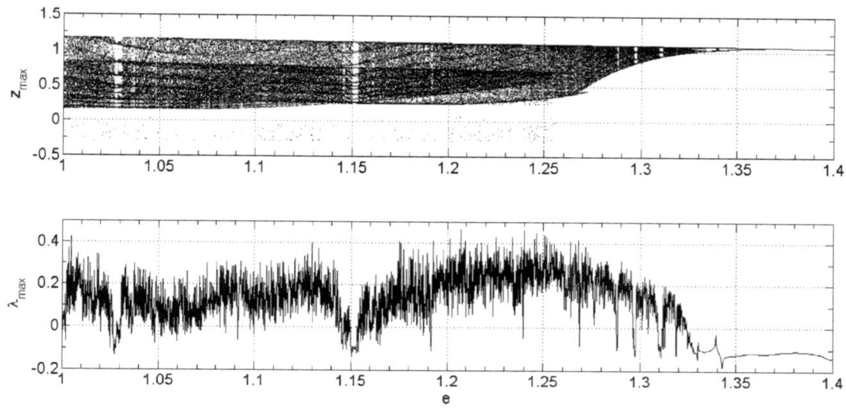

FIGURE 8.8 Bifurcation diagram depicting the local maxima of $z(t)$ and corresponding graph of LLE ($\lambda_{\max}$). (b) versus the parameter e for $a = 17.8571$; $b = 28$; $c = 37$; $d = 0.01$; $\alpha = 1.3$, and $\beta = 0.1$.

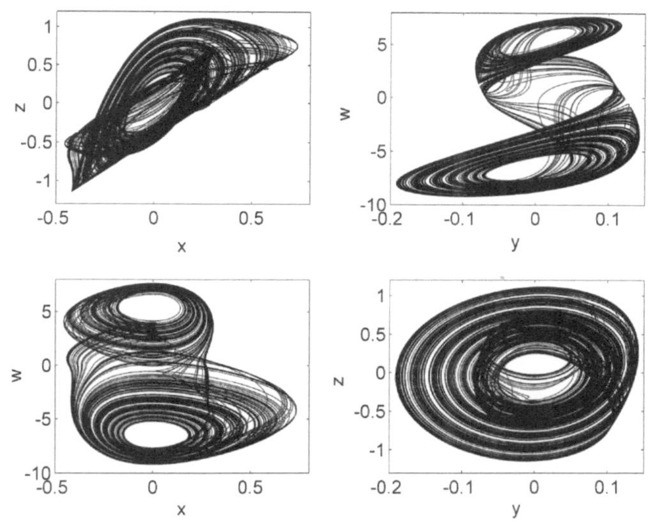

FIGURE 8.9 Phase portraits of system (8.4) for $d = 0.01$ and $e = 1.2$. The initial conditions are $(x(0), y(0), z(0), w(0)) = (0.1, 0.1, 0.1, 0.1)$.

The chaotic hidden attractor is depicted in Fig. 8.4. It is important to note that system (8.4) has the feature of partial amplitude control. This is due to the fact that its state variable w(t) appears only in the first equation of autonomous system (8.4) and its amplitude can be changed by inserting a boosting controller k into system (8.4) through the transformation, $w \to w + k$. The resulting trans-

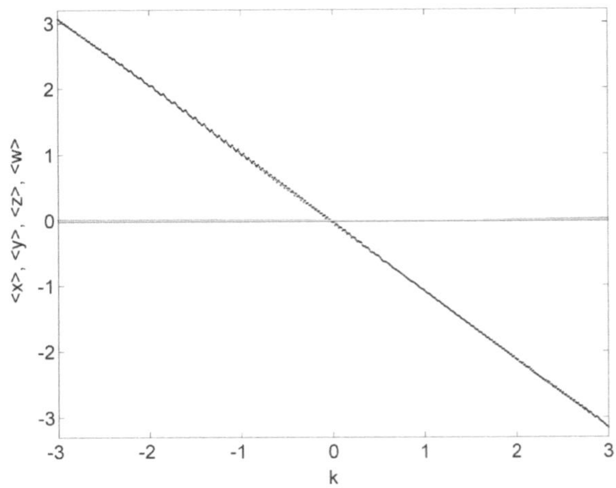

FIGURE 8.10 The average values of the state variables x (green), y (blue), z (red) and w (black), versus boosting controller k for $a = 17.8571$; $b = 28$; $c = 37$; $d = 0.01$; $e = 1.2$; $\alpha = 1.3$, and $\beta = 0.1$. For the interpretation of the references to color in this figure, the reader is referred to the web version of this book chapter.

formed system is as follows:

$$\frac{dx}{dt} = a(y - x + x(\alpha - \beta|\mathrm{w} + k|)), \tag{8.11a}$$

$$\frac{dy}{dt} = x - ey - z, \tag{8.11b}$$

$$\frac{dz}{dt} = b(y - dz), \tag{8.11c}$$

$$\frac{dw}{dt} = -cx, \tag{8.11d}$$

System (8.11) has a line equilibrium given by $E_1 = \{(x, y, z, w) \in R^4 \ / \ x = y = z = 0, w = w^*\}$. The local stability of the line equilibrium reveals that the stability of line equilibrium E_1 is independent the boosting controller k. In order to check the partial amplitude control of chaotic system (8.11), the average values of the state variables x, y, z, and z versus boosting controller k are plotted in Fig. 8.10.

It is shown in Fig. 8.10 that the average of the state variable w decreases and other three state variables (x, y, and z) remain unchanged when boosting controller k is varied. The phase portraits and time series of the state variable w of system (8.11) are depicted in Fig. 8.11 for different of values of boosting controller k.

The reader can notice that the attractor is moved in the w-axis for different values of the offset booting k as shown in Fig. 8.11.

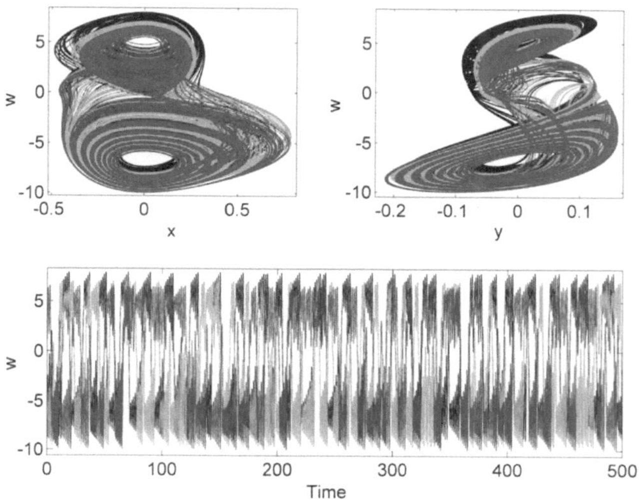

FIGURE 8.11 Phases portraits illustrating amplitude control feature of the system (8.3) in the planes (x, w), (y, w) and the time trace series of signal of the state variable $w(t)$ for different values of the control parameter k: $k = -0.1$ (red), $k = 0.2$ (blue) and $k = 0.35$ (black). The rest of the parameters are those in the stability analysis section. For the interpretation of the references to color in this figure, the reader is referred to the web version of this book chapter.

System (8.4) can also display the feature of total amplitude control. Indeed, the amplitude of the state variable of system (8.4) can be boosted by inserting the transformation $x \to x/m$, $y \to y/m$, $z \to z/m$, $\omega \to \omega/m$, and $z \to z/m$ in system (8.4). The parameter m remains in the absolute terms as shown in the following system (8.12):

$$\frac{dx}{dt} = a[y - x + x(\alpha - \beta\left|\frac{w}{m}\right|)], \tag{8.12a}$$

$$\frac{dy}{dt} = x - ey - z, \tag{8.12b}$$

$$\frac{dz}{dt} = b(y - dz), \tag{8.12c}$$

$$\frac{d\mathrm{w}}{dt} = -cx. \tag{8.12d}$$

System (8.12) has a line equilibrium given by $E_2 = \{(x, y, z, w) \in R^4 \ / \ x = y = z = 0, w = w^*\}$. The local stability of the line equilibrium reveals that the stability of line equilibrium E_2 is independent of the boosting controller m. The phase portrait of the system (8.12) showing the amplitude boosting behavior is presented in Fig. 8.12 for some value of m.

As shown in Fig. 8.12, the amplitude of the chaotic signals x, y, z, and w are adjusted simultaneously by the control parameter m. Small amplitudes are

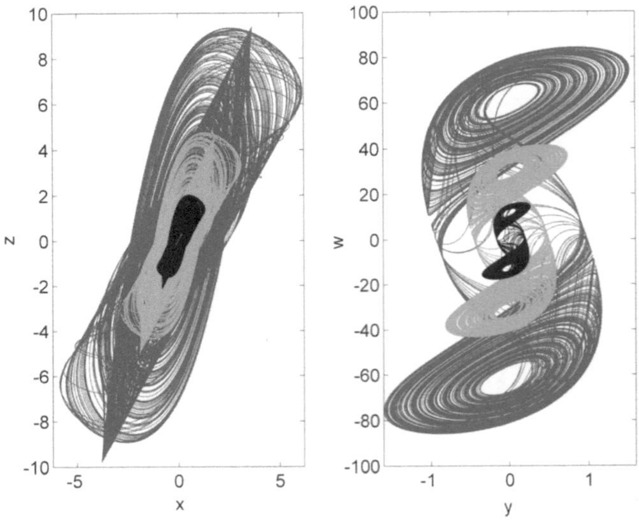

FIGURE 8.12 Phase portraits of the interesting total amplitude control dynamics of the x, y, z, w variables in the system (8.3) in the (x, z) and (y, w) planes for different values of the control parameter m: $m = 2$ (red), $m = 5$ (blue) and $m = 10$ (black). The rest of the parameters are those in the stability analysis section. For the interpretation of the references to color in this figure, the reader is referred to the web version of this book chapter.

achieved by small values of m ($m = 1$) while large amplitudes are obtained by a higher value of m ($m = 10$).

8.3 Electronic implementation of an absolute memristor autonomous Van der Pol–Duffing circuit

In this section, PSIM software package is used to implement the proposed circuit in order to check the numerical results obtained in the previous section. In Table 8.1, the electronic component values of the circuits in Fig. 8.1 are presented.

The characteristic voltage-intensity of the absolute memristor is presented in Fig. 8.13.

Fig. 8.13 shows three fingerprints which identify the absolute memristor. The frequency-dependent pinched hysteresis loops of the absolute memristor are obtained by a sinusoidal voltage source V_1 considered as $V_1 = v_1 \sin(2\pi F)$, where v_1 and F are the stimulus amplitude and frequency, respectively. For the stimulus frequency $F = 150$ Hz, the decrease of the stimulus amplitudes v_1 (from 0.1 V, to 0.09 V and to 0.05 V) leads to a decrease of the pinched hysteresis loop as seen Fig. 8.13(a). While for the stimulus amplitude $v_1 = 0.05$ V, the decrease of stimulus frequency Γ (from 200 Hz to 150 Hz and to 50 Hz), the hysteresis loops are pinched at the origin and the lobe area decreases with the frequencies,

TABLE 8.1 Electronic components values of the circuits in Fig. 8.1.

Parameters	Signification	Values
R_{01}, R_{02}	Resistor	10 kΩ
R_1	Resistor	1 kΩ
R_2	Resistor	14 kΩ
R_3	Resistor	1.21 kΩ
R_0	Resistor	5.26 kΩ
R	Resistor	1.583 kΩ
R_L	Resistor	14.28 kΩ
R_P	Potentiometer	6.34 kΩ
L	Inductor	5.66 mH
C_0	Capacitor	5 nF
C_a	Capacitor	3.92 nF
C_b	Capacitor	70 nF
g_1	Gain of multiplier AD633JN chip	1
g_2	Gain of multiplier AD33JN chip	0.23
OP_1 to Op_4	Operational amplifier	IC chip of TL 084

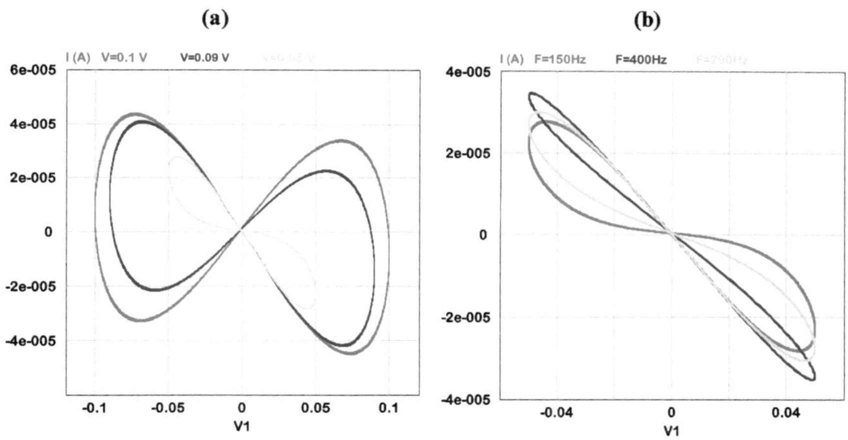

FIGURE 8.13 Voltage–current characteristic ($v1 - i$) of the circuit of the flux-controlled memristor in Fig. 8.1 obtained in PSIM: (a) $F = 150$ Hz for different stimulus voltages and (b) $V = 0.05$ V for different stimulus frequencies, illustrating the fingerprint of the memristor. For the interpretation of the colors in this figure, the reader is referred to the web version of this book chapter.

and the hysteresis loop shrinks into a single-value function at zero frequency as illustrated in Fig. 8.13(b).

The first confirming result is the extreme multistability predicted during numerical findings in Section 8.2. The circuit is tested in PSIM software package and oscilloscopes are connected to measure voltage across capacitors C_A and C_B. The values of the electronic components are set as in Table 8.1. Only the

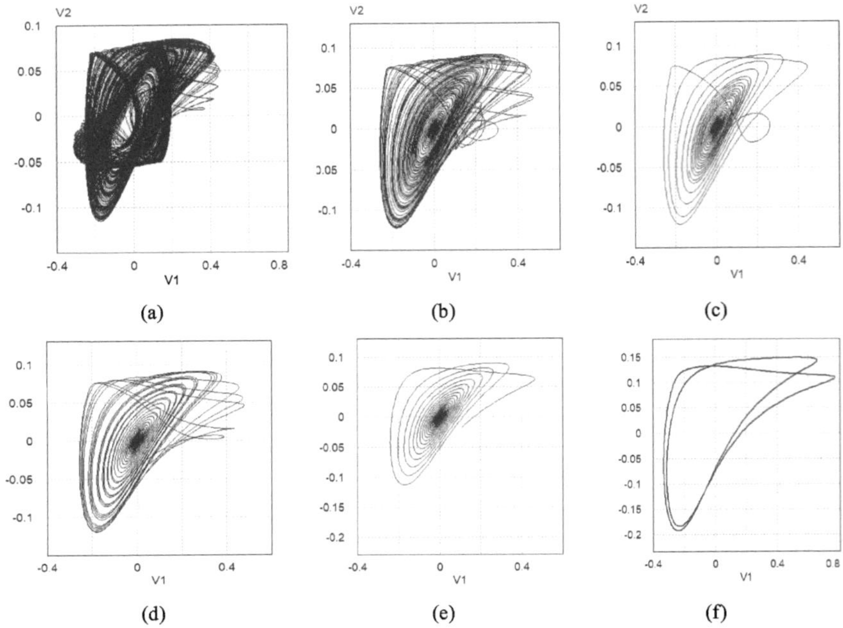

FIGURE 8.14 Phase portraits of the circuit captured in PSIM in the (V_2, V_1) plane displaying rare feature of extreme multistability of the circuit in Fig. 8.1: (a) $V_0 = 0.0$ V chaos; (b) $V_0 = 2.9$ V another chaos; (c) $V_0 = 2.92$ V periodic cycle; (d) $V_0 = 2.92$ V another periodic cycle; (d) $V_0 = 2.9205$ V another periodic cycle; (f) $V_0 = -1.33$ V period-2. The resistors, inductor, and capacitor values of the circuit are kept constant. Initial voltage of the other capacitors is $V_1(0) = V_2(0) = 0.1$ V and the initial inductor current $I_L(0) = 0.1$ mA.

voltage across capacitor C_0 is varied. The phase portrait of the voltage V_2 versus V_1 is displayed in Fig. 8.14.

As we can see, the phase portraits in Fig. 8.14, resembles the one predicted in the numerical computation on Fig. 8.4.

The second confirming result is the chaotic behavior of the circuit predicted by numerical computation illustrated in Fig. 8.9 is displayed in Fig. 8.15.

It is easy to see that the phase portraits captured from the electronic implementation reproduce those in Fig. 8.9, confirming the numerical computation findings.

8.4 Conclusion

The present paper dealt with the analysis and electronic implementation of an absolute memristor Van der Pol–Duffing circuit. By using the Routh–Hurwitz stability criteria, the stability of the line equilibrium points of the system describing the proposed circuit was carried out. The proposed circuit can display the striking behavior of extreme multistability, periodic and hidden chaotic at-

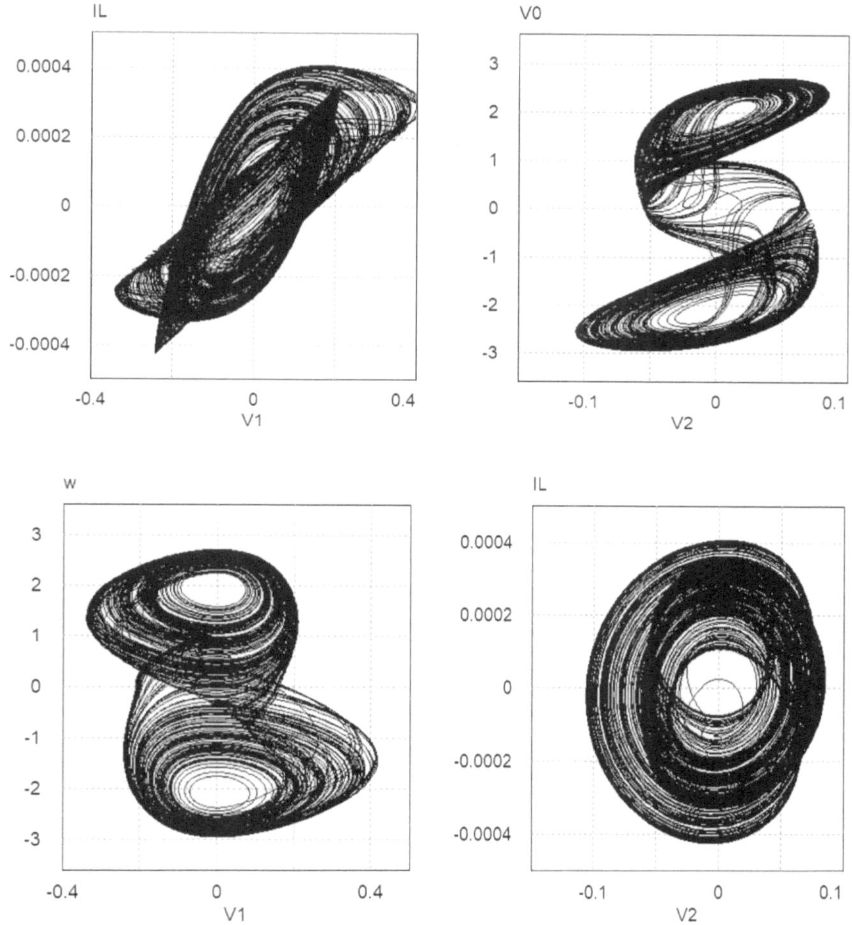

FIGURE 8.15 Phase portraits of the different capacitor voltages and inductor current captured from PSIM in different planes simulating the circuit of Fig. 8.1. The resistors are $R_L = 14.2\ \Omega$; $R_P = 6.3\ \mathrm{k}\Omega$. The initial voltages of the capacitors are $V_1(0) = V_2(0) = V_0(0) = 0.1$ V and $I_L(0) = 0.1$ mA. The rest of the circuit components values are in Table 8.1 of this section.

tractors. By adding two new parameters in the system describing the proposed circuit, a flexible chaotic absolute memristor Van der Pol–Duffing circuit with partial or total amplitude control was achieved. The proposed circuit was implemented and tested using the PSIM software to verify the numerical simulations results. Comparison of the results obtained from the analog circuit and numerical simulations showed good qualitative agreement. It could be used as for neuromorphic circuits with artificial intelligence applications owing to its mem-element.

References

[1] G.P. King, S.T. Gaito, Bistable chaos. I. Unfolding the usp, Physical Review A 46 (1992) 3092–3099.

[2] L.O. Chua, G.-N. Lin, Canonical realization of Chua's circuit family, IEEE Transactions on Circuits and Systems 37 (1990) 885–902.

[3] H.B. Fotsin, P. Woafo, Adaptive synchronization of a modified and uncertain chaotic van der Pol–Duffing oscillator based on parameter identification, Chaos, Solitons and Fractals 24 (2005) 1363–1371.

[4] M. Lakshmanan, K. Murali, Chaos in Nonlinear Oscillators: Controlling and Synchronization, World Scientific, Singapore, 1996.

[5] H.A. Haus, Waves and Fields in Optoelectronics, Prentice-Hall, Englewood Cliffs, NJ, 1984.

[6] B.J.A. Zielinska, D. Mukamel, V. Steinberg, S. Fishman, Physical Review A 32 (1985) 702.

[7] P.J. Probert, J.E. Carrol, Proceedings of the IEEE 134 (295) (1987), 136, 22 (1989).

[8] S. Rajasekar, M. Lakshmanan, Physica D 32 (1988) 146.

[9] J. Guckenheimer, P. Holmes, Nonlinear Oscillations and Bifurcations of Vector Fields, Springer-Verlag, New York, 1983.

[10] E. Atlee Jackson, Perspectives of Nonlinear Dynamics, Cambridge University Press, New York, 1991.

[11] L. Kengne, J.R. Mboupda Pone, H.T.K. Tagne, J. Kengne, Dynamics, control and symmetry breaking aspects of a modified van der Pol–Duffing oscillator, and its analog circuit implementation, Analog Integrated Circuits and Signal Processing (2020) 1–21.

[12] N. Inaba, T. Tsubone, Nested mixed-mode oscillations, Part II: experimental and numerical study of a classical Bonhoeffer–van der Pol oscillator, Physica D: Nonlinear Phenomena (2020) 132493.

[13] A.E. Matouk, H.N. Agiza, Bifurcations, chaos and synchronization in ADVP circuit with parallel resistor, Journal of Mathematical Analysis and Applications 341 (2008) 259–269.

[14] A.M.A. El-Sayed, A. Elsaid, H.M. Nour, A. Elsonbaty, Dynamical behavior, chaos control and synchronization of a memristor-based advp circuit, Communications in Nonlinear Science and Numerical Simulation 18 (2013) 148–170.

[15] B.-C. Bao, J.-P. Xu, Z. Liu, Initial state dependent dynamical behaviors in a memristor based chaotic circuit, Chinese Physics Letters 27 (2010) 070504, 070506.

[16] V.A. Botta, C. Nespoli, M. Messias, Mathematical analysis of a third order memristor based Chua's oscillator, Tendencias em Matematica Aplicada e Computacional 12 (2011) 9199.

[17] V.A. Slipko, Y.V. Pershin, M. Di Ventra, Changing the state of a memristive system with white noise, Physical Review E 87 (2013) 042103, 042109.

[18] L. Teng, Herbert H.C. Iu, X. Wang, X. Wang, Chaotic behavior in fractional order memristor-based simplest chaotic circuit using fourth degree polynomial, Nonlinear Dynamics 77 (2014) 231–241.

[19] S. Sabarathinam, C.K. Volos, K. Thamilmaran, Implementation and study of the nonlinear dynamics of a memristor-based Duffing oscillator, Nonlinear Dynamics 87 (2017) 37–49.

[20] Y. Lu, X. Huang, S. He, D. Wang, B. Zhang, Memristor based van der Pol oscillation circuit, Nonlinear Dynamics 24 (2014) 1450154–1450168.

[21] G. Martinsen, S. Grimnes, C.A. Lütken, G.K. Johnsen, Memristance in human skin, Journal of Physics Conference Series 224 (2010) 012071.

[22] O. Sporns, G. Tononi, R. Kötter, The human Connectome: a structural description of the human brain, PLoS Computational Biology 1 (2005), 0245-0251.

[23] D.B. Strukov, G.S. Snider, D.R. Stewart, R.S. Williams, The missing memristor found, Nature 453 (2008) 80–83.

[24] L. Luo, Z. Dong Duan, L. Shukai, S. Chun, Memristor-based stateful logic gates for multi-functional logic circuit, IET Circuits, Devices & Systems (20 April 2020), https://doi.org/10.1049/iet-cds.2019.0422.

[25] G.M. Mahmoud, A.A. Farghaly, T.M. Abed-Elhameed, S.A. Aly, A.A. Arafa, Dynamics of distributed-order hyperchaotic complex van der Pol oscillators and their synchronization and control, The European Physical Journal Plus 135 (1) (2020) 32.
[26] K. Rajagopal, J.R. Mboupda Pone, S.T. Kingni, S. Arun, A. Karthikeyan, Analysis and electronic implementation of an absolute memristor autonomous van der Pol–Duffing circuit, The European Physical Journal Special Topics 228 (2019) 2287–2299.
[27] C. Yakopi, T.M. Taha, G. Subramanyam, R.E. Pino, Generalized memristive device spice model and its application in circuit design, IEEE Transactions on Computer-Aided Design of Integrated Circuits and Systems 32 (2013) 1201–1214.
[28] N. Bao, H. Wang, B.C. Bao, M. Chen, P. Jin, G. Wang, Initial condition-dependent dynamics and transient period in memristor-based hypogenetic jerk system with four line equilibria, Communications in Nonlinear Science and Numerical Simulation 57 (2018) 264–275.
[29] H. Fotsin, S. Bowong, J. Daafouz, Adaptive synchronization of two chaotic systems consisting of modified van der Pol Duffing and Chua oscillators, Chaos, Solitons and Fractals 26 (2005) 215–229.
[30] G.J. Fodjouong, H.B. Fotsin, P. Woafo, Synchronizing modified van der Pol Duffing oscillators with offset terms using observer design: application to secure communications, Physica Scripta 75 (2007) 638–644.
[31] E. Miranda, J. Suñé, Memristors for Neuromorphic Circuits and Artificial Intelligence Applications, Materials (Basel, Switzerland) 13 (4) (2020) 938, https://doi.org/10.3390/ma13040938.

Chapter 9

Memristor-based novel 4D chaotic system without equilibria

Analysis and projective synchronization

Piyush Pratap Singh[a], Binoy Krishna Roy[b], and Christos Volos[c]
[a]*Department of Electrical Engineering, NIT Meghalaya, Shillong, Meghalaya, India,* [b]*Department of Electrical Engineering, NIT Silchar, Cachar, Assam, India,* [c]*Laboratory of Nonlinear Systems, Circuits and Complexity (LaNSCom), Department of Physics, Aristotle University of Thessaloniki, Thessaloniki, Greece*

Chaos is the score upon which reality is written.

Sir Henry Miller

9.1 Introduction

Chaos, the nonlinear phenomenon, has been growing significantly in the interdisciplinary field of nonlinear dynamics over the past three decades. Chaos, its control and applications encompass almost all branches of science such as mathematics, physics, chemistry, biology, and engineering [1–13]. Simultaneously, several investigations on the development and design of novel chaotic systems [14–20] and their special behavior such as hidden chaotic attractors have been reported in [15,21–27].

In 1971, the memristor, the fourth fundamental electrical circuit element was investigated by Leon O. Chua [28]. After three decades, HP team fabricated a memristor element in the laboratory [29]. Only a handful of organizations and companies have memristors since they do not have a reliable experimental results. That is why theoretical analysis and simulations of memristors are more common than their experimental studies. The memristors are useful in various applications such as image encryption and signal processing, bio-system, neural network, complex neural network, and filter design [30–38] due to their chaotic and complex behaviors. Therefore, analysis of a memristor-based novel chaotic system with unique behavior becomes the primary motivation of this work.

Mem-elements for Neuromorphic Circuits with Artificial Intelligence Applications
https://doi.org/10.1016/B978-0-12-821184-7.00017-7

In 2008, the memristor-based chaotic circuit was derived from the canonical Chua's circuit [39] for the first time in the literature. Chua circuit was redesigned via a memristor element and a technical report on the detail study of autonomous memristor-based chaotic circuits was archived by Muthuswamy in [40] and [41], respectively. After the pioneering work by Muthuswamy, the memristor was considered as a promising circuit element for circuit design. Various research results on memristor-based chaotic systems or circuits have been reported in the literature and are summarized here based on the memristor circuit design and implementation [42–48], its control and synchronization [49–61], and design of novel chaotic system/circuits with unique behaviors [54,62–82].

9.1.1 Literature survey

Practical implementation of a memristor element using micro controller emulation and off-the-shelf components based chaotic circuit on a breadboard was reported in [42]. A simple memristor-based autonomous chaotic system was studied in [43]. In [43], three energy storage elements: a linear passive inductor and a linear passive capacitor and a nonlinear active memristor were used in series and circuit topology was studied. A memristor-based circuit was implemented using SPICE emulator in [44], off-the-shelf solid state components in [45], FPGA implementation in [47], to develop real world memristor circuit applications. A scroll chaotic system containing the HP memristor model and triangular wave sequence was proposed in [46]. The memristor-based Chua chaotic circuit was implemented by replacing the nonlinear resistor with negative conductance in [48].

An adaptive synchronization of a class of memristor-based neural chaotic systems using a novel adaptive backstepping approach was reported in [49]. Synchronization between memristor-based complex Lorenz systems was achieved in [50]. Control and synchronization of Chua's memristor-based chaotic system were studied in a terminal sliding mode [51], and by finite time impulsive control [52] and feedback control in [53]. The study of dynamical behaviors, control and synchronization of a new chaotic model with complex variables and cubic nonlinear terms was done in [54]. Chaos and Hopf bifurcation control in a memristor-based chaotic system with time delay was reported in [55]. A fractional order memristor chaotic system with no equilibrium and its synchronization using adaptive sliding mode control [56], backstepping control [57] and based on the novel conformable Adomian decomposition method in [58], were reported. A hyperchaotic memristor oscillator with fuzzy based chaos control and LQR based chaos synchronization was reported [59]. A complex novel 4D memristor hyperchaotic system and its synchronization using adaptive sliding mode control was reported [60]. Memristor design of a Sprott-A chaotic system was reported in [61] and a simple intermittent control scheme with adaptive mechanism was developed to achieve complete synchronization. It is found that the memristor-based system under investigation exhibits fruitful dynamic behaviors such as coexisting bifurcation, multistability, transient chaos, and transient quasiperiods.

The design and implementation of memristor chaotic circuits were studied with complex dynamics [62], without equilibrium [63,66,71,75], complex variables and cubic nonlinear terms [54], a line of equilibria [64], with a linear passive capacitor type of energy-storage elements [65], multi-scroll attractors [67]. A new charge-controlled memristor-based on the simplest chaotic circuit was studied and simulation was done in Multisim [68]. A four-wing 4D hyperchaotic attractor with a line equilibrium was reported in [69]. The design of memristor-based 4D hyperchaotic systems with multi-scroll attractors was reported in [70] and [72], respectively. In Ref. [70], a new memristor-based multi-scroll hyperchaotic system was designed. The proposed memristor-based system possesses multiple complex dynamic behaviors and various coexisting and hidden coexisting attractors. A new 4D dynamical system is proposed to set a chaotic circuit composed of memristor and Josephson junction [73]. A novel time-delay system with no equilibrium points was proposed [74]. A new memristor hyperchaotic oscillator was derived using exponential memductance and discontinuous memductance functions in [76].

Hidden extreme multistability in a novel memristive 6D hyperchaotic autonomous system was studied in [77]. A novel memristor-based chaotic system and its applications in image encryption and digital watermarking were investigated in [78] and [79], respectively. A non-ideal flux-controlled, absolutely nonlinear active memristor model is introduced in [81]. The proposed system has the phenomenon of a multi-type quasi-periodic limit cycle and multi-attractor chaotic attractors with various topologies, which indicates that the system has numerous hidden attractors. A novel memristor-based complex-valued chaotic system and its projective synchronization were investigated in [80] and [82], respectively. In short, various novel memristor-based chaotic and hyperchaotic systems are reported in the literature with different behaviors such as no equilibria, fixed equilibria, line of equilibria, line and plane of equilibria, and these are summarized in Table 9.1.

Various research results on memristor-based circuits and their applications have been reported in the literature and are summarized here, based on the design of non-volatile memristor memory system, digital and analog systems; and neuromorphic systems [39,40,42,83–96].

9.1.2 Application of memristor and memristive circuit

Application of memristor may be divided into three important classes, namely: 1) design of non-volatile memristor memory system; 2) digital and analog systems; and 3) neuromorphic systems. Chaos applications of memristors have a large implementation space ranging from secure communication to medical purposes like seizure detection. Some important research efforts in this area attempt to use memristor properties to invoke a chaotic response. In Ref. [83], the memristor connected to a power source exhibits chaos by using the second-order effects of a memristor. The second-order properties of the memristor are reproduced to design chaotic circuits [80,82]. Other nonlinear circuits with memristors are also reported by adapting a modified Chua circuit that exhibit chaos. Chaos is observed by replacing the Chua diode with the memristor [39,40,84]. An experimental demonstration of chaos with analog components used to build the memristor is presented in [42]. An image encryption application using piecewise linear memristors is presented in [85].

Several research have been displayed on the use of memristors in biometric and neuromorphic circuits [86–98]. In the neuromorphic circuit approach, observable biological behavior or processing is aimed to be replicated. A simple way to build neuromorphic circuit using memristors is to build processing elements or neuron circuits with standard CMOS while the adaptive synapses are implemented using the memristor crossbar. The synapse plays an important role in signal exchange and information encoding between neurons. Electric and chemical synapses are often used to investigate the synchronization in electrical activities of neurons. The memristor is used to connect two neurons and the phase synchronization in electrical activities is discussed in [98]. Memristors have revived the design of neuromorphic circuit area because they have the potential of synaptic integration, which is lacking in CMOS circuits. In the literature, various groups have demonstrated through simulation how to achieve STDP with memristors [87,90,92].

In order to have overall perspectives of numerous applications of memristors, we have categorized the application into two categories; one is discrete device applications and the other is array devices applications. The discrete device applications use memristors in a manner that takes advantage of their nonlinear characteristics and their controllable resistance changes to enhance the performance of the desired application. Array device applications on the other hand not only depend on memristor properties but are also coupled with the ongoing trend to increase the device density by reducing the width of the wires in nanocrossbar structure [97]. The taxonomy of applications for memristors in both crossbar array and discrete device form is shown in Fig. 9.1. The comprehensive review of papers provides a glimpse on various applications of memristors and computer-aided analysis of memristor circuits and memristive systems during the past few years.

Many researchers have introduced their memristor-based chaotic and hyperchaotic systems without equilibria and these are listed in Table 9.2. It is evident that memristor-based systems have chaotic/hyperchaotic behaviors based on significantly large numbers of nonlinear terms out of their total numbers of terms in the dynamics. Therefore, to investigate a novel memristor-based simple chaotic system without equilibria and with unique properties and behaviors is critical and of major concern from the application point of view [30–34]. Also, it is difficult to reproduce the accurate dynamics of the memristor-based chaotic system, which is highly advantageous in encryption and communication [61].

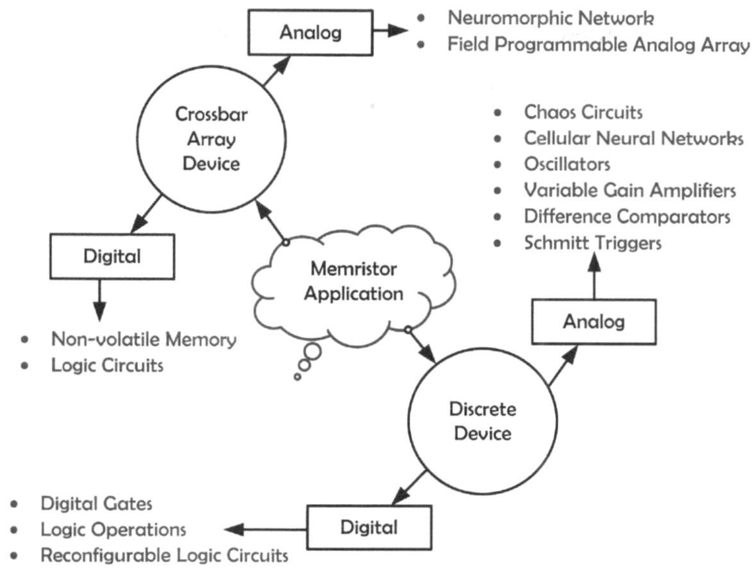

FIGURE 9.1 Taxonomy of memristor applications in different fields.

Motivated from the literature review and issues, the following objectives are defined for the present work:

1. To explore and analyze a novel memristor-based simple chaotic system without equilibria.
2. To achieve projective synchronization between the memristor-based proposed chaotic systems using a nonlinear active control technique.

The following points explain the interesting and unique properties of the proposed novel 4D memristor-based chaotic system:

- The proposed memristor-based chaotic system without equilibria is simple and rare in the literature.
- The system has two nonlinear terms and a total of nine terms. Although the 4D chaotic system in [57] has same number of total terms, but it has four nonlinear terms (refer to Table 9.2). Thus, the system is the simplest compared with similar available chaotic or hyperchaotic systems.
- The proposed system has periodic, 2-torus quasi-periodic, chaotic and chaotic 2-torus natures of Lyapunov exponents. Such a memristor-based dissipative chaotic system is not available in the literature.
- The proposed chaotic system has thumb and parachute shapes of Poincaré map. Such types of unique Poincaré map behavior in the no equilibria class of memristor-based chaotic systems are not reported in the literature.

The above-mentioned interesting and unique properties reflect the novelty and contributions of this work.

TABLE 9.1 Reported memristor-based hyperchaotic/chaotic systems with different nature of equilibria.

Memristor-based chaotic/ hyperchaotic system	Complex valued	Nature/number of equilibria	Total no. of terms	References
4D Chaotic system	No	No	9	[57]
4D Chaotic system	No	Line	8	[70], [67]
4D Hyperchaotic system	No	Line	9	[64], [72]
4D Hyperchaotic system	No	Line	10	[69]
6D Chaotic system	Yes	Line	16	[50]
6D Hyperchaotic system	Yes	Fixed	18	[54]
6D Hyperchaotic system	Yes	Fixed	17, 19	[60]
6D Hyperchaotic system	No	Line and plane	14	[77]
6D Chaotic system	Yes	Line and plane	14	[82]

TABLE 9.2 Reported memristor-based no equilibrium chaotic, hyperchaotic systems with no. of nonlinear terms and total number of terms.

Memristor-based systems	No. of equilibrium points	Nonlinear terms	Total no. of terms	References
3D Chaotic	No	3	7	[66]
3D Chaotic	No	3	6	[61]
4D Hyperchaotic	No	5	12	[63]
4D Chaotic	No	6	12	[56]
4D Chaotic	No	4	9	[57]
4D Hyperchaotic	No	5	11	[75]
4D Chaotic	**No**	**2**	**9**	**This work**
5D Chaotic	No	3	10	[71]
5D Chaotic	No	6	13	[81]

The rest of the paper is organized as follows: a brief description of memristor models and the novel memristor-based 4D chaotic system are presented in Section 9.2. In Section 9.3, different properties and behaviors of the memristor-based chaotic system are discussed. The projective synchronization using nonlinear active control technique is discussed in Section 9.4. In Section 9.5, numerical simulation results are presented and discussed. Finally, conclusions are summarized and future scopes are pointed out in Section 9.6.

9.2 Brief introduction to flux- and charge-controlled memristor models and novel chaotic system

In this section, a brief introduction to flux- and charge-controlled memristor models is presented. Furthermore, the flux-controlled memristor-based novel chaotic system with real state variables is discussed.

9.2.1 Introduction to flux- and charge-controlled memristor models

A memristor element is known as a nonlinear resistor with memory, and defined by $f(\varphi, q) = 0$ which represents flux- and charge-controlled memristors. The flux-controlled and charge-controlled current and voltage relationship can be written as

$$\begin{cases} W(\varphi) = \frac{dq(\varphi)}{d\varphi} \\ i(t) = W(\varphi)v(t) \\ v(t) = \frac{d\varphi(t)}{dt} \end{cases} \tag{9.1}$$

The charge-controlled current and voltage relationship can be written as

$$\begin{cases} M(q) = \frac{d\varphi(q)}{dq} \\ v(t) = M(q)i(t) \\ i(t) = \frac{dq(t)}{dt} \end{cases} \tag{9.2}$$

where φ and q represent the magnetic flux and charge, respectively. The current and voltage terminals of the memristor device are denoted as i and v. $W(\varphi)$ and $M(q)$ are known as memductance and memristance, respectively. In this paper, a flux-controlled memristor with cubic nonlinear characteristics is considered as $q(\varphi) = \gamma\varphi + \beta\varphi^3$. Then the memductance $W(\varphi)$ is calculated as

$$W(\varphi) = \frac{dq(\varphi)}{d\varphi} = \gamma + 3\beta\varphi^2 \tag{9.3}$$

9.2.2 Flux-controlled memristor-based novel chaotic system

The dynamics of the flux-controlled memristor-based new chaotic system is defined in (9.4).

$$\begin{cases} \dot{x} = -y - z \\ \dot{y} = x - [W(w)]y \\ \dot{z} = a - ay^2 - bz \\ \dot{w} = y \end{cases} \tag{9.4}$$

where x, y, z, w are the state variables. $W(w) = \gamma + 3\beta w^2$ is the memductance and γ, β are positive constants. The system shows chaotic be-

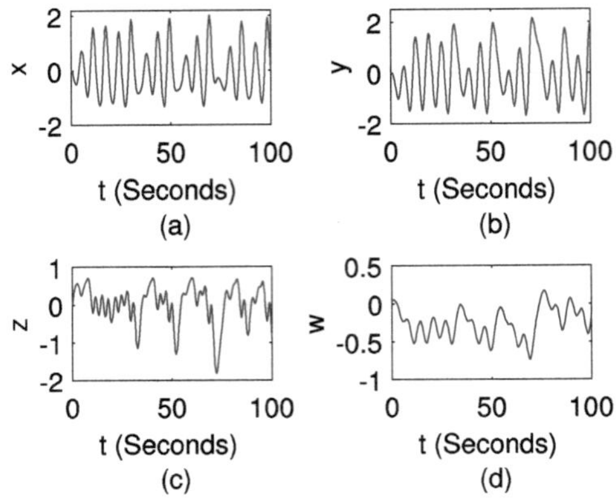

FIGURE 9.2 Time series behaviors of state variables show chaotic nature of the proposed memristor-based system.

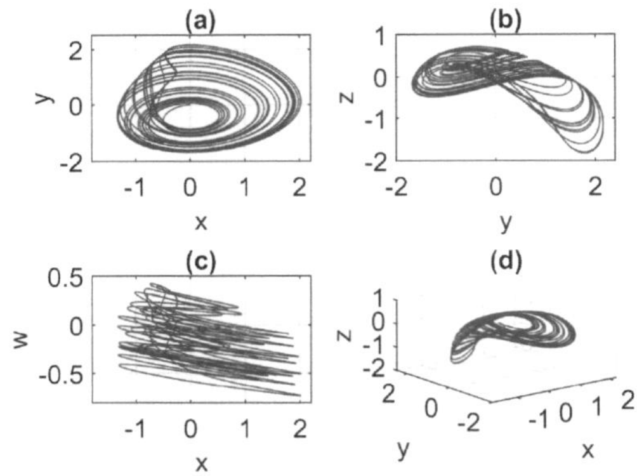

FIGURE 9.3 Phase plane behaviors show chaotic nature of the proposed memristor-based system.

havior for the parameters values $a = 0.5$, $b = 0.5$, $\gamma = 0.00067$, and $\beta = 0.00002$. The initial condition is considered as $[x(0),\ y(0),\ z(0),\ w(0)] = [0.001,\ 0.001,\ 0.001,\ 0.01]$. The time series behaviors and the corresponding phase plane behaviors of the state variables are shown in Figs. 9.2 and 9.3, respectively. The Lyapunov exponents of the proposed memristor-based chaotic system (9.4) are $L_1 = 0.0583$, $L_2 = 0$, $L_3 = -0.0072$, and $L_4 = -0.6630$.

By solving $\dot{x} = \dot{y} = \dot{z} = \dot{w} = 0$, the system in (9.4) has no equilibria. The generalized Jacobian matrix $J(x, y, z, w)$ is calculated as

$$J(x, y, z, w) = \begin{bmatrix} 0 & -1 & -1 & 0 \\ 1 & -(\gamma + 3\beta w^2) & 0 & -6\beta y w \\ 0 & -2ay & -b & 0 \\ 0 & 1 & 0 & 0 \end{bmatrix}$$

9.3 Properties and behaviors of memristor-based novel chaotic system

In this section, different properties and dynamical behaviors of memristor-based novel chaotic system are explored.

9.3.1 Symmetry and invariance

The system in (9.4) is symmetrical about the y-axis because it is invariant under the transformation from $(x,\ y,\ z,\ w)$ to $(-x,\ y,\ -z,\ -w)$.

9.3.2 Dissipation

The divergence of system (9.4) is calculated as

$$\nabla V = \frac{\delta \dot{x}}{\delta x} + \frac{\delta \dot{y}}{\delta y} + \frac{\delta \dot{z}}{\delta z} + \frac{\delta \dot{w}}{\delta w} = -(\gamma + 3\beta w^2) - b \tag{9.5}$$

The system is dissipative, i.e. $\nabla V < 0$ for all positive values of parameters.

9.3.3 Lyapunov spectrum

The Lyapunov spectra are plotted by varying one parameter at a time and keeping others fixed. Lyapunov spectra are explored for all parameters a and b. The Lyapunov spectrum of the memristor-based chaotic system with a as a varying parameter is shown in Fig. 9.4 where $b = 0.6$ and $c = 0.1$ parameters are kept fixed. The Lyapunov spectrum is plotted when a is varied between $0 \leq a \leq 0.510$. The system in (9.4) exhibits periodic, 2-torus quasi-periodic, chaotic and chaotic 2-torus behaviors for different range of a parameter. The behavior is periodic when $a \in [0,\ 0.26) \cup [0.435,\ 0.445] \cup [0.475,\ 0.485]$ (approximately). The behavior is 2-torus quasi-periodic for $a \in [0.26,\ 0.435)$. The 2-torus quasi-periodic behavior is considered when the nature of the Lyapunov exponents is $(0,\ 0,\ -,\ -)$. The behavior is chaotic for $a \in (0.445,\ 0.475) \cup (0.490,\ 0.505]$. The system reflects chaotic 2-torus behavior when $a \in (0.505,\ 0.510]$. Chaotic 2-torus behavior is considered when the nature of Lyapunov exponents is $(+,\ 0,\ 0,\ -)$. Fig. 9.4 reveals that the memristor-based novel chaotic system (9.4) without equilibria shows chaotic behavior along with chaotic 2-torus behavior for a parameter. The chaotic 2-torus behavior is unique for the memristor-based chaotic system without equilibria.

Similarly, the Lyapunov spectrum of the memristor-based chaotic system in (9.4) with b as a varying parameter is explored and shown in Fig. 9.5. Parameter b ranges between $0.518 \leq b \leq 0.85$ and shows chaotic behavior when $b \in [0.518,\ 0.622] \cup [0.647,\ 0.70)$. The behaviors are periodic and 2-torus quasi-periodic for b range as (0.622, 0.647) and [0.70, 0.85], respectively.

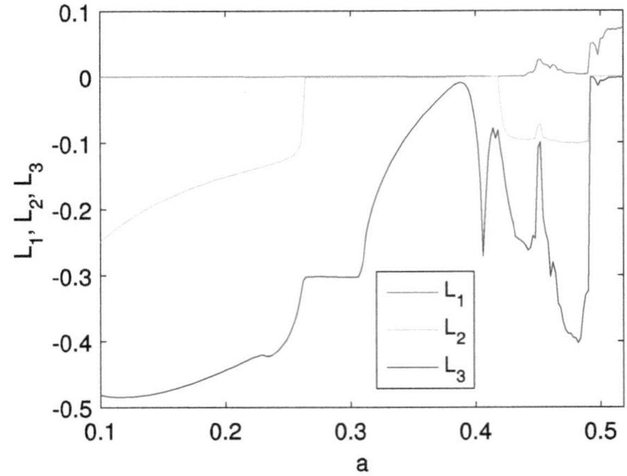

FIGURE 9.4 Lyapunov spectrum of the memristor-based novel chaotic system with the variation of parameter $a \in [0,\ 0.51]$. (For interpretation of the colors in the figure, the reader is referred to the web version of this chapter.)

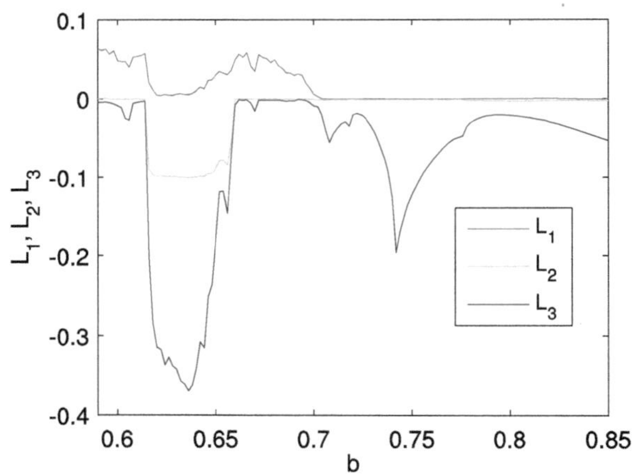

FIGURE 9.5 Lyapunov spectrum of the proposed chaotic system with respect to b parameter. (For interpretation of the colors in the figure, the reader is referred to the web version of this chapter.)

9.3.4 Bifurcation diagram

Bifurcation diagrams are analyzed by varying one parameter at a time and keeping others fixed. The bifurcation diagrams are explored for the parameters a and b. The bifurcation diagram of the memristor-based chaotic system with a as a varying parameter is shown in Fig. 9.6 where $b = 0.6$ and $c = 0.1$ parameters are kept fixed. Bifurcation diagram is plotted when a is varied between $0.25 \leq a \leq 0.51$. Similarly, the bifurcation diagram of the memristor-based chaotic system in (9.4) with b as a varying parameter is explored and shown

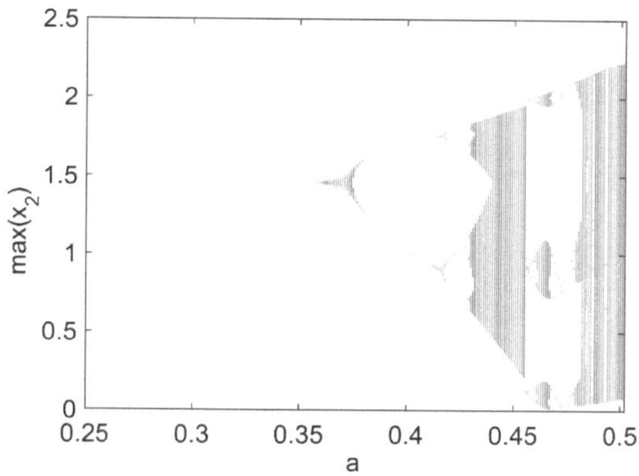

FIGURE 9.6 Bifurcation diagram of the memristor-based novel chaotic system when parameter $a \in [0.25,\ 0.51]$.

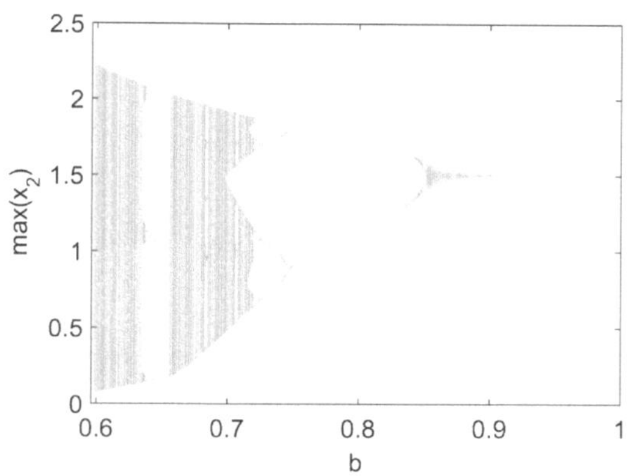

FIGURE 9.7 Bifurcation diagram of the proposed chaotic system when parameter $b \in [0.622,\ 1]$.

in Fig. 9.7. The bifurcation plots correspond to similar behaviors obtained via Lyapunov spectra in Figs. 9.4 and 9.5.

9.3.5 Kaplan–Yorke dimension

Dimensions are the static characterization of attractor. Nevertheless the attractor is formed by the dynamical system and so it is interesting to look for a connection between dimensions and dynamical characteristics. Kaplan and Yorke proposed a dimension based on the Lyapunov exponents, known as the Kaplan–Yorke dimension D_{KY} [99] and defined as

$$D_{KY} = k + \frac{1}{|\lambda_{k+1}|}\sum_{i=1}^{k}\lambda_i \qquad (9.6)$$

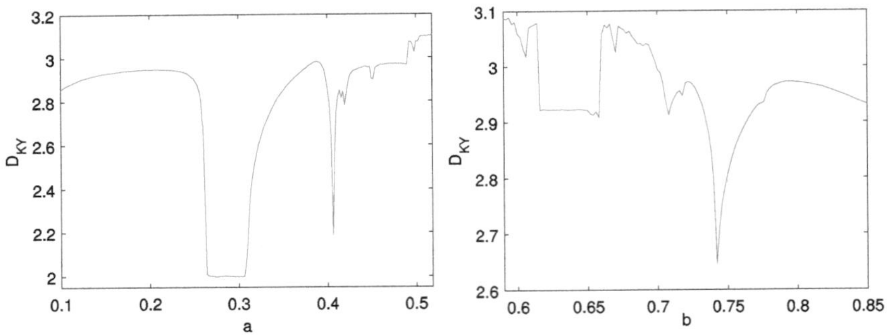

FIGURE 9.8 Shows spectrum of Kaplan–Yorke dimension (D_{KY}) for a and b parameters.

where k is the largest integer for which the sum of k Lyapunov exponents is positive. Using (9.6), the Kaplan–Yorke dimension of the memristor-based proposed chaotic system, for the parameters values $a = 0.5$, $b = 0.5$, $\gamma = 0.00067$, and $\beta = 0.00002$, is calculated as

$$D_{KY} = 3 + \frac{0.0511}{0.6630} = 3.0771 \tag{9.7}$$

$D_{KY} = 3.0771$ is the estimate of the dimension of the volume that neither grows nor decays. The spectrum of the Kaplan–Yorke dimension for the range of a and b parameters are shown in Fig. 9.8.

9.3.6 Poincaré section

Poincaré sections are analyzed between different planes with $x = 5$, $z = 0$, $y = -2$, and $w = 8$, and are shown in Figs. 9.9 to 9.12, respectively. In Figs. 9.9 and 9.10 the Poincaré sections are placed radially with respect to $x = 5$ and $z = 0$ axes, respectively, and illustrate the stretch and fold action of the complex-valued chaotic system flow and reflects the chaotic behavior with parachute and thumb shapes of the Poincaré map. Thus one claims the unique chaotic behaviors of the proposed memristor-based chaotic system.

9.4 Projective synchronization between the memristor-based chaotic systems

In this section, projective synchronization between the identical memristor-based chaotic systems (9.4) is discussed. The system in (9.4) is considered as

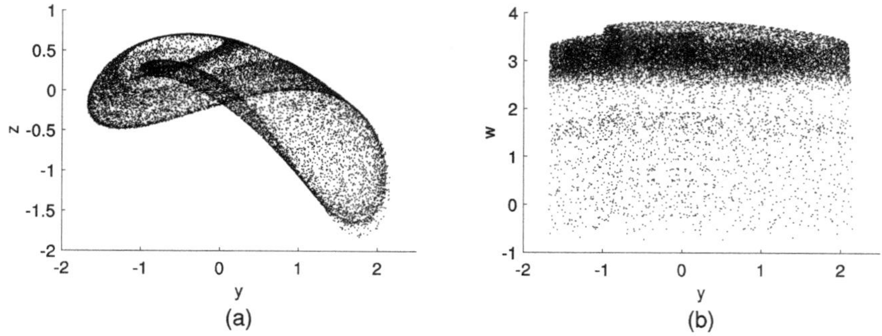

FIGURE 9.9 Poincaré sections between different planes at $x = 5$.

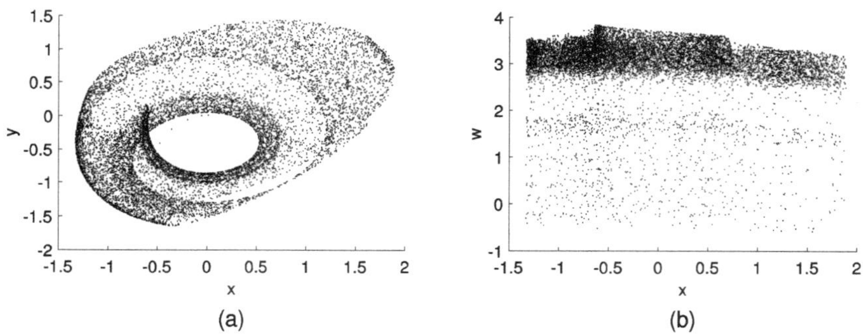

FIGURE 9.10 Poincaré sections between different planes at $z = 0$.

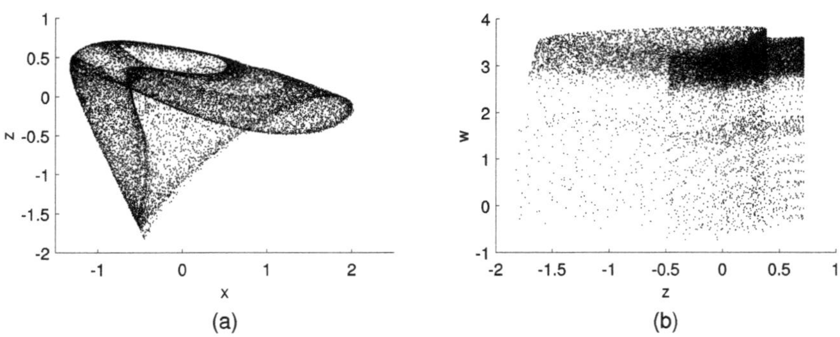

FIGURE 9.11 Poincaré sections between different planes at $x = -2$.

the drive system and defined in (9.8). We have

$$\begin{cases} \dot{x}_d = -y_d - z_d \\ \dot{y}_d = x_d - (\gamma + 3\beta w_d^2) y_d \\ \dot{z}_d = a - a y_d^2 - b z_d \\ \dot{w}_d = y_d \end{cases} \tag{9.8}$$

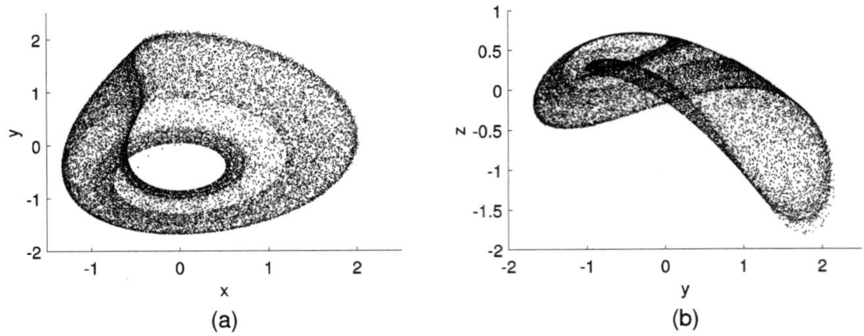

FIGURE 9.12 Poincaré sections between different planes at $w = 8$.

where x_d, y_d, z_d, and w_d are the state variables of the drive systems (9.8).

The chaotic system in (9.4) with added control inputs is considered as the response system and represented in (9.9):

$$\begin{cases} \dot{x}_r = -y_r - z_r + u_1 \\ \dot{y}_r = x_r - (\gamma + 3\beta w_r^2)y_r + u_2 \\ \dot{z}_r = a - ay_r^2 - bz_r + u_3 \\ \dot{w}_r = y_r + u_4 \end{cases} \tag{9.9}$$

where x_d, y_d, z_d, and w_d are the state variables of the drive systems (9.9). Added control inputs defined as u_1, u_2, u_3, and u_4 are required in the design.

Let the projective synchronization error ($e \in R^4$) between the memristor-based drive (9.8) and response (9.9) chaotic systems be defined as

$$\begin{cases} e_1 = x_d - k_1 x_r \\ e_2 = y_d - k_2 y_r \\ e_3 = z_d - k_3 z_r \\ e_4 = w_d - k_4 w_r \end{cases} \tag{9.10}$$

where k_i for $i = 1, 2, 3, 4$ are the projective factors between each state, respectively, of the memristor-based drive (9.8) and response (9.9) chaotic systems. Then the error dynamics is written as

$$\begin{cases} \dot{e}_1 = -y_d - z_d - k_1(-y_r - z_r) - k_1 u_1 \\ \dot{e}_2 = x_d - (\gamma + 3\beta w_d^2)y_d - k_2(x_r - (\gamma + 3\beta w_r^2)y_r) - k_2 u_2 \\ \dot{e}_3 = a - ay_d^2 - bz_d - k_3(a - ay_r^2 - bz_r) - k_3 u_3 \\ \dot{e}_4 = y_d - k_4 y_r - k_4 u_4 \end{cases} \tag{9.11}$$

Theorem 1. *The designed control inputs in* (9.12) *ensure the projective synchronization between the drive* (9.8) *and response* (9.9) *chaotic systems asymp-*

totically.

$$\begin{cases} u_1 = \frac{1}{k_1}[-y_d + k_1 y_r - z_d + k_1 z_r + me_1] \\ u_2 = \frac{1}{k_2}[x_d - 3\beta w_d^2 y_d - k_2(x_r - 3\beta w_r^2 y_r) + ne_2] \\ u_3 = \frac{1}{k_3}[a - ay_d^2 - k_3(a - ay_r^2) + pe_3] \\ u_4 = \frac{1}{k_4}[y_d - k_4 y_r + qe_4] \end{cases} \tag{9.12}$$

where m, n, p, and q are positive parameters used by designer.

Proof. Let a Lyapunov function candidate, a continuously differentiable function, be defined as

$$V(e) = \frac{1}{2}\sum_{i=1}^{4} e_i^2 \tag{9.13}$$

$\dot{V}(e)$ is calculated as

$$\dot{V}(e) = \sum_{i=1}^{4} e_i \dot{e}_i \tag{9.14}$$

$$\begin{aligned} \dot{V}(e) &= e_1\dot{e}_1 + e_2\dot{e}_2 + e_3\dot{e}_3 + e_4\dot{e}_4 \\ &= e_1[-y_d - z_d - k_1(-y_r - z_r) - k_1 u_1] + e_2[x_d - (\gamma + 3\beta w_d^2)y_d \\ &\quad - k_2(x_r - (\gamma + 3\beta w_r^2)y_r) - k_2 u_2] + e_3[a - ay_d^2 - bz_d \\ &\quad - k_3(a - ay_r^2 - bz_r) - k_3 u_3] + e_4[y_d - k_4 y_r - k_4 u_4] \end{aligned}$$

Using the designed control inputs (9.12), $\dot{V}(e_w)$ is obtained:

$$\dot{V}(e) = -me_1^2 - (n+\gamma)e_2^2 - (p+b)e_3^2 - qe_4^2 \tag{9.15}$$

$\dot{V}(e) < 0 \ \forall\ e \neq 0$, i.e. as per Lyapunov stability theorem [100], the error (e) tends to zero as $t \to \infty$. Therefore, the memristor-based response chaotic system in (9.9) is asymptotically synchronized with the memristor-based drive chaotic system (9.8). In addition, the Lyapunov function candidate $V(e)$ is radially unbounded, i.e. $V(e) \to \infty$ as $t \to \infty$, thus, the error dynamics is globally asymptotically stable.

Therefore, the states of the memristor-based response system (9.9) are synchronized with the chaotic states of the memristor-based drive system (9.8). □

Remark. Unlike the trivial available nonlinear active control technique, which cancels the nonlinear terms, the designed control technique does not require the cancellation of all nonlinear terms and involves external forcing parameters m, n, p, and q to achieve projective synchronization. Moreover, due to the achieved projective synchronization, the behavior of the state variables changes and may be useful in communication security.

In Section 9.5, simulation results are presented to verify the design of non-linear active control technique to achieve the projective synchronization.

9.5 Simulation results and discussion

Simulation is performed in MATLAB® environment and solver ode45 is used for the simulation. The simulation runs for 60 seconds with step time 0.001. The initial condition for the states of memristor-based drive (9.8) and response (9.9) systems are considered as $[x_d(0),\ y_d(0),\ z_d(0),\ w_d(0)] = [0.001,\ 0.001,\ 0.001,\ 0.001]$ and $[x_r(0),\ y_r(0),\ z_r(0),\ w_r(0)] = [1,\ 1,\ 1,\ 1]$, respectively. The projective factors are $k_1 = 2$, $k_2 = -2$, $k_3 = 1$, $k_4 = -1$. Positive and negative values of k_i mean synchronization and anti-synchronization. The design parameters are considered as $m = n = p = q = 2$.

The time series plots and phase plane behaviors of the proposed memristor-based system are shown in Figs. 9.2 and 9.3, revealing the chaotic nature of the system. The projective synchronized state variables of the memristor-based drive and response chaotic systems using nonlinear active control technique are shown in Fig. 9.13. It is apparent from Fig. 9.13 that each state is synchronized as per the desired projective factor.

The designed control laws and the projective synchronization errors, to achieve projective synchronization between the memristor-based drive (9.8) and response (9.9) chaotic systems, are depicted in Figs. 9.14 and 9.15, respectively. Figs. 9.13 and 9.15 reveal that the synchronization between the state variables is achieved in a very small settling time (approximately 2.5 seconds), i.e. the objective is achieved. Compared to the conventional synchronization of chaotic systems, the modified projective synchronization can achieve high unpredictability to improve the security of information signal in communication.

Therefore, the states of the memristor-based response system (9.9) are synchronized with the chaotic states of the drive system (9.8). Moreover, the results presented here suggest that the behavior of the state variables changes due to projective synchronization and may be useful in communication security.

9.6 Conclusions and future scope

In this paper, dynamic behaviors of flux-controlled memristor-based novel 4D chaotic system without equilibrium are analyzed. The proposed memristor-based chaotic system is simple because it has two parameters, two nonlinear terms and a total of nine terms. The properties and behaviors of memristor-based novel chaotic system are analyzed by different qualitative and quantitative tools such as time series, phase plane, Lyapunov exponent, Lyapunov spectrum and Poincaré map. The memristor-based proposed chaotic system has some unique

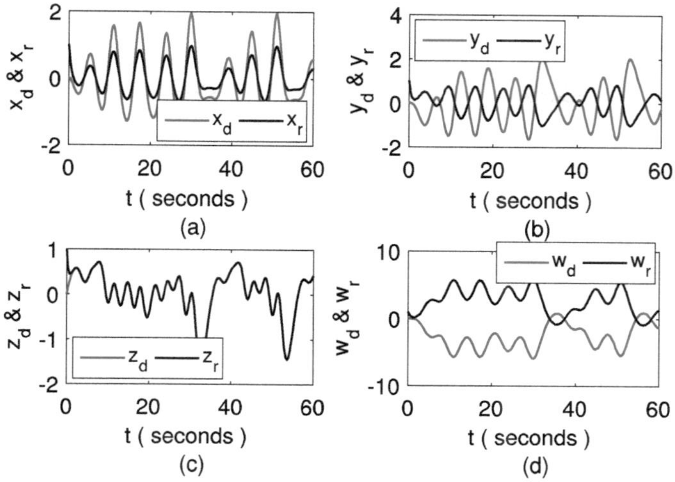

FIGURE 9.13 Synchronized state of the proposed memristor-based chaotic system using nonlinear active control technique. (For interpretation of the colors in the figure, the reader is referred to the web version of this chapter.)

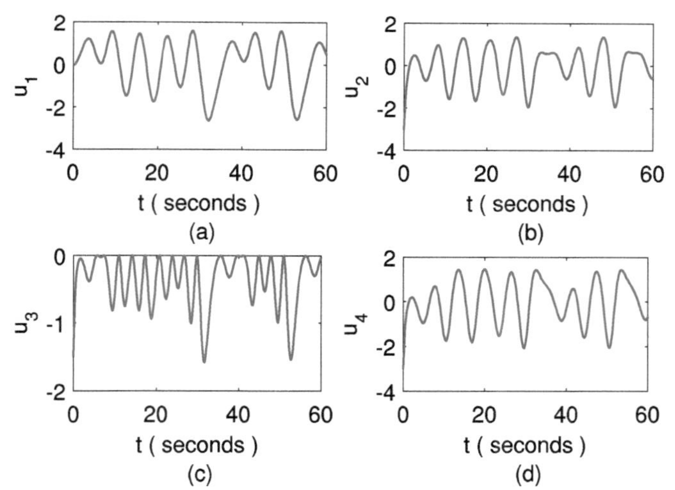

FIGURE 9.14 Designed nonlinear active control laws to achieve projective synchronization between the states of memristor-based drive and response chaotic systems.

and interesting behaviors such as being periodic, 2-torus quasi-periodic, and being chaotic and of chaotic 2-torus type. The proposed chaotic system has thumb and parachute shapes of the Poincaré map. A memristor-based chaotic system with such type of behaviors is not available in the literature and reflects the unique and interesting features.

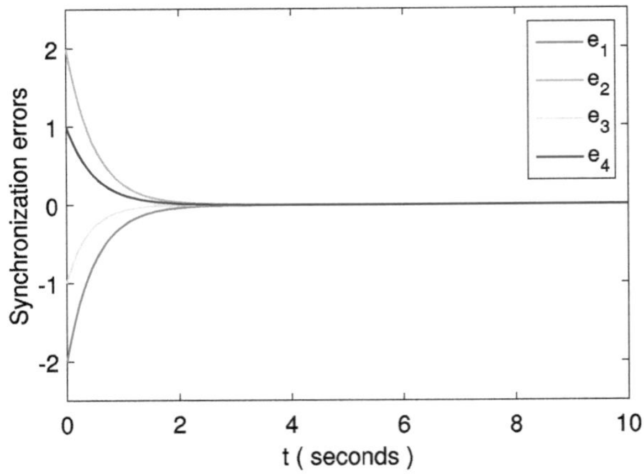

FIGURE 9.15 Projective synchronization errors between the states of memristor-based drive and response chaotic systems. (For interpretation of the colors in the figure, the reader is referred to the web version of this chapter.)

In addition, the projective synchronization between the two identical memristor-based chaotic systems using nonlinear active control technique is achieved. It is worth noting that projective synchronization between the memristor-based chaotic systems can be applied to secure communication using active control [7,9–11]. The proposed memristor-based chaotic system can provide unique state variables to improve the efficiency of information transmission, and projective synchronization can improve the communication security.

Finally, the inter and intra layer network synchronization of memristor-based chaotic system in the complex dynamical network is an open research problem nowadays. The proposed memristor-based chaotic system and its projective synchronization approach can be exploited as application to the complex dynamical network as future work.

References

[1] G. Chen, X. Dong, From Chaos to Order: Methodologies, Perspectives and Applications, World Scientific, Singapore, 1998, pp. 311–387.

[2] P.P. Singh, K.M. Singh, B.K. Roy, Chaos control in biological system using recursive backstepping sliding mode control, The European Physical Journal Special Topics 227 (2018) 731–746.

[3] P.P. Singh, B.K. Roy, Comparative performances of synchronisation between different classes of chaotic systems using three control techniques, Annual Reviews in Control 45 (2018) 152–165.

[4] S. Banerjee, L. Rondoni, M. Mitra, Applications of Chaos and Nonlinear Dynamics in Science and Engineering – Vol. 3, Springer, 2015, pp. 371–392.

[5] S. Mukherjee, S.K. Palit, S. Banerjee, M.R.K. Ariffin, L. Rondoni, D.K. Bhattacharya, Can complexity decrease in Congestive Heart failure?, Physica A: Statistical Mechanics and its Applications 439 (2015) 93–102.

[6] S. Banerjee, J. Kurths, Chaos and cryptography: a new dimension in secure communications, The European Physical Journal Special Topics 223 (2014) 1441–1445.
[7] P.P. Singh, J.P. Singh, B.K. Roy, NAC-based synchronisation and anti-synchronisation between hyperchaotic and chaotic systems, its analogue circuit design and application, IETE Journal of Research 63 (2017) 853–869.
[8] J.P. Singh, B.K. Roy, Second order adaptive time varying sliding mode control for synchronization of hidden chaotic orbits in a new uncertain 4-D conservative chaotic system, Transactions of the Institute of Measurement and Control 40 (2017) 3573–3586.
[9] P.P. Singh, J.P. Singh, B.K. Roy, SMC based synchronization and anti-synchronization of chaotic systems for secure communication and analog circuit realization, International Journal of Control Theory and Applications 9 (2016) 171–183.
[10] P.P. Singh, J.P. Singh, B.K. Roy, Implementation for synchronization of Chen and Liu-Yang chaotic systems using SMC and active control schemes, International Journal of Control Theory and Applications 9 (2016) 159–169.
[11] P.P. Singh, J.P. Singh, B.K. Roy, Synchronization of chaotic systems using NAC and its application to secure communication, International Journal of Control Theory and Applications 8 (2015) 995–1004.
[12] P.P. Singh, J.P. Singh, B.K. Roy, Adaptive control scheme for anti-synchronization of identical Bhalekar–Gejji chaotic systems, Research and Reviews: Journal of Physics 3 (2014) 1–7.
[13] P.P. Singh, J.P. Singh, B.K. Roy, Synchronization and anti-synchronization of Lu and Bhalekar–Gejji chaotic systems using nonlinear active control, Chaos, Solitons and Fractals 69 (2014) 31–39.
[14] J.P. Singh, B.K. Roy, S. Jafari, New family of 4-D hyperchaotic and chaotic systems with quadric surfaces of equilibria, Chaos, Solitons and Fractals 106 (2018) 243–257.
[15] J.P. Singh, B.K. Roy, Coexistence of asymmetric hidden chaotic attractors in a new simple 4-D chaotic system with curve of equilibria, Optik – International Journal for Light and Electron Optics 145 (2017) 209–217.
[16] P.P. Singh, J.P. Singh, M. Borah, B.K. Roy, On the construction of a new chaotic system, in: IFAC-PapersOnLine: 4th Conference on Advances in Control and Optimization of Dynamical Systems (ACODS), vol. 49, 2016, pp. 522–525.
[17] S. Jafari, V.T. Pham, T. Kapitaniak, Multiscroll chaotic sea obtained from a simple 3D system without equilibrium, International Journal of Bifurcation and Chaos 26 (2016) 1650031–1650036.
[18] S. Jafari, J.C. Sprott, F. Nazarimehr, Recent new examples of hidden attractors, The European Physical Journal Special Topics 224 (2015) 1469–1476.
[19] S. Jafari, J.C. Sprott, Simple chaotic flows with a line equilibrium, Chaos, Solitons and Fractals 57 (2013) 79–84.
[20] S. Jafari, J.C. Sprott, S.M.R.H. Golpayegani, Elementary quadratic chaotic flows with no equilibria, Physics Letters A 337 (2013) 699–702.
[21] G.A. Leonov, N.V. Kuznetsov, V.I. Vagaitsev, Localization of hidden Chua attractors, Physics Letters A 375 (2011) 2230–2233.
[22] G.A. Leonov, N.V. Kuznetsov, V.I. Vagaitsev, Hidden attractor in smooth Chua systems, Physica D 241 (2012) 1482–1486.
[23] Z. Wei, I. Moroz, J.C. Sprott, A. Akgul, W. Zhang, Hidden hyperchaos and electronic circuit application in a 5D self-exciting homopolar disc dynamo, Chaos 27 (2017) 033101.
[24] Z. Wei, I. Moroz, J.C. Sprott, Z. Wang, W. Zhang, Detecting hidden chaotic regions and complex dynamics in the self-exciting homopolar disc dynamo, International Journal of Bifurcation and Chaos 27 (2017) 1730008–1730015.
[25] Z. Wei, P. Yu, W. Zhang, M. Yao, Study of hidden attractors, multiple limit cycles from Hopf bifurcation and boundedness of motion in the generalized hyperchaotic Rabinovich system, Nonlinear Dynamics 82 (2015) 131–141.

[26] Z. Wei, W. Zhang, Hidden hyperchaotic attractors in a modified Lorenz–Stenflo system with only one stable equilibrium, International Journal of Bifurcation and Chaos 24 (2014) 1450127–1450132.
[27] Z. Wei, W. Zhang, Z. Wang, M. Yao, Hidden attractors and dynamical behaviors in an extended Rikitake system, International Journal of Bifurcation and Chaos 25 (2015) 1550028–1550036.
[28] L.O. Chua, Memristor – the missing circuit element, IEEE Transactions on Circuit Theory 18 (1971) 507–519.
[29] D.B. Strukov, G.S. Snider, D.R. Stewart, R.S. Williams, The missing memristor found, Nature 453 (2008) 80–83.
[30] L. Wang, T. Dong, M.F. Ge, Finite-time synchronization of memristor chaotic systems and its application in image encryption, Applied Mathematics and Computation 347 (2019) 293–305.
[31] K. Usha, P.A. Subha, Hindmarsh-Rose neuron model with memristors, Biosystems 178 (2019) 1–9.
[32] Y. Cao, Y. Cao, S. Wen, T. Huang, Z. Zeng, Passivity analysis of delayed reaction–diffusion memristor-based neural networks, Neural Networks 109 (2019) 159–167.
[33] Z. Li, C. Zhou, M. Wang, Symmetrical coexisting attractors and extreme multistability induced by memristor operating configurations in SC-CNN, International Journal of Electronics and Communications (AEU) 100 (2019) 127–137.
[34] M.E. Sahin, Z.G. Cam Taskiran, H. Guler, S.E. Hamamci, Simulation and implementation of memristive chaotic system and its application for communication systems, Sensors and Actuators A, Physical 290 (2019) 107–118.
[35] M.F. Tolba, W.S. Sayed, M.E. Fouda, H. Saleh, M. Al-Qutayri, M. Baker, A.G. Radwan, Digital emulation of a versatile memristor with speech encryption application, IEEE Access 7 (2019) 174280–174297.
[36] J. Sun, Q. Yang, Y. Wang, Dynamical analysis of novel memristor chaotic system and DNA encryption application, Iranian Journal of Science and Technology (2019), https://doi.org/10.1007/s40998-019-00239-x.
[37] O. Krestinskaya, A. James, L.O. Chua, Neuro-memristive circuits for edge computing: a review, IEEE Transactions on Neural Networks and Learning Systems 31 (2020) 4–23.
[38] B. Bao, N. Wang, Q. Xu, H. Wu, Y. Hu, A simple third-order memristive band pass filter chaotic circuit, IEEE Transactions on Circuits and Systems II, Express Briefs 64 (2017) 977–981.
[39] M. Itoh, L.O. Chua, Memristor oscillators, International Journal of Bifurcation and Chaos 18 (2008) 3183–3206.
[40] B. Muthuswamy, P.P. Kokate, Memristor-based chaotic circuits, IETE Technical Review 26 (2014) 1–16.
[41] B. Muthuswamy, Contributions to the Study of Autonomous Chaotic Circuits and Cellular Automata, Technical Report No. UCB/EECS-2009-164, Electrical Engineering and Computer Sciences, University of California, Berkeley, 2009.
[42] B. Muthuswamy, Implementing memristor based chaotic circuits, International Journal of Bifurcation and Chaos 20 (2010) 1335–1350.
[43] B. Muthuswamy, L.O. Chua, Simplest chaotic circuit, International Journal of Bifurcation and Chaos 20 (2010) 1567–1580.
[44] D. Batas, H. Fiedler, A memristor SPICE implementation and a new approach for magnetic flux-controlled memristor modeling, IEEE Transactions on Nanotechnology 10 (2011) 250–255.
[45] H. Kim, M.Pd. Sah, C. Yang, S. Cho, L.O. Chua, Memristor emulator for memristor circuit applications, IEEE Transactions on Circuits and Systems I, Regular Papers 59 (2012) 2422–2431.
[46] H. Li, L. Wang, S. Duan, A memristor-based scroll chaotic system-design, analysis and circuit implementation, International Journal of Bifurcation and Chaos 24 (2014) 1450099–1450110.

[47] X. Ya-Ming, W. Li-Dan, D. Shu-Kai, A memristor-based chaotic system and its field programmable gate array implementation, Acta Physica Sinica 65 (2016) 120503–120512.
[48] M. Abdirash, I. Dolzhikova, A.P. James, Implementation of Chua's chaotic oscillator with an HP memristor. CoRR, abs/1805.08081, 2018.
[49] X. Hu, S. Duan, Adaptive synchronization of memristor-based chaotic neural systems, Journal of Engineering Science and Technology Review 8 (2015) 17–23.
[50] S. Wang, X. Wang, Y. Zhou, A memristor-based complex Lorenz system and its modified projective synchronization, Entropy 17 (2015) 7628–7644.
[51] A.H. Abolmasoumi, S. Khosravinejad, Chaos control in memristor-based oscillators using intelligent terminal sliding mode controller, International Journal of Computer Theory and Engineering 8 (2016) 506–511.
[52] W. Xiong, J. Huang, Finite-time control and synchronization for memristor-based chaotic system via impulsive adaptive strategy, Advances in Difference Equations 26 (2016) 101–109.
[53] K. Ding, On designing feedback controllers for master-slave synchronization of memristor-based Chua's circuits, Complexity (2018) 5431619.
[54] E.E. Mahmoud, M.A. Al-Adwani, Dynamical behaviors, control and synchronization of a new chaotic model with complex variables and cubic nonlinear terms, Results in Physics 7 (2017) 1346–1356.
[55] D. Ding, X. Qian, W. Hu, N. Wang, D. Liang, Chaos and Hopf bifurcation control in a fractional-order memristor-based chaotic system with time delay, The European Physical Journal Plus 132 (2017) 447–459.
[56] K. Rajagopal, L. Guessas, A. Karthikeyan, A. Srinivasan, G. Adam, Fractional order memristor no equilibrium chaotic system with its adaptive sliding mode synchronization and genetically optimized fractional order PID synchronization, Complexity (2017) 1892618.
[57] P. Prakash, J.P. Singh, B.K. Roy, Fractional-order memristor-based chaotic jerk system with no equilibrium point and its fractional-order backstepping control, IFAC PapersOnLine 51 (2018) 1–6.
[58] J. Ruan, K. Sun, J. Mou, S. He, L. Zhang, Fractional-order simplest memristor-based chaotic circuit with new derivative, The European Physical Journal Plus 133 (2018) 3–13.
[59] K. Rajagopal, H. Jahanshahi, M. Varan, I. Bayir, V.T. Pham, S. Jafari, A. Karthikeyan, A hyperchaotic memristor oscillator with fuzzy based chaos control and LQR based chaos synchronization, International Journal of Electronics and Communications (AEU) 94 (2018) 55–68.
[60] K. Rajagopal, S. Vaidyanathan, A. Karthikeyan, A. Srinivasan, Complex novel 4D memristor hyperchaotic system and its synchronization using adaptive sliding mode control, Alexandria Engineering Journal 57 (2018) 683–694.
[61] H.M. Li, Y.F. Yang, Y. Zhou, C.L. Li, K. Qian, Z.Y. Li, J.R. Du, Dynamics and synchronization of a memristor-based chaotic system with no equilibrium, Complexity (2019) 4647608.
[62] B. Bao, Z. Ma, J. Xu, Z. Liu, Q. Xu, A simple memristor chaotic circuit with complex dynamics, International Journal of Bifurcation and Chaos 21 (2011) 2629–2645.
[63] V.T. Pham, C. Volos, L.V. Gambuzza, A memristive hyperchaotic system without equilibrium, Scientific World Journal (2014) 368986.
[64] Q. Li, S. Hu, S. Tang, G. Zeng, Hyperchaos and horseshoe in a 4D memristive system with a line of equilibria and its implementation, International Journal of Circuit Theory and Applications 42 (2014) 1172–1188.
[65] Z. Hrubos, T. Gotthans, Analysis and synthesis of chaotic circuits using memristor properties, Journal of Electrical Engineering 65 (2014) 129–136.
[66] A.T. Azar, C. Volos, N.A. Gerodimos, G.S. Tombras, V.T. Pham, A.G. Radwan, S. Vaidyanathan, A. Ouannas, J.M. Munoz-Pacheco, A novel chaotic system without equilibrium: dynamics, synchronization, and circuit realization, Complexity (2017) 7871467.
[67] C. Wang, X. Liu, H. Xia, Multi-piecewise quadratic nonlinearity memristor and its 2N-scroll and 2N+1-scroll chaotic attractors system, Chaos 27 (2017) 033114.

[68] Z. Chen, H. Tang, Z. Wang, Q. Zhang, J. Han, Design and circuit implementation for a novel charge-controlled chaotic memristor system, Journal of Applied Analysis and Computation 5 (2015) 251–261.
[69] J. Ma, Z. Chen, Z. Wang, Q. Zhang, A four-wing hyper-chaotic attractor generated from a 4-D memristive system with a line equilibrium, Nonlinear Dynamics 81 (2015) 1275–1288.
[70] F. Yuan, G. Wang, X. Wang, Extreme multistability in a memristor-based multi-scroll hyper-chaotic system, Chaos 26 (2016) 073107.
[71] X. Hu, C. Liu, L. Liu, Y. Yao, G. Zheng, Multi-scroll hidden attractors and multi-wing hidden attractors in a 5-dimensional memristive system, Chinese Physics B 26 (2017) 110502.
[72] C. Wang, H. Xia, L. Zhou, Implementation of a new memristor-based multiscroll hyper-chaotic system, Pramana Journal of Physics 88 (2017) 34–41.
[73] J. Ma, P. Zhou, B. Ahmad, G. Ren, C. Wang, Chaos and multi-scroll attractors in RCL-shunted junction coupled Jerk circuit connected by memristor, PLoS ONE 13 (2018) e0191120, https://doi.org/10.1371/journal.pone.0191120.
[74] V.T. Pham, S. Vaidyanathan, C.K. Volos, S. Jafari, N.V. Kuznetsov, T.M. Hoang, A novel memristive time-delay chaotic system without equilibrium points, The European Physical Journal Special Topics 225 (2016) 127–136.
[75] K. Rajagopal, A. Bayani, A.J.M. Khalaf, H. Namazi, S. Jafari, V.T. Pham, A no-equilibrium memristive system with four-wing hyperchaotic attractor, International Journal of Electronics and Communications (AEU) 95 (2018) 207–215.
[76] K. Rajagopal, S. Arun, A. Karthikeyan, P. Duraisamy, A. Srinivasan, A hyperchaotic memristor system with exponential and discontinuous memductance function, International Journal of Electronics and Communications (AEU) 95 (2018) 249–255.
[77] B.A. Mezatio, M.T. Motchongom, B.W. Tekam, R. Kengne, R. Tchitnga, A. Fomethe, A novel memristive 6D hyperchaotic autonomous system with hidden extreme multistability, Chaos, Solitons and Fractals 120 (2019) 100–115.
[78] B. Wang, F.C. Zou, J. Cheng, A memristor-based chaotic system and its application in image encryption, Optik 154 (2018) 538–544.
[79] B. Wang, F.C. Zou, Y. Zhang, New memristive chaotic system and the application in digital watermark, Optik 172 (2018) 873–878.
[80] E.E. Mahmoud, O.A. Althagafi, A new memristive model with complex variables and its generalized complex synchronizations with time lag, Results in Physics 15 (2019) 102619.
[81] L. Liu, C. Du, L. Liang, X. Zhang, A high spectral entropy (SE) memristive hidden chaotic system with multi-type quasi-periodic and its circuit, Entropy 21 (2019), https://doi.org/10.3390/e21101026.
[82] P.P. Singh, B.K. Roy, Memristor-based novel complex-valued chaotic system and its projective synchronisation using nonlinear active control technique, The European Physical Journal Special Topics 228 (2019) 2197–2214.
[83] T. Driscoll, Y. Pershin, D. Basov, M. Di Ventra, Chaotic memristor, Applied Physics A, Materials Science & Processing 102 (2011) 885–889.
[84] B.C. Bao, Z. Liu, J.P. Xu, Steady periodic memristor oscillator with transient chaotic behaviours, Electronics Letters 46 (2010) 237–238.
[85] Z.H. Lin, H.X. Wang, Image encryption based on chaos with PWL memristor in Chua's circuit, in: Proc. Int. Conf. Commun. Circuits Syst., 2009, pp. 964–968.
[86] A. Afifi, A. Ayatollahi, F. Raissi, Implementation of biologically plausible spiking neural network models on the memristor crossbar-based CMOS/nano circuits, in: Proc. Eur. Conf. Circuit Theory Design, 2009, pp. 563–566.
[87] I. Ebong, P. Mazumder, Memristor based STDP learning network for position detection, in: Proc. Int. Conf. Microelectron, 2010, pp. 292–295.
[88] S.H. Jo, T. Chang, I. Ebong, B.B. Bhadviya, P. Mazumder, W. Lu, Nanoscale memristor device as synapse in neuromorphic systems, Nano Letters 10 (2010) 1297–1301.
[89] G.S. Snider, Self-organized computation with unreliable, memristive nanodevices, Nanotechnology 18 (2007) 365202–365215.

[90] S. Kyungah, K. Insung, S. Jung, M. Jo, S. Park, J. Park, J. Shin, K.P. Biju, J. Kong, K. Lee, B. Lee, H. Hwang, Analog memory and spike-timing-dependent plasticity characteristics of a nanoscale titanium oxide bilayer resistive switching device, Nanotechnology 22 (2011) 254023–254030.

[91] K. Likharev, A. Mayr, I. Muckra, O. Türel, CrossNets: high-performance neuromorphic architectures for CMOL circuits, Annals of the New York Academy of Sciences 1006 (2003) 146–163.

[92] J.A. Perez-Carrasco, C. Zamarreno-Ramos, T. Serrano-Gotarredona, B. Linares-Barranco, On neuromorphic spiking architectures for asynchronous STDP memristive systems, in: Proc. IEEE Int. Symp. Circuits Syst., 2010, pp. 1659–1662.

[93] S.J. Choi, G.B. Kim, K. Lee, K.H. Kim, W.Y. Yang, S. Cho, H.J. Bae, D.S. Seo, S.I. Kim, K.J. Lee, Synaptic behaviours of a single metal-oxide-metal resistive device, Applied Physics A 102 (2011) 1019–1025.

[94] G.D. Howard, E. Gale, L. Bull, B.D.L. Costello, A. Adamatzky, Evolving spiking networks with variable memristors, in: Proc. 13th Annu. Conf. Genetic Evol. Comput., Ireland, 2011, pp. 1275–1282.

[95] D. Querlioz, O. Bichler, C. Gamrat, Simulation of a memristor-based spiking neural network immune to device variations, in: Proc. Int. Joint Conf. Neural Netw., 2011, pp. 1775–1781.

[96] F. Merrikh-Bayat, S.B. Shouraki, A. Rohani, Memristor crossbar-based hardware implementation of the IDS method, IEEE Transactions on Fuzzy Systems 19 (2011) 1083–1096.

[97] M. Hu, H. Li, Y. Chen, Q. Wu, G.S. Rose, R.W. Linderman, Memristor crossbar-based neuromorphic computing system: a case study, IEEE Transactions on Neural Networks and Learning Systems 104 (2014) 1864–1877.

[98] Y. Xu, Y. Jia, J. Ma, A. Alsaedi, B. Ahmad, Synchronization between neurons coupled by memristor, Chaos, Solitons and Fractals 104 (2017) 435–442.

[99] P. Frederickson, J. Kaplan, E. Yorke, J. Yorke, The Lyapunov dimension of strange attractors, Journal of Differential Equations 49 (1983) 185–207.

[100] J.E. Slotine, W. Li, Applied Nonlinear Control, Prentice Hall Inc., Englewood Cliffs, New Jersey, 1991, pp. 100–154, 276–311.

Chapter 10

Memristor Helmholtz oscillator: analysis, electronic implementation, synchronization and chaos control using single controller

Cyrille Ainamon[a], André Chéagé Chamgoué[b], Sifeu Takougang Kingni[c], René Yamapi[d], Jean Bio Chabi Orou[a], and Paul Woafo[e]

[a]*Institute of Mathematics and Physical Sciences, University of Abomey-Calavi, Porto Novo, Benin,* [b]*School of Geology and Mining Engineering, University of Ngaoundere, Meiganga, Cameroon,* [c]*Department of Mechanical, Petroleum and Gas Engineering, Faculty of Mines and Petroleum Industries, University of Maroua, Maroua, Cameroon,* [d]*Fundamental Physics Laboratory, Department of Physics, Faculty of Science, University of Douala, Douala, Cameroon,* [e]*Laboratory of Modelling and Simulation in Engineering, Biomimetics and Prototypes (LaMSEBP), Department of Physics, Faculty of Science, University of Yaoundé I, Yaoundé, Cameroon*

10.1 Introduction

A memristor is a nonlinear two-terminal electronic memory component which was first introduced by Chua in 1971 as the fourth fundamental passive circuit element following the resistor, the capacitor and the inductor [1]. The possibility of modifying its resistance as a function of the current density passing through it, and keeping these changes in memory, makes memory components ideal candidates to act as synapses in artificial neural networks [2].

This passive circuit element links the fundamental quantities of electric charge and magnetic flux linkage. After its physical realization by Stanley William's group from Hewlett–Packard Laboratory [3], the memristor has attracted wide attention in the scientific community and has been investigated in novel applications [4–10]. For example, the memristor can be useful for low-power computation and storage to store data without the need of power using [10] and so on. It is a nonlinear device with memory characteristics, which can be used to design chaotic circuits, memory devices and neural networks [11].

Mem-elements for Neuromorphic Circuits with Artificial Intelligence Applications
https://doi.org/10.1016/B978-0-12-821184-7.00018-9

On the other hand, chaotic circuits with memristors have been investigated [12–20] for their applications related to secure communications with chaos [21]. Several nonlinearities for characterizing the pinched hysteresis loop of memristors, such as a HP memristor model [22], non-smooth piecewise linearity [21–23] smooth cubic nonlinearity [24–26], smooth piecewise-quadratic nonlinearity [13,27,28], practical circuit emulators (made of resistors, capacitors, operational amplifiers, transistors and analog multipliers) [29–31] and absolute value nonlinearity [19,20,27,32], were used to emulate the relation between magnetic flux and electric charge. In Refs. [19,20,27,32], circuits which possess absolute value nonlinearity characteristics are employed to realize the memristor-based chaotic circuit. In this chapter, a Memristor Helmholtz (**MH**) oscillator is designed by adding a flux-controlled memristor with absolute nonlinearity value in the circuit of an autonomous Helmholtz jerk oscillator [33]. The autonomous jerk oscillator has been built by converting the second-order Helmholtz oscillator into a three dimensional oscillator using the jerk model. The Helmholtz oscillator is a two-dimensional oscillator with a quadratic nonlinearity. The Helmholtz oscillator known to naval architects as the Helmholtz–Thompson equation has been investigated in the literature [34–39].

The rest of this chapter is structured as follows. In Section 10.2, the **MH** oscillator is presented and its rate-equations are derived and investigated. Section 10.3 deals with electronic implementation of MH oscillator. In Section 10.4, the feedback controller method is used for achieving synchronization in unidirectional coupled identical chaotic memristor Helmholtz oscillators. In Section 10.5, a single simple controller is employed for achieving the control chaotic behavior found in memristor Helmholtz oscillator. Some concluding remarks end this chapter.

10.2 Design and analysis of the proposed memristor Helmholtz oscillator

An ideal and active flux-controlled **MH** oscillator is presented in Fig. 10.1.

The circuit of Fig. 10.1(a) is designed by adding an ideal and active flux-controlled memristor [27,32] in the circuit of an autonomous Helmholtz jerk oscillator [33]. This circuit includes five operational amplifiers, one analog multiplier, ten resistors, three capacitors and, an ideal and active flux-controlled memristor. The circuit of an ideal and active flux-controlled memristor with an absolute memristance function (characterized by smooth piecewise-quadratic nonlinearity [27,32]) is shown in Fig. 10.1(b). The details of the flux-controlled memristor described are given in Ref. [20]. The memductance function of the flux-controlled memristor is expressed as

$$I = kM(V_0)\,V_X = k\left(\frac{1}{R_\alpha} - \frac{g_1 g_2}{R_\beta}\,|V_0|\right)V_X, \tag{10.1a}$$

$$\frac{dV_0}{dt} = -\frac{1}{R_C C_0} V_M - \frac{1}{R_d C_0} V_0, \tag{10.1b}$$

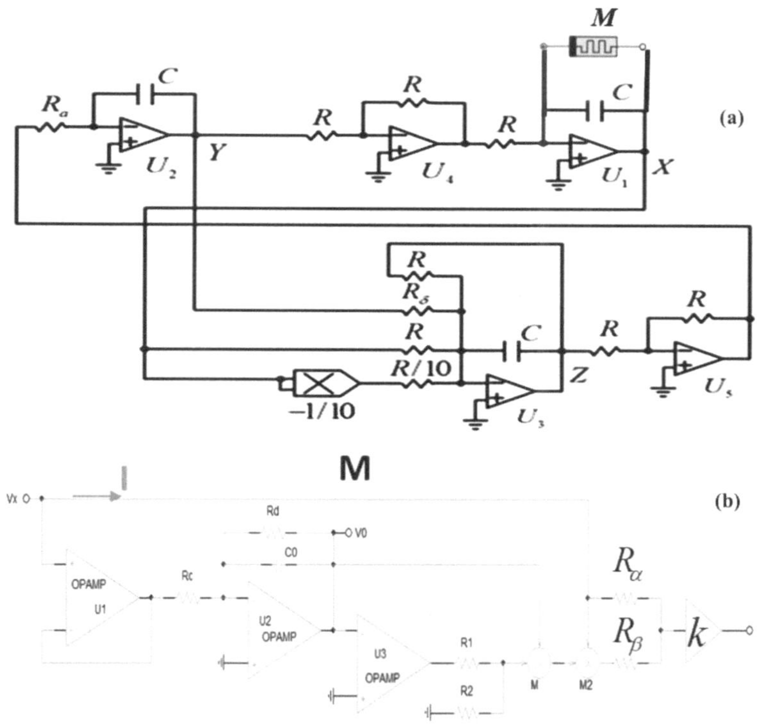

FIGURE 10.1 Schematic diagrams of: (a) memristor Helmholtz oscillator and (b) the ideal and active flux-controlled memristor M used in (a).

where V_0 is the voltage across the capacitor C_0 in the memristor circuit of Fig. 10.1(b), $M(V_0) = \left(\frac{1}{R_\alpha} - \frac{g_1 g_2}{R_\beta}|V_0|\right)$ is the memristance and g_1, g_2 are the gains of multipliers M_1 and M_2 of Fig. 10.1(b), respectively. From Fig. 10.1(a), it is easy to derive the equations of the **MH** oscillator based on Kirchhoff's laws as follows:

$$C\frac{dV_X}{dt'} = \frac{V_Y}{R} - kM(V_0)V_X, \tag{10.2a}$$

$$C\frac{dV_Y}{dt'} = \frac{V_Z}{R_a}, \tag{10.2b}$$

$$C\frac{dV_Z}{dt'} = -\frac{V_Z}{R} - \frac{V_Y}{R_\delta} - \frac{V_X}{R} + \frac{V_X^2}{R_k K}, \tag{10.2c}$$

$$\frac{dV_0}{dt'} = -\frac{1}{R_C C_0}V_X - \frac{1}{R_d C_0}V_0, \tag{10.2d}$$

where V_X, V_Y, V_Z are the voltage at the outputs of three operational amplifiers (U_1, U_2, U_3).

For possible dynamical analysis, a scale transformation is often presented to map the set of Eq. (10.2) into a dimensionless dynamical system by setting

$$t = t'/(RC),\ x = V_X/V_{ref},\ y = V_Y/V_{ref},\ z = V_Z/V_{ref},\ a = R/R_a,$$
$$\delta = R/R_\delta,\ V_{ref} = 1\text{v},\ w = V_0/V_{ref},\ R_k = R/10; \tag{10.3}$$
$$c = RC/R_C C_0,\ d = RC/R_d C_0.$$

The dynamical equations are approached as follows:

$$\frac{dx}{dt} = y - k(\alpha - \beta\,|\mathrm{w}|)\,x, \tag{10.4a}$$
$$\frac{dy}{dt} = az, \tag{10.4b}$$
$$\frac{dz}{dt} = -z - \delta y - x + x^2, \tag{10.4c}$$
$$\frac{dw}{dt} = -cx - dw, \tag{10.4d}$$

where $\alpha = \frac{R}{R_\alpha}$ and $\beta = \frac{R g_1 g_2}{R_\beta}\,|V_0|$ are the memductance parameters. In the following subsections, the **MH** oscillator is analytically and numerically investigated using linear stability analysis and the fourth order Runge–Kutta methods [40], respectively.

10.2.1 Equilibrium points and their stabilities

The equilibrium points of the **MH** oscillator system can be derived by setting $\frac{dx}{d\tau} = \frac{dy}{d\tau} = \frac{dz}{d\tau} = \frac{dw}{d\tau} = 0$. Then, by solving the obtained system of nonlinear equations, the equilibrium points of system (10.4) are $O\,(0,0,0,0)$ and $E\,(\bar{x}, k\bar{x}\,(\alpha - \beta\,|c\bar{x}/d|)\,, 0, -c\bar{x}/d)$ with $\bar{x}$, satisfied the equation $\frac{\delta k \beta c}{d}\,|\bar{x}| + \bar{x} - (1 + \alpha\delta k) = 0$, which gives $\bar{x} = \frac{d(1+\alpha\delta k)}{d-\delta k\beta c} < 0$ with $d < \delta k \beta c$ and $\bar{x} = \frac{d(1+\alpha\delta k)}{d+\delta k\beta c} > 0$. The Jacobian matrix of system (10.4) at equilibrium point O is given by

$$J = \begin{pmatrix} -k\alpha & 1 & 0 & 0 \\ 0 & 0 & a & 0 \\ -1 & -\delta & -1 & 0 \\ -c & 0 & 0 & -d \end{pmatrix}. \tag{10.5}$$

The characteristic equation of system (10.4) evaluated at the equilibrium point O is

$$\lambda^4 + \delta_1\lambda^3 + \delta_2\lambda^2 + \delta_3\lambda + \delta_4 = 0, \tag{10.6}$$

where the expressions for δ_i $(i = 1, 2, 3, 4)$ are given by

$$\begin{aligned}\delta_1 &= \alpha\mathrm{k} + \mathrm{d} + 1,\\ \delta_2 &= \alpha\mathrm{dk} + \mathrm{a}\delta + \alpha\mathrm{k} + \mathrm{d},\\ \delta_3 &= \mathrm{a}\alpha\delta\mathrm{k} + \mathrm{ad}\delta + \alpha\mathrm{dk} + \mathrm{a},\\ \delta_4 &= \mathrm{da}\left(\alpha\delta\mathrm{k} + 1\right).\end{aligned} \tag{10.7}$$

Using the Routh–Hurwitz conditions, Eq. (10.6) has all roots with negative real parts if and only if

$$\begin{cases}\delta_i > 0, \ i = 1, 2, 3, 4\\ \delta_1\delta_2\delta_3 > \delta_3^2 + \delta_1^2\delta_4\end{cases} \tag{10.8}$$

The Jacobian matrix of system (10.4) at equilibrium point $E\left(\bar{x}, \bar{y}, \bar{z}, \bar{w}\right)$ is given by

$$J = \begin{pmatrix} -k\left(\alpha - \beta\left|\bar{w}\right|\right) & 1 & 0 & k\beta\bar{x}sign\left(\bar{w}\right)\\ 0 & 0 & a & 0\\ 2\bar{x} - 1 & -\delta & -1 & 0\\ -c & 0 & 0 & -d \end{pmatrix}. \tag{10.9}$$

The characteristic equation of system (10.4) evaluated at the equilibrium point E is

$$\lambda^4 + \delta_1\lambda^3 + \delta_2\lambda^2 + \delta_3\lambda + \delta_4 = 0, \tag{10.10}$$

where the expressions for δ_i $(i = 1, 2, 3, 4)$ are given by

$$\begin{aligned}\delta_1 &= d - k\beta\left|\bar{w}\right| + k\alpha + 1,\\ \delta_2 &= ck\bar{x}\beta\mathrm{sgn}\left(\bar{w}\right) - \beta kd\left|\bar{w}\right| - \beta k\left|\bar{w}\right| + \alpha dk + a\delta + k\alpha + d,\\ \delta_3 &= ck\bar{x}\beta\mathrm{sgn}\left(\bar{w}\right) - \beta ka\delta\left|\bar{w}\right| - \beta kd\left|\bar{w}\right| + a\alpha\delta k + ad\delta + \alpha dk - 2a\bar{x} + a,\\ \delta_4 &= a\beta c\delta k\bar{x}\mathrm{sgn}\left(\bar{w}\right) - a\beta d\delta k\left|\bar{w}\right| + a\alpha d\delta k - 2ad\bar{x} + ad.\end{aligned} \tag{10.11}$$

Using the Routh–Hurwitz conditions, Eq. (10.11) has all roots with negative real parts if and only if

$$\begin{cases}\delta_i > 0, \ i = 1, 2, 3, 4\\ \delta_1\delta_2\delta_3 > \delta_3^2 + \delta_1^2\delta_4\end{cases} \tag{10.12}$$

Using the parameters values $c = 37$, $d = 12$, $\alpha = 1.3$, $\beta = 0.1$, $k = 0.05$ and by varying the parameter δ from 0.0 to 1.7 and a from 0.0 to 14.0, the stability boundaries of the equilibrium point O are shown in Fig. 10.2.

In Fig. 10.2, the conditions of stabilities (10.8) are met in gray zone while in the black zone the equilibrium point O is unstable. From Fig. 10.2, we can observe that increasing the value of the parameter k decreases the unstability

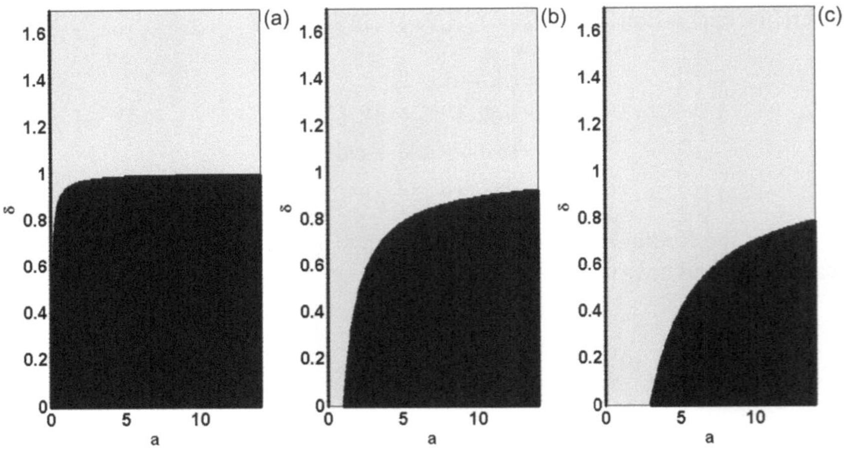

FIGURE 10.2 Stability boundaries of the equilibrium point O versus the parameters δ and a. The other parameters are $c = 37$, $d = 12$, $\alpha = 1.3$ and $\beta = 0.1$. (a): $k = 0.05$, (b): $k = 0.5$, (c): $k = 1.0$.

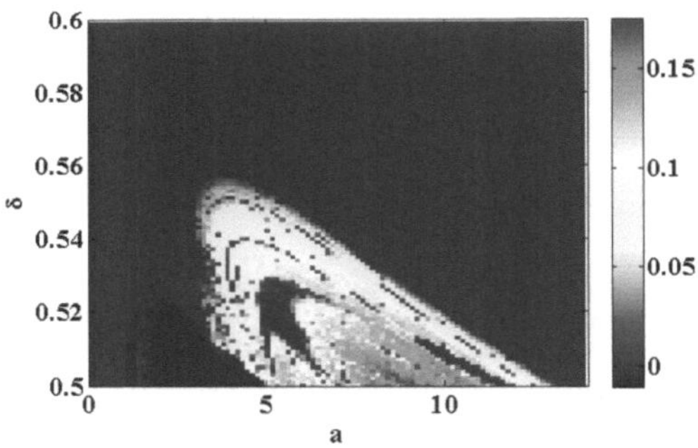

FIGURE 10.3 Two parameters LLE diagrams in the parameter space a and δ. The other parameters are $c = 37$, $d = 12$, $k = 0.05$, $\alpha = 1.3$, and $\beta = 0.1$. (For interpretation of the colors in the figure, the reader is referred to the web version of this chapter.)

zone. It is also noticed that with the same parameters and by varying in the same range the parameters δ and a the conditions of stabilities (10.12) are not met. Therefore, the equilibrium point E is unstable.

10.2.2 Dynamical analysis of memristor Helmholtz oscillator

The dynamical behavior of the proposed oscillator is investigated by plotting the bifurcation diagrams, Lyapunov exponents, phase portraits and basin of attraction plots. The two parameters (a, δ), the largest Lyapunov exponent (LLE), are illustrated in Fig. 10.3.

In Fig. 10.3, steady state behaviors are in dark blue regions, periodic oscillations are in blue regions and chaotic behaviors are located in light blue, red, green and yellow regions. For $\delta = 0.5$, the bifurcation diagram showing the lo-

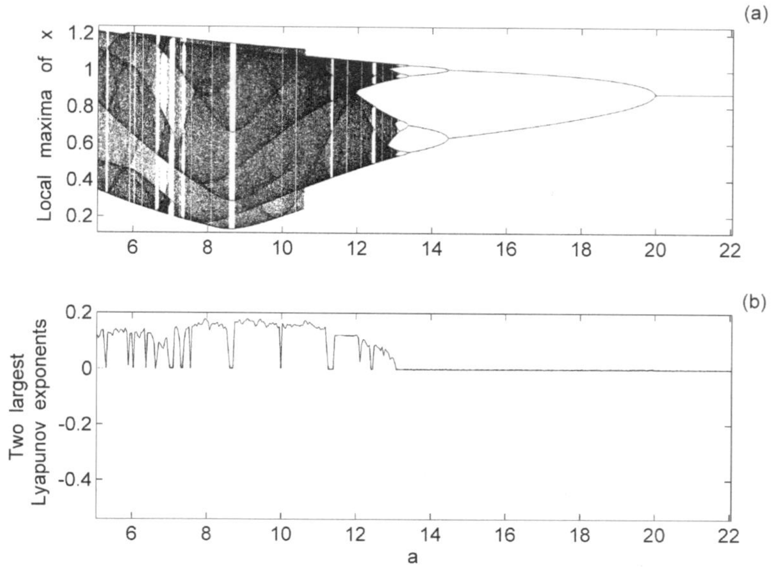

FIGURE 10.4 Bifurcation diagram depicting the maxima of $x(t)$ (a) and the corresponding two LLE (b) versus the parameter a for $\delta = 0.5$, $c = 37$, $d = 12$, $k = 0.05$, $\alpha = 1.3$ and $\beta = 0.1$.

cal maxima of state variable $x(t)$ and the two LLE versus the parameter a is plotted in Fig. 10.4.

The bifurcation diagram of the output $x(t)$ in Fig. 10.4(a) displays a chaotic region interspersed with periodic windows followed by a reverse period-doubling bifurcation to period-1-oscillations. The bifurcation diagram in Fig. 10.4(a) is confirmed by the two LLE shown in Fig. 10.4(b). For $a = 5.05$, the bifurcation diagram of $x(t)$ and the two LLE versus the parameter δ are plotted in Fig. 10.5.

The bifurcation diagram of the output $x(t)$ in Fig. 10.5(a) displays a chaotic region interspersed with periodic windows followed by a reverse period-doubling bifurcation to period-1-oscillations. By comparing the two set of data [for increasing (black) and decreasing (red) the parameter δ] used to plot Fig. 10.5(a), it is revealed that the proposed oscillator displays coexistence between period-3-oscillations and chaotic behavior in the range $0.5293 < \delta < 0.5301$. The bifurcation diagram in Fig. 10.5(a) is confirmed by the two LLE shown in Fig. 10.5 (b). The phase portraits of chaotic behaviors found in Figs. 10.4(a) and 10.5(a) are illustrated in Fig. 10.6 for specific values of δ and a.

Two different shapes of one-scroll chaotic attractors are presented in Fig. 10.6. The coexisting attractors found in Fig. 10.5 are illustrated in Fig. 10.7 which depicts the phase portrait of the resulting attractors of the proposed oscillator in the plane (x, y) for specific values of parameter δ and initial conditions.

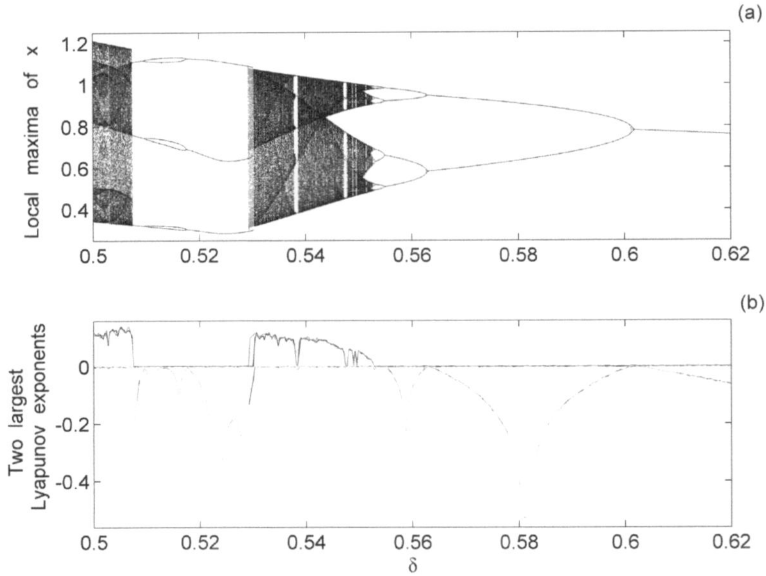

FIGURE 10.5 Bifurcation diagram depicting the maxima of $x(t)$ (a) and the two LLE (b) versus the parameter δ for $a = 5.05$, $c = 37$, $d = 12$, $k = 0.05$, $\alpha = 1.3$, and $\beta = 0.1$. Bifurcation diagrams are obtained by scanning the parameter δ upwards (black) and downwards (red). (For interpretation of the colors in the figure, the reader is referred to the web version of this chapter.)

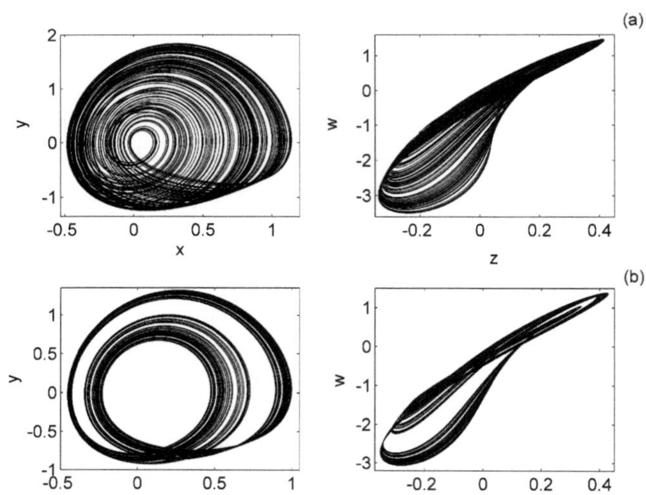

FIGURE 10.6 Phase portrait of system (10.4) in planes (x, y) and (z, w) for specific values of δ and a: (a) $\delta = 0.5$, $a = 9$ and (b) $\delta = 0.55$, $a = 5.05$ with the initial conditions $(x(0), y(0), z(0), \mathrm{w}(0)) = (0.005, 0, 0, 0)$. The other parameters are $c = 37$, $d = 12$, $k = 0.05$, $\alpha = 1.3$, and $\beta = 0.1$.

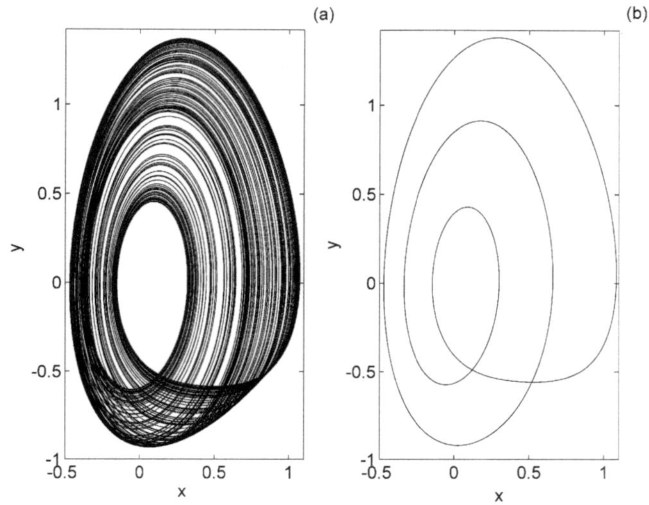

FIGURE 10.7 Coexisting attractors for $\delta = 0.53$ and specific values of initial conditions: (a) $(x(0), y(0), z(0), w(0)) = (0.005, 0, 0, 0)$ and (b) $(x(0), y(0), z(0), w(0)) = (0, 0.1, 0, 0)$. The other parameters are $a = 5.05$, $c = 37$, $d = 12$, $k = 0.05$, $\alpha = 1.3$, and $\beta = 0.1$.

At $a = 5.05$, the **MH** oscillator displays chaotic attractor and period-3 oscillations for two specific initial conditions as shown in Figs. 10.7(a) and (b), respectively. The coexistence of attractors shown in Fig. 10.7 is further detailed in Fig. 10.8, which presents the basin of attraction of the proposed oscillator in the plane $z = 0$ and $w = 0$ for $\delta = 0.53$.

From Fig. 10.8, one can see that system (10.4) can display not only periodic and chaotic behaviors but it can also lead to unbounded orbits depending on the initial conditions.

10.3 Electronic circuit simulations of the proposed memristor Helmholtz oscillator

The numerical results found in Section 10.2 are verified in this section by the electronic implementation of the proposed oscillator by using ORCAD-Pspice software. Table 10.1 presents the electronic component values of the circuits in Fig. 10.1.

The frequency-dependent pinched hysteresis loops of the ideal and active flux-controlled memristor are illustrated in Fig. 10.9 which shows its characteristic voltage–current. To plot Fig. 10.9, a sinusoidal voltage source is considered as $V_1 = V_{1m} \cos(2\pi f t)$ where V_{1m} and f are the stimulus amplitude and frequency, respectively.

For the stimulus frequency $f = 2.5$ kHz, the decrease of the stimulus amplitudes V_{1m} (from 3.0v to 2.0v passing through 2.5v) leads to a decrease of the pinched hysteresis loop as shown in Fig. 10.9(a). While for the stimulus

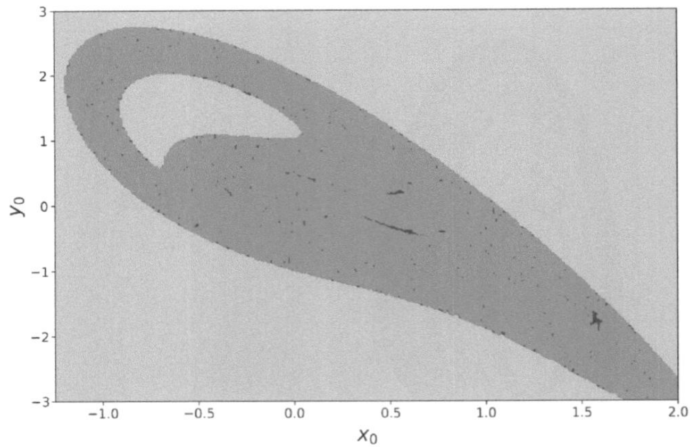

FIGURE 10.8 Cross section of the basin of attraction of system (10.4) in the plane (x, y) at $z = 0$ and $w = 0$ for $\delta = 0.53$, $a = 5.05$, $c = 37$, $d = 12$, $k = 0.05$, $\alpha = 1.3$, and $\beta = 0.1$. The values of initial conditions chosen in cyan regions lead to unbounded orbits as well as the red and blue regions represents chaotic and periodic attractors, respectively. (For interpretation of the colors in the figure, the reader is referred to the web version of this chapter.)

TABLE 10.1 Electronic components values of the circuits in Fig. 10.1.

Parameters	Signification	Values
R_α	Resistor	153.8462 kΩ
R_β	Resistor	200 kΩ
R_C	Resistor	0.720 kΩ
R_d	Resistor	0.833 kΩ
R	Resistor	10 kΩ
R_a, R_δ	Resistors	Values are given in the text
C, C_0	Capacitors	100 nF
g_1	Gain of multiplier	1.0
g_2, g_3	Gain of multiplier	0.10

amplitude $V_{1m} = 1.5$v, the increase of stimulus frequency f (from 2.0 kHz to 3.0 kHz passing through 2.5 kHz) leads to a decrease of the pinched hysteresis loop as shown in Fig. 10.9(b). The phase portraits captured in ORCAD-PSpice oscilloscope are presented in Figs. 10.10 and 10.11.

The two different shapes of one-scroll chaotic attractors obtained from the PSPICE-oscilloscope are presented in Fig. 10.10. Meanwhile in Fig. 10.11, the coexistence between one-scroll chaotic and period-3 attractors is obtained from the PSpice-oscilloscope. One can note that there is a good qualitative agreement between the PSpice results (Figs. 10.10 and 10.11) and the numerical simulations results (Figs. 10.6 and 10.7).

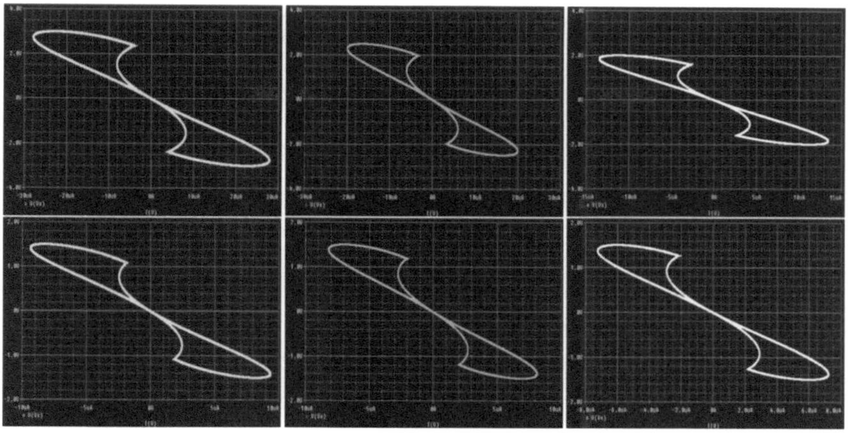

FIGURE 10.9 The characteristic voltage–current of the ideal and active flux-controlled memristor for $f = 2.5$ kHz with different stimulus amplitudes (first line) and for $V_{1m} = 1.5$ v with different stimulus frequencies (second line).

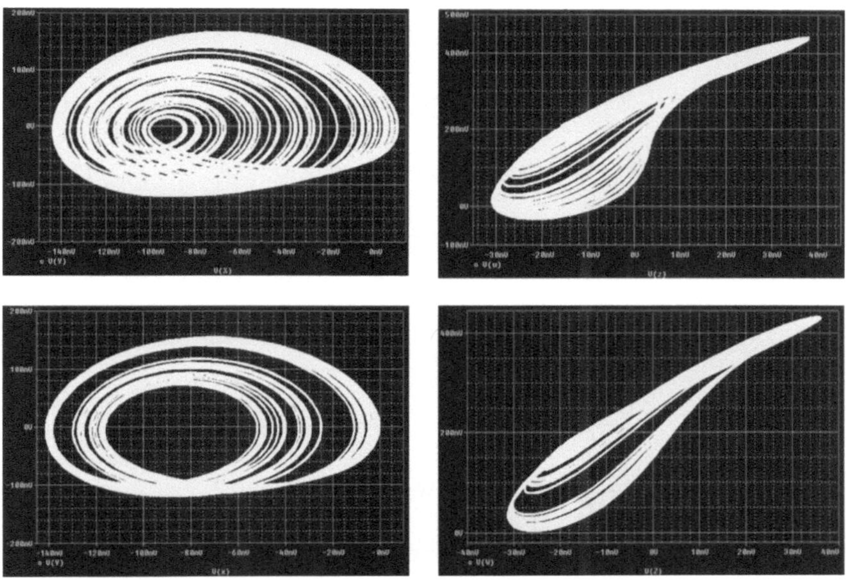

FIGURE 10.10 Chaotic attractors in the planes (V_x, V_y) and (V_z, V_w) observed on Orcard-PSPICE. The values of resistors R_a and R_δ are: (a) $R_a = 1.111$ kΩ, $R_\delta = 20.90$ kΩ and (b) $R_a = 1.111$ kΩ, $R_\delta = 20$ kΩ.

10.4 Chaos synchronization of unidirectional coupled identical chaotic memristor Helmholtz oscillators

In this section, a feedback control method is used to synchronize unidirectional coupled identical chaotic proposed oscillators. The drive and the response

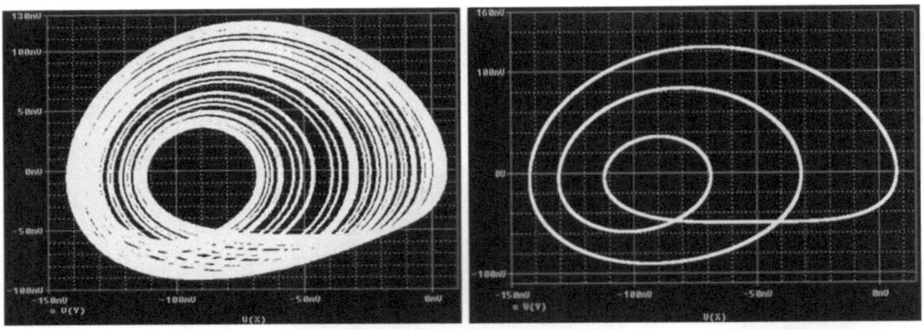

FIGURE 10.11 Coexisting attractors in the plane (V_x, V_y) observed on Orcard-PSPICE. The values of resistors R_a and R_δ are: (a) $R_\delta = 19.8$ kΩ, $R_a = 1.9801980$ kΩ and (b) $R_\delta = 20$ $k\Omega$, $R_a = 1.9801980$ kΩ.

chaotic proposed oscillators are expressed, respectively, as follows:

$$\frac{dx_1}{dt} = -y_1 - k\left(\alpha - \beta\left|\mathrm{w}_1\right|\right)x_1, \tag{10.13a}$$

$$\frac{dy_1}{dt} = az_1, \tag{10.13b}$$

$$\frac{dz_1}{dt} = -z_1 - \delta y_1 - x_1 + x_1^2, \tag{10.13c}$$

$$\frac{d\mathrm{w}_1}{dt} = -cx_1 - d\mathrm{w}_1, \tag{10.13d}$$

$$\frac{dx_2}{dt} = -y_2 - k\left(\alpha - \beta\left|\mathrm{w}_2\right|\right)x_2 + u_1, \tag{10.14a}$$

$$\frac{dy_2}{dt} = az_2 + u_2, \tag{10.14b}$$

$$\frac{dz_2}{dt} = -z_2 - \delta y_2 - x_2 + x_2^2, \tag{10.14c}$$

$$\frac{d\mathrm{w}_2}{dt} = -cx_2 - d\mathrm{w}_2, \tag{10.14d}$$

where u_1 and u_2 are the controllers to be determined in order to achieve synchronization between the systems (10.13) and (10.14). The synchronization errors are defined as $e_1 = x_2 - x_1$, $e_2 = y_2 - y_1$, $e_3 = z_2 - z_1$, $e_4 = \mathrm{w}_2 - \mathrm{w}_1$. Its derivatives are given by

$$\frac{de_1}{dt} = e_2 - k\alpha e_1 + k\beta\left(e_1\left|\mathrm{w}_2\right| + x_1 e_4\right) + u_1, \tag{10.15a}$$

$$\frac{de_2}{dt} = ae_3 + u_2, \tag{10.15b}$$

$$\frac{de_3}{dt} = -e_3 - \delta e_2 - e_1 + e_1\left(x_2 + x_1\right), \tag{10.15c}$$

$$\frac{de_4}{dt} = -ce_1 - de_4, \tag{10.15d}$$

where the controllers are chosen as $u_1 = -e_2 - k\beta(|w_2| e_1 + x_1 e_4)$ and $u_2 = -e_2 - ae_3$. By substituting the expressions of the controllers u_1 and u_2 into the system (10.15) it becomes

$$\frac{de_1}{dt} = -k\alpha e_1, \tag{10.16a}$$

$$\frac{de_2}{dt} = -e_2, \tag{10.16b}$$

$$\frac{de_3}{dt} = -e_3 - \delta e_2 - e_1 + e_1(x_2 + x_1), \tag{10.16c}$$

$$\frac{de_4}{dt} = -ce_1 - de_4. \tag{10.16d}$$

The solutions of Eqs. (10.16a) and (10.16b) are $e_1(t) = e_1(0)e^{-k\alpha t}$ and $e_2(t) = e_2(0)e^{-t}$, respectively. Therefore, one can write $\lim_{t\to\infty} e_1(t) = 0$ and $\lim_{t\to\infty} e_2(t) = 0$. Thus, the system (10.16) can be reduced as follows:

$$\frac{de_3}{dt} = -e_3, \tag{10.17a}$$

$$\frac{de_4}{dt} = -de_4. \tag{10.17b}$$

The solutions of system (10.17) are $e_3(t) = e_3(0)e^{-t}$ and $e_4(t) = e_4(0)e^{-dt}$. One can write $\lim_{t\to\infty} e_3(t) = 0$ and $\lim_{t\to\infty} e_4(t) = 0$. Therefore the following theorem is given.

Theorem 1. *The controllers* $u_1 = -e_2 - k\beta(|w_2| e_1 + x_1 e_4)$ *and* $u_2 = -e_2 - ae_3$ *can synchronize the drive and response systems* (10.13) *and* (10.14).

Proof. The proof is obvious, so we omit it. □

For numerical simulations, the initial conditions of the drive and response systems (10.13) and (10.14) are $(x_1(0), y_1(0), z_1(0), w_1(0)) = (0.005, 0, 0, 0)$ and $(x_2(0), y_2(0), z_2(0), w_2(0)) = (0.005, 0, 0, 0.001)$, respectively. The dynamics of the synchronization errors for $\delta = 0.5$ and $a = 9$ is depicted in Fig. 10.12.

In Fig. 10.12, the synchronization error variables converge to zero with exponentially asymptotical speed when the time is great than 60.0.

10.5 Chaos control of memristor Helmholtz oscillator using single controller

In this section, the chaotic behavior found in memristor Helmholtz oscillator is controlled to the equilibrium point O behavior by using a single simple con-

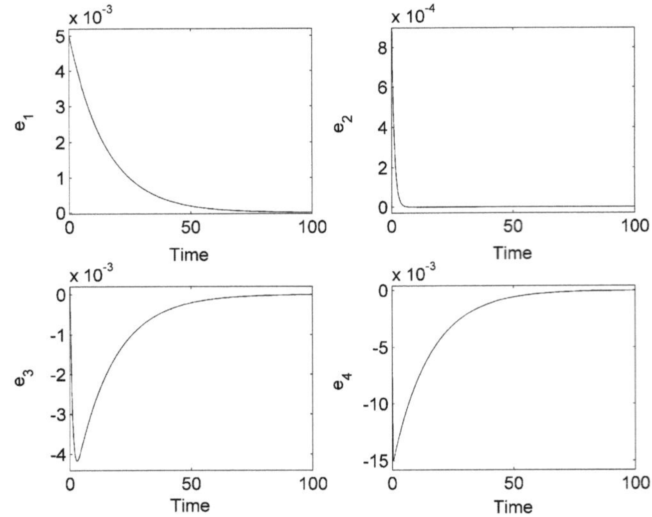

FIGURE 10.12 Time series of synchronization error variables for $\delta = 0.5$ and $a = 9$. The other parameters are $c = 37$, $d = 12$, $k = 0.05$, $\alpha = 1.3$, and $\beta = 0.1$.

troller. The simple and single controller is mathematically designed based on the principle of Lyapunov's method for asymptotic global stability [41]. The controller u_3 is added to Eq. (10.4a), the system (10.4) becomes

$$\frac{dx}{dt} = y - k(\alpha - \beta|\mathrm{w}|)x + u_3, \tag{10.18a}$$

$$\frac{dy}{dt} = az, \tag{10.18b}$$

$$\frac{dz}{dt} = -z - \delta y - x + x^2, \tag{10.18c}$$

$$\frac{d\mathrm{w}}{dt} = -cx - d\mathrm{w}, \tag{10.18d}$$

where $u_3 = -y - k\beta x|\mathrm{w}|$. By substituting the expression of the controller u_3 into the controlled system (10.18) and it becomes

$$\frac{dx}{dt} = -k\alpha x, \tag{10.19a}$$

$$\frac{dy}{dt} = az, \tag{10.19b}$$

$$\frac{dz}{dt} = -z - \delta y - x + x^2, \tag{10.19c}$$

$$\frac{d\mathrm{w}}{dt} = -cx - d\mathrm{w}. \tag{10.19d}$$

The solution of Eq. (10.19a) is $x(t) = x(0)e^{-k\alpha t}$. Therefore, one can write $\lim_{t \to \infty} x(t) = 0$. Thus, the system (10.19) can be reduced as follows:

$$\frac{dy}{dt} = az, \tag{10.20a}$$

$$\frac{dz}{dt} = -z - \delta y, \tag{10.20b}$$

$$\frac{dw}{dt} = -dw. \tag{10.20c}$$

The solution of Eq. (10.20c) is $w(t) = w(0)e^{-dt}$. That is yield $\lim_{t \to \infty} w(t) = 0$. Thus, the system (10.13) can be reduced as follows:

$$\begin{pmatrix} \frac{dy}{dt} \\ \frac{dz}{dt} \end{pmatrix} = \begin{pmatrix} 0 & a \\ -\delta & -1 \end{pmatrix} \begin{pmatrix} y \\ z \end{pmatrix} = A \begin{pmatrix} y \\ z \end{pmatrix}. \tag{10.21}$$

For the set of parameters $a = 9$ and $\delta = 0.5$, the eigenvalues of A at the equilibrium point ($y = 0, z = 0$) are $\lambda_{1,2} = -0.5 \pm 2.06155280883j$ with $j^2 = -1$. Based on the Routh–Hurwitz criteria, the subsystem (10.14) is asymptotically stable.

Theorem 2. *The chaotic behavior found in **MH** oscillator can be controlled using the controller* $u_3 = -y - k\beta x\,|w|$.

Proof. The proof is obvious, so we omit it. □

The time series of the state responses and the output of the controller are shown in Fig. 10.13.

From Fig. 10.13, the controllers u_3 are activated at $t \geq 1150$. It is noted that Theorem 2 is effective.

10.6 Conclusion

This chapter was devoted to the analysis, electronic implementation, synchronization and chaos control of the MH oscillator. The equilibrium points of the mathematical model describing the proposed oscillator were determined and their stabilities were analyzed referring to the Routh–Hurwitz criteria. For specific parameters, the proposed oscillator exhibited two different shapes of one-scroll chaotic attractors and coexistence between period-3 limit cycle and one-scroll chaotic attractor. The electronic implementation of the proposed oscillator was performed on ORCAD-PSpice software. Orcard-PSpice results confirmed the numerical simulations results. Moreover, the synchronization of unidirectional coupled identical chaotic memristor Helmholtz oscillators was achieved using simple controllers. Finally, based on the Routh–Hurwitz criteria a proposed simple and single controller was designed to suppress the chaotic behavior

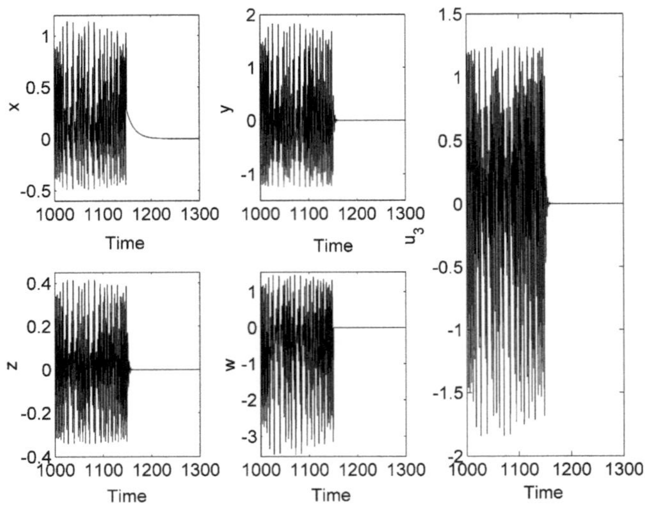

FIGURE 10.13 Time series of x, y, z, w and the output of the controller u_3 for $\delta = 0.5$ and $a = 9$. The initial conditions are $(x(0), y(0), z(0), \mathrm{w}(0)) = (0.005, 0, 0, 0)$. The other parameters are $c = 37$, $d = 12$, $k = 0.05$, $\alpha = 1.3$, and $\beta = 0.1$.

found in MH oscillator. Future work will concern the embedded cryptographic applications using chaotic memristor Helmholtz oscillators and investigating the possibility of using memristors as synapses for neuron miming. Also one might consider memristors based on Helmholtz oscillators for neuromorphic implementations.

References

[1] L. Chua, IEEE Trans. Circuit Theory 18 (1971) 507–519.
[2] S.N. Truong, K.S. Min, J. Semicond. Technol. Sci. (2014).
[3] D.B. Strukov, G.S. Snider, D.R. Stewart, R.S. Williams, Nature 453 (2008) 80–83.
[4] A.A. Zakhidov, B. Jung, J.D. Slinker, H.D. Abruña, G.G. Malliaras, Org. Electron. (2010).
[5] G. Martinsen, S. Grimnes, C.A. Lütken, G.K. Johnsen, J. Phys. Conf. Ser. (2010).
[6] I. Valov, E. Linn, S. Tappertzhofen, S. Schmelzer, J. Van Den Hurk, F. Lentz, R. Waser, Nat. Commun. (2013).
[7] O. Sporns, G. Tononi, R. Kötter, PLoS Comput. Biol. 1 (2005) 0245.
[8] X.Y. Wang, A.L. Fitch, H.H.C. Iu, W.G. Qi, Phys. Lett. Sect. A Gen. At. Solid State Phys. (2012).
[9] S. Shin, K. Kim, S.M. Kang, IEEE Trans. Nanotechnol. 10 (2011) 266–274.
[10] M. Di Ventra, Y.V. Pershin, L.O. Chua, Proc. IEEE (2009).
[11] C. Wang, H. Lin, J. Sun, L. Zhou, C. Zhou, Q. Deng, Dianzi Yu Xinxi Xuebao/J. Electron. Inf. Technol. 42 (2020) 795–810.
[12] I. Petráš, IEEE Trans. Circuits Syst. II, Express Briefs (2010).
[13] B.C. Bao, J.P. Xu, Z. Liu, Chin. Phys. Lett. (2010).
[14] V.A. Botta Pirani, C. Néspoli, M. Messias, TEMA – Tendências Em Matemática Apl. e Comput. (2011).
[15] L. Teng, H.H.C. Iu, X. Wang, X. Wang, Nonlinear Dyn. (2014).

[16] S. Sabarathinam, C.K. Volos, K. Thamilmaran, Nonlinear Dyn. (2017).
[17] Y. Lu, X. Huang, S. He, D. Wang, B. Zhang, Int. J. Bifurc. Chaos (2014).
[18] A.M.A. El-Sayed, A. Elsaid, H.M. Nour, A. Elsonbaty, Commun. Nonlinear Sci. Numer. Simul. (2013).
[19] H. Bao, N. Wang, B. Bao, M. Chen, P. Jin, G. Wang, Commun. Nonlinear Sci. Numer. Simul. (2018).
[20] K. Rajagopal, J.R. Mboupda Pone, S.T. Kingni, S. Arun, A. Karthikeyan, Eur. Phys. J. Spec. Top. (2019).
[21] B. Muthuswamy, P. Kokate, IETE Tech. Rev. (Institution Electron. Telecommun. Eng. India) (2009).
[22] G.Y. Wang, J.L. He, F. Yuan, C.J. Peng, Chin. Phys. Lett. (2013).
[23] M. Itoh, L.O. Chua, Int. J. Bifurc. Chaos (2008).
[24] B. Muthuswamy, Int. J. Bifurc. Chaos (2010).
[25] H.H.C. Iu, D.S. Yu, A.L. Fitch, V. Sreeram, H. Chen, IEEE Trans. Circuits Syst. I, Regul. Pap. (2011).
[26] N. Henry Alombah, H. Fotsin, E.B. Megam Ngouonkadi, T. Nguazon, Int. J. Bifurc. Chaos (2016).
[27] B.C. Bao, J.P. Xu, G.H. Zhou, Z.H. Ma, L. Zou, Chin. Phys. B (2011).
[28] F. Yuan, G. Wang, X. Wang, Chaos (2016).
[29] B. Bao, J. Yu, F. Hu, Z. Liu, Int. J. Bifurc. Chaos (2014).
[30] Y. Babacan, A. Yesil, F. Kacar, AEU – Int. J. Electron. Commun. (2017).
[31] U.E. Ayten, S. Minaei, M. Sağbaş, AEU – Int. J. Electron. Commun. (2017).
[32] B.C. Bao, F.W. Hu, Z. Liu, J.P. Xu, Chin. Phys. B (2014).
[33] C. Ainamon, S.T. Kingni, V.K. Tamba, J.B.C. Orou, P. Woafo, J. Control Autom. Electr. Syst. (2019).
[34] E. Del Río, A. Rodríguez-Lozano, M.G. Velarde, Rev. Sci. Instrum. (1992).
[35] J.A. Gottwald, L.N. Virgin, E.H. Dowell, J. Sound Vib. (1995).
[36] I.S. Kang, L.G. Leal, J. Fluid Mech. 218 (1990) 41.
[37] K.J. Spyrou, B. Cotton, B. Gurd, J. Sh. Res. (2002).
[38] J.M.T. Thompson, Chaotic Phenomena Triggering the Escape from a Potential Well, 1991.
[39] J.M.T. Thompson, Appl. Mech. Rev. (1997).
[40] W.H. Press, S.A. Teukolsky, W.T. Vetterling, B.P. Flannery, Numerical Recipes in Fortran 90, 2010.
[41] X. Liao, P. Yu, Absolute Stability of Nonlinear Control Systems, Springer, Netherlands, Dordrecht, 2008.

Chapter 11

Design guidelines for physical implementation of fractional-order integrators and its application in memristive systems

Ciro Fabian Bermúdez-Márquez and Jesus Manuel Muñoz-Pacheco
Faculty of Electronics Sciences, Meritorious Autonomous University of Puebla, Puebla, Mexico

11.1 Introduction

Memristor and fractional-order derivatives can be feasible options for constructing new systems with complex dynamics. In 1971 the memristor was postulated theoretically by Chua as the fourth element in electronic circuits [1]. The first practical implementation was delivered until 2008 by HP Labs [2]. Due to the non-volatility, re-programmability, and analog storage properties, memristors have several applications in different areas such as neural networks, secure communications, neuromorphic computation, memristor-based chaotic oscillators [3–8]. As is well known, the memristor has been addressed as an elemental unit for developing the most promising applications such as low-energy, high-density memories, and neuromorphic hardware [10]. Regarding memristor-based artificial neural networks it promises high-impact artificial intelligence applications, i.e., low-power, brain-like computing, image processing, filtering, and classification, and so forth. These relatively new neuromorphic circuits arise and are strongly encouraged by big companies: Intel, IBM, and Google [10]. One of the first applications of memristors was the non-volatile memory since the fundamental feature of memristors is that they "remember" data without need of power supplies as a resistive state [11]. Memristors would improve the reconfigurable digital circuits, for instance, field-programmable gate array (FPGA) technology, and generic logic circuits to perform digital computation [12]. Lastly, memristors are applied for neuromorphic engineering and neuromorphic hardware, which encompass the design of microelectronic circuits that emulate biological neural networks, i.e., the neurons and their interactions as they occur in the brain or the nervous system [13–15]. Along this line, much

Mem-elements for Neuromorphic Circuits with Artificial Intelligence Applications
https://doi.org/10.1016/B978-0-12-821184-7.00019-0

interesting work has been reported. An et al. proposed a novel neuromorphic computing based cloud robotics (NC-robotics) system. In order to implement this NC-robotics system, neurons and memristor-based (RRAM) synaptic array are stacked vertically, and its concise equivalent circuit model is created, evaluated, and analyzed through SPICE simulations [16]. Also, An et al. reported a novel 3D neuromorphic IC architecture that combines monolithic 3D integration and a synaptic array based on vertical resistive random-access memory structure (V-RRAM), also known as memristor memory [17]. It was determined that applying the monolithic 3D integration technology on neuromorphic computing system hardware implementation can reduce the power consumption by 50%. Filippov et al. developed an original biomorphic neuron, which differs from conventional IT models by a more complex neuron structure and biological models by accounting mean fire rates of neurons [18]. They evaluated experimentally memristor-based synaptic connections and found the memristor conductance changes when the process of the neural link strengthening. Liu et al. proposed a memristor-based design of bidirectional transmission excitation/inhibition synapses and implemented a neuromorphic computing system based on their proposed synapse designs [19]. Also, they evaluated the robustness of the approach by considering the process variation at the physical level. Getachew et al. develop a SPICE modeling for a nanoscale memristor device using modified piece-wise linear (PWL) functions. In this manner, a binary-weighted memristor-based artificial synaptic circuit can be designed [20]. Yan et al. studied a new training approach that enables prevailing neural network training techniques to be applied for memristor-based neuromorphic networks. Their simulation results validate the accuracy and fast speed of the proposed circuit [21]. Dias et al. reviewed the neuromorphic properties of memristors, comparing them with the key players of neuronal computations, synapses, and neurons. Next, their findings were extended to multiple computing units that can be appended in networks to get more complex dynamics [22]. Slavova et al. analyzed the hysteresis cellular nonlinear network (HCNN) with memristor synapses. They determined the edge of the chaos regime for the HCNN model based on local activity theory [23]. Li et al. studied a small Hopfield neural network with a memristive synaptic weight. By using a memristor as one weight, the previous stable network evolves to rich, complex dynamics, such as quasi-periodic orbits, chaos, and hyperchaos [24]. Wang et al. showed the synchronization of two coupled hyperbolic-type Hopfield neural networks with a memristive synaptic connection. Two different cases of single and multiple time-delays were analyzed, where the synchronization error using two time-delays reduced [25]. In [26], the chaotic dynamics in a three-neuron-based Hopfield neural network (HNN) with stimulation of electromagnetic radiation was investigated. In particular, a flux-controlled memristor was incorporated into the neural network to model electromagnetic field effects. Infinitely many hidden attractors were found when the initial conditions were varied. Zhang et al. proposed a memristive autapse involving magnetic coupling and found that the memristive system is expected to be used to mimic biological synapses for advances in neuromorphic computing [27]. Chaotic dynamics in nanoscale NbO 2 Mott memristors for analog computing were investigated by Kumar et al. [28]. They recognized that a controllable chaotic behavior might be critical to a neural-inspired circuit. The experimental realization of chaotic dynamics was also shown. Neural networks have been analyzed from different perspectives. In [29] proposed a simplified memristor-based fractional-order neural network using a discontinuous memductance function, whereas in Ref. [30] proposes a new cryptographic algorithm using chaotic neural networks based on Hermite and Chebyshev polynomials to incorporate a memristor conductance.

However, the memristor, similar to the resistor, inductor, and capacitor, is a real electrical element [2]. As a result, the nonlinear constitutive relation between the device voltage and current has been more effectively explained using the arbitrary-order version of its mathematical model [9,32]. In this manner, the fractional calculus theory can help us to describe the memory (storage) of memristive systems due to the analytical definitions of the fractional-order derivative are based on the hereditary and memory properties of their kernels. Therefore, the link between memristors and fractional order is straightforward. The conceptual framework of the *fracmemristor*, which joins the concepts underlying the fractional-order circuit elements and the memristor, has been recently introduced [31–39].

In that scenario, one of the remaining goals is implementing the fractional-order derivative physically according to the required design specifications. The main idea of this calculus is the proper exploitation of the fractional Laplacian operator s^{α}. From an electronics circuit point of view, the transfer function of such circuits is given by

$$H(s) = (\tau s)^{j}, \tag{11.1}$$

where $j = \pm\alpha$, with $0 < \alpha < 1$ being the order of differentiator/integrator, respectively. When $j = -\alpha$ represents a fractional-order integrator and τ means its time-constant. However, as fractional-order calculus is a new field of interest, there are no commercially available devices with these characteristics. Therefore, we can find in literature several works on how to obtain integer-order approximations for fractional-order integrators. Among these approximations, the most frequently used are the continued fraction expansions (CFEs) and interpolation techniques such as the rational approximation method [40] and Carlson method [41]. Other techniques use curve fitting or identification formulas such as the Oustaloup approximation method [42]. Finally, an approach that combines CFE and fitting techniques is the Matsuda approximation [43].

Once the approximation is generated, the next step consists of synthesizing the transfer function with electronic circuits. In this regard, there are two main paths; the former is based on full passive RC networks. That is the easiest way for implementing emulators of these elements, but this solution suffers from a simple adjustment to modify the required fractional order. The latter uses active topologies where distinct amplifiers (OpAmps, OTAs, CFOAs, CCIIs, etc.) have been explored as proper solutions. Among those active-device-based topologies, a less studied approach is the programmable analog non-embedded hardware, such as the Field Programmable Analog Array (FPAA). As well know an FPAA is an integrated circuit device containing computational analog blocks and interconnects between these blocks offering field-programmability. Indeed, the active topologies could be a suitable approach to designing programmable fractional-order integrators, but the FPAAs with discrete-time operation mode requires an appropriate master clock frequency.

In this chapter, the CFE approach is applied to get the approximated integer-order transfer function of $(s)^{-1}$. Then, FPAAs are used to implement the fractional-order integrators. The contributions of this work rely on proposing a metric performance defined as the rate of the desired error regarding the fractional order of the integer-order rational approximation. Additionally, a figure of merit (FOM) relating the fractional order with the FPAA clock frequency is introduced.

11.2 Fractional-order calculus preliminaries

On the other hand, the fractional calculus is a mathematical discipline concerned with studying differential and integral operators of real or even complex orders. Moreover, the integer-order operators are a small subspace of the available fractional orders. In recent years, the fractional calculus has received much attention due to the fractional derivatives provide more accurate models than the integer-order counterpart. Many examples have been found in different interdisciplinary fields [44], ranging from a description of anomalous viscoelastic diffusion in complex liquids, D-decomposition technique for control problems, chaotic systems, to macroeconomic models with dynamic memory, and forecast of the trend of complex systems [45–53]. This work has demonstrated that fractional derivatives provide an excellent approach for describing the memory and hereditary properties of real physical phenomena.

The three most frequently used definitions for the general fractional-order operator are Riemann–Liouville (RL), Grünwald–Letnikov (GL), and Caputo (C) [49]. From the point of view of applied science, the Grünwald–Letnikov fractional derivatives seem to be a natural way for generalizing the notion of derivative. They have a substantial meaning for any real or complex order, and we recover the classic results for integer orders [54].

11.2.1 Grünwald–Letnikov definition

The general formula for n-derivative of the function f for $n \in \mathbb{N}$ and $j > n$ is given by

$$f^{(n)}(t) = \frac{d^n f}{dt^n} = \lim_{h \to 0} \frac{1}{h^n} \sum_{j=0}^{n} (-1)^j \binom{n}{j} f(t - jh) \tag{11.2}$$

where $\binom{n}{j}$ are the binomial coefficients defined as

$$\binom{n}{j} = \frac{n!}{j!(n-j)!} \tag{11.3}$$

In the case of negative value of n we have

$$\binom{-n}{j} = \frac{-n(-n-1)(-n-2)\cdots(-n-j+1)}{j!} = (-1)^j \begin{bmatrix} n \\ j \end{bmatrix} \tag{11.4}$$

where $\begin{bmatrix} n \\ j \end{bmatrix}$ is defined as

$$\begin{bmatrix} n \\ j \end{bmatrix} = \frac{2(n+1)\cdots(n+j-1)}{j!} \tag{11.5}$$

Generalizing (11.2) we can write the fractional-order derivative definition of order α, ($\alpha \in \mathbb{R}$) as follows:

$$D_t^{\alpha} f(t) = \lim_{h \to 0} \frac{1}{h^{\alpha}} \sum_{j=0}^{\infty} (-1)^j \binom{\alpha}{j} f(t - jh) \tag{11.6}$$

For binomial coefficients calculation, we can use the relation between Euler's Gamma function and the factorial, defined as

$$\binom{\alpha}{j} = \frac{\alpha!}{j!(\alpha - j)!} = \frac{\Gamma(\alpha+1)}{\Gamma(j+1)\Gamma(\alpha - j + 1)} \tag{11.7}$$

where Euler's Gamma function with $r > 0$ is $\Gamma(r) = \int_0^{\infty} t^{r-1} e^{-t} dt$.

11.3 Fractional-order memristive systems

A memristor represents the relation between the integral of the current $I(t)$ and the integral of the voltage $V(t)$. By the duality principle, one can define the memristor device as charge-controlled $\phi(t) = h(q(t))$ or flux-controlled $q(t) = f(\phi(t))$. The previous concepts are not necessarily limited to resistance (or conductance) but can be generalized not only to capacitative and inductive systems but also to the fractional-order domain defining a general class of αth-order u-controlled dynamical systems called fractional-order memristive systems [31–33], given by

$$\begin{aligned} D^{\alpha} w(t) &= F(w, u, t), \\ v(t) &= G(w, u, t)u(t), \end{aligned} \tag{11.8}$$

where u, v are the input and output respectively, w is the internal state of the fractional-order memristive system, F is a function continuously differentiable,

G is a function continuous scalar, and D^q denotes the fractional-order derivative operator in the sense of Grünwald–Letnikov (11.6).

From a physical point of view, a solution of the fractional-order differential equation (11.8) in the frequency domain is obtained using fractional-order integrator circuits $1/s^{\alpha}$. As already mentioned in the introduction, since there are no fractional-order capacitors, integer-order transfer functions are proposed as an approximation of the fractional order. In this chapter, we use the continued fraction expansion to get first- and second-order approximations.

11.4 Continued fraction expansion (CFE)

An expression of the form

$$a_1 + \cfrac{b_1}{a_2 + \cfrac{b_2}{a_3 + \cfrac{b_3}{a_4 + \ddots}}} \tag{11.9}$$

is called a continued fraction. In general, the numbers $a_1, a_2, a_3, \cdots, b_1, b_2, b_3$ may be any real or complex numbers, and the number of terms may be finite or infinite. We can also write (11.9) as

$$a_1 + \frac{b_1}{a_2 +}\,\frac{b_2}{a_3 +}\,\frac{b_3}{a_4 + \cdots} \tag{11.10}$$

Both notations are very similar and can be switched. From (11.10) we can generate the fractions

$$c_1 = \frac{a_1}{1}, \quad c_2 = a_1 + \frac{b_1}{a_2}, \quad c_3 = a_1 + \frac{b_1}{a_2 +}\,\frac{b_2}{a_3}, \quad \cdots \tag{11.11}$$

obtained by truncating the expansion to the first, second, third $\cdots$ steps, respectively. These fractions are also called the first, second, third, $\cdots$ convergents of the continued fraction (11.10). In this manner, the nth convergent can be given by

$$c_n = a_1 + \frac{b_1}{a_2 +}\,\frac{b_2}{a_3 +} \cdots \frac{b_{n-1}}{+ \ a_n} \tag{11.12}$$

In 1776 Lagrange obtained the continued fraction expansion (CFE) for $(1+x)^{\alpha}$ as follows

$$(1+x)^{\alpha} = \cfrac{1}{1 - \cfrac{\alpha x}{1 + \cfrac{\frac{1(1+\alpha)}{1\cdot 2}x}{1 + \cfrac{\frac{1(1-\alpha)}{2\cdot 3}x}{1 + \cfrac{\frac{2(2+\alpha)}{3\cdot 4}x}{1 + \cfrac{\frac{2(2-k)}{4\cdot 5}x}{1 + \cfrac{\frac{3(3+\alpha)}{5\cdot 6}x}{1 + \ddots}}}}}}} \tag{11.13}$$

and written in a more compact way

$$(1+x)^{\alpha} = \frac{1}{1-}\,\frac{\alpha x}{1+}\,\frac{\frac{1(1+\alpha)}{1\cdot 2}x}{1\quad +}\,\frac{\frac{1(1-\alpha)}{2\cdot 3}x}{1\quad +}\,\frac{\frac{2(2+\alpha)}{3\cdot 4}x}{1\quad +}\,\frac{\frac{2(2-\alpha)}{4\cdot 5}x}{1\quad +}\cdots \tag{11.14}$$

Eq. (11.14) can be conveniently recast by multiplying m in the numerator and denominator as shown below

$$(1+x)^{\alpha} = \frac{1}{1-}\,\frac{\alpha x}{1+}\,\frac{2\cdot\frac{1(1+\alpha)}{1\cdot 2}x}{2\cdot 1\quad +}\,\frac{3\cdot 2\cdot\frac{1(1-\alpha)}{2\cdot 3}x}{3\cdot 1\quad +}\,\frac{3\cdot\frac{2(2+\alpha)}{3\cdot 4}x}{1\quad +}\cdots \tag{11.15}$$

It is important to notice that the denominator is composed of 2 terms, this can be clearly seen in (11.13). In order to maintain the similarity, it is necessary to multiply m in three different places. In this case $m_1 = 2$, $m_2 = 3$, $m_3 = 2$, ..., thus

$$(1+x)^{\alpha} = \frac{1}{1-}\,\frac{\alpha x}{1+}\,\frac{(1+\alpha)x}{2\quad +}\,\frac{(1-\alpha)x}{3\quad +}\,\frac{(2+\alpha)x}{2\quad +}\,\frac{(2-\alpha)x}{5\quad +}\cdots \tag{11.16}$$

Eq. (11.14) is the one most appropriate to compute the nth term applying the following formula:

$$\frac{\psi(n)\,[\psi(n) + (-1)^n \alpha]}{(n-1)n} \tag{11.17}$$

TABLE 11.1 Rational approximations for $\frac{1}{s^{0.5}}$.

Order	No. of terms	Rational approximation
1	2	$\frac{s+3}{3s+1}$
2	4	$\frac{s^2+10s+5}{5s^2+10s+1}$
3	6	$\frac{s^3+21s^2+35s+7}{7s^3+35s^2+21s+1}$
4	8	$\frac{s^4+36s^3+126s^2+84s+9}{9s^4+84s^3+126s^2+36s+1}$
5	10	$\frac{s^5+55s^4+330s^3+462s^2+165s+11}{11s^5+165s^4+462s^3+330s^2+55s+1}$

where the function $\psi(x)$ with $x \geq 2$, $x \in \mathbb{Z}^+$ is defined as[1]: $\psi(x) = \lfloor \frac{x}{2} \rfloor$, while Eq. (11.17) can be used recursively from the nth term to the second term being the first term $\frac{1}{1-}\frac{\alpha x}{1}$.

By replacing $x = s - 1$, we can get the rational approximation for s^α or $\frac{1}{s^\alpha}$. In this manner, the CFE based-approximation for a fractional integrator $\frac{1}{s^\alpha}$ using the first two terms results in a first order transfer function as shown by

$$\frac{1}{s^\alpha}_{(c_2)} \approx \frac{(1-\alpha)s+(1+\alpha)}{(1+\alpha)s+(1-\alpha)}, \tag{11.18}$$

whereas the second order approximation has the form of

$$\frac{1}{s^\alpha}_{(c_4)} \approx \frac{(\alpha^2-3\alpha+2)s^2+(8-2\alpha^2)s+(\alpha^2+3\alpha+2)}{(\alpha^2+3\alpha+2)s^2+(8-2\alpha^2)s+(\alpha^2-3\alpha+2)}. \tag{11.19}$$

One of the advantages of using the CFE approximation is that we systematically convert the problem of fractional order to an integer order problem. For instance, for a fractional-order integrator with $\alpha = 0.5$, the integer-order approximations are shown in Table 11.1, and their corresponding Bode plots are given in Fig. 11.1. By analyzing Fig. 11.1, we observe whether the order of the approximation increases, the bandwidth (10^{-1} rad/s up to 10^1 rad/s) also increases, and the error compared to the ideal integrator decreases. A detailed error analysis is conducted in the next section.

---

1 $\lfloor x \rfloor$ is the round down to the nearest integer function.

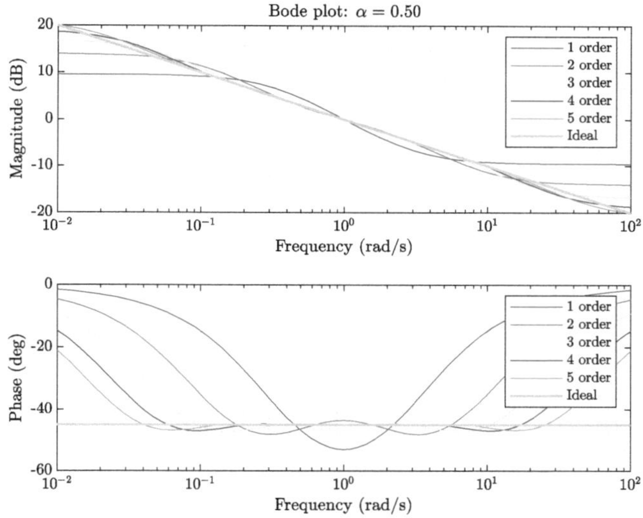

FIGURE 11.1 Bode plots of the fractional integrator $\alpha = 0.5$ approximated with integer-order transfer functions from first to fifth order. (For interpretation of the colors in the figure, the reader is referred to the web version of this chapter.)

11.4.1 CFE error analysis

In CFE approach, we can notice that the smaller α the larger the approximation error compared to an ideal fractional integrator. This is considering a two-decades bandwidth To quantify the error, we define the unnormalized error and the normalized error. On the one side, the unnormalized error of the magnitude in dB can be calculated using

$$\text{error}_{\text{dB}} = 20\log_{10}\left|\frac{H(j\omega)}{G(j\omega)}\right| \tag{11.20}$$

where $H(j\omega)$ is the ideal fractional-order integrator $\frac{1}{s^\alpha}$ and $G(j\omega)$ is the approximate transfer function of the integrator using CFE $_{(c_n)}\frac{1}{s^\alpha}$. The unnormalized error of the phase in degrees is obtained by

$$\text{error}_{\text{deg}} = \angle H(j\omega) - \angle G(j\omega) \tag{11.21}$$

Because an ideal fractional integrator is essentially a pole at the origin power of α, it has a behavior of -20α dB/decade in magnitude and -90α degrees in phase. To normalize the error, we consider that the magnitude of the ideal integrator at 0.1 rad/s and 10 rad/s for any α is 20α dB and -20α dB, respectively. Similarly, the phase remains constant at -90α degrees for any α. Therefore, the

TABLE 11.2 Maximum absolute magnitude error in dB varying α and transfer function order (unnormalized errors).

α/Order	1st	2nd	3rd	4th	5th
0.1	0.4587	0.3333	0.2427	0.0620	0.0220
0.2	0.8931	0.6524	0.4665	0.1172	0.0423
0.3	1.2792	0.9426	0.6527	0.1596	0.0591
0.4	1.5942	1.1889	0.7838	0.1848	0.0710
0.5	1.8155	1.3748	0.8449	0.1904	0.0767
0.6	1.9193	1.4807	0.8259	0.1767	0.0755
0.7	1.8769	1.4676	0.7232	0.1458	0.0669
0.8	1.6464	1.2410	0.5415	0.1021	0.0510
0.9	1.1460	0.7507	0.2939	0.0513	0.0284

TABLE 11.3 Maximum absolute phase error in degrees varying α and transfer function order (unnormalized errors).

α/Order	1st	2nd	3rd	4th	5th
0.1	6.7092	2.8727	0.6548	0.5632	0.2796
0.2	13.2833	5.5250	1.2598	1.0838	0.5338
0.3	19.5614	7.7297	1.7677	1.5208	0.7390
0.4	25.3200	9.2514	2.1362	1.8358	0.8760
0.5	30.2099	9.8632	2.3294	1.9960	0.9311
0.6	33.6307	9.3936	2.3198	1.9760	0.8975
0.7	34.4722	7.8155	2.0880	1.7611	0.7756
0.8	30.6494	5.3543	1.6237	1.3492	0.5739
0.9	19.0601	2.5243	0.9255	0.7528	0.3080

normalized error for both magnitude and phase is

$$\text{error}_{\text{norm mag}} = \frac{20\log_{10}\left|\frac{H(j\omega)}{G(j\omega)}\right|}{20\alpha} = \frac{\log_{10}\left|\frac{H(j\omega)}{G(j\omega)}\right|}{\alpha} \tag{11.22}$$

$$\text{error}_{\text{norm phase}} = \frac{\angle H(j\omega) - \angle G(j\omega)}{90\alpha} \tag{11.23}$$

Tables 11.2 and 11.3 show the maximum unnormalized absolute errors for the magnitude and phase regarding to α and the order of the transfer function. However, Table 11.2 and Table 11.3 only give us information about the total error but not of its percentage. Therefore, the normalized errors (1.22) and (1.23) are used instead. Please see Fig. 11.2 and Fig. 11.3 for a detailed set of simulation results.

Tables 11.4 and 11.5 show the percentages of the normalized maximum absolute error for the magnitude and phase, respectively. We found that **the higher the error rate the smaller the fractional order α.**

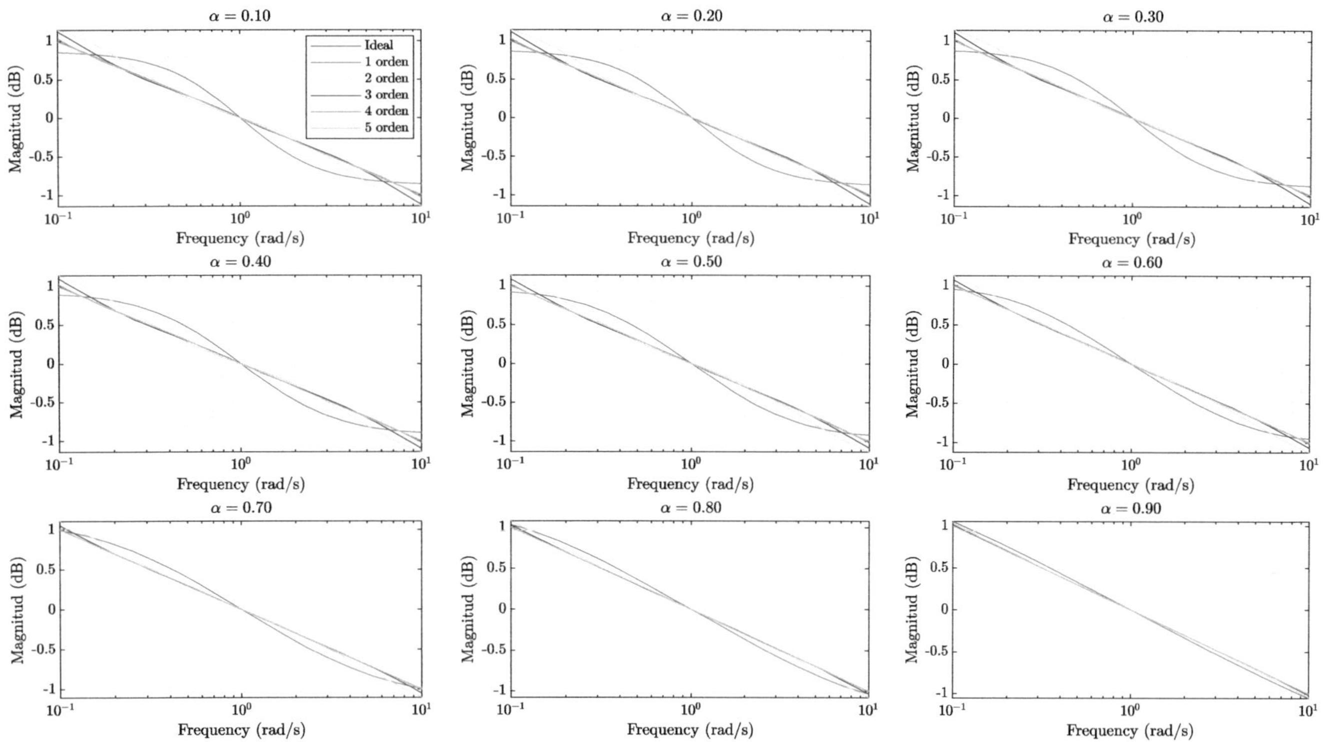

FIGURE 11.2 Normalize magnitude bode plots of fractional integrator approximations. (For interpretation of the colors in the figure, the reader is referred to the web version of this chapter.)

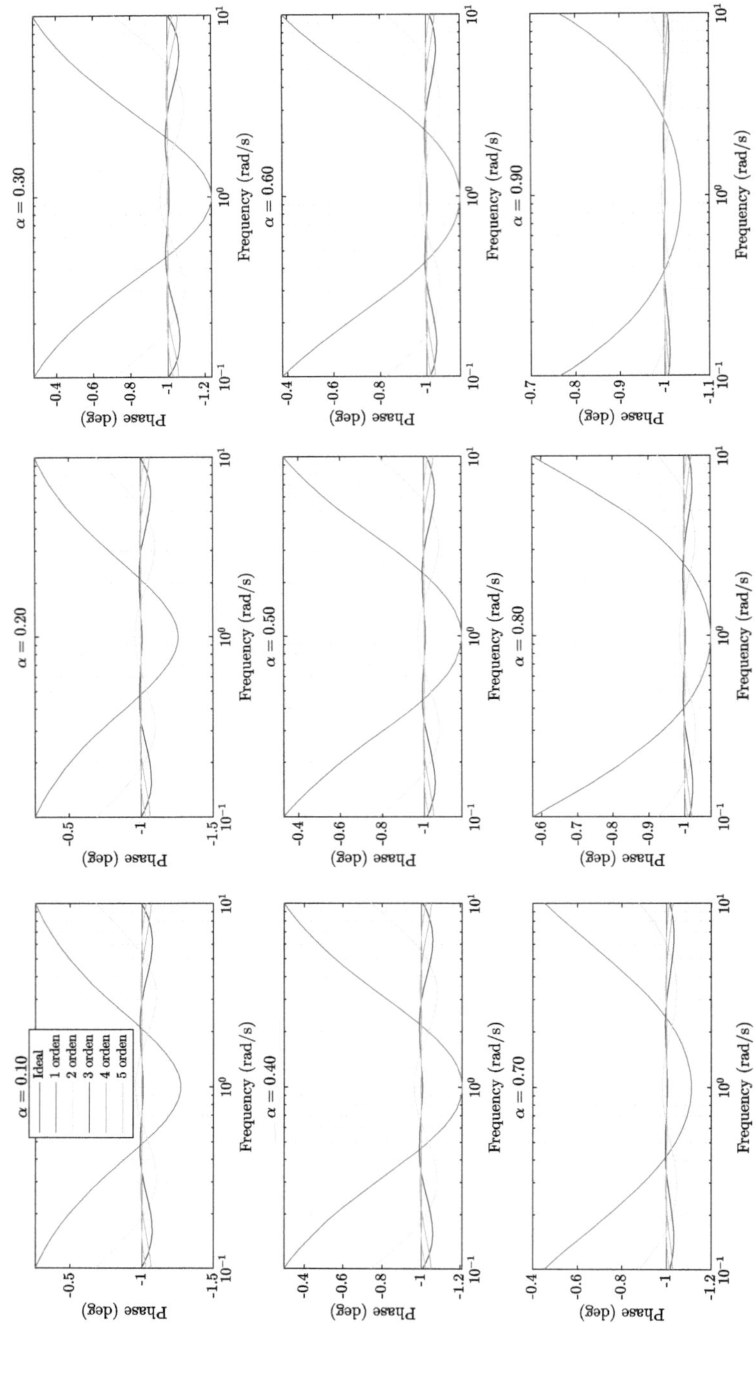

FIGURE 11.3 Normalize phase bode plots of fractional integrator approximations. (For interpretation of the colors in the figure, the reader is referred to the web version of this chapter.)

TABLE 11.4 Maximum absolute magnitude error in % varying α and transfer function order (normalized error).

α/Order	1st	2nd	3rd	4th	5th
0.1	22.9364	16.6661	12.1365	3.1006	1.1022
0.2	22.3267	16.3092	11.6624	2.9299	1.0576
0.3	21.3204	15.7107	10.8780	2.6595	0.9850
0.4	19.9281	14.8618	9.7974	2.3094	0.8869
0.5	18.1555	13.7481	8.4493	1.9044	0.7669
0.6	15.9941	12.3390	6.8823	1.4724	0.6290
0.7	13.4063	10.4826	5.1656	1.0415	0.4779
0.8	10.2900	7.7565	3.3847	0.6380	0.3190
0.9	6.3665	4.1706	1.6328	0.2848	0.1578

TABLE 11.5 Maximum absolute normalized phase error in % varying α and transfer function order (normalized error).

α/Order	1st	2nd	3rd	4th	5th
0.1	74.5462	31.9184	7.2756	6.2574	3.1070
0.2	73.7962	30.6945	6.9988	6.0212	2.9653
0.3	72.4496	28.6284	6.5472	5.6324	2.7370
0.4	70.3334	25.6983	5.9338	5.0995	2.4334
0.5	67.1331	21.9182	5.1765	4.4355	2.0692
0.6	62.2790	17.3955	4.2958	3.6593	1.6620
0.7	54.7178	12.4056	3.3143	2.7954	1.2312
0.8	42.5686	7.4365	2.2552	1.8739	0.7971
0.9	23.5310	3.1165	1.1426	0.9293	0.3802

Since the maximum error occurs in a very short frequency range, we also analyze the average error. Table 11.6 and Table 11.7 give the normalized average error. This type of error has the same behavior as the maximum normalized error. We observe the error decreases by 5.86% for $\alpha = 0.1$ and a 2.96% for $\alpha = 0.9$ in the magnitude response using a second-order approximation. In phase, the change is even more noticeable, the error decreases by 26.71% for $\alpha = 0.1$, and a 5.61% for $\alpha = 0.9$, respectively.

11.5 Implementation of fractional-order integrators using FPAAs

The next step is to propose design rules for the electronic implementation of fractional-order integrators. As well known, the FPAA conforms by CAMs (Configurable Analog Modules). The CAMs settings depend on the clock frequencies that we select for each FPAA. Each FPAA has two system clock frequency sources Sys1 and Sys2, and six chip clock frequencies, from Clock 0

TABLE 11.6 Average absolute normalized magnitude error in % varying α and transfer function order.

α/Order	1st	2nd	3rd	4th	5th
0.1	13.4397	7.5755	2.4058	0.5407	0.2754
0.2	13.1089	7.3398	2.2960	0.5155	0.2639
0.3	12.5614	6.9432	2.1189	0.4751	0.2454
0.4	11.8034	6.3812	1.8827	0.4217	0.2203
0.5	10.8454	5.6502	1.5991	0.3580	0.1897
0.6	9.7075	4.7507	1.2822	0.2875	0.1548
0.7	8.4275	3.6945	0.9478	0.2133	0.1168
0.8	6.8606	2.5119	0.6123	0.1387	0.0772
0.9	4.2249	1.2563	0.2916	0.0667	0.0377

TABLE 11.7 Average absolute normalized phase error in % varying α and transfer function order.

α/Order	1st	2nd	3rd	4th	5th
0.1	34.9335	8.2174	2.6873	1.3313	0.3667
0.2	34.0430	7.8450	2.5975	1.2737	0.3498
0.3	32.5319	7.2389	2.4512	1.1805	0.3226
0.4	30.3547	6.4219	2.2473	1.0556	0.2864
0.5	27.4424	5.4308	1.9840	0.9042	0.2431
0.6	23.6947	4.3175	1.6640	0.7329	0.1948
0.7	18.9857	3.1492	1.2933	0.5488	0.1439
0.8	13.2152	2.0009	0.8821	0.3598	0.0929
0.9	6.5510	0.9388	0.4450	0.1742	0.0441

to Clock 5, which are subdivisions of any of the system clock sources. The FPAA has a primary clock frequency source called ACLK or f_C, whose frequency is equal to 16 MHz and cannot be changed. Sys1 and Sys2 depend on this frequency and can be modified using the equations

$$\text{Sys1} = \frac{f_C}{m}, \qquad \text{Sys2} = \frac{f_C}{m}, \tag{11.24}$$

where $m \in [1, 510]$ and is an even integer number. The parameters Sys1 and Sys2 control the chip clock sources using the following equations:

$$\text{Clock}\, h = \frac{\text{Sys1}}{n}, \qquad \text{Clock}\, h = \frac{\text{Sys2}}{n}, \tag{11.25}$$

where $n \in [1, 510]$ and is an even number and $h \in [0, 5]$.

11.5.1 Fractional-order integrator based on a first order transfer function

For the sake of simplicity, we show how to implement a fractional-order integrator based on a first-order transfer function. Still, the same considerations are valid for any other approximation. In this manner, the *FilterBilinear CAM* was used to synthesize the first order CFE approximation. By considering the Pole and Zero operation mode, the CAM transfer function is as follows:

$$\frac{V_{\text{out}}(s)}{V_{\text{in}}(s)} = -\frac{G_H(s + 2\pi f_z)}{s + 2\pi f_p}, \tag{11.26}$$

where G_L is defined as $G_L = \frac{f_z}{f_p} G_H$, and represents the gain in DC, G_H is the high frequency gain, f_p is the frequency of the pole and f_z is the frequency of the zero. On the other hand, the first-order transfer function (1.18) can be rewritten as

$$\frac{1}{{}_{(c_2)}\, s^\alpha} \approx \frac{As + 1}{s + A}, \tag{11.27}$$

with A being $A = \frac{1-\alpha}{1+\alpha}$. However, the bandwidth of Eq. (11.27) is from 10^{-1} rad/s to 10^1 rad/s, which is a very low frequency range. Then, we perform a frequency scaling to Eq. (11.27) using the following steps.

1. We set $s = p$, which represents the normalized frequency:

$$N(p) = \frac{Ap + 1}{p + A} \tag{11.28}$$

2. Next, the substitution $p = \dfrac{s}{k_f}$ is applied:

$$N(s) = \frac{Ask_f^{-1} + 1}{sk_f^{-1} + A} \tag{11.29}$$

3. We recast the previous equation:

$$\frac{1}{{}_{(c_2)}\, s^\alpha} \approx \frac{As + k_f}{s + Ak_f} \tag{11.30}$$

As a result, the scaling factor k_f can be chosen according to the required bandwidth. Fig. 11.4 shows the frequency scaling when $k_f = 2\pi 1000$. The new bandwidth is from 100 Hz to 10 kHz, which is within the operating range of the CAMs.

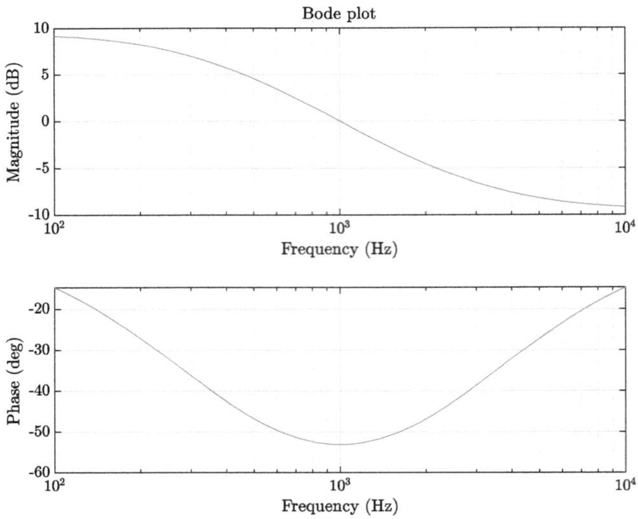

FIGURE 11.4 Frequency response of Eq. (11.30) for a fractional integrator with $k_f = 2\pi 1000$ and fractional order $\alpha = 0.5$.

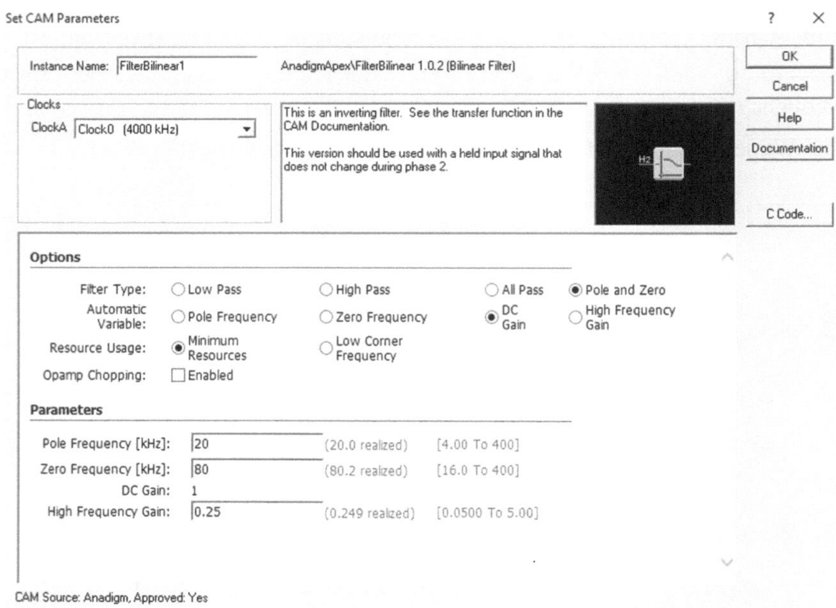

FIGURE 11.5 Configuration parameters for the CAM: FilterBilinear.

By equating Eqs. (11.30) and (11.26), the values of the parameters G_H, G_L, f_z and f_p for a certain α order can be determined:

$$\frac{G_H(s + 2\pi f_z)}{s + 2\pi f_p} = \frac{As + k_f}{s + Ak_f} \tag{11.31}$$

TABLE 11.8 Synthesis of the FilterBilinear CAM in FPAA to design fractional-order integrators.

α	f_p [kHz]	f_z [kHz]	G_L	G_H
0.10	0.818182	1.222222	1.222222	0.818182
0.15	0.739130	1.352941	1.352941	0.739130
0.20	0.666667	1.500000	1.500000	0.666667
0.25	0.600000	1.666667	1.666667	0.600000
0.30	0.538462	1.857143	1.857143	0.538462
0.35	0.481481	2.076923	2.076923	0.481481
0.40	0.428571	2.333333	2.333333	0.428571
0.45	0.379310	2.636364	2.636364	0.379310
0.50	0.333333	3.000000	3.000000	0.333333
0.55	0.290323	3.444444	3.444444	0.290323
0.60	0.250000	4.000000	4.000000	0.250000
0.65	0.212121	4.714286	4.714286	0.212121
0.70	0.176471	5.666667	5.666667	0.176471
0.75	0.142857	7.000000	7.000000	0.142857
0.80	0.111111	9.000000	9.000000	0.111111
0.85	0.081081	12.333333	12.333333	0.081081
0.90	0.052632	19.000000	19.000000	0.052632
0.95	0.025641	39.000000	39.000000	0.025641

TABLE 11.9 Absolute frequency range dependent on F_c and the value of n.

F_c [kHz]	n	min = $F_c/1000$ [kHz]	max = $F_c/10$ [kHz]
16000.0000	1.0000	16.0000	1600.0000
8000.0000	2.0000	8.0000	800.0000
4000.0000	4.0000	4.0000	400.0000
⋮	⋮	⋮	⋮
31.6206	506.0000	0.0316	3.1621
31.4961	508.0000	0.0315	3.1496
31.3725	510.0000	0.0314	3.1373

with $G_H = A$, $f_p = \frac{Ak_f}{2\pi}$, $f_z = \frac{k_f}{2A\pi}$, and $G_L = \frac{1}{A}$. Table 11.8 is obtained when $k_f = 2\pi 1000$ and the fractional order varies from 0.1 to 0.95. Nevertheless, the *FilterBilinear CAM* depends directly on the clock frequency being selected in the parameter Clock 0 (Fig. 11.5), and therefore certain values of Table 11.8 cannot be designed.

The absolute limits of the values that can be entered for the frequencies f_z and f_p are $\left[\frac{F_{clock}}{1000}, \frac{F_{clock}}{10}\right]$ where F_{clock} is the clock frequency that is selected for the parameter Clock 0 (Fig. 11.5). The range for Gain parameter is $\left[0.01\frac{V}{V}, 1000.0\frac{V}{V}\right]$. It is important to remark that F_{clock} could be set by the other

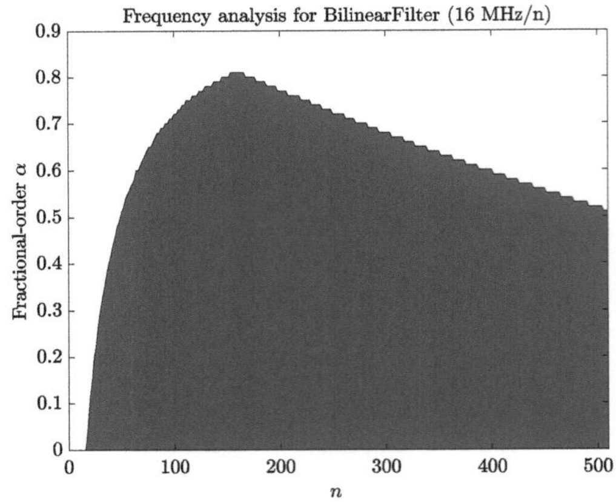

FIGURE 11.6 Figure of Merit (FOM) to design fractional-order integrators which relates the fractional order α and n-clock divisions in Eq. (11.25).

clocks of FPAA as given in Eq. (11.25). In this manner, we compute all possible frequencies for F_{clock} depending on the subdivisions given by n in Eq. (11.25) as shown in Table 11.9.

However, the process of finding the quasi-optimal n is challenging since it is a multi-parametric function. We propose a figure of merit to designing fractional-order integrators, which relates the fractional order α and the value n. Fig. 11.6 shows the surface where the fractional integrators could be implemented. As a first result, we found the maximum α we can develop in the FPAA using a first-order approximation is $\alpha \in [0.01, 0.81]$. Besides, the quasi-optimal n is 160.

11.6 Electronic implementation of a fractional-order memristive system

The design guidelines for fractional-order integrators previously introduced are herein applied to a fractional-order memristive system, defined by

$$\begin{aligned} D^{\alpha} w &= k_2 u, \\ v &= (w-1)u, \end{aligned} \tag{11.32}$$

with u and v being the input and output respectively, w represents the internal state of fracmemristor, and $k_2 = 60$. The fingerprint of the fracmemristor is characterized by an external periodic signal, $u = A\sin(\omega t)$ with A the amplitude, and f the frequency being applied. As is well known, an integer-order memristor may evolve into three cases. First, the hysteresis loop of the memristive system is a Lissajous curve pinched at the origin. Second, if the frequency of the periodic input signal increases, then the hysteresis loop becomes smaller and shrinks continuously. Finally, if the frequency tends to infinity, then the fig-

TABLE 11.10 Synthesis of the *FilterBilinear CAM* in FPAA to design fractional-order integrators with $k_f = 1250$.

q	f_p [kHz]	f_z [kHz]	G_L	G_H
0.10	0.162772	0.243153	1.222222	0.818182
0.15	0.147045	0.269159	1.352941	0.739130
0.20	0.132629	0.298416	1.500000	0.666667
0.25	0.119366	0.331573	1.666667	0.600000
0.30	0.107124	0.369467	1.857143	0.538462
0.35	0.095788	0.413191	2.076923	0.481481
0.40	0.085262	0.464202	2.333333	0.428571
0.45	0.075461	0.524488	2.636364	0.379310
0.50	0.066315	0.596831	3.000000	0.333333
0.55	0.057758	0.685250	3.444444	0.290323
0.60	0.049736	0.795775	4.000000	0.250000
0.65	0.042200	0.937877	4.714286	0.212121
0.70	0.035108	1.127348	5.666667	0.176471
0.75	0.028421	1.392606	7.000000	0.142857
0.80	0.022105	1.790493	9.000000	0.111111

ure of the loop of the memristive system tends to a straight line. For the case of a fractional-order memristive system, we have an additional degree of freedom (DOF). It means that the hysteresis loop is also a function of the fractional-order derivative.

To implement the memristive system (11.32), we first design a fractional-order integrator following the steps in Eq. (11.28) to Eq. (11.31). By considering $k_f = 1250$, we obtain Table 11.10.

In this manner, the memristive system (11.32) is implemented on an FPAA, as shown in Fig. 11.7. In particular, it is given the implementation for a fractional order $\alpha = 0.5$. Then, the fingerprint of the fracmemristor is analyzed, setting the external periodic signal with $A = 0.6$, and $\omega = 1250$ rad/s. We implement several fractional orders ($\alpha = 0.5$, $\alpha = 0.6$, $\alpha = 0.7$, and $\alpha = 0.8$) to observe the trade-off between the fractional order and the hysteresis loop, as illustrated in Fig. 11.8.

11.7 Conclusions

As is well known, the memristor-based artificial synapses have demonstrated great potential for bioinspired neuromorphic computing in recent years. Also, neural networks implemented with traditional hardware face the inherent limitation of memory latency. It has motivated memristor-based neuromorphic computing. Regarding memristive systems, there have been considerable advances in using memristors to propose artificial neural networks with chaotic behavior as well as synchronization techniques. Hardware implementations of memristors are needed to get insights into diverse future applications. However, this is

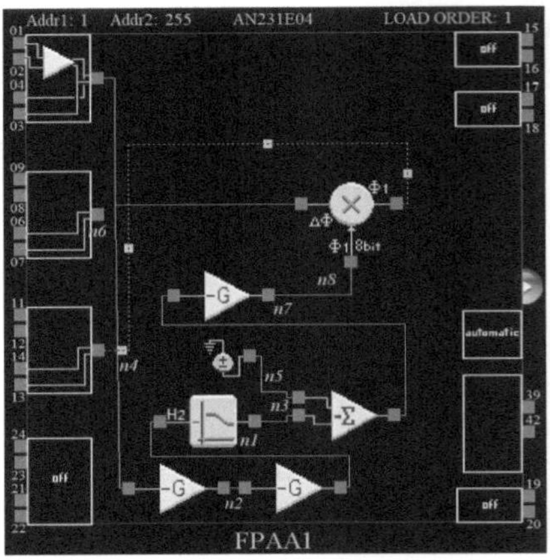

Name	Options	Parameters	Clocks
GainInv1 (GainInv v1.0.1) Anadigm (Approved)		Gain 4.00	ClockA 20.833 kHz (Chip Clock 0)
FilterBilinear1 (FilterBilinear v1.0.2) Anadigm (Approved)	Filter Type: Pole and Zero; Automatic Variable: DC Gain; Resource Usage: Minimum Resources	Pole Frequency [kHz] 0.0664; Zero Frequency [kHz] 0.594; DC Gain 3.00; High Frequency Gain 0.335	ClockA 20.833 kHz (Chip Clock 0)
Voltage2 (Voltage v1.0.1) Anadigm (Approved)	Polarity: Negative (-2V)		
GainInv2 (GainInv v1.0.1) Anadigm (Approved)		Gain 15.0	ClockA 20.833 kHz (Chip Clock 0)
Multiplier2 (Multiplier v1.0.2) Anadigm (Approved)	Sample and Hold: Off	Multiplication Factor 1.00	ClockA 20.833 kHz (Chip Clock 0); ClockB 333.333 kHz (Chip Clock 1)
SumInv1 (SumInv v1.0.1) Anadigm (Approved)	Input 3: Off	Gain 1 (UpperInput) 0.500; Gain 2 (LowerInput) 1.00	ClockA 20.833 kHz (Chip Clock 0)
GainInv3 (GainInv v1.0.1) Anadigm (Approved)		Gain 1.00	ClockA 20.833 kHz (Chip Clock 0)

FIGURE 11.7 Implementation of fracmemristor (11.32) using AnadigmDesigner2 and FPAAs.

not a trivial task. The first step would be to characterize the novel architectures fully. In this manner, distinct emulator circuits can be helpful.

Along this line, the design guidelines for the physical implementation of fractional-order integrators, which are highly required for fractional-order memristive systems, have been presented. More particularly, we discuss a synthesis methodology based on continued fraction expansion to get transfer functions of nth order. A metric performance defined as the rate of the desired error between the fractional-order and the integer-order rational approximation was first introduced. We found that as the fractional order reduces, the integer order of the approximation increases to limit the error of amplitude and phase. However, first-order approximations for fractional orders higher than 0.85 are suitable. Next, we propose a physical realization using FPAAs. Here, we discovered that there is a substantial trade-off between the fractional order of the integrator and the frequency of the FPAA clocks. This figure of merit can be extended to other clocks frequencies and will be a designed guideline for fractional-order systems in FPAAs. Experimental results of a fractional-order memristive system showed the usefulness of the proposed approach. In particular, we showed the hysteresis loop is in agreement with the theoretical findings. The results presented in this chapter could be beneficial for diverse engineering applications, such as artificial intelligence based on memristors.

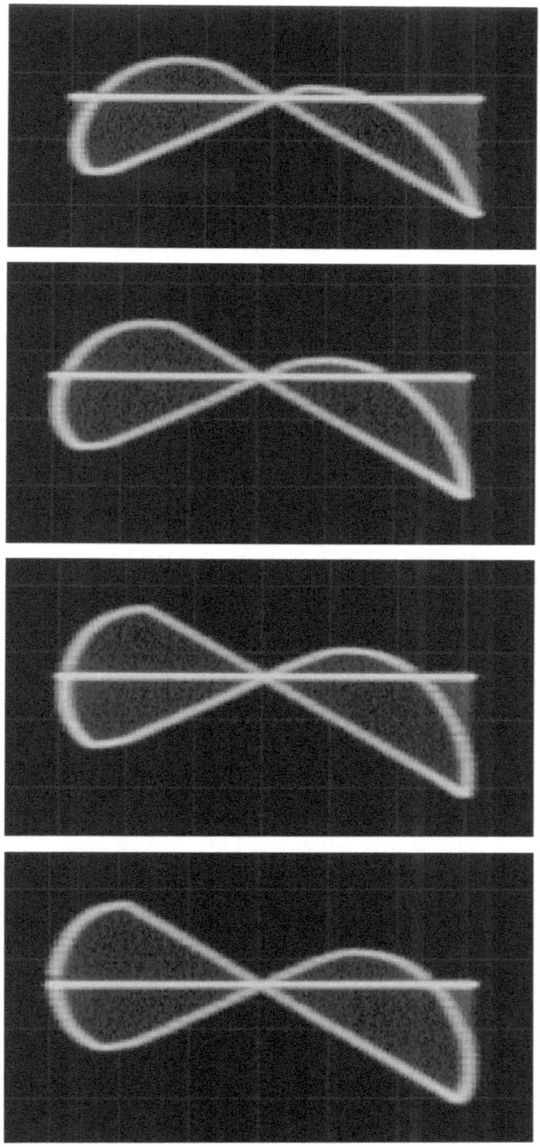

FIGURE 11.8 Experimental results of the hysteresis loop of memristive system (11.32) for a) $\alpha = 0.5$, b) $\alpha = 0.6$, c) $\alpha = 0.7$, and d) $\alpha = 0.8$.

Acknowledgments

J.M. Muñoz-Pacheco acknowledges CONACYT/MEXICO for the financial support through Project No. 258880 (Proyecto Apoyado por el Fondo Sectorial de Investigacion para la Educacion). C.F. Bermudez-Marquez also thanks CONACYT/MEXICO for the engineering scholarship. Both authors thank VIEP-BUAP 2020 project. The authors thankfully acknowledge computer resources, technical advice and support provided by Laboratorio Nacional de Supercómputo del Sureste de México (LNS), a member of the CONACYT national laboratories, with project No. 202001025C.

References

[1] L. Chua, IEEE Transactions on Circuit Theory 18 (1971) 507.

[2] D.B. Strukov, G.S. Snider, D.R. Stewart, R.S. Willians, Nature 453 (2008) 80.

[3] Q. Hong, J. Zhao, X. Wang, Neurocomputing 330 (2019) 11.

[4] A.H. Fouad, A.G. Radwan, International Journal of Electronics and Communications (AEU) 98 (2019) 123.

[5] B. Bodo, J.S. Armand Eyebe Fouda, A. Mvogo, S. Tagne, Chaos, Solitons and Fractals 115 (2018) 190.

[6] Y. Xu, Y. Jia, J. Ma, A. Alsaedi, B. Ahmad, Chaos, Solitons and Fractals 104 (2017) 435.

[7] S. Shin, K. Kim, S.-M. Kang, IEEE Transactions on Nanotechnology 10 (2011) 266.

[8] S.H. Jo, T. Chang, I. Ebong, B.B. Bhadviya, P. Mazumder, W. Lu, Nano Letters 10 (2010) 1297.

[9] F. Corinto, M. Forti, IEEE Transactions on Circuits and Systems I, Regular Papers 63 (2016) 1997.

[10] M.A. Trefzer, Memristor in a nutshell, in: E. Miranda (Ed.), Guide to Unconventional Computing for Music, Springer, Cham, 2017.

[11] L. Chua, Resistance switching memories are memristors, Journal of Applied Physics 102 (4) (2011) 765.

[12] I. Kazi, P. Meinerzhagen, P.E. Gaillardon, D. Sacchetto, A. Burg, G. De Micheli, A ReRAM-based non-volatile flip-flop with sub-V T read and CMOS voltage-compatible write, in: IEEE 11th International Conference on New Circuits and Systems (NEWCAS), IEEE, 2013, pp. 1–4.

[13] M.A. Trefzer, A.M. Tyrrell, Evolvable Hardware: From Practice To Application, Springer, Berlin, 2015.

[14] F. Corinto, S.M. Kang, A. Ascoli, Memristor-based neural circuits, in: IEEE international symposium on circuits and systems (ISCAS), IEEE, 2013, pp. 1832–1835.

[15] L.O. Chua, The fourth element, Proceedings of the IEEE 100 (6) (2012) 1920–1927.

[16] H. An, J. Li, Y. Li, X. Fu, Y. Yi, Three dimensional memristor-based neuromorphic computing system and its application to cloud robotics, Computers & Electrical Engineering 63 (2017) 99–113.

[17] H. An, M.A. Ehsan, Z. Zhou, F. Shen, Y. Yi, Monolithic 3D neuromorphic computing system with hybrid CMOS and memristor-based synapses and neurons, Integration 65 (2019) 273–281.

[18] V.A. Filippov, A.N. Bobylev, A.N. Busygin, A.D. Pisarev, S.Y. Udovichenko, A biomorphic neuron model and principles of designing a neural network with memristor synapses for a biomorphic neuroprocessor, Neural Computing and Applications 32 (7) (2020) 2471–2485.

[19] B. Liu, Y. Chen, B. Wysocki, T. Huang, Reconfigurable neuromorphic computing system with memristor-based synapse design, Neural Processing Letters 41 (2) (2015) 159–167.

[20] M.N. Getachew, R. Priyadarshini, R.M. Mehra, SPICE model of HP-memristor using PWL window function for neuromorphic system design application, Materials Today: Proceedings (2020).

[21] Z. Yan, J. Chen, R. Hu, T. Huang, Y. Chen, S. Wen, Training memristor-based multilayer neuromorphic networks with SGD, momentum and adaptive learning rates, Neural Networks (2020).

[22] C. Dias, J. Ventura, P. Aguiar, Memristive-based neuromorphic applications and associative memories, in: Advances in Memristors, Memristive Devices and Systems, Springer, Cham, 2017, pp. 305–342.

[23] A. Slavova, G. Bobeva, Determination of edge of chaos in hysteresis CNN model with memristor synapses, in: 2017 European Conference on Circuit Theory and Design (ECCTD), IEEE, 2017, September, pp. 1–4.

[24] Q. Li, S. Tang, H. Zeng, T. Zhou, On hyperchaos in a small memristive neural network, Nonlinear Dynamics 78 (2) (2014) 1087–1099.

[25] Z. Wang, F. Parastesh, K. Rajagopal, I.I. Hamarash, I. Hussain, Delay-induced synchronization in two coupled chaotic memristive Hopfield neural networks, Chaos, Solitons and Fractals 134 (2020) 109702.

[26] H. Lin, C. Wang, Y. Tan, Hidden extreme multistability with hyperchaos and transient chaos in a Hopfield neural network affected by electromagnetic radiation, Nonlinear Dynamics 99 (3) (2020) 2369–2386.

[27] G. Zhang, D. Guo, F. Wu, J. Ma, Memristive autapse involving magnetic coupling and excitatory autapse enhance firing, Neurocomputing 379 (2020) 296–304.

[28] S. Kumar, J.P. Strachan, R.S. Williams, Chaotic dynamics in nanoscale NbO 2 Mott memristors for analogue computing, Nature 548 (7667) (2017) 318–321.

[29] Y. Fan, X. Huang, Z. Wang, Y. Li, Nonlinear dynamics and chaos in a simplified memristor-based fractional-order neural network with discontinuous memductance function, Nonlinear Dynamics 93 (2) (2018) 611–627.

[30] A.R. Bevi, S. Tumu, N.V. Prasad, Design and investigation of a chaotic neural network architecture for cryptographic applications, Computers & Electrical Engineering 72 (2018) 179–190.

[31] J.M. Munoz-Pacheco, Infinitely many hidden attractors in a new fractional-order chaotic system based on a fracmemristor, The European Physical Journal Special Topics 228 (10) (2019) 2185–2196.

[32] Y-F. Pu, X. Yuan, IEEE Access 4 (2016) 1872.

[33] L. Xu, G. Huang, Y-F. Pu, Journal of Circuits, Systems, and Computers 27 (2018) 1850227.

[34] N. Yang, C. Xu, C. Wu, R. Jia, International Journal of Bifurcation and Chaos 27 (2017) 1750199.

[35] C. Coopmans, I. Petras, Y.Q. Chen, in: ASME 2009 International Design Engineering Technical Conferences & Computers and Information in Engineering Conference, California, 2009, p. 1127.

[36] Y. Yu, Z. Wang, International Journal of Bifurcation and Chaos 28 (2018) 1850091.

[37] I. Petras, IEEE Transactions on Circuits and Systems II, Express Briefs 57 (2010) 975.

[38] G. Si, L. Diao, J. Zhu, Nonlinear Dynamics 87 (2017) 2625.

[39] C. Sanchez-Lopez, V.H. Carbajal-Gomez, M.A. Carrasco-Aguilar, I. Carro-Perez, Complexity (2018) 2806976.

[40] S.D. Roy, On the realization of a constant-argument immittance or fractional operator, IEEE Transactions on Circuit Theory 14 (3) (1967) 264–274.

[41] G. Carlson, C. Halijak, Approximation of fractional capacitors $(1/s)\hat{}(1/n)$ by a regular Newton process, IEEE Transactions on Circuit Theory 11 (2) (1964) 210–213.

[42] D. Valerio, J.S. Da Costa, Introduction to single-input, single-output fractional control, IET Control Theory & Applications 5 (8) (2011) 1033–1057.

[43] K. Matsuda, H. Fujii, H (infinity) optimized wave-absorbing control – analytical and experimental results, Journal of Guidance, Control, and Dynamics 16 (6) (1993) 1146–1153.

[44] H. Sun, Y. Zhang, D. Baleanu, W. Chen, Y. Chen, A new collection of real world applications of fractional calculus in science and engineering, Communications in Nonlinear Science and Numerical Simulation 64 (2018) 213–231.

[45] J.A. Tenreiro Machado, A.M. Lopes, Complex and fractional dynamics, Entropy 19 (2017) 62.

[46] W.C. Chen, Nonlinear dynamics and chaos in a fractional-order financial system, Chaos, Solitons and Fractals 36 (2008) 1305–1314.

[47] J. Machado, M.E. Mata, A.M. Lopes, Fractional state space analysis of economic systems, Entropy 17 (2015) 5402–5421.

[48] E. Zambrano-Serrano, E. Campos-Canton, J.M. Munoz-Pacheco, Strange attractors generated by a fractional order switching system and its topological horseshoe, Nonlinear Dynamics 83 (2016) 1629–1641.

[49] I. Petras, Fractional-Order Nonlinear Systems: Modeling, Analysis and Simulation, Springer Science & Business Media, 2011.

[50] B.S.T. Alkahtani, A. Atangana, Chaos on the Vallis Model for El Niño with Fractional Operators, Entropy 18 (2016) 100.

[51] S. He, K. Sun, H. Wang, Complexity Analysis and DSP Implementation of the Fractional-Order Lorenz Hyperchaotic System, Entropy 17 (2015) 82998311.

[52] E. Zambrano-Serrano, J.M. Munoz-Pacheco, E. Campos-Canton, Chaos generation in fractional-order switched systems and its digital implementation, International Journal of Electronics and Communications (AEU) 79 (2017) 43–52.

[53] O.I. Tacha, J.M. Munoz-Pacheco, E. Zambrano-Serrano, I.N. Stouboulos, V-T. Pham, Determining the chaotic behavior in a fractional-order finance system with negative parameters, Nonlinear Dynamics 94 (2018) 1303–1317.

[54] M.D. Ortigueira, J.T. Machado, What is a fractional derivative?, Journal of Computational Physics 293 (2015) 4–13.

Chapter 12

Control of bursting oscillations in memristor based Wien-bridge oscillator

S. Dinesh Vijay and K. Thamilmaran
Department of Nonlinear Dynamics, School of Physics, Bharathidasan University, Tiruchirappalli, India

12.1 Introduction

The memristor theory was initiated by Simmons and Verderber in the year 1967, they proposed new conduction and reversible memory phenomena in insulating thin films [1]. A novel two-terminal circuit element namely memristor is proposed by Leon O. Chua in the year 1971 [2]. It is characterized by the relationship between the charge (q) and flux (ϕ), also it has non-volatile memory behavior. William's group physically realized nanoscale two-terminal memristor from HP laboratory in 2008 [3]. The memristors can retain memory states, and data, in power-off modes. Non-volatile random access memory, or NVRAM, is pretty much the first to-market memristor application we shall be seeing. There are already 3 nm memristors in fabrication now. Crossbar latch memory developed by Hewlett Packard is reportedly currently about one-tenth of the speed of DRAM. The fab prototypes resistance is read with an alternating current so that the stored value remains unaffected [4]. The non-volatile memory behavior of memristor has great potential applications such as DRAM, neuronal synapse, neuromorphic systems, artificial intelligence devices, system controls, and signal processing [3–6,18,19]. Dynamical studies on memristor based systems are added to by the development of memristor emulators. One has realized this by a diode bridge based memristors [7–10], op-amps with multiplier [11]. Memristor emulators are more convenient to modify, which is constructed with a diode bridge with LC, L, RC networks [8–10]. Most of the memristor based oscillators focused on the neuronal behaviors [12–14]. Bursting oscillations (BOs) is one of the most important neuronal behaviors. Two neurons communicate by sharing an action potential between them, the action potential is known as a spike. The continuous spikes are called a spiking train, and one or more spikes with a quiescent state form a burst [15,16]. This consists of repetitive active and quiescent (rest) states in its time trajectory [15–17]. A burst may have one or

Mem-elements for Neuromorphic Circuits with Artificial Intelligence Applications
https://doi.org/10.1016/B978-0-12-821184-7.00020-7

more active states. If a time series has periodically repetitive bursts with quiescent states it is known as periodic BOs, if the envelope consists of non-periodic trajectories it is known as chaotic BOs.

In the past few decades, the memristor emulator based oscillators have gained great attention for its simplicity and its rich dynamics. From this, numerical and experimental studies in memristor based oscillators such as the self-excited attractors and hidden attractors [7,24], chaotic and periodic bursting phenomena [8], quasi-periodic and chaotic BOs [9], extremely slow passage effect [18] chaotic beats [19], complex transient dynamics [19–23], transient chaos [22], and multistability [25,26] have been reported. However, most of the memristor based emulators are focused on the complex dynamics and other rich dynamics. But the control of oscillations in dynamical systems only leads to application points to various fields. Chaos has the property of sensitively as regards dependence on the initial conditions. So, the idea of chaos control is, when we have trajectory transportation to stochastically considered periodic orbit implanted in the attractor, a small perturbation to stabilize such an orbit. If changing the perturbation, the periodic orbit to be stabilized also the trajectory moves its neighborhood. This statement indicates the critical sensitivity of complex systems. The initial conditions may affect the trajectories in the complex systems [27]. According to [27] the control of complex dynamics has been reported numerically and experimentally [27–30]. On the other hand, studies on neuronal dynamics control are reported in various aspects. We have for example thalamocortical neuronal oscillation control [31], oscillation, and frequency control in neural networks [32], irregular bursting oscillation control [33], chaotic oscillation tuning [34], and control of oscillations in higher dimensional neuronal networks [35]. Previous studies focused on controlling dynamics in a particular way. This means control of the global transitions. This chapter focuses on control of the bursting oscillations (BOs) in all the aspects. Namely, amplitude, rest time, number of bursts and, also the nature (periodic/chaotic) of a burst.

We report controlling of bursting phenomena using emulated second-order diode bridge based memristor with an op-amp based Wien-bridge oscillator. To vary the system parameter R, we have observed the periodic and chaotic bursting oscillations at a particular set of parameters. Further increasing the parameter we have to control the amplitude, the number of spikes, and quiescent time in the bursting oscillations. On the other hand, we need to vary the negative feedback gain of the memristive Wien-bridge oscillator. It produced the self-excited oscillations which control the amplitude of bursting oscillations in another set of fixed parameters of the system.

The chapter organized as follows. In Section 12.2 the mathematical model of a diode bridge based second-order memristor nonlinearity and its functions are analyzed. The memristive Wien-bridge oscillator and its circuit equations are investigated in Section 12.3. The definitions and differences between chaotic and periodic BOs are characterized in Section 12.4. Amplitude control BOs, spike control BOs, quiescent state control BOs are investigated with numerical results in Section 12.5. The amplitude control BOs by negative feedback gain is discussed in Section 12.6. Finally, the numerically obtained results are addressed in Section 12.7.

12.2 Mathematical model of LC network based diode bridge memristor

The memristor is a fourth fundamental two terminal electronic element alongside the basic elements resistor (R), capacitor (C), and inductor (L). These are having the well-defined (v–i), (q–v), and (ϕ–i) relationships, respectively [1]. The missing relationship is (q–ϕ) and also there is a missing element. A passive two terminal element, namely memristor, is characterized by the relationship between charge (q) and flux (ϕ) proposed by Chua in 1971 [2]. The second-order generalized memristor is simplified by replacing the second-order memristive RLC filter proposed in [3]. The second-order LC network based diode bridge memristor is used as nonlinearity, it is shown in Fig. 12.1. This is formed by using an inductor (L), a capacitor (C), and four diodes (D_1, D_2, D_3, D_4). Denote by v and i the input voltage and current of the memristor emulator. The voltage and current relation of the second-order memristor is expressed as

$$i = 2I_s e^{-\rho v_c} \sinh(\rho v_c) \tag{12.1}$$

$$\frac{dv_c}{dt} = \frac{1}{C} 2I_s[e^{(-\rho v_c)} \cosh(\rho v_1) - 1] - i_L \tag{12.2}$$

$$\frac{di_L}{dt} = \frac{v_c}{L} \tag{12.3}$$

where $\rho = 1/(2nv_T)$, I_S, n, and v_T for the reverse saturation current, emission coefficient, and the thermal voltage of the diode, respectively. The passive floating generalized diode bridge memristor is fabricated with easily available components. The emulated memristor models are useful for further modifications of future memristor theory.

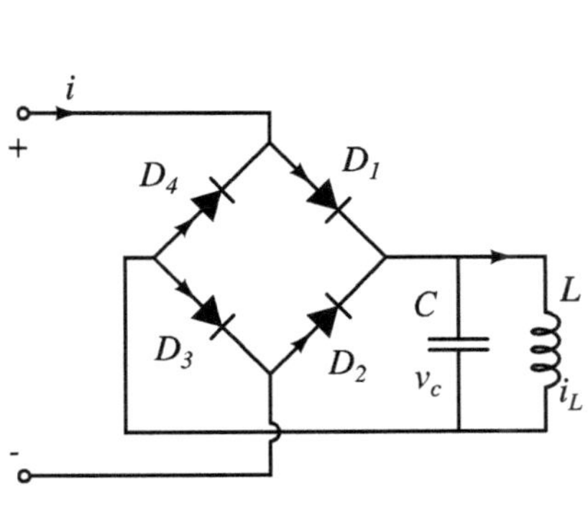

FIGURE 12.1 LC network based diode bridge memristor nonlinearity.

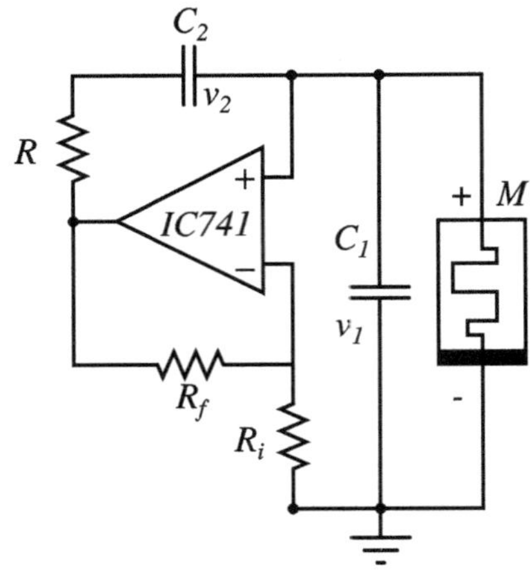

FIGURE 12.2 Memristive Wien-bridge oscillator.

12.3 Memristive Wien-bridge oscillator

The Wien bridge oscillator was developed by Max Wien in the year 1981. The Wien bridge oscillator is based on the bridge circuit; it consists of four resistors and two capacitors and it is used for the measurement of impedance. A huge frequency is produced by the Wien bridge oscillator. A feedback circuit is used by the Wien bridge oscillator and the circuit consists of a series RC circuit which is connected to a parallel RC circuit. The components of the circuit have the same values, which give the phase delay and phase advance circuit with the help of the frequency. The two stage RC circuit amplifier circuit has a high quality of resonant frequency, low distortion, and also sufficient tuning. Considering the very simple sine wave oscillator used by the RC circuit and place in the conventional LC circuit, the construct of the output of the sinusoidal waveform is called a Wien bridge oscillator, as shown in Fig. 12.2.

The memristive Wien-bridge oscillator as the same topology as the classic Wien-bridge oscillator. To apply Kirchhoff's voltage and current laws from the Fig. 12.2, the circuit equations can be obtained:

$$\frac{dv_c}{dt} = \frac{1}{C} 2I_s[e^{(-\rho v_c)}\cosh(\rho v_1) - 1] - i_L$$

$$\frac{di_L}{dt} = \frac{v_c}{L}$$

$$\frac{dv_1}{dt} = \frac{1}{C_1}\frac{kv_1}{R} - \frac{v_2}{R} - 2I_s e^{(-\rho v_c)}\sinh(\rho v_1)$$

$$\frac{dv_2}{dt} = \frac{kv_1}{RC_2} - \frac{v_2}{RC_2} \tag{12.4}$$

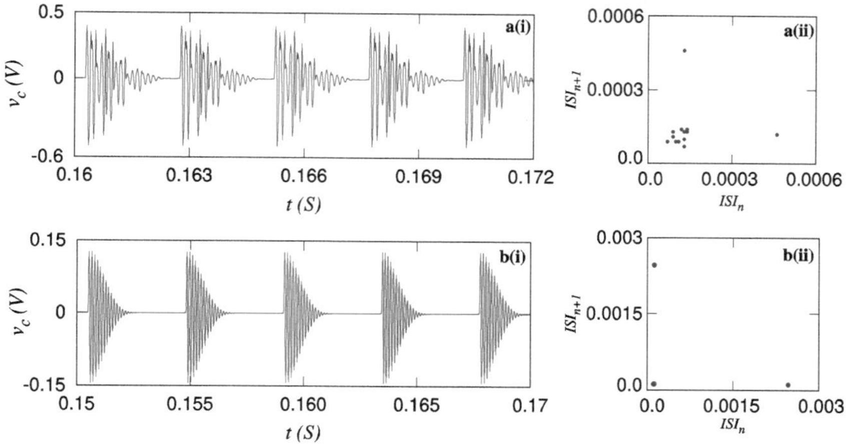

FIGURE 12.3 (i) Time series of $v_C(t)$ plane and corresponding (ii) inter spike interval (ISI) return map in the (ISI_n–ISI_{n+1}) plane. (a) Chaotic BOs and (b) periodic BOs for parameter $R = 10$ kΩ and $R = 20$ kΩ, respectively.

The current through the inductor (i_L) and voltage across the capacitors v_c, v_1, and v_2 are chosen as the state variables. The proposed memristive Wien-bridge oscillator has two important parts, with positive and negative feedback. $k = R_f/R_i$ is described as the negative feedback gain. It takes care of the self-excited oscillations, which helps to control the amplitude of the oscillations [5]. From this, to modify the negative feedback gain, we have to control the oscillations. We used the fourth-order Runge–Kutta algorithm to study the dynamics of Eq. (12.4) through a numerical simulation.

12.4 Chaotic and periodic bursting oscillations (BOs)

The successive sequence of continuous active states with a quiescent (rest) state in a time trajectory is known as bursting oscillations (BOs). The periodic BOs consists of periodic active states as well as chaotic BOs consists of chaotic active states [Ref. 6]. These are obtained numerically from Eq. (12.4). We have the fixed system parameters $R_f = 6$ kΩ, $R_i = 2$ kΩ, $C = 4.7$ nF, $C_1 = 10$ nF, $C_2 = 100$ nF, and $L = 25$ mH. To vary the control parameter R, we have observed the time series of chaotic BOs ($R = 10$ $k\Omega$) and periodic BOs ($R = 20$ kΩ) as shown in Fig. 12.3(i), for the initial conditions (0, 0, 0.0001, 0). The time series 12.3a(i) has a chaotic envelope in its active states and b(i) has periodic active states. The two oscillations are characterized by a ISI return map to confirm the chaotic and periodic nature of the time series. In Fig. 12.3a(ii) we have randomly distributed points in the (ISI_n–ISI_{n+1}) plane, it represents the chaotic nature of the time series 12.3a(i). As well as in the Fig. 12.3b(ii) have periodically distributed points in the (ISI_n–ISI_{n+1}) plane, it represents the periodic nature of the time series 12.3b(i).

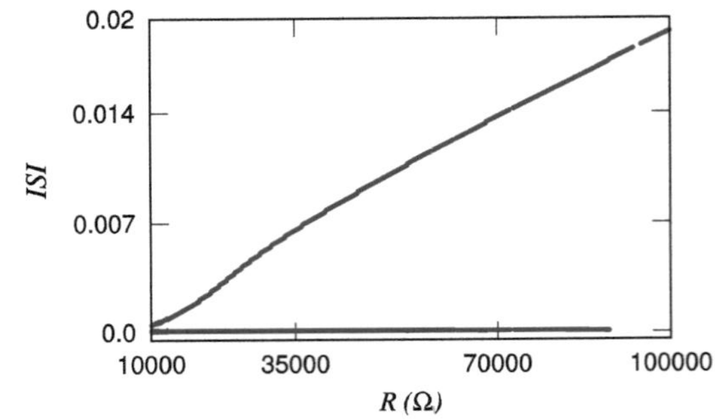

FIGURE 12.4 Inter spike interval (ISI) bifurcation diagram in the (R–ISI) plane for the fixed system parameters $R_i = 2$ kΩ, $R_f = 6$ kΩ, $C_1 = 10$ nF, $C_2 = 100$ nF, $C = 4.7$ nF, $L = 25$ mH.

12.5 Control of active states and quiescent states in BOs

In this section, the active and quiescent state control is examined as described in the previous section. The variation of the positive feedback resistance (R) in Fig. 12.2 controls the active and quiescent states. For the fixed system parameters $R_f = 6$ kΩ, $R_i = 2$ kΩ, $C = 4.7$ nF, $C_1 = 10$ nF $C_2 = 100$ nF, and $L = 25$ mH. The control parameter R to varying (10 kΩ - 100 kΩ). We have observed the amplitude decreasing and quiescent time increasing behaviors in BOs shown in Figs. 12.4 and 12.5. The increment of the time difference between each two successive active states is confirmed by the ISI bifurcation diagram in the plane (R–ISI) as shown in Fig. 12.4. At the same time, the increment of the quiescent time is also confirmed by the time series $v_c(t)$ as shown in Fig. 12.5(i). In parallel, the variation of the control parameter (R) tends to decrease the amplitude of the BOs, as clearly shown in Fig. 12.5.

The variation of the control parameter (R) controls the amplitude of BOs, and the quiescent time between each successive active state also the number of active states in a burst. From this, we have plotted the relationship between control parameter and amplitude of BOs (R–v_c), control parameter, and the number of spikes in a burst (R–n), control parameter, and quiescent time between two successive bursts (R–r_t). In Fig. 12.6(a) shows the relation between (R–v_c). When we increase the control parameter (R) value (30 kΩ–100 kΩ), the amplitude of BOs (v_c) gets decrease (0.0602 V–0.0038 V), it indicates the control parameter (R) value is inversely proportional to the amplitude of BOs (v_c). In the same way, when the control parameter (R) is increasing (30 kΩ–100 kΩ), the number of active states in a burst (n) gets decreased (8–1). On the other hand, when the control parameter (R) increases (40 kΩ–100 kΩ), the quiescent time between two successive bursts (r_t) gets an increase (0.004 S–0.0163 S). It is shown in Fig. 12.7 and it obeys a linear fit, which indicates that R is directly proportional to r_t.

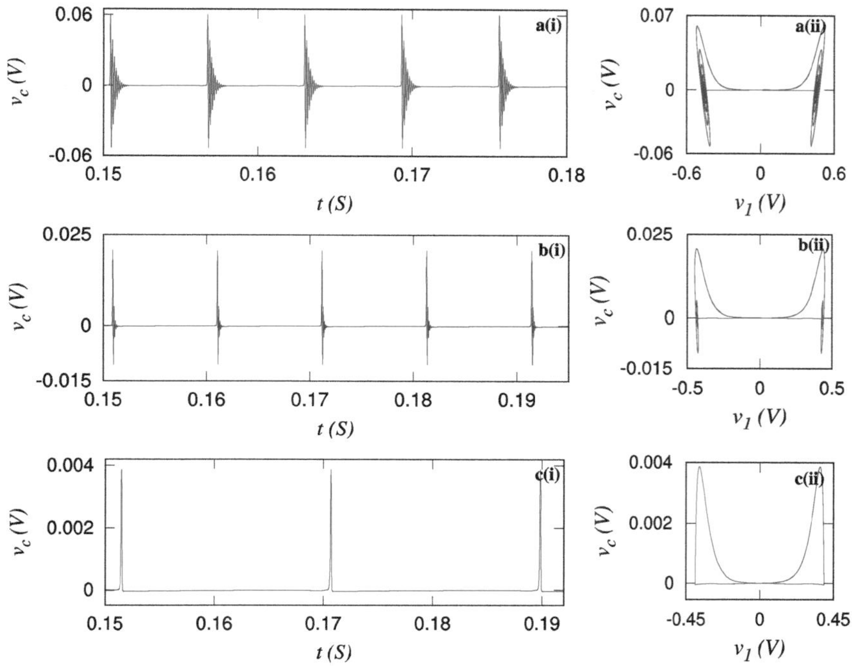

FIGURE 12.5 (i) Time series of $v_C(t)$ plane and corresponding (ii) Phase portrait in the $(v_1–v_C)$ plane. (ac) Spike decreasing as well as quiescent state increasing BOs for parameter $R = 30\ \text{k}\Omega, 50\ \text{k}\Omega, 100\ \text{k}\Omega$, respectively.

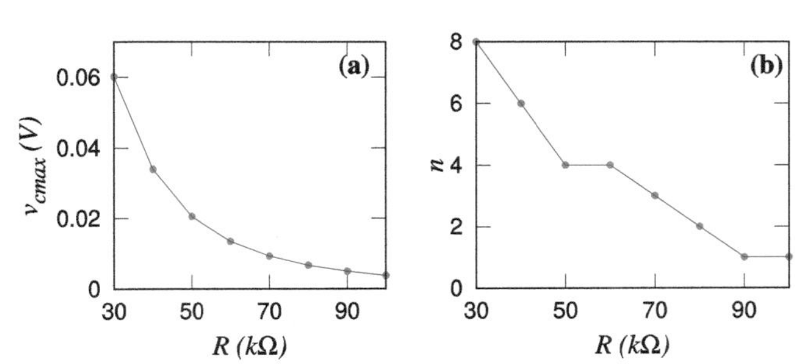

FIGURE 12.6 (a) Amplitude control relationship between control parameter (R) and amplitude of BOs (v_C) in the $(R–v_C)$ plane. (b) Spike control relationship between control parameter (R) and number of spikes in a burst (n) in the $(R–n)$ plane.

12.6 Control of amplitude in BOs

Control of the amplitude in a chaotic signal is an important application of chaos. It turns out that the amplitude control technique yields the expected amplification without any extra circuitry. The signum piecewise linear function switches from −1 (negative) to +1 (positive). If it is the only nonlinearity and without con-

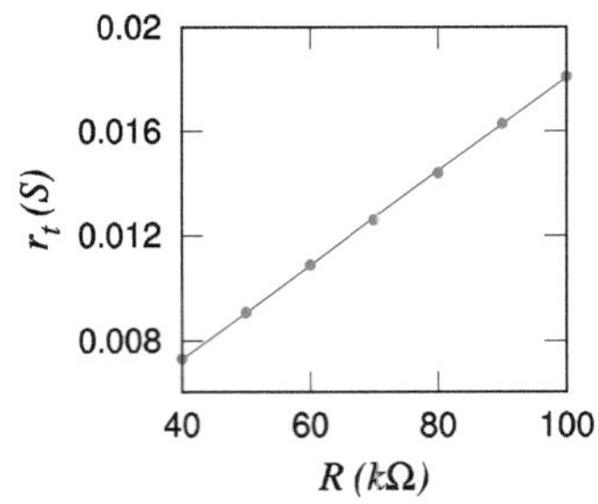

FIGURE 12.7 Linear fit for relationship between control parameter (R) and quiescent time between two successive bursts (r_t) in the (R–r_t) plane.

stant terms, then its coefficient can provide amplitude control since it uniquely determines the scale of the variables. If the signum is multiplied by one of the variables, the signum determines only the sign of the corresponding term and not its magnitude, then a constant term is also needed to provide amplitude control. In this section, the amplitude of BOs controls due to the variation of negative feedback have a gain ratio (R_f/R_i) in hysteresis nonlinearity based oscillator. When increasing the negative feedback gain ratio to vary R_f, the amplitude of BOs is increasing, and vice versa the amplitude decreases for R_i varying from higher to lower values.

12.6.1 Amplitude increasing in BOs

The amplitudes of BOs (v_c) get increase to varying the control parameter (R_f) value (2 kΩ–5.9 kΩ), for the fixed system parameters are $R = 47$ kΩ, $R_i =$ 2 kΩ, $C_1 = 1$ nF, $C_2 = 10$ nF, $C = 4.7$ nF, $L = 25$ mH. Increment of bursting amplitude as shown in Fig. 12.8 in the (R_f–v_c) plane. It is also confirmed by the time series of ($v_c(t)$) and its corresponding phase portraits in the ($v_1 - v_c$) plane as shown in Fig. 12.9. The scale of time series Fig. 12.9 a(i) has lower amplitude compare than Fig. 12.9 c(i) as same as the phase portraits Fig. 12.9 a(ii) and c(ii). Additionally, a curve plotted for the relationship between control parameter and amplitude of BOs in the (R_f–v_c) plane. This curve is calculated manually from the time series in Fig. 12.9 (i), and it clearly shows that the amplitude of the BOs (v_c) get increases (0.0169 V–0.3142 V), when the control parameter is increase. Further increasing the control parameter value chaotic oscillation has occurred, as shown in Fig. 12.10.

12.6.2 Amplitude decreasing in BOs

In this case, again the ratio of negative feedback gain is varied, but the control parameter is R_i. It is inversely proportional to the negative feedback gain (k). When increase the control parameter (R_i) value (2 kΩ–8 kΩ) the amplitude of the BOs gets decreasing and the (k) also decrease. The fixed system parameters

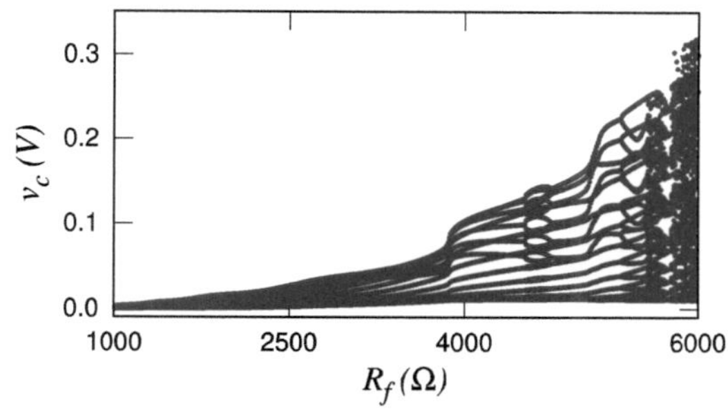

FIGURE 12.8 Bifurcation diagram in the (R_f–v_c) plane for the fixed system parameters $R = 47$ kΩ, $R_i = 2$ kΩ, $C_1 = 1$ nF, $C_2 = 10$ nF, $C = 4.7$ nF, $L = 25$ mH.

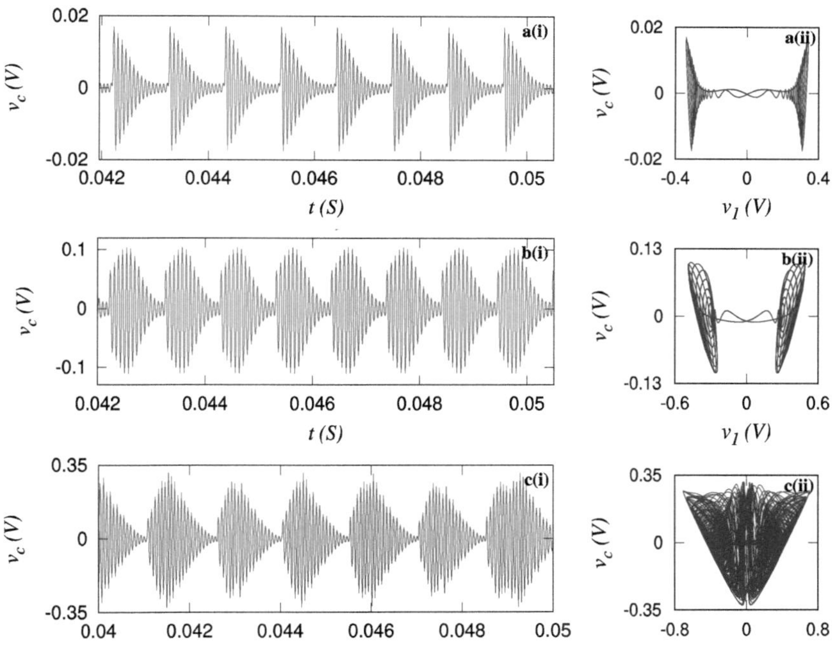

FIGURE 12.9 (i) Time series of $v_c(t)$ plane and corresponding (ii) phase portrait in the (v_1–v_c) plane. (a)–(c) Amplitude increasing BOs for the control parameter $R_f = 2$ kΩ, 4 kΩ, 5.9 kΩ, respectively.

are $R = 47$ kΩ, $R_f = 5$ kΩ, $C_1 = 1$ nF, $C_2 = 10$ nF, $C = 4.7$ nF, $L = 25$ mH. In Fig. 12.11 is shown that the amplitude is decreasing of BOs and also the time series $v_c(t)$ and so are its corresponding phase portraits (v_1–v_c) of amplitude decreasing BOs, as shown in Fig. 12.12. The manually plotted, amplitude and parameter relationship curve shown in Fig. 12.13. It clearly shows the inversely proportional relationship between the ($R_i - v_c$).

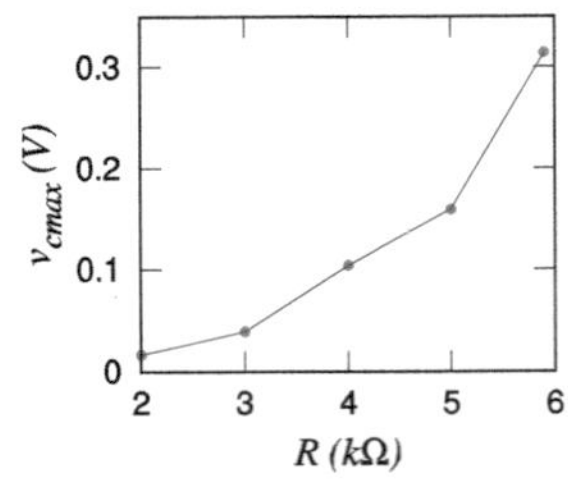

FIGURE 12.10 Amplitude increasing relationship between control parameter and amplitude of BOs in $(R_f - v_c)$ plane.

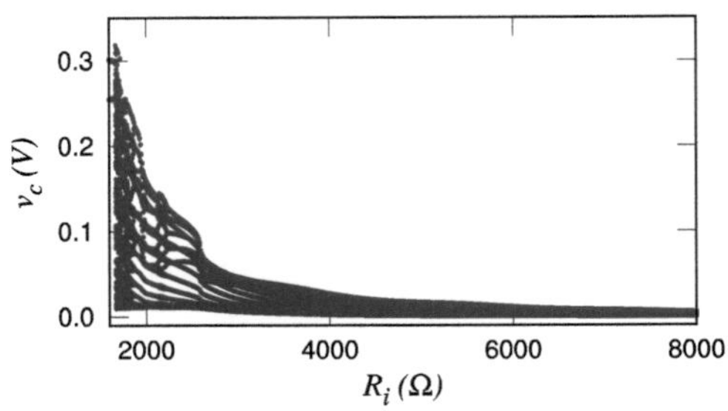

FIGURE 12.11 Bifurcation diagram in the $(R_i - v_c)$ plane for the fixed system parameters $R = 47$ kΩ, $R_f = 5$ kΩ, $C_1 = 1$ nF, $C_2 = 10$ nF, $C = 4.7$ nF, $L = 25$ mH.

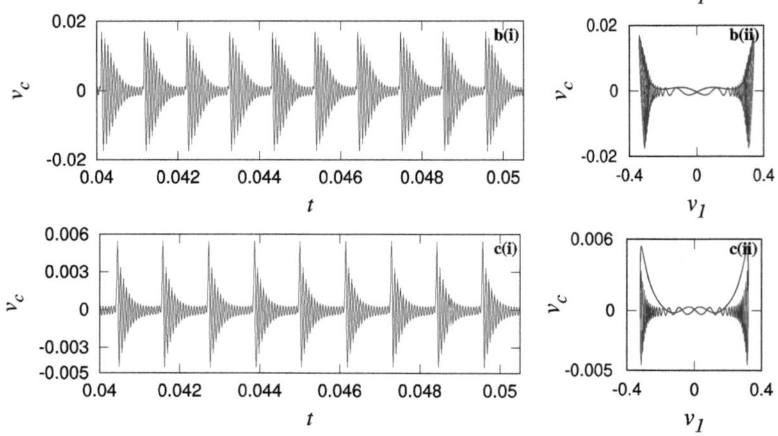

FIGURE 12.12 (i) Time series of $v_c(t)$ plane and corresponding (ii) phase portrait in the $(v_1 - v_c)$ plane. (a)–(c) Amplitude decreasing BOs for the control parameter $R_i = 2$ kΩ, 5 kΩ, 8 kΩ, respectively.

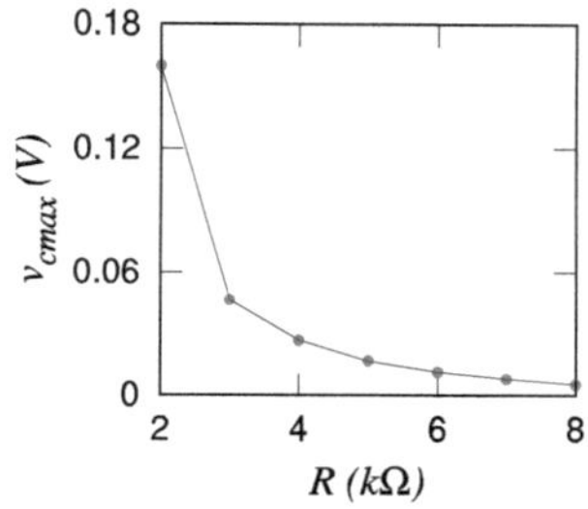

FIGURE 12.13 Amplitude decreasing relationship between control parameter and amplitude of BOs in the $(R_i – v_c)$ plane.

12.7 Conclusion

We have investigated the control of bursting oscillations (BOs) using the well-known Wien-bridge oscillator with second-order diode bridge memristor. The amplitude, spike, and quiescent time are controlled by different parameters of the memristive Wien-bridge oscillator. The quiescent time is increasing between the two successive bursts, as confirmed by the ISI bifurcation diagram. The relationships between control parameter versus amplitude, number of spikes, quiescent time are studied by manually plotted curves. In another set of parameters, we have studied the control amplitude of BOs to changing the negative feedback gain. For the first time, in these studies, we have controlled most of the possible dynamics of BOs.

Acknowledgments

S. Dinesh Vijay acknowledges Bharathidasan University for providing a University Research Fellowship. K. Thamilmaran acknowledges DST-PURSE, Govt. of India for financial support.

References

[1] J.G. Simmons, R.R. Verderber, New conduction and reversible memory phenomena in thin insulating films, Proceedings of the Royal Society of London Series A, Mathematical and Physical Sciences 301 (1967) 77–102.
[2] L.O. Chua, Memristor – the missing circuit element, IEEE Circuit Theory 18 (1971) 507–519.
[3] D.B. Strukov, G.S. Snider, D.R. Stewart, R. Stanley Williams, The missing memristor found, Nature 453 (2008) 80–83.
[4] C. Xu, X. Dong, N.P. Jouppi, Y. Xie, Design implications of memristor-based RRAM cross-point structures, IEEE, 2011, pp. 1–6.
[5] B. Mouttet, Proposal for memristors in signal processing, in: International Conference on Nano-Networks, Springer, 2008, pp. 11–13.
[6] A. Wu, Z. Zeng, Dynamic behaviors of memristor-based recurrent neural networks with time-varying delays, Neural Networks 36 (2012) 1–10.
[7] M.O. Chen, M. Li, Q. Yu, M.O. Chen, B. Bao, Q. Xu, J. Wang, Dynamics of self-excited attractors and hidden attractors in generalized memristor-based Chua's circuit, Nonlinear Dynamics 81 (2015) 215–226.

[8] H. Wu, B. Bao, Z. Liu, Q. Xu, P. Jiang, Chaotic and periodic bursting phenomena in a memristive Wien-bridge oscillator, Nonlinear Dynamics 83 (2016) 893–903.

[9] B.C. Bao, P.Y. Wu, H. Bao, Q. Xu, M. Chen, Numerical and experimental conformations of quasi-periodic behavior and chaotic bursting in third oder memristive oscillator, Chaos, Solitons and Fractals 106 (2018) 161–170.

[10] F. Ascoli, Memristive diode bridge with LCR filter, Electronics Letters 48 (2012) 824–825.

[11] S. Sabarathinam, C.K. Volos, K. Thamilmaran, Implementation and study of the nonlinear dynamics of a memristor-based Duffing oscillator, Nonlinear Dynamics 87 (2017) 37–49.

[12] Z. Guo, J. Wang, Z. Yan, Global exponential synchronization of two memristor-based recurrent neural networks with time delays via static or dynamic coupling, IEEE Transactions on Systems, Man, and Cybernetics: Systems 45 (2014) 235–249.

[13] I.E. Bong, J. Wang, P. Mazumder, CMOS and memristor-based neural network design for position detection, Proceedings of the IEEE 100 (2011) 2050–2060.

[14] A. Thomas, Memristor-based neural networks, Journal of Physics D, Applied Physics 46 (2013) 093001.

[15] S.D. Vijay, S.L. Kingston, K. Thamilmaran, Different transitions of bursting and mixed-mode oscillations in Liénard system, International Journal of Electronics and Communications (AEU) 111 (152898) (2019) 1–13.

[16] E.M. Izhikevich, Neural excitability, spiking and bursting, International Journal of Bifurcation and Chaos 10 (2000) 117–266.

[17] G. Innocenti, A. Morelli, R. Genesio, A. Torcini, Dynamical phases of the Hindmarsh–Rose neuronal model: studies of the transition from bursting to spiking chaos, Chaos 17 (043128) (2007) 1–10.

[18] H. Wu, Y. Ye, M.O. Chen, Q. Xu, B. Bao, Extremely slow passages in low-pass filter-based memristive oscillator, Chaos 97 (2019) 2339–2353.

[19] A.I. Ahamed, K. Srinivasan, K. Murali, M. Lakshmanan, Observation of chaotic beats in a driven memristive Chua's circuit, International Journal of Bifurcation and Chaos 21 (2011) 737–757.

[20] B.C. Bao, P. Jiang, H.G. Wu, F.W. Hu, Complex transient dynamics in periodically forced memristive Chua's circuit, Nonlinear Dynamics 79 (2015) 2333–2343.

[21] A. Buscarino, L. Fortuna, M. Frasca, L.V. Gambuzza, Chaotic circuit based on Hewlett–Packard memristor, Chaos 22 (023136) (2012) 1–9.

[22] B.C. Bao, Z. Liu, J.P. Xu, Steady periodic memristor oscillator with transient chaotic behaviors, Electronics Letters 46 (2010) 228–230.

[23] B.C. Bao, Z.H. Ma, J.P. Xu, Z. Liu, Q. Xu, Memristor chaotic circuit with complex dynamics, International Journal of Bifurcation and Chaos 21 (2011) 2629–2645.

[24] G.A. Leonov, N.V. Kuznetsov, T.N. Mokaev, Homoclinic orbits, and self-excited and hidden attractors in a Lorenz-like system describing convective fluid motion, The European Physical Journal Special Topics 224 (2015) 1421–1458.

[25] C. Li, F. Min, Q. Jin, H. Ma, Extreme multistability analysis of memristor-based chaotic system and its application in image decryption, AIP Advances 7 (25204) (2017) 1–15.

[26] E.B.M. Ngouonkadi, H.B. Fotsin, P.L. Fotso, V.K. Tamba, H.A. Hilda, Bifurcations and multistability in the extended Hindmarsh–Rose neuronal oscillator, Chaos, Solitons and Fractals 85 (2016) 151–163.

[27] A.L. Fradkov, A.Y. Pogromsky, Introduction to Control of Oscillations and Chaos, World Scientific, 1998, 35.

[28] W.L. Ditto, S.N. Rauseo, M.L. Spano, Experimental control of chaos, Physical Review Letters 65 (1990) 3211–3215.

[29] B. Stefanos, G. Celso, Y.C. Lai, H. Mancini, D. Maza, The control of chaos: theory and applications, Physics Reports 329 (2000) 103–197.

[30] S. Eckehard, H.L.K. Sabine, H. Philipp, Control of Self-Organizing Nonlinear Systems, Springer, 2016.

[31] H.C. Pape, R. Mager, Nitric oxide controls oscillatory activity in thalamocortical neurons, Neuron 9 (1992) 441–448.

[32] F.K. Skinner, N. Kopell, E. Marder, Mechanisms for oscillation and frequency control in reciprocally inhibitory model neural networks, Journal of Computational Neuroscience 1 (1994) 69–87.

[33] R.C. Elson, R. Huerta, H.D.I. Abarbanel, M.I. Rabinovich, A.I. Selverston, Dynamic control of irregular bursting in an identified neuron of an oscillatory circuit, Journal of Neurophysiology 82 (1999) 115–122.

[34] N. Faisal, Q. Thomas, M. Louai, V. Horia, N. Jay, L. Jinzhi, G. Leon, Tunable oscillations and chaotic dynamics in systems with localized synthesis, Physical Review E 85 (2012) 046210.

[35] D. Niccolo, S.H. Jan, R. Deep, Controlling oscillations in high-order Discontinuous Galerkin schemes using artificial viscosity tuned by neural networks, Journal of Computational Physics 409 (2020) 109304.

Part II

Applications of mem-elements

Chapter 13

Memristor, mem-systems and neuromorphic applications: a review

Viet-Thanh Pham[a,b,c], Christos Volos[d], and Tomasz Kapitaniak[c]
[a]*Faculty of Electrical and Electronic Engineering, Phenikaa Institute for Advanced Study (PIAS), Phenikaa University, Hanoi, Vietnam,* [b]*Phenikaa Research and Technology Institute (PRATI), A&A Green Phoenix Group, Hanoi, Vietnam,* [c]*Division of Dynamics, Lodz University of Technology, Lodz, Poland,* [d]*Laboratory of Nonlinear Systems, Circuits & Complexity (LaNSCom), Department of Physics, Aristotle University of Thessaloniki, Thessaloniki, Greece*

13.1 Introduction

The memristor was introduced a long time ago [26,28]. There has been an increasing interest in memristor and its applications since the real memristor was implemented at Hewlett–Packard Labs [131,137]. Pershin et al. investigated memristive model of amoeba learning [109]. Itoh and Chua introduced memristor cellular automata and memristor discrete time cellular neural networks [64]. Memristive adaptive filters were studied in [39]. Pershin and Ventra demonstrated associative memory with memristive neural networks experimentally [110]. Memristor applications for programmable analog ICs were presented by Shin et al. [126]. Interestingly, Chua proofed that resistance switching memories were memristors [27]. Corinto et al. discovered memristive elements' current–voltage characteristics for pattern recognition systems [33]. The memristor bridge synapse-based neural network and its learning were reported in [3], while memristive devices for computing were discussed in [153]. Applications in secure communications of memristive systems were proposed in [84,122].

Published works in memristor field boomed last years, leading to the appearance of useful review works. Prodromakis and Toumazou wrote a short review on memristive devices and applications in 2010 [119]. The authors focused on the capacity of memristive networks on non-volatile memory, dynamic load, neuromorphic systems, and image processing. In 2018, recent advances in memristive materials for artificial synapses were presented in [73]. Various materials for memristor-based artificial synapses were reviewed, especially organic materials, two-dimensional materials as well as emerging materials, such as halide perovskites and low-dimensional materials. Li el al. reviewed memristor devices in neuromorphic computing from the viewpoints of materials sciences and

Mem-elements for Neuromorphic Circuits with Artificial Intelligence Applications
https://doi.org/10.1016/B978-0-12-821184-7.00022-0

device challenges [82]. Based on material system engineering, different memristors were summarized, for example oxide based memristors, phase change memristors, and other memristors (using heterogeneous materials, organic materials, electrolytic materials, chalcogenides). The authors explained two materials challenges of the memristor dielectric and electrode: stochasticity, and CMOS compatibility. In addition, the authors also discussed two challenges and strategies for memristor integration: three-dimension stacking, and scaling [82]. Sung et al. presented a review on memristive hardware for neuromorphic computation [132]. The authors analyzed the competitiveness of the memristor-based neuromorphic device with the aim to find the appropriate position of the memristor in the future artificial intelligence ecosystem. By combining industrial trends and academic pursuits, authors showed that publications (both papers and patents) on memristor-based neuromorphic computation concentrated on main topics such as neuron, synapse, neural network, training and learning, and neural processor. Ravichandran et al. provides an overview of various artificial neural networks based on memristive devices. The authors analyzed three memristor-based neuromorphic architectures: in-situ analog arithmetic in crossbars (ISAAC) [124], processing-in-memory solution (PRIME) [24], area efficient and power efficient (AEPE) [135]. Implementations of the memristor neural networks were illustrated via one transistor one memristor (1T1R) structure [154] and biologically plausible network [108,144]. Khalid surveyed various memristor models, characteristics as well as potential applications [71]. In particular, nonlinear ionic drift model and tunnel barrier memristive model were described. It is interesting that neuromemristive circuits for edge computing were reviewed by Krestinskaya et al. [76]. Numerous neuromorphic CMOS-memristive architectures were provided in order to integrate into edge computing devices. Although there were various review papers related to the memristor field, recent advanced results about memristive systems, especially for neuromorphic applications, should still be surveyed.

In this chapter, we provide an overview of recent researches related to memristor, and mem-systems and their neuromorphic applications. The main aim of this chapter is to show the neuromorphic circuits/architectures and artificial intelligence applications.

13.2 Memristor and mem-systems

13.2.1 Memristor

The definition of a memristor was introduced by Professor L.O. Chua in 1971 [26]. As illustrated in Fig. 13.1, the three classical elements are the resistor, the inductor, and the capacitor, while memristor is considered as the fourth basic circuit element [137].

Let consider a memristor in Fig. 13.2, where v_M is the voltage across the memristor, and i_M is the current via the memristor. The memristor is defined in terms of a relationship between the charge q and the magnetic flux φ. Therefore,

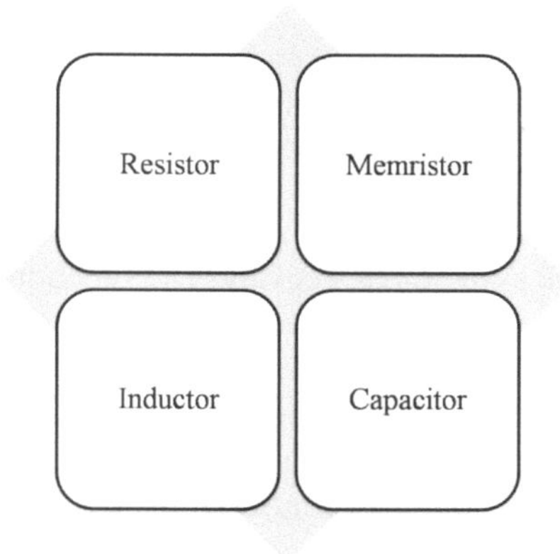

FIGURE 13.1 Four fundamental electrical elements: resistor, inductor, capacitor, and memristor.

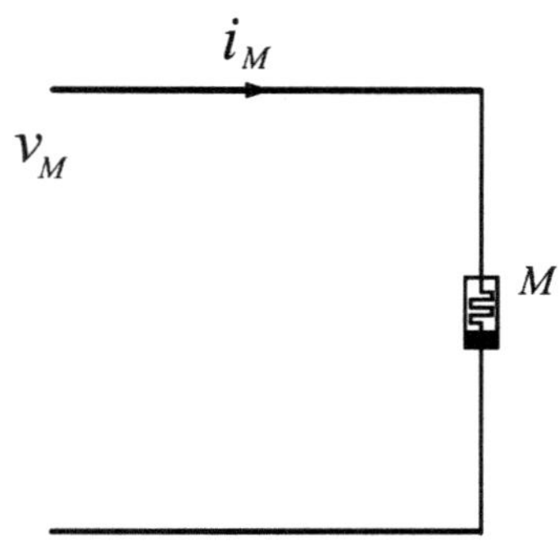

FIGURE 13.2 Memristor M with the voltage v_M, and the current i_M.

a charge-controlled memristor is given by

$$v_M = M(q)\, i_M \tag{13.1}$$

with the memristance

$$M(q) = \frac{d\varphi(q)}{dq} \tag{13.2}$$

The flux-controlled memristor is given by

$$i_M = W(\varphi)\, v_M \tag{13.3}$$

with the memductance

$$W(\varphi) = \frac{dq(\varphi)}{d\varphi} \tag{13.4}$$

It is worth noting that in practical there are memristive systems, which were generalized from the definition of a memristor [26], described by the general form

$$\begin{cases} \dot{x} = f(x, u, t) \\ y = g(x, u, t)\, u \end{cases} \tag{13.5}$$

In memristive system (13.5), u, y, and x present the input, the output and the state of the memristive system, respectively. The function f is a continuous n-dimensional vector function. The function g is a continuous scalar function.

Different memristor models have been constructed to support simulation and design processes [14,90,94,97,127]. Commonly used memristor models are the linear dopant drift model [116,131], nonlinear dopant drift model [15,69,118], TEAM model [78], modified Biolek's model [13], and data driven model [95]. It is worth noting that some memristor models are suitable for fast and large-scale simulations [13,95]. Such memristor models have been also compared in [76].

A great deal of previous research into memristor has focused on its implementation [51]. Previous studies have explored various memeristors, such as titanium dioxide memristor, polymeric memristor, ferroelectric memristor, and the spintronic memristor [22,23,36,125,150]. Some studies have assessed the efficacy of memristor materials compatible with the CMOS fabrication process [42]. Issues related to memristor implementation are still open problems.

13.2.2 Memristive systems

Memristive systems found potential applications in various areas [136]. In this section, we only consider briefly four main areas with the presence of memristive systems: nonlinear circuits, storage technology, communications, and neuromorphic systems (see Fig. 13.3).

The intrinsic nonlinear characteristic of the memristor could be exploited in nonlinear circuits [6,7,32,38]. There is a large volume of published studies describing the role of the memristor in memristor oscillators [63,65]. Especially, novel memristor-based oscillators displayed complex dynamics like chaos [8,66,103]. Muthuswamy implemented memristor-based chaotic circuits with analog circuit components [101]. Based on cellular nonlinear networks, Buscarino realized memristive chaotic circuits [18]. Muthuswamy and Chua introduced simplest chaotic circuits with memristors [102]. Chaos was generated in a memristive circuit with time-delayed feedback [156,157]. Simple memristive time-delayed chaotic systems were built in [111], while a novel memristive time-delay chaotic system without equilibrium points was proposed in [112]. In addition, memristors were applied to design hyperchaotic oscillators [16,44]. Fitch et al. combined a memristor with cubic nonlinear characteristics and a modified canonical Chua circuit to obtain hyperchaos [44]. The authors in [17] presented a six-dimensional hyperchaotic oscillator. Li et al. developed four-dimensional hyperchaotic memristive systems [80,81]. Memristor-based hyperchaotic systems without equilibrium were also reported [113,115]. For a

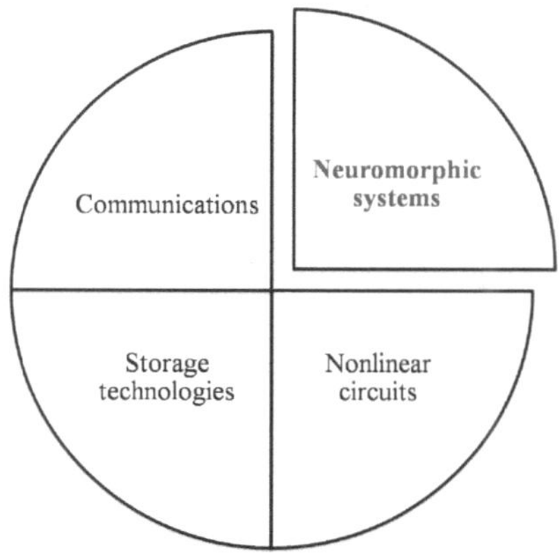

FIGURE 13.3 Four main areas with the presence of memristive systems.

comprehensive literature review on research trends for implementing chaotic systems with memristors, the reader can refer to the published chapter [114].

As have been known, silicon-based semiconductor memories belong to two categorizes: volatile memory and nonvolatile memory [5,12,92]. Volatile memory needs power to maintain the stored information. In contrast to volatile memory, the stored information in nonvolatile memory is retained when the power supply is turned off. It is now well established from a variety of studies that memristor is a potential candidate for emerging nonvolatile memory technologies [92]. The authors found that power consumption of the memristors were smaller compared to the existing memory devices [42,54]. Taherinejad et al. explored the possible storage of multi-bit data in a single memristor [133]. The memristor-based memory cell had less noise margins and stored non-binary data [151]. Discovering application of the memristor for developing emerging memories is still an attractive research trend.

More recent attention has focused on memristor as an emerging tool in communications and networking [76]. Modern networks such as Internet of Things (IoT) and Unmanned Aerial Vehicle (UAV) require numerous sensors. Different kinds of memristor sensors, especially memristive-biosensors, were introduced [20,52,53,121,143]. Memristive-biosensors were used for early detection of prostate cancer [139], and PSA-IgM detection [140]. Gao et al. introduced emerging physical unclonable functions [47]. Interestingly, the memristive crypto primitive was fundamental for building highly secure physical unclonable functions [48]. The authors proposed the block diagram of the memristive-based key generation scheme [1]. Moreover, memristor-based security approach was developed for secure IoT conference communication in which keys were generated using unique memristor devices. Security characteristics of such an approach were analyzed via mutual authentication, confidentiality, integrity, authorization, and replay attacks [1]. In addition, applications in secure communications of memristive systems were reported in other published works [84,122]. The memristor with nanoscale features will overcome various problems and challenges of communications and networking areas.

Neuromorphic system has been the subject of many classic studies [129]. There are opportunities for electronic implementation of neuromorphic system because of the rapid development in modern microelectronics [91]. Neuromorphic computing is an important research direction having significant influence on the development of artificial intelligence [98,160]. Many applications of neuromorphic computing in practice were found in computer vision, pattern recognition, natural language processing, and robotics. The memristor has been introduced as a new breakthrough technique for realizing neuromorphic computing [141]. Memristor-based neuromorphic computing is expected to solve current computational bottleneck. Memristive neuromorphic system exhibits advanced features, such as requiring much lower power supply voltage, consuming lower power consumption, having nanoscale size. Therefore, a very promising memristive era is opening for future practical systems [46,75,161]. Memristive neuromorphic systems are discussed further in Section 13.3.

13.3 Neuromorphic systems

Fig. 13.4 presents briefly two simplified neurons and their connection. As can be seen in Fig. 13.4, a neuron includes dendrites, soma, and axon [67]. A neuron can have various dendrites, which collect signals from different neurons. The soma is an important part to perform nonlinear processing. In general, an output signal is generated at a certain threshold and is passed to axon. We suppose that the neuron i (the presynaptic cell) transmits signal to the neuron j (the postsynaptic cell) through a junction, called synapse.

Inspired from biological concepts, artificial neuron models have been proposed. A simple linear neuron model [70] is shown in Fig. 13.5, in which weighted connections described synapses. The inputs are x_i while the output is y. By applying the weighted summation and a threshold function $f()$, the output can be obtained by

$$y = f\left[\sum_{i=1}^{N} w_i x_i\right] \tag{13.6}$$

Artificial neuron network has been constructed by connecting many neurons [55,86]. As illustrated in Fig. 13.6, this artificial neuron network example includes layers: an input layer, three hidden layers, and an output layer. It is considered as a deep neural network because there are multiple layers between the input and output layers [50]. The exact connections between neurons does not drawn, but they are expressed by weight matrices. From the view point of the memristor, there are three main research areas related to neuromorphic systems (see Fig. 13.7). In the next sections, we discuss further each of these research areas.

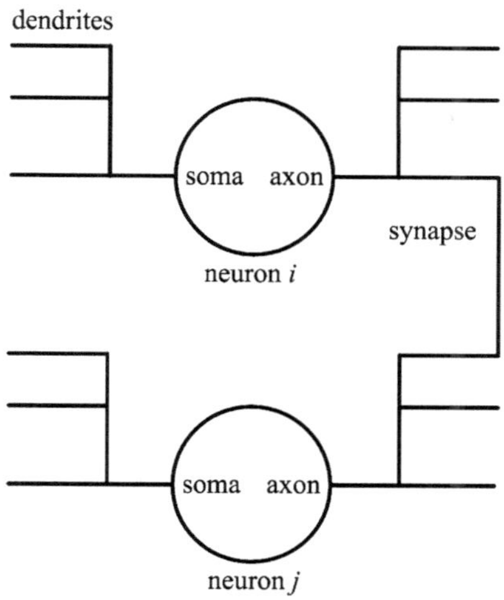

FIGURE 13.4 Simplified neurons with synapse [67].

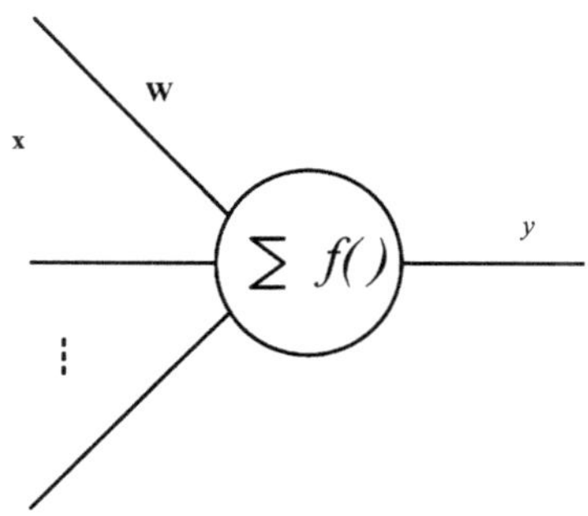

FIGURE 13.5 Artificial neuron with weighted synapse [67].

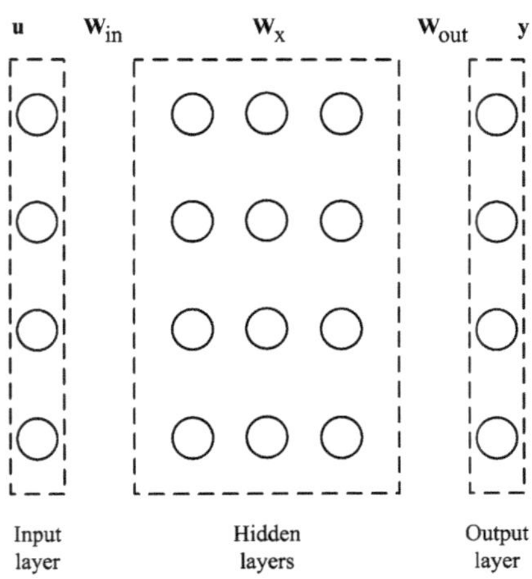

FIGURE 13.6 Artificial neuron network including various layers [67].

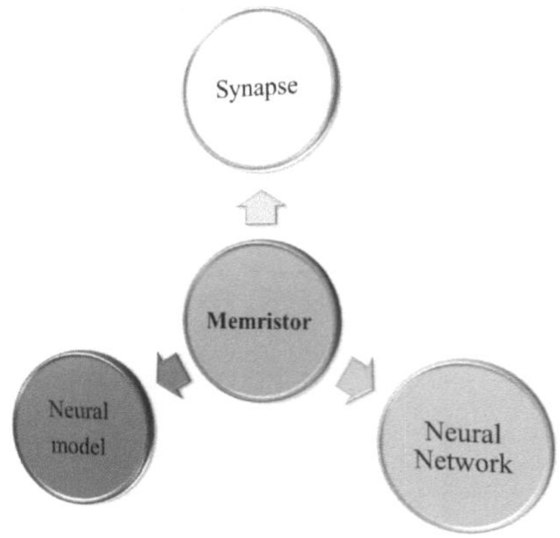

FIGURE 13.7 Three main research areas of the memristor related to neuromorphic systems.

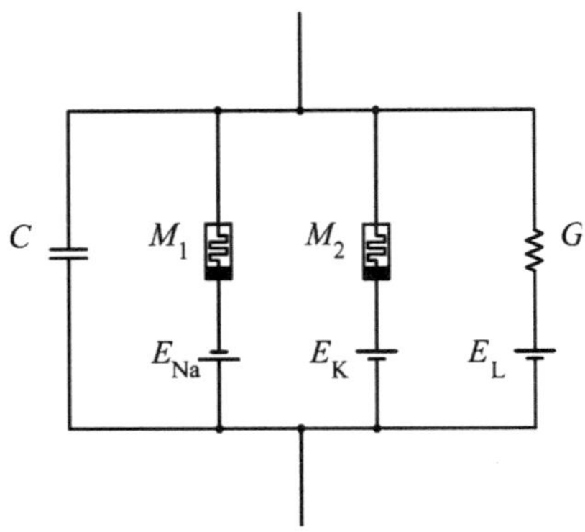

FIGURE 13.8 Circuit presentation of memristive Hodgkin–Huxley axon [25].

13.3.1 Neuron model

Hodgkin and Huxley introduced a model of the squid axon, which is important in neuroscience [57–59]. Hodgkin and Huxley model was presented by electrical circuit with seven circuit elements. The memristive Hodgkin–Huxley axon was made by Chua et al. by using two memristors, a capacitor, a resistor, and three batteries as illustrated in Fig. 13.8 [25]. Two memristors were a sodium ion-channel memristor (M_1) and a potassium ion-channel memristor (M_2).

By simplifying the Hodgkin and Huxley model, the FitzHugh–Nagumo model was obtained [45,104,120]. The FitzHugh–Nagumo model was given by

$$\begin{aligned} \dot{x} &= x - \tfrac{1}{3}x^3 - y + I_{ext} \\ \dot{y} &= \tfrac{1}{c}\left(x - by + a\right) \end{aligned} \tag{13.7}$$

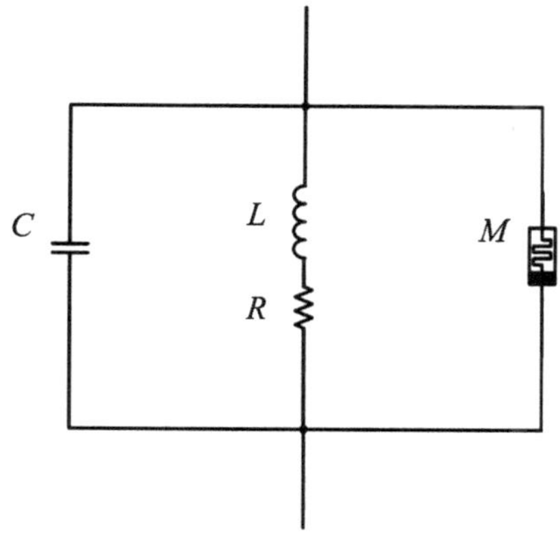

FIGURE 13.9 Presentation of memristive FitzHugh–Nagumo circuit [155].

where x presents the membrane potential, and y is a recovery variable. Here a, b, c are model's parameter and I_{ext} is a stimulus current. It is interesting that Nagumo et al. introduced an equivalent circuit with a tunnel diode [104]. The employment of the memristor could lead to memristor-based FitzHugh–Nagumo circuit [11,155]. As shown in Fig. 13.9, the memristor-based FitzHugh–Nagumo circuit includes a capacitor, an inductor, a resistor and a memristor [155]

By using three coupled first order differential equations, Hindmarsh and Rose described a model by

$$\begin{aligned}\dot{x} &= y - ax^3 + bx^2 - z + I_{ext}\\ \dot{y} &= c - y - dx^2\\ \dot{z} &= r\left[s\left(x + 1.6\right) - z\right]\end{aligned} \tag{13.8}$$

In Eq. (13.8), x, y, z are the membrane potential, slow current for recovery variable, and adaption current, respectively [56]. The parameters in the model are a, b, c, r, and s while I_{ext} is the external forcing current. Hindmarsh–Rose neuron model displays numerous behaviors of neuron. When studying the effect of electromagnetic induction in the biological system, Lv et al. built an improved Hindmarsh–Rose neuron model by adding a variable as magnetic flux which adjusted the membrane potential via a memristor [9,81,101]. The improved Hindmarsh–Rose neuron model has the form

$$\begin{aligned}\dot{x} &= y - ax^3 + bx^2 - z + I_{ext} - k_1 x\rho\left(\phi\right)\\ \dot{y} &= c - y - dx^2\\ \dot{z} &= r\left[s\left(x + 1.6\right) - z\right]\\ \dot{\phi} &= x - k_2\phi\end{aligned} \tag{13.9}$$

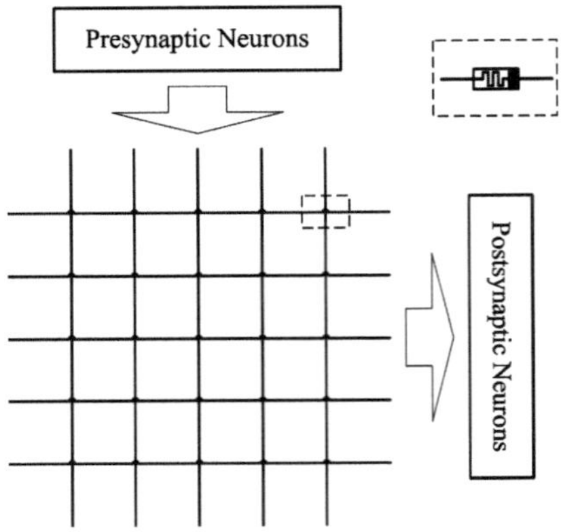

FIGURE 13.10 General crossbar array architecture with memristive synapses [132].

It is noted that the magnetic flux across membrane is ϕ while the memory conductance of a magnetic flux-controlled memristor is

$$\rho(\phi) = \alpha + 3\beta\phi^2 \tag{13.10}$$

with two parameters (α, β) [81]. A complete dynamical analysis of such a neuron under magnetic flow effect was reported in [106].

Wang et al. used diffusive memristors as leaky integrate-and-fire neurons [144]. A diffusive memristor includes a SiO_xN_y:Ag layer between two Pt electrodes. By considering a leaky integrate and fire neuron with noise, stochastic memristor-based neuron model was presented in [105]. Yakopcic et al. combined a memristor with two operational amplifiers to realize a neuron [152].

13.3.2 Synapse

A considerable amount of literature has been published on implementation of the synapse with memristor and related issues [75,141]. A typical implementation is a memristive cross bar array in which a memristor describes a synapse [158,159]. As illustrated in Fig. 13.10, memristors connect the input neurons (presynaptic ones) and the output neurons (postsynaptic ones). As a result, and a memristor-based synaptic weight crossbar is obtained.

Based on the common crossbar array architecture of memristor synapses, there were different versions of the synapse implementation. Prezioso et al. used the synapse with two memristors [117] while a compact memristor-based dynamic synapse with two memristors and one resistor was introduced in [61]. Synapse with one transistor and one memristor (named 1T1M) solved sneak path problems [146,154]. Other work reporting kinds of memristive synapses was [2,72,74,130]. Memristor synaptic weights face three challenging issues: nonvolatility, linearity, and multilevel [132]. Furthermore, the learning mechanism should be determined when designing memristive synapses.

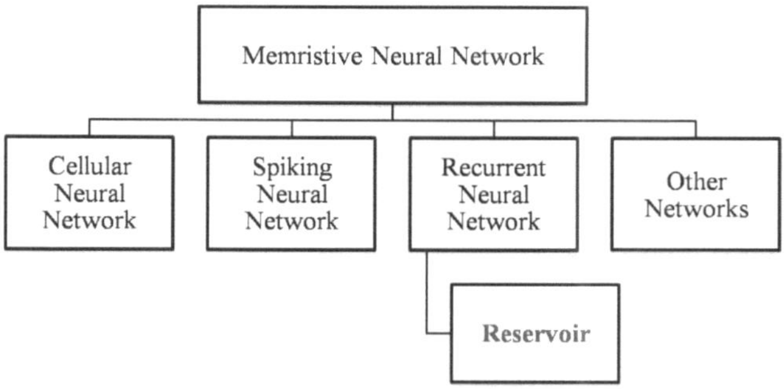

FIGURE 13.11 Taxonomy of memristive neural network architectures.

13.3.3 Neural network

Numerous memristive architectures have proposed for neuromorphic computing purposes. In this section, we focus on some fundamental memristive neural network architectures (as shown in Fig. 13.11).

Cellular Neural/Nonlinear Networks (CNNs) were invented by Chua in the 1980s [30,31]. Previous researches established that the CNN architecture includes of a number of cells. Each cell was designed by using linear capacitors, linear resistors, linear and nonlinear voltage-controlled current sources and independent sources. Therefore CNN architecture matched for VLSI implementation and was useful for signal processing, pattern formation, and bio-inspired robotic visions [29,30]. Memristive synapses were applied in new memristive-CMOS architectures, which used for image filtering, and edge detection [41,62,89,138].

Considering as the third generation of neural network models [87], spiking neural networks mimic natural brain processing [49]. It is worth noting that spike timing dependent plasticity permits un-supervised learning supporting real-time adaptive systems [34]. Spiking neural network hardware implementation provided high parallelism and low power consumption [21]. The memristor was suitable for building spiking neural networks [4,100,107,108,123]. Memristive spiking neural network showed good performance for handwritten digits recognition, letter recognition, and pattern recognition [35,148].

Recurrent neural network was an attractive architecture for VLSI integration [79,88,142]. Mathematical analyses of memristive recurrent neural networks were presented in [10,83]. However, realization of memristive recurrent neural networks was still in first steps. Xavier et al. proposed a continuous-time neural network using memristors [149]. The behavioral and electrical analysis of Xavier's memristive recurrent neural network was done in Simulink-SPICE simulation. Deng et al. applied a bio-plausible recurrent memristive network for the learning of various timing patterns as well as complex spatiotemporal pattern of human motor [37]. The authors fabricated iron oxide memristor-based synapses.

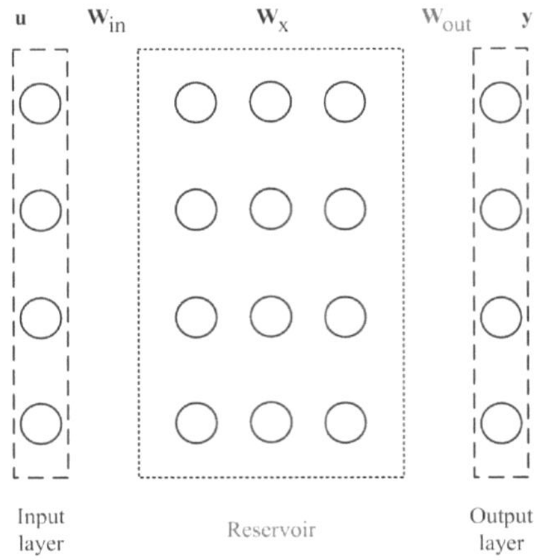

FIGURE 13.12 Reservoir computing system [134]. It is noted that only the output layer weights must be trained.

In recent years, there has been an increasing interest in other emerging memristor-based neuromorphic architectures, such as long short term memory network [128], hierarchical temporal memory architecture [43], and reservoir computing [145]. Reservoir computing is presented further in the next section. Exploring efficient memristive neuromorphic computing paradigm is still an interesting research topic.

13.4 Reservoir computing

Previous studies have explored the capability of recurrent neural networks for processing temporal data [60,147]. Compared with conventional feedforward neural networks, recurrent neural networks obtained better performance for discovering temporal correlations in the data. In RRNs, outputs depend on not only current inputs but also previous states of neurons. However, there are still some problems, such as efficient learning algorithms, and complexity architectures. In order to solve such problems, reservoir computing [68,85] is proposed as a neural network-based computing paradigm (see a conceptually illustration in Fig. 13.12). In the reservoir computing system, there is a vital part called "reservoir" connecting with the input [134]. It is worth noted that the connections in the reservoir are fixed and never changed during the testing phases of the network. As a result, training does not be required. However, the temporal inputs are mapped into a high-dimensional space through evolution of neurons. The internal reservoir states are processed at the output. Therefore, the training occurs only for the weights in the readout function, that reduces the training cost significantly.

Two different approaches for memristor simulation models were theoretically studied in [19]. They were an extension of Strukov's model and an equivalent Wiener model approximation. Delay and binary operator tasks were presented. Ability of the implementation of reservoir computing in memristor networks was discussed. The authors indicated that these models could lead to useful memristor-based reservoir computing systems but their computational performance could be different. Therefore, they suggested that experimental modeling research is required to develop accurate volatile memristor models [19].

In [93], a hardware reservoir was implemented with bi-stable memristive synapses. The authors designed two synapse circuits: an excitatory memristive synapse circuit and an inhibitory memristive synapse circuit. Each of the synapse circuits had two bi-stable memristors in parallel, two diode-connected transistors, and a current mirror. It is simple to see that the weights of the synapses were determined by their memristor conductances. The inhibitory synapse and excitatory synapse were similar to a GABAergic synapse and a glutamatergic synapse in a biological brain [93]. In addition, a neuron circuit was built with a MOSFET differential pair and a current mirror realizing sigmoid activation function. By using the hardware reservoir, the epileptic seizure signal was detected with the accuracy 85%.

Kudithipudi et al. developed a neuromemristive reservoir computing architecture with a novel readout layer [77]. The authors used a memristor crossbar to implement the synaptic weights. In the other word, trainable weights were introduced via the crossbar. It is noted that linear output neuron was a transistor-based opamp circuit. There were two memristors for each reservoir output (one memristor inhibits the output and one excites the output). However, both reservoir neuron circuits and reservoir synapse circuits were based on transistors. Hyperbolic tangent neuron circuit improved reservoir dynamics supporting better classification feature.

A reservoir computing system was implemented with a dynamic memristor array [40]. The memristor-based reservoir computing system processed effectively temporal information tasks such as handwritten digit recognition and signal prediction. Memristive reservoir was constructed by tungsten oxide WO_x memristors with short term dynamics, which were designed during device fabrication. In this work, the short term memory effect of the memristor was considered as a time constant around 50 ms. Reservoir computing system has a readout network, which can be trained to generate the desired output. Similarly, a memristor-based reservoir computing system was applied to recognize the speech signal and forecast chaotic time-series [99]. The hardware implementation used a 32×32 WOx memristor array.

In the work of Midya et al., authors introduced a physical memristive reservoir computing system [96]. By combining diffusive memristor-based reservoir and drift memristor-based readout layer, the system was tested via handwritten digit classification task. The layer diffusive memristor-based reservoir was built by diffusive memeristors with short term dynamics. The readout layer was developed in a memristor-based 1T1R crossbar array.

Based on recent memristive reservoir computing systems, it seems that they can provide a generic architecture for signal processing applications. However, important challenges and open issues should require future research.

13.5 Conclusion

This chapter summarized recent results of the memristor and memeristive system for neuromorphic aims. The recent progress in implementation of memristive systems for neuromorphic applications was also surveyed. Practical memristive neuromorphic application is still an attractive research area. By using neuromorphic computing, the drawbacks of current computing systems will be overcome gradually in future studies. We believe that memristive neuromorphic architectures will appear widely in different areas ranging from deep reinforcement learning, Internet of Things, robotics, to unmanned aerial vehicle (UAV) networks.

Acknowledgment

This work has been supported by the Polish National Science Centre, MAESTRO Programme (Project No. 2013/08/A/ST8/00780) and the OPUS Programme (No. 2018/29/B/ST8/00457)

References

[1] H. Abunahla, B. Mohammad, Memristor Technology: Synthesis and Modeling for Sensing and Security Applications, Springer, Switzerland, 2018.

[2] S.P. Adhikari, H. Kim, R.K. Budhathoki, C. Yang, L.O. Chua, A circuit-based learning architecture for multilayer neural networks with memristor bridge synapses, IEEE Trans. Circuits Syst. I, Regul. Pap. 62 (2015) 215–223.

[3] S.P. Adhikari, C. Yang, H. Kim, L.O. Chua, Memristor bridge synapse-based neural network and its learning, IEEE Trans. Neural Netw. Learn. Syst. 23 (2012) 1426–1435.

[4] M. Al-Shedivat, R. Naous, G. Cauwenberghs, K.N. Salama, Memristors empower spiking neurons with stochasticity, IEEE J. Emerg. Sel. Top. Circuits Syst. 5 (2015) 242–253.

[5] S. Aritome, Advanced flash memory technology and trends for files storage application, in: Proc. IEEE IEDM Tech. Dig., 2000, pp. 763–766.

[6] A. Ascoli, F. Corinto, Memristor models in a chaotic neural circuit, Int. J. Bifurc. Chaos 23 (2013) 1350052.

[7] A. Ascoli, F. Corinto, V. Senger, R. Tetzlaff, Memristor model comparison, IEEE Circuits Syst. Mag. 13 (2013) 89–105.

[8] B. Bao, Z. Liu, B. Xu, Dynamical analysis of memristor chaotic oscillator, Acta Phys. Sin. 59 (2010) 3785–3793.

[9] B.C. Bao, Z. Liu, J.P. Xu, Steady periodicmemristor oscillator with transient chaotic behaviors, Electron. Lett. 46 (2010) 228–230.

[10] G. Bao, Z. Zeng, Y. Shen, Region stability analysis and tracking control of memristive recurrent neural network, Neural Netw. 98 (2018) 51–58.

[11] H. Bao, W. Liu, M. Chen, Hidden extreme multistability and dimensionality reduction analysis for an improved non-autonomous memristive FitzHugh–Nagumo circuit, Nonlinear Dyn. 96 (2019) 1879–1894.

[12] R. Bez, E. Camerlenghi, A. Modelli, A. Visconti, Introduction to flash memory, Proc. IEEE 91 (2003) 489–502.

[13] D. Biolek, Z. Kolka, V. Biolkova, Z. Biolek, Memristor models for spice simulation of extremely large memristive networks, in: Proc. IEEE Int. Symp. Circuits Syst. (ISCAS'2016), 2016, pp. 389–392.
[14] D. Biolek, Z. Kolka, V. Biolkova, Z. Biolek, M. Potrebic, D. Tosic, Modeling and simulation of large memristive networks, Int. J. Circuit Theory Appl. 46 (2018) 50–65.
[15] Z. Biolek, D. Biolek, V. Biolkova, SPICE model of memristor with nonlinear dopant drift, Radioengineering 18 (2009) 210–214.
[16] A. Buscarino, L. Fortuna, M. Frasca, L.V. Gambuzza, A chaotic circuit based on Hewlett–Packard memristor, Chaos 22 (2012) 023136.
[17] A. Buscarino, L. Fortuna, M. Frasca, L.V. Gambuzza, A gallery of chaotic oscillators based on hp memristor, Int. J. Bifurc. Chaos 22 (2012), 1330015–14.
[18] A. Buscarino, L. Fortuna, M. Frasca, L.V. Gambuzza, G. Sciuto, Memristive chaotic circuits based on cellular nonlinear networks, Int. J. Bifurc. Chaos 22 (2012) 1250070–1–13.
[19] J.P. Carbajal, J. Dambre, M. Hermans, B. Schrauwen, Memristor models for machine learning, Neural Comput. 27 (2015) 725–747.
[20] S. Carrara, D. Sacchetto, M.-A. Doucey, C. Baj-Rossi, G.D. Micheli, Y. Leblebici, Memristive-biosensors: a new detection method by using nanofabricated memristors, Sens. Actuators B, Chem. 171 (2012) 449–457.
[21] S. Carrillo, J. Harkin, L.J. McDaid, F. Morgan, S. Pande, S. Cawley, B. McGinley, Scalable hierarchical network-on-chip architecture for spiking neural network hardware implementations, IEEE Trans. Parallel Distrib. Syst. 24 (2013) 2451–2461.
[22] A. Chanthbouala, V. Garcia, R.O. Cherifi, K. Bouzehouane, S. Fusil, X. Moya, S. Xavier, H. Yamada, C. Deranlot, N.D. Mathur, M. Bibes, A. Barthelemy, J. Grollier, A ferroelectric memristor, Nat. Mater. 11 (2012) 860–864.
[23] A. Chanthbouala, R. Matsumoto, J. Grollier, V. Cros, A. Anane, A. Fert, A.V. Khvalkovskiy, K.A. Zvezdin, K. Nishimura, Y. Nagamine, H. Maehara, K. Tsunekawa, A. Fukushima, S. Yuasa, Vertical-current-induced domain-wall motion in MgO-based magnetic tunnel junctions with low current densities, Nat. Phys. 7 (2011) 626–630.
[24] P. Chi, S. Li, C. Xu, T. Zhang, J. Zhao, Y. Liu, Y. Wang, Y. Xie, PRIME: a novel processing-in-memory architecture for neural network computation in reRAM-based main memory, in: Proc. ACM/IEEE 43rd Annual International Symposium on Computer Architecture (ISCA'2016), 2016, pp. 27–39.
[25] L. Chua, V. Sbitnev, H. Kim, Hodgkin–Huxley axon is made of memristors, Int. J. Bifurc. Chaos 22 (2012) 1230011.
[26] L.O. Chua, Memristor – the missing circuit element, IEEE Trans. Circuit Theory 18 (1971) 507–519.
[27] L.O. Chua, Resistance switching memories are memristors, Appl. Phys. A 102 (4) (2011) 765–783.
[28] L.O. Chua, S.M. Kang, Memristive devices and system, Proc. IEEE 64 (1976) 209–223.
[29] L.O. Chua, T. Roska, Cellular Neural Networks and Visual Computing, Cambridge University Press, Cambridge, 2002.
[30] L.O. Chua, L. Yang, Cellular neural networks: application, IEEE Trans. Circuits Syst. 35 (1988) 1273–1290.
[31] L.O. Chua, L. Yang, Cellular neural networks: theory, IEEE Trans. Circuits Syst. 35 (1988) 1257–1272.
[32] F. Corinto, A. Ascoli, M. Gilli, Nonlinear dynamics of memristive oscillators, IEEE Trans. Circuits Syst. I, Regul. Pap. 58 (2011) 1323–1336.
[33] F. Corinto, A. Ascoli, M. Gilli, Analysis of current–voltage characteristics for memristive elements in pattern recognition systems, Int. J. Circuit Theory Appl. 40 (2012) 1277–1320.
[34] E. Covi, S. Brivio, A. Serb, T. Prodromakis, M. Fanciulli, S. Spiga, Analog mem-ristive synapse in spiking networks implementing unsupervised learning, IEEE J. Emerg. Sel. Top. Circuits Syst. 10 (2016) 482.

[35] E. Covi, S. Brivio, A. Serb, T. Prodromakis, M. Fanciulli, S. Spiga, Hfo_2-based memristors for neuromorphic applications, in: Proc. IEEE Int. Symp. Circuits Syst. (ISCAS), 2016, pp. 393–396.
[36] M. Crupi, L. Pradhan, S. Tozer, Modelling neural plasticity with memristors, IEEE Canadian Rev. 68 (2012) 10–14.
[37] L. Deng, G. Li, N. Deng, D. Wang, Z. Zhang, W. He, H. Li, J. Pei, L. Shi, Complex learning in bio-plausible memristive networks, Sci. Rep. 5 (2015) 10684.
[38] T. Driscoll, Y.V. Pershin, D.N. Basov, M.D. Ventra, Chaotic memristor, Appl. Phys. A 102 (2011) 885–889.
[39] T. Driscoll, J. Quinn, S. Klien, H.T. Kim, B.J. Kim, Y.V. Pershin, M.D. Ventra, D.N. Basov, Memristive adaptive filters, Appl. Phys. Lett. 97 (2010) 093502.
[40] C. Du, F. Cai, M.A. Zidan, W. Ma, S.H. Lee, W.D. Lu, Reservoir computing using dynamic memristors for temporal information processing, Nat. Commun. 8 (2017) 2204.
[41] S. Duan, X. Hu, Z. Dong, L. Wang, P. Mazumder, Memristor-based cellular nonlinear/neural network: design analysis and applications, IEEE Trans. Neural Netw. Learn. Syst. 26 (2015) 1202–1213.
[42] K. Eshraghian, K.-R. Cho, O. Kavehei, S.-K. Kang, D. Abbott, S.-M.S. Kang, Memristor MOS content addressable memory (MCAM): hybrid architecture for future high performance search engines, IEEE Trans. Very Large Scale Integr. Syst. 19 (2011) 1407–1417.
[43] D. Fan, M. Sharad, A. Sengupta, K. Roy, Hierarchical temporal memory based on spin-neurons and resistive memory for energy-efficient brain-inspired computing, IEEE Trans. Neural Netw. Learn. Syst. 27 (2016) 1907–1919.
[44] A.L. Fitch, D. Yu, H.H.C. Iu, V. Sreeram, Hyperchaos in a memristor-based modified canonical Chua's circuit, Int. J. Bifurc. Chaos 22 (2012), 1250133–8.
[45] R. Fitzhugh, Impulses and physiological states in theoretical models of nerve membrane, Biophys. J. 1 (1961) 445–466.
[46] J. Fu, Z. Liao, J. Wang, Memristor-based neuromorphic hardware improvement for privacy-preserving ANN, IEEE Trans. Very Large Scale Integr. Syst. 27 (2019) 2745–2754.
[47] Y. Gao, D.C. Ranasinghe, S.F. Al-Sarawi, O. Kavehei, D. Abbott, Memristive crypto primitive for building highly secure physical unclonable functions, Sci. Rep. 5 (2015) 12785.
[48] Y. Gao, D.C. Ranasinghe, S.F. Al-Sarawi, O. Kavehei, D. Abbott, Emerging physical unclonable functions with nanotechnology, IEEE Access 5 (2016) 61–80.
[49] W. Gerstner, W.M. Kistler, Spiking Neuron Models: Single Neurons, Populations, Plasticity, Cambridge University Press, Cambridge, 2002.
[50] I. Goodfellow, Y. Bengio, A. Courville, Deep Learning, MIT Press, USA, 2016.
[51] L. Guckert, E.E. Swartzlander Jr., System Design with Memristor Technologies, The Institution of Engineering and Technology, London, UK, 2018.
[52] N.S.M. Hadis, A.A. Manaf, S.H. Herman, Trends of deposition and patterning techniques of tio_2 for memristor based bio-sensing applications, Microsyst. Technol. 19 (2013) 1889–1896.
[53] N.S.M. Hadis, A.A. Manaf, S.H. Herman, S.H. Ngalim, Roff/ron ratio of nano-well fluidic memristor sensor towards hydroxide based liquid detection, in: Proc IEEE 15th International Conference on Nanotechnology (IEEE-NANO'2015), 2015, pp. 1078–1081.
[54] Y. Halawani, B. Mohammad, D. Homouz, M. Al-Qutayri, H. Saleh, Modeling and optimization of memristor and STT-RAM-based memory for low-power applications, IEEE Trans. Very Large Scale Integr. Syst. 24 (2016) 1003–1014.
[55] S.O. Haykin, Neural Networks and Learning Machines, Pearson, USA, 2008.
[56] J.L. Hindmarsh, R. Rose, A model of neuronal bursting using three coupled first order differential equations, Proc. R. Soc. Lond. B, Biol. Sci. 221 (1984) 87–102.
[57] A.L. Hodgkin, A.F. Huxley, B. Katz, Ionic currents underlying activity in giant axon of the squid, Arch. Sci. Physiol. 3 (1949) 129–150.
[58] A.L. Hodgkin, A.F. Huxley, B. Katz, A quantitative description of membrane current and its application to conduction and excitation in nerve, J. Physiol. 117 (1952) 500–544.

[59] A.L. Hodgkin, R.D. Keynes, Experiments on the injection of substances into squid giant axons by means of microsyringe, J. Physiol. London 131 (1956) 592.
[60] J.J. Hopfield, Neural networks and physical systems with emergent collective computational abilities, Proc. Natl. Acad. Sci. USA 79 (1982) 2554–2558.
[61] M. Hu, Y. Chen, J.J. Yang, Y. Wang, H.H. Li, A compact memristor-based dynamic synapse for spiking neural networks, IEEE Trans. Comput.-Aided Des. Integr. Circuits Syst. 36 (2017) 1353–1366.
[62] X. Hu, G. Feng, S. Duan, L. Liu, A memristive multilayer cellular neural network with applications to image processing, IEEE Trans. Neural Netw. Learn. Syst. 28 (2017) 1889–1901.
[63] M. Itoh, L.O. Chua, Memristor oscillators, Int. J. Bifurc. Chaos 18 (2008) 3183–3206.
[64] M. Itoh, L.O. Chua, Memristor cellular automata and memristor discrete time cellular neural networks, Int. J. Bifurc. Chaos 19 (2009) 3605–3656.
[65] M. Itoh, L.O. Chua, Memristor Hamiltonian circuits, Int. J. Bifurc. Chaos 21 (2011) 2395–2425.
[66] H.H.C. Iu, D.S. Yu, A.L. Fitch, V. Sreeram, H. Chen, Controlling chaos in a memristor based circuit using a twin–T notch filter, IEEE Trans. Circuits Syst. I, Regul. Pap. 58 (2011) 1337–1344.
[67] E.M. Izhikevich, Dynamical Systems in Neuroscience the Geometry of Excitable and Bursting, MIT Press, USA, 2007.
[68] H. Jaeger, H. Haas, Harnessing nonlinearity: predicting chaotic systems and saving energy in wireless communication, Science 304 (2004) 78–80.
[69] Y.N. Joglekar, S.J. Wolf, The elusive memristor: properties of basic electrical circuits, Eur. J. Phys. 30 (2009) 661.
[70] J. Kacprzyk, W. Pedrycz, Handbook of Computational Intelligence, Springer, Berlin, Germany, 2015.
[71] M. Khalid, Review on various memristor models, characteristics, potential applications, and future works, Trans. Electr. Electron. Mater. 20 (2019) 289–298.
[72] H. Kim, M.P. Sah, C. Yang, T. Roska, L.O. Chua, Memristor bridge synapses, Proc. IEEE 100 (2012) 2061–2070.
[73] S.G. Kim, J.S. Han, H. Kim, S.Y. Kim, H.W. Jang, Recent advances in memristive materials for artificial synapses, Adv. Mater. Technol. 3 (2018) 1800457.
[74] Y.-S. Kim, K.-S. Min, Synaptic weighting circuits for cellular neural networks, in: Proc. Int. Workshop Cellular Nanosc. Netw. Appl., 2012, pp. 1–6.
[75] R. Kozma, R.E. Pino, G.E. Pazienza, Advances in Neuromorphic Memristor Science and Applications, Springer, Germany, 2012.
[76] O. Krestinskaya, A.P. James, L.O. Chua, Neuromemristive circuits for edge computing: a review, IEEE Trans. Neural Netw. Learn. Syst. 31 (2020) 4–23.
[77] D. Kudithipudi, Q. Saleh, C. Merkel, J. Thesing, B. Wysocki, Design and analysis of a neuromemristive reservoir computing architecture for biosignal processing, Front. Neurosci. 9 (2016) 502.
[78] S. Kvatinsky, E.G. Friedman, A. Kolodny, U.C. Weiser, TEAM: threshold adaptive memristor model, IEEE Trans. Circuits Syst. I, Regul. Pap. 60 (2013) 211–221.
[79] S.-W. Lee, H.-H. Song, A new recurrent neural-network architecture for visual pattern recognition, IEEE Trans. Neural Netw. 8 (1997) 331–340.
[80] Q. Li, S. Hu, S. Tang, G. Zeng, Hyperchaos and horseshoe in a 4D memristive system with a line of equilibria and its implementation, Int. J. Circuit Theory Appl. 42 (2014) 1172–1188.
[81] Q. Li, H. Zeng, J. Li, Hyperchaos in a 4D memristive circuit with infinitely many stable equilibria, Nonlinear Dyn. 79 (2015) 2295–2308.
[82] Y. Li, Z. Wang, R. Midya, Q. Xia, J.J. Yang, Review of memristor devices in neuromorphic computing: materials sciences and device challenges, J. Phys. D, Appl. Phys. 51 (2018) 503002.
[83] H. Liu, Z. Wang, B. Shen, Discrete-time memristive recurrent neural networks with time-varying delays: exponential stability analysis, in: Proc. 35th Chin. Control Conf. (CCC), 2016, pp. 3584–3589.

[84] P. Liu, R. Xi, P. Ren, J. Hou, X. Li, Analysis and implementation of a new switching memristor scroll hyperchaotic system and application in secure communication, Complexity 2018 (2018) 3497640.
[85] M. Lukosevicius, H. Jaeger, Reservoir computing approaches to recurrent neural network training, Comput. Sci. Rev. 3 (2009) 127–149.
[86] W.W. Lytton, From Computer to Brain: Foundations of Computational Neuroscience, Springer, Germany, 2002.
[87] W. Maass, Networks of spiking neurons: the third generation of neural network models, Neural Netw. 10 (1997) 1659–1671.
[88] Y. Maeda, M. Wakamura, Simultaneous perturbation learning rule for recurrent neural networks and its FPGA implementation, IEEE Trans. Neural Netw. 16 (2005) 1664–1672.
[89] M.D. Marco, M. Forti, L. Pancioni, Memristor standard cellular neural networks computing in the flux–charge domain, Neural Netw. 93 (2017) 152–164.
[90] P. Mazumder, S.M. Kang, R. Waser, Memristors: devices models and applications, Proc. IEEE 100 (2012) 1911–1919.
[91] C. Mead, Neuromorphic electronic systems, Proc. IEEE 78 (1990) 1629–1636.
[92] J.S. Meena, S.M. Sze, U. Chand, T.-Y. Tseng, Overview of emerging nonvolatile memory technologies, Nanoscale Res. Lett. 9 (2014) 526.
[93] C. Merkel, Q. Saleh, C. Donahue, D. Kudithipudi, Memristive reservoir computing architecture for epileptic seizure detection, Proc. Comput. Sci. 41 (2014) 249–254.
[94] F. Merrikh-Bayat, B. Hoskins, D.B. Strukov, Phenomenological modeling of memristive devices, Appl. Phys. A Solids Surf. 118 (2014) 779–786.
[95] I. Messaris, A. Serb, S. Stathopoulos, A. Khiat, S. Nikolaidis, T. Prodromakis, A data-driven verilog—a ReRAM model, IEEE Trans. Comput.-Aided Des. Integr. Circuits Syst. 37 (2018) 3151–3162.
[96] R. Midya, Z. Wang, S. Asapu, X. Zhang, M. Rao, W. Song, Y. Zhuo, N. Upadhyay, Q. Xia, J.J. Yang, Reservoir computing using diffusive memristors, Adv. Intell. Syst. 1 (2019) 1900084.
[97] V. Mladenov, Advanced Memristor Modeling: Memristor Circuits and Networks, MDPI, Basel, Switzerland, 2019.
[98] D. Monroe, Neuromorphic computing gets ready for the (really) big time, Commun. ACM 57 (2014) 13–15.
[99] J. Moon, W. Ma, J.H. Shin, F. Cai, C. Du, S.H. Lee, W.D. Lu, Temporal data classification and forecasting using a memristor-based reservoir computing system, Nat. Electron. 2 (2019) 480–487.
[100] H. Mostafa, C. Mayr, G. Indiveri, Beyond spike-timing dependent plasticity in memristor crossbar arrays, in: Proc. IEEE Int. Symp. Circuits Syst. (ISCAS), 2016, pp. 926–929.
[101] B. Muthuswamy, Implementing memristor based chaotic circuits, Int. J. Bifurc. Chaos 20 (2010) 1335–1350.
[102] B. Muthuswamy, L.O. Chua, Simplest chaotic circuits, Int. J. Bifurc. Chaos 20 (2010) 1567–1580.
[103] B. Muthuswamy, P.P. Kokate, Memristor-based chaotic circuits, IETE Tech. Rev. 26 (2009) 415–426.
[104] J. Nagumo, S. Arimoto, S. Yoshizawa, An active pulse transmission line simulating nerve axon, Proc. IRE 50 (1962) 2061–2070.
[105] R. Naous, M. AlShedivat, E. Neftci, G. Cauwenberghs, K.N. Salama, Memristor-based neural networks: synaptic versus neuronal stochasticity, AIP Adv. 6 (2016) 111304.
[106] S. Panahi, S. Jafari, A.J.M. Khalaf, K. Rajagopal, V. Pham, F.E. Alsaadi, Complete dynamical analysis of a neuron under magnetic flow effect, Chin. J. Phys. 56 (2018) 2254–2264.
[107] N. Panwar, B. Rajendran, U. Ganguly, Arbitrary spike time dependent plasticity (STDP) in memristor by analog waveform engineering, IEEE Electron Device Lett. 38 (2017) 740–743.
[108] G. Pedretti, V. Milo, S. Ambrogio, R. Carboni, S. Bianchi, A. Calderoni, N. Ramaswamy, A.S. Spinelli, D. Ielmini, Memristive neural network for on-line learning and tracking with brain-inspired spike timing dependent plasticity, Sci. Rep. 7 (2017) 5288.

[109] Y.V. Pershin, S.L. Fontaine, M.D. Ventra, Memristive model of amoeba learning, Phys. Rev. E 80 (2009) 021926.
[110] Y.V. Pershin, M.D. Ventra, Experimental demonstration of associative memory with memristive neural networks, Neural Netw. 23 (2010) 881–886.
[111] V.T. Pham, A. Buscarino, L. Fortuna, M. Frasca, Simple memristive time-delayed chaotic systems, Int. J. Bifurc. Chaos 23 (2013) 1350073.
[112] V.T. Pham, S. Vaidyanathan, C.K. Volos, S. Jafari, N.V. Kuznetsov, T.M. Hoang, A novel memristive time-delay chaotic system without equilibrium points, Eur. Phys. J. Spec. Top. 225 (2016) 127–136.
[113] V.-T. Pham, C. Volos, L.V. Gambuzza, A memristive hyperchaotic system without equilibrium, Sci. World J. 2014 (2014), 368986–9.
[114] V.-T. Pham, C.K. Volos, T.D. Son, New research trends for implementing chaotic systems with memristors, in: New Research Trends in Nonlinear Circuits: Design, Chaotic Phenomena and Applications, Nova Science Publishers, USA, 2014, pp. 171–198.
[115] V.T. Pham, C.K. Volos, S. Vaidyanathan, T.P. Le, V.Y. Vu, A memristor-based hyperchaotic system with hidden attractors: dynamics, synchronization and circuital emulating, J. Eng. Sci. Technol. Rev. 8 (2015) 205–214.
[116] M.D. Pickett, D.B. Strukov, J.L. Borghetti, J.J. Yang, G.S. Snider, D.R. Stewarta, R.S. Williams, Switching dynamics in titanium dioxide memristive devices, J. Appl. Phys. 106 (2009) 074508.
[117] M. Prezioso, F. Merrikh-Bayat, B.D. Hoskins, G.C. Adam, K.K. Likharev, D.B. Strukov, Training and operation of an integrated neuromorphic network based on metal-oxide memristors, Nature 521 (2015) 61–64.
[118] T. Prodromakis, B.P. Peh, C. Papavassiliou, C. Toumazou, A versatile memristor model with nonlinear dopant kinetics, IEEE Trans. Electron Devices 58 (2011) 3099–3105.
[119] T. Prodromakis, C. Toumazou, A review on memristive devices and applications, in: Proc. 17th IEEE International Conference on Electronics, Circuits and Systems (ICECS'2010), 2010, pp. 934–937.
[120] C. Rocsoreanu, A. Georgescu, N. Giurgiteanu, The FitzHugh–Nagumo Model: Bifurcation and Dynamics, Kluwer Academic Publishers, USA, 2000.
[121] D. Sacchetto, M.-A. Doucey, G.D. Micheli, Y. Leblebici, S. Carrara, New insight on biosensing by nano-fabricated memristors, BioNanoScience 1 (2011) 1–3.
[122] M.E. Sahin, Z.G.C. Taskiran, H. Guler, S.E. Hamamci, Simulation and implementation of memristive chaotic system and its application for communication systems, Sens. Actuators A, Phys. 209 (2019) 107–118.
[123] T. Serrano-Gotarredona, T. Prodromakis, B. Linares-Barranco, A proposal for hybrid memristor-cmos spiking neuromorphic learning systems, IEEE Circuits Syst. Mag. 13 (2013) 74–88.
[124] A. Shafiee, A. Nag, N. Muralimanohar, R. Balasubramonian, J.P. Strachan, M. Hu, R.S. Williams, V. Srikumar, ISAAC: a convolutional neural network accelerator with in-situ analog arithmetic in crossbars, in: Proc. ACM/IEEE 43rd Annual International Symposium on Computer Architecture (ISCA'2016), 2016, pp. 14–26.
[125] A.M. Sheri, H. Hwang, M. Jeon, B.-G. Lee, Neuromorphic character recognition system with two PCMO memristors as a synapse, IEEE Trans. Ind. Electron. 61 (2014) 2933–2941.
[126] S. Shin, K. Kim, S.M. Kang, Memristor applications for programmable analog ICs, IEEE Trans. Nanotechnol. 410 (2011) 266–274.
[127] J. Singh, B. Raj, Comparative analysis of memristor models and memories design, J. Semicond. 39 (2018) 074006.
[128] K. Smagulova, O. Krestinskaya, A.P. James, A memristor-based long short term memory circuit, Analog Integr. Circuits Signal Process. 95 (2018) 467–472.
[129] L.S. Smith, Neuromorphic systems: past, present and future, in: Brain Inspired Cognitive Systems, Springer, New York, 2010, pp. 167–182.

[130] D. Soudry, D.D. Castro, A. Gal, A. Kolodny, S. Kvatinsky, Memristor-based multilayer neural networks with online gradient descent training, IEEE Trans. Neural Netw. Learn. Syst. 26 (2015) 2408–2421.
[131] D.B. Strukov, G.S. Snider, D.R. Stewart, R.S. Williams, The missing memristor found, Nature 453 (2008) 80–83.
[132] C. Sung, H. Hwang, I.K. Yoo, Perspective: a review on memristive hardware for neuromorphic computation, J. Appl. Phys. 124 (2018) 151903.
[133] N. Taherinejad, P.D.S. Manoj, A. Jantsch, Memristors' potential for multi-bit storage and pattern learning, in: Proc. IEEE European Modelling Symposium (EMS'2015), 2015, pp. 450–455.
[134] G. Tanaka, T. Yamane, J.B. Heroux, R. Nakane, N. Kanazawa, S. Takeda, H. Numata, D. Nakano, A. Hirose, Recent advances in physical reservoir computing: a review, Neural Netw. 115 (2019) 100–123.
[135] S. Tang, S. Yin, S. Zheng, P. Ouyang, F. Tu, L. Yao, J. Wu, W. Cheng, L. Liu, S. Wei, AEPE: an area and power efficient RRAM crossbar-based accelerator for deep CNNs, in: Proc. IEEE 6th Non-Volatile Memory Systems and Applications Symposium (NVMSA'2017), 2017, pp. 1–6.
[136] R. Tetzlaff, Memristors and Memristive Systems, Springer, New York, USA, 2014.
[137] J.M. Tour, T. He, The fourth element, Nature 453 (2008) 42–43.
[138] S.N. Truong, S. Shin, J. Song, H.-S. Mo, F. Corinto, K.-S. Min, Memristor-based cellular nanoscale networks: theory circuits and applications, in: Proc. IEEE Int. Symp. Circuits Syst. (ISCAS), 2015, pp. 1134–1137.
[139] I. Tzouvadaki, N. Madaboosi, R. Soares, V. Chu, J. Conde, G.D. Micheli, S. Carrara, Biofunctionalization study of memristive-biosensors for early detection of prostate cancer, in: Proc. Conference on Ph.D. Research in Microelectronics and Electronics (PRIME'2015), 2015, pp. 17–20.
[140] I. Tzouvadaki, C. Parrozzani, A. Gallotta, G.D. Micheli, S. Carrara, Memristive biosensors for PSA-IgM detection, BioNanoScience 5 (2015) 189–195.
[141] I. Vourkas, G.C. Sirakoulis, Memristor-Based Nanoelectronic Computing Circuits and Architectures, Springer, Germany, 2015.
[142] J. Wang, Analysis and design of an analog sorting network, IEEE Trans. Neural Netw. 6 (1995) 962–971.
[143] X. Wang, Y. Chen, Y. Gu, H. Li, Spintronic memristor temperature sensor, IEEE Electron Device Lett. 31 (2010) 20–22.
[144] Z. Wang, S. Joshi, S. Savelev, W. Song, R. Midya, Y. Li, M. Rao, P. Yan, S. Asapu, Y. Zhuo, H. Jiang, P. Lin, C. Li, J.H. Yoon, N.K. Upadhyay, J. Zhang, M. Hu, J.P. Strachan, M. Barnell, Q. Wu, H. Wu, R.S. Williams, Q. Xia, J.J. Yang, Fully memristive neural networks for pattern classification with unsupervised learning, Nat. Electron. 1 (2018) 137–145.
[145] Z. Wang, S. Joshi, S.E. Savelev, H. Jiang, R. Midya, P. Lin, M. Hu, N. Ge, J.P. Strachan, Z. Li, Q. Wu, M. Barnell, G.-L. Li, H.L. Xin, R.S. Williams, Q. Xia, J.J. Yang, Memristors with diffusive dynamics as synaptic emulators for neuromorphic computing, Nat. Mater. 16 (2017) 101–108.
[146] Z. Wang, W. Zhao, W. Kang, Y. Zhang, J.-O. Klein, C. Chappert, Ferroelectric tunnel memristor-based neuromorphic network with 1T1R crossbar architecture, in: Proc. Int. Joint Conf. Neural Netw. (IJCNN'2014), 2015, pp. 29–34.
[147] P.J. Werbos, Backpropagation through time: what it does and how to do it, Proc. IEEE 78 (1990) 1550–1560.
[148] X. Wu, V. Saxena, K. Zhu, Homogeneous spiking neuromorphic system for real-world pattern recognition, IEEE J. Emerg. Sel. Top. Circuits Syst. 5 (2015) 254–266.
[149] G.M.T. Xavier, F.G. Castañeda, L.M.F. Nava, J.A.M. Cadenas, Memristive recurrent neural network, Neurocomputing 273 (2018) 281–295.
[150] S. Xiao, X. Xie, S. Wen, Z. Zeng, T. Huang, J. Jiang, GST-memristor-based online learning neural networks, Neurocomputing 272 (2017) 677–682.

[151] P. Yadav, B.K. Das, Memristor-based memory cell with less noise margins and storing non-binary data, in: Physics of Semiconductor Devices, Springer, Cham, 2014, pp. 183–187.
[152] C. Yakopcic, M.Z. Alom, T.M. Taha, Memristor crossbar deep network implementation based on a convolutional neural network, in: Proc. Int. Joint Conf. Neural Netw. (IJCNN), 2016, pp. 963–970.
[153] J.J. Yang, D.B. Strukov, D.R. Stewart, Memristive devices for computing, Nat. Nanotechnol. 8 (2013) 13–24.
[154] P. Yao, H. Wu, B. Gao, S.B. Eryilmaz, X. Huang, W. Zhang, Q. Zhang, N. Deng, L. Shi, H.-S.P. Wong, H. Qian, Face classification using electronic synapses, Nat. Commun. 8 (2017) 15199.
[155] J.H. Zhang, X.F. Liao, Synchronization and chaos in coupled memristor-based FitzHugh–Nagumo circuits with memristor synapse, AEU Int. J. Electron. Commun. 75 (2017) 82–90.
[156] W. Zhang, Y. Yu, Chaos generator based on a memristive circuit with time-delayed feedback, in: Proc IEEE Int.Conference on Communication, Circuit and Systems (ICCCAS'10), Chengdu, China, Sep. 2010, pp. 817–819.
[157] X. Zhang, Z. Cui, Y. Zhu, Synchronization and circuit experiment simulation of chaotic time-delay systems, in: Proc. Pacific–Asia Conf. Circuits, Commun. Syst., Chengdu, China, Sep. 2009, pp. 781–784.
[158] Y. Zhang, Y. Li, X. Wang, E.G. Friedman, Synaptic characteristics of Ag/AgInSbTe/Ta-based memristor for pattern recognition applications, IEEE Trans. Electron Devices 64 (2017) 1806–1811.
[159] Y. Zhang, X. Wang, E.G. Friedman, Memristor-based circuit design for multilayer neural networks, IEEE Trans. Electron Devices 65 (2017) 677–686.
[160] W.S. Zhao, G. Agnus, V. Derycke, A. Filoramo, J.P. Bourgoin, C. Gamrat, Nanotube devices based crossbar architecture: toward neuromorphic computing, Nanotechnology 21 (2010) 175202.
[161] M.A. Zidan, J.P. Strachan, W.D. Lu, The future of electronics based on memristive systems, Nat. Electron. 1 (2018) 22.

Chapter 14

Guidelines for benchmarking non-ideal analog memristive crossbars for neural networks

Alex James
Indian Institute of Information Technology and Management, Kerala, India

14.1 Introduction

Smarter sensors and higher communication bandwidth imply that the volume, velocity, and variety of data will increase further in the next decade[1–3]. The rush to make devices faster requires faster intelligent processing of the sensor data, which can be implemented with edge computing technologies [4–7]. Adding a co-processor next to the sensors, and performing intelligent computational tasks makes this an edge AI solution [8–10]. To make edge AI solutions, energy efficiency is essential to ensure that sensors can work a long period with a limited supply of electricity. One of the emerging approaches to building low power AI chips that can act as co-processors is by using memristor crossbar arrays [8,11–14].

The memristor crossbar arrays make use of memristors and transistor switches to emulate the neurons and artificial neural networks [15–18]. The heart of the analog artificial neural network is the dot-product computation or multiple-accumulate operations. The crossbar takes voltage as inputs through the rows and provides currents as its outputs at the column. The conductance of the memristors acts as the weights in the neural network layer. The weight summation of input voltages is represented as the column currents, and each crossbar emulates one layer of the neural network.

The simplicity of the crossbar arrays in performing analog dot-product computations at the cost of low area and power makes it an attractive option to build analog neural networks. However, the scaling up of analog neural networks with crossbar arrays is not an easy problem, as it involves solving the issues of sneak path currents and memristor non-idealities. While there are many successful ways to elevate the issue of sneak path currents [19–21], the problems of non-idealities are hardware to address. In this chapter, a guideline is provided to bring focus on the most critical issues to address when designing and simulating analog memristive neural networks.

Mem-elements for Neuromorphic Circuits with Artificial Intelligence Applications
https://doi.org/10.1016/B978-0-12-821184-7.00023-2

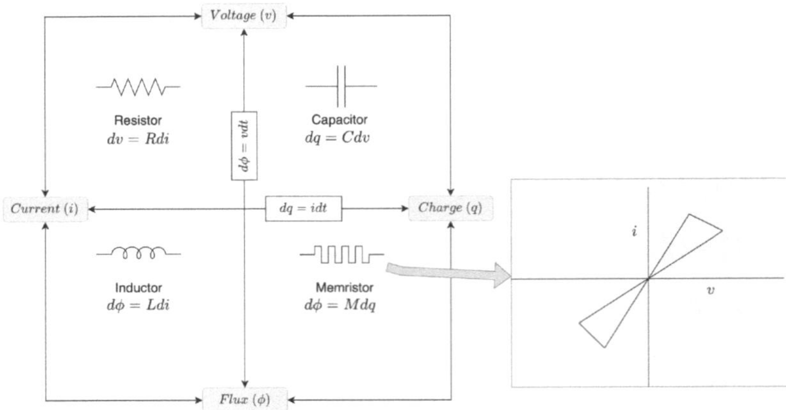

FIGURE 14.1 The memristor element, nonlinear memristor symbol, and IV characteristics.

14.2 Basic concepts

14.2.1 Memristor simplified

The memristor, as of now, is considered as a class of two-terminal resistive devices, that shows the behavior of a non-linear resistor and shows non-volatile or volatile memory properties. The memristor can be programmed using a series of voltage pulses such that its resistance changes from one value to another [22,23]. This change in resistance is a crucial characteristic of memristors.

Fig. 14.1 shows the three frequent representations of a memristor, as a fundamental element, as a nonlinear circuit element and by its signature pinched hysteresis characteristic. The Φ-q relationship that memristors offer can be linear or nonlinear. The nonlinear memristor symbol is more often used than the linear memristor symbol. It will be more appropriate to use the nonlinear symbol as the nonlinear behavior is more often reflective of real devices. The memristor being a fundamental circuit element has been under several scrutinies in the recent years [24,25]. The fact that flux–charge relationships are challenging to validate has prompted researchers to question its fundamentality. It is argued that the memristor is nothing but only a nonlinear resistor. There have been arguments against the ideal or generalization of memristor to cover a broad range of devices, contradicted by evidence of non-passive nano battery effect [26] observable in some of the resistive switching memories that are classified as memristors.

While these arguments on the fundamentality exist, the quest for having an ideal or generalized memristor [27,28] has led to the rapid progress of several resistive switching devices that can have numerous practical applications. It has come a long way from being a lab-based concept to be introduced into some of the mainstream semiconductor industry applications that can be integrated along with CMOS devices. As far as practical use and application are concerned, the fundamentality of the memristor is the least concerning factor.

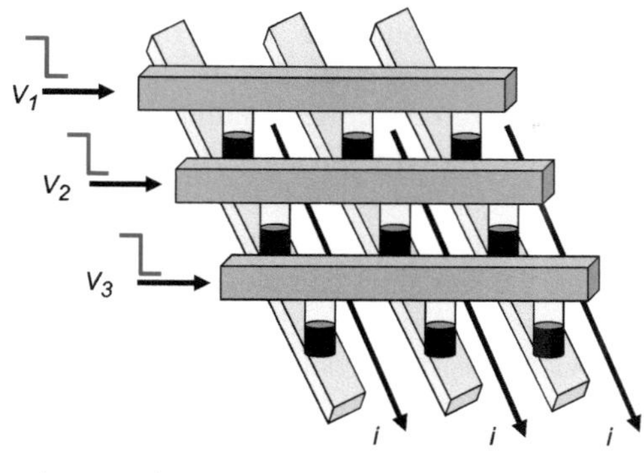

FIGURE 14.2 A memristor crossbar arrangement.

It should be further clarified that the memristor originally is designed as a device to fill the gap in the relationship between magnetic flux linkage $\Phi_m(t)$ and electric charge $q(t)$, i.e., having a characteristic relation $f(\Phi_m(t), q(t)) = 0$. It should be noted that magnetic flux linkage does not show a magnetic field here, and is represented as an integral of voltage over time generalized from the characteristic of an inductor. The memristor as an element will have memristance that can be described as $M(q(t)) = d\Phi_m/dq$, where the unit of the memristance is that of resistance (Wb / C, or ohm).

14.2.2 Crossbar simplified

The crossbar or crosspoint arrangement of memristors is one of the most useful circuit arrangements of memristors [22,29–32]. Fig. 14.2 shows the arrangement of memristors in the crossbar configuration, i.e., the two-terminal memristor being placed in between two metal lines. The memristor can be combined along with a selector device such as a transistor to turn ON or OFF specific memristor crossbar nodes. More often, when a transistor is used along with the memristor, the nodes are called 1T1M or 1T1R or 1S1R or IS1M configuration, with T implying a transistor, S implying any selector device, R implying any resistive device, and M indicating a memristor device.

When the selectors are not used, the currents from one column leaks to the other column, causing sneak path current errors in the column readouts. The use of the selector will allow for the targeted selection of columns and nodes. Such a collection of memristor nodes is useful for the memristor to be used as a memory array and for implementing memristor-based logic circuits. A large number of threshold logic systems have been proposed in recent years based on the sequenced programming of memristor nodes such as RTL [33], IMPLY [34], and MAGIC [35].

As the size of the crossbar increases, the sneak path current increases, and it introduces increased errors in the column currents. Other than the selector switching, another possible configuration for scaling is the use of the modular crossbar or tiled crossbar architecture. In the modular architecture, several small crossbars are used in parallel, such that the currents from one column of the crossbar is summed up with the corresponding column in another crossbar. As a smaller crossbar has lower errors, this modular summation keeps the overall errors lower. This can be further extended to 3D crossbar [19,35,36] as well, where the readouts and controls can be done across the column or along the vertical lines. However, the signaling complexity of the 3D crossbar increases with the density and requires extensive testing for signal integrity. The 2D crossbar, while lower in device density, poses lower complexity in designing the data paths and is easier to realize.

14.2.3 Analog neural networks simplified

The crossbar forms the heart of most of the modern analog neural networks. The voltage signals v_i are applied across the rows and the currents i_i are read across the columns. Each node in the crossbar has a conductance $g(i, j)$, which means along with the columns the current $i_j = \sum g(i, j)v_i$, i.e., the output current is the weighted summation of input voltages, where the conductance of each node represents the weights. This weighted summation of inputs is the signature characteristic of neural networks and provides a highly simplified circuit configuration for implementing a layer of neural network.

The weighted summation of input is followed by an activation function for implementing the neural network layer. Each column readout can be used to represent a single neuron. The simplest way to create the activation function is by using an Op-Amp buffer circuit that can implement a ReLu function or using a comparator circuit for implementing a sigmoid function.

An analog-only neural network implementation would require both the weighted summation of inputs and the processing circuits between the layers to be in the analog domain. This means there will be a need to design analog amplifiers and circuits that consume low area and power. Often, switching mechanisms are used to lower the power, and configurations such as based on spiking neural networks help with this.

Another approach is to keep the weighted summation of inputs (or multiply and accumulate or dot-product computations) in analog, while the processing circuits to be mixed-signal circuits. The output of the crossbar is converted to the digital domain and passed through the memories, before moving the data to the next crossbar. This can help with the ease of implementation, however, will increase the overall chip area.

14.3 Non-idealities of memristors

14.3.1 Aging

Memristor devices contain multiple conductance states [37–42]. The more the number of states it has, the better its use to mimic the neuron weight in analog

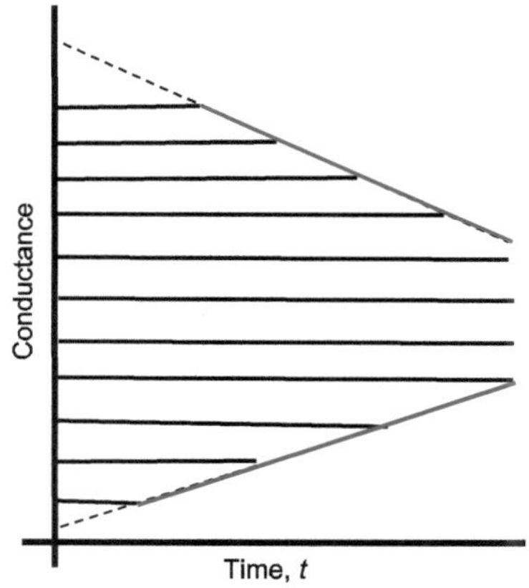

FIGURE 14.3 A simplified view of the aging issue in memristor.

neural network implementations. Aging of the memristor devices that occurs as a result of electrical stress on the devices during writing, erase and reading stages. The increased number of these cyclic operations results in the shifting of conductive states such that over a period of time, several of the conductive states become overlapping. Also, aging makes devices electrically damaged, resulting in devices getting stuck at a conductance level. The assessment of endurance and conductance levels of the memristors over some time is an effective way to study aging.

Fig. 14.3 shows an example of the disappearing conductive state of a hypothetical memristor device over some time. As the cycle time increases, i.e. the number of times the memristor is run in the full cycle of write, read and erase, the conductive state of the memristor starts to disappear and merge with the nearby conductive states. Each conductance state for a type of memristor device has variability on top of this, which makes analog conductance behavior and usage a challenging problem.

The memristor in crossbar configuration for neural networks requires a CMOS-memristor hybrid circuit configuration. The CMOS circuits in the crossbar array consist of the read-write circuitry to change the states of memristors, and also for accumulating the weighted summation results from the crossbar columns. The activation function and timing control between the neural network layers are also usually implemented with CMOS circuits. Hence, with aging, the impact of CMOS circuits involving switching and computational operations is equally important to be considered. The technology-specific issues of the CMOS process and reliability of the circuits become essential with aging.

Guideline 1. *The design of analog neural network needs to consider the failure of the devices, loss of conductance states in the memristor, and CMOS reliability issues.*

14.3.2 Electromagnetics and signal integrity

The crossbar circuits offer an advantage in analog neural network computation if the packing density is high [29,43,44]. The longer the signal path between the neurons from a layer to another, the higher the chance of having signal loss and interference. The use of 3D technologies can help pack more memristor neurons. However, the subsequent processing is usually done in analog CMOS circuits, which can be an energy-hungry proposition [36,45,46]. The optimization of the CMOS circuits, such as the amplifiers and control circuits, would require careful physical design. The electromagnetics issues become an essential aspect of being considered in the optimization of the physical design and placement of the crossbar-CMOS circuits.

The design challenges can significantly increase with the number of crossbar arrays and the size of the crossbar. Typically, many of the deep neural networks contain more than 20 layers and have the ability to process a large number of inputs. In an analog neural network, this introduces considerable wiring complexity, which can add signal path delays and cross-coupling effects. The high-speed signals can create challenging issues in the analog circuits which require innovations in architecture for the analog neural net implementation. The more scalable the design needs to be, the problems related to signal integrity take prominence as it can directly affect the overall working of the system.

Guideline 2. *Scalability of practical analog memristive neural networks requires architecture-specific criteria to be laid for dealing with signal integrity issues.*

14.3.3 Conductance variabilities

The variation of memristor conductance can be mainly due to the process variations and device manufacturing issues [14,16,22,23,47]. In any new technology, the device to device variability can be much higher than a mature technology. As most of the memristor technologies are new to the market, and the process technologies are not fully mature for full-blown commercial use, they exhibit variabilities in its conductance state. This can be a major problem in analog memristive neural networks, where the weights are analog, and any change in the weight values results in the change in the network performance during inference tasks.

The robustness of the memristive neural networks with conductance variabilities in the crossbar is essential to assess for ensuring the proper functioning of the analog neural networks. Fine-tuning and learning strategies have proven to be useful to reduce the impact of variabilities during the training. The use of dropouts and regularization strategies also helps to provide robustness, along with solving the issue of overtraining.

Guideline 3. *The incorporation of different learning strategies is essential to compensate for the conductance variabilities.*

14.3.4 Noise effects

In general, the electrical performance of the semiconductor devices gets affected by temperature and signal noise [48–50]. The heating of devices occurs due to excessive electrical stress on the devices, which in turn has a further impact on the circuit performance. In neural networks, the noises to the signals can shift the currents in the crossbar, which can lead to wrong dot-product computations. The crossbars in neural networks are used as dot-product computational engines, which, even with a small amount of noise can lead to incorrect inference results.

Not all neural architectures are robust to signal noise, which can occur at the input source side, within the crossbar or across the devices. The amount of noise and the type of noise has an impact on the range of errors that can occur either at the output of the crossbar or that at the output of the activation function. It can also be noted that training the memristor becomes challenging under noisy conditions. At the same time, the noise can be taken into the advantages of building stochastic neural networks, and also, the signal noise can be used to create adversarial attacks.

Guideline 4. *The noise has several origins and can ultimately impact the performance of the analog memristive neural network. Analyzing the robustness of the neural architecture to signal and hardware noise is essential to ensure stable system performance.*

14.3.5 System implementations

The large analog neural network implementations consist of millions of memristive devices [13,14,22]. It is not practically feasible to test the large design for a wide range of conditions using conventional circuit design tools. The use of scripting languages such as Python comes in handy for analysis of the system performance. The high-level languages can be used to export the circuit characteristics into the architecture, and use traditional deep learning libraries for simulating large networks. The non-ideality issues can be incorporated into these libraries and also incorporate various learning strategies to stabilize the system performance.

The existing electronic design tools for analog circuit simulations use extensive computational resources to run large designs. In most neural network designs, one will be required to simulate millions of units in parallel, which is obviously a sizeable computational task. The scaling up of the neural networks requires architecture designs that can help perform modular simulations yet preserve the reliability of the design. The use of machine learning techniques to optimize electronic design is another strategy to avoid human-in-the-loop from analog circuit design. However, this is a challenging task as the number of parameters to consider increases significantly with an increase in the network size. At the same time, the use of machine learning and automation possibly is the only way ahead to build extremely large-scale analog neural networks that work.

Guideline 5. *Scaling up studies of neural networks requires simulations on an extremely large number of circuit components to assess the practical feasibility of the design. Higher-level languages and automation of the analysis are essential to ensure speed and accuracy of hardware design.*

14.4 Applications

14.4.1 Programmable logic

The memristor crossbar can be used to build programmable logic gates. There are several ways in which programmable logic can be built with a crossbar. The best-known approach is by switching the resistance of the crossbar node to represent the data. There are several configurations that use this IMPLY [51], MAGIC [35], CNIMP [52], MRL [53], MAD [54], RTL [14], Scouting Logic [55] etc. They all can be used in a crossbar and often are assessed based on logic gate area, energy, and data latency. Almost, the majority of the memristor logic gates tend to outperform the CMOS-based logic gates. Among these gates, the threshold logic gates make use of neuron models for implementing the logic functions.

The memristor devices have much more variability than CMOS counterparts. This makes it necessary to test the performance of the memristor logic against the variability of the conductance. The device to device as well as the endurance becomes a significant issue for this. Another problem with many of the logic designs with memristors is the frequent need to change the resistance. The higher number of cycles are required to tune the memristor-based gates; their overall reliability becomes lower. The threshold logic gates have a high level of tolerance to device variability due to the ability to tolerate the variations in the weight values.

14.4.2 Neuromorphic accelerators

There are numerous neural networks in use today. Many are extremely power-hungry as one needs a large amount of memory and computational cost of implementing parallel dot-product computations, among many others. The analog neural networks that can be used as a way to accelerate the neural computations during inferences operations, especially for near sensor computations.

The general-purpose analog neural processor is an extremely challenging task, as dealing with real-time signals and analog memories does not concern trivial problems to address. On the other hand, it offers the possibility to drastically change the landscape of neural computing in programmable edge devices.

There are several early attempts to build analog neural network hardware in the recent past with memristor arrays. Some of the examples include Long short-term memory [56,57], generative adversarial networks [58], and convolutional neural networks [59–61]. These works are only indicative, and there are several other neural networks that are increasingly being used. The spiking neural networks are considered to be the most go-to approach in the last decade for real-time implementations [62–68]. The popularity and attraction of spiking neural networks have been their biological resemblance to biological neuron models, and the ability to reduce the energy consumption per neuron. Most of the practical implementations of the memristor crossbar-based neural networks are not purely analog, instead, it is a mix of digital and analog circuits. The analog implementations will be more suitable for use in edge AI sensors, such as for real-time computer vision or biomedical applications. In contrast, the digital deployments will be more suitable for general-purpose neural computing applications.

14.5 Conclusions

The deep learning neural networks in use today are representative of a weak AI system, where the neural network is application-specific. The vast majority of the analog memristive neural networks proposed in the literature are all focusing on weak AI system development. These weak AI systems find applications in face recognition, speech recognition, language recognition, image recognition, and character recognition.

Most of the deep learning system consists of say 10 million neurons, which corresponds to roughly ~30 million semiconductor devices if implemented in memristive hardware. This is a relatively small size compared to the complexity of digital chips. However, for analog circuit implementation of such neural networks, it is still a challenging task to ensure low power, compact and robust design. In most cases, the design would need to be precisely tuned to the applications, and often loses the generality of the neural network established in the first place. The presented guidelines must be used to benchmark the system performance across different technologies to ensure community-wide coherence of reporting the performance of memristive analog neural network designs.

References

[1] S. Nihtianov, A. Luque, Smart Sensors and MEMS: Intelligent Sensing Devices and Microsystems for Industrial Applications, Woodhead Publishing, 2018.

[2] G. van der Horn, J. Huijsing, Integrated Smart Sensors: Design and Calibration, Springer Science & Business Media, 2012.

[3] G. Meijer, K. Makinwa, M. Pertijs, Smart Sensor Systems: Emerging Technologies and Applications, John Wiley & Sons, 2014.

[4] M. Alnemari, N. Bagherzadeh, Efficient deep neural networks for edge computing, in: 2019 IEEE International Conference on Edge Computing (EDGE), 2019, https://doi.org/10.1109/edge.2019.00014.

[5] M. Caprolu, R. Di Pietro, F. Lombardi, S. Raponi, Edge computing perspectives: architectures, technologies, and open security issues, in: 2019 IEEE International Conference on Edge Computing (EDGE), 2019, https://doi.org/10.1109/edge.2019.00035.

[6] K. Dolui, S.K. Datta, Comparison of edge computing implementations: fog computing, cloudlet and mobile edge computing, in: 2017 Global Internet of Things Summit (GIoTS), 2017, https://doi.org/10.1109/giots.2017.8016213.

[7] J. Cao, Q. Zhang, W. Shi, Edge Computing: A Primer, Springer, 2018.
[8] J. Zeng, C. Li, L.-J. Zhang, A face recognition system based on cloud computing and AI edge for IOT, in: Edge Computing – EDGE 2018, 2018, pp. 91–98, https://doi.org/10.1007/978-3-319-94340-4_7.
[9] X. Wang, Y. Han, C. Wang, Q. Zhao, X. Chen, M. Chen, In-edge AI: intelligentizing mobile edge computing, caching and communication by federated learning, IEEE Network 33 (5) (2019) 156–165, https://doi.org/10.1109/mnet.2019.1800286.
[10] E. Li, L. Zeng, Z. Zhou, X. Chen, Edge AI: on-demand accelerating deep neural network inference via edge computing, IEEE Transactions on Wireless Communications 19 (1) (2020) 447–457, https://doi.org/10.1109/twc.2019.2946140.
[11] R. Kozma, R.E. Pino, G.E. Pazienza, Are memristors the future of AI?, Advances in Neuromorphic Memristor Science and Applications (2012) 9–14, https://doi.org/10.1007/978-94-007-4491-2_2.
[12] A.P. James, A hybrid memristor–CMOS chip for AI, Nature Electronics 2 (7) (2019) 268–269, https://doi.org/10.1038/s41928-019-0274-6.
[13] E. Miranda, J. Suñé, Memristors for neuromorphic circuits and artificial intelligence applications, Materials 13 (4) (Feb. 2020), https://doi.org/10.3390/ma13040938.
[14] A.P. James, Deep Learning Classifiers with Memristive Networks: Theory and Applications, Springer, 2019.
[15] S. Vaidyanathan, C. Volos, Advances in Memristors, Memristive Devices and Systems, Springer, 2017.
[16] R. Tetzlaff, Memristors and Memristive Systems, Springer Science & Business Media, 2013.
[17] I. Vourkas, G.C. Sirakoulis, Memristor-Based Nanoelectronic Computing Circuits and Architectures: Foreword by Leon Chua, Springer, 2015.
[18] A. Adamatzky, Chaos, CNN, Memristors and Beyond: A Festschrift for Leon Chua, World Scientific, 2013.
[19] D. Mikhailenko, C. Liyanagedera, A.P. James, K. Roy, M2CA: modular memristive crossbar arrays, in: 2018 IEEE International Symposium on Circuits and Systems (ISCAS), 2018, https://doi.org/10.1109/iscas.2018.8351112.
[20] S.H. Jo, K.-H. Kim, W. Lu, High-density crossbar arrays based on a Si memristive system, Nano Letters 9 (2) (2009) 870–874, https://doi.org/10.1021/nl8037689.
[21] G.P.I. Vourkas, G.C. Sirakoulis, A. Rubio, Heterogeneous memristive crossbar for in-memory computing, in: 2015 International Conference on Memristive Systems (MEMRISYS), 2015.
[22] L. Chua, G.C. Sirakoulis, A. Adamatzky, Handbook of Memristor Networks, Springer Nature, 2019.
[23] D. Ielmini, R. Waser, Resistive Switching: From Fundamentals of Nanoionic Redox Processes to Memristive Device Applications, John Wiley & Sons, 2015.
[24] I. Abraham, The case for rejecting the memristor as a fundamental circuit element, Scientific Reports 8 (1) (Jul. 2018) 10972.
[25] J. Kim, Y.V. Pershin, M. Yin, T. Datta, M. Di Ventra, An experimental proof that resistance-switching memory cells are not memristors, Advanced Electronic Materials 6 (7) (2020) 2000010, https://doi.org/10.1002/aelm.202000010.
[26] I. Valov, et al., Nanobatteries in redox-based resistive switches require extension of memristor theory, Nature Communications 4 (2013) 1771.
[27] L. Chua, Resistance switching memories are memristors, Memristor Networks (2014) 21–51, https://doi.org/10.1007/978-3-319-02630-5_3.
[28] D. Ielmini, Universal set/reset characteristics of metal-oxide resistance switching memories, ECS Transactions 35 (4) (2019) 581–596, https://doi.org/10.1149/1.3572306.
[29] I. Vourkas, G.C. Sirakoulis, Modeling memristor-based circuit networks on crossbar architectures, Memristor Networks (2014) 505–535, https://doi.org/10.1007/978-3-319-02630-5_23.
[30] R. Hasan, T.M. Taha, Memristor crossbar based unsupervised training, in: 2015 National Aerospace and Electronics Conference (NAECON), 2015, https://doi.org/10.1109/naecon.2015.7443091.

[31] Y. Zhao, G. Shi, A circuit implementation method for memristor crossbar with on-chip training, in: 2018 IEEE Asia Pacific Conference on Circuits and Systems (APCCAS), 2018, https://doi.org/10.1109/apccas.2018.8605612.
[32] I. Vourkas, G.C. Sirakoulis, Modeling memristor–based circuit networks on crossbar architectures, Handbook of Memristor Networks (2019) 973–1004, https://doi.org/10.1007/978-3-319-76375-0_34.
[33] A.P. James, V. Linu Rose, D.S. Kumar, Resistive threshold logic, IEEE Transactions on Very Large Scale Integration (VLSI) Systems 22 (1) (2014) 190–195, https://doi.org/10.1109/tvlsi.2012.2232946.
[34] S. Kvatinsky, A. Kolodny, U.C. Weiser, E.G. Friedman, Memristor-based IMPLY logic design procedure, in: 2011 IEEE 29th International Conference on Computer Design (ICCD), 2011, https://doi.org/10.1109/iccd.2011.6081389.
[35] B. Hoffer, V. Rana, S. Menzel, R. Waser, S. Kvatinsky, Experimental demonstration of memristor-aided logic (MAGIC) using valence change memory (VCM), IEEE Transactions on Electron Devices 67 (8) (2020) 3115–3122, https://doi.org/10.1109/ted.2020.3001247.
[36] N.G. Murali, P. Satya Vardhan, F. Lalchhandama, K. Datta, I. Sengupta, Mapping of boolean logic functions onto 3D memristor crossbar, in: 2019 32nd International Conference on VLSI Design and 2019 18th International Conference on Embedded Systems (VLSID), 2019, https://doi.org/10.1109/vlsid.2019.00106.
[37] Advanced Memristor Modeling, 2019, https://doi.org/10.3390/books978-3-03897-103-0.
[38] Y. Nishi, Advances in Non-volatile Memory and Storage Technology, Elsevier, 2014.
[39] A. James, Memristor and Memristive Neural Networks, BoD – Books on Demand, 2018.
[40] R. Kozma, R.E. Pino, G.E. Pazienza, Advances in Neuromorphic Memristor Science and Applications, Springer Science & Business Media, 2012.
[41] L. Chua, If it's pinched it's a memristor, in: Handbook of Memristor Networks, 2019, pp. 15–88.
[42] A. Slavova, R. Tetzlaff, Mathematical analysis of memristor CNN, in: Memristors – Circuits and Applications of Memristor Devices [Working Title], 2019, https://doi.org/10.5772/intechopen.86446.
[43] R. Hasan, T.M. Taha, C. Yakopcic, A fast training method for memristor crossbar based multi-layer neural networks, Analog Integrated Circuits and Signal Processing 93 (3) (2017) 443–454, https://doi.org/10.1007/s10470-017-1051-y.
[44] A.G. Ruotolo, M. Ottavi, S. Pontarelli, F. Lombardi, A novel write-scheme for data integrity in memristor-based crossbar memories, in: Proceedings of the 2012 IEEE/ACM International Symposium on Nanoscale Architectures – NANOARCH'12, 2012, https://doi.org/10.1145/2765491.2765521.
[45] B. Chakrabarti, et al., Corrigendum: a multiply-add engine with monolithically integrated 3D memristor crossbar/CMOS hybrid circuit, Scientific Reports 7 (Jul. 2017) 46874.
[46] A. Pisarev, A. Busygin, S. Udovichenko, O. Maevsky, 3D memory matrix based on a composite memristor-diode crossbar for a neuromorphic processor, Microelectronic Engineering 198 (2018) 1–7, https://doi.org/10.1016/j.mee.2018.06.008.
[47] A.G. Radwan, M.E. Fouda, On the Mathematical Modeling of Memristor, Memcapacitor, and Meminductor, Springer, 2015.
[48] K.M. Sundqvist, L.B. Kish, Memristors and thermal noise: is the memristor indeed the missing passive circuit element?, in: 2017 International Conference on Noise and Fluctuations (ICNF), 2017, https://doi.org/10.1109/icnf.2017.7985980.
[49] C. Yakopcic, T.M. Taha, G. Subramanyam, R.E. Pino, Impact of memristor switching noise in a neuromorphic crossbar, in: 2015 National Aerospace and Electronics Conference (NAECON), 2015, https://doi.org/10.1109/naecon.2015.7443090.
[50] A. Adamatzky, L. Chua, Memristor Networks, Springer Science & Business Media, 2013.
[51] J. Borghetti, G.S. Snider, P.J. Kuekes, J.J. Yang, D.R. Stewart, R.S. Williams, 'Memristive' switches enable 'stateful' logic operations via material implication, Nature 464 (7290) (Apr. 2010) 873–876.

[52] Q. Chen, X. Wang, H. Wan, R. Yang, A logic circuit design for perfecting memristor-based material implication, IEEE Transactions on Computer-Aided Design of Integrated Circuits and Systems 36 (2) (2017) 279–284, https://doi.org/10.1109/tcad.2016.2578881.

[53] S. Kvatinsky, N. Wald, G. Satat, A. Kolodny, U.C. Weiser, E.G. Friedman, MRL — memristor ratioed logic, in: 2012 13th International Workshop on Cellular Nanoscale Networks and their Applications, 2012, https://doi.org/10.1109/cnna.2012.6331426.

[54] L. Guckert, E.E. Swartzlander, MAD gates—memristor logic design using driver circuitry, IEEE Transactions on Circuits and Systems II, Express Briefs 64 (2) (2017) 171–175, https://doi.org/10.1109/tcsii.2016.2551554.

[55] L. Xie, et al., Scouting logic: a novel memristor-based logic design for resistive computing, in: 2017 IEEE Computer Society Annual Symposium on VLSI (ISVLSI), 2017, https://doi.org/10.1109/isvlsi.2017.39.

[56] C. Li, et al., Long short-term memory networks in memristor crossbar arrays, Nature Machine Intelligence 1 (1) (2019) 49–57, https://doi.org/10.1038/s42256-018-0001-4.

[57] K. Smagulova, O. Krestinskaya, A.P. James, A memristor-based long short term memory circuit, Analog Integrated Circuits and Signal Processing 95 (3) (2018) 467–472, https://doi.org/10.1007/s10470-018-1180-y.

[58] O. Krestinskaya, B. Choubey, A.P. James, Memristive GAN in analog, Scientific Reports 10 (1) (Apr. 2020) 5838.

[59] K. Smagulova, O. Krestinskaya, A. James, Who is the winner? Memristive-CMOS hybrid modules: CNN-LSTM versus HTM, IEEE Transactions on Biomedical Circuits and Systems 14 (2) (Apr. 2020) 164–172.

[60] P. Yao, et al., Fully hardware-implemented memristor convolutional neural network, Nature 577 (7792) (Jan. 2020) 641–646.

[61] C. Yakopcic, M.Z. Alom, T.M. Taha, Extremely parallel memristor crossbar architecture for convolutional neural network implementation, in: 2017 International Joint Conference on Neural Networks (IJCNN), 2017, https://doi.org/10.1109/ijcnn.2017.7966055.

[62] A.M. Zyarah, N. Soures, D. Kudithipudi, On-device learning in memristor spiking neural networks, in: 2018 IEEE International Symposium on Circuits and Systems (ISCAS), 2018, https://doi.org/10.1109/iscas.2018.8351813.

[63] M. Shahsavari, P. Devienne, P. Boulet, Spiking neural computing in memristive neuromorphic platforms, Handbook of Memristor Networks (2019) 691–728, https://doi.org/10.1007/978-3-319-76375-0_25.

[64] L.A. Camuñas-Mesa, B. Linares-Barranco, T. Serrano-Gotarredona, Neuromorphic spiking neural networks and their memristor-CMOS hardware implementations, Materials 12 (17) (Aug. 2019), https://doi.org/10.3390/ma12172745.

[65] Q. Duan, et al., Spiking neurons with spatiotemporal dynamics and gain modulation for monolithically integrated memristive neural networks, Nature Communications 11 (1) (Jul. 2020) 3399.

[66] S. Saïghi, et al., Plasticity in memristive devices for spiking neural networks, Frontiers in Neuroscience 9 (Mar. 2015) 51.

[67] E. Gale, B. de Lacy Costello, A. Adamatzky, Spiking in memristor networks, Memristor Networks (2014) 365–387, https://doi.org/10.1007/978-3-319-02630-5_17.

[68] S. Spiga, A. Sebastian, D. Querlioz, B. Rajendran, Memristive Devices for Brain-Inspired Computing: From Materials, Devices, and Circuits to Applications – Computational Memory, Deep Learning, and Spiking Neural Networks, Woodhead Publishing, 2020.

Chapter 15

Bipolar resistive switching in biomaterials: case studies of DNA and melanin-based bio-memristive devices

Gayatri M. More[a], Arpita P. Tiwari[b], Kiran D. Pawar[a], Tukaram D. Dongale[a,c], and Tae Geun Kim[c]

[a]*School of Nanoscience and Biotechnology, Shivaji University, Kolhapur, India,* [b]*Centre of Interdisciplinary Research, D.Y. Patil University, Kolhapur, India,* [c]*School of Electrical Engineering, Korea University, Seoul, Republic of Korea*

15.1 Introduction

Computing devices that play very vital roles in many applications are now a days ubiquitous part of human life. The rise of the internet of things, big data, and artificial intelligence has paved the way for the development of highly efficient and intelligent computing devices. As storing and providing on-demand data to the processing units is one of the key aspects of efficient computing devices, a large amount of data processing and storing is required. The high speed and error-free data storing and retrieving operations are decisive factors of any computing system. Given this, various kinds of nonvolatile memory devices are being utilized. On the commercial front, flash-based nonvolatile memories are now part of the many computing systems. However, further scaling down of flash-based memories in the range of < 10 nm gives rise to various technological challenges such as the increase in the leakage current, non-reliability in the data storing and retrieving, cell-to-cell cross-talk, reduction in memory window, etc. [1]. Because of this, various technological solutions are being researched to fulfill future computing demands. These solutions include, but are not limited to non-conventional nonvolatile memory devices, exotic materials, and breakthrough device structures.

One of the nanoscale phenomena that have attracted a lot of research interest in recent years is resistive switching in which external bias can control the resistance state of the device. Simply by controlling the resistance state of the device, one can store '1' and '0' and also retrieve stored data. This simplest mechanism can offer huge advantages over its other counterparts such as two-terminal

https://doi.org/10.1016/B978-0-12-821184-7.00024-4

structures, complementary metal-oxide-semiconductor (CMOS) compatibility, high-density integration, etc. [2]. In the past, the nanoscale resistive switching effect was used to realize the memristor/memristive device and considered as a fundamental circuit element, similar to resistor, inductor, and capacitor [3]. The resistive random access memory (ReRAM) devices based on memristive or resistive switching effect are considered as a disruptive technology, which can be utilized for the next-generation smart computing devices. The memristive devices are gaining a lot of attention due to their intrinsic properties such as nonvolatility, non-linearity, and passivity. These properties can be exploited for the resistive memory [4], brain-inspired computing [5], and sensor applications [6].

The memristive effect can be understood by the resistive switching property of the device and this can be achieved by properly tuning the material properties. In recent years, inorganic, organic, and hybrid materials have been explored for the resistive switching/memristive devices fabrication. The inorganic materials such as TiOx, NiOx, HfOx, AlOx, TaOx, ZnOx, etc. are promising materials for the nonvolatile memory application [7–9]. In addition to this, organic materials consisting of several traditional polymers, synthetic polymers, and small molecules are also utilized for memristive device development [10]. In the case of hybrid materials, an organic–inorganic combination of perovskites, polymer-oxides composite, graphene oxide-polymers composite, etc. are gaining a lot of attention for nonvolatile memory application [11,12]. The memory devices based on inorganic and hybrid materials have demonstrated the excellent nonvolatile memory effect, high speed, scalability, and low power consumption properties [11,12]. Nevertheless, inorganic and hybrid materials have some disadvantages like non-biodegradability and toxicity. Furthermore, they are the main source of electronic waste and other environmental issues. Thus, the development of biocompatible and biodegradable memory devices is a need of the hour. For this reason, many research groups are focusing on biomaterial-based memory devices. Biomaterials like proteins, biomolecules, natural polymers, pigments, polysaccharides, etc. can be used as a resistive switching medium for resistive memory application. Furthermore, these materials provide additional degrees of freedom such as biocompatibility, biodegradability, renewability, and environmental friendliness [13]. Due to these properties, bio-memristive devices can be used for secure memory devices, implantable medical devices, bio-inspired electronic systems, and medical diagnosis devices in the near future.

15.2 Brief overview of resistive switching and memristive devices

The resistive switching-based devices undergo two or multiple state variations, in which resistance of the device changes its resistance from its original resistance state to another resistance state. The change could be abrupt or gradual, which also decides the application of the resistive switching effect. In general, an abrupt change from low resistance state (LRS) to high resistance state (HRS) could be used for the nonvolatile memory application, while synaptic properties can be mimicked with the help of gradual state change [14]. Moreover, multiple intermediate resistance states can also be engineered for high-density memory applications [15,16]. It is interesting to note that the origin and mechanism of the resistive switching effect are unique for each resistive switching devices and can be changed (device-to-device) according to active switching material, device structure, top and bottom electrode and external stimulus (voltage, current, light, etc.). In the case of switching materials, inorganic, organic, hybrid, polymers, and biomaterials have been used [1,2,8–10,12,13]. The device structure of ReRAM is preferably vertical type, however, for some cases; planer device structures are also being used. It is now well understood that the top and bottom electrodes influence the resistive switching properties. For instance, the silver or copper-based active metals can undergo the redox reaction due to voltage stress [17] that can cause the formation and breaking of conductive filament. This effect is very different from the oxygen vacancies supported formation and breaking of the conductive filament [18]. In addition to this, the external stimulus plays a key role in resistive switching and is used to trigger the resistive switching process. The external voltage or current is commonly utilized to get the resistive switching; however, a photonic stimulus is also used for some novel applications [19,20].

The resistive switching effect is generally classified in terms of unipolar or bipolar, however, in some cases, the co-existence of both types was also observed [21]. In the case of unipolar, the resistive switching occurs at the same voltage polarity whereas dissimilar voltage polarity is a prerequisite for bipolar resistive switching effect. Apart from this, understanding of the resistive switching mechanism is very important to develop reliable memory devices for practical applications. Two types of resistive switching mechanisms, filamentary and homogeneous are generally observed in the resistive memory devices. However, the co-existence of the filamentary and homogeneous resistive switching has also been reported for some devices [22,23]. In the case of the filamentary type resistive switching, a conductive filament is formed and broken between the top and bottom electrodes, which decreases and increases the resistance of the device, respectively. Thus, the reproducible LRS and HRS can be achieved by forming and breaking the conductive filament. On the other hand, interface effects dominate in the homogeneous resistive switching and give rise to the repeatable LRS and HRS. Several conduction models have been reported to investigate the resistive switching phenomena. The filamentary conduction is accomplished due to the hopping, space-charge-limited conduction (SCLC), Ohmic conduction, Poole–Frenkel emission, and trap-assisted tunneling while the homogeneous or interfacial conduction is due to Schottky emission, Fowler–Nordheim tunneling and direct tunneling [24]. Additionally, depending upon switching material and electrode material, the ReRAM cells are classified as cation-based witching device and anion-based witching device. A resistive switching device based on cation migrations is known as an electrochemical metallization cell (ECM) or sometimes called programmable metallization cell (PCM) or conductive bridge random access memory (CBRAM). On the other hand, devices based on anion migrations are called valence change memory cells (VCMs) [25].

The ideal memristor is a nonvolatile, non-linear, and passive circuit element predicted by Prof. L. Chua in 1971 based on the symmetric relationship between magnetic flux and charge [26]. Other circuit elements such as a resistor, capacitor, and inductor follow the relationship between current–voltage, voltage-charge, and current-magnetic flux, respectively. In 2008, the HP group announced the first practical realization of the memristor device based on the $Pt/TiO_2/Pt$ sandwich structure with the help of Chua's theory [3]. In addition to this, there is a class of devices that possesses the non-ideal properties of the memristor device and popularly referred to as memristive devices [27]. It is very hard to get the ideal memristor properties, therefore, we will refer our devices as memristive devices in this chapter, hereafter.

15.3 Materials for resistive switching application

The resistive switching devices are made up of metal–insulator–metal (MIM) structure in which the middle layer acts as an active switching medium. There are different types of inorganic, organic, hybrid, and bio-materials reported for the resistive switching memory application, as shown in Fig. 15.1. Inorganic materials include metal oxides, chalcogenides, nitrides and halides [7–9], whereas the organic materials consist of many traditional polymers, synthetic polymers, small molecules, etc. [28], while the hybrid materials are made up of organic–inorganic combinations such as hybrid perovskite, polymer-oxides composites, graphene oxide and reduced graphene oxide composites with polymers [29,30]. These inorganic and hybrid materials are popular choices for many applications and show maximum nonvolatile memory property with high speed, scalability, and low power consumption characteristics. Nevertheless, these materials are non-biodegradable, toxic, and costly [31]. Therefore, the development of biocompatible, biodegradable, and transient memory devices is not possible with inorganic and hybrid materials [32]. The efforts have been focused on the development of biomaterial-based memory devices. Biomaterials like proteins, natural polymers, pigments, polysaccharides, etc. can be used for nonvolatile memory applications [33]. Table 15.1 summarizes the recent advance in the biomaterial-based resistive switching memory devices. Most of the reported devices are based on the bipolar resistive switching effect. The survey suggested that only a few devices show excellent endurance and retention memory properties, comparable to oxide-based memory devices. However, low operating voltage and a good memory window are key aspects of many devices. Therefore, these devices can be an excellent choice for the low power embedded memory

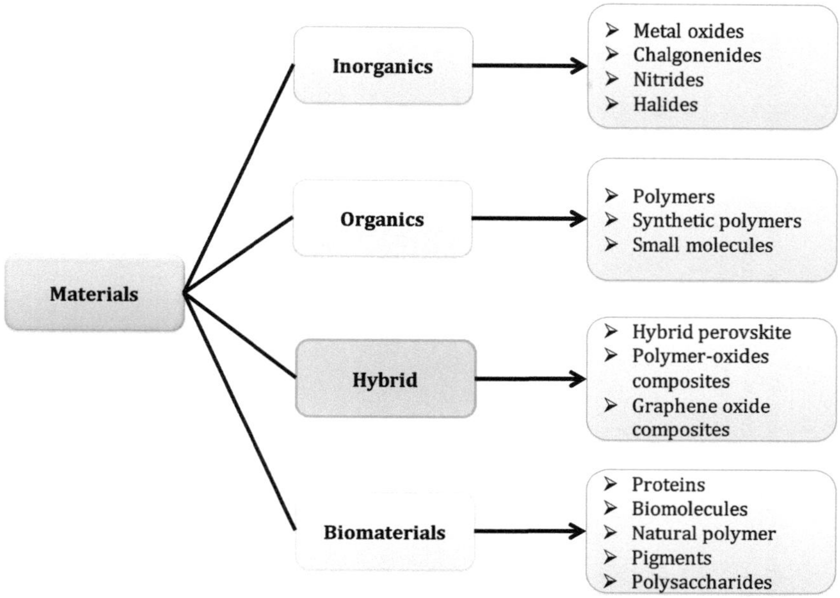

FIGURE 15.1 Active switching layer materials for resistive switching memory application.

application. The literature survey suggested that the biomaterials have great potential to be prospective candidates for transient and implantable electronics. In particular, transient biomaterials can be used for the secure memory application, whereas implantable memory devices can be developed using biocompatible biomaterials.

15.4 Biomaterial-based memristive devices

15.4.1 Case study 1: DNA-based memristive device

As an alternative to inorganic material-based devices, biomaterial-based memory devices have become an active area of research owing to their low cost, biodegradability, biocompatibility, and excellent memory properties [17,59]. In recent years, polymer materials are extensively studied for nonvolatile memory and neuromorphic computing application [60]. However, traditional polymers based organic memory devices require complex chemical synthesis processes that add to the cost of production and therefore make faster marketing difficult. In the light of these difficulties, great efforts have been made to develop the organic memory devices using various biomaterials such as protein, deoxyribonucleic acid (DNA), chitosan, cellulose, gelatin and many more [61]. Among these, the DNA has attracted a lot of attention and has been utilized as an active layer for resistive memory applications. Besides, a recent study suggests that

TABLE 15.1 Biomaterial-based resistive switching memory devices.

Biomaterials	Device structure	Switching mode (unipolar or bipolar)	Endurance (# of Cycles)	Retention (s)	Memory window (HRS/LRS)	Resistive switching voltage (V) (SET/RESET)	References
Melanin	Ag/melanin/SS	Bipolar	2×10^4	10^3	10	±0.6	[34]
DNA	Au/CuO-DNA-Al/Au/Si	Bipolar	100	500	50	±2.25	[35]
	Pt/Cu^{2+} doped DNA/FTO	Bipolar	10×10^4	10^4	10^3	−3.5/+2.5	[36]
	Ag/DNA-CTMA/ITO	Bipolar	200	10^4	10^3	+0.65/−1.25	[37]
	Au/DNA_{10}/Au	Bipolar	100	$\sim 10^6$	20	+0.73/−0.85	[38]
Cellulose	Ag/Cellulose fibres/Al	Bipolar	6×10^3	300	10	±1V	[17]
	Ag/cellulose fibres/ITO	Bipolar	30	10^4	10^7	+1/−0.5	[39]
Egg Albumen	Al/Egg albumen /ITO	Bipolar	500	10^4	10^3	−0.3/+3.6	[40]
	Ag/ Egg albumin/ ITO/PET	Bipolar	50	10^4	10^2	+0.6/−0.7	[41]
Gelatin	Al/Gelatin/ITO	Bipolar	120	10^5	10^6	−0.7/+2.4	[42]
	ITO/Al-chelated gelatin/ITO	Bipolar	60	10^5	10^4	−2/+2.5	[43]
Lotus Leaf	Ag/lotus leaves/ITO	Bipolar	75	10^3	40	+5/−6.5	[44]
Aloe Vera	Ag/Aloe vera/ITO	Bipolar	100	10^5	10^4	+0.87/−1.15	[45]
	Ag/Aloe vera/ITO	Bipolar	100	10^4	10^3	±0.6	[46]
	Al/Aloe vera/ITO	Bipolar	21	10^4	10^3	−0.9/+3.1	[47]
Pectin	Ag/Pectin/FTO	Bipolar	100	500	450	+3.3/−4.5	[48]
	Ag/ Pectin/ITO	Bipolar	500	10^4	25	+1/−0.5	[49]

continued on next page

TABLE 15.1 *(continued)*

Biomaterials	Device structure	Switching mode (unipolar or bipolar)	Endurance (# of Cycles)	Retention (s)	Memory window (HRS/LRS)	Resistive switching voltage (V) (SET/RESET)	References
Protein	Al/silk fibroin protein/ITO	Bipolar	120	10^3	11	+10.4/−11.5	[50]
	Al/S-layer protein/ITO/PET	Bipolar	500	4.5×10^3	5.9	±8	[51]
	Al/silkworm hemolymoh protein/ITO	Bipolar	500	10^4	10^3	−1.2/+3.5	[52]
	Ag/Sericin/Au	Bipolar	21	10^3	10^6	+2.5/−0.8	[53]
Chitosan	Mg/Ag doped chitosan/Mg	Bipolar	60	10^4	10^2	+1.63/−0.82	[54]
	Ag/Ag doped chitosan/Pt	Bipolar	100	10^4	10^5	+0.5/−0.25	[55]
Starch	Au/starch-chitosan/ITO/PET	Bipolar	–	10^4	100	+3/−4	[56]
Enzymes	Ag/lysozyme/poly-styrene sulfonate/Pt	Bipolar	200	10^4	10^2	+1/−1.3	[57]
	Ag/poly(allylamine hydrochloride/Ferritin/Pt	Bipolar	300	10^4	10^2	−1.5/+1.5	[58]

the DNA is a strong candidate for bio-memristor applications [38] which can be used as a basic building block for the transient resistive memory application with inherent biodegradability, biocompatibility and flexibility properties. To achieve comparable performance with the inorganic resistive memory device, the DNA-based resistive memory should possess excellent memory properties such as long terms endurance and retention, uniformity in resistive switching states, and electroforming free process with low voltage resistive switching operation. The earlier reported DNA-based memory devices required higher resistive switching voltage [35], showed poor memory window [38] and switching instability [62]. In the present work, we have developed a DNA based organic memory device using a solution-processable technique and demonstrated the uniform and excellent nonvolatile memory properties.

15.4.1.1 Materials and methods

All the reagents and chemicals used in the study were of analytical grade (Sigma-Aldrich, Mumbai). Reagents and washing solutions used in the experiments were prepared with phosphate buffer saline (PBS, 10 mM, pH 7.4). *Escherichia coli* (ATCC 25922 strain) was used for bacterial DNA purification. All reagents were prepared under aseptic conditions and by using deionized water. For the present study, bacterial genomic DNA (gDNA) was isolated from *Escherichia coli* using a Qiagen spin column kit (Qiagen, New Delhi) as described elsewhere [63]. To this end, *E. coli* cells were grown overnight in Nutrient Broth medium at 37 °C, harvested by centrifugation at 8000 rpm for 10 minutes, and then suspended in lysis buffer to disrupt the cells and release the nucleic acids and proteins into the solution. This was then followed by a digestion step to remove the protein and RNA in the solution using proteinase and RNase enzymes, respectively. The digested proteins and RNAs were removed by centrifugation using the DNA binding spin column. The gel documentation system (Bio-Rad) was used to investigate the purity of the isolated gDNA. The DNA was eluted from the column in 100 μL of deionized water and the quality of isolated DNA was analyzed by agarose gel (1% w/v) electrophoresis. The optical spectra were recorded using a UV-Vis spectrophotometer (Cary-60, Agilent Technology). The purity and concentration of the isolated gDNA were also measured spectrophotometrically by recording optical density at 260 and 280 nm and estimated using the A260/A280 ratio. The concentration of gDNA was calculated using a formula, $C = A/e * l$, where, C is the concentration of the nucleic acid, in μg/mL, A is the absorbance at 260 nm, l is the width of the cuvette (1 cm) and e is extinction coefficient (1 OD260 Unit = 50 μg/mL for double-stranded DNA).

To prepare the memory device, the fluorine-doped tin oxide (FTO) substrates (resistance: 7 Ω/sq.) were thoroughly cleaned first by double distilled water followed by acetone and iso-polypropanol and finally, the slides were dried to remove any water and other residues. 20 μl of the gDNA solution was then applied onto the cleaned FTO substrate using the drop-casting method in the layer-by-layer manner. Ten consecutive layers of gDNA were drop casted on FTO. The entire preparation and fabrication processes were undertaken in a dust-free environment (the temperature at 21 °C and relative humidity of 70 to 80%). The top silver (Ag) contacts were created using a vacuum deposition technique. The surface structure of the deposited material was studied by atomic force microscopy (AFM) (Nanoman, Veeco). All electrical measurements were carried out using a memristor characterization platform (ArC ONE).

15.4.1.2 Results and discussion

The UV-Vis spectrum of isolated gDNA is shown in Fig. 15.2a. The A260/A280 value of 1.92 indicated that the gDNA isolated from a bacterium was highly pure and contamination-free. The isolation of DNA from *E. coli* yielded 188 μg/mL of gDNA with high purity which was evident by agarose gel electrophoretogram as shown in Fig. 15.2b. The presence of a clear and discrete band in lane 2 confirmed the presence of gDNA (Fig. 15.2b). Such concentrated, high purity DNA is an important requisite for memory application. Fig. 15.2c represents the atomic force microscopy (AFM) image of the gDNA active layer. The DNA molecules were found distributed on the FTO substrate and were mutually aggregated with each other. The root mean square roughness calculated over the entire imaged area was found to be 39.75 nm. Fig. 15.2d represents the typical structure of the developed DNA memory device. In the present case, Ag acted as a top electrode whereas FTO worked as a bottom electrode. The active DNA layers were sandwiched between these two electrodes and responsible for the resistive switching effect.

The resistive switching (RS) effect is considered to be a new paradigm shift in the nonvolatile memory device technology. In the RS effect, the active material changes the resistance state due to external electrical stress and device shuffles in two or more resistive states. These states are generally known as high resistance state (HRS), intermediate resistance state (IRS), and low resistance state (LRS). In general, the change of resistance is nonvolatile and reversible, which could be exploited for the development of high-performance memory devices. Fig. 15.3a represents typical hysteresis-like current–voltage (I–V) characteristics of the Ag/DNA/FTO device and switching direction is indicated by arrows. The hysteresis loop in I–V mode is one of the fingerprint properties of a memristor device. The device showed the bipolar resistive switching effect under very low (±0.5 V) resistive switching voltage. Initially, the device was in HRS at 0 V. As the external electrical stress was increased, the device started to conduct and switch its resistance state from HRS to LRS at +0.5 V. The device was in the LRS during +0.5 V → 0 V → −0.5 V and changed resistance state again to HRS at −0.5 V. It is interesting to note that the device was free from the electroforming process and current compliance protocol. In the case of resistive memory devices, the electroforming free process reduces the complexity of memory architecture and unit cost of the memory cell [64]. In addition to this, the external current limiter circuit is eliminated in the case of self-compliance

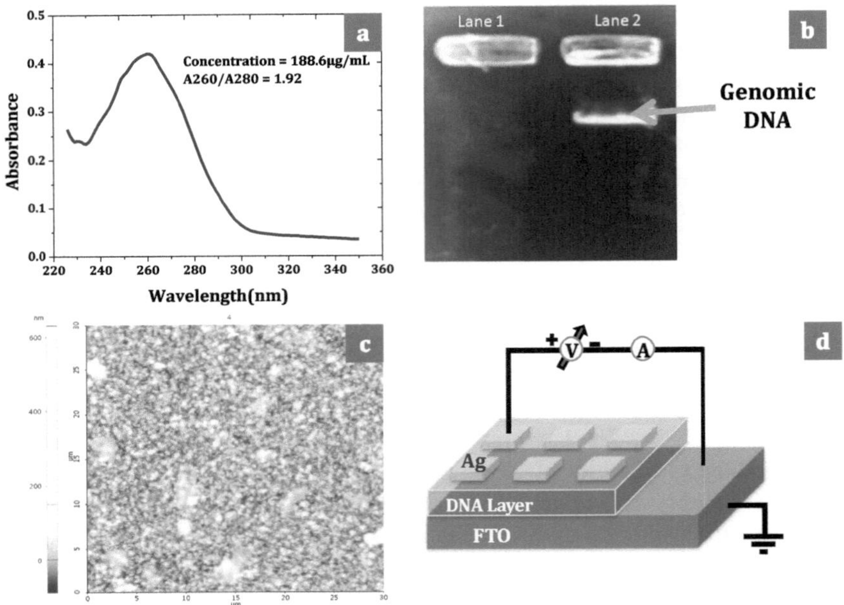

FIGURE 15.2 (a) UV-Vis spectrum of isolated gDNA of *E. coli* used in the study; (b) agarose gel electrophoretogram; Lane 1: Buffer solution, Lane 2: gDNA in Buffer solution (10 µL); (c) AFM image of DNA active layer, and (d) typical Ag/DNA/FTO memory device structure. (For interpretation of the colors in the figure, the reader is referred to the web version of this chapter.)

effect [65]. Instead of a linear change in the current, the high non-linearity was observed in the current switching. In some voltage range, the stepwise switching was also observed. The ideal memristor is characterized in terms of the single-valued charge-magnetic flux relation; however, it is hard to get the ideal memristor characteristics. Instead of single-valued charge-magnetic flux relation, the practical devices show a double-valued relation [66,67], therefore, they can be considered as extended memristor devices or memristive devices [68]. To categorize the DNA memory device either memristor or memristive device, we have calculated the device charge and flux using an earlier reported method [69]. The time-domain flux, time-domain charge, and charge-magnetic flux properties of DNA memristive devices are shown in Fig. 15.3b, c, and d, respectively. The resistive switching states of the device such as initial, half-period, final-period and turning point are represented by A_1, B_{CW}, A_2, B_{CW}^N, respectively. In the present case, the time-domain flux was found to be symmetric, whereas quasi-asymmetric behavior was observed for the device charge characteristic. The device showed a gradual state change for time-domain flux characteristic, whereas the quasi-abrupt transition was observed for time-domain charge characteristic, as shown in Fig. 15.3b and c. As against the ideal memristor definition, double-valued charge-magnetic flux characteristic was observed for the

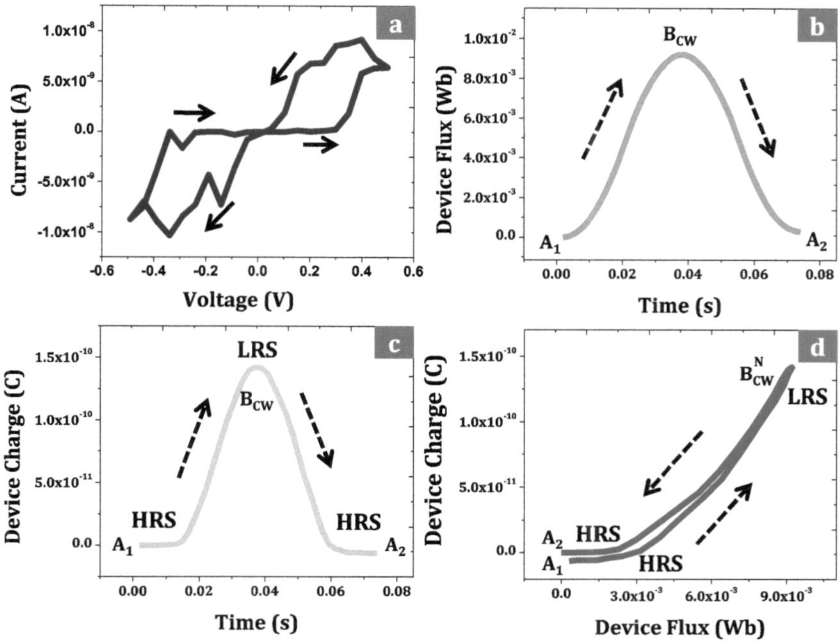

FIGURE 15.3 (a) I–V, (b) time-domain flux, (c) time-domain charge, and (d) charge-magnetic flux characteristics of DNA memristive device. The arrow indicates the direction of resistive switching.

DNA-based device, as shown in Fig. 15.3d. Given this, the DNA memory device acted like an extended memristor device or memristive device.

To confirm the nonvolatile resistive switching properties of the DNA memristive device, the memory cell using endurance and retention measurements was also tested. The room temperature endurance and retention characteristics of the Ag/DNA/FTO device are shown in Fig. 15.4a and b. For the nonvolatile memory measurements, series of write pulses (pulse magnitude: ±0.5 V and pulse duration: 500 μs) were applied and resistance of the test cell was measured at 0.2 V. The DNA memristive device showed the bipolar resistive switching effect without any gradual SET/RESET switching or current compliance approach, as shown in Fig. 15.3a. In the present study, two distinct and well-resolved resistance states were observed viz. LRS and HRS, as shown in Fig. 15.4a. The results indicate that the DNA memristive device switches error-free between LRS and HRS up to 200 cycles without degradation in resistance states and memory window (HRS/LRS: ~1.75). The retention characteristic of the DNA memristive device (Fig. 15.4b) showed the stability of storing data with respect to time. The retention characteristic further confirmed the bi-level resistive switching and demonstrated the stability of two resistance states over the 10^3 seconds. The endurance and retention memory results asserted the high reliability of DNA memristive device. The cycle to cycle switching uniformity

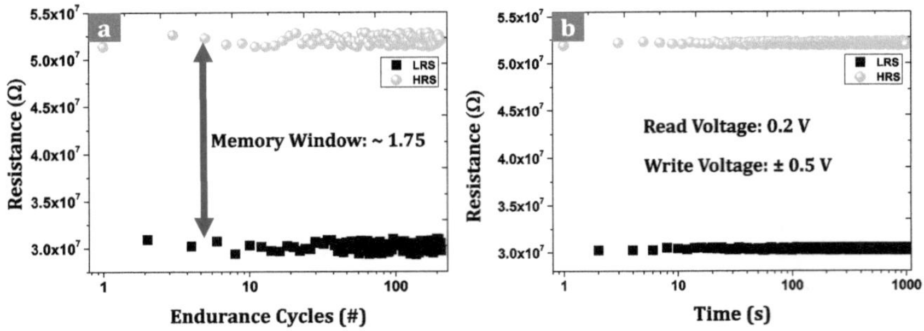

FIGURE 15.4 (a) Endurance and (b) retention memory properties of DNA memristive device.

was quite good, as evident from the endurance result. This suggested the good reliability and stable switching behavior of DNA-based memristive device.

The conduction mechanism of Ag/DNA/FTO memristive device could be understood by plotting I–V curves in the log-log scale, as shown in Fig. 15.5a and b. The positive and negative biased I–V data had two slopes in low voltage and high voltage regions. The magnitude of the slope was calculated to be 3.99 and 3.43 for low voltage region for positive bias data and negative bias data, respectively. In the same way, slope values of high voltage region for positive bias and negative bias were 6.99 and 3.09, respectively. To confirm the possible conduction mechanism involved in the resistive switching process, different conduction mechanisms were investigated. In the present case, Child's square law fitted well to the experimental HRS data with good-adjusted R^2 values (Fig. 15.5c to f). This suggested that the current of the device was directly proportional to the square of the applied voltage.

The resistive switching characteristics and nonvolatile memory results (endurance and retention) of DNA memristive device showed the abrupt transition of current during I–V measurements and resistive states, respectively. In general, this kind of abrupt transition happens due to the filamentary resistive switching effect. The possible bipolar resistive switching effect is depicted in Fig. 15.6a and b. In our case, Ag which is known as an electrochemically active metal worked as a top electrode. When a positive voltage was applied to the Ag electrode, the Ag ions were oxidized at the interface and started to migrate through the DNA active layer towards the bottom FTO electrode where they reduced (Ag → Ag+ + e-). This would have formed the highly conductive filament, as shown in Fig. 15.6a and device switched to LRS. When negative bias was applied to the Ag electrode, the electrochemical dissolution of conductive filament started. When voltage reached to V_{RESET}, the complete breaking of filament took place due to oxidation of Ag (Ag+ + e- → Ag), which drove the system into HRS, as represented in Fig. 15.6b. Given this, the Ag/DNA/FTO memristive device repeatedly switched between LRS and HRS as a bipolar resistive switching device.

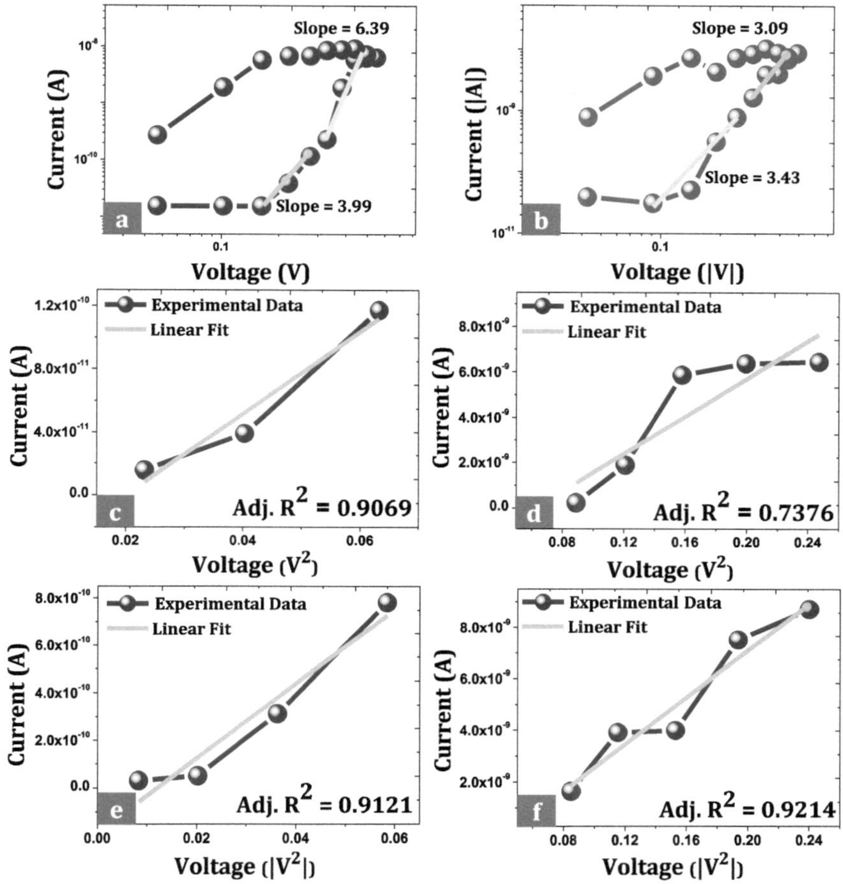

FIGURE 15.5 Log-Log I–V characteristics of DNA memristive device in (a) positive and (b) negative bias, respectively. Child's square law conduction mechanism fitting to the positive and negative biased (c and e) low and (d and f) high voltage region, respectively.

15.4.1.3 Summary

In summary, we have developed an organic Ag/DNA/FTO thin film memristive device and demonstrated its nonvolatile resistive switching effect. The bipolar resistive switching with a pinched hysteresis loop was observed at very low resistive switching voltage (±0.5 V). The device shows gradual state change for time-domain flux characteristics, whereas the quasi-abrupt transition was observed for time-domain charge characteristics. Furthermore, double-valued charge-magnetic flux characteristic was observed which suggests the memristive device like nature of DNA-based device. The DNA memory device switched between LRS and HRS up to 200 cycles without degradation in resistance states. Furthermore, data retention capability of DNA memristive device was found to

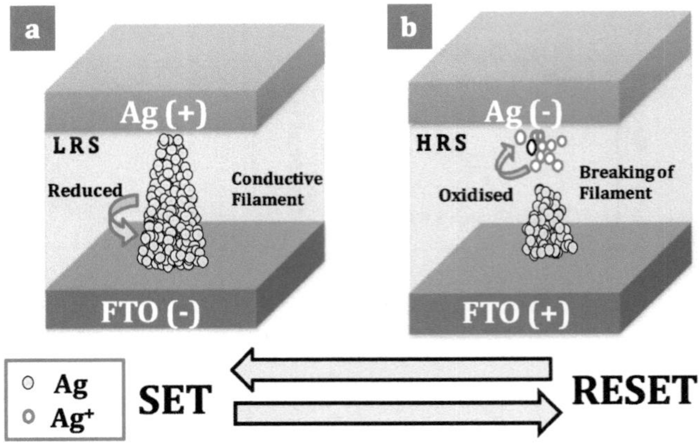

FIGURE 15.6 Possible bipolar resistive switching mechanism of Ag/DNA/FTO memristive device during (a) LRS and (b) HRS.

be 10^3 seconds with good uniformity during the cycle to cycle variations. The analysis of resistive switching results suggested that Child's square law was responsible for the device conduction mechanism and the resistive switching effect was due to the formation and breaking of conductive Ag filament.

15.4.2 Case study 2: melanin-based memristive device

As another case study, we demonstrate the fabrication and characterization of bio-memristive devices using synthetic melanin. To understand the switching dynamics, melanin was deposited on different bottom electrodes such as FTO, $TiN/SiO_2/Si$ and flexible indium tin oxide coated PET substrates with aluminum (Al) as a top electrode. It was observed that the switching property of the Al/melanin/FTO device was superior to other devices. In the case of nonvolatile memory property, the endurance of the Al/melanin/FTO device was stable over 10^3 cycles. Furthermore, LRS and HRS resistance states were retained over 10^3 s. The present study reveals that melanin is a promising biomaterial for nonvolatile memory applications.

Melanin is one of the interesting classes of natural pigments found in most organisms which can be used for the resistive switching memory application due to its tunable electrical properties. Furthermore, excellent biocompatibility and low-cost synthesis make melanin a promising choice for green electronics. Melanin is a natural biological pigment found in living organisms and can be formed by either tyrosine or dopamine [70]. The most common form of melanin is eumelanin, obtained from the copolymerization of derivatives of 5, 6-dihydroxy-indole-2-carboxylic acid (DHICA), and 5, 6-dihydroxy-indole (DHI). Melanin has some unique physical-chemical properties such as broad absorbance spectra throughout the UV region, strong non-radiative relaxation of photo-excited electronic states, hybrid ionic–electronic conduction, and biocompatibility [71]. It is interesting to note that the electrical properties of the melanin are similar to the organic semiconductors with energy band gap values 0.6 and 1.1 eV for natural and synthetic melanin, respectively [72]. This suggested that the melanin can be used as an active switching layer for resistive memory devices. In the past, few successful attempts were undertaken by various research groups to study the resistive switching properties of melanin [34,73,74]. In the present case study, we have synthesized the synthetic melanin by using chemical oxidation of 3, 4-dihydroxy-DL-phenylalanine (DL-DOPA). The prepared melanin was deposited on different substrates and their resistive switching properties were investigated.

15.4.2.1 Materials and methods

3,4-Dihydroxy-DL-phenylalanine (DL-DOPA), potassium permanganate ($KMnO_4$), (30% NH_3), ammonia solution (NH_4OH), FTO, indium tin oxide coated PET and SiO_2/Si substrates were purchased from the Sigma Aldrich (Mumbai, India). The melanin was synthesized by the oxidation-polymerization reaction of DL-DOPA with $KMnO_4$. In the typical process, 10 mM of DL-DOPA was dissolved in 20 mL of distilled water and heated at 50 °C. Under vigorous stirring, 0.6 mL of 0.1 M $KMnO_4$ was added to this solution. It was observed that the color of the solution changed from colorless to dark purple and gradually changed to black. The reaction was kept under continuous stirring for 6 h at 50 °C and pH was maintained at 7. After completion of the reaction, the solution was centrifuged at 15,000 rpm for 30 minutes. The black pellets of melanin were collected as product and further used for device fabrication and characterizations [75]. In the stock solution of synthetic melanin, 2–3 drops of 30% ammonia were added before deposition. The solution was stirred at room temperature for 1 hour and then ultra-sonicated for a further 1 hour. The substrates were cleaned thoroughly by deionized water, triton, ethanol, acetone, and iso-propyl alcohol and finally, the slides were dried to remove water and other residues. The final solution was spin-coated on different substrates at 4000 rpm for 1 minute [76]. For uniform melanin layer formation, the spin coating was repeated for 3 times. We have developed the three devices: (i) Al/melanin/FTO; (ii) Al/melanin/TiN/SiO_2/Si, and (iii) Al/melanin/ITO coated PET. The aluminum (Al) and titanium nitride (TiN) layers were deposited using vacuum deposition techniques. The optical spectra were recorded using a UV-Vis spectrophotometer (Cary-60, Agilent Technology). The Fourier-transform infrared spectroscopy (FTIR) (FTIR-4100, JASCO) was utilized to understand the functional groups present in the synthetic melanin. The surface structure of the deposited material was studied by atomic force microscopy (AFM) (Nanoman, Veeco). All electrical measurements were carried out using a memristor characterization platform (ArC ONE). During the electrical measurements, the top Al electrode was biased while the bottom electrode was grounded.

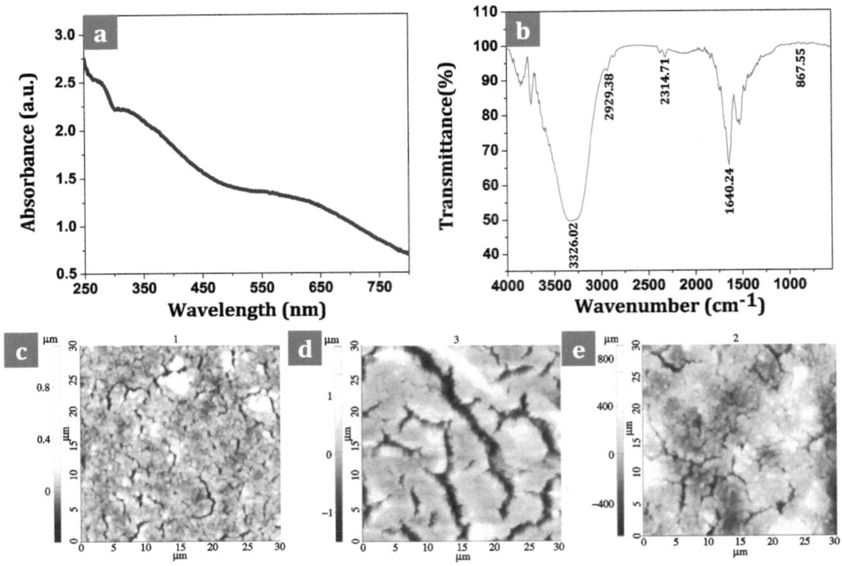

FIGURE 15.7 (a) UV-Vis and (b) FT-IR spectra of synthetic melanin. The AFM images of melanin deposited on (c) FTO, (d) TiN, and (e) flexible PET substrates. (For interpretation of the colors in the figure, the reader is referred to the web version of this chapter.)

15.4.2.2 Results and discussion

The black melanin pigments were synthesized from the oxidation-polymerization reaction of DL-DOPA with $KMnO_4$. The absorption spectra of synthetic melanin at room temperature are shown in Fig. 15.7a. The spectra of synthetic melanin showed a broad absorption in UV-Vis. region and found that the absorbance intensity decreases as wavelength increases. Such behavior is in good agreement with the photo-protection function of melanin [72]. The FT-IR spectroscopy was used to analyze the functional groups present in the synthetic melanin. The FT-IR spectra of synthesized melanin shown in Fig. 15.7b and verified by the absorbance bands at 3326, 2929, 2314, 1640, 867 cm^{-1}. The FT-IR spectrum of synthetic melanin showed a broad absorption band at 3326.02 cm^{-1} which demonstrated the presence of the hydroxyl group (OH). The stretching vibrations for aliphatic C-H bonding appeared at 2929.38 cm^{-1}. The characteristic strong band at 1640.24 cm^{-1} could be attributed to $C = O$ stretching vibrations of amide-I. The FT-IR spectrum of synthetic melanin ranging from 3326 to 867 cm^{-1} has a good degree of similarity to earlier reported melanin spectra [77]. The imaging of morphologies of melanin on FTO, TiN, and flexible PET films by AFM are shown in Fig. 15.7c, d, and e, respectively. The AFM images suggested that the synthetic melanin was deposited uniformly with a compact structure. The root mean square roughness calculated over the imaged area (30 × 30 μm) was found to be 106, 427, and 181.8 nm for FTO, $TiN/SiO_2/Si$ and flexible PET, respectively. In the present case, melanin de-

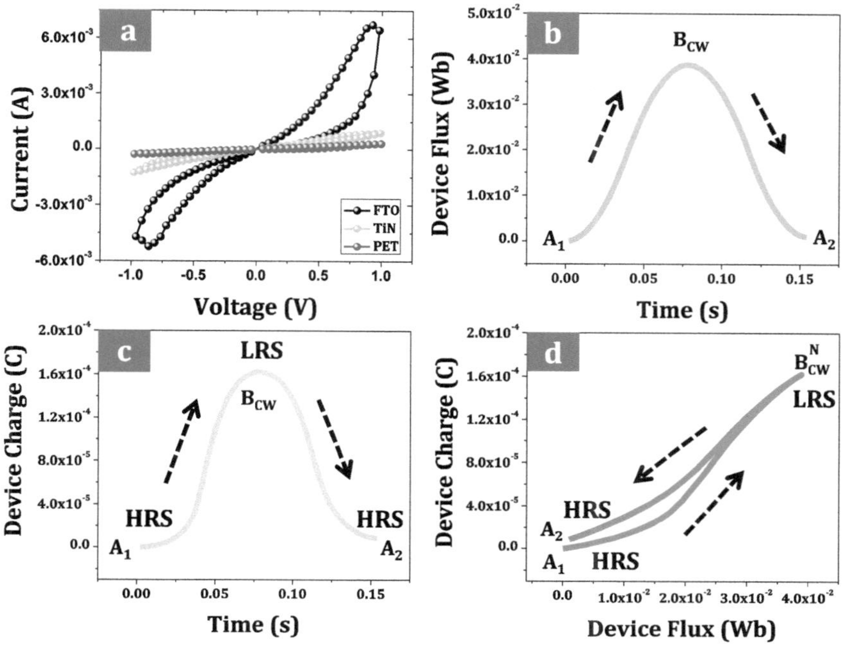

FIGURE 15.8 (a) I–V characteristic of synthetic melanin-based devices. (b) Time-domain flux, (c) time-domain charge, and (d) charge-magnetic flux characteristics of Al/melanin/FTO device. The arrow indicates the direction of resistive switching. (For interpretation of the colors in the figure, the reader is referred to the web version of this chapter.)

posited on FTO showed good surface morphology with low roughness. On the other hand, melanin deposited on TiN showed a lot of cracks and high roughness.

Fig. 15.8a represents the I–V characteristic of the synthetic melanin on different substrates. The I–V results suggested the developed devices demonstrate the typical bipolar resistive switching behavior. It is observed that the devices show good resistive switching at ± 1 V. During the electrical measurements, devices possessed HRS at the initial stage, then switched to LRS after the application of suitable electrical stress, called a SET voltage. Moreover, devices were switched again to HRS state at RESET voltage. These observations suggested that the devices possess two resistance states viz. LRS ($+1$ V $\rightarrow$ 0 V $\rightarrow -1$ V) and HRS (0 V to $+1$ V and 0 V to -1 V). Interestingly, all devices were found operational under electro-forming free mode. In the electro-forming assisted switching, a very high electrical bias, usually greater than SET and RESET voltages are applied to the device. The initial electro-forming process switches the device into the operational mode. However, a high electrical bias could increase the power consumption and circuit complexity of memory architecture. Forming-free memory devices can minimize power consumption

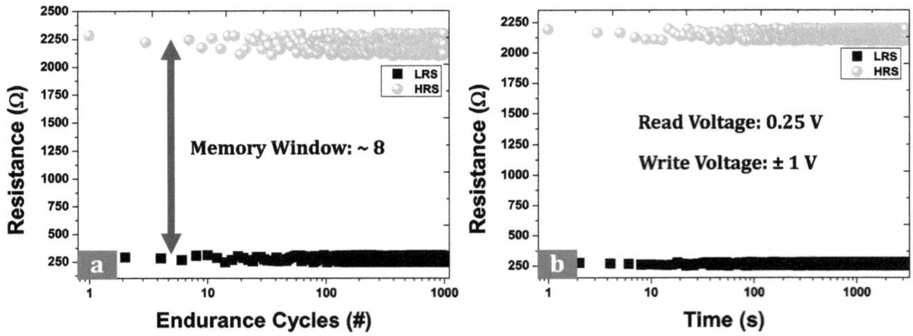

FIGURE 15.9 (a) Endurance and (b) retention memory properties of Al/melanin/FTO-based memristive device.

and circuit complexity issues [78,79]. In view of this, fabricated synthetic melanin devices could be useful for low power memory applications. I–V results suggested that the Al/melanin/FTO device had good resistive switching properties than other devices (Al/melanin/TiN/SiO_2/Si and Al/melanin/PET). The time-domain flux, time-domain charge, and charge-flux characteristics of the Al/melanin/FTO device are shown in Fig. 15.8b, c, and d, respectively. In the present case, the time-domain flux was found to be symmetric, whereas quasi-asymmetric behavior was observed for the charge characteristic. In addition to this, a double-valued charge-magnetic flux characteristic was observed for the Al/melanin/FTO device. In the nutshell, Al/melanin/FTO device shows non-ideal memristor properties and can be referred to as a memristive device.

The melanin-based memristive devices exhibited a typical bipolar resistive switching effect. To evaluate the nonvolatile memory performance of the Al/melanin/FTO-based memristive device, endurance, and retention tests were performed. Fig. 15.9a represents the endurance characteristic of the Al/melanin/FTO memristive device at room temperature. The endurance test showed two discrete and well-resolved resistance states (HRS and LRS) with a good memory window (~8). The device switched effectively in two states up to 10^3 cycles without noticeable degradation in the LRS and HRS. The retention characteristic of the Al/melanin/FTO memristive device is shown in Fig. 15.9b. The retention characteristic demonstrates the data storing capability of the device for an extended time. The LRS and HRS values of the device were stable for 3×10^3 seconds. A good uniformity during a cycle-to-cycle switching was observed, which suggested the good reliability and stable switching behavior of the device. This further asserts the reliable operation of the Al/melanin/FTO memristive device.

To understand the conduction mechanism of the Al/melanin/FTO memristive device, we have plotted the positive and negative bias I–V characteristics on a log-log scale, as shown in Fig. 15.10a and b, respectively. The slope values were calculated by dividing the positive and negative bias I–V data into low and high voltage regions. The slope values are summarized in Fig. 15.10a and b. For

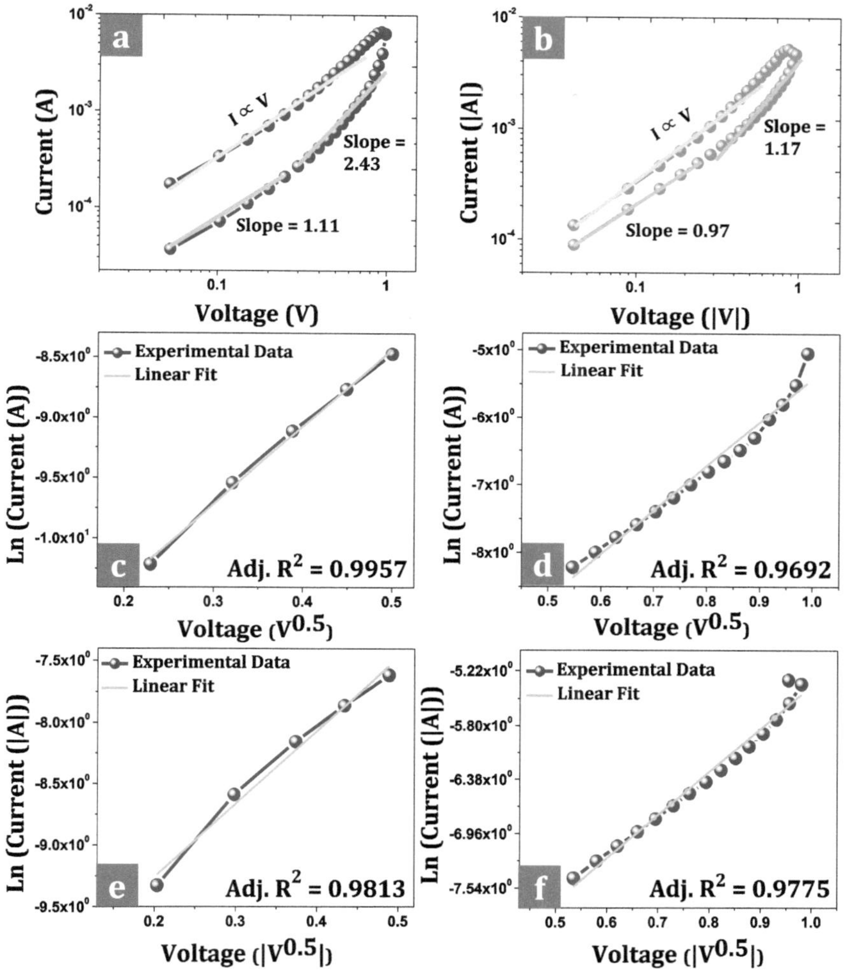

FIGURE 15.10 Log-Log I–V characteristics of Al/melanin/FTO memristive device in (a) positive and (b) negative bias, respectively. The Schottky plot of I Vs $V^{0.5}$ for the positive (c and d) and negative (e and f) biased low and high voltage region, respectively.

finding the conduction mechanism of the Al/melanin/FTO memristive device, we have investigated different types of conduction mechanisms such as Poole–Frenkel, Schottky, space charge limited current, trap-assisted tunneling, hopping conduction, Ohmic conduction, etc. In the present case, the Schottky emission model is well fitted to the HRS data with good-adjusted R^2 values, as shown in Fig. 15.10c to f. On the other hand, LRS data follows the Ohmic conduction model.

The melanin is a hybrid electronic-ionic conductor in which charge transportation occurs due to the free radicals (electrons), hydronium ions (protons), positive (H^+), and negative (OH^-) charged ions [34]. The electrical behavior of melanin is mainly affected by the water (H^+ and OH^-) content. However, Rosenberg's theory suggested that the electrons are dominantly responsible for charge transport in the device at lower water content [80]. Considering these facts and experimental conditions, a Schottky barrier may be formed between the Al/melanin interface. In addition to this, a quasi-analog I–V nature of melanin memristive device suggested that the interfacial resistive switching could be dominant in the developed device. In this case, a large amount of electron starts to extricate from the interface with respect to the increase in the voltage. Once the magnitude of the applied voltage reaches a sufficient value (SET voltage), a Schottky barrier height could be lowered down and the device switches to the LRS [81]. Therefore, modulation of external voltage could lead to switching from HRS to LRS and vice versa.

15.4.2.3 Summary

In summary, we have investigated the bottom electrode dependent bipolar resistive switching characteristic of synthetic melanin. I–V characteristic of synthetic melanin clearly shows the hysteresis loop with non-ideal memristor properties. Among tested devices, Al/melanin/FTO memristive device shows better resistive switching properties than other devices. The nonvolatile study reveals that the Al/melanin/FTO memristive device possesses good endurance and retention properties. The conduction model fitting results suggested that the Al/melanin interface plays an important role in resistive switching. The Schottky emission is responsible for the conduction in the device during HRS, whereas Ohmic conduction is dominated during LRS.

15.5 Conclusion and future outlook

In the present chapter, we have overviewed the resistive switching effect, materials for resistive switching, and the importance of the biomaterials for resistive switching memory applications. In particular, we have focused on synthesis, characterization, and resistive switching characteristics of DNA and synthetic melanin-based biomaterials. The developed memory devices show typical bipolar resistive switching characteristics at low voltage, indicating the possibility of these devices for low power memory applications. The detailed analyses of the I–V characteristic of both devices are indicating the presence of non-ideal memristor or memristive properties. The endurance and retention tests suggested that the DNA and synthetic melanin can be used for the nonvolatile memory application. However, further research is needed for these materials to compete with CMOS standard materials. In addition to this, control over resistive switching and device reliability (temporal as well as spatial) is major challenges in front of the bio-memristive devices.

Acknowledgments

This work was supported by a National Research Foundation of Korea (NRF) grant, funded by the Korean government (Ministry of Science, ICT & Future Planning, No. 2016R1A3B1908249). Dr. Kiran D. Pawar acknowledges the DBT BUILDER-SUK programme sanctioned to Shivaji University, Kolhapur.

References

[1] J.S. Meena, S.M. Sze, U. Chand, T.Y. Tseng, Overview of emerging nonvolatile memory technologies, Nanoscale Research Letters 9 (1) (2014) 526.

[2] H. Akinaga, H. Shima, ReRAM technology: challenges and prospects, IEICE Electronics Express 9 (8) (2012) 795–807.

[3] D.B. Strukov, G.S. Snider, D.R. Stewart, R.S. Williams, The missing memristor found, Nature 453 (7191) (2008) 80.

[4] P. Rabbani, R. Dehghani, N. Shahpari, A multilevel memristor–CMOS memory cell as a ReRAM, Microelectronics Journal 46 (12) (2015) 1283–1290.

[5] S.S. More, P.A. Patil, K.D. Kadam, H.S. Patil, S.L. Patil, A.V. Pawar, S.S. Kanapally, D.V. Desai, S.M. Bodake, R.K. Kamat, S. Kim, Resistive switching and synaptic properties modifications in gallium-doped zinc oxide memristive devices, Results in Physics 12 (2019) 1946–1955.

[6] A.V. Pawar, S.S. Kanapally, K.D. Kadam, S.L. Patil, V.S. Dongle, S.A. Jadhav, S. Kim, T.D. Dongale, MemSens: a new detection method for heavy metals based on silver nanoparticle assisted memristive switching principle, Journal of Materials Science: Materials in Electronics 30 (12) (2019) 11383–11394.

[7] A. Sawa, Resistive switching in transition metal oxides, Materials Today 11 (6) (2008) 28–36.

[8] D.B. Strukov, H. Kohlstedt, Resistive switching phenomena in thin films: materials, devices, and applications, MRS Bulletin 37 (2) (2012) 108–114.

[9] D.S. Jeong, R. Thomas, R.S. Katiyar, J.F. Scott, H. Kohlstedt, A. Petraru, C.S. Hwang, Emerging memories: resistive switching mechanisms and current status, Reports on Progress in Physics 75 (7) (2012) 076502.

[10] T. Lee, Y. Chen, Organic resistive nonvolatile memory materials, MRS Bulletin 37 (2) (2012) 144–149.

[11] X. Zhu, J. Lee, W.D. Lu, Iodine vacancy redistribution in organic–inorganic halide perovskite films and resistive switching effects, Advanced Materials 29 (29) (2017) 1700527.

[12] C. Wang, P. Gu, B. Hu, Q. Zhang, Recent progress in organic resistance memory with small molecules and inorganic–organic hybrid polymers as active elements, Journal of Materials Chemistry C 3 (39) (2015) 10055–10065.

[13] H. Wang, F. Meng, B. Zhu, W.R. Leow, Y. Liu, X. Chen, Resistive switching memory devices based on proteins, Advanced Materials 27 (46) (2015) 7670–7676.

[14] T.D. Dongale, N.B. Mullani, V.B. Patil, R.S. Tikke, P.S. Pawar, S.V. Mohite, A.M. Teli, A.A. Bagade, K.K. Pawar, K.V. Khot, S.S. Shinde, Mimicking the biological synapse functions of analog memory, synaptic weights, and forgetting with ZnO-based memristive devices, Journal of Nanoscience and Nanotechnology 18 (11) (2018) 7758–7766.

[15] C. He, Z. Shi, L. Zhang, W. Yang, R. Yang, D. Shi, G. Zhang, Multilevel resistive switching in planar graphene/SiO_2 nanogap structures, ACS Nano 6 (5) (2012) 4214–4221.

[16] K.K. Pawar, D.V. Desai, S.M. Bodake, H.S. Patil, S.M. More, A.S. Nimbalkar, S.S. Mali, C.K. Hong, S. Kim, P.S. Patil, T.D. Dongale, Highly reliable multilevel resistive switching in a nanoparticulated In_2O_3 thin-film memristive device, Journal of Physics D, Applied Physics 52 (17) (2019) 175306.

[17] A.P. Rananavare, S.J. Kadam, S.V. Prabhu, S.S. Chavan, P.V. Anbhule, T.D. Dongale, Organic non-volatile memory device based on cellulose fibers, Materials Letters 232 (2018) 99–102.

[18] H.D. Lee, B. Magyari-Köpe, Y. Nishi, Model of metallic filament formation and rupture in NiO for unipolar switching, Physical Review B 81 (19) (2010) 193202.
[19] J. Park, S. Lee, K. Yong, Photo-stimulated resistive switching of ZnO nanorods, Nanotechnology 23 (38) (2012) 385707.
[20] T.S. Bhat, A.S. Kalekar, D.S. Dalavi, C.C. Revadekar, A.C. Khot, T.D. Dongale, P.S. Patil, Hydrothermal synthesis of nanoporous lead selenide thin films: photoelectrochemical and resistive switching memory applications, Journal of Materials Science: Materials in Electronics (2019) 1–10.
[21] D. Choi, C. Soo Kim, Coexistence of unipolar and bipolar resistive switching in Pt/NiO/Pt, Applied Physics Letters 104 (19) (2014) 193507.
[22] G.U. Kamble, N.P. Shetake, S.D. Yadav, A.M. Teli, D.S. Patil, S.A. Pawar, M.M. Karanjkar, P.S. Patil, J.C. Shin, M.K. Orlowski, R.K. Kamat, Coexistence of filamentary and homogeneous resistive switching with memristive and meminductive memory effects in Al/MnO_2/SS thin film metal–insulator–metal device, International Nano Letters 8 (4) (2018) 263–275.
[23] K.P. Biju, X. Liu, S. Kim, S. Jung, J. Park, H. Hwang, Coexistence of filamentary and homogeneous resistive switching in graded WOx thin films, Physica Status Solidi (RRL)—Rapid Research Letters 5 (3) (2011) 89–91.
[24] C. Wang, H. Wu, B. Gao, T. Zhang, Y. Yang, H. Qian, Conduction mechanisms, dynamics and stability in ReRAMs, Microelectronic Engineering 187 (2018) 121–133.
[25] Y. Yang, W. Lu, Nanoscale resistive switching devices: mechanisms and modeling, Nanoscale 5 (21) (2013) 10076–10092.
[26] L. Chua, Memristor – the missing circuit element, IEEE Transactions on Circuit Theory 18 (5) (1971) 507–519.
[27] L.O. Chua, S.M. Kang, Memristive devices and systems, Proceedings of the IEEE 64 (2) (1976) 209–223.
[28] J.C. Scott, L.D. Bozano, Nonvolatile memory elements based on organic materials, Advanced Materials 19 (11) (2007) 1452–1463.
[29] J. Choi, J.S. Han, K. Hong, S.Y. Kim, H.W. Jang, Organic–inorganic hybrid halide perovskites for memories, transistors, and artificial synapses, Advanced Materials 30 (42) (2018) 1704002.
[30] B. Hwang, J.S. Lee, Recent advances in memory devices with hybrid materials, Advanced Electronic Materials 5 (1) (2019) 1800519.
[31] M. Irimia-Vladu, E.D. Głowacki, G. Voss, S. Bauer, N.S. Sariciftci, Green and biodegradable electronics, Materials Today 15 (7–8) (2012) 340–346.
[32] Z. Lv, Y. Zhou, S.T. Han, V.A.L. Roy, From biomaterial-based data storage to bio-inspired artificial synapse, Materials Today 21 (5) (2018) 537–552.
[33] N. Raeis-Hosseini, J.S. Lee, Resistive switching memory using biomaterials, Journal of Electroceramics 39 (1–4) (2017) 223–238.
[34] S.T. Gurme, T.D. Dongale, S.N. Surwase, S.D. Kumbhar, G.M. More, V.L. Patil, P.S. Patil, R.K. Kamat, J.P. Jadhav, An organic bipolar resistive switching memory device based on natural melanin synthesized from aeromonas sp. SNS, Physica Status Solidi A 215 (24) (2018) 1800550.
[35] B. Sun, L. Wei, H. Li, X. Jia, J. Wu, P. Chen, The DNA strand assisted conductive filament mechanism for improved resistive switching memory, Journal of Materials Chemistry C 3 (46) (2015) 12149–12155.
[36] Y. Abbas, S.R. Dugasani, M.T. Raza, Y.R. Jeon, S.H. Park, C. Choi, The observation of resistive switching characteristics using transparent and biocompatible $Cu2^+$-doped salmon DNA composite thin film, Nanotechnology 30 (33) (2019) 335203.
[37] H.Y. Jeng, T.C. Yang, L. Yang, J.G. Grote, H.L. Chen, Y.C. Hung, Non-volatile resistive memory devices based on solution-processed natural DNA biomaterial, Organic Electronics 54 (2018) 216–221.
[38] S. Qin, R. Dong, X. Yan, Q. Du, A reproducible write–$(read)_n$–erase and multilevel biomemristor based on DNA molecule, Organic Electronics 22 (2015) 147–153.

[39] U. Celano, K. Nagashima, H. Koga, M. Nogi, F. Zhuge, G. Meng, Y. He, J. De Boeck, M. Jurczak, W. Vandervorst, T. Yanagida, All-nanocellulose nonvolatile resistive memory, NPG Asia Materials 8 (9) (2016) e310.

[40] Y.C. Chen, H.C. Yu, C.Y. Huang, W.L. Chung, S.L. Wu, Y.K. Su, Nonvolatile bio-memristor fabricated with egg albumen film, Scientific Reports 5 (2015) 10022.

[41] J.X. Zhu, W.L. Zhou, Z.Q. Wang, H.Y. Xu, Y. Lin, W.Z. Liu, J.G. Ma, Y.C. Liu, Flexible, transferable and conformal egg albumen based resistive switching memory devices, RSC Advances 7 (51) (2017) 32114–32119.

[42] Y.C. Chang, Y.H. Wang, Resistive switching behavior in gelatin thin films for nonvolatile memory application, ACS Applied Materials & Interfaces 6 (8) (2014) 5413–5421.

[43] Y.C. Chang, Y.H. Wang, Solution-processed Al-chelated gelatin for highly transparent nonvolatile memory applications, Applied Physics Letters 106 (12) (2015) 34_1.

[44] Y. Qi, B. Sun, G. Fu, T. Li, S. Zhu, L. Zheng, S. Mao, X. Kan, M. Lei, Y. Chen, A nonvolatile organic resistive switching memory based on lotus leaves, Chemical Physics 516 (2019) 168–174.

[45] Z.X. Lim, S. Sreenivasan, Y.H. Wong, F. Zhao, K.Y. Cheong, Filamentary conduction in aloe vera film for memory application, Procedia Engineering 184 (2017) 655–662.

[46] Z.X. Lim, S. Sreenivasan, Y.H. Wong, F. Zhao, K.Y. Cheong, Effects of electrode materials on charge conduction mechanisms of memory device based on natural Aloe vera, MRS Advances 1 (36) (2016) 2513–2518.

[47] Z.X. Lim, K.Y. Cheong, Effects of drying temperature and ethanol concentration on bipolar switching characteristics of natural Aloe vera-based memory devices, Physical Chemistry Chemical Physics 17 (40) (2015) 26833–26853.

[48] B. Sun, X. Zhang, G. Zhou, P. Li, Y. Zhang, H. Wang, Y. Xia, Y. Zhao, An organic nonvolatile resistive switching memory device fabricated with natural pectin from fruit peel, Organic Electronics 42 (2017) 181–186.

[49] J. Xu, X. Zhao, Z. Wang, H. Xu, J. Hu, J. Ma, Y. Liu, Biodegradable natural pectin-based flexible multilevel resistive switching memory for transient electronics, Small 15 (4) (2019) 1803970.

[50] M.K. Hota, M.K. Bera, B. Kundu, S.C. Kundu, C.K. Maiti, A natural silk fibroin protein-based transparent bio-memristor, Advanced Functional Materials 22 (21) (2012) 4493–4499.

[51] A. Moudgil, N. Kalyani, G. Sinsinbar, S. Das, P. Mishra, S-layer protein for resistive switching and flexible nonvolatile memory device, ACS Applied Materials & Interfaces 10 (5) (2018) 4866–4873.

[52] L. Wang, D. Wen, Nonvolatile bio-memristor based on silkworm hemolymph proteins, Scientific Reports 7 (1) (2017) 17418.

[53] H. Wang, F. Meng, Y. Cai, L. Zheng, Y. Li, Y. Liu, Y. Jiang, X. Wang, X. Chen, Sericin for resistance switching device with multilevel nonvolatile memory, Advanced Materials 25 (38) (2013) 5498–5503.

[54] N.R. Hosseini, J.S. Lee, Biocompatible and flexible chitosan-based resistive switching memory with magnesium electrodes, Advanced Functional Materials 25 (35) (2015) 5586–5592.

[55] N. Raeis Hosseini, J.S. Lee, Resistive switching memory based on bioinspired natural solid polymer electrolytes, ACS Nano 9 (1) (2014) 419–426.

[56] N. Raeis-Hosseini, J.S. Lee, Controlling the resistive switching behavior in starch-based flexible biomemristors, ACS Applied Materials & Interfaces 8 (11) (2016) 7326–7332.

[57] H. Baek, C. Lee, K.I. Lim, J. Cho, Resistive switching memory properties of layer-by-layer assembled enzyme multilayers, Nanotechnology 23 (15) (2012) 155604.

[58] Y. Ko, Y. Kim, H. Baek, J. Cho, Electrically bistable properties of layer-by-layer assembled multilayers based on protein nanoparticles, ACS Nano 5 (12) (2011) 9918–9926.

[59] F. Meng, L. Jiang, K. Zheng, C.F. Goh, S. Lim, H.H. Hng, J. Ma, F. Boey, X. Chen, Protein-based memristive nanodevices, Small 7 (21) (2011) 3016–3020.

[60] Y. Chen, G. Liu, C. Wang, W. Zhang, R.W. Li, L. Wang, Polymer memristor for information storage and neuromorphic applications, Materials Horizons 1 (5) (2014) 489–506.

[61] Y. Yang, J. Ouyang, L. Ma, R.H. Tseng, C.W. Chu, Electrical switching and bistability in organic/polymeric thin films and memory devices, Advanced Functional Materials 16 (8) (2006) 1001–1014.
[62] Y. Wang, X. Yan, R. Dong, Organic memristive devices based on silver nanoparticles and DNA, Organic Electronics 15 (12) (2014) 3476–3481.
[63] M.H. Haque, M.N. Islam, F. Islam, V. Gopalan, N.T. Nguyen, A.K. Lam, M.J. Shiddiky, Electrochemical detection of FAM134B mutations in oesophageal cancer based on DNA-gold affinity interactions, Electroanalysis 29 (5) (2017) 1359–1367.
[64] M.M. Rehman, G.U. Siddiqui, Y.H. Doh, K.H. Choi, Highly flexible and electroforming free resistive switching behavior of tungsten disulfide flakes fabricated through advanced printing technology, Semiconductor Science and Technology 32 (9) (2017) 095001.
[65] H. Wu, X. Li, F. Huang, A. Chen, Z. Yu, H. Qian, Stable self-compliance resistive switching in $AlO_\delta/Ta_2O_{5-x}/TaO_y$ triple layer devices, Nanotechnology 26 (3) (2014) 035203.
[66] S. Patil, M. Chougale, T. Rane, S. Khot, A. Patil, O. Bagal, S. Jadhav, A. Sheikh, S. Kim, T. Dongale, Solution-processable ZnO thin film memristive device for resistive random access memory application, Electronics 7 (12) (2018) 445.
[67] V.S. Dongle, A.A. Dongare, N.B. Mullani, P.S. Pawar, P.B. Patil, J. Heo, T.J. Park, T.D. Dongale, Development of self-rectifying ZnO thin film resistive switching memory device using successive ionic layer adsorption and reaction method, Journal of Materials Science: Materials in Electronics 29 (21) (2018) 18733–18741.
[68] L. Chua, Five non-volatile memristor enigmas solved, Applied Physics A 124 (8) (2018) 563.
[69] N. Du, Y. Shuai, W. Luo, C. Mayr, R. Schüffny, O.G. Schmidt, H. Schmidt, Practical guide for validated memristance measurements, Review of Scientific Instruments 84 (2) (2013) 023903.
[70] M. d'Ischia, K. Wakamatsu, A. Napolitano, S. Briganti, J.C. Garcia-Borron, D. Kovacs, P. Meredith, A. Pezzella, M. Picardo, T. Sarna, J.D. Simon, Melanins and melanogenesis: methods, standards, protocols, Pigment Cell & Melanoma Research 26 (5) (2013) 616–633.
[71] E.S. Bronze-Uhle, J.V. Paulin, M. Piacenti-Silva, C. Battocchio, M.L.M. Rocco, C.F.D.O. Graeff, Melanin synthesis under oxygen pressure, Polymer International 65 (11) (2016) 1339–1346.
[72] V.I.T.O. Capozzi, G. Perna, P. Carmone, A. Gallone, M. Lastella, E. Mezzenga, G. Quartucci, M. Ambrico, V. Augelli, P.F. Biagi, T. Ligonzo, Optical and photoelectronic properties of melanin, Thin Solid Films 511 (2006) 362–366.
[73] M. Ambrico, A. Cardone, T. Ligonzo, V. Augelli, P.F. Ambrico, S. Cicco, G.M. Farinola, M. Filannino, G. Perna, V. Capozzi, Hysteresis-type current–voltage characteristics in Au/eumelanin/ITO/glass structure: towards melanin based memory devices, Organic Electronics 11 (11) (2010) 1809–1814.
[74] M. Ambrico, P.F. Ambrico, A. Cardone, T. Ligonzo, S.R. Cicco, R.D. Mundo, V. Augelli, G.M. Farinola, Melanin layer on silicon: an attractive structure for a possible exploitation in bio-polymer based metal–insulator–silicon devices, Advanced Materials 23 (29) (2011) 3332–3336.
[75] D.J. Kim, K.Y. Ju, J.K. Lee, The synthetic melanin nanoparticles having an excellent binding capacity of heavy metal ions, Bulletin of the Korean Chemical Society 33 (11) (2012) 3788–3792.
[76] J.P. Bothma, J. de Boor, U. Divakar, P.E. Schwenn, P. Meredith, Device-quality electrically conducting melanin thin films, Advanced Materials 20 (18) (2008) 3539–3542.
[77] N.M. Raman, S. Ramasamy, Genetic validation and spectroscopic detailing of DHN-melanin extracted from an environmental fungus, Biochemistry and Biophysics Reports 12 (2017) 98–107.
[78] Y.S. Chen, H.Y. Lee, P.S. Chen, T.Y. Wu, C.C. Wang, P.J. Tzeng, F. Chen, M.J. Tsai, C. Lien, An ultrathin forming-free HfOx resistance memory with excellent electrical performance, IEEE Electron Device Letters 31 (12) (2010) 1473–1475.
[79] A.C. Khot, N.D. Desai, K.V. Khot, M.M. Salunkhe, M.A. Chougule, T.M. Bhave, R.K. Kamat, K.P. Musselman, T.D. Dongale, Bipolar resistive switching and memristive properties of

hydrothermally synthesized TiO_2 nanorod array: effect of growth temperature, Materials & Design 151 (2018) 37–47.
[80] M.M. Jastrzebska, H. Isotalo, J. Paloheimo, H. Stubb, Electrical conductivity of synthetic DOPA-melanin polymer for different hydration states and temperatures, Journal of Biomaterials Science, Polymer Edition 7 (7) (1996) 577–586.
[81] X.B. Yin, Z.H. Tan, X. Guo, The role of Schottky barrier in the resistive switching of $SrTiO_3$: direct experimental evidence, Physical Chemistry Chemical Physics 17 (1) (2015) 134–137.

Chapter 16

Nonvolatile memristive logic: a road to in-memory computing

Ling Yang, Yi Li, Long Cheng, and Xiangshui Miao
Wuhan National Laboratory for Optoelectronics, School of Optical and Electronic Information, Huazhong University of Science and Technology, Wuhan, China

16.1 Introduction

For the past 10 years, starting with the conceptual connection between the predicted missing memristor and TiO_X-based resistive switching devices [1], memristors and memristive devices have been widely investigated. This interest has led to a new wave of research on the known resistive random access memories based on valence change and electrometallization mechanisms [2,3], and the research of alternative in-memory computing (IMC) paradigms for the "More than Moore" era [4,5].

In traditional von Neumann computing architecture, the performance gap between the physically separated computing engine and memory is becoming greater. The increasingly severe data transfer bottleneck, not the processor speed, limits system performance, especially for data-intensive workloads, such as machine learning, data mining, and gene sequencing on massive datasets. Additionally, the physical movement of the data consumes more energy and time than the actual processor computation itself.

IMC is the concept of *in situ* executing computing in the physical location where data are stored. Instead of "near-memory computing" in which more memory is simply moved closer to the logic elements, true IMC can integrate the memory and computation at the device and circuit level, reducing the time and power consumed by constantly moving data back and forth across the system [6,7].

One such an IMC approach (based on the memristor) is the brain-inspired neuromorphic computing [8]. Firstly, analogue memristors in a crossbar configuration can naturally execute vector-matrix multiplication, the fundamental computation in an artificial neural network, based on Ohm's law and Kirchhoff's law in a highly parallel manner with $O(1)$ computation complexity. This means that a hardware neural network accelerator can be implemented with a high degree of energy efficiency. Secondly, studies have demonstrated spike-timing dependent plasticity in analogue memristors, which provide a compact and effi-

Mem-elements for Neuromorphic Circuits with Artificial Intelligence Applications
https://doi.org/10.1016/B978-0-12-821184-7.00025-6

cient way to build spiking neural networks. However, to achieve a performance improvement in a memristor-based neuromorphic system, the requirement for controllable analogue behaviors, including linear and symmetric conductance tuning, high conductance precision (>5–6 bits), large dynamic range, and small variation, still poses real challenges in the development of memristive materials and devices that can enable this approach.

Another subset of IMC is the digital logic-in-memory approach, which executes Boolean logic using nonvolatile memories, such as RRAM, PCRAM, MRAM, and flash memory. Compared with neuromorphic computing, which can tolerate device variation, randomness, and errors to a certain extent, the logic-in-memory approach is based on deterministic computing, which compares favorably with non-deterministic methods (such as stochastic, probability approaches) when maintaining constant precision and reliability.

In this chapter, we do not discuss memristive neuromorphic computing, which has been studied extensively; the reader is referred to several comprehensive reviews [9–11]. We also note another excellent reference review, which examines both analogue and digital IMC schemes and offers a critical viewpoint by Ielmini et al. [12]. Here we focus on the illustration of the principles and implementations of memristive Boolean logic and their potential applications. Whenever possible, recent work on memristive logic will be presented.

16.2 Memristive logic gates in crossbar array

Generally, memristors that are used to build logic gates are bipolar resistive switches (BRS) which have two stable resistance states corresponding to logic one and logic zero, respectively, as shown in Fig. 16.1 (a). The switching between the high resistance state (HRS) and the low resistance state (LRS) is triggered by the external operating voltages, termed the SET and RESET voltages, of opposite polarities [13], as shown in Fig. 16.1 (b). As a two-terminal device, the memristor is easily integrated into a crossbar array, which is able to achieve the highest integration $4F^2$ in two-dimensional circuits, where F (feature size) refers to the half-pitch of the bitline (BL) at a certain technology node [14]. If the memristors are integrated into an n-layer-stacked 3D array, the integration density that can be achieved is $4\ F^2/n$ [15]. The memristive logic gates are usually executed based on a row of devices connected to a wordline (WL) in parallel, whereas the other terminal is connected to a BL with other devices in the same column, as shown in Fig. 16.1 (c).

Presently, the various reported memristive logic methods can be classified into four categories in terms of the input and output variables and physical operations: R-R logic (sometimes referred to as stateful logic) whose input and output are both represented as a resistance, V-R logic (sometimes referred to as sequential logic), all or part of whose inputs are represented as a voltage whereas the output is represented as a resistance, V-V logic whose input and output are both represented as a voltage [12], and R-V logic whose input and output are represented by a resistance and a voltage, respectively.

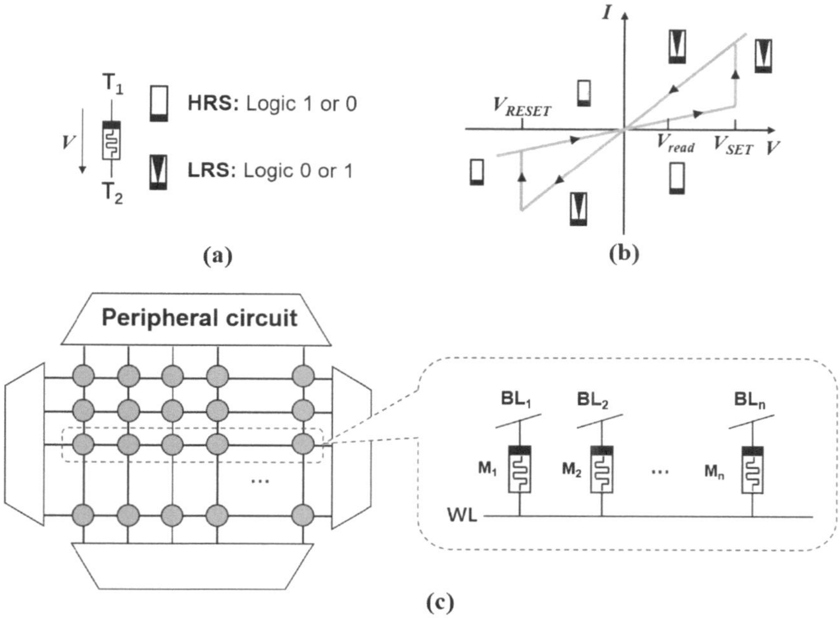

FIGURE 16.1 (a) A two-terminal bipolar binary memristive device. (b) I–V characteristic. (c) Integration structure.

Although there are many reported memristive logic methods, in this chapter we only focus on R-R logic and V-R logic, which demonstrate IMC properties represented by an output resistance, and introduce various typical logic gates of each type, since they have similar mechanisms. Fig. 16.2 shows the state machine of the logic operation. In the device initialization step, it is common to write an HRS via the RESET operation and write an LRS via the SET operation. The readout is executed by applying a low voltage pulse to the cell one wants to read to ensure that the state of the cell is not destroyed by the V_{read}.

16.3 R-R logic gate

16.3.1 Material implication logic (IMPLY or IMP)

IMP, which is one of the 16 two-input Boolean logic relationships (Fig. 16.3 (a)), was the first logic gate proposed in a memristive crossbar array [16]. It was a landmark work that first experimentally demonstrated memristors that could perform Boolean logic operations. IMP and FALSE operations were proposed that could be combined as building blocks to construct functionally complete logic gates and complex computation functions.

Technically, IMP is performed on a circuit containing two memristors (M_1 and M_2) and a load resistor (R_G) in a row. Next, V_{DD} and V_{COND} are applied to

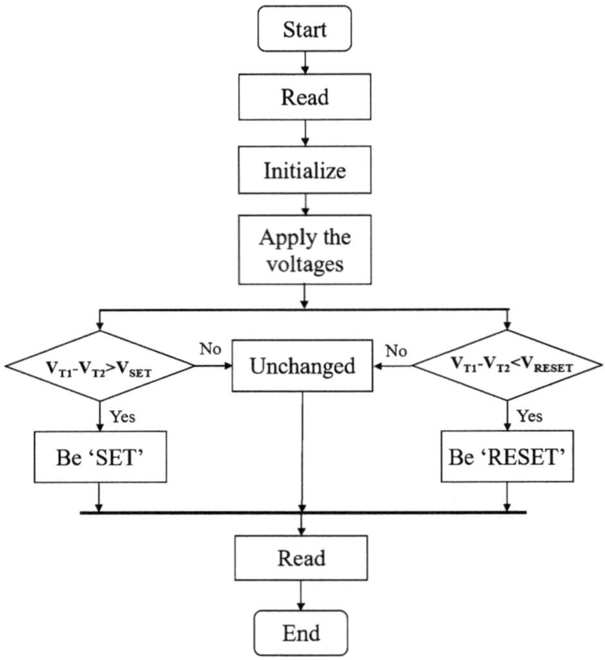

FIGURE 16.2 State machines of R-R logic gate and V-R logic gate.

M_2 and M_1 simultaneously with R_G grounded, as shown in Fig. 16.3 (b). Note that the amplitude of V_{DD} is higher than that of V_{SET}, while V_{COND} is lower than V_{SET}. The value of R_G is generally chosen as $\sqrt{R_H \cdot R_L}$, guaranteeing that R_G is lower than R_H (the resistance of the HRS) and higher enough than R_L (resistance of LRS). In this case, M_2 will be programmed into the HRS or the LRS depending on the combination of the initial state of the two devices. The logic output q' is stored in M_2 as its final resistance state. At the same time, the state of M_1 will not alter during the logic computation process throughout.

Specifically, if M_1 is in the HRS ($p = 0$), V_{COND} has little influence on the V_{WL}, and thus the voltage crossing M_2 is sufficient to set it into the LRS ($q' \leftarrow 1$); if M_1 is in the LRS ($p = 1$), the electrical potential of the WL will be pulled up by V_{COND}, leaving insufficient voltage drop on M_2 to set it. Hence, M_2 remains in its initial state. These behaviors of this circuit correspond exactly with the truth table of the material implication logic gate.

Based on IMP and FALSE (reset operation), we can realize other Boolean logic gates. Here, we introduce the operations of the universal NAND gate. The relationship between the NAND gate and the IMP gate is shown in Eq. (16.1). Fig. 16.4 shows the circuit and operating steps. According to Eq. (16.1), three steps are required to construct a NAND gate using an IMP gate. During the first cycle, M_3 is initialized to the HRS (0) via a reset operation. During the second

Inputs		Output
p (M_1)	q (M_2)	q'(M_2)
0	0	1
0	1	1
1	0	0
1	1	1

(a)

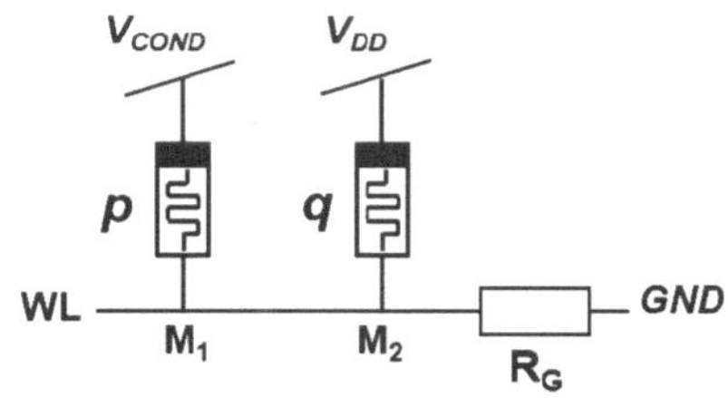

(b)

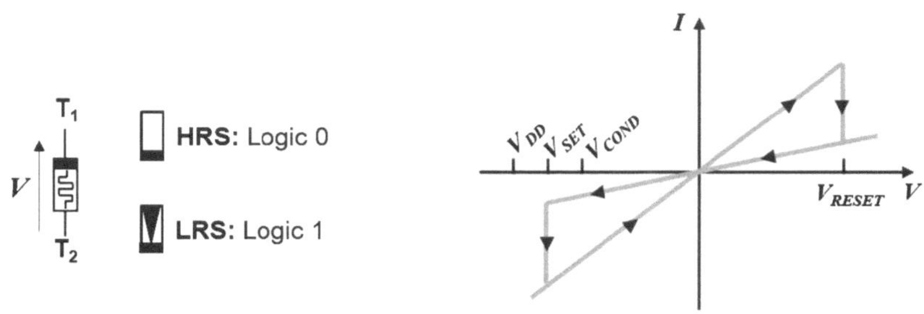

(c) (d)

FIGURE 16.3 Memristive IMP logic gate. (a) The truth table for $q' \leftarrow p$ IMP q logic operation. (b) Schematic: circuit structure and operating voltages applied to the corresponding terminals. (c) Logic value and corresponding resistant state. The HRS is defined as logic 0 and the LRS is defined as logic 1. The positive direction of the voltage potential: from T_2 to T_1. (d) I–V characteristics and values of operating voltage.

cycle, the first IMP operation is carried out with p and s and the result is s'. Lastly, the second IMP operation is performed on q and s', the result s'' is the final output of pNANDq and is stored in M_3 in the form of a resistance.

$$
\begin{aligned}
s' &= pIMPs = \overline{p\bar{s}} \\
s'' &= qIMPs' = \overline{q\overline{s'}} = \overline{qp\bar{s}} \\
s &= 0 \Rightarrow s'' = \overline{pq}
\end{aligned}
\tag{16.1}
$$

16.3.2 Variants of IMP — NOR logic gate

Several similar memristive logic gates have been reported since the demonstration of IMP logic. Here, we introduce an efficient NOR gate proposed by Lu et al. [17]. Fig. 16.5 shows the schematic of this method. The operating steps are executed as follows:

(1) Initialize the output cell M_3 into the HRS; Write A and B into the input cells M_1 and M_2, respectively.

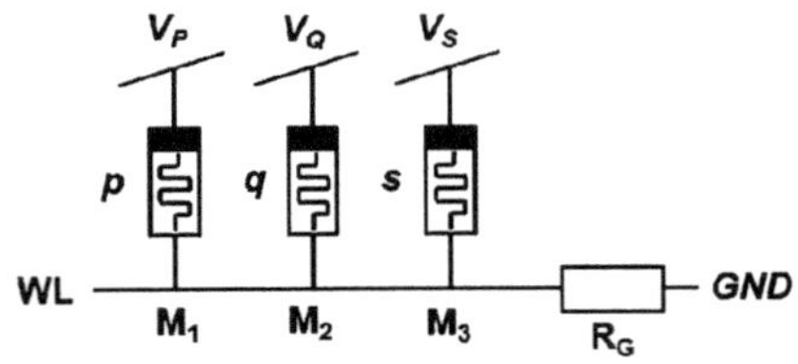

(a)

Cycle	V_P	V_Q	V_S
1			$-V_{DD}$
2	V_{COND}		V_{DD}
3		V_{COND}	V_{DD}

(b)

Cycle 1	Cycle 2			Cycle 3			Cycle 1,2, 3		
s=0	*s'←pIMPs*			*s"←qIMPs'*			*s"←pNANDq*		
s	*p*	*s*	*s'*	*q*	*s'*	*s"*	*p*	*q*	*s"*
0	0	0	1	0	1	1	0	0	1
0	0	0	1	1	1	1	0	1	1
0	1	0	0	0	0	1	1	0	1
0	1	0	0	1	0	0	1	1	0

(c)

FIGURE 16.4 Memristive NAND gate built by an IMP gate. (a) Schematic of the NAND gate. (b) The sequence of voltage operations. (c) Logic operation in each cycle.

(2) Apply $V_{DD}/2$ and V_{DD} simultaneously to the input cells and the output cell, respectively.

(3) Apply a read voltage V_{read} to M_3 to obtain the resistance state as the output of the NOR gate.

The state of M_3 will be programmed into either the HRS or the LRS depending on the combination of the two inputs. If any one of the input cells is in the LRS (logic 1), the electrical potential of the WL (V_{WL}) will be pulled up and close to $V_{DD}/2$. Hence, the voltage drop on M_3 is insufficient to trigger the SET operation and the outputs of these cases are 0. However, if both the input cells are in the HRS (logic 0), V_{WL} is approximately 0 and the voltage drop on M_3 is $\sim V_{DD}$. It is higher than V_{SET} and sufficient to set the device into the LRS. Fig. 16.6 shows an abstract distribution of the voltage drop in the two cases mentioned above (represented by '10' and '00', respectively). Consequently, the output of this combination is logic 1. The behaviors of this circuit conform to the truth table of the NOR logic gate as shown in Fig. 16.5 (a).

It is worth noting that, ignoring the wire resistance while adding more (finite) input cells, the same operating method is also available in this k-input NOR logic circuit, as shown in Fig. 16.7 [18]. However, the computing error rate may increase with more input cells in the gate. Device variations (cycle-to-cycle and device-to-device) inevitably lead to unpredicted voltage division. We will discuss these non-ideal effects in detail in Section 16.5.

Inputs		Output
A (M_1)	B (M_2)	C (M_3)
0	0	1
0	1	0
1	0	0
1	1	0

(a)

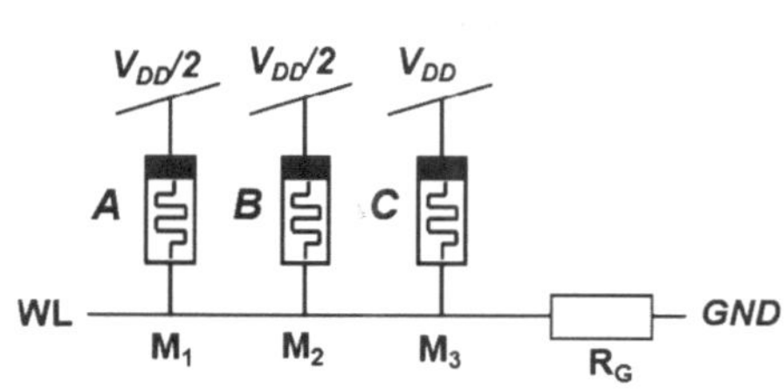

(b)

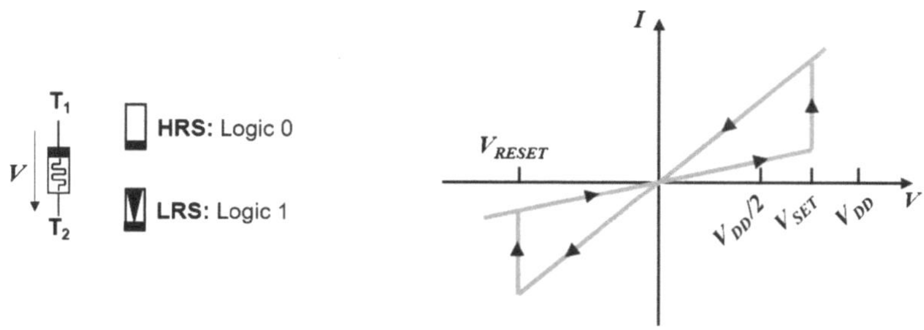

(c)

(d)

FIGURE 16.5 Memristive NOR gate. (a) The truth table of the NOR logic gate ($C = \overline{A + B}$). (b) Schematic: circuit structure and the voltages applied to the corresponding terminals. (c) Logic value and corresponding resistant state. The HRS is defined as logic 0 and the LRS is defined as logic 1. The positive direction of the potential: from T_1 to T_2. (d) I–V characteristics and values of operating voltages. Here, V_{COND} is chosen to be $V_{DD}/2$.

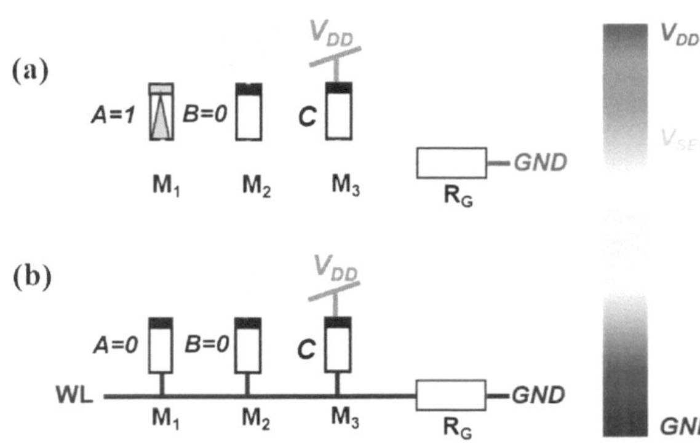

FIGURE 16.6 Distribution of the voltage drop in the case of '10' (a) and '00' (b). (For interpretation of the colors in the figure, the reader is referred to the web version of this chapter.)

16.3.3 Variants of IMP – NAND and AND logic gates

Besides the IMC properties, the memristive logic gate has also been praised as a "software-defines-hardware" reconfigurable methodology, which means that diverse logic functions can be realized in the same physical structure by merely applying varied operating signals. For example, in the NOR scheme described

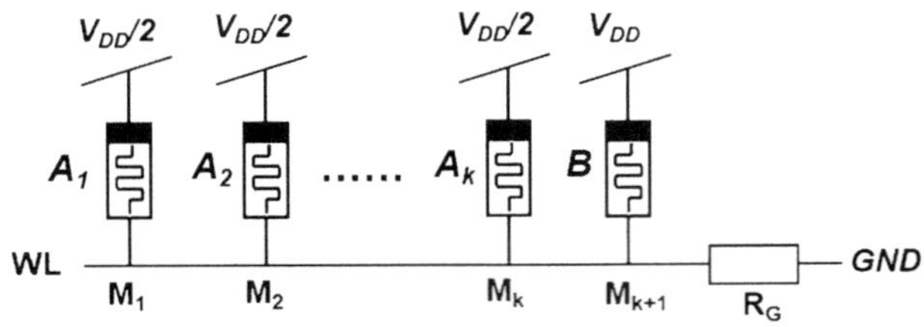

FIGURE 16.7 Schematic of a k-input NOR operation.

Inputs		Output
A (M_1)	B (M_2)	C (M_3)
0	0	1
0	1	1
1	0	1
1	1	0

(a)

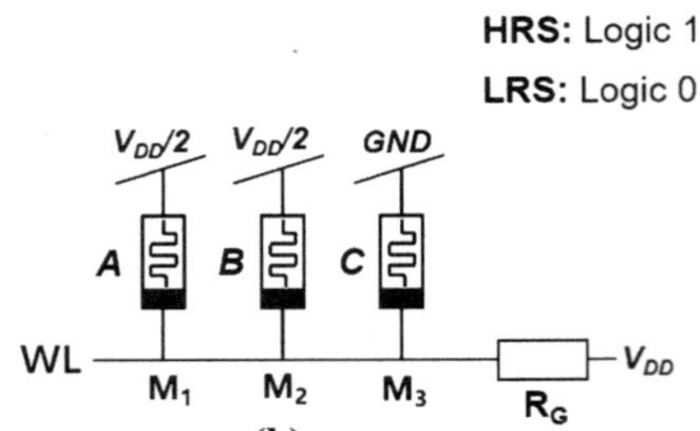

(b)

Inputs		Output
A (M_1)	B (M_2)	C (M_3)
0	0	0
0	1	0
1	0	0
1	1	1

(c)

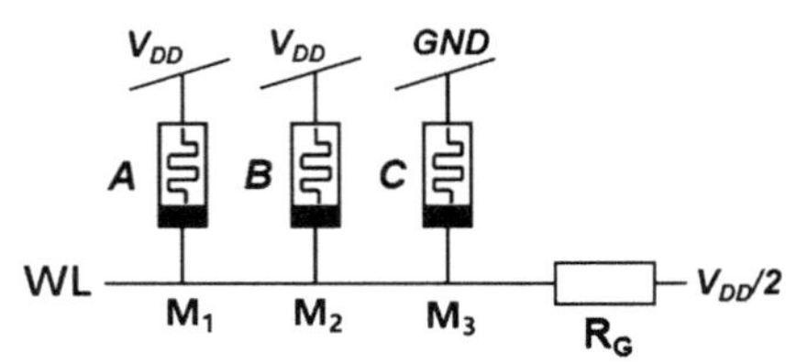

(d)

FIGURE 16.8 Memristive NAND and AND gate. (a) The truth table of the NAND gate. (b) Schematic for the memristive NAND gate. (c) The truth table of the AND gate. (d) Schematic for the memristive AND gate.

in Section 16.3.2, if the HRS represents logic 1 and the LRS represents 0, the logic function of the identical circuit will correspond to a NAND gate, instead of a NOR gate [19].

Distinct from the methods mentioned above, this method defines the HRS and the LRS as 1 and 0, respectively. As shown in Fig. 16.8 (b) and (d), the positive polarities of the three devices share the same WL. The operating steps are similar to the other methods:

(1) Initialize the output cell M_3 into the HRS; A and B are written to the input cells M_1 and M_2, respectively.

(2) $V_{DD}/2$ and V_{DD} are applied to the input cells and R_G, respectively, with the output cell grounded.

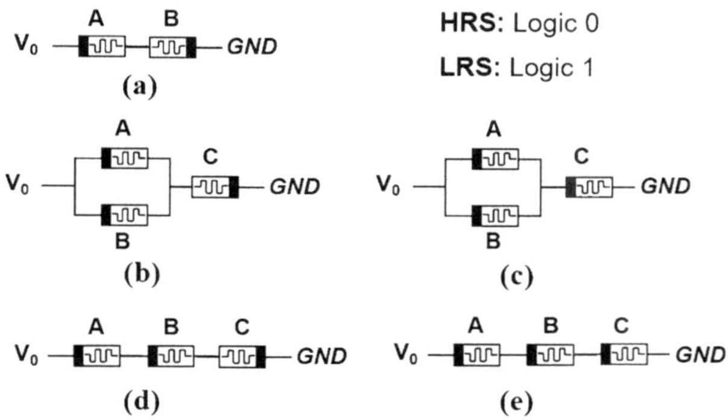

FIGURE 16.9 MAGIC logic family. (a) NOT gate. (b) NOR gate. (c) OR gate. (d) NAND gate. (e) AND gate.

(3) A read voltage is applied to M3 to obtain the resistance state as the output of the NAND gate.

If there is an input cell in the LRS (0), the voltage potential of the WL will be pulled down to about $V_{DD}/2$, leaving an insufficient voltage drop on M3 to perform the SET operation. Hence, M3 is unchanged, with an output of logic 1. If both the input cells are in the HRS, the paths from $V_{DD}/2$ to the WL are closed. Hence, V_{WL} is pulled up to approximately V_{DD}, which leaves a sufficient voltage drop to programme M3 into the LRS (0).

In the AND gate, V_{DD} and $V_{DD}/2$ are exchanged. Since the operation and the mechanism are similar to the NAND gate, we only give the schematic diagram shown in Fig. 16.8 (c) and (d).

16.3.4 Memristor-aided logic (MAGIC)

All of the methods mentioned above require at least two different voltage drivers to trigger the logic gate. To simplify the circuit structure, a memristor-aided logic (MAGIC) method was proposed [20]. The MAGIC logic circuit is constructed using only memristors and is driven by a single operating voltage. Fig. 16.9 shows five representative logic gates of the MAGIC family, where a logic family is that variety of operations that can be executed by using the same computing element flexibly [21].

Note that the MAGIC NOR and NOT gate could be mapped in a crossbar array, as shown in Fig. 16.10, whereas other MAGIC gates cannot be realized in a crossbar due to their special topological connections. Here, we introduce the operating principle of the MAGIC NOR gate as an example.

Compared to IMP and its variants, the load resistor R_G is not required in the MAGIC logic family. However, the MAGIC NOR gate is also executed through voltage division. The operating steps are:

Inputs		Output
A (M_1)	B (M_2)	C (M_3)
0	0	1
0	1	0
1	0	0
1	1	0

(a)

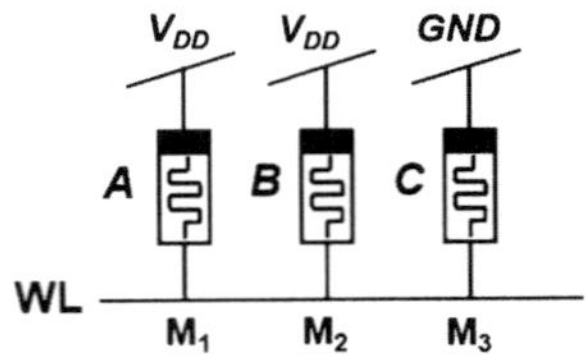

(b)

FIGURE 16.10 MAGIC NOR gate. (a) Truth table. (b) Schematic of the gate circuit.

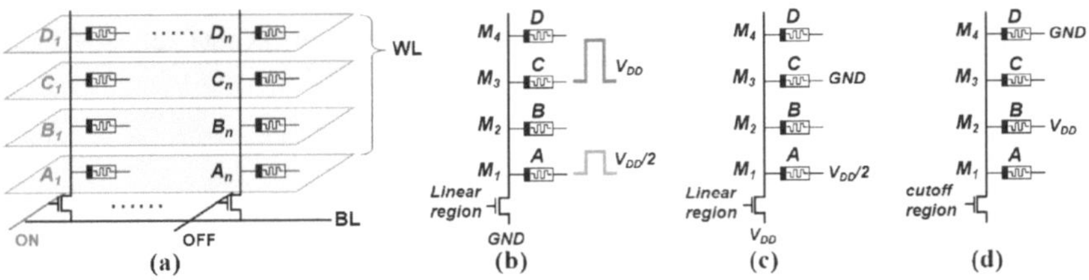

FIGURE 16.11 Schematic of the logic operations in a 3D vertical RRAM. (a) 3D-LUT scheme. (b) Half-V_{DD} scheme. (c) A variant of the half-V_{DD} scheme. (d) V_{DD}/GND scheme.

(1) Initialize the output cell M_3 into the LRS; A and B are written to the input cells M_1 and M_2, respectively.
(2) Apply V_{DD} to M_1 and M_2 simultaneously with M_3 grounded.
(3) Read the state of M_3.

If anyone of the input devices is in the LRS (logic 1), V_{WL} is pulled up to approximately V_{DD}, leaving sufficient voltage drop on M_3 to perform a RESET. Hence, the output is logic 0. By contrast, if both the inputs are logic 0, the voltage drop across M_3 is low and cannot RESET the device into the HRS. Hence, it remains in the LRS (logic 1).

The NOT gate obeys the same operating principles when applying the voltages to only M_1 and M_3.

16.3.5 Hyperdimensional computing in 3D RRAM

When R_G is replaced by a FinFET device, the logic circuit structure can be extended from 2D to 3D [22]. Fig. 16.11 shows how the logic gate is implemented in a vertical pillar structure array.

The 3D-LUT (lookup table) scheme shows the method of reading data stored in vertical pillars: by turning on the FET to select the pillar in which the data are stored, while V_{read} is applied to the correct WL. Fig. 16.11 (b) shows a schematic of the half-V_{DD} scheme. In the half-V_{DD} scheme, the circuit contains a transistor in the linear region and three memristors (including an output cell M_3 and two input cells (M_1, M_2)). V_{DD} and $V_{DD}/2$ are applied to the input cell

	Cycle1	Cycle2
C	V_{DD}	V_{DD}
B		$V_{DD}/2$
A	$V_{DD}/2$	

(a)

	Cycle1	Cycle2
C	GND	GND
B		V_{DD}
A	V_{DD}	

(b)

FIGURE 16.12 Operations for AND and OR. (a) Execute AND with the half-V_{DD} scheme in two cycles ($C = A$ AND B). (b) Execute OR with V_{DD}/GND scheme in two cycles ($C = A$ OR B).

and output cells, respectively. However, $V_{DD}/2$ is applied to M_1 and M_2 in two different cycles, instead of the same cycle. In this case, the transistor serves as a resistor to construct a voltage diving structure. Fig. 16.11 (c) shows a variant of the half-V_{DD} scheme, a pair of V_{DD} and $V_{DD}/2$ pulses are applied to the source of the FET and the input cell. Distinct from the half-V_{DD} scheme, the transistor is turned off in the same circuit. V_{DD} is applied to the input cell with the output cell grounded.

This work designs consecutive operating procedures to realize several logic functions such as AND and OR. Fig. 16.12 shows the operations used to execute the AND and OR functions with the half-V_{DD} scheme and V_{DD}/GND, respectively. In the AND scheme, the output cell is initialized into the LRS (logic 1). If any one of the input cells is logic 0 (HRS) in the corresponding cycle, the potential of the pillar electrode is pulled down, leaving sufficient voltage drop across M_3, programming M_3 into the HRS. However, if both the input cells are in the LRS, the potential of the pillar electrode is influenced by $V_{DD}/2$ and is pulled up. As a result, the voltage drop across M_3 is insufficient to trigger the state switching, and M_3 remains in the LRS (logic 1).

As for the OR scheme, V_{DD} is applied to the WL of the input cell and the output cell is initialized into the HRS. If any one of the input cells is in the LRS, the path from V_{DD} to the pillar electrode is opened and the voltage potential of the pillar electrode is pulled up to approximately V_{DD}, leaving sufficient voltage drop across M_3 to perform the SET operation. However, if both the inputs are in the HRS, V_{DD} has a slight influence on the pillar electrode. Hence, the output cell remains in the initial state HRS.

16.3.6 R-R logic based on neural networks

Sun et al. formulated a logic computing method based on the concept of neural networks [23]. This method is a universal framework to perform complete Boolean logic functions in the same circuit structure. In particular, linearly separable functions could be realized in one step, such as AND and NAND, whereas linearly nonseparable functions were unable to be realized in one step: for instance, the XOR function requires two steps. Fig. 16.13 and Fig. 16.14 show the operations of NAND and XOR with neural networks, respectively.

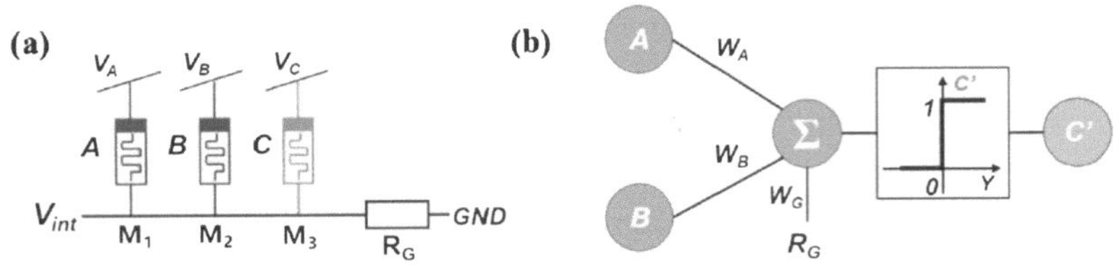

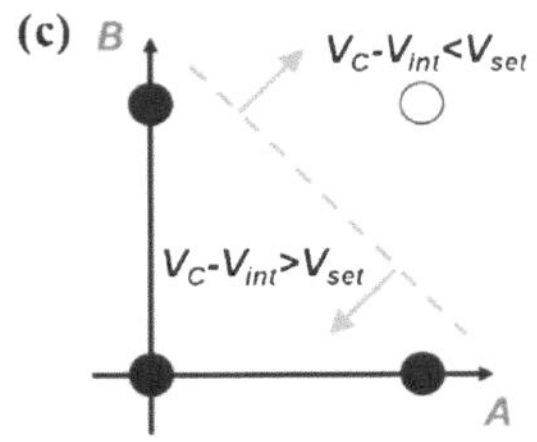

(d)

	A	B	C	C'
Logic value	0	0	0	1
	0	1	0	1
	1	0	0	1
	1	1	0	0
Weight	V_A=0.7Vset	V_B=0.7Vset	V_C=1.35V_{set}	

FIGURE 16.13 Mechanism of neural network logic. (a) Memristive circuit. (b) Neural network model. (c) Input and output characteristics of NAND. (d) The truth table of NAND and the value of weight voltages.

Fig. 16.13 (a) illustrates the circuit structure of neural network logic, consisting of three memristive devices (A, B, and C), and a grounded resistor, where A and B are the input cells, and C is the output device. Voltages (V_A, V_B, and V_C) applied to the corresponding BL represent the weights W_A, W_B, and W_C, respectively, as shown in Fig. 16.13 (b).

Fig. 16.13 (c) illustrates the input–output characteristics of a NAND function that is a linearly separable function which can be realized within a single-step operation. During the operation, the values of the voltages correspond closely to the values of the weights in the neural networks as shown in Fig. 16.13 (d).

By contrast, XOR is a linearly nonseparable function that requires at least two steps. Fig. 16.14 (a) shows the method to build the XOR function by cascading a NIMP function and a C-INMP function.

Performing Boolean logic using neural networks in a memristive array is a creative idea. However, this scheme requires that the voltage values correspond closely to the values of the weights, which is problematic as the device resistance variation easily influences the computing result and reduces robustness.

16.3.7 Conclusion of memristive R-R logic

In this section, several typical R-R logic methods have been introduced. Memristive R-R logic is a truly IMC method, where all the elements participating in the computing functions are represented as resistance. Computing is entirely implemented within the memory array as virtually no data moves in and out of memory during the computing process [13]. However, to date, only a few of the 16 two-input Boolean logic relationships can be realized in a single step under a standard operating voltage mode. Other logic gates can be constructed

(a)

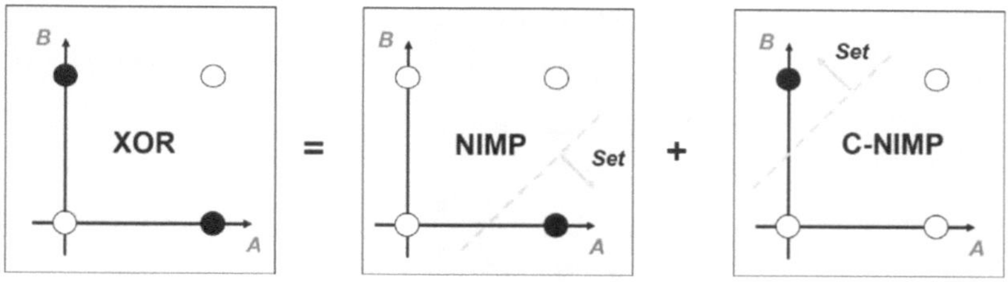

(b)

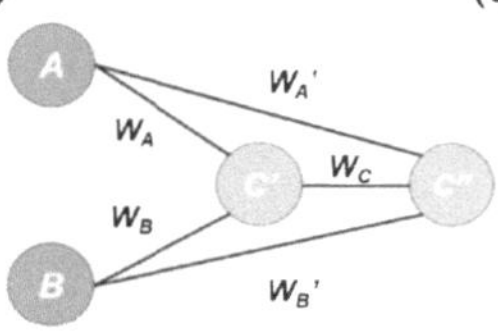

(c)

	A	B	C	C'	C''
Logic value	0	0	0	0	0
	0	1	0	0	1
	1	0	0	1	1
	1	1	0	0	0
Weight(W)	V_A=-0.5Vset	V_B=0.5Vset	$V_C=0.75V_{set}$		
W'	V_A=0.5Vset	V_A=-0.5Vset		$V_C=0.75V_{set}$	

FIGURE 16.14 Schematic of neural-network-based XOR. (a) Input/output characteristics of XOR. (b) Neural network model of XOR. (c) The truth table of XOR and the value of weight voltages.

by the cascading of these single-step logic gates, in particular XOR and XNOR, which are complex in either CMOS logic or memristive logic implementations. Consequently, R-R logic could be used for NAND/AND/NOR/OR intensive computations to decrease the delays and power consumption caused by the effective movement of the data.

16.4 Memristive V-R logic

16.4.1 Memristive sequential logic based on a single device

As shown in Fig. 16.1 (b), switching between the HRS and the LRS of a bipolar resistive switch (BRS) is dependent on the voltage polarity. In this chapter, if it is not pointed out, the positive direction is implicit for the polarity of the SET voltage and the HRS is implicit for logic 0. Furthermore, analogous to the CMOS logic circuit, the high voltage pulse is implicit for logic 1, whereas the low voltage pulse or zero potential corresponds to logic 0. Fig. 16.15 shows the logic operations of a BRS cell.

Similar operations can be conducted in a complementary resistive switch (CRS) device based on two anti-serially connected bipolar devices. Fig. 16.16 shows the logic operations of a CRS cell [24].

As for the CRS, the HRS/LRS state (B/A) is defined as logic 1 and the LRS/HRS state is defined as logic 0. There is an intermediate state LRS/LRS defined as 'ON'. Distinct from the BRS, the read operation is a destructive readout method since the resistances of '0' and '1' are the same. Hence, a spike-read scheme is used to read out the state of the CRS, as shown in Fig. 16.16 (b). A read voltage pulse, the amplitude of which is higher than V_{th2} is applied to the device. In this case, if the device is in the LRS/HRS state, then switching will

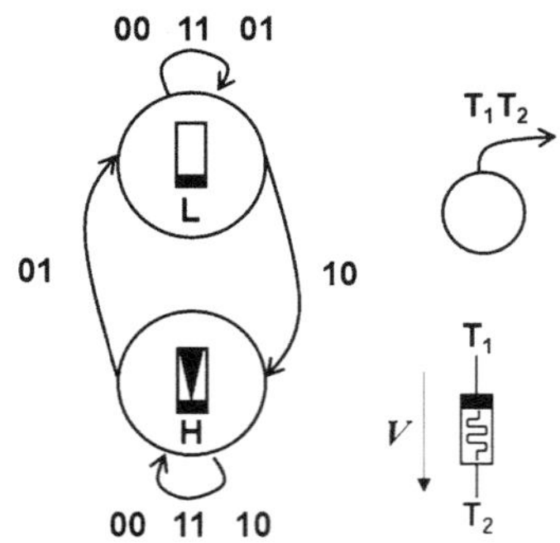

FIGURE 16.15 Operations of a BRS cell.

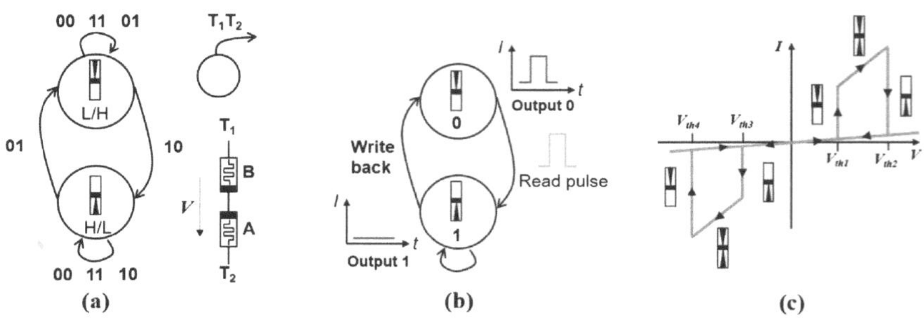

FIGURE 16.16 Operations of a CRS cell. (a) State machine. (b) Read operation. (c) I–V characteristics of the CRS.

be triggered, leaving a current spike that corresponds to logic 0 output. After the readout, the state must be written back into its initial state due to the destructive readout model.

In this section, the two inputs are represented as voltage pulses applied to the two terminals, respectively. The final result is related to the initial state Z'. The relationship between the final state Z and initial state Z' is defined by the equation

$$Z = (T_1 RIMP T_2) \cdot Z' + (T_1 RIMP T_2) \cdot \left(notZ'\right) \tag{16.2}$$

Fig. 16.17 shows the four cases of Eq. (16.2). Regardless of the initial state Z', the finial state Z is 0 if T_1 is 0 and T_2 is 1. By contrast, if T_1 is 1 and T_2 is 0, the final state Z will be 1, since the two cases are a negative pulse and a positive pulse, respectively.

Based on the above equation, 14 Boolean logic functions can be realized within three steps, except the XOR and XNOR logic functions. Fig. 16.18 illustrates the two logic functions in detail.

Later, You et al. showed that all 16 Boolean logic functions could be realized in three logic cycles with a single symmetric bipolar resistive switching BFTO/BFO bilayer structure cell [25]. With this method, the reading process

T_1	T_2	Initial state	Final state BRS(CRS)	Readout
0	0	Z'	Z=Z'	Z
0	1	Z'	H(L/H)	0
1	0	Z'	L(H/L)	1
1	1	Z'	Z=Z'	Z

FIGURE 16.17 Calculation of the state variable Z.

	Cycle 1	Cycle 2
T_1	0	q
T_2	1	p

(a)

p	q	RS BRS (CRS)		Readout
		Cycle 1	Cycle 2	
0	0	H (L/H)	H (L/H)	0
0	1	H (L/H)	H (L/H)	0
1	0	H (L/H)	L (H/L)	1
1	1	H (L/H)	H (L/H)	0

(b)

	Cycle 1	Cycle 2	Cycle 3
T_1	1	p	q
T_2	0	1	1

(c)

p	q	RS BRS (CRS)			Readout
		Cycle 1	Cycle 2	Cycle 3	
0	0	L (H/L)	H (L/H)	H (L/H)	0
0	1	L (H/L)	H (L/H)	H (L/H)	0
1	0	L (H/L)	L (H/L)	H (L/H)	0
1	1	L (H/L)	L (H/L)	L (H/L)	1

(d)

FIGURE 16.18 Operation of (a) (b) RNIMP and (c) (d) AND.

can be used as an additional logic cycle since the device resistance depends on the polarity of the read voltage. In other words, the resistance state is inverted by reversing the polarity of the reading bias. Based on this novel device and methodology, the final output is defined by the four input variables together, including T_1, T_2, the initial state S' and the read voltage (r). The relationship is summarized by the following equation:

$$S=\left(T_1+\overline{T_2}\right)\cdot S'\cdot r+\left(\overline{T_1}\cdot T_2\right)\cdot S'\cdot\overline{r}+\left(T_1\cdot\overline{T_2}\right)\cdot\overline{S'}\cdot r+\left(\overline{T_1}+T_2\right)\cdot\overline{S'}\cdot\overline{r} \quad (16.3)$$

According to Eq. (16.3), S' can be programmed into a certain state 1 or 0 via the application of a writing pulse ($T_1=1$, $T_2=0$) or ($T_1=0$, $T_2=1$), respectively. Consequently, Eq. (16.3) can be simplified to

$$S=\left(T_1+\overline{T_2}\right)\cdot r+\left(\overline{T_1}\cdot T_2\right)\cdot\overline{r} \quad (16.4)$$

or

$$S=\left(T_1\cdot\overline{T_2}\right)\cdot r+\left(\overline{T_1}+T_2\right)\cdot\overline{r} \quad (16.5)$$

Based on this method, XOR and XNOR can be realized using equations:

$$S=(1\cdot\overline{p})\cdot q+(0+p)\cdot\overline{q}=\overline{p}q+p\overline{q} \quad (16.6)$$

$$S=(p\cdot 1)\cdot q+(\overline{p}+0)\cdot\overline{q}=pq+\overline{pq} \quad (16.7)$$

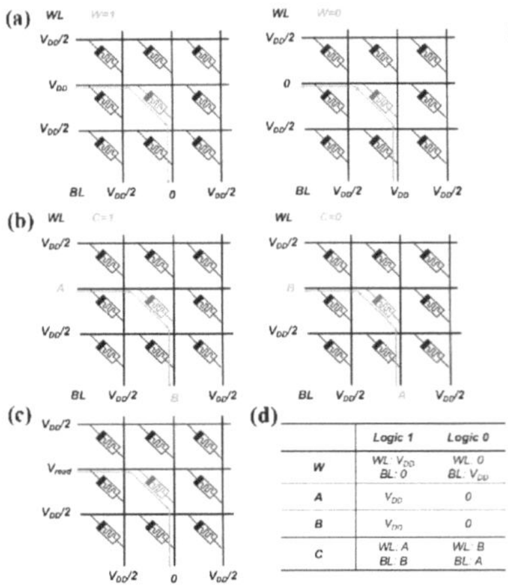

(c)

Function	name	W(R)	A(V)	B(V)	C(state)
0	null	0	p	p	q
1	identity	1	p	p	q
p	transfer	q	1	0	p
q	transfer	p	1	0	q
$\bar{p}$	complement	q	0	1	p
q	complement	p	0	1	q
pq	AND	0	p	0	q
$\overline{pq}$	NAND	1	0	p	q
$p+q$	OR	q	0	p	0
$\overline{p+q}$	NOR	0	p	1	q
$p+q$	implication (IMP)	1	p	q	0
$p+\bar{q}$	implication (IMP)	1	p	q	1
$\bar{p}q$	inhibition	0	p	q	0
pq	inhibition	0	p	q	1
$pq+pq$	XOR	0	p	q	p
$pq+\bar{p}\bar{q}$	XNOR	1	p	q	q

FIGURE 16.19 Schematic of the four-variable logic method in a crossbar array. (a) Step 1: Initialization. (b) Step 2: Writing. (c) Step 3: Readout. (d) Four-input variables and the corresponding physical state. (e) Logic operations of 16 Boolean logic functions.

As shown in Eq. (16.6), XOR is executed by (S' $=0$, $T_1=1$, $T_2=p$) with $r=q$ based on Eq. (16.5). XNOR is executed by (S' $=0$, $T_1=p$, $T_2=0$) with $r=q$ based on Eq. (16.5) too.

16.4.2 Four-variables methods

16.4.2.1 A four-variable method using a single BRS in a crossbar array

In Section 16.4.1, the operations based on a single symmetric BRS is one of the four-variable methods. In this section, we introduce several additional four-variable methods using other physical media.

Fig. 16.19 shows a method for realizing the 16 Boolean logic functions within three logic steps using a bipolar device integrated into a crossbar array [23]. In the first step, the device is programmed into a certain state corresponding to W via applying V_{DD} on the WL (logic 1) or the BL (logic 0). In the second step, voltages corresponding to A and B are applied to certain terminals defined by C, as shown in Fig. 16.19 (d). If C is 1, A will be applied to the WL, and B to the BL, whereas, if C is 0, A will be applied to the BL, and B to the WL. Fig. 16.19 (c) shows the method of reading a cell in a crossbar array.

The final state of the cell is defined by W, A, B, and C together. The relationship between the output and the four input variables is summarized as

$$L=\left(A+\overline{B}\right)\cdot C\cdot W+\left(\overline{A}+B\right)\cdot\overline{C}\cdot W+A\cdot\overline{B}\cdot C\cdot\overline{W}+\overline{A}\cdot B\cdot\overline{C}\cdot\overline{W} \quad (16.8)$$

All of the 16 Boolean logic functions are contained in Eq. (16.8), according to the variable assignment in Fig. 16.19 (e).

16.4.2.2 A four-variable method using a one-transistor–one-resistor (1T1R) cell

As shown in Fig. 16.14, the CRS can mitigate the impact of the leakage current in a sneak path. However, the destructive readout model results in an additional writing-back cycle which leads to extra delay and power consumption. To further address this limitation, studies have found that in a single 1T1R cell, all of the 16 Boolean logic functions could be realized within two logic steps, including the initialization and application of the operating voltages [26].

In the scheme, the four variables include the voltages applied to the three terminals (G, T_1, and T_2) of the cell and the initial state (I) of the memristor, as shown in Fig. 16.20. The relationship between the inputs and output can be defined by the equation

$$R = I\overline{G} + IGT_1 + IG\overline{T_1}\,\overline{T_2} + \overline{I}GT_1\overline{T_2} \tag{16.9}$$

Here, we illustrate the principle of the method with an example of the NAND gate. To realize the NAND gate, there are two steps required:

(1) Initialize the memristor into the LRS (logic 1) via the applications of a writing pulse (G=1, $T_1 = 1$, $T_2 = 0$).
(2) Application of the voltage pulses corresponding to p and q on the gate of the transistor (G) and T_2, respectively, with T_1 grounded (logic 0).

In the case of ($p = 1$, $q = 1$), the transistor is opened, and a sufficient negative voltage is applied to the memristor to trigger a RESET operation, leaving the memristor programmed into the HRS (logic 0). If any one of p and q are 0, the switching is triggered and the memristor remains in the initial state of logic 1.

16.4.3 Other memristive V-R logic methods

Siemon et al. conceived of methods for executing all 16 Boolean logic functions using the threshold switching (TS) device and the CRS or complementary switching (CS) device [27]. Zhou et al. demonstrated experimentally a reconfigurable Boolean logic method via a flexible configuration of the BRS and the CRS structures in an array [28]. Gao et al. presented a solution to realize complete Boolean logic functions based on a single unsymmetrical CRS [29]. These studies enriched the family of V-R logic methods, expanding their range of application.

16.4.4 Conclusion of the memristive V-R logic

In this section, we have introduced some typical V-R logic (sequential logic) methods. In V-R logic, all or part of the inputs are represented as a voltage state, and the final result is stored in the output cell as a resistance state. All of the methods mentioned in this section are implemented in a cell containing one

(a) (b)

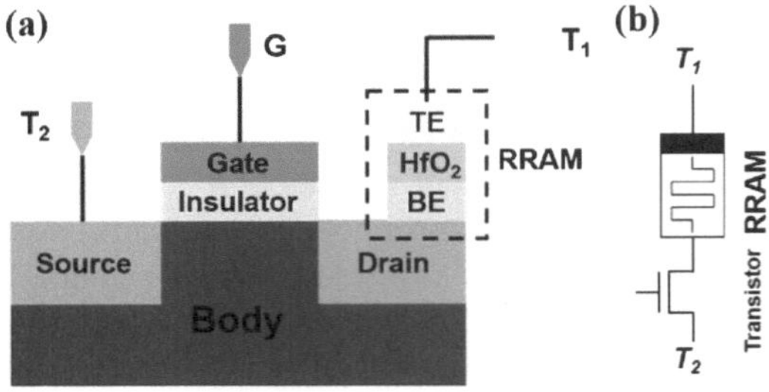

(c)

Function	name	$I(R)$	$G(V)$	$T_1(V)$	$T_2(V)$
0	null	0	0	p	q
1	identity	1	0	p	q
p	transfer	p	1	0	q
q	transfer	q	1	0	p
p	complement	p	0	0	q
q	complement	q	0	p	0
pq	AND	0	p	q	0
$\overline{pq}$	NAND	1	p	0	q
$p+q$	OR	p	1	q	0
$\overline{p+q}$	NOR	q	p	q	1
$p+q$	implication	1	p	q	1
$p+\bar{q}$	implication	1	1	p	q
$\bar{p}q$	inhibition	q	1	0	p
$p\bar{q}$	inhibition	p	1	0	q
$\bar{p}q+p\bar{q}$	XOR	p	q	$\bar{p}$	p
$pq+\bar{p}\bar{q}$	XNOR	q	p	q	$\bar{q}$

FIGURE 16.20 Schematic of the four-variable method based on a 1T1R cell. (a) Integration structure of the device. (b) Circuit schematic and terminals. (c) Operations of 16 Boolean logic functions.

or two devices. Compared with R-R logic executing the same logic function, V-R logic requires fewer memristive cells and fewer steps, which reduces the device area and the power consumption of the logic circuit. However, since the inputs and output are physically inhomogeneous [12], it is difficult to realize a cascade of V-R logic gates. Moreover, there is data movement between the memory array and peripheral circuit in the computing process. Consequently, we think that V-R logic is not wholly IMC. It could, however, be used in special scenarios to increase computing speed.

16.5 Challenges and outlooks

Memristive based IMC has opened new opportunities for "beyond CMOS" computing architecture for data-intensive computation applications. In this chapter, we have introduced the basic operational principles of memristive logic computing and summarized the recent advances in this area, with a particular focus on two methodologies, namely R-R and V-R logic. Finally, we discuss some key challenges and opportunities in this fast-growing field.

If we want to develop a Memristive Processing Unit (MPU) for true IMC, we believe there should be long-term and short-term goals to consider. The long-term goal is to build a memristive universal computing system for various computational tasks. Several pieces of literature have reported progress in the architectural design and hardware implementation of varied types of MPU [19]. Cheng et al. demonstrated a fully functional memristive arithmetic logic unit (MemALU) design based on four reconfigurable stateful logic functions in a crossbar array: IMP, destructive OR, NOR, and non-destructive OR. Moreover, efficient arithmetic computation, increment, decrement, and left/right shift functions have also been designed and realized in a MemALU [30]. However, the further improvement of MPUs calls for more systematic studies and the overall optimization of devices, circuits, algorithms, and architectures. By contrast, the short-term goal is to construct a mixed architecture by integrating a specific MPU within a mature transistor-based processing unit to address specific logic-intensive edge computing or embedded computing. For instance, the intensive hamming distance calculation in gene sequencing could be implemented based on memristive computation in a crossbar array with high parallelism and low energy cost [31]. The MPU would be dedicated to executing simple but repetitive computations with low precision, whereas the CMOS processing unit would execute the complicated and precise computations. By taking advantage of the two means, the computing capability would be increased. This has been explored in a number of studies [32].

Although this IMC technology has exhibited appealing properties, it is still quite far from commercialization. In its current phase, the proposed logic methods mostly rely on the assumption that all device operating conditions are ideal. However, practically, the non-ideal characteristics of the devices and circuits need to be considered to ensure high computing accuracy.

Firstly, the memristor performance determines the computing performance. For a memristive logic system, such as R-R logic realized by voltage dividing, a memristive device with a high on/off ratio (R_H/R_L) is recommended for correct logic operations. Moreover, the computation accuracy is restricted by the device parameter variation, in terms of resistance state, switching threshold voltage, and switching time, from cycle-to-cycle (C2C) and device-to-device (D2D). Fig. 16.21 shows the general device parameter requirements for error-free and energy-efficient logic computation.

Secondly, for the logic circuit, the non-ideal effects, such as wire resistance, parasitic capacitance, and sneak path, seriously degrade the logic implementation. The degradation is more evident when considering high-speed logic operations in a large-scale crossbar array. The computation errors caused by these non-ideal effects may be accumulated in each operating cycle, and finally lead to an incorrect result in the sequential operation. To avoid those obstacles,

Indicators	Relevant performances	Requirements of Memristive logic
On-off ratio	Accuracy	High
Variation of switching threshold voltages		Small
Variations of the resistance (C2C and D2D)		Small
Switch speed	Speed	High
Resistance value	Power consumption	High
Threshold voltage value		Compatible with the work voltage of CMOS
Ability to suppress the leakage current	Accuracy Power consumption	Required in large scale crossbar
Endurance	Reliability	High

FIGURE 16.21 Device-level requirements for memristive logic.

circuit optimization and fault tolerance design are of fundamental importance [33].

Lastly, most research has been focused on the materials, devices, and logic algorithms. However, there have been fewer studies focused on the intermediate part between the hardware implementation and the software instruction design. To build a new memristive computer distinct from classical von Neumann architecture, it is important to develop a corresponding new instruction set and compiler.

In summary, in this chapter we have tried to cover some of the important progress made in non-volatile, reconfigurable, and efficient memristive logic methods that have significant prospects of being used in "beyond CMOS" IMC systems. The challenges and outlooks have been discussed at the device, circuit, and system levels. We believe that this chapter will stimulate more research efforts in this exciting field. We believe that significant changes unexpectedly will come about following the exploration of brand-new routes.

Acknowledgments

This work was supported by the National Key Research and Development Plan of MOST of China (Grant No. 2019YFB2205100), by the National Natural Science Foundation of China (Grant No. 61841404, 61874164, 51732003), by Hubei Key Laboratory of Advanced Memories, and by Hubei Engineering Research Center on Microelectronics.

References

[1] D.B. Strukov, G.S. Snider, D.R. Stewart, R.S. Williams, The missing memristor found, Nature 453 (7191) (2008) 80–83.

[2] R. Waser, R. Dittmann, G. Staikov, K. Szot, Redox-based resistive switching memories – nanoionic mechanisms, prospects, and challenges, Advanced Materials 21 (25–26) (2009) 2632–2663.
[3] J.J. Yang, D.B. Strukov, D.R. Stewart, Memristive devices for computing, Nature Nanotechnology 8 (1) (2013) 13.
[4] Z. Wang, H. Wu, G.W. Burr, C.S. Hwang, K.L. Wang, Q. Xia, J.J. Yang, Resistive switching materials for information processing, Nature Reviews Materials (2020) 1–23.
[5] Q. Xia, J.J. Yang, Memristive crossbar arrays for brain-inspired computing, Nature Materials 18 (4) (2019) 309–323.
[6] N. Talati, A.H. Ali, R.B. Hur, N. Wald, R. Ronen, P.-E. Gaillardon, S. Kvatinsky, Practical challenges in delivering the promises of real processing-in-memory machines, in: 2018 Design, Automation & Test in Europe Conference & Exhibition (DATE), IEEE, 2018, pp. 1628–1633.
[7] A. Haj-Ali, R. Ben-Hur, N. Wald, R. Ronen, S. Kvatinsky, Not in name alone: a memristive memory processing unit for real in-memory processing, IEEE MICRO 38 (5) (2018) 13–21.
[8] J. Tang, F. Yuan, X. Shen, Z. Wang, M. Rao, Y. He, Y. Sun, X. Li, W. Zhang, Y. Li, Bridging biological and artificial neural networks with emerging neuromorphic devices: fundamentals, progress, and challenges, Advanced Materials 31 (49) (2019) 1902761.
[9] H. Jeong, L. Shi, Memristor devices for neural networks, Journal of Physics D: Applied Physics 52 (2) (2018) 023003.
[10] D. Ielmini, S. Ambrogio, Emerging neuromorphic devices, Nanotechnology 31 (9) (2019) 092001.
[11] Y. Zhang, Z. Wang, J. Zhu, Y. Yang, M. Rao, W. Song, Y. Zhuo, X. Zhang, M. Cui, L. Shen, Brain-inspired computing with memristors: challenges in devices, circuits, and systems, Applied Physics Reviews 7 (1) (2020) 011308.
[12] D. Ielmini, H.-S.P. Wong, In-memory computing with resistive switching devices, Nature Electronics 1 (6) (2018) 333–343.
[13] P.-E. Gaillardon, L. Amarú, A. Siemon, E. Linn, R. Waser, A. Chattopadhyay, G. De Micheli, The programmable logic-in-memory (PLiM) computer, in: 2016 Design, Automation & Test in Europe Conference & Exhibition (DATE), IEEE, 2016, pp. 427–432.
[14] L. Hai, Y. Chen, An overview of non-volatile memory technology and the implication for tools and architectures, in: Design, Automation and Test in Europe, DATE 2009, Nice, France, April 20–24, 2009, 2009.
[15] Y.-C. Chen, H. Li, W. Zhang, R.E. Pino, The 3-D stacking bipolar RRAM for high density, IEEE Transactions on Nanotechnology 11 (5) (2012) 948–956.
[16] J. Borghetti, G.S. Snider, P.J. Kuekes, J.J. Yang, D.R. Stewart, R.S. Williams, 'Memristive' switches enable 'stateful' logic operations via material implication, Nature 464 (7290) (2010) 873–876.
[17] B. Chen, F. Cai, J. Zhou, W. Ma, P. Sheridan, W.D. Lu, Efficient in-memory computing architecture based on crossbar arrays, 2015, pp. 17.5.1–17.5.4.
[18] S. Shin, K. Kim, S.-M. Kang, Reconfigurable stateful nor gate for large-scale logic-array integrations, IEEE Transactions on Circuits and Systems II, Express Briefs 58 (7) (2011) 442–446.
[19] P. Huang, J. Kang, Y. Zhao, S. Chen, R. Han, Z. Zhou, Z. Chen, W. Ma, M. Li, L. Liu, X. Liu, Reconfigurable nonvolatile logic operations in resistance switching crossbar array for large-scale circuits, Advanced Materials 28 (44) (2016) 9758–9764.
[20] S. Kvatinsky, D. Belousov, S. Liman, G. Satat, N. Wald, E.G. Friedman, A. Kolodny, U.C. Weiser, MAGIC—memristor-aided logic, IEEE Transactions on Circuits and Systems II, Express Briefs 61 (11) (2014) 895–899.
[21] J. Reuben, R. Benhur, N. Wald, N. Talati, A.H. Ali, P. Gaillardon, S. Kvatinsky, Memristive logic: a framework for evaluation and comparison, in: Power and Timing Modeling Optimization and Simulation, 2017, pp. 1–8.
[22] H. Li, T.F. Wu, S. Mitra, H.-S.P. Wong, Resistive RAM-centric computing: design and modeling methodology, IEEE Transactions on Circuits and Systems I, Regular Papers 64 (9) (2017) 2263–2273.

[23] Z. Sun, E. Ambrosi, A. Bricalli, D. Ielmini, Logic computing with stateful neural networks of resistive switches, Advanced Materials 30 (38) (2018) e1802554.

[24] E. Linn, R. Rosezin, S. Tappertzhofen, U. Böttger, R. Waser, Beyond von Neumann—logic operations in passive crossbar arrays alongside memory operations, Nanotechnology 23 (30) (2012) 305205.

[25] T. You, Y. Shuai, W. Luo, N. Du, D. Bürger, I. Skorupa, R. Hübner, S. Henker, C. Mayr, R. Schüffny, T. Mikolajick, O.G. Schmidt, H. Schmidt, Exploiting memristive $BiFeO_3$ Bilayer structures for compact sequential logics, Advanced Functional Materials 24 (22) (2014) 3357–3365.

[26] Z.-R. Wang, Y.-T. Su, Y. Li, Y.-X. Zhou, T.-J. Chu, K.-C. Chang, T.-C. Chang, T.-M. Tsai, S.M. Sze, X.-S. Miao, Functionally complete Boolean logic in 1T1R resistive random access memory, IEEE Electron Device Letters 38 (2) (2017) 179–182.

[27] A. Siemon, T. Breuer, N. Aslam, S. Ferch, W. Kim, J. van den Hurk, V. Rana, S. Hoffmann-Eifert, R. Waser, S. Menzel, E. Linn, Realization of Boolean logic functionality using redox-based memristive devices, Advanced Functional Materials 25 (40) (2015) 6414–6423.

[28] Y.-X. Zhou, Y. Li, Y.-T. Su, Z.-R. Wang, L.-Y. Shih, T.-C. Chang, K.-C. Chang, S.-B. Long, S.M. Sze, X.-S. Miao, Nonvolatile reconfigurable sequential logic in a HfO_2 resistive random access memory array, Nanoscale 9 (20) (2017) 6649–6657.

[29] S. Gao, F. Zeng, M. Wang, G. Wang, C. Song, F. Pan, Implementation of complete Boolean logic functions in single complementary resistive switch, Scientific Reports 5 (2015) 15467.

[30] L. Cheng, Y. Li, K.S. Yin, S.Y. Hu, Y.T. Su, M.M. Jin, Z.R. Wang, T.C. Chang, X.S. Miao, Functional demonstration of a memristive arithmetic logic unit (MemALU) for in-memory computing, Advanced Functional Materials 29 (49) (2019) 1905660.

[31] N. Ge, J.H. Yoon, M. Hu, E.J. Merced-Grafals, N. Davila, J.P. Strachan, Z. Li, H. Holder, Q. Xia, R.S. Williams, X. Zhou, J.J. Yang, An efficient analog Hamming distance comparator realized with a unipolar memristor array: a showcase of physical computing, Scientific Reports 7 (1) (2017).

[32] M. Le Gallo, A. Sebastian, R. Mathis, M. Manica, H. Giefers, T. Tuma, C. Bekas, A. Curioni, E. Eleftheriou, Mixed-precision in-memory computing, Nature Electronics 1 (4) (2018) 246–253.

[33] Y. Li, Y. Zhou, Z. Wang, X. Miao, Memcomputing: fusion of memory and computing, Science China (Information Sciences) 61 (6) (2018) 159–161.

Chapter 17

Implementation of organic RRAM with ink-jet printer: from design to using in RFID-based application

Toan Dao Thanh[a], Viet-Thanh Pham[b,c], and Christos Volos[d]

[a]*Faculty of Electrical-Electronic Engineering, University of Transport and Communications, Hanoi, Vietnam,* [b]*Faculty of Electrical and Electronic Engineering, Phenikaa Institute for Advanced Study (PIAS), Phenikaa University, Hanoi, Vietnam,* [c]*Phenikaa Research and Technology Institute (PRATI), A&A Green Phoenix Group, Hanoi, Vietnam,* [d]*Laboratory of Nonlinear Systems, Circuits & Complexity (LaNSCom), Department of Physics, Aristotle University of Thessaloniki, Thessaloniki, Greece*

17.1 Introduction

It is now well established from a variety of studies that memories is generally classified into two fundamental categorizes: volatile memory and nonvolatile memory [2,3,17]. In general, volatile memory needs power to maintain the stored information. In contrast, the stored information is retained in nonvolatile memory when the power supply is turned off. Fig. 17.1 shows a taxonomy of memories. Volatile memory can be categorized into static random-access memory (SRAM) and dynamic random-access memory (DRAM) while nonvolatile memory may be divided into several groups (see Fig. 17.1). In numerous electronic devices (portable devices, sensors in wireless sensor nodes, embedded systems etc.), a matter of considerable concern is battery life [16]. A great deal of previous research into memory has focused on how to enlarge the operational life time of such electronic devices [20,38,42].

In recent years, there has been an increasing amount of literature on memristor, a potential candidate for emerging memory technologies [9,10,17,21,31]. Chua suggested that resistance switching memories were memristors [5]. Compared to the existing memory devices, the power consumption of a memristor was smaller [9,10]. Taherinejad et al. explored the possible storage of multi-bit data in a single memristor [30]. A memristor-based memory cell had less noise margins and stored non-binary data [39]. Discovering applications of the

Mem-elements for Neuromorphic Circuits with Artificial Intelligence Applications
https://doi.org/10.1016/B978-0-12-821184-7.00026-8

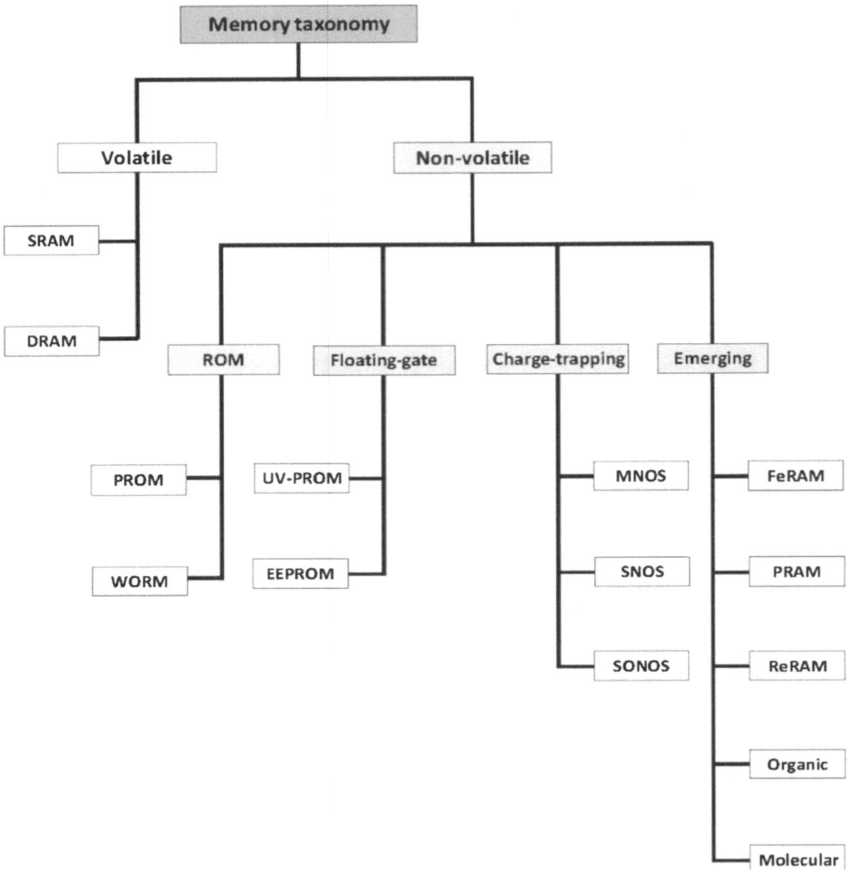

FIGURE 17.1 Taxonomy of memories including two main categorizes: volatile memory and non-volatile memory.

memristor for developing emerging memories is still an attractive research trend [7,18,25,34,35].

The development of the 4.0 era has received considerable critical attention. In recent years, there has been an increasing interest in portable devices, smart phones, smart homes, and smart cities. As a result, the demand for memories has increased significantly [6,11,26,29]. Embedded memory, emerging memory technologies as well as in-memory computing have received considerable attention recently. Resistive RAM (RRAM, or ReRAM) is a potential candidate for emerging nonvolatile memory technologies [1,14,24,37,40]. Compared with current RAM or read-only memory (ROM), resistive RAM is suitable for faster computing power and higher intensity [4,13,33,36,41].

From the view point of manufacturing, it is simple to see that printed electronics is an emerging fabrication method because of its lithography-free or vacuum-free processing [24]. Moreover, screen and ink-jet printing technologies have been investigated and used in electronic manufacturing [1,14]. It is noted that ink-jet method has been applied to fabricate electronic devices because such a method provides advanced features such as having lowest price and being easy to use [1,14,24]. Recently, there has been renewed interest in electronics with organic materials [19,27]. The key aspects of advancements can be easily listed as follows: low temperature process, low-cost, and mechanical flexibility [4,13,15,23,33]. Therefore, organic RRAM is promising for novel storage and/or processing information technologies. Organic RRAM promotes emerging applications of flexible electronics. Synthesis of Au nanoparticle and its application to fabricate organic RRAM device were reported in [12]. Dao introduced a high-performance organic resistive device to illustrate the application of the Au nanoparticle for RRAM array [8].

This chapter summarizes the implementation of organic RRAM with ink-jet printer. Design process is presented in Section 17.2 while fabrication process is reported in Section 17.3. Section 17.4 introduced a real application of the fabricated Organic RRAM for RFID.

17.2 Design process

In the first step, a cross bar array architecture is designed as shown in Fig. 17.2. It is worth noting that the cross bar array architectures were commonly applied in a variety of work [22,28,32,43,44]. As can be seen from Fig. 17.2, there are 16 word-lines and 16 bit-lines.

Fig. 17.3 shows the illustrated shape of organic RRAM in this study. A crossbar chip architecture, which is simply stacked by a word-line layer/active layer/bit-line layer, was selected because it is suitable to fabricate by a process of printing or spin-coating [1,14]. A capacity of 32 bytes could be expressed by $8 \times 8 \times 4$ bits. Under the crossbar rule, the RRAM is compatible to the Quad in-line as presented in Fig. 17.3.

Fig. 17.4 shows the designed layout on 25 mm $\times$ 25 mm substrate. The factor was set to be 300 μm in order to adapt to ability of our fabrication facilities which are presented in Section 17.3.

17.3 Fabrication process

The fabrication processes of the RRAM array are described as follows. Fig. 17.5 shows experiential setup to print electrode layer of RRAM array, where an inkjet printer of Epson T60 was used due to its popularity and easy modification. Ag Nanoink was provided from AgIC Inc (Japan) with the size of 20 nm, at concentration of 15 in ethylene glycol, ethanol, and water multi-solvent. The cartridge of the printer was cleaned using acetone and pure water, then dried at 60 °C for

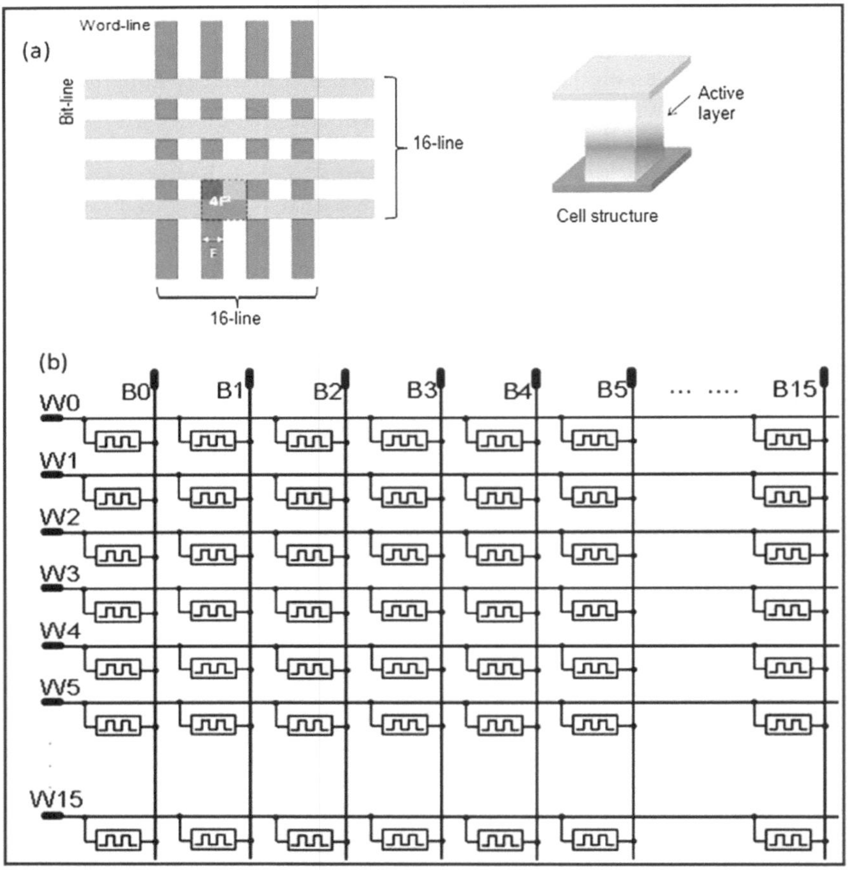

FIGURE 17.2 Theoretical design showing a cross bar array architecture.

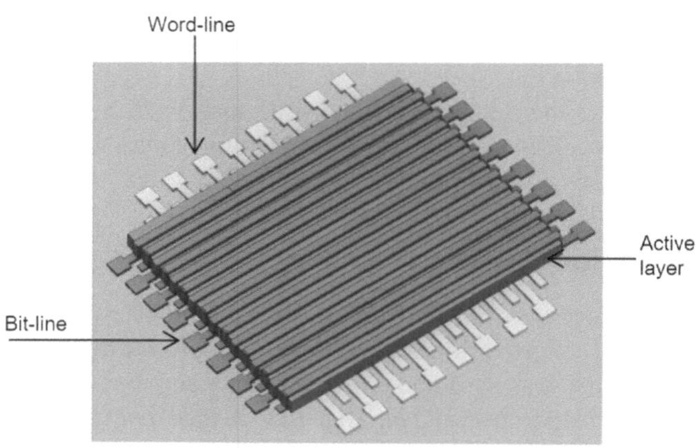

FIGURE 17.3 Illustration of quad in-line 32-byte-RRAM used in this work.

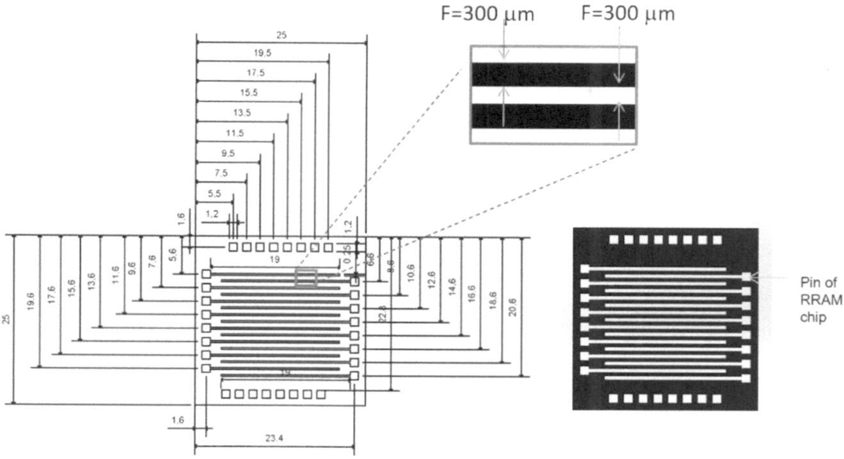

FIGURE 17.4 Detailed layout of RRAM array.

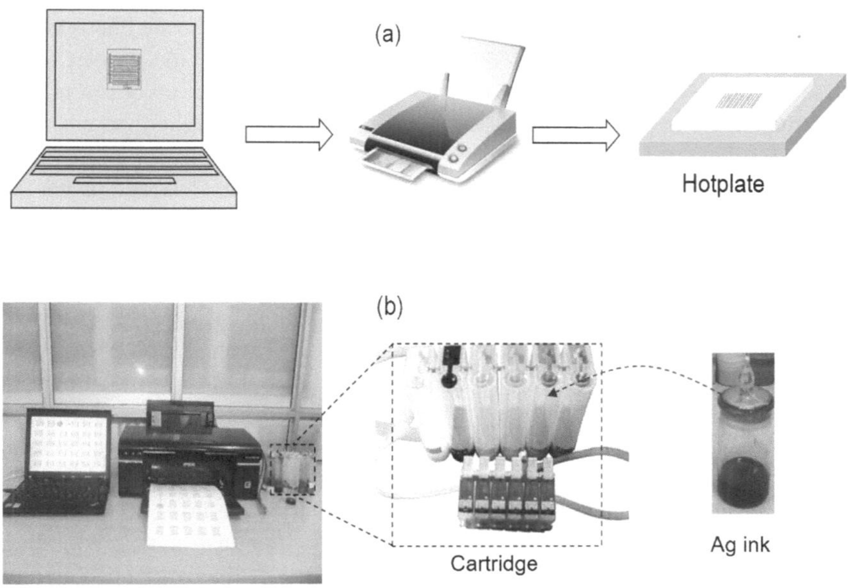

FIGURE 17.5 Experiment system to fabricate electrode.

about 4 hours. After that, nanoink was carefully loaded into the cartridge using a syringe needle. The 16-line per substrate was drawn with a Altium Designer CAD and then printed on a A4-size Canon photo paper and dried maturely at room temperature for a period of 60 minutes. The bit-line exhibits well conductive. The sheet resistance value was estimated to be 0.22 W/sq, the sheet resistance of the printed line is relatively larger than that of pure Ag metal, but

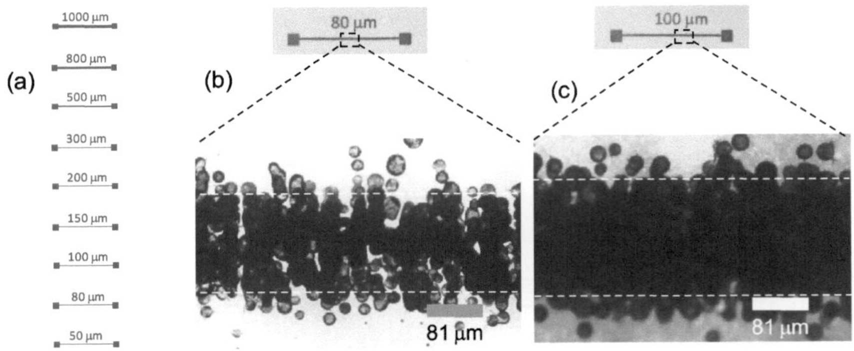

FIGURE 17.6 Optical images of printed electrodes with different line width values.

it can be used as electrodes of electronic devices [4,13,41]. There is no crosstalk between two different lines since they are completely insulated from each other. Fig. 17.6 shows the optical images of printed lines at different width values. At the designed width less than 300 μm, the actual width of electrode line is larger than that by designing. This is due to several factors those the droplet tends to spread out when approaching the substrate and the resolution of the printer together with the repeated processes [1,14]. At 300 μm or higher, the error between the designed and fabricated lines significantly reduces. In this experiment, the designed width of 300 μm was chosen to fabricate the electrode layers for organic RRAM chip.

For the active layer, Au nanoparticle synthesis and characterization were presented in our previous work [12]. These Au NP have diameters in the range 2–6 nm and are well dispersed in the poly(3-hexylthiophene) (P3HT) (Aldrich, Mw = 100,000) host matrix with a weight ratio of 20:1 were dissolved in cyclohexanone which was used in order to avoid the removal of the printed bottom Ag electrode layer. The solution was stirred and ultrasonically treated in order to obtain a well-dispersed composite. A ~250 nm active layer of Au NP and P3HT polymer was prepared by spin-coating of the solution on the printed Ag bottom electrode layer at 1000 rpm for 40 s, and dried the layer at 80 °C for 2 hours. Finally, the organic RRAMs were formed by lamination or evaporation of an Ag top electrode. The RRAM cell was defined by the overlapped area of the bottom and top electrodes, leading to the 32 bytes (16 × 16 bit cells) memory array as shown in Fig. 17.7 and Fig. 17.8. The electrical characteristics of the electrode and the RRAM array chip were measured with a Keithley 4200 semiconductor parameter analyzer and a Tektronix TDS 2022B digital oscilloscope at room temperature.

To explore a memory behavior, a cell was randomly chosen to characterize. Fig. 17.9 shows the current–voltage (IV) characteristics of the memory cell. As can be seen, the voltage sweep started at 0 V, continued to −6 V, then was returned back to 6 V and ended again at 0 V. The device was initially in high

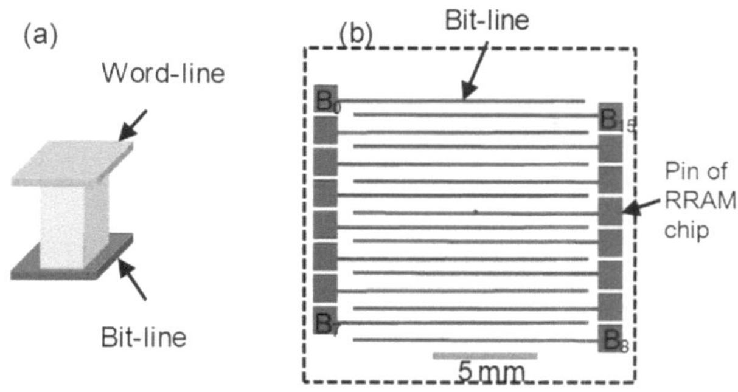

FIGURE 17.7 Structure of a memory cell.

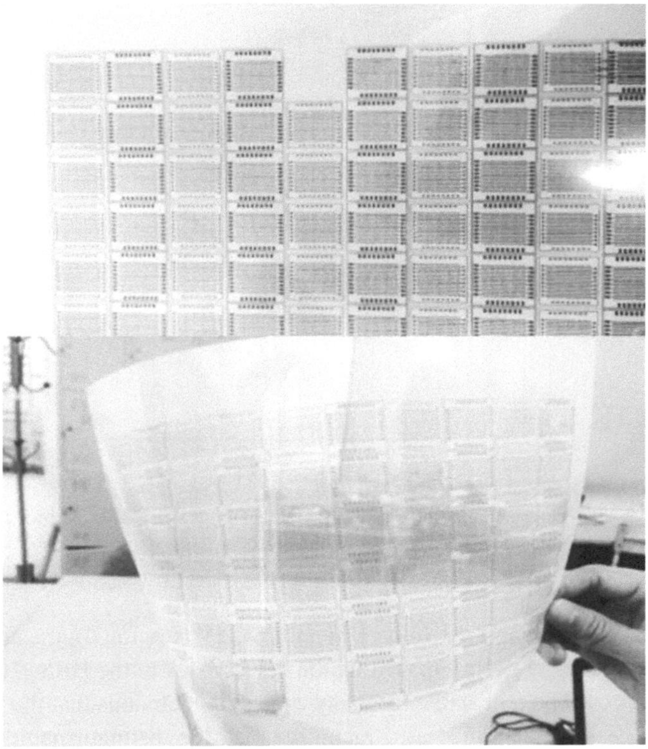

FIGURE 17.8 Image of printed electrodes in an A4 size paper.

resistivity state (HRS). The HRS state was maintained below about −4 V (Region 1), and then the current rapidly increased at a threshold voltage of −4 V, changing to low resistivity state (LRS). Once the transition from HRS to LRS took place, the state of the device did not change under subsequent negative (Region 2) or positive (Region 3) voltages, which suggests a non-volatile prop-

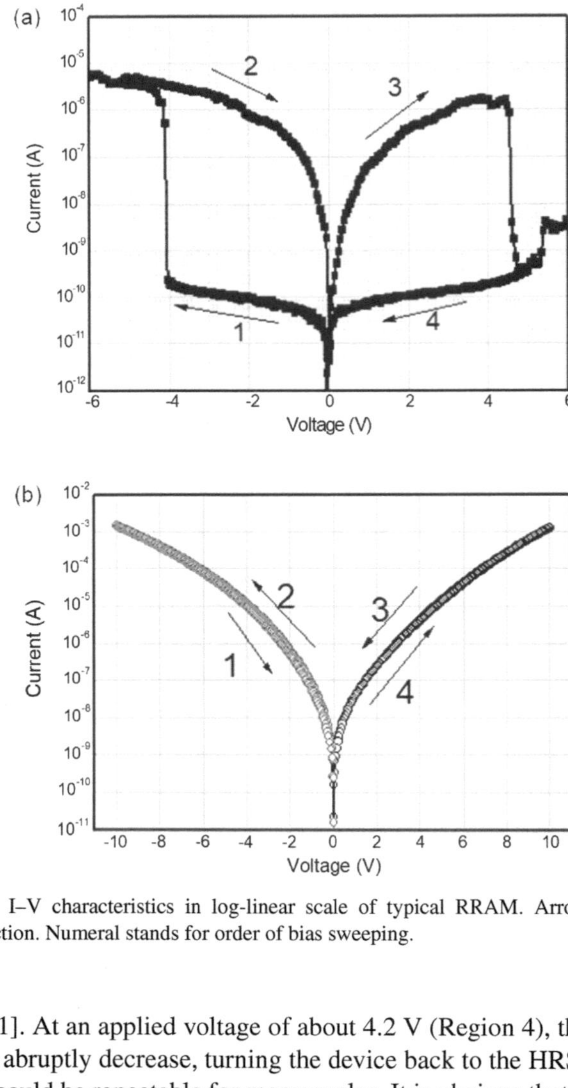

FIGURE 17.9 I–V characteristics in log-linear scale of typical RRAM. Arrows indicate bias sweeping direction. Numeral stands for order of bias sweeping.

erty [4,13,41]. At an applied voltage of about 4.2 V (Region 4), the current was observed to abruptly decrease, turning the device back to the HRS. The I–V relationships could be repeatable for many cycles. It is obvious that the fabricated device array was functioned as a memory since the distinguishable values of HRS and LRS can represent "0" and "1" logic states. The operating mechanism of an organic RRAM device can be understood the charge trapping by Au NP which has been presented in our previous work [12]. At a read voltage VR of 0.5 V, the on/off current ratio was estimated to be ~2.5×10^3, which is similar to that in our previous device with an evaporated electrode [12], indicating that the printed electrode does strongly affect device performance.

Beside the above-mentioned parameters, the retention time is one of the most important parameters in a memory. To examine the retention time, the mem-

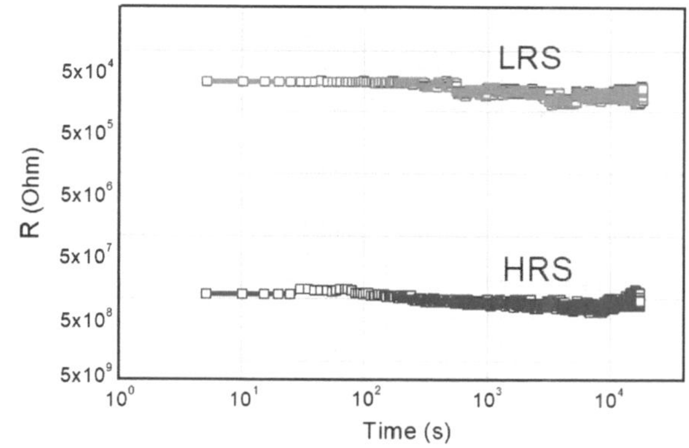

FIGURE 17.10 Retention time characteristics of typical RRAM at V_R of 0.5 V.

ory cell was programmed at −5 V for 1 μs then LRS current was continuously measured at $V_R = 0.5$ V. Subsequently, the device was erased by applying 5 V for 1 ms and the current in the HRS state was also continuously observed at $V_R = 0.5$ V. The retention time characteristics of the memory cell are shown in Fig. 17.10. As can be seen in Fig. 17.10, the LRS currents slightly decreased while the HRS currents almost unchanged, suggesting that the non-destructive readout and highly reliable non-volatile storage were realized in our organic RRAM device.

17.4 A practical application example

In order to demonstrate an application of the memory array in data storage, a readout circuit to code the fabricated resistive memory was designed based on the reference work by Andersson et al. [1]. Fig. 17.11 and Fig. 17.12 display the schematic of the readout circuit and the pictures of the encoder system when the memory element is being read. The memory array in the socket can be accessed using two analog switches to select the row and column.

The output of the selected column is then connected to Arduino Nano with an 8-bit analog to digital converter. The computer serially communicates with the Arduino Nano over USB to receive data.

A computer program with C language is developed to encode and display the state of each memory cell as illustrated in Fig. 17.13. The binary of "1" or "0" is defined to be less than 500 kΩ and more than 1 MΩ, respectively. That method also allows for defining a "dead cell" (i.e. infinite resistance) as a "0" bit.

To visually observe values, "1" or "0" is presented as red (light gray in print version) or "0" blue (gray in print version) squares, respectively. ASCII code is obtained by combining 8 bits as shown in Fig. 17.13, so that the system visualized letters corresponding to the ASCII codes. For example, as can be seen

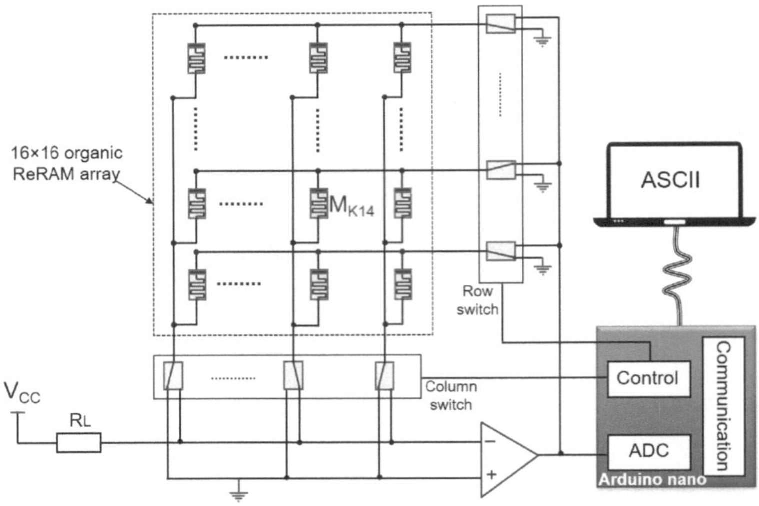

FIGURE 17.11 Equivalent diagram of encoder system. Memory chip embedded in a lab-made socket.

FIGURE 17.12 Equivalent diagram of encoder system. Memory chip embedded in a lab-made socket.

in Fig. 17.14, the letters of "ô" and "d" can be encoded to be "10010011" and "01100100", respectively, in the fabricated memory chip.

17.5 Conclusion

In this chapter, design and fabrication processes for implementing organic RRAM The organic RRAM array was based on an Au NP:P3HT composite on a paper substrate. Interesting that we generated the electrode layer by using a commercial Epson T60 inkjet printer. We believe that the described method is useful for investigating a low-cost organic RRAM array in smart electronic devices.

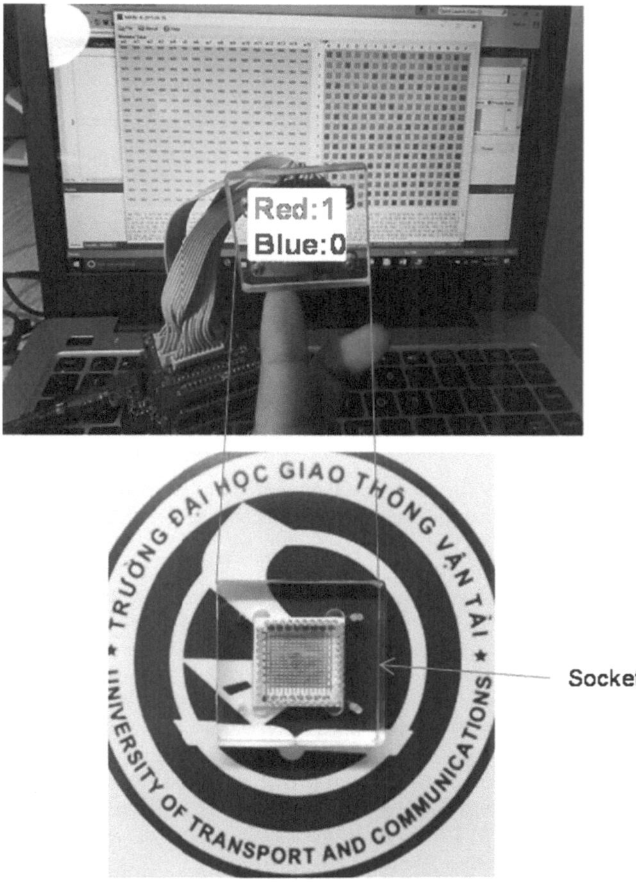

FIGURE 17.13 Developed computer program to encode and display the state of each memory cell.

References

[1] H. Andersson, A. Rusu, A. Manuilskiy, S. Haller, S. Ayoz, H.-E. Nilsson, System of nano-silver inkjet printed memory cards and PC card reader and programmer, J. Microelectron. 42 (2011) 21–27.

[2] S. Aritome, Advanced flash memory technology and trends for files storage application, in: Proc. IEEE IEDM Tech. Dig., 2000, pp. 763–766.

[3] R. Bez, E. Camerlenghi, A. Modelli, A. Visconti, Introduction to flash memory, Proc. IEEE 91 (2003) 489–502.

[4] B. Cho, S. Song, Y. Ji, H.-G. Choi, H.C. Ko, J.-S. Lee, G.-Y. Jung, T. Lee, Demonstration of addressable organic resistive memory utilizing a pc-interface memory cell tester, IEEE Electron Device Lett. 34 (2013) 51–53.

[5] L.O. Chua, Resistance switching memories are memristors, Appl. Phys. A 102 (4) (2011) 765–783.

[6] L.O. Chua, The fourth element, Proc. IEEE 100 (2012) 1920–1927.

[7] E. Covi, S. Brivio, A. Serb, T. Prodromakis, M. Fanciulli, S. Spiga, Analog mem-ristive

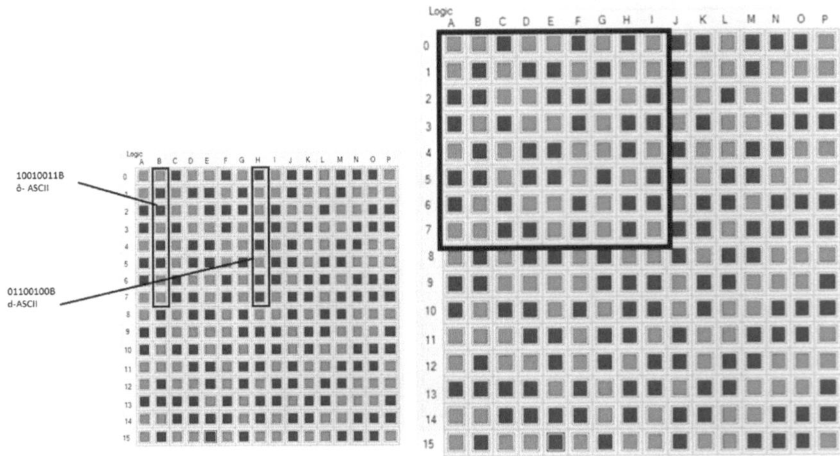

FIGURE 17.14 Binary of "1" or "0" in red (light gray in print version) or blue (gray in print version) color and ASCII format appearing on screen of computer.

synapse in spiking networks implementing unsupervised learning, IEEE J. Emerg. Sel. Top. Circuits Syst. 10 (2016) 482.

[8] T.T. Dao, An organic RRAM chip fabricated using ink-jet printer and its readout circuit for data storage, in: Proc. International Conference on Recent Advances in Signal Processing, Telecommunications & Computing (SigTelCom'2019), 2019, pp. 140–144.

[9] K. Eshraghian, K.-R. Cho, O. Kavehei, S.-K. Kang, D. Abbott, S.-M.S. Kang, Memristor MOS content addressable memory (MCAM): hybrid architecture for future high performance search engines, IEEE Trans. Very Large Scale Integr. Syst. 19 (2011) 1407–1417.

[10] Y. Halawani, B. Mohammad, D. Homouz, M. Al-Qutayri, H. Saleh, Modeling and optimization of memristor and STT-RAM-based memory for low-power applications, IEEE Trans. Very Large Scale Integr. Syst. 24 (2016) 1003–1014.

[11] C.H. Ho, Rram technology and its embedded potential on iot applications, in: Proc International Symposium on VLSI Technology Systems and Application (VLSI-TSA'2017), 2017, pp. 24–27.

[12] M.H. Hoang, T.T. Dao, N.T.T. Trang, P.H.N. Nguyen, T.T. Ngo, Synthesis of gold nanoparticles capped with quaterthiophene for transistor and resistor memory devices, J. Chemistry 2016 (2016) 12471754.

[13] Y. Ji, D.F. Zeigler, D.S. Lee, H. Choi, A.K.-Y. Jen, H.C. Ko, T.-W. Kim, Flexible and twistable non-volatile memory cell array with all-organic one diode–one resistor architecture, Nat. Commun. 4 (2013) 2707.

[14] Y. Kawahara, S. Hodges, B.S. Cook, C. Zhang, G.D. Abowd, Instant inkjet circuits: lab-based inkjet printing to support rapid prototyping of UbiComp devices, in: Proc. ACM International Joint Conference on Pervasive and Ubiquitous Computing (UbiComp'2013), 2013, pp. 363–372.

[15] Y. Kim, D. Yoo, J. Jang, Y. Song, H. Jeong, K. Cho, W.-T. Hwang, W. Lee, T.-W. Kim, T. Lee, Characterization of PI: PCBM organic nonvolatile resistive memory devices under thermal stress, Org. Electron. 33 (2016) 48–54.

[16] G. Mathur, P. Desnoyers, P. Chukiu, D. Ganesan, P. Shenoy, Ultralow power data storage for sensor networks, ACM Trans. Sens. Netw. 5 (2009) 33.

[17] J.S. Meena, S.M. Sze, U. Chand, T.-Y. Tseng, Overview of emerging nonvolatile memory technologies, Nanoscale Res. Lett. 9 (2014) 526.

[18] H. Mostafa, C. Mayr, G. Indiveri, Beyond spike-timing dependent plasticity in memristor crossbar arrays, in: Proc. IEEE Int. Symp. Circuits Syst. (ISCAS), 2016, pp. 926–929.
[19] S. Nau, C. Wolf, K. Popovic, A. Blümel, F. Santoni, A. Gagliardi, A. di Carlo, S. Sax, E.J.W. List-Kratochvil, Inkjet-printed resistive switching memory based on organic dielectric materials: from single elements to array technology, Adv. Electron. Mater. 1 (2015) 1400003.
[20] Y. Nishi, Advances in Non-volatile Memory and Storage Technology, Woodhead Publishing, Cambridge, UK, 2014.
[21] Y.V. Pershin, M.D. Ventra, Experimental demonstration of associative memory with memristive neural networks, Neural Netw. 23 (2010) 881–886.
[22] M. Prezioso, F. Merrikh-Bayat, B.D. Hoskins, G.C. Adam, K.K. Likharev, D.B. Strukov, Training and operation of an integrated neuromorphic network based on metal-oxide memristors, Nature 521 (2015) 61–64.
[23] M.M. Rehman, B.-S. Yang, Y.-J. Yang, K.S. Karimov, K.H. Choi, Effect of device structure on the resistive switching characteristics of organic polymers fabricated through all printed technology, Curr. Appl. Phys. 17 (2017) 533–540.
[24] P. Rosa, A. Camara, C. Gouveia, The potential of printed electronics and personal fabrication in driving the internet of things, Open J. Internet of Things 1 (2015) 16–36.
[25] T. Serrano-Gotarredona, T. Prodromakis, B. Linares-Barranco, A proposal for hybrid memristor-cmos spiking neuromorphic learning systems, IEEE Circuits Syst. Mag. 13 (2013) 74–88.
[26] S. Shirinzadeh, R. Drechsler, In-Memory Computing: Synthesis and Optimization, Springer, Germany, 2019.
[27] Y. Song, J. Jang, D. Yoo, S.-H. Jung, S. Hong, J.-K. Lee, T. Lee, 4k-bit and microlithographic integration of organic nonvolatile resistive memory devices, Org. Electron. 17 (2015) 192–197.
[28] C. Sung, H. Hwang, I.K. Yoo, Perspective: a review on memristive hardware for neuromorphic computation, J. Appl. Phys. 124 (2018) 151903.
[29] M. Suri, Applications of Emerging Memory Technology: Beyond Storage, Springer, Germany, 2020.
[30] N. Taherinejad, P.D.S. Manoj, A. Jantsch, Memristors' potential for multi-bit storage and pattern learning, in: Proc. IEEE European Modelling Symposium (EMS'2015), 2015, pp. 450–455.
[31] S. Tang, S. Yin, S. Zheng, P. Ouyang, F. Tu, L. Yao, J. Wu, W. Cheng, L. Liu, S. Wei, AEPE: an area and power efficient RRAM crossbar-based accelerator for deep CNNs, in: Proc. IEEE 6th Non-Volatile Memory Systems and Applications Symposium (NVMSA'2017), 2017, pp. 1–6.
[32] I. Vourkas, G.C. Sirakoulis, Memristor-Based Nanoelectronic Computing Circuits and Architectures, Springer, Germany, 2015.
[33] R. Wang, S. Chen, L. Feng, Q. Li, W. Hu, W. Liu, X. Guo, Stable fully-printed polymer resistive read-only memory and its operation in mobile readout system, Org. Electron. 27 (2015) 259–265.
[34] Z. Wang, S. Joshi, S. Savelev, W. Song, R. Midya, Y. Li, M. Rao, P. Yan, S. Asapu, Y. Zhuo, H. Jiang, P. Lin, C. Li, J.H. Yoon, N.K. Upadhyay, J. Zhang, M. Hu, J.P. Strachan, M. Barnell, Q. Wu, H. Wu, R.S. Williams, Q. Xia, J.J. Yang, Fully memristive neural networks for pattern classification with unsupervised learning, Nat. Electron. 1 (2018) 137–145.
[35] Z. Wang, S. Joshi, S.E. Savelev, H. Jiang, R. Midya, P. Lin, M. Hu, N. Ge, J.P. Strachan, Z. Li, Q. Wu, M. Barnell, G.-L. Li, H.L. Xin, R.S. Williams, Q. Xia, J.J. Yang, Memristors with diffusive dynamics as synaptic emulators for neuromorphic computing, Nat. Mater. 16 (2017) 101–108.
[36] C. Wolf, S. Nau, S. Sax, Y. Busby, J.-J. Pireaux, E.J.W. List-Kratochvil, Resistive switching based on filaments in metal/PMMA/metal thin film devices, Jpn. J. Appl. Phys. 54 (2015) 120301.
[37] C.D. Wright, P. Hosseini, J.A.V. Diosdado, Beyond von Neumann computing with nanoscale phase-change memory devices, Adv. Funct. Mater. 23 (2013) 2248–2254.

[38] Y. Xie, Emerging Memory Technologies: Design, Architecture, and Applications, Springer, Germany, 2013.

[39] P. Yadav, B.K. Das, Memristor-based memory cell with less noise margins and storing non-binary data, in: Physics of Semiconductor Devices, Springer, Cham, 2014, pp. 183–187.

[40] J.J. Yang, D.B. Strukov, D.R. Stewart, Memristive devices for computing, Nat. Nanotechnol. 8 (2013) 13–24.

[41] D. Yoo, Y. Song, J. Jang, W.-T. Hwang, S.-H. Jung, S. Hong, J.-K. Lee, T. Lee, Vertically stacked microscale organic nonvolatile memory devices toward three-dimensional high integration, Org. Electron. 21 (2015) 198–202.

[42] H. Yu, Y. Wang, Design Exploration of Emerging Nano-Scale Non-volatile Memory, Springer, Germany, 2002.

[43] Y. Zhang, Y. Li, X. Wang, E.G. Friedman, Synaptic characteristics of Ag/AgInSbTe/Ta-based memristor for pattern recognition applications, IEEE Trans. Electron Devices 64 (2017) 1806–1811.

[44] Y. Zhang, X. Wang, E.G. Friedman, Memristor-based circuit design for multilayer neural networks, IEEE Trans. Electron Devices 65 (2017) 677–686.

Chapter 18

Neuromorphic vision networks for face recognition

Akshay Kumar Maan[a] and Alex James[b]

[a]*Griffith University, Brisbane, QLD, Australia,* [b]*Indian Institute of Information Technology and Management, Kerala, India*

18.1 Introduction

The real-time automatic detection and recognition of image patterns such as faces [35,37] pose the problems of low utilization of imaging sampling rates [10,40,42,43], speed of processing [14], and low speed of matching with increased number of classes. The earliest attempts of pattern matching hardware [4,6,13,25] focused on developing hardware circuits to learn patterns in static images [19], [20]. The neuromorphic vision circuits such as retina pixel array [23,41] are limited by low spatial resolution. Circuits in the literature for retina pixel arrays exhibit complexities when scaling to higher resolution. The human visual system, on the other hand, is highly efficient and accurate on similar tasks such as face recognition. It can also be noted that higher resolution of images and faster sampling rates increases the number of available features in face recognition [27,31], [3], and usually recognition algorithms favor the probability of detecting discriminative features [28,34].

The neuromorphic circuits need to be able to adapt to the changes and lack of consistency introduced due to partial obstructions, illumination variations, pose differences, and changes in facial features due to aging in a face-recognition application. [2,18,33], occlusions [1,18,21,22,36,44], pose variations [7,9,39,45], and age related changes. In real-time implementations, we require higher resolution both spatial and temporal so that the neuromorphic system can extract finer features. In real systems though, the limitations of low bandwidth, high algorithmic complexity and low available memory make it hard to realize high speed high resolution face-recognition systems. This chapter presents a hardware neuromorphic face-recognition system that memorizes the feature in an image in a network of memristive threshold logic circuits and then performs template matching from the array of sensors. The added advantage of the proposed system is its ability to be implemented in a layered structure of pixel array planes and in analog-to-digital converters.

Mem-elements for Neuromorphic Circuits with Artificial Intelligence Applications
https://doi.org/10.1016/B978-0-12-821184-7.00027-X

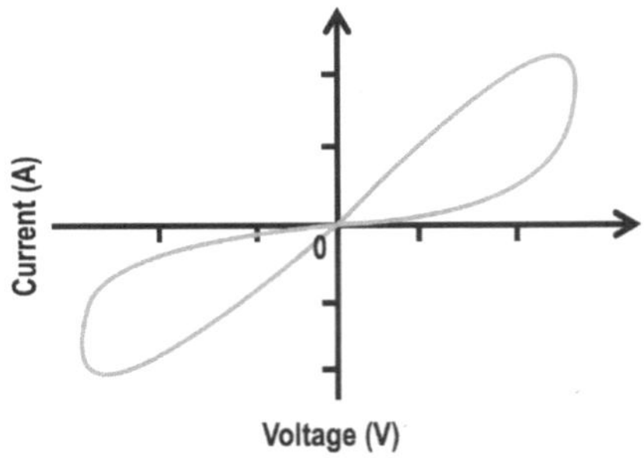

FIGURE 18.1 A pinched hysteresis loop in the v–i characteristic is a distinctive feature of memristors.

To build neuromorphic systems in hardware, memristors, which are highly dense, non-volatile, and naturally amenable to be operated analogously to the biological synapse, will be to build a threshold logic cell. This cell is programmable and is the building block of memory networks that can be reconfigured to perform face-recognition tasks. The said memory networks also serve to shift focus from memory being a storage-only unit to being a computational unit as well, just as in the human brain. Existing neuromorphic systems are mostly software-based, and therefore present computational bottlenecks when faced with high speed, high resolution input and real-time output. Fully hardware implementations of reconfigurable memory networks to deal with the finer space-time input in a real-time setting are shown. The networks are highly parallel and linearly scalable. Sizeable improvements in power consumption, area-on-chip and leakage power as compared to CMOS-only circuits have been achieved.

The key element of the presented CMOS inverter/memristor circuits is a memristive [11,12,17] device that exhibits resistance-switching [15,24,32,38] give a review and applications of such devices. A memristive device consists of an active thin film layer of some doped material sandwiched between two metal (platinum) electrodes. When a relatively large electrical field is applied across the electrodes, the resistivity (memory state) of the doped material changes due to the modulation of the width of the doped region depending on the amount of electric charge passed through the memristor. With electric current passing in a given direction, the boundary between the two regions (doped and undoped) moves in the direction of current flow, thus leading to the resistive switching phenomenon where an external electric field coupled with the previous resistance state of the device determines the new resistance state [8]. The memristance is the slope of the flux-charge curve, which has two different slopes (Fig. 18.1). This means there are two different values for memristance which makes it suitable for binary logic, and that the memristors are purely passive two-terminal devices, unlike MOSFETs.

18.2 Preliminaries

A threshold logic gate has the following transfer function:

$$f(x_1, \ldots, x_n) = \begin{cases} 1\text{ V}, & \text{if } \sum_{i=1}^{n} w_i x_i \geq T \\ 0\text{ V}, & \text{otherwise} \end{cases} \tag{18.1}$$

where x_i is a Boolean input variable, w_i is an integer weight of the corresponding input i, and the threshold T is an integer number [6]. In the design presented here the weights w_i using are implemented the resistive switching property of the memristor device, x_i are the inputs available to the gates and the threshold T is implemented using a CMOS inverter [17]. On applying an initial threshold voltage, memristive devices are brought into the ON state upon which they show linear conductance with a configurable slope [17]. Thus they are especially suitable for implementing weights in the proposed threshold logic circuits. The memristor model used for simulation is the spice model that was proposed by [8]. This device has a large R_{OFF}/R_{ON} ratio (10^2-10^3), while still retaining a relatively low switching time (about 10 ns).

From Fig. 18.2 it can be seen that when voltages 3.5 V and −3.5 V are applied across the positive and negative terminals of the memristor, we get a high resistance state (R_H) and if the reverse voltages (−3.5 V and 3.5 V) are applied across its positive and negative terminals, we get low resistance state (R_L). These voltage levels +/−3.5 V, are used to set high and low resistance states of memristor and are represented as $+/-V_{TR}$, the training voltage [7]. If we use any voltage in-between the training voltages, the resistance state of the memristor will not change. Hence, we can use these voltage levels for the working of the cell and will refer to them as the testing voltage V_{TE} which can be 1 V for logic high and 0 V for logic low.

18.3 Model description

The system has two-stage process: (1) template formation, and (2) matching and recognition. In the template formation stage the idea is to store only the most stable information across a sufficient number of frames in a video while accommodating only the very significant changes in pixel intensity when they occur. This is done by implementing an averaging circuit (Fig. 18.3, Block 1) that fires when the change in the intensities of the corresponding pixels in the arriving n frames is above/below a predefined threshold. A parallel circuit for each of the bits per pixel is implemented. If the average of the incoming n bits is more than 0.5 V, the output of the averaging circuits is 0 V, otherwise 1 V. If output of the averaging circuit is 0 V, it sets the resistance (Fig. 18.3, Block 2) to R_L. If, on the other hand, the output of the averaging circuit is 1 V, it sets the resistance (Fig. 18.3, Block 2) to R_H. This configuration ensures that the template output state is changed only when there is a significant change in the input frames.

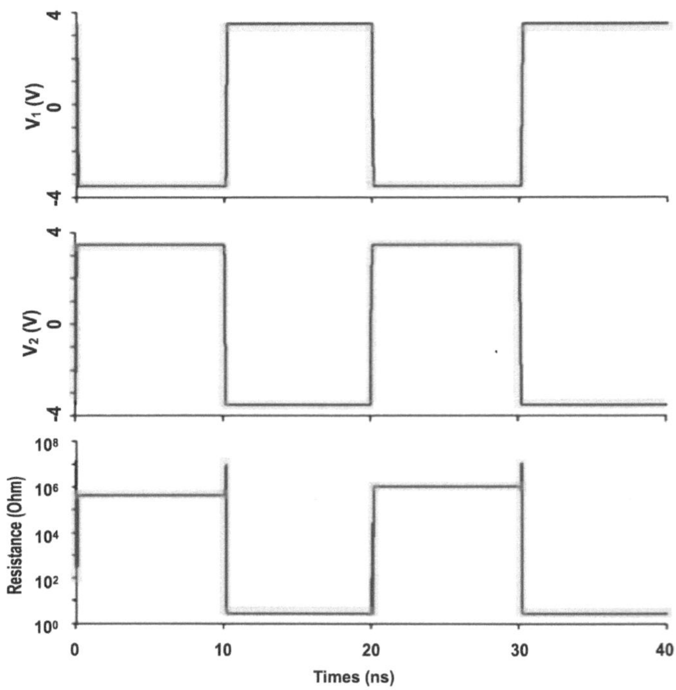

FIGURE 18.2 Simulation result showing the switching of resistance state of the memristor when voltages V_1 and V_2 applied across it. V_1 and V_2 are training voltages for the memristor having values +/−3.5 V. We see that when voltages 3.5 V and −3.5 V are applied across the positive and negative terminals of the memristor, we get a high resistance state (R_H) and if the reverse voltages (−3.5 V and 3.5 V) are applied across its positive and negative terminals, we get a low resistance state (R_L).

In the recognition phase of Fig. 18.4, the template is used to set the values of resistors that memorize the features in a set of parallel circuits that we call threshold logic (also, cognitive cell) cells and the network so formed threshold logic network (also, cognitive network). The resistors are set to R_L or R_H so that output of each cell is logic 1. This is accomplished by the following logic: if the input to a cell is greater than the average of the inputs to the cell, the corresponding resistor is set to R_H, otherwise to R_L. In the case the inputs are equal, the default setting is R_H. Thus our network is trained for each bit of each pixel of the template image in a parallel configuration. The output of all the trained cells is averaged through a simple averaging circuit at the end. Now during the testing phase, a test image is applied to the trained network and is ranked based on how close its score is to the template score of 1. The switching of the appropriate resistances is done as described in the preceding section.

18.4 Template formation

Template formation is part of the training stage in the face-recognition system. It comprises two blocks, namely Block 1 and Block 2. The main function of Block 1 is to perform a temporal average of a certain number of grayscaled and normalized image frames coming from the video at sensor. The corresponding bit in each frame is used to take the mean. Once this mean is calculated, it is applied to an inverting circuit. The normalization happens with respect to the features that are stable in the incoming frames. The physical location of the mouth, eyes and nose constitute the stable features. For a fixed number of frames, the average is performed for each bit corresponding to each pixel on a temporal basis. For an n-input averaging circuit with input voltages $V = \{V_1, V_2, \ldots, V_n\}$, the output voltage V_{block1} is given by

$$V_{block1} = \begin{cases} 1, & \text{if } \left(\frac{1}{n}\right) \sum_{i=1}^{n} V_i \le V_{th} \\ 0, & \text{otherwise} \end{cases} \tag{18.2}$$

where V_{th} is the threshold of the inverter. Consider the example of the three consecutive frames from a video coming into Block 1 for Fig. 18.3. As explained above, the average is performed for each bit corresponding to each pixel via a memristive averaging configuration and is fed to the inverting amplifier, which serves as a NOT logic. The memristor resistance state is of the same value in all three memristors ending in a common node. The common node represents the voltage as the mathematical mean of the input pixel voltages. For example, let us suppose one set of the three consecutive bits in the incoming image slices is given as [0; 1; 1]. When these are applied to the memristive averaging circuit followed by the inverter, we get 0:67 V as the common node voltage. The subsequent inverter turns it to 0 V. Likewise, inputs of [0; 0; 0] and [1; 0; 1] go through the averaging circuit to give 0 V and 0.67 V, respectively, and are changed to 1 V and 0 V at the output. Block 1 of Fig. 18.3 thus acts as a pre-processor detecting the significant pixel intensity variations in the incoming images frames.

Thus the state change owing to significant changes in the pixel density in consecutive frames) is represented by the output that changes from 0 to 1 or from 1 to 0. The meristors will then be switched to High Resistance State (HRS) and a value of RH or Low Resistance State (LRS) and a value of RL if the output is 1 or 0, respectively, which is how memorization occurs in the proposed circuit:

$$R_{block2} = \begin{cases} R_H, & \text{if } V_{block1} = 1 \\ R_L, & \text{if } V_{block1} = 0 \end{cases} \tag{18.3}$$

where R_H is a resistance of very high value ($10^9 \Omega$) and to R_L is a resistance of very low value ($10^6 \Omega$). The voltage divider circuit is formed by memristors having states of R_H or R_L which depend on the output of Block 1 (V_{block1})

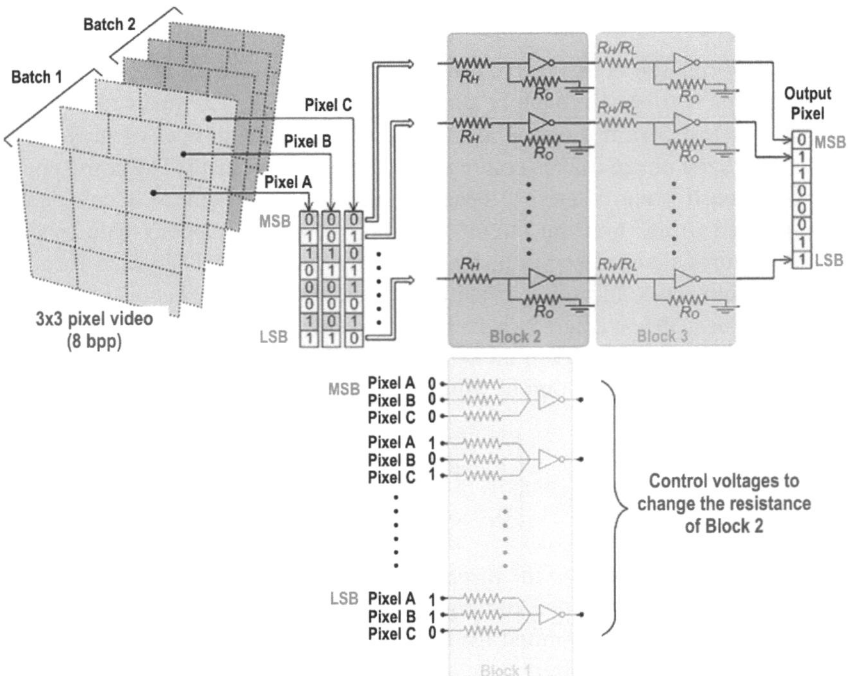

FIGURE 18.3 The illustration of how a template pixel is formed by temporal processing using memory cells. The main function of Block 1 is to perform a temporal average of a certain number of grayscaled and normalized image frames coming from the video at sensor. The corresponding bit in each frame is used to take the mean. Once this means is calculated, it is applied to an inverting circuit. The normalization happens with respect to the features that are stable in the incoming frames. The physical location of the mouth, eyes and nose constitute the stable features. For a fixed number of frames, the average is performed for each bit corresponding to each pixel on a temporal basis. Input digital stream continues to be received at the input of Block 1 while the input to Block 2 V_{block2} is kept at logic 1 irrespective of the input changes. This input of logic 1 is applied to a voltage divider circuit followed by an inverter of Block 2. The voltage divider circuit is composed of R_H or to R_L depending on the voltage on the output node of Block 1 (V_{block1}) which defines how the resistors in Block 2 are set, and ground resistance Ro (107). Output V_{block2} is then given by Eq. (18.4) at end of Block 2 where all the bits are rearranged to the corresponding pixels and a template image is formed [26].

needed to set the memristor resistance in Block 2, and earthed resistance Ro ($10^7\,\Omega$) so that the output V_{block2} of Block 2 is given by

$$V_{block2} = \begin{cases} 1, & \text{if } \frac{V/R_H}{1/R_H + 1/R_L} < 0 \\ 0, & \text{otherwise} \end{cases} \tag{18.4}$$

This is where the training of the circuit to form a template is achieved according to Eq. (18.3). The block output V_{block2} is then given by Eq. (18.4). At the output

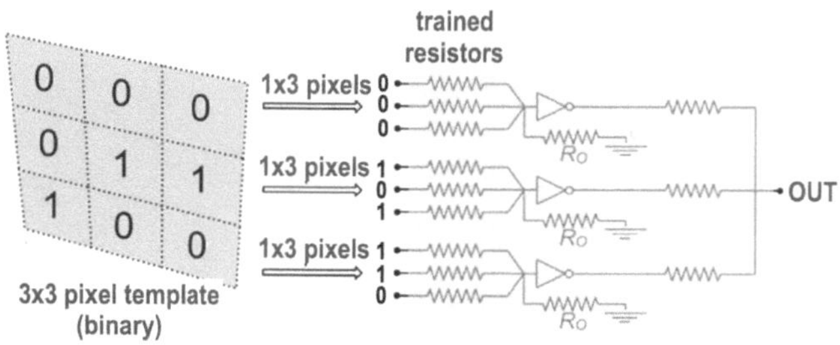

FIGURE 18.4 Example showing template matching with a trained network. Template image form the template formation phase is applied to the circuit as input. Each group of pixels is applied to a separate threshold logic cell in parallel. Each cell contains memristors trainable to either R_L or R_H so that output of each cell is logic 1. Similar to a neuron firing mechanism, the cognitive cell would fire when the voltage at V_O reaches a threshold expected by the inverter. Firing of a cognitive cell refers to a change in the cell output, V_{OUT} and that indicates a significant change at the input side of the cell [26].

all the bits are rearranged to the corresponding pixels and a template image is formed. The circuit shown in Fig. 18.4 is for each corresponding bit of the incoming frames so that parallel implementation for all bits can be done for real-time, fast template formation.

18.5 Face recognition

Fig. 18.4 shows the face-recognition circuitry that is trained by the template formed in the template formation phase and then applied to test images for recognition. The template is used to set the values of resistors that memorize the features in a set of parallel circuits that we call memristive threshold logic cells (MTLs) (also, cognitive cells) and the network so formed memristive threshold logic memory network (also, cognitive network). Based on the input voltages, each cell contains memristors trainable to either R_L or R_H so that output of each cell is logic 1. Similar to a neuron firing mechanism, the cognitive cell would fire when the voltage at V_o reaches a threshold expected by the inverter. Firing of a cognitive cell refers to a change in the cell output, V_{OUT} and that indicates a significant change at the input side of the cell. This property of the memristive threshold logic cell is used for face recognition.

18.6 Memristive threshold logic (MTL)

Fig. 18.5 illustrates the working of a memristive threshold logic (MTL) cell. The input voltages V_1, V_2 and V_3 are applied to the memristor resistance R_1, R_2 and R_3, respectively. To process the full image spatially, the MTL cells of Fig. 18.5 are replicated in parallel to cover the entire image. The memristance values are set either to LHS or RHS, which is guided by the digital pixel intensity at the input. In general, for an n-input cognitive cell, the memristive values $R = \{R_1, R_2, R_3, \ldots, R_N\}$ for a corresponding set of input values $V = \{V_1, V_2, V_3, \ldots, V_N\}$ are set based on

$$R_i = \begin{cases} R_H, & \text{if } V_i > V_t = (\frac{1}{n}) \sum_{i=1}^{n} V_i \\ R_L, & \text{otherwise} \end{cases} \tag{18.5}$$

where V_t denotes the threshold of the mean of the inputs applied. V_O which is the output of the potential divider of a cell with n inputs thus becomes $V_O = \frac{\sum_{i=1}^{n} V_i/R_i}{\frac{1}{R_O} + \sum_{i=1}^{n} V_i/R_i}$. Thus the output V_{OUT} of the inverter is given as

$$V_{OUT} = \begin{cases} 1, & \text{if } V_O > V_{th} \\ x, & \text{otherwise} \end{cases} \tag{18.6}$$

where V_{th} is the threshold of the inverter. To realize Eq. (18.6) in hardware, we use a logic function. Another stage of averaging circuit then combines all input averages to perform a circuit-wide mean of the features. This forms the template image which switches on the appropriate cells.

18.7 Edge detection with memristive threshold logic (MTL) cells

The cognitive cells described here can be gainfully configured to detect edges in the images. Edge detection is the most commonly used operation in image processing applications like face recognition, segmentation and pattern analysis. A higher level of feature processing is possible if we can integrate edge detection in the process of face recognition. Almost all of the edge detection methods are algorithm oriented and pay only a very little attention on the hardware implementation. As in the approach proposed in [30] a hardware-based edge detection method is introduced which detects the edge features of the available images. Fig. 18.5 shows edge detection process. Since edges are local discontinuities in the image intensities, a new method is developed for edge detection which elegantly takes advantage of the configurable memristive threshold logic cells. We note that (in grayscale images) an edge appears whenever there is transition from black to white pixels or from white to black pixels. Therefore, two MTL cells are built: black-to-white detection cell that detects a transition in an image from black to white pixels, and white-to-black detection cell that detects a transition in an image from white to black pixels. When the corresponding output of black-to-white and white-to-black networks are applied to a XOR gate, the edges of the input image are obtained. Color images are converted to gray scale images and normalized before applying to the network. The intermediate output is then arrived at by feeding the normalized input image, taking 2×2 pixel blocks a time, to the black-to-white and white-to-black networks. We choose the inverter threshold voltage V_{th} as 0.3 V. When $V_i < V_{th}$ meaning the input

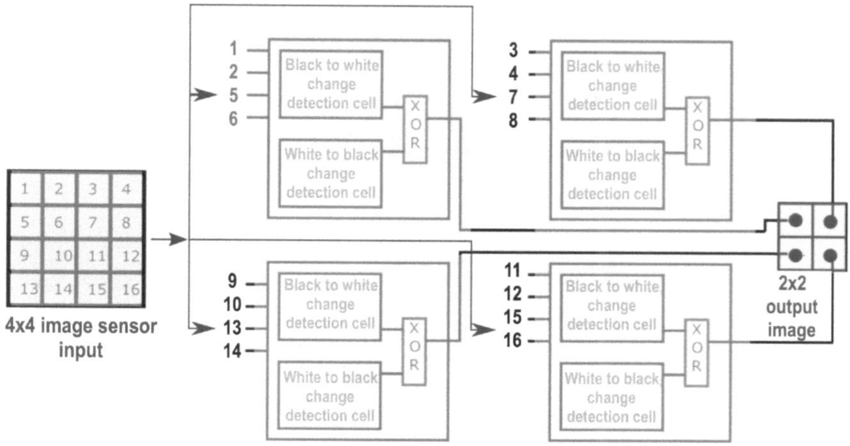

FIGURE 18.5 Figure shows how black-to-white network and white-to-black network are fed by 2×2 pixel blocks of the input image, whose outputs are XORed to form output pixels detecting the edges in the input image [26].

voltage is a logic low, the input resistance of the network cell is set at HRS. On the other hand, it changes to LHS if input voltage is logic high i.e. $V_i > V_{th}$, which forms the white-to-black cell. Before applying to black-to-white network the input image is inverted. In like manner, the input resistance of the threshold logic cell is set at HRS when the input voltage is a logic high state, while it is at LHS when the input image is a logic low, forming the corresponding black-to-white network. The final image showing the edges in the input frame is then constructed by XOR'ing output voltages.

In the illustration, a block size of 2×2 is shown that configures the threshold logic cells. The block size is an important threshold which has implication on the density of the cognitive networks thus created. Finally, the edge detected images are applied to the template formation network to form template images for both training and test sets.

Fig. 18.6 summarizes the operation of the edge detection cells. The MTL network cell is used with n inputs similar to the conventional logic of a moving window operation on the image with a size of m pixels, resulting in m binary outputs. In such an operation, each of the m pixels are processed using the n input cell, where the n inputs represents a $n \times n$ neighborhood of the ith pixel in the image. Initially the memristances of the cell are trained based on the conditions specified for a black-to-white or a white-to-black detection cell. For the black-to-white detection cell, input memristance is set to R_H if input V_i is greater than the inverter threshold V_{th}, and to R_L otherwise. For the white-to-black detection cell, the input image is inverted and input memristance is set to R_H if input V_i is less than the inverter threshold V_{th}, and to R_H otherwise. The succeeding neighborhood of input voltages are applied to the memory network

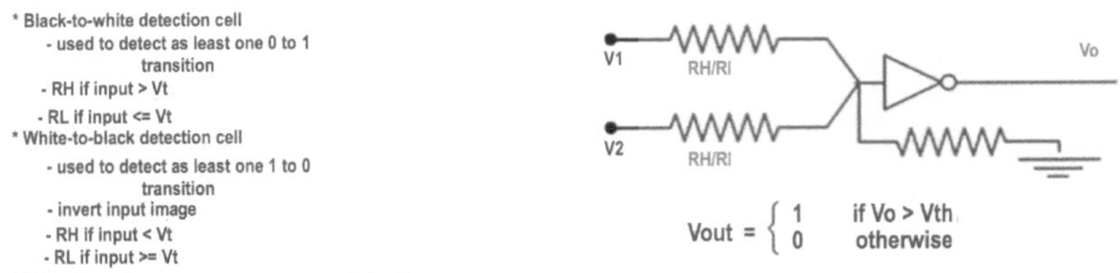

0-to-1

	V1	V2	R1	R2	Vo
Initial state	0	1	RL	RH	1
	0	0	RL	RL	1
	1	0	RH	RL	0
1/21 →	1	1	RH	RH	0

1-to-0

V1	V2	R1	R2	Vo
0	1	RH	RL	0
0	0	RH	RH	1
1	0	RL	RH	1
1	1	RL	RL	0

FIGURE 18.6 White-to-black and black-to-white detection. The memristances of the Memristive Threshold Logic Cell are trained based on the conditions specified for a black-to-white or a white-to-black detection cell. The succeeding neighborhood of input voltages are applied to the memory network whose input resistances are set according to the immediately preceding input voltages. Detection of transition happens when the first state change happens. For instance, a 0 to 1 transition is detected when the output changes its state (here from 0 to 1) whenever there is a change in inputs from initial state of 0 to 1. The big arrows alongside the table show the output changing as soon as the inputs change from 0 to 1 (in the case of a black-to-white detection cell) or 1 to 0 (in the case of a white-to-black detection cell).

whose input resistances are set according to the immediately preceding input voltages. Detection of a transition happens when the first state change occurs. For e.g. 0-to-1 transition is detected when the output changes its state (here from 0 to 1) whenever there is a change in inputs from initial state of 0 to 1. The big arrows alongside the table show the output changing as soon as the inputs change from 0-to-1 (in the case of a black-to-white detection cell) or 1-to-0 (in the case of a white-to-black detection cell). The resulting outputs are XOR'ed to detect the corresponding edges.

Once the resistor values in Fig. 18.4 have been set by the template image, test images can be applied to these circuits. Any image of a different class (face of a different person) will have heavily differing pixel intensities (and hence bit values) which will flip the output of corresponding cells to 0 or switch them off. The more the number of cells that are switched off, the greater is mismatch between the template and test images. This score is collected at the averaging output that basically gives a weight to this degree of mismatch. An image of the same class (face of the same person) will switch off much fewer cells (or none) and thus have the average score much closer to the 1 of the template.

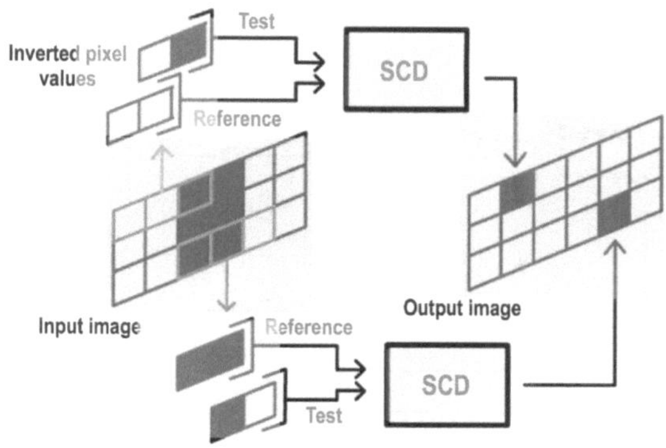

FIGURE 18.7 Spatial Change Detection unit of Block 1 in Fig. 18.3. A sliding window of 1 × 2 pixels sets the memristance of the MTL memory cells in the SCD units. The next window of pixels is applied to the input resistances set by previous window, which in turn train the memristors for the next window. Lower block detects black-to-white changes and the pixels are inverted and applied to upper block to detect white-to-black changes. The output from both SCD units are XOR'ed to arrive at final edges in the image.

As in the template formation network, this threshold logic network of the recognition phase is implemented in a parallel architecture enabling real-time face recognition from data arriving immediately from the sensors.

18.8 Circuit realization

18.8.1 Edge detection and template formation

As described above, template formation, which comprises two blocks: Block 1 and Block 2, can be integrated with edge detection for a higher level of feature processing. In this section the two processes are implemented in hardware for the circuits of described above. Block 1 of Fig. 18.3.1 detects the edges in each frame. The incoming picture frames come directly from sensor data and are grayscaled, and normalized to get the stable features in the images. Based on this image Block 1 will detect the edges which is then used by Block 2 to generate the facial feature template.

Fig. 18.7 shows the working of Block 1, which comprises a Spatial Change Detection (SCD) unit, used to detect spatial change (edge) in a group adjacent pixels. SCD contains two 2-input memristive threshold logic (MTL) cells, one for detecting black-to-white changes and the other to detect white-to-black changes. The output of the cell is in accordance to Eq. (18.3). Initially a window size of 1 × 2 pixels is selected which will take the two adjacent pixel values of an image frame and move horizontally on the image to detect horizontal spatial changes. The window values are then applied as voltages to a two input MTL cell. The window values are reference values used to set the resistor values of corresponding memristors in the memory cell. The memristance of the mem-

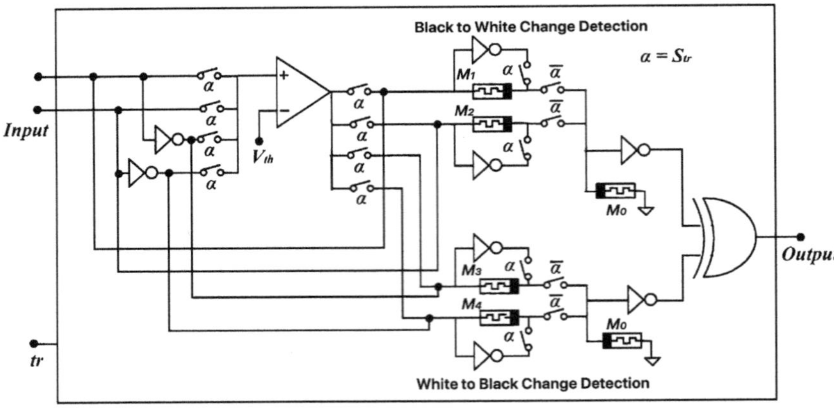

FIGURE 18.8 SCD detailed diagram. The inputs are the two pixel values and one control signal tr for synchronizing the training and testing operations, and the Output is the response of SCD. Two 2input MTL cells, one for black-to-white change detection and other for white-to-black change detection are trained using Switches S_{tr} and $\underline{S_{tr}}$. S_{tr} selects training phase and S_{tr} select testing phase. The op-amp generates training voltages $+/-V_{tr}$, to train the memristors and input pixels are applied to trained memristors in the testing phase to detect spatial changes (edges).

ristors in the cell are set based on the Eq. (18.5). After the memristor values are set, the window will move to the next two pixels. These pixel values will serve as test values, which will be fed to the same cell whose memristances were just trained by applying the reference pixel values in the previous stage. If there is any change between the window values, the cell will detect and write a low ('0') at the first pixel value in the window's position in the result image. If there is no change, the cell will not record any change. Now the second window value will become reference pixel values and set the memristor values for the next window of pixels. Likewise, the window is moved to the next pixel values, which is applied to the memory network detecting the horizontal spatial change. Since the cell is only able to detect dark to light changes, another memory cell is used whose input pixel values are the inverted values of those applied to the first memory cell. This will detect light-to-dark changes. The output of the two cells is XOR'ed to get the final output of an SCD unit. This procedure is applied on the all pixels of the image in order to find the horizontal spatial change of the image. If the window movement is in vertical direction, vertical edges can be detected and combining the horizontal and vertical spatial changes the process will give all edges in the image.

Fig. 18.8 is the internal diagram of the SCD unit. Here the inputs are the two pixel values and one control signal tr for synchronizing the training and testing operations, and the $Output$ is the response of SCD. Training and testing phases are controlled by the actions of the switches S_{tr} based on the control signal tr. In order to train the memristors (M_1, M_2, M_3, M_4) in the MTL cells based on Eq. (18.5), $+/-V_{tr}$ is applied across the memristors. Recall that $+/-V_{tr}$ are

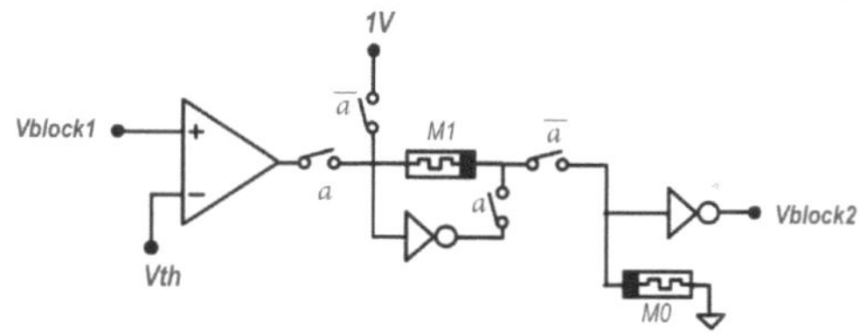

FIGURE 18.9 MTL Cell realization diagram of Block 2 of Fig. 18.3. Corresponding to each pixel of the image from Block 1 the memristance of the cell will change during the training phase, as in the case of Block 1. Input to each cell is a fixed voltage of 1 V.

used to train memristors to their R_H and R_L values. The memristor M_O of the cell is always fixed to an intermediate resistance value in between R_H and R_L. The training voltage V_{tr} is higher than the input pixel voltages V_i. In order to achieve that, an op-amp is used as a comparator and a group of switches S_{tr}. The reference voltage of the op-amp is set to V_{th} which is equal to 0.5 V. Based on the Eq. (18.5) the op-amp will generate the training voltage which is applied to the positive terminal of memristor and the inverted voltage to the negative terminal. A $+V_{tr}$ at the positive terminal and a $-V_{tr}$ at the negative terminal of the memristor will set memristance to R_L and the opposite condition will give R_H. Based on the input pixel values the memristances of MTL cell for black-to-white changes are set and based on the inverted pixel values the memristances of the white-to-black cell are set during the training phase. During the testing phase ($\underline{tr}$, signal) switches S_{tr} will open and switches $\underline{S_{tr}}$ will close. In this phase, the input pixel values will directly apply to the cells. Since these values are in the range of [0, 1] and are smaller than V_{tr}, they will not change the memristor state. If there is any change with the memorised states, the cell will detect it and give an output as low ('0') as explained earlier. By combining the outputs from black-to-white and white-to-black cells using an XOR gate spatial changes or edges in the image are detected.

Note that the SCD units can be used either in a normal windowing operation or as a parallel processing unit, wherein each unit will work on two pixels each. In the latter approach, all memristor values will be set in parallel during training phase. In testing phase, applying one bit-shifted input image horizontally or vertically will detect the corresponding edges. The second approach is faster and suitable in the case of a hardware model as presented in this work.

In Block 2 of Fig. 18.4 a network of single input MTL cells corresponding to each pixel of the image from Block 1 is used to generate template images. Fig. 18.9 shows the architecture of the cells in Block 2. Corresponding to each pixel of the image from Block 1 the memristance of the cell will change based on Eq. (18.3) during the training phase, as in the case of Block 1. But in the testing phase, different from the case of Block 1, the input to each cell is a fixed voltage of 1 V. The goal of this configuration is to keep the non-changing features of an image even when there is variation in lighting and in the presence

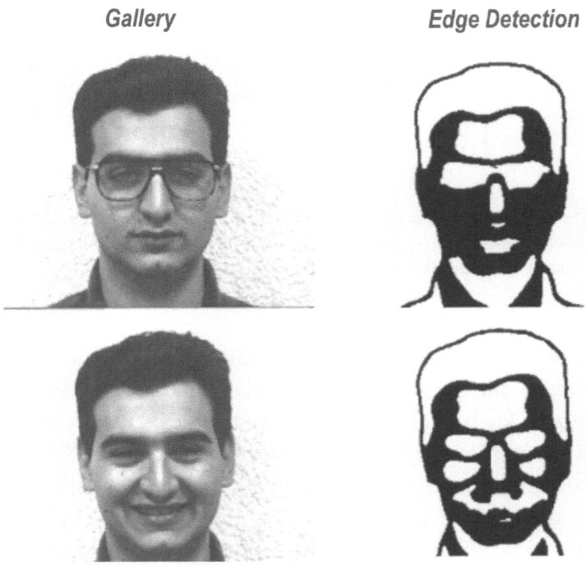

FIGURE 18.10 Edge detection illustration of the edge detection MTL Cell.

of occlusions. The output response of the cell is in accordance with Eq. (18.4). Fig. 18.10 shows an illustration of edge detection on two gallery images.

18.9 Experimental setup

The varying-resistance part of the circuit was realized by a memristor, which is a device that can vary its resistance according to the voltage applied across it. The simulations were performed in SPICE with a SPICE model of memristor in [8] using feature size of 0:25 μm TSMC process, BSIM models and HP memristor model. The template formation phase of the system was tested on three face-recognition databases – AR ("ARDb", [5]), ORL ("ORLDb", [29]), and faces94 ("faces94Db", [16]) using MATLAB®. AR database contains 100 classes each containing 26 images. ORL database has 40 different classes each containing 10 images. For the faces94 database 20 classes were tested on each containing 20 images. Equal and random distribution is applied to each database to form train and test sets.

18.10 Results and discussion

Fig. 18.11 shows how the changes in resolution of the input images affects the recognition accuracy across the databases. Accuracy generally increases with increase in resolution until it hits a maximum after which it decreases. It shows that there is a built-in resolution threshold in the system which only requires an

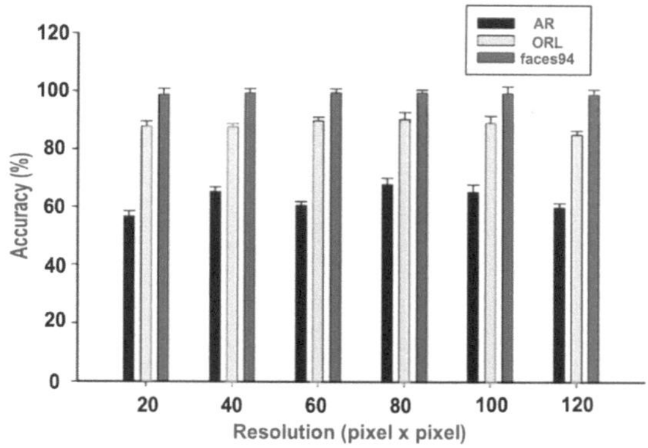

FIGURE 18.11 Recognition accuracy variation with increase in pixel × pixel resolution for train and test images for the three databases AR, ORL and faces94. Accuracy generally increases with increase in resolution until it hits a maximum after which it decreases.

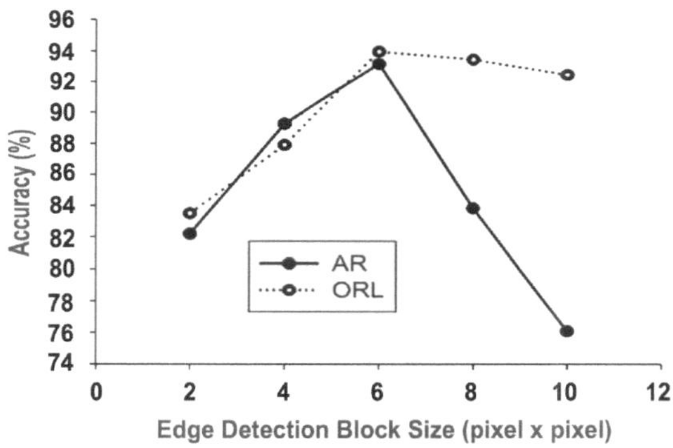

FIGURE 18.12 Recognition accuracy variation with changing the cell size of edge detection circuit for (pixel × pixel) resolution for train and test images for databases AR, ORL. It can be noted that a 6 × 6 cell size is optimum for maximum accuracy.

optimum amount of resolution for maximum accuracy. In our experiments an image resized to 80 × 80 gives the best results for all the three databases. This is a useful result in that it indicates the concentration of useful information within fewer pixels and thus tackles the need for large amounts of data for recognition, resulting in higher efficiency in terms of speed of operation. Another important result is the maximum block size of pixels during the edge detection with the threshold logic network gives maximum accuracy. It was found that a block size of 6 × 6 gave the best accuracy on a random 10 class AR and 40 class ORL databases. This thresholding result is depicted in Fig. 18.12. An accuracy increase of up to an optimal block size value of (18.6) was observed, after which it started to decline. The reason for this decrease is that as the window size increases the number of inputs per cell increases for the edge detection circuit so that the cell needs to set the memristances and detect changes (from 0-to-1

TABLE 18.1 Performance analysis of the circuit for template formation.

Dimension	Area (m^2)	Power dissipation (W)	Leakage power (W)
3 × 3	677.37 μ	145.87 n	333.45 p
10 × 10	7500 μ	1.62 μ	3.71 n
100 × 100	0.75 m	162.08 μ	370.50 n
1000 × 1000	75.26 m	16.21 m	37.05 μ

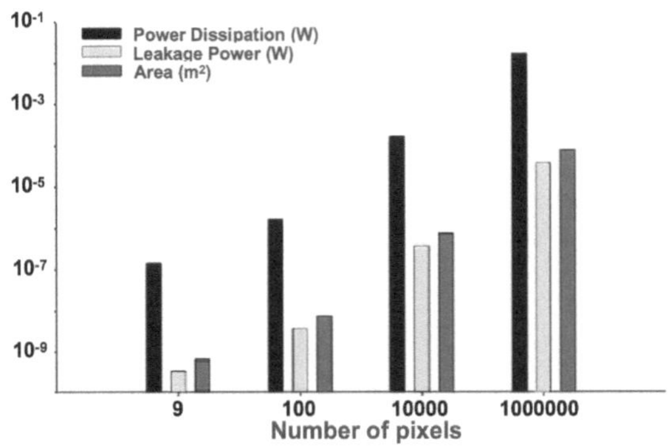

FIGURE 18.13 Variation of on-chip area, power dissipation, leakage power shown as a function of the number of pixels processed in the template formation stage. The fact that all bits of all pixels have their individual circuit, the power dissipation, area, delay and leakage power for images of any dimension scale linearly.

or 1-to-0) of a more closely spaced pixel values. Also increasing the number of inputs to a cell will decrease the number of parallel edge detection cells, so limiting the inputs to an optimal size is in accordance with the circuits' parallel architecture. Area density and power dissipation are important implications of this result of the threshold logic network.

Among the databases, faces94 database shows the best results, followed by ORL and AR. AR is arguably the more difficult of the databases owing to its design for unconstrained face recognition under varying occlusion and lighting conditions.

Table 18.1 shows how the template formation part of the circuit performs. A template circuit of Fig. 18.3 is made for each one of the 8 bits corresponding to each pixel. For this single bit circuit the area, power dissipation and leakage power and delay have been found to be resp. 9.41 μm^2, 2.03 nW and 4.60 pW and 2.55 ns. The linear scaling of the template formation circuit is demonstrated by the circuits ability to perform in parallel as shown in Fig. 18.13.

Tables 18.2 and 18.3 show the performance analysis of the circuits for edge detection and template formation with edge detection circuits, respectively. Again it is noted that the area on-chip and power dissipation vary linearly with pixel resolution.

TABLE 18.2 Performance analysis of the circuit for edge detection.

Dimension	Area (m^2)	Power dissipation (W)
3 × 3	884.7 μ	0.625 m
10 × 10	9830 μ	6.9 m
100 × 100	0.98 m	0.69
1000 × 1000	98.30 m	69.46

TABLE 18.3 Performance analysis of the circuit for template formation with edge detection.

Dimension	Area (m^2)	Power dissipation (W)
3 × 3	402.75 μ	0.285 m
10 × 10	447 μ	3.168 m
100 × 100	0.45 m	0.32
1000 × 1000	44.75 m	31.68

TABLE 18.4 Performance analysis of the circuit for face recognition.

Dimension	Area (m^2)	Power dissipation (W)	Leakage power (W)
3 × 3	18.80 μ	15.17 n	9.4 p
10 × 10	117.52 μ	75.51 n	58.7 p
100 × 100	11751.5 μ	7.55 μ	5.88 n
1000 × 1000	1.175 m	0.76 m	0.59 μ

For the face-recognition part of the network, the performance analysis is shown in Table 18.4. A 4-input threshold logic cell constitutes the basic cell. The area, power dissipation and leakage power of this basic cell are 4.7 μm^2, 3.88 nW and 2.35 pW, respectively. The linear scaling of the circuits is demonstrated by Fig. 18.14.

18.11 Future research directions

The next logical step to the current work would be to build the systems presented here in hardware. Memristors are not yet available as off-the-shelf commercial devices and much of their fabrication and layout vis-a-vis other components such as CMOS and interconnects is still under development, which presents a challenge to realizing much of the systems researched in the literature. The presented research draws heavily from the interdisciplinary work in neurobiology and neuroscience wherein the synaptic activity in pre-synaptic and postsynaptic neurons has been modeled into a memory network of threshold logic memristive cells. It is believed that a deeper understanding of how a large but not

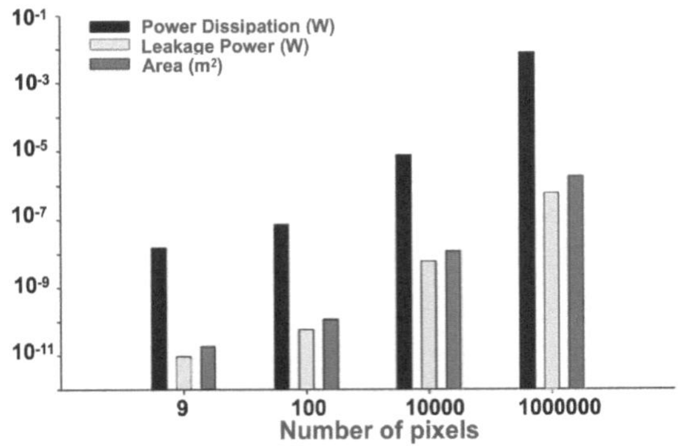

FIGURE 18.14 Variation of power dissipation, leakage power and area with the number of pixels processed in the face-recognition stage. As in the template formation stage linear scalability is achieved as the number of the pixels processed increases.

overwhelmingly large connectivity of neurons, that enables both storage and computation in biological brain, is achieved, and the electrical, chemical and feedback processes involved. The feedback mechanism calibrating the sensory input to the brain is an activity, which if incorporated in the current work, can improve the performance of the cognitive systems in hardware. This research builds memristive threshold logic memory networks for vision part of the human cognitive system. A next step in this direction would be to build MTL circuits for speech recognition.

18.12 Conclusion

The proposed reconfigurable memristor threshold logic circuit for face recognition tackles the problem of high speed, high resolution processing in real time in parallel, scalable hardware, as opposed to algorithmic approach that suffers from these challenges. The approach makes use of the signification pixel variations between consecutive frames temporally and spatially. The primitive cell proposed can handle a large number of inputs but only one output, which makes it amenable to be used in a network of cells. The memristive threshold logic is used to then memorize these significant pixel variations by setting cell parameters as per the input. Face recognition is then accomplished by calculating the overall change in the test image and comparing with the memorized template image. Through the presented approach, it may be noted that the advantages offered by a truly hardware approach include the high speed and the ability to utilize all the available image frames from the imaging sensors without compromising highly relevant and accurate object tracking in real-time images. The hardware nature of the cell architecture makes it a futuristic alternative to algorithmic approaches and highly attractive to the practical utilization of high frame rates that could lead to near-continuous real-time object tracking, and can surpass human object tracking ability. In terms of hardware implementation, this needs resistive memories that can do large number of cycles, smaller area and long data retention time which can be made possible with emerging memory technologies such as MRAM and QsRAM. Furthermore, the primitive processing element, namely the memristive threshold logic cell, requires only a small area on the chip.

18.13 Key terms and definitions

Leakage power: Power dissipated during the reverse biasing of diffusion regions and wells (for e.g., p-type diffusion vs. n-well), wells and substrate (for e.g., n-well vs. p-substrate) in CMOS.

Memristor: Theoretically, a fundamental electrical device analogous to a resistor, capacitor and inductor which links flux and charge flowing through it. For example, a class of nanoscale solid-state devices (a.k.a. memristive switching devices) formed by doping a thin layer of an insulator (for e.g. TiO_2) between two metal electrodes (for e.g. platinum), that change its memristance according to the history of applied voltage.

Neuromorphic computing: A term coined by Carver Mead in his seminal 1990 paper to describe the use of very-large-scale integration (VLSI) systems to mimic neuro-biological architectures present in the nervous system.

Template matching: is a technique used in digital image processing applications and more generally in pattern recognition tasks for finding parts of an image (pattern) which match a template image.

Testing: Testing is a broad term used for the many ways an unknown label of a set of data points is determined by applying them to a function learned from the labeled set of points.

Threshold logic: Login in which an output signal is activated when the weight of the inputs reach a threshold.

Training: The process of realizing an appropriate function from labeled set of data points.

References

[1] A.F. Abate, M. Nappi, D. Riccio, G. Sabatino, 2D and 3D face recognition: a survey, Pattern Recognition Letters 28 (14) (2007) 1885–1906.

[2] Y. Adini, Y. Moses, S. Ullman, Face recognition: the problem of compensating for changes in illumination direction, IEEE Transactions on Pattern Analysis and Machine Intelligence 19 (7) (1997) 721–732.

[3] T. Ahonen, A. Hadid, M. Pietikainen, Face description with local binary patterns: application to face recognition, IEEE Transactions on Pattern Analysis and Machine Intelligence 28 (12) (2006) 2037–2041.

[4] M. Aldwairi, T. Conte, P. Franzon, Configurable string matching hardware for speeding up intrusion detection, ACM SIGARCH Computer Architecture News 33 (1) (2005) 99–107.

[5] ARDb, Retrieved from http://www2.ece.ohiostate.edu/aleix/ARdatabase.html, 2014, May 1.

[6] Z.K. Baker, V.K. Prasanna, Time and area efficient pattern matching on FPGAs, in: Proceedings of the 2004 ACM/SIGDA 12th International Symposium on Field Programmable Gate Arrays, ACM, 2004, February, pp. 223–232.

[7] P.N. Belhumeur, J.P. Hespanha, D.J. Kriegman, Eigenfaces vs. fisherfaces: recognition using class specific linear projection, IEEE Transactions on Pattern Analysis and Machine Intelligence 19 (7) (1997) 711–720.

[8] Z. Biolek, D. Biolek, V. Biolkova, SPICE model of memristor with nonlinear dopant drift, Radioengineering 18 (2) (2009) 210–214.

[9] V. Blanz, T. Vetter, Face recognition based on fitting a 3D morphable model, IEEE Transactions on Pattern Analysis and Machine Intelligence 25 (9) (2003) 1063–1074.

[10] Z. Cao, Q. Yin, X. Tang, J. Sun, Face recognition with learning-based descriptor, in: Computer Vision and Pattern Recognition (CVPR), 2010 IEEE Conference on, IEEE, 2010, June, pp. 2707–2714.

[11] L.O. Chua, Memristor-the missing circuit element, IEEE Transactions on Circuit Theory 18 (5) (1971) 507–519.

[12] L. Chua, Resistance switching memories are memristors, Applied Physics A 102 (4) (2011) 765–783.

[13] C.R. Clark, D.E. Schimmel, Scalable pattern matching for high speed networks, in: Field-Programmable Custom Computing Machines, 2004. FCCM 2004. 12th Annual IEEE Symposium on, IEEE, 2004, April, pp. 249–257.

[14] S.M. Crouzet, H. Kirchner, S.J. Thorpe, Fast saccades toward faces: face detection in just 100 ms, Journal of Vision 10 (4) (2010) 16.

[15] K. Eshraghian, O. Kavehei, K.R. Cho, J.M. Chappell, A. Iqbal, S.F. Al-Sarawi, D. Abbott, Memristive device fundamentals and modeling: applications to circuits and systems simulation, Proceedings of the IEEE 100 (6) (2012) 1991–2007.

[16] faces94, Retrieved from http://www.face-rec.org/databases/, 2014, May 1.

[17] L. Gao, F. Alibart, D.B. Strukov, Programmable CMOS/memristor threshold logic, IEEE Transactions on Nanotechnology 12 (2) (2013) 115–119.

[18] A.S. Georghiades, P.N. Belhumeur, D.J. Kriegman, From few to many: illumination cone models for face recognition under variable lighting and pose, IEEE Transactions on Pattern Analysis and Machine Intelligence 23 (6) (2001) 643–660.

[19] G. Indiveri, Neuromorphic analog VLSI sensor for visual tracking: circuits and application examples, IEEE Transactions on Circuits and Systems. 2, Analog and Digital Signal Processing 46 (11) (1999) 1337–1347.

[20] S.H. Jo, T. Chang, I. Ebong, B.B. Bhadviya, P. Mazumder, W. Lu, Nanoscale memristor device as synapse in neuromorphic systems, Nano Letters 10 (4) (2010) 1297–1301.

[21] J. Kim, J. Choi, J. Yi, M. Turk, Effective representation using ICA for face recognition robust to local distortion and partial occlusion, IEEE Transactions on Pattern Analysis and Machine Intelligence 27 (12) (2005) 1977–1981.

[22] K.C. Lee, J. Ho, M.H. Yang, D. Kriegman, Video-based face recognition using probabilistic appearance manifolds, in: Computer Vision and Pattern Recognition, 2003. Proceedings. 2003 IEEE Computer Society Conference on, Vol. 1, IEEE, 2003, June, I-313 pp.

[23] P. Lichtsteiner, C. Posch, T. Delbruck, A 128×128 120 dB 15 μs latency asynchronous temporal contrast vision sensor, IEEE Journal of Solid-State Circuits 43 (2) (2008) 566–576.

[24] K.K. Likharev, Hybrid CMOS/nanoelectronic circuits: opportunities and challenges, Journal of Nanoelectronics and Optoelectronics 3 (3) (2008) 203–230.

[25] C.H. Lin, C.T. Huang, C.P. Jiang, S.C. Chang, Optimization of regular expression pattern matching circuits on FPGA, in: Design, Automation and Test in Europe, 2006. DATE'06. Proceedings, Vol. 2, IEEE, 2006, pp. 1–6.

[26] A.K. Maan, D.S. Kumar, A.P. James, Memristive threshold logic face recognition, Procedia Computer Science 41 (2014) 98–103.

[27] H. Moon, P.J. Phillips, Computational and performance aspects of PCA-based face-recognition algorithms, Perception-London 30 (3) (2001) 303–322.

[28] H. Murase, S.K. Nayar, Visual learning and recognition of 3-D objects from appearance, International Journal of Computer Vision 14 (1) (1995) 5–24.

[29] ORLDb, Retrieved from http://www.cl.cam.ac.uk/research/dtg/attarchive/facedatabase.html, 2014, May 1.
[30] A. Pappachen James, A. Pachentavida, S. Sugathan, Edge detection using resistive threshold logic networks with CMOS flash memories, International Journal of Intelligent Computing and Cybernetics 7 (1) (2014) 79–94.
[31] P.J. Phillips, H. Moon, S. Rizvi, P.J. Rauss, The FERET evaluation methodology for face-recognition algorithms, IEEE Transactions on Pattern Analysis and Machine Intelligence 22 (10) (2000) 1090–1104.
[32] D.B. Strukov, H. Kohlstedt, Resistive switching phenomena in thin films: materials, devices, and applications, MRS Bulletin 37 (02) (2012) 108–114.
[33] X. Tan, B. Triggs, Enhanced local texture feature sets for face recognition under difficult lighting conditions, IEEE Transactions on Image Processing 19 (6) (2010) 1635–1650.
[34] S. Thorpe, D. Fize, C. Marlot, Speed of processing in the human visual system, Nature 381 (6582) (1996) 520–522.
[35] J. Touryan, L. Gibson, J.H. Horne, P. Weber, Real-time measurement of face recognition in rapid serial visual presentation, Frontiers in Psychology 2 (2011).
[36] M. Turk, A.P. Pentland, Face recognition using eigenfaces, in: IEEE Computer Society Conference on Computer Vision and Pattern Recognition, 1991, Proceedings CVPR'91, IEEE, 1991, June, pp. 586–591.
[37] M. Van den Bergh, L. Van Gool, Combining RGB and ToF cameras for real-time 3D hand gesture interaction, in: Applications of Computer Vision (WACV), 2011 IEEE Workshop on, IEEE, 2011, January, pp. 66–72.
[38] R. Waser, R. Dittmann, G. Staikov, K. Szot, Redox-based resistive switching memories–nanoionic mechanisms, prospects, and challenges, Advanced Materials 21 (2009) 2632–2663.
[39] L. Wiskott, J.M. Fellous, N. Kuiger, C. Von Der Malsburg, Face recognition by elastic bunch graph matching, IEEE Transactions on Pattern Analysis and Machine Intelligence 19 (7) (1997) 775–779.
[40] J. Wright, A.Y. Yang, A. Ganesh, S.S. Sastry, Y. Ma, Robust face recognition via sparse representation, IEEE Transactions on Pattern Analysis and Machine Intelligence 31 (2) (2009) 210–227.
[41] C.Y. Wu, C.T. Chiang, A low-photocurrent CMOS retinal focal-plane sensor with a pseudo-BJT smoothing network and an adaptive current Schmitt trigger for scanner applications, IEEE Sensors Journal 4 (4) (2004) 510–518.
[42] Y. Xu, D. Zhang, J. Yang, J.Y. Yang, A two-phase test sample sparse representation method for use with face recognition, IEEE Transactions on Circuits and Systems for Video Technology 21 (9) (2011) 1255–1262.
[43] M. Yang, L. Zhang, Gabor feature based sparse representation for face recognition with Gabor occlusion dictionary, in: Computer Vision–ECCV 2010, Springer Berlin Heidelberg, 2010, pp. 448–461.
[44] W. Zhang, S. Shan, W. Gao, X. Chen, H. Zhang, Local Gabor binary pattern histogram sequence (lgbphs): a novel non-statistical model for face representation and recognition, in: Computer Vision, 2005. ICCV 2005. Tenth IEEE International Conference on, Vol. 1, IEEE, 2005, October, pp. 786–791.
[45] W. Zhao, R. Chellappa, P.J. Phillips, A. Rosenfeld, Face recognition: a literature survey, ACM Computing Surveys 35 (4) (2003) 399–458.

Chapter 19

Synaptic devices based on HfO_2 memristors

M.B. González[a], M. Maestro-Izquierdo[a], S. Poblador[a], M. Zabala[a], F. Campabadal[a], G. González-Cordero[b], S. Aldana[b], D. Maldonado[b], F. Jiménez-Molinos[b], and J.B. Roldán[b]

[a]*Institute of Microelectronics of Barcelona, IMB-CNM (CSIC), Bellaterra, Spain,* [b]*Electronics and Computer Technology Department, Science Faculty, Granada University, Granada, Spain*

19.1 Introduction

Artificial intelligence (AI) systems based on brain-inspired (neuromorphic) computing will be able to replicate in the future brain functions, such as learning from experience, reasoning or decision-making mechanisms [48], [122]. These systems exhibit a high potential for applications in cognitive tasks as visual pattern recognition, medical sensors and self-driving cars. In electronic AI systems, artificial neurons can be typically implemented using CMOS technology. However, there are no industrial solutions so far for the electronic synapses to connect pre- and post-neurons. Recently, it has been shown that nanodevices based on resistive switching structures (composed of Metal-Insulating-Metal structures, MIM) could be suitable devices to mimic biological synapses due to the analog control of the device resistance (synaptic strength), their scaling capability, and their low power operation [6,48,53]. Besides that, resistive switching devices should satisfy the synaptic biological learning rules, being the processes that adjust the connection strengths (known as weights in artificial neural networks) between neurons based on the relative timing of an individual neuron output and input action potentials (or spikes) [99,118,120]. Currently, the most suitable materials to mimic the biological synapses are being explored; among the different alternatives, high-k dielectrics have shown highly promising synaptic properties due to their excellent ability to modulate their electrical resistance and learning capabilities [6,18,48,69].

In this chapter, the conductance tunability of HfO_2-based memristors, which can be attributed to field and temperature assisted oxygen vacancy/ion migration processes, will be investigated. For this purpose, a dedicated electrical characterization in combination with physical simulations will be presented, and the influence of several parameters such as the pulse amplitude and the number of pulses on the device conductance will be shown. The results indicate that the

Mem-elements for Neuromorphic Circuits with Artificial Intelligence Applications
https://doi.org/10.1016/B978-0-12-821184-7.00028-1

intrinsic series resistance and the applied electric field are two key parameters to control the resistive switching dynamics and the conductance modulation capability of these devices. Next, the reliability of HfO_2-based synaptic devices will be discussed. The energy consumption per synaptic event and the non-ideal asymmetry related to the linearity of the synaptic weight update will be evaluated. Furthermore, different physical simulation and modeling approaches for the devices analyzed here will be introduced, where Kinetic Monte Carlo algorithms are considered and several compact models with different degrees of complexity are studied. These models include the redox reactions that control resistive switching and the thermal effects. Finally, the appearance of current fluctuations due to random telegraph noise and several numerical techniques available for the analysis of this type of noise signals will also be assessed.

19.2 HfO_2-based resistive switching structures

Nanodevices based on CMOS compatible HfO_2-based resistive switching (or memristive) structures are being widely investigated as promising candidates for non-volatile resistive random access memory (RRAM) applications, due to their potential device scaling, fast switching speed, low power operation, and high density 3D integration [5,37,46,85]. In the last years, these devices that exhibit a typical pinched hysteresis current-voltage loop, are also being explored as a potential technology to mimic biological synapses in neuromorphic networks [48,55,100]. This is due to their scalable size within a crossbar array, large on/off ratio and the capability to tune their resistance state (synaptic weight in biological synapses). However, at the same time, ideal synaptic characteristics would require an improvement in the stability of the resistive state, a low programming energy consumption and the linearity tuning of the synaptic weight update [48,118].

Resistive switching structures are mostly built by Metal-Insulator-Metal (MIM) or Metal-Insulator-Semiconductor (MIS) devices that show the resistive switching (RS) behavior. The RS phenomenon consists in a nonvolatile sudden change of the electrical resistance as a result of the application of a voltage or electric current. The main physical mechanism responsible for the RS behavior is the formation and partial disruption of a conductive filament connecting the top and bottom electrodes [12]. In the case of filamentary resistive switching structures based on the Valence Change Mechanism (VCM), the RS is triggered by electrochemical reduction-oxidation (redox) processes and electric-field-assisted oxygen anion migration [7,19,79]. These processes induce a local change in the stoichiometry of the dielectric layer and lead to the creation of an oxygen deficient conductive nanofilament. Among various alternatives, VCM RS devices based on HfO_2 have shown highly reliable resistive switching properties [37].

A typical description of HfO_2-based resistive switching devices is shown below. MIM devices consisting of TiN/10 nm-Ti/10 nm-HfO_2/W structures were

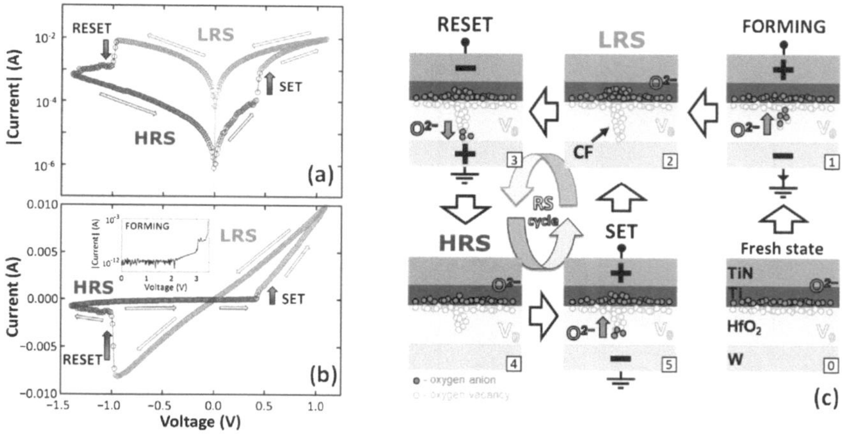

FIGURE 19.1 Typical current-voltage resistive switching characteristics of TiN/Ti/HfO_2/W VCM-based memristor devices in log (a) and linear (b) scales. The inset in (b) shows the typical current-voltage characteristics of the forming process in fresh devices. (c) Illustration of the physical mechanism responsible for the forming and RS processes.

fabricated on silicon wafers [87,88]. The 10 nm-thick HfO_2 layer was deposited by atomic layer deposition (ALD) at 225 °C using TDMAH and H_2O as precursors and N_2 as carrier and purge gas. The 10 nm-Ti capping layer acts as an oxygen scavenging layer on top of HfO_2. Both electrodes were deposited by magnetron sputtering and patterned by photolithography. The fabricated devices are square cells with sizes of $5 \times 5\ \mu m^2$ and $15 \times 15\ \mu m^2$.

The RS assessment is performed by applying sequential voltage sweeps or trains of pulses. Fig. 19.1 shows the typical current-voltage (I-V) characteristics of the forming process and the resistive switching behavior by applying the voltage sweeps to the top electrode with the bottom electrode grounded. Fig. 19.1c shows a schematic representation of the physical mechanisms behind these processes and how the Ti capping layer, with high oxygen affinity, enhances the RS capability [21,79]. After device processing (in the fresh state), oxygen vacancies (V_O) are present in the adjacent region of the Ti oxygen scavenging layer. Prior to the resistive switching behavior, the fresh devices need to be formed in order to create a V_O rich filamentary path [77]. The forming process is performed by applying a positive voltage ramp, leading the device to the low resistance state (LRS). In this step, the current is limited with a compliance value to prevent a hard dielectric breakdown of the insulator. After that, the device can be switched to a high resistance state (HRS) by applying a negative voltage ramp (RESET). The RESET process is attributed to the drift of oxygen anions by the electric field and the recombination of oxygen vacancies at the tip of the conductive filament (CF). Finally, a positive voltage ramp is applied (SET), and the filamentary path is restored, leading the device again to the LRS. This sequence constitutes a complete RS cycle. Fig. 19.2 shows 3000 sequential RS

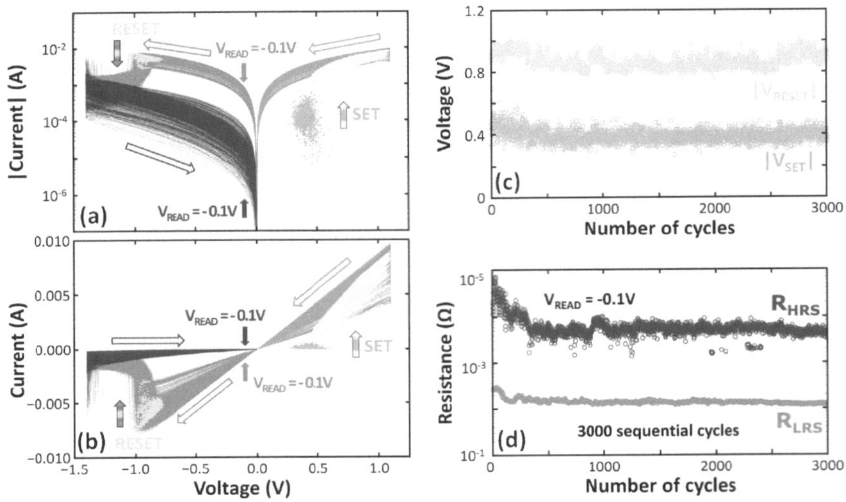

FIGURE 19.2 Current-voltage characteristics of 3000 sequential bipolar sweep cycles in logarithmic (a) and linear (b) scales. The SET/RESET events of each cycle are highlighted in orange/green (light gray/mid gray in print version). (c) Evolution of the SET and RESET voltages as a function of the cycle number. (d) Resistance values measured at -0.1 V in the HRS (blue, dark gray in print version) and LRS (red, gray in print version) of the devices versus the number of cycles.

cycles in logarithmic and linear scales by applying voltage sweeps from -1.4 V to $+1.1$ V. The SET/RESET transitions are associated to the field and temperature assisted oxygen vacancy generation, recombination and ion drift processes (Fig. 19.1c) [7,19,79]. As shown in Fig. 19.2b, the current is linearly dependent on voltage in the LRS, indicating an ohmic conduction [17] due to the metallic behavior of the fully formed CF. In Fig. 19.2c, the evolution of the SET/RESET voltages as a function of the number of cycles is represented showing a low cycle-to-cycle variability. In Fig. 19.2d, the evolution of the resistance in the LRS and HRS measured at low field, $V_{READ} = -0.1$ V, is represented. It can be observed that the cycle-to-cycle variability is larger in the HRS than in the LRS. This fact can be explained by the cyclical variations of the gap distances between the CF filament tip and the metal electrode in the HRS [63,73].

The RS behavior obtained by applying trains of sequential pulses is also proved in Ti/HfO_2 VCM-based resistive switching devices (Fig. 19.3). The current-voltage characteristics of one pulse sequence are represented in Fig. 19.3a. The maximum and minimum voltages applied to the device in pulse train operation are programmed to be the same as for a sweep cycle, being in the studied devices $+1.1$ V and -1.4 V, respectively. In Fig. 19.3b, the evolution of R_{HRS} and R_{LRS} at $V_{READ} = -0.1$ V for more than 10^5 sequential cycles under pulse train operation is given. Notice that in pulse mode, the resistance values for both resistance states are similar to those in the sweep mode, while variability is slightly reduced. The low cycle-to-cycle variability of the switch-

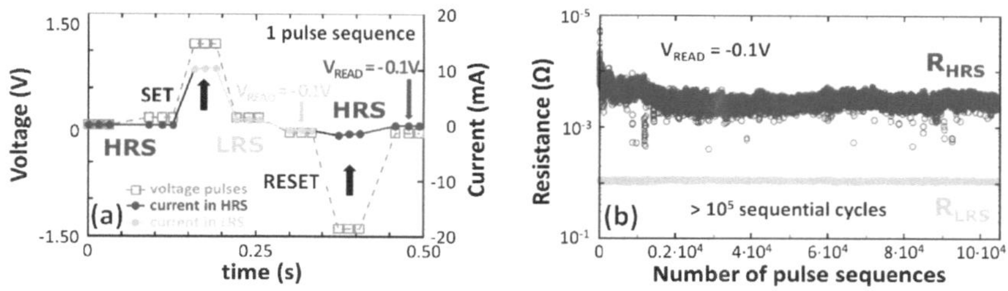

FIGURE 19.3 Applied voltage pulse scheme including one SET and one RESET operations versus time and the corresponding measured current in TiN/Ti/HfO_2/W VCM-based memristors. A read process is performed before and after the RESET pulse. (b) Resistance values measured at $V_{READ} = -0.1$ V at the HRS (blue, dark gray in print version) and LRS (red, gray in print version) of the devices for more than 10^5 sequential cycles.

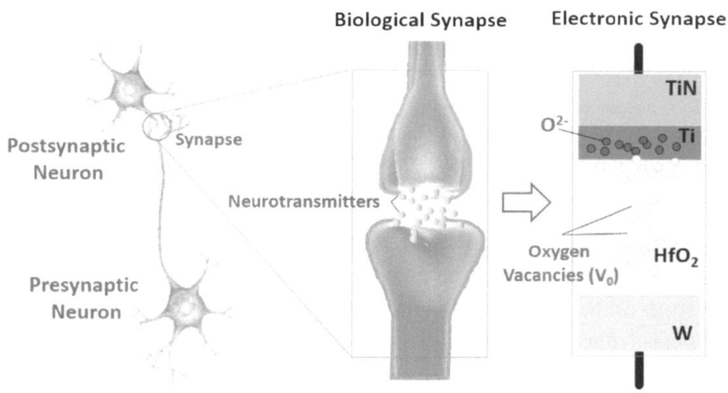

FIGURE 19.4 Illustration of the biological synapse that acts as a connection between two neurons and the VCM memristors studied to mimic the synaptic functionality.

ing parameters and resistances, and the large R_{HRS}/R_{LRS} ratio of more than one decade indicate the potential of these devices to be studied as electronic synapses in neuromorphic circuits.

19.3 Demonstration of learning rules in memristor devices to mimic biological synapses

Several publications have investigated the suitability of resistive switching devices processed with different metal/dielectric combinations [6,48,53] and specifically those based on HfO_2 [18,23,69,80,84] to act as artificial synapses. Fig. 19.4 shows an illustration of the analogy between a biologic synapse and a TiN/Ti/HfO_2/W VCM-based memristor in terms of the electrochemical reactions taking place. In biological systems, a neuron transmits and processes information through electrical signals (also called action potentials or spikes). The transmission of an action potential from the pre-synaptic neuron is weighted

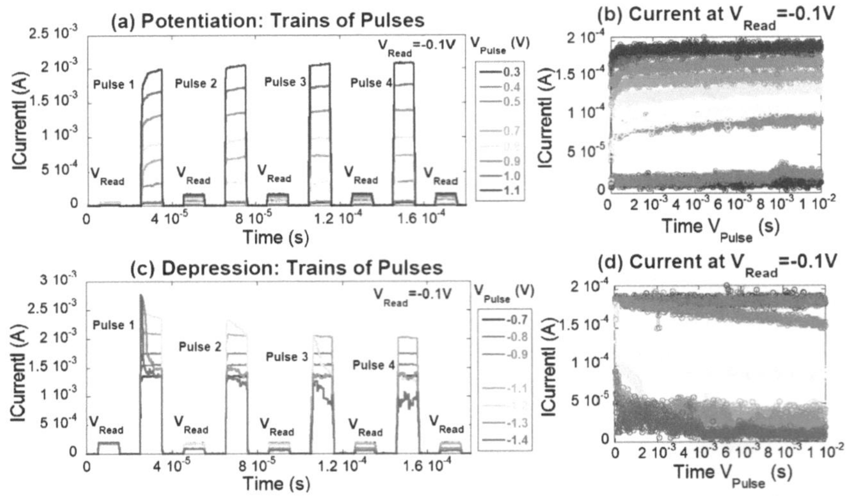

FIGURE 19.5 Measured current during sequential trains of (a) SET and (c) RESET pulses with different amplitudes. (b) Potentiation and (d) depression characteristics measured at $V_{READ} = -0.1$ V as a function of time at V_{Pulse} for 1000 trains of pulses of different amplitudes.

by the synaptic weight strength and converted into a chemical signal through the release of neurotransmitters. The post-synaptic neuron integrates all the signals received from all the connected synapses, and if the result exceeds a threshold, then the post-synaptic neuron will fire and transmit a post-synaptic action potential. The amount of neurotransmitters released into the synapse is quantified by the synaptic weight that can be strengthened (potentiated) or weakened (depressed) when the pre-synaptic and post-synaptic neurons fire in short relative timings. The capability of a synapse to adjust the connection force between neurons (or synaptic weight) is known as synaptic plasticity. In HfO_2 memristors the device resistance can be modulated by the applied voltage due to oxygen ion migration process between reactive Ti oxygen scavenging layer and HfO_2 switching layer (Fig. 19.4). However, although Ti/HfO_2-based memristors are currently being intensively investigated to evaluate their synaptic properties, the implementation of these devices in neuromorphic circuits is still a challenge even at device level, where several reliability issues need to be addressed, such as the linearity of the weight update and the programming energy consumption.

In order to evaluate the analog behavior and plasticity of the devices, trains of SET and RESET pulses with different amplitudes were applied to memristors left at the HRS (Fig. 19.5a), and at the LRS (Fig. 19.5c). The evolution of the current measured at $V_{READ} = -0.1$ V as a function of the accumulated stressing time in the V_{pulse} is shown in Fig. 19.5b for the potentiation case (positive pulse amplitudes) and in Fig. 19.5d for the depression case (negative pulse amplitudes). Notice that the capability of the memristors to potentiate or depress are essential features to mimic the synaptic functionality in biological-inspired

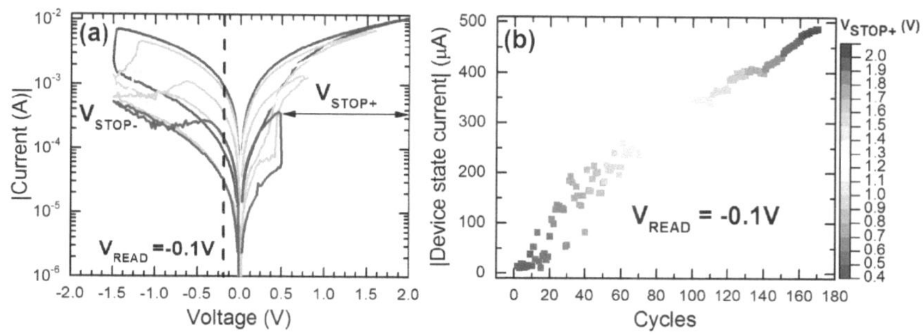

FIGURE 19.6 (a) Typical experimental RS characteristics of the studied memristors using several V_{STOP+} voltages in successive cycles in one device. The results show a progressive and self-limiting SET switching. (b) Current extracted at −0.1 V from the I-V curves in (a) for several V_{STOP+} conditions. A continuous conductance modulation capability of the devices with the applied voltage is obtained.

neuromorphic networks [23]. A total of 1000 pulses were assessed for each bias value and polarity. The results show that the higher the pulse amplitude, the lower the number of pulses required to modify the device conductance. Moreover, during the first pulses there is a strong dependence of the device conductance with time, indicating that most of the ion migration processes take place in the first pulses. Next, a saturation of the current is observed [121]. The level of saturation is also dependent on the voltage, indicating that pulses with higher pulse amplitude lead to a larger morphological and stoichiometric changes of the CF [80]. The results have shown long term potentiation (LTP) and depression (LTD) characteristics of HfO_2-based memristors and their capability to reach different conductive states. However, the results in Fig. 19.5 b, d point out that the curve between the device current and the number of identical programming pulses is not linear. In ideal electronic synapses, there should be a linearity of the synaptic weight update with time [118]. This is an undesired feature since the resistance variations of the device depend on the resistance state, or in other words, the changes in the weight state depend on its history. This non-linearity results in a low accuracy of the learning process in neuromorphic circuits [11], [16]. In addition, symmetric LTD and LTP characteristics are not typically observed in resistive switching devices [108] due to the different behavior during SET and RESET transitions (Fig. 19.1).

The capability of the fabricated devices to reach multilevel conductance states, has also been investigated by varying linearly the voltage values applied to the memristor device through voltage ramps. Fig. 19.6a shows RS characteristics of devices subjected to double sweeps by increasing each cycle 0.1 V the stop positive voltage (V_{STOP+}) from 0.5 V to 2 V and fixing the stop negative voltage (V_{STOP}) at −1.5 V. The experimental results demonstrate that intermediate states can be achieved by tuning the voltage amplitude. Fig. 19.6b shows the current measured at $V_{READ} = -0.1$ V with the number of cycles, indicating a continuously modulation capability of the conductance state for more than two decades with the applied voltage instead of the fast saturation trend observed under pulse schemes (Fig. 19.5). The reported results show the improved capability of HfO_2-based devices to control their conductivity by linearly varying the applied voltage [65,88].

The synaptic plasticity of resistive switching devices according to spike timing has been widely studied to emulate the learning mechanisms of biological synapses [54,89,97]. Among the synaptic weight update rules, the most studied is the spike-timing dependent plasticity (STDP), which has been experimentally demonstrated in HfO_2-based resistive memristors [18,65,69].

This mechanism can be tested in these devices as the memristor conductivity can be electrically tuned by the applied bias. The STDP mechanism depends on the arrival time of the spikes from pre and post-synaptic neurons. The STDP rule adjusts the synaptic weight between neurons according to the relative timing ($\Delta t = t_{post} - t_{pre}$) between the output and input spikes of a neuron. Fig. 19.7a shows the waveforms employed during STDP testing for different relative timings (Δt) in the range of ms (as biological synapses). The pre-spike shape is indicated in blue (dark gray in print version) and the corresponding ahead and delayed post-spikes are indicated in green and orange (mid gray and light gray in print version), respectively. Notice that the resultant square pulses applied to the memristors are negative (Fig. 19.7b, d) for $\Delta t < 0$ (post-spike preceding the pre-spike) resulting in the depression of the device state. In the case of $\Delta t > 0$ (pre-spike preceding the post-spike) the resultant square pulses are positive (Fig. 19.7c, e) and the device state will be potentiated. As shown in Fig. 19.7b-e, the shorter the $|\Delta t|$, the larger the resultant pulse amplitude.

The obtained experimental spike-time dependent plasticity (STDP) behavior is represented in Fig. 19.7f. The relative conductance variations of the device state as a function of the relative time are represented by the ratio between the relative change in conductance to the initial conductance state ($\Delta G/G_{\text{INITIAL}}$), being $\Delta G = G_{\text{FINAL}} - G_{\text{INITIAL}}$. The experimental STDP in resistive switching devices can be analyzed according to the biological trend that considers the exponential behavior for the synaptic weight update following Eq. (19.1) [99, 120]

$$\begin{cases} \dfrac{\Delta G}{G_{\text{INITIAL}}} = +A_+ e^{-\Delta t/\tau_+} \text{ for } \Delta t > 0 \\ \dfrac{\Delta G}{G_{\text{INITIAL}}} = -A_- e^{+\Delta t/\tau_-} \text{ for } \Delta t < 0 \end{cases} \tag{19.1}$$

where τ_+ and τ_- are the time parameters determining the range of relative times in which potentiation and depression take place, and A_+ and A_- are the scaling parameters that indicate the maximum synaptic weight variations during potentiation and depression processes, respectively. The results and the exponential fitting trend show a slight asymmetry in the STDP, due to the different increase in conductance variations between the potentiation and depression processes. This asymmetry can be tuned by the initial conductive state of the device [89].

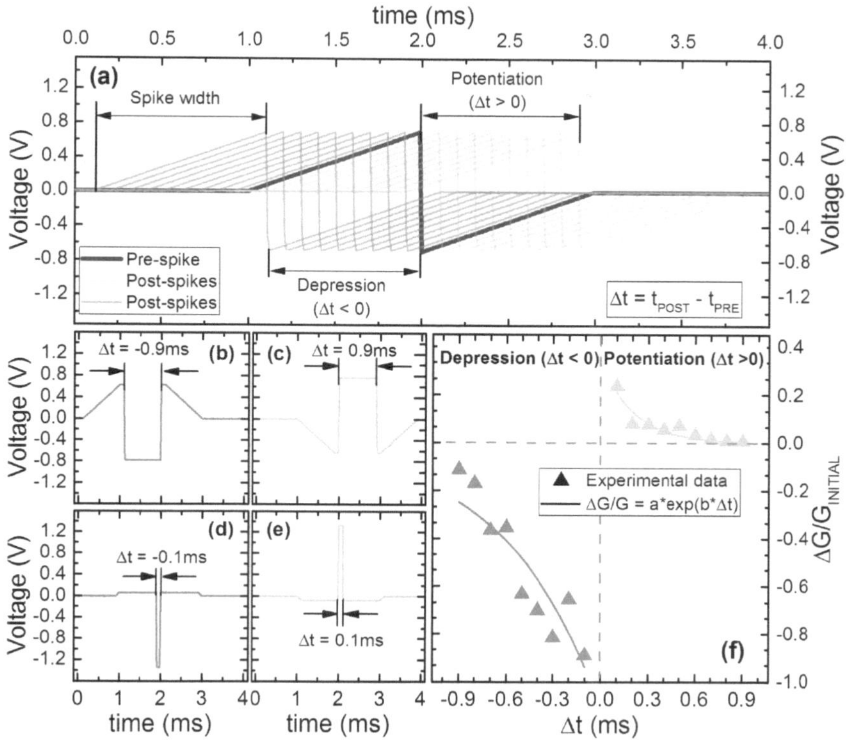

FIGURE 19.7 (a) Waveforms employed during STDP testing for different relative timings (Δt) between spikes. (b)-(e) Characteristic voltages applied to the device during (b), (d) depression (Δt < 0) and (c), (e) potentiation (Δt > 0) processes of the device state. (f) Experimental STDP data (symbols) for the studied TiN/Ti/HfO_2/W VCM-based memristors using the spike shapes shown in (a). Lines indicate the exponential fitting trend according to Eq. (19.1).

In addition, the impact of the different time scales of the spikes on the STDP behavior has been investigated from 10 ms to 100 ns [65]. The experimental results demonstrate the successful implementation of the STDP in these devices at lower time scales, allowing to minimize the energy consumption per synaptic event. Notice that the spike time and shape have to be carefully designed as these parameters can strongly impact the programming energy consumption and STDP behavior [13,120]. Another reported approach is to consider two memristors per synapse [11] or synaptic configurations based on hybrid CMOS-RRAM [109].

19.4 Stability and reliability issues of resistive synaptic devices

As synaptic elements, memristors have some critical challenges related to reliability concerns such as the stability of the synaptic state with time, the pro-

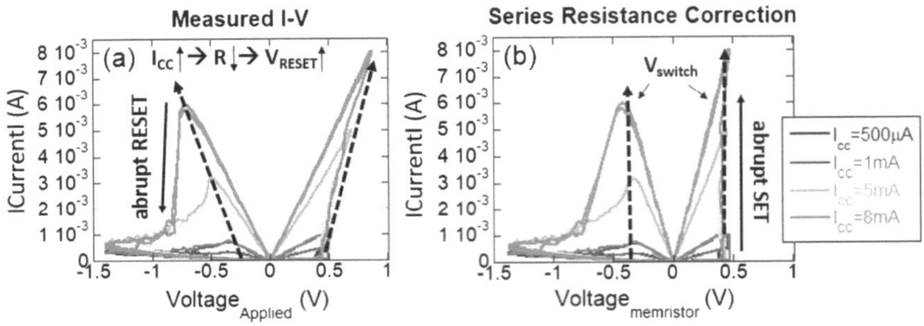

FIGURE 19.8 (a) Typical RS characteristics of a Ti/HfO_2-based memristor measured using several current limitations during the SET process and (b) extracted intrinsic RS characteristics of the studied memristors after series resistance correction. The intrinsic characteristic shows a $|V_{switch}| \sim 0.4$ V for the measured quasi-static voltage ramps.

gramming energy consumption, and the presence of parasitic effects such as the intrinsic series resistance (R_{series}). In biological synapses, the estimated energy consumption is around 10 fJ [118]. However, in resistive synaptic devices, the large programming voltages (~1 V) and programming currents involved in the RS phenomenon, significantly increase the energy consumption per synaptic event. Therefore, further optimization in the operating conditions and device characteristics is desired. The reduction of the pulse width is a powerful technique to reduce the programming energy consumption [26,44,45,72,115]. However, the intrinsic series resistance present in memristor devices and crosspoint structures could significantly affect their resistive switching dynamics [26,42].

Fig. 19.8a shows typical RS characteristics of a Ti/HfO_2-based memristor measured using several current limitations (I_{cc}) during the SET sweep process. As in the case of applying different V_{STOP+} biases (Fig. 19.6), the I_{CC} value employed influences the resistance after the SET process (Fig. 19.8a). Notice that the results reveal a gradual nature of the RESET transition at low I_{CC} values and consequently at high resistance. This behavior is typical of RS devices based on the VCM mechanism [113]. However, at high I_{cc} values, and therefore, lower resistance, the RESET transition becomes more abrupt and higher voltages are necessary to reset the device. In addition, a progressive and self-limiting SET process is observed in the measured I-V characteristics of these devices (also shown in Fig. 19.6). This behavior can be attributed to the presence of a series resistance, R_{series} [26,42] in the memristor structure. Notice that this self-limiting behavior allows the multilevel control of the resistance states by varying the voltage amplitude (Fig. 19.6). Without a R_{series}, the SET process would be expected to be more abrupt, and therefore, varying the voltage would not be able to tune the memristor resistance.

Fig. 19.8b shows the extracted intrinsic RS characteristics of the studied memristors obtained by subtracting the voltage drop across R_{series}. The results confirm that the RS characteristics after R_{series} correction present an abrupt

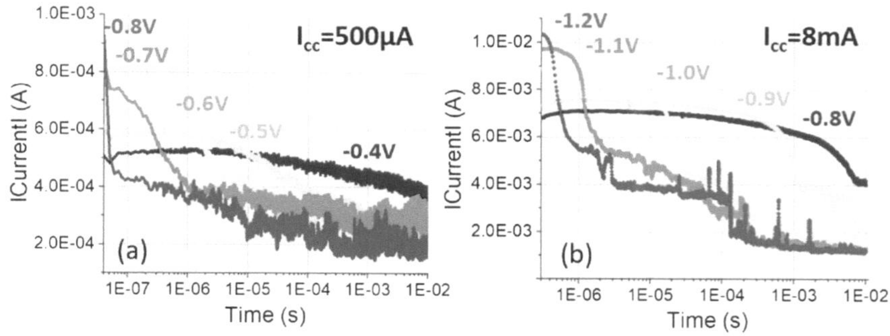

FIGURE 19.9 Current evolution as a function of time measured under constant voltage for different biases with a 10 ns sampling interval. Current transients were measured for two different initial states defined by a current compliance of (a) 500 μA and (b) 8 mA in the previous SET process.

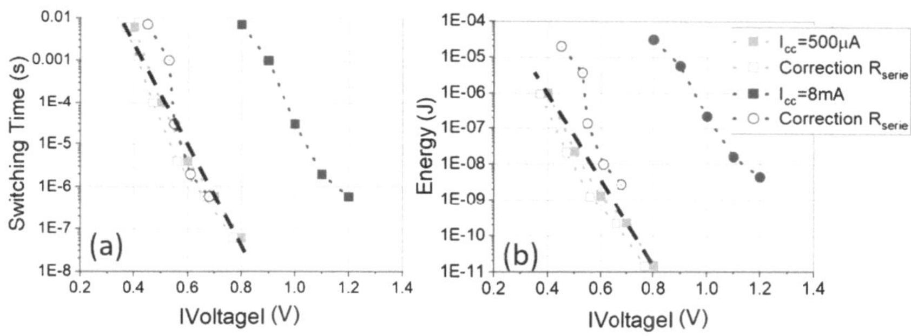

FIGURE 19.10 (a) RESET switching time and (b) RESET energy versus voltage for different initial states with (empty symbols) and without (full symbols) R_{series} correction. Initial states were defined by the current compliance of the previous SET process.

and non-self-limiting SET process with the characteristic snapback and snap-forward effects [22,42,75,112,113]. In addition, the intrinsic RS characteristics present symmetrical switching voltages (V_{switch}) for the SET and RESET processes with values $|V_{switch}| \sim 0.4$ V. Previously published work reported similar values for V_{switch} for HfO_2-based memristors under quasi-static voltage ramps [22,26,45,112]. Fig. 19.8b also reveals that $|V_{switch}|$, and therefore intrinsic switching dynamics, are not significantly dependent on the current level in the range studied.

To further understand how R_{series} affects the measured RESET voltages, current transient characteristics were recorded at several constant voltages (Fig. 19.9). The measurements were done in the same device departing from two different initial resistance states, which were defined by the I_{CC} (500 μA and 8 mA) of the previous SET process. It is observed that by applying higher voltages the time required to RESET the device exponentially decreases. This trend is important in terms of energy consumption (Fig. 19.10) [26,44,45,72,115] and can be explained by the fact that some of the main physical processes involved

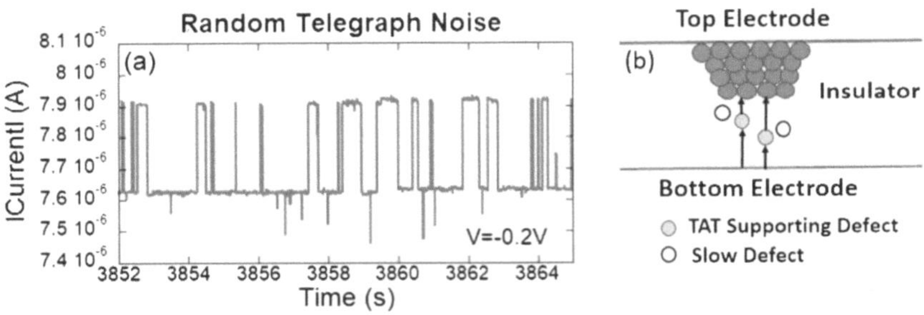

FIGURE 19.11 (a) Typical current fluctuations versus time observed in the studied Ti/HfO_2-based devices. (b) Schematic of the physical mechanism responsible for RTN fluctuations in the HRS of the devices.

in RS processes, such as oxygen ions migration and the injection of oxygen ions into the dielectric from the Ti layer, are triggered by the applied electric field by reducing their corresponding activation energy. However, it can be seen in Figs. 19.9 and 19.10 that, as a result of R_{series}, the lower the resistance of the initial state, the higher the applied voltages required to RESET the device. Fig. 19.10 shows that when a R_{series} correction is applied, the exponential dependence of the switching time and energy of the curves corresponding to different initial resistive states tend to merge at lower biases with an experimental rate of ~0.05 V/dec. This trend corresponds to the kinetics of the slowest physical processes responsible for the reversible CF formation.

Furthermore, it should be pointed out that HfO_2 VCM-based devices can suffer from some reliability problems, such as retention at high temperatures, endurance issues and stability problems [48,118]. The stability of the resistance states in RS devices has been widely studied for RRAM applications. It has been reported that the presence of reversible random telegraph noise fluctuations (RTN) can affect the stability of the resistance state [25,90,91]. This phenomenon (illustrated in Fig. 19.11) is associated with fluctuations between electronic states of defects located in the vicinity of the CF. In addition, irreversible current changes owing to trap density variations inside or near the CF may occur [25,103]. These irreversible fluctuations can be attributed to oxygen vacancy generation or recombination processes, and drift of oxygen ions, which are field-driven processes, thus influencing the shape, size, or length of the filamentary path and their stoichiometry.

19.5 Physical simulation of memristors

19.5.1 Simulation schemes for the description of memristor physics

For the study and simulation of a group of atoms –a low number of them–, different techniques can be employed such as density functional theory (DFT) and molecular dynamics (MD). These are powerful tools and allow a very accurate representation of the system physics; nevertheless, they are computationally costly [111]. In this kind of approaches, the interaction between particles is described by a force field, the accuracy of the method depends strongly on the selection of this force field. However, these methods present a clear limitation, very short steps are required, in the order of the atomic vibration ($\sim 10^{-15}$ s). Therefore, some physical processes, such as a complete resistive switching cycle in a memristor simulation, cannot be dealt with since a long computing time is needed [107].

In the context of device simulation, the opposite case to the methodologies explained above is represented by continuum models that include the self-consistent solution of partial differential equations. These equations (continuity, Poisson and heat equations, among others) describe the device at a macroscopic level in a previously determined computational domain. In this case different parameters are used to characterize the materials. To discretize the simulation domain and solve the partial differential equations there are conventional numerical techniques based on the finite element method (FEM), the finite difference method (FDM) or the finite volume method (FVM). An intermediate solution between the accurate procedures, named as *ab initio* since they are based on first principles, with a high computational cost, and continuum models with low computational burden and a less rigorous physical description can be found in the kinetic Monte Carlo (kMC) techniques [107]. This latter methodology reproduces the stochastic behavior of the system using random numbers and relatively simple rules to reproduce the evolution of a system through its states.

In the kMC simulation of memristors, such as the devices described in the previous sections, the majority of the physical processes involved take place on a long-time scale that DFT and MD cannot achieve. This time scale can be modeled easily within a kMC algorithm assuming transitions between equilibrium states with inactivity periods [107]. In this way, a kMC algorithm can achieve longer time scales. Memristors show a stochastic nature linked to the rupture and creation of the conductive filaments as explained above [86,96]. Kinetic Monte Carlo algorithms can reproduce this behavior in a natural manner since they are based on the random reproduction of the physical processes behind resistive switching operation. The Poisson and heat equations are solved in the simulation domain to obtain the electric potential and temperature distributions needed to calculate the probability of the processes included in the kMC algorithm. This is quite important since the temperature is responsible for the failure of the real devices and affects the mechanisms behind RS that are thermally activated; the electric field is also key since it determines the migration of charged particles and the generation of oxygen vacancies linked to RS. The kMC methodology can provide information of internal variables that cannot be obtained by experiments. Furthermore, the simulators are able to reproduce complete cycles of RS and even short RS series under ramped and pulsed voltage signals with a good correspondence to experimental data [1–4,19,38,41,70,71,79,81,82].

In this simulation approach, for the ions movement we consider that each state corresponds to a single energy valley (in terms of the ion energy states within the dielectric) and the long time between transitions arises because the particle must exceed the energy barrier to get to the adjacent energy valley. Although only one or a few atoms change their states, the entire system evolves to a new state and the latter transition is what matters from the simulation viewpoint. As the ions stay in the same state for a time long enough in comparison with the atom vibrational time, they "forget" their previous history. Therefore, the probability of moving from one state to another will not depend on its previous history, what it is known as a Markov walk. In the literature we can find some memristor kMC simulation tools [1–3,19,38,41,70,72,79,81,82]. An important feature to take into account is the simulation domain, it can be two-dimensional (2D) [38,41,71,82], three-dimensional (3D) [1–3,79] or even a hybrid 2D-3D as in [19]; this latter approach was made to optimize the computational burden. The choice of the simulation domain dimensionality implies a trade-off between accuracy and computational load. The manner Poisson and heat equations are solved and included in the simulation flow can also make a difference in these tools, even for non-kMC approaches [1–3,38,41,79,104,105]. Some simulators are prepared to deal with complete RS cycles [1–3,38,79], while others are focused in one or two processes, such as the forming and SET [41,81]. Regarding the charge conduction determination through the dielectric we also find different perspectives. Some tools make use of trap-assisted tunneling (TAT), taking into consideration charge conduction through the defects distributed in the dielectric [19,38,41,79]. Others employ a macroscopic approach (with a lower computational cost) making use of a Poole-Frenkel emission with effective parameters [1,2] to account for the possibility of different mechanisms that might happen simultaneously depending on the voltage regime: thermionic emission, direct tunneling, Fowler-Nordheim... Once the CF is fully formed, different charge transport schemes can be considered. It is assumed in this case that the main conduction mechanism is ohmic, through the CF [2,3,19,74,104,105]; in some cases the contribution of the set-up and the Maxwell resistances [1,2,104,105] in an equivalent circuit that models the complete device is considered. For certain devices that show non-linearity in the I-V curves at low voltages, the Quantum Point Contact (QPC) model is introduced [1,95,105].

19.5.2 Kinetic Monte Carlo simulation approach

For the implementation of the kMC algorithm all the rate constants of the processes involved in the study of the system evolution have to be calculated. The Transition State Theory (TST) is used [41]. The transition rates can be calculated as follows [1,107]:

$$\Gamma = \nu \exp\left(-\frac{E_A}{k_B T}\right) \tag{19.2}$$

where ν stands for the vibration constant, E_A is the energy barrier height, k_B the Boltzmann constant and T the temperature. This value corresponds to the inverse of the time needed for a determined event to occur. Once we have all the transition rates, we can weigh each process statistically and calculate an optimized iteration time t, the average time that takes at least one event to occur [1,41]:

$$t = -\frac{\ln(1 - randm)}{\sum \Gamma} \tag{19.3}$$

where *randm* is a random number between 0 and 1. The time scale depends on the rate constants, and they are linked to the energy barriers that characterize the physical processes considered in the simulation. Some of these barriers can be affected by the electric field and, obviously, by the temperature. This implies that the dominant physical mechanism and, so, the time scale of the simulation approach can change as the external device voltage and the temperature evolve.

Since the temperature and the electric field distributions are important, it is essential an accurate description of these variables. The solution of Poisson and heat equations in a 3D simulation domain is recommended where both the dielectric and the electrodes are included; however, the computational cost of each section of the simulator (in this case the Poisson and heat equations) has to be taken into consideration to build an efficient tool. The solution of these differential equations needs of realistic and physical coherent boundary conditions. For both cases, Neumann (for lateral faces) and Dirichlet (fixing the voltage and the temperature at the electrode-dielectric interface) boundary conditions can be used. A fixed temperature of the electrode interface assumes the consideration of the electrodes as infinite heat sinks, a null normal temperature derivative corresponds to adiabatic conditions at the lateral faces.

The simulation of the processes connected with the creation and destruction of the percolation paths within the dielectric is crucial for the description of resistive switching. By reproducing the evolution of conductive filaments we can study the progressive device resistance change in an analog manner. These features allow the analysis of memristors in the role to emulate neural synapses. In this respect, as highlighted above, the change of device conductance can mimic the biological plasticity of synapses [34]. The percolation paths made of metallic ions in conducting bridge random access memories (CBRAMs) and oxygen vacancies in VCMs are formed or ruptured in the cycles that characterize the resistive switching operation. The percolation path evolution can be described accurately in the kMC approach, taking into consideration the stochasticity of the physical processes behind. Algorithms to calculate percolation paths, such as the Hoshen-Kopelman, have to be employed in kMC simulation tools.

A brief summary of the kMC algorithm is depicted in the flow-graph in Fig. 19.12.

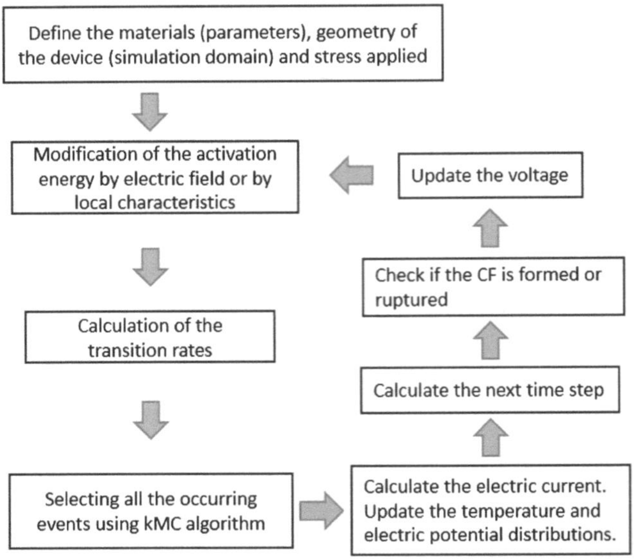

FIGURE 19.12 Simulation flow-graph of a general kMC algorithm for RRAM simulation.

19.5.3 Kinetic Monte Carlo simulation of conductive-bridge RAMs

One of the more successful types of memristors or resistive memories is the electrochemical metallization memory (ECM) also known as CBRAM. Their operation is based on the oxidation and reduction of atoms from the active electrode that move in the dielectric; these ions group together to form percolation paths (conductive filaments) with metallic-like behavior. The formation and rupture of conductive filaments can be simulated in a kMC approach that accounts for the inherent stochasticity and variability of the devices [1,2,41,114]. In this sort of devices at least one of the electrodes must be chemically active. The forming and SET processes are based on the generation of cations by oxidizing active electrode atoms. Then these cations travel through the dielectric medium field-driven until they get reduced, usually when they reach the inert electrode. This process can help to make a cluster of reduced atoms grow to form a percolation path that bridges the electrodes [1,2,64,81,101]. If the cation mobility in the dielectric is low enough a layer close to the active electrode (in contact with it) is formed, this layer can be treated as an extension of the electrode and it is called virtual electrode [1,2,110]. The CF remnants in contact with the electrodes left in the RESET processes at different RS cycles are also considered virtual electrodes.

Concerning ECMs, two types of devices depending on the polarity of their operation are distinguished: unipolar or bipolar. In the unipolar case the RESET process is dominated by the thermal effects that lead to higher oxidation and ion diffusion rates [1,82,105]. However, for bipolar devices, thermal phenomena

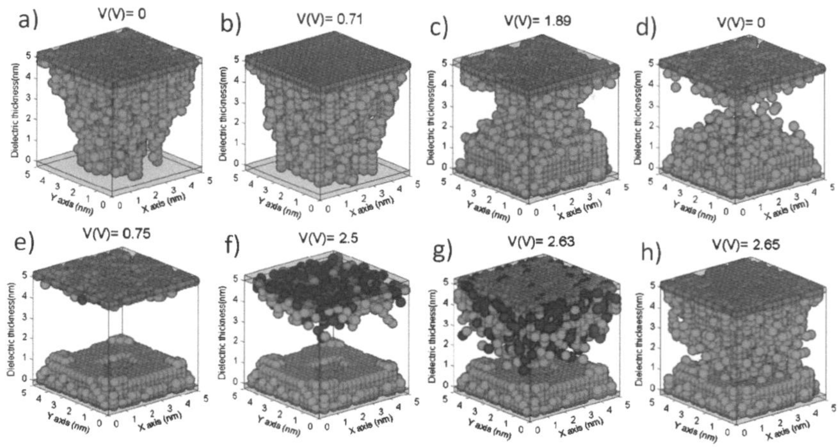

FIGURE 19.13 kMC simulation of a HfO_2-based unipolar ECM [1]. Plots a-d correspond to a RESET process, plots e-h correspond to the SET process that takes place over using the CF remnants of the previous RESET process. Red (gray in print version) balls represent Ni atoms, blue (dark gray in print version) ones represent Ni cations; once the filament is formed, the only elements shown are the atoms within the percolation path.

are not so important since the electric field is more determinant [2,57,82,106]. In Fig. 19.13 we plot the evolution of the conductive filament in a RS cycle for unipolar ECMs. The kMC simulation allows to plot all the atoms that form the percolation path that switches the devices.

19.5.4 Valence change memories kinetic Monte Carlo simulation

As highlighted above, other important group of resistive memories are valence change memories, whose conductive filaments are formed by oxygen vacancies. A metal layer that performs as an oxygen ion reservoir (such a Ti) is usually needed [19]. RS in VCMs has filamentary nature [59]. The dielectric breakdown is produced by means of the electric field that can induce a current. The electric field and the device temperature lead to the breakdown of molecule bonds generating Frenkel pairs (an oxygen vacancy and an oxygen ion) that forms defect-rich regions that lead the devices to operate in the LRS [7,19,79]. In these bipolar devices, the RESET process needs a change in the voltage polarity to inject the stored oxygen ions from the reservoir layer to the dielectric. The oxygen ions cross the dielectric field-driven till they recombine with the vacancies created in the previous SET process. In doing so, the CF can be destroyed [79]. It is also important to take into consideration that oxygen ions diffuse more efficiently than vacancies; therefore, in many cases, for modeling purposes, oxygen vacancies are assumed to be immobile [7,43,79,92,102]. The operation of these devices can be simulated under the kMC approach. In Fig. 19.14 we show the CF evolution in a RS cycle for a VCM.

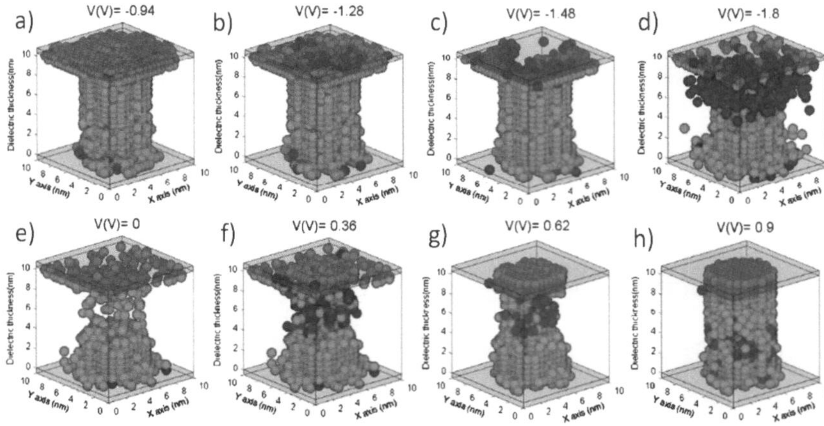

FIGURE 19.14 kMC simulation of a VCM. Plots a-d correspond to a RESET process, plots e-h correspond to the SET process that takes place over using the CF remnants of the previous RESET process. The red (gray in print version) balls represent oxygen vacancies, the blue (dark gray in print version) ones symbolize oxygen ions and the purple (darker gray in print version) ones represent grid points where both an oxygen vacancy and an oxygen ion (not recombined) are found.

19.6 Memristor compact modeling

Modeling of memristors is a key issue in technology computer-aided design (TCAD). Different modeling approaches provide a variety of degrees of accuracy and physical description at the simulation level [49,58,83,106]. In general, a more physically detailed description implies more computational cost, as it is the case of *ab initio* [47,78] or kMC [1,116] tools, as those described in the previous section. In some applications, in which the number of devices to simulate is large (in a Kbit or Mbit memory cross-bar architecture, for instance), device analytical simple models are required. These models must have a low computational cost and they should be able to work on standard circuit simulators describing the devices accurately. These models are named compact models [47,58,106] and, in general, work with currents and voltages as input/output variables, although they usually manage internal variables for describing the memristor state. To do so, simplified 1D differential equations are solved in reasonable computing times. Although they do not provide a detailed description of the device physics (such as the conductive filament configuration or temperature distribution), they are able to reproduce the device behavior in terms of switching characteristics: SET/RESET voltages or LRS and HRS resistances [47,58].

Usually, compact models are classified in two groups: empirical or phenomenological and physically based [49,58,83,106,119]. We will focus on the latter, for which the model equations rely on a physical but simplified description of the device operation.

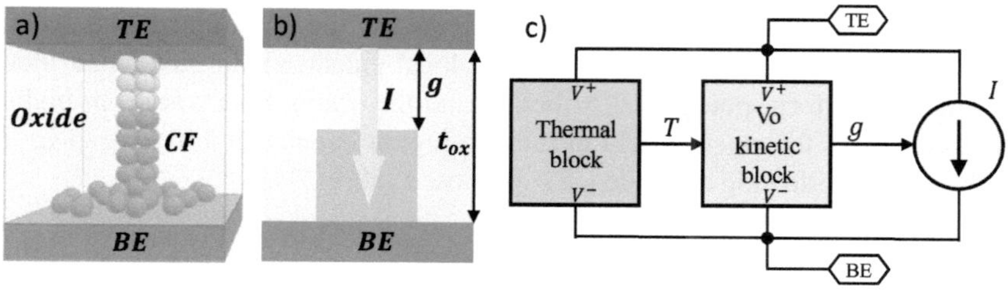

FIGURE 19.15 a) Three-dimensional view of the SU-VCM modeling structure with an indication of different device areas (TE, oxide, CF and BE), b) geometrical parameters in the model. The gap (g) between the TE and the filament tip is one of the state variables, the other one is the temperature (T), c) sub-circuit for the model implementation.

19.6.1 VCM compact modeling

In this section we deal with two of the most important VCM models described in the literature and present the main features of a new model based on them (the UGR-VCM model).

The unidimensional Compact Model for VCM (SU-VCM) [39,40,50,51,117], developed initially at Stanford University, describes the RS mechanisms in terms of generation/recombination processes of oxygen vacancies in the oxide layer. The gap between the top electrode and the conductive filament tip acts as an internal state variable. The conduction mechanism is assumed to be a general tunneling current dependent on the gap and the applied voltage. The temperature dependence is modeled by means of a thermal resistance in most related papers.

The SU-VCM assumes that the RS operation can be described by generation/recombination of oxygen vacancies (V_o) and oxygen ions that allow the formation/disruption of a conductive filament (a region or path deficient in oxygen atoms) that can short the electrodes. Due to the differences in the activation energies linked to the migration of oxygen vacancies and oxygen ions, the vacancies are supposed to be immobile while oxygen ions diffuse in the dielectric [7].

The simplified structure of the SU-VCM model is shown in Fig. 19.15a, the gap (g) is the main internal state variable that controls RS operation. The gap variation (dg/dt) is calculated as follows [39,40,50,117], accounting for the thermally activated RS physical mechanisms [40]:

$$\frac{dg}{dt} = \upsilon_0 \cdot e^{-\frac{E_{a,m}}{k_b T}} \cdot \sinh\left(\frac{q \cdot \gamma \cdot a_0 \cdot V}{L \cdot k_b \cdot T}\right), \quad g \geq g_{min} \tag{19.4}$$

where υ_0 is a velocity factor depending on the attempt-to-escape frequency multiplied by the atomic hopping distance, E_a is the activation energy for vacancy generation that activates the SET process, E_m is the barrier for oxygen ion migration that activates the RESET process, k_b is the Boltzmann constant, q is the electron charge, a_0 is the atomic hopping distance, V is the applied voltage in the gap, L is the oxide layer thickness, g_{min} is the minimum gap distance, T is the device temperature and γ is the local enhancement factor that models the nonuniform potential distribution and the strong polarization in high-k dielectrics. As the electric field in the gap region is enhanced if the gap (g) is reduced, the following dependence is proposed [50,117]:

$$\gamma = \gamma_0 - \beta \cdot g^3 \tag{19.5}$$

where γ_0 and β are fitting parameters. Other expressions for γ are also employed: references [39,40] use a linear dependence ($\gamma = \gamma_0 - \beta \cdot g$), in reference [51] an exponential relationship is used ($\gamma = \gamma_0 - \beta \cdot g^\alpha$, where α is a fitting parameter) and in reference [15] a normalized exponential function is assumed ($\gamma = \gamma_0 - \beta \cdot (g/g_1)^3$, where g_1 is a reference gap value).

For a circuital simulation approach, the model is split into three subcircuits or blocks (see Fig. 19.15c). The kinetic block calculates the evolution of the gap size g (by means of Eq. (19.4)). Its inputs are the applied voltage and the temperature.

The current in the dielectric layer is assumed to be linked to a tunneling mechanism in the gap region. Current components connected to trap-assisted tunneling, Poole-Frenkel tunneling, Fowler-Nordheim tunneling or direct tunneling could be feasible. The following gap depending expression is proposed in order to account for a generalized tunneling current [15,39,40,50,51,117]:

$$I = I_0 \cdot e^{-g/g_0} \cdot \sinh(V/V_0) \tag{19.6}$$

where I_0, g_0 and V_0 are fitting parameters. Eq. (19.6) is implemented by the current source I (Fig. 19.15c).

As generation/recombination of oxygen vacancies are drastically affected by the local temperature [40], a thermal model has to be implemented in order to calculate the temperature (thermal block in Fig. 19.15c). The temperature is obtained making use of a simple effective thermal resistance [40,50,51,117],

$$T = T_0 + V(t) \cdot I(t) \cdot R_{th} \tag{19.7}$$

where R_{th} is the effective thermal resistance of the entire device, T_0 is the absolute room temperature and $V(t)I(t)$ is the power dissipated by means of Joule heating. Thermal inertia can be also taken into account if a capacitor is included in the thermal subcircuit [15,39].

Notice that the gap would be obtained by integration of Eq. (19.4). However, the high resistance state normally shows a prominent variability from cycle-to-cycle [60]. This is mainly due to the RS stochasticity and the spatial variation of the average gap size in the case of multiple filaments. In order to include variability in the model, a transient noise signal is added to the average gap distance [39,50,51,117], which results in a random perturbation of the device resistance,

$$g|_{t+\Delta t} = \int \left(\frac{dg}{dt} + \delta_g(T) \cdot \chi(t) \right) dt \tag{19.8}$$

where $\chi(t)$ is a Gaussian noise sequence randomly generated with zero-mean and unitary root mean square generated at each simulation time step and $\delta_g(T)$ is a sigmoid function dependent on the temperature that can be expressed as follows:

$$\delta_g(T) = \frac{\delta_g^0}{1 + e^{\frac{T_{crit}-T}{T_{smth}}}} \tag{19.9}$$

where δ_g^0 is a fitting parameter, T_{crit} is an activation temperature (for high temperatures random fluctuations are triggered) and T_{smth} is other fitting parameter that determines the transition smoothness for activation of the random fluctuations.

Fig. 19.16 shows simulation results obtained by the SU-VCM model implemented in SPICE when a ramped voltage signal is employed. The initial state is assumed to be HRS (the current is low and the g value is close to 3 nm). When the voltage increases the SET process begins, the temperature rises, the gap g reaches a given minimum value of 0.25 nm (the approximated size of an oxygen vacancy), the device state changes to LRS and the current increases. Then the ramped voltage signal changes the polarity and a RESET process begins. In this process the current drops off and the gap g increases.

The bidimensional Compact Model for VCMs (PU-VCM) could be seen as a natural evolution of the SU-VCM model and it was developed at Peking University [43,61,62]. This model introduces the CF radius as a second state variable.

The model assumes that a device in the lowest resistance state is represented by a cylindrical CF between the top and bottom electrodes. The RESET process reduces the length of the cylinder (thus, the gap between the CF tip and the top electrode increases) until the gap reaches the maximum allowed value (see Fig. 19.17) when the RESET is fulfilled. According to this model, the set process is modeled following two consecutive steps. In the first step, the growth of a narrow cylindrical filament of radius r takes place from the tip of the main body of the filament toward the top electrode (see Fig. 19.17). The second step of the set process begins when the gap reaches a minimum value ($g = 0$). In this stage, the radius of the small conductive cylindrical filament (r) increases [43,61,62].

The set process is mainly controlled by the generation of oxygen vacancies. Therefore, according to the previously described schema for the CF evolution, the generation of oxygen vacancies is translated to the change of the CF geometry (described by variables g and r) by means of Eqs. (19.10) or (19.11) (according to the set stage) [43,61,62]

$$\frac{dg}{dt} = a_0 \cdot f \cdot e^{-\frac{E_a - \alpha_a \cdot Z \cdot q \cdot E}{k_b \cdot T}} \tag{19.10}$$

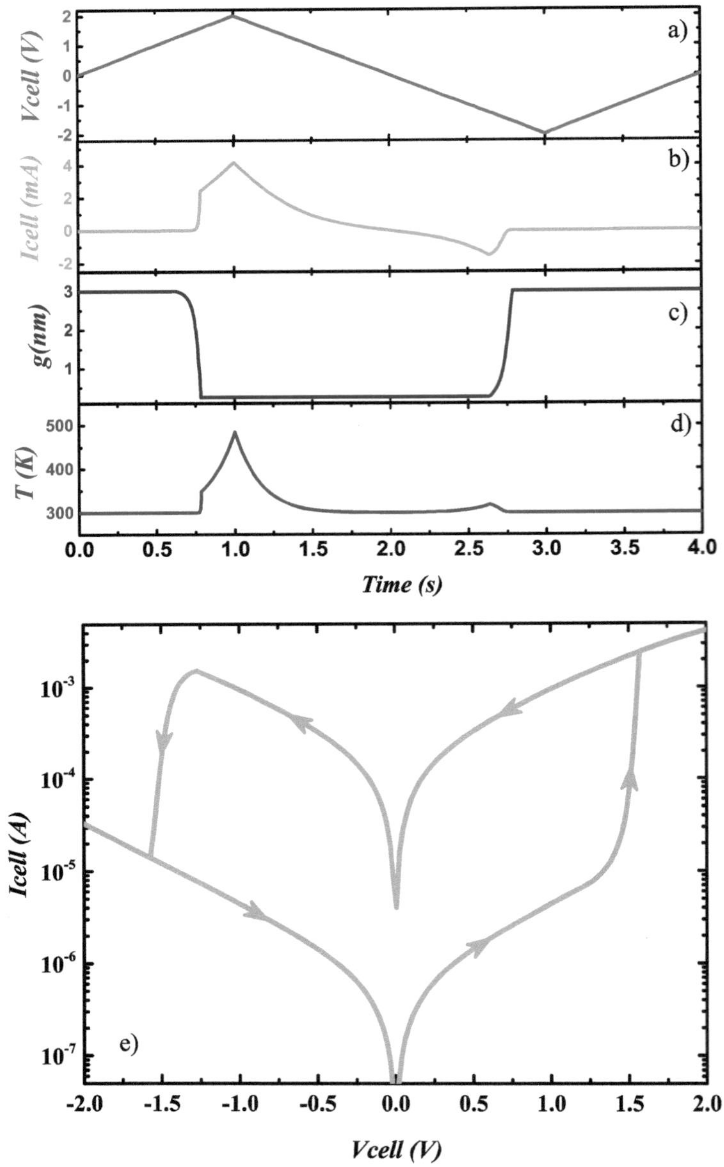

FIGURE 19.16 a) Ramped voltage signal applied to a memristor. Time evolution of the main model magnitudes: b) current, c) CF to electrode gap, d) device temperature. e) Current versus voltage curve.

$$\frac{dr}{dt} = \left(\Delta r + \frac{\Delta r^2}{2r}\right) f \cdot e^{-\frac{E_a - \alpha_a \cdot Z \cdot q \cdot E}{k_b \cdot T}} \tag{19.11}$$

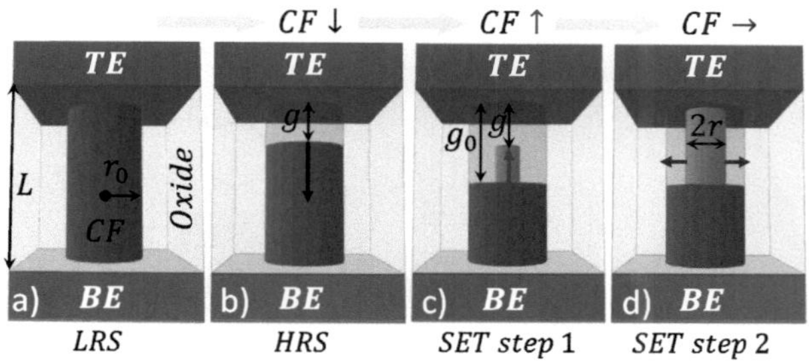

FIGURE 19.17 Diagram of PU-VCM model conductive filament evolution. a) The device is initially in the LRS (the gap g is closed) and is modeled by a cylindrical CF that shorts the electrodes. The CF has an ohmic resistance R_{CF} with radius r_0. b) When the RESET process is over the gap is non-zero; a maximum value is allowed, g_0. c) The first step in the set process consists in the growth of a small cylindrical CF, from the point of rupture of the CF to the electrode. d) The set second step begins when the gap is zero, the CF expands radially increasing its radius, r [43,61,62].

where a_0 is the atomic hopping distance, f is the vibration frequency of the oxygen atom, E_a is the activation energy for oxygen vacancies generation, Z is the charge number of oxygen ions, q is the elementary electron charge, α_a is an enhancement factor of the electric field across the gap (E), k_b is the Boltzmann constant, T is the temperature and Δr is the effective CF expanding width [43].

The RESET process is controlled by three mechanisms: release of oxygen ions by the electrode (Eq. (19.12)), oxygen ion hopping in the oxide layer (Eq. (19.13)) and recombination between oxygen vacancies and oxygen ions (Eq. (19.14)) [43,61,62]. We have

$$\frac{dg}{dt} = a_0 \cdot f \cdot e^{-\frac{E_i - \gamma \cdot Z \cdot q \cdot V}{k_b \cdot T}} \tag{19.12}$$

$$\frac{dg}{dt} = a_0 \cdot f \cdot e^{-\frac{E_h}{k_b \cdot T}} \cdot \sinh\left(\frac{\alpha_h \cdot Z \cdot q \cdot E}{k_b \cdot T}\right) \tag{19.13}$$

$$\frac{dg}{dt} = a_0 \cdot f \cdot e^{-\frac{\Delta E_r}{k_b \cdot T}} \tag{19.14}$$

where E_h, E_i and ΔE_r are the oxygen ion migration activation energy, the energy barrier for the oxygen ion electrode release processes and the relaxation energy in the recombination process, respectively. The increase of g, (dg/dt), is controlled by the slowest of the three mechanisms (Li et al. [62]).

The charge transport in the CF is assumed to be ohmic and the corresponding resistance is calculated according to the CF geometry (determined by r and g) [43,61,62]. In addition, tunneling or hopping currents (I_{hop}) are also considered following Eq. (19.6). For the HRS, two hopping components are included: a current path from the main CF body to the top electrode and other current path from the filament tip to the top electrode. Furthermore, the latter depends on the

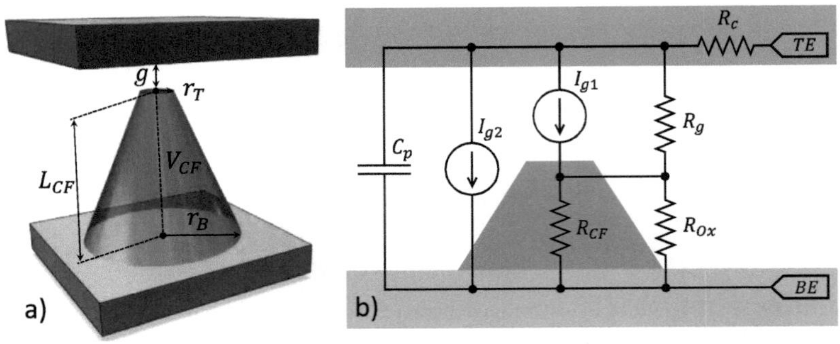

FIGURE 19.18 a) Three-dimensional geometrical representation of the CF in the UGR-VCM model, b) equivalent circuit for calculating the current and voltage drop in the modeled RRAMs.

CF geometry (r, g) [43,61,62]:

$$I_{hop} = I_0\left(\pi r^2/4\right)e^{-g/g_T} \cdot \sinh\left(\frac{V_{gap}}{V_T}\right) \tag{19.15}$$

where V_{gap} is the voltage in the gap; I_0, g_T and V_T are the fitting parameters of the hopping current model.

Finally, the thermal model is similar to that of the SU-VCM model. Variability is also included in a similar way (introducing random variations in g, and in r). Random variations of the activation energies values E_i and E_h are also included [62].

Based on the previous models, we have developed the UGR-VCM model with the following set of characteristics [30–32]:

- Truncated-cone shaped filaments are employed with metallic-like transport characteristics. Its electrical conductivity is temperature dependent. Hopping currents from the CF to the electrodes are also considered.
- The differential equation that determines the resistive switching operation is established in terms of the conductive filament volume, instead of using unidimensional variables (as the CF radius or the gap).
- The resistances connected with the metal electrodes are considered as well as the device capacitance.
- The temperature is obtained by solving the heat equation including Joule heating effects in the CF and the CF lateral heat dissipation to the surrounding dielectric.
- Cycle-to-cycle variability is considered and included in the model equations.

A more detailed description of the UGR-VCM model is given below. Fig. 19.18 shows a UGR-VCM model scheme. Truncated-cone shaped CFs were assumed for the description of RS operation, following previous experimental results [56]. The distance between the CF tip and the electrode (g) determines the hopping transport processes [43]. The electrodes and CF ohmic resistances were also included (R_C and R_{CF}, respectively). Two additional resistances (R_{ox} and R_g) model the role of the dielectric surrounding the CF and the dielectric region filling the gap (for the sake of simplicity, an ohmic dependence was assumed for them).

For this model, we assumed the same physics than in the PU-VCM model to describe SET/RESET mechanisms. However, we have formulated the transient variation of the CF shape in terms of its total CF volume, instead of using the gap or CF radius, as in the PU-VCM model (Eqs. (19.10) and (19.11)). Therefore, the volume variations for the SET and RESET processes can be expressed as:

$$\frac{dV_{TC}}{dt} = \upsilon_0 \exp\left(-\frac{E_a - \alpha_a ZqE}{k_b T}\right) \qquad (set) \qquad (19.16)$$

$$\frac{dV_{TC}}{dt} = -\upsilon_0 \exp\left(-\frac{E_r - \alpha_r(g) ZqE}{k_b T}\right) \qquad (reset) \qquad (19.17)$$

where υ_0 is a constant equal to $a_0^3 \cdot f$. The remaining parameters in Eq. (19.16) are the same than those in Eq. (19.10). For the RESET process, E_r is an average energy that accounts for the processes involved in the RESET process (see Eqs. (19.12)–(19.14)) and $\alpha_r(g) = \gamma g + \alpha_h$, accounts for the electric field enhancement factor that lowers the hopping barrier (α_h) and for the external voltage enhancement factor (γ) during oxygen ion release.

Having assumed a truncated-cone shaped CF, its state is determined by three variables (g, r_T and r_B; see Fig. 19.18). In order to get a compact model, we have assumed that the SET process follows these consecutive steps: firstly, the increase in the volume is located in the area of the gap (g), which is narrowed down until its minimum value is reached; then the minor radius, r_T, is increased from its minimum value until its maximum allowed value; and finally, the bottom radius r_B is increased. A reverse order is followed during the RESET process: firstly, a reduction of r_B is applied until it reaches its minimum value. Then the top radius, r_T, decreases until its minimum value and, finally, the gap is increased until its maximum value.

The thermal block in this model is based on the solution of the following heat equation, which takes into account lateral heat dissipation from the CF to the dielectric:

$$\sigma_{CF}\xi^2 = -k_{th}\frac{\partial^2 T(x)}{\partial x^2} + 2h\frac{T(x) - T_{ox}}{r_{CF}(x)} \qquad (19.18)$$

σ_{CF} is the CF electric conductivity (a linear temperature dependence is also considered), r_{CF} is the CF radius at each position x, ξ is the average electric field in the CF, k_{th} the CF thermal conductivity and the parameter h is the lateral heat transfer coefficient. As Eq. (19.18) cannot be solved analytically for a truncated-cone CF, we have assumed that a truncated-cone shaped CF with constant conductivity is equivalent to a cylindrical CF and a conductivity dependent on the position along the CF main axis. In this case, the cylinder maximum temperature can be analytically calculated and it is assumed to represent the CF

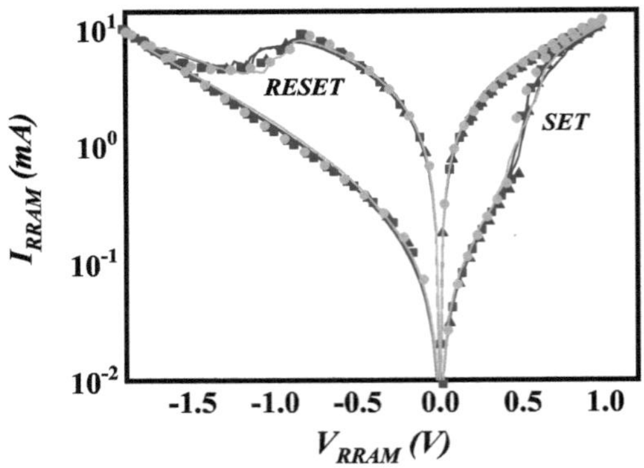

FIGURE 19.19 Experimental (solid lines) and modeled (symbols) I-V curves of RS in TiN/Ti/10 nm-HfO_2/W devices for several cycles in a long RS series.

temperature for evaluating Eqs. (19.16) and (19.17) [30]:

$$T = T_0 + \frac{\sigma_{CF} r_g r_B \xi^2}{r_T h} \left[\frac{1}{2} - \frac{e^{\frac{\alpha}{2}}}{e^{\alpha} + 1} \right] \tag{19.19}$$

where the α parameter is

$$\alpha = L_{CF} \sqrt{\frac{2h}{k_{th} r_g}} \tag{19.20}$$

Finally, variability is introduced by including a random component added to the CF volume, as is done for g in SU-VCM and PU-VCM models.

Once the CF state is known at each simulation time step, the current is calculated by means of the electrical model shown in Fig. 19.18. We have considered the contributions to the hopping currents from the top and lateral sides of the truncated-cone shaped CF (I_{g1} and I_{g2} current sources, respectively). In both cases, the calculation of the current is based on Eq. (19.6). For the lateral component (I_{g2}), an average distance between the CF surface and the top electrode is calculated.

The UGR-VCM model has been tested with hundreds of experimental I-V RS cycles of different technologies based on the following stacks: TiN/Ti/HfO_2/W, Pt/TiO_2/Al_2O_3/TiO_2/RuO_x/TiN and Au/Ti/TiO_2/SiO_x/Si-n^{++} [30–32]. Fig. 19.19 shows an example of experimental I-V curves of RS in TiN/Ti/10 nm-HfO_2/W devices, which are accurately reproduced by the model.

19.6.2 Compact modeling of CBRAMs

Regarding CBRAMs, several macroscopic simulators have been developed considering RS based on redox reactions and diffusion processes [8,104,105]. In

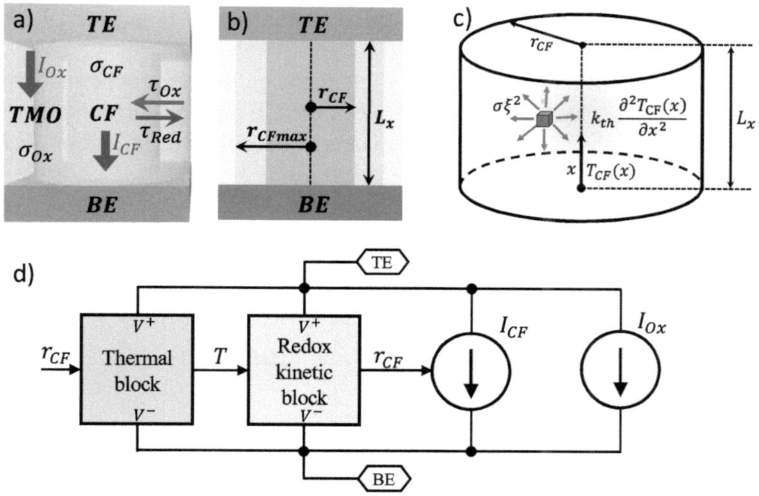

FIGURE 19.20 a) Three-dimensional view of the cylindrical structure of the CF assumed by the IM2NP model. The creation/destruction of the CF in the dielectric layer depends on the kinetics of the redox processes. b) Geometrical parameters involved in the model. c) Thermal representation of the CF. d) Electrical sub-circuit used for implementing the IM2NP model.

these cases, temperature, reaction rates and currents are calculated in a self-consistent procedure that includes the numerical resolution of the heat equation (19.18). These modeling tools provide a detailed CF description (temperature and the CF radius are obtained as a function of the position along the dielectric). However, simplified versions have been developed for compact modeling in a circuit simulation context. This is the case of models presented in [9,10] and [27,28], which will be briefly described below.

In the model presented in [9,10] (we name it as IM2NP), the CF is represented by a metallic-like cylinder that bridges the electrodes through the insulator (see Fig. 19.20). Its radius, r_{CF}, is an internal variable that determines the CF state and it is changed according to the redox processes rates. A maximum CF size (defined by the maximum radius, r_{CFmax}) is fixed. The conduction in the CF is assumed to be ohmic, as well as in the hollow cylinder of surrounding oxide with radius r_{CFmax} that wraps the CF. The CF and surrounding oxide electrical conductivities are σ_{CF} and σ_{Ox}, respectively. A generalized function to include a trap-assisted tunneling current through the oxide, I_{Ox}, is implemented following the next equation [9]:

$$I_{OX} = A_{HRS} \cdot S_{Cell} \cdot \left(\frac{V_{cell}}{L_x}\right)^{\alpha_{HRS}} \tag{19.21}$$

Here A_{HRS} and α_{HRS} are fitting parameters, S_{Cell} is the area of the device, V_{Cell} is the applied voltage between the top and the bottom electrodes and L_x is the oxide thickness. This current component is dominant in the HRS, but negligible in the LRS. In the circuital representation of the model (see Fig. 19.20d), it corresponds to the current source, I_{OX}.

The evolution of the CF radius (r_{CF}) depends on the electrochemical oxidation/reduction rates at the interface between the CF and the surrounding oxide. It is calculated in the kinetic block by means of the following equation [9,10]:

$$\frac{dr_{CF}(t)}{dt} = \frac{r_{CFmax} - r_{CF}(t)}{\tau_{Red}} - \frac{r_{CF}(t)}{\tau_{Ox}} \tag{19.22}$$

where the time constants for the reduction and oxidation processes (τ_{Red}, τ_{Ox}) are calculated according to the following equations [9,10]:

$$\frac{1}{\tau_{Red}} = A_{RedOx} \cdot e^{-\frac{E_a - q \cdot \alpha_{Red} \cdot V_{cell}}{k_b T}} \tag{19.23}$$

$$\frac{1}{\tau_{Ox}} = A_{RedOx} \cdot e^{-\frac{E_a + q \cdot \alpha_{Ox} \cdot V_{cell}}{k_b T}} \tag{19.24}$$

In these equations, τ_{Red} (τ_{Ox}) is the time constant for the electrochemical reduction (oxidation) rate, A_{RedOx} is the nominal redox rate, E_a is the activation energy, q is the electron charge, α_{Red} and α_{Ox} are the transfer coefficients (with values between 0 to 1), k_b is the Boltzmann constant, T is the device temperature and V_{cell} is the applied voltage. Note that electrochemical reduction/oxidation rates depend on the applied voltage (V_{cell}) and the local CF temperature (T). The temperature inside the cylinder-shaped CF is calculated by solving the heat equation (19.18) without lateral heat transfer ($h = 0$) [9,10], with boundary conditions $T(0) = T(L_x) = T_0$ (T_0 is the room temperature). An analytical expression for the maximum temperature is obtained and then used [9,10].

The IM2NP model is fully compatible with electrical simulators based on SPICE. Fig. 19.20 graphically summarizes the model and a possible subcircuit structure for a SPICE implementation. Finally, Fig. 19.21 shows the simulation results obtained with the IM2NP model implemented in SPICE when a ramped voltage is applied to the device.

At the University of Granada we have developed a physically based model for unipolar RRAMs [104,105] in which the evolution of conductive filaments (whose radii depend on the position along the CF) is calculated taking into account metallic species diffusion and redox reactions in the SET/RESET processes. Ohmic and quantum conduction through a constriction (by means of the Quantum Point Contact model, QPC [73,105]) are also considered.

To make circuit simulation easier and faster, we have implemented a SPICE model based on the simulation tool previously developed [104,105]. This means that the physical principles and model parameters are similar; however, the RRAM is treated following a discrete approach, using lumped circuital elements. In addition, the thermal behavior is modeled by equivalent thermal resistances accounting also for the heat loss from the CF to the surrounding oxide. A first version of this circuital model was introduced in [52] and a simplified in a more compact version in [28]. The latter is described below.

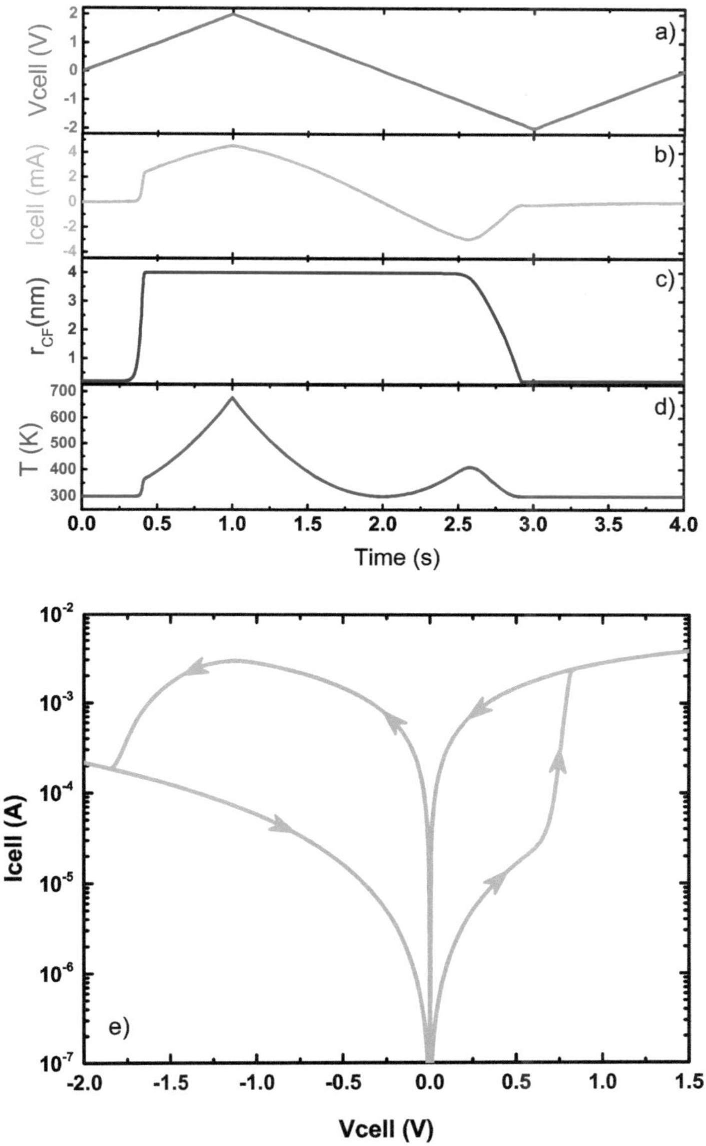

FIGURE 19.21 a) Ramped voltage signal applied. The IM2NP model implemented in SPICE has been used in order to calculate: b) current versus time, c) transient evolution of internal variable r_{CF}, d) time evolution of the CF temperature, and e) semi logarithmic *I-V* curve.

The analysis of a RESET event allows to distinguish that only a small part of the filament is deeply heated and narrowed until the filament disruption happens. However, the CF main body temperature is much lower and its shape

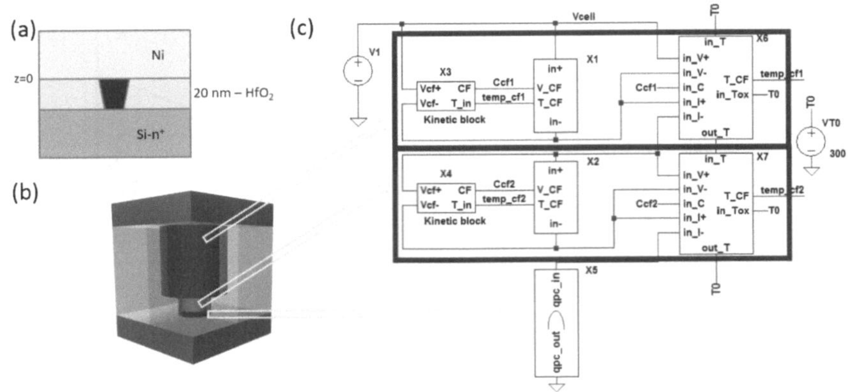

FIGURE 19.22 a) Sketch of the device structure, with a truncated-cone shaped conductive filament. b) The CF is represented by two different cylinders. Furthermore, a QPC barrier is also considered. c) LTspice subcircuit hierarchy built in order to simulate the device. Each cylinder is modeled by means of a subcircuit that includes a kinetic subcircuit (for the description of the cylinders evolution), a thermal subcircuit (for the temperature calculation) and, finally, a subcircuit for the modeling of the ohmic conduction through the filament. A special subcircuit is added in order to reproduce the conduction through the constriction by means of the QPC model [105].

hardly changes. Therefore, we have chosen two cylinders for representing the CF. The big one (top cylinder, Fig. 19.22) accounts for the CF main body, while the other one (bottom cylinder) models the hottest region where the breakdown takes place.

To model the devices, we considered different cylinders as CF building blocks. Each cylinder is represented by a circuit (shown inside the black rectangles in Fig. 19.22). Each circuit block is composed of three subcircuits: an ohmic subcircuit, a thermal subcircuit and a kinetic subcircuit. This block contains a capacitor whose charge is modified by current sources depending on the diffusion and redox rates [20,52]. Therefore, the cylinder radius is stored as the capacitor voltage. The ohmic block includes a voltage controlled resistor for the conduction through the CF and depends on the cylinder radius and on its temperature. In the thermal block, the temperature is calculated by means of an equivalent electric circuit that includes longitudinal and transversal thermal resistances.

Ni/HfO_2/Si-n^{++} RRAMs fabricated at IMB-CNM have been simulated [24] with this model. These samples show a non-linear behavior at low voltages that has been related to quantum conduction through a barrier. This effect was included in macroscopic simulators by means of the Quantum Point Contact model [105]. In order to account for this in our circuital model, a new subcircuit is added for implementing the I-V relationship imposed by the QPC model [73]. See that a perfect agreement has been achieved in both operation regimes, constant voltage (I-t plots) and ramped voltage (I-V curves), using the same physical parameters and cylinder sizes for both types of simulations (Fig. 19.23).

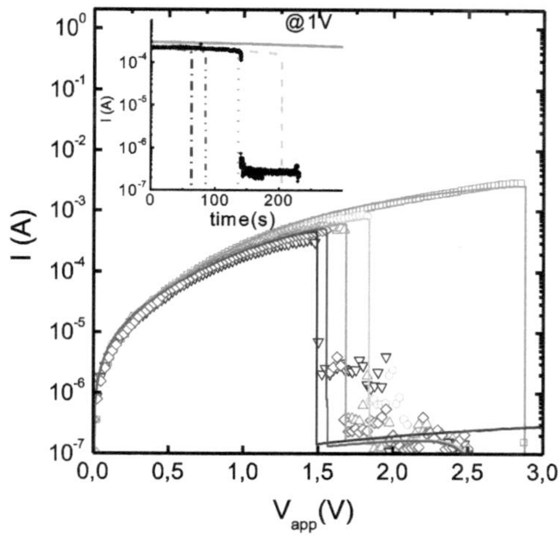

FIGURE 19.23 Experimental (symbols) and simulated (lines) I-V curves obtained by applying a voltage ramp (absolute value is employed for the applied voltage). Five SET-RESET cycles were measured (only the RESET portion is shown). The inset shows the experimental data (black symbols) obtained when the same sample is subjected to a constant voltage of 1 V. Lines show the simulation results obtained with the same model parameters and cylinder sizes as those used for the simulations in the main plot. Therefore, the model is useful to reproduce the device behavior under different operation conditions.

19.7 Memristor random telegraph noise

Random Telegraph Noise (RTN) signals are well known in the electron device reliability landscape [67,90,98]. RTN is a major reliability problem for MOS-like transistors (Gate-all-around, FinFETs, Double-gate FETs, ...). For memristors and, in particular, for resistive memories, RTN is also an important issue to deal with. In this respect, the physics behind this type of signals has to be studied.

In the operation of RRAMs as memories, RTN can affect multilevel nonvolatile applications by reducing the read margin between different resistance levels. For applications concerning neuromorphic circuits, the temporal changes in the device resistance state can alter the mimicking process of synaptic weights when memristors act as artificial synapses within neural network circuits [34,35]. Nevertheless, RTN fluctuations can also be used for good. They allow the implementation of entropy sources of random number generators that can be key component in cryptography applications such as secret key generation, stochastic spiking signals and stochastic neurons [14,93]. As commented in the previous section, the accurate modeling of these devices is essential for circuit design. In particular, RTN signals have to be considered from this viewpoint for circuit simulation [29]. These signals are related to the emission and capture of electrons by defects close to the conductive filament in devices with

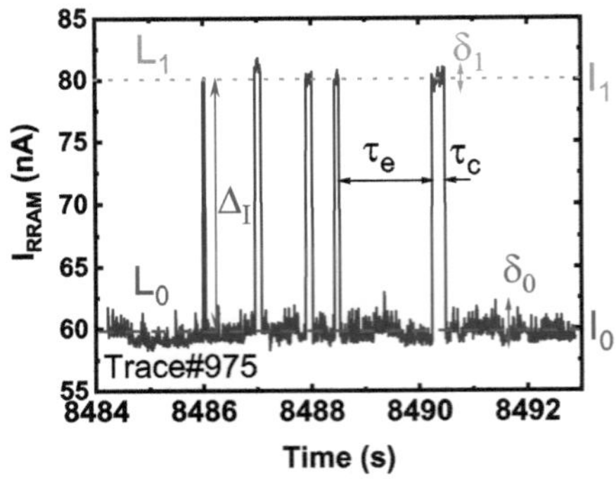

FIGURE 19.24 Current versus time trace measured in a Ni/HfO_2/Si-n^{++} device (trace #975, 1000 samples).

filamentary conduction. Hence, RTN features are connected with the number of active traps and their mutual interaction [25,33,92,93].

19.7.1 Numerical procedures to analyze random telegraph signals

As it was highlighted in the introduction, random telegraph noise signals can be observed in some memristors, both at the LRS and the HRS. See Fig. 19.24, which shows a (I-t) trace obtained for Ni/HfO_2/Si-n^{++} devices in the HRS.

Several techniques have been developed to scrutinize the physical processes behind RTN signals and the statistical dependence of the measurements. We review some of them here and highlight their main features.

19.7.1.1 Current versus time trace representation

The current versus time traces are employed to analyze the data. This simple representation plots N points (I_i, t_i). N is the number of sampled data and i is an integer ($1 \leq i \leq N$). See Fig. 19.24, there are two clear current levels (labeled L_0 and L_1); the corresponding currents values for these levels are the following: $I_0 = 60$ nA and $I_1 = 80$ nA. Δ_I stands for the difference of these current levels and the variation of each of these current levels is represented as δ_0 and δ_1. The emission (capture) times τ_e (τ_c) are also shown. When we deal with a short RTN trace, this methodology could be useful. Nevertheless, for longer traces it is useless. For this reason, other approaches have been introduced, as shown below.

19.7.1.2 Time lag plot (TLP) representation

The TLP consists of a graph where the sampled current $i + 1$ is plotted versus the current sample i; see Fig. 19.25. The plot has $N - 1$ points (I_{i+1}, I_i) of RTN sampled data [76]. The stable current levels are labeled as L_0 and L_1 in the TLP, they can be seen as a cluster of points on the main diagonal (described math-

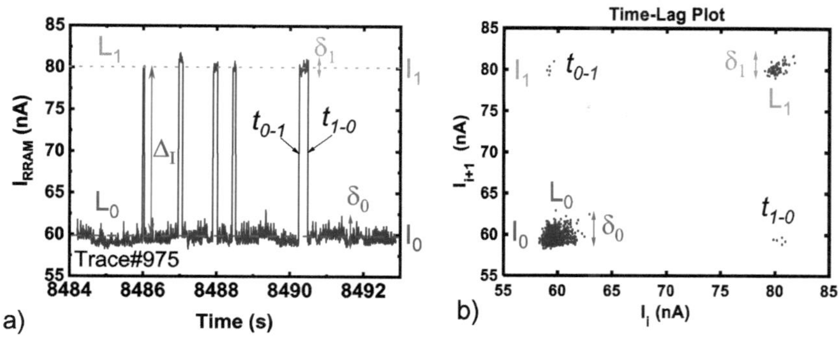

FIGURE 19.25 a) Current versus time measured in Ni/HfO_2/Si-n^{++} devices (trace #975, 1000 samples), b) Time lag plot (TLP) of the same trace [76].

ematically by $I_{i+1} = I_i$). In the opposite diagonal the points can be related to transitions between the stable current levels, they are labeled t_{0-1} (t_{1-0}), which stand for transitions from I_0 to I_1 (I_1 to I_0).

The current levels variability (δ_0, δ_1) is also seen in the TLP. The relative probability of each current level can be clearly observed, in this case, see the different cluster sizes linked to L_1 and L_0; however, this issue is not clear with this representation technique in all the cases.

The approximate number of traps (N_T) behind the RTN signal could be estimated by the equation [76,94]

$$N_T = \text{Ceiling}[\log_2(N_L)] \tag{19.25}$$

where N_L is the number of local maxima in the TLP main diagonal of TLP. The ceiling function (in the previous equation, Ceiling [x]) maps x to the least integer greater than or equal to x.

19.7.1.3 Color code time lag plot (CCTLP)

The TLP allows to extract information for short RTN traces, but for long RTN traces it does not work correctly. The proximity of stable current levels, the background noise and the consequent overlapping obfuscate and hide the information that can be extracted by using this procedure [33,36,66–68,90]. The color code time lag plot (CCTLP) [90] consist of a TLP graph that discriminates the probability occurrence of points (I_{i+1}, I_i) making use of a color code (high probability: black, medium probability: yellow and low probability: red). A clearer picture can be obtained in this manner.

19.7.1.4 Radius time lag plot (RTLP)

The RTLP also represents the occurrence probability of a point (I_i, I_{i+1}) in the conventional TLP. It employs a circular region of radius r to obtain the number of counts of each (I_i, I_{i+1}) pair [66]. The details of the equations that describe this methodology are given in [66].

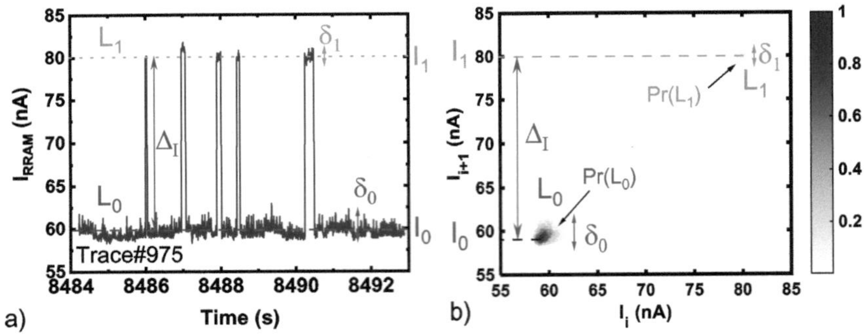

FIGURE 19.26 a) Current versus time plot for a Ni/HfO_2/Si-n^{++} device (trace #975, 1000 samples), b) RTLP of the same trace #975 [66].

A radius r of a circular region was calculated as $10^{-6}\sigma$, where σ stands for the standard deviation of the current fluctuations in the RTN sequence [66]. Fig. 19.26 shows the same points that the TLP plotted in Fig. 19.25. However, a color map helps to represent the number of points in each region, assuming a relative normalized count, where the most probable state is L_0 ($Pr(L_0) \approx 1$, and Pr is the normalized count of L_0). Within this normalized count, the relative count $Pr(L_1)$ is close to 0.2. In this case a more refined calculation of the occurrence probability is obtained.

This methodology is computationally expensive; for each point, the distance and counts added by the other neighbor points must be determined by an algorithm whose complexity depends on N^2.

19.7.1.5 Weighted time lag plot (WTLP)

The WTLP allows to obtain information similar to the RTLP in order to solve the problems of the TLP. The occurrence probability in a given TLP position is obtained by means of a bidimensional Gaussian distribution (BGD) [67,68].

The data space (I_{i+1}, I_i) of a trace is translated to a $M \times M$ matrix ($M < N$). For each position (x, y), the probability can be calculated as a weighted sum of all the sampled data (I_{i+1}, I_i). The mathematical methodology employs Eq. (19.26) [67],

$$\psi(x, y) = \frac{k}{2\pi\alpha^2} \sum_{i=1}^{N-1} \exp\left(\frac{-\left[(I_i - x)^2 + (I_{i+1} - y)^2\right]}{2\alpha^2}\right) \tag{19.26}$$

The values of x and y are obtained by considering the current values and the numbers of points in the matrix accordingly to Eq. (19.27).

$$(x, y) = \left(\frac{\hat{x}}{M-1}(I_{max} - I_{min}) + I_{min}, \frac{\hat{y}}{M-1}(I_{max} - I_{min}) + I_{min}\right) \tag{19.27}$$

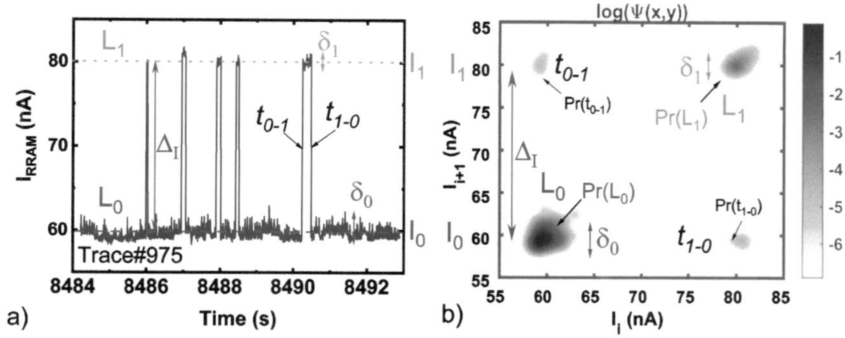

FIGURE 19.27 Current versus time plot for a Ni/HfO_2/Si-n^{++} device (1000 experimental samples were employed, trace #975). b) WTLP ($M = 200$ points, $\alpha = 0.1$) for trace #975 [67].

where $\hat{x}$ and $\hat{y}$ are the indices of the values ($\hat{x} = 0, 1, \ldots, M-1$, $\hat{y} = 0, 1, \ldots, M-1$), and I_{max} and I_{min} are the maximum and minimum currents of the RTN trace.

Trace #975 is analyzed by means of a WTLP in Fig. 19.27 (logarithmic scale and $M = 200$). The region of highest occurrence probability is at the main diagonal, as expected. Both techniques (RTLP and WTLP) describe the occurrence probability through a color code and circumvent the lacks of a simple TLP. The price paid in both cases is linked to the computational burden to be assumed.

19.7.1.6 Locally weighted time lag plot (LWTLP)

This technique provides a plot similar to the WTLP although the computation time needed is lower. The plotting space is changed from (I_{i+1}, I_i) pairs to $M \times M$ points (x, y) [33], and, as in the previous technique, the procedure employs the coordinates $(\hat{x}, \hat{y})$, as described by the following equation:

$$(\hat{x}, \hat{y}) = \left(int \left(M \frac{I_i - I_{min}}{I_{max} - I_{min}} \right), int \left(M \frac{I_{i+1} - I_{min}}{I_{max} - I_{min}} \right) \right) \tag{19.28}$$

After the determination of $(\hat{x}, \hat{y})$, the matrix is updated, considering this pair of values and the surrounding coordinates, weighted by a function that depends on the distance. For the calculation of the occurrence probability, submatrices of different dimensions can used; i.e., the region considered for the occurrence probability estimation corresponds to a 3×3 or 5×5 matrix, as explained in [33].

The LWTLP for trace #975 is shown in Fig. 19.28, calculated for $M = 200$ and a 5×5 sub-matrix. In the main diagonal are represented the different current levels, as in previous methodologies. This method is more computationally efficient that the previous ones, on average 750 times faster than WTLP or RTLP [33], while the information given is enough for the RTN analysis.

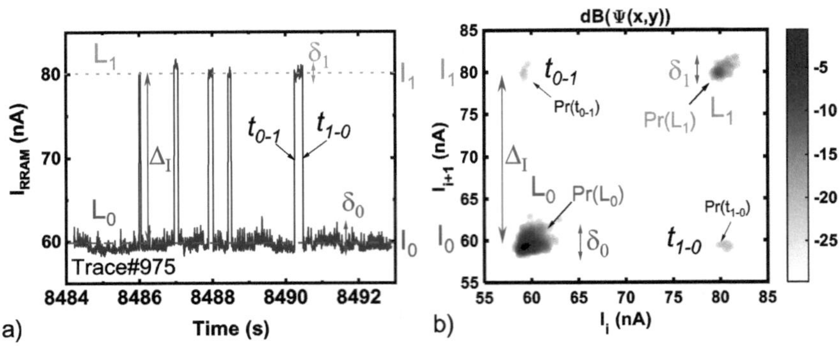

FIGURE 19.28 a) Current versus time plot for a Ni/HfO_2/Si-n^{++} device (trace #975 including 1000 samples). b) LWTLP ($M = 200$ points, M 5 × 5, $\sigma_f = 1.5$) of the same trace [33].

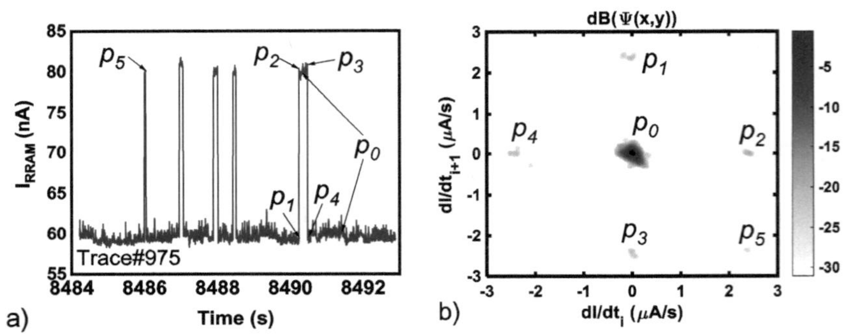

FIGURE 19.29 a) Current versus time plot for a Ni/HfO_2/Si-n^{++} device (trace #975 including 1000 samples). b) DLWTL Plot ($M = 200$, M 5 × 5 matrix, $\sigma_f = 1.5$) of the trace #975 [35].

19.7.1.7 Differential locally weighted time lag plot (DLWTLP)

For this methodology we employ the current numerical derivative with respect to time (Eq. (19.29)) [35]. The derivatives are obtained and represented in a LWTLP plot. The current derivative corresponding to sample $i + 1$ is plotted versus the current derivative of sample i. The current stable levels are not analyzed with this technique; instead, the temporal sequence of current values are studied; i.e., the occurrence probability of changes between current states (p_1, p_2, p_3 and p_4) and the presence of spikes (p_5), as described in [35].

$$\frac{dI_i}{dt} \approx \frac{I_{i+1} - I_i}{t_{i+1} - t_i}, \; i = 1, \ldots, n - 1. \tag{19.29}$$

The DLWTLP of trace #975 is shown in Fig. 19.29. The occurrence of p_5 indicates the presence of spikes in the RTN signal from state 0-1-0. The marks p_1 and p_2 indicate the transitions from state 0 to state 1, and the presence of marks p_3 and p_4 indicates the transitions from state 1 to state 0.

19.8 Conclusion

In this chapter, the resistive switching properties of HfO_2 VCM-based memristors have been described, and their potential to emulate the synaptic functionality in biological-inspired neuromorphic networks has been discussed. The results presented show that the spike-timing dependent plasticity learning rule of biological synapses has been validated in these devices, indicating their capability to tune the synaptic weight update in neuromorphic systems. However, the spike time and shape have to be carefully designed as these parameters can strongly influence the programming energy consumption and STDP behavior. Furthermore, some reliability concerns such as the stability of the synaptic state with time, the programming energy consumption, and the presence of parasitic effects such as the intrinsic series resistance have been presented. The experimental results indicate that the intrinsic series resistance and the applied electric field are two key parameters that determine the energy consumption and the transient characteristics of these devices.

In addition, simulation tools for different types of memristors have been introduced. In particular, kinetic Monte Carlo simulation schemes have been described and their most representative results have been shown. The stochastic nature of the formation and rupture processes linked to conductive filaments has been explained in-depth. The kinetic Monte Carlo procedures allow to study resistive switching at the atomic level and this feature gives us access to internal variables such as the device temperature, the electric potential distribution and the geometric features of the conductive filaments. In addition, memristor compact modeling has been described in detail. Different models have been presented in terms of the mathematical description of the redox reactions controlling the conductive filaments evolution and the thermal effects. Finally, random telegraph noise signals have been analyzed by means of different numerical techniques. Their stochastic features and the physics behind the behavior of these noise signals have been studied from the fundamental and the modeling viewpoints.

Acknowledgments

This work has been partially funded by the Spanish Ministry of Science, Innovation and Universities and the ERDF program through projects TEC2017-84321-C4-1-R, TEC2017-84321C4-3-R. The support of FENDECYT project, reference 3180130, is also acknowledged.

References

[1] S. Aldana, P. García-Fernández, A. Rodríguez-Fernández, R. Romero-Zaliz, M.B. González, F. Jiménez-Molinos, J.B. Roldán, A 3D kinetic Monte Carlo simulation study of resistive switching processes in Ni/HfO_2/Si-n^+-based RRAMs, Journal of Physics. D, Applied Physics 50 (33) (2017) 335103.

[2] S. Aldana, J.B. Roldán, P. García-Fernández, J. Suñe, R. Romero-Zaliz, F. Jiménez-Molinos, S. Long, F. Gómez-Campos, M. Liu, An in-depth description of bipolar resistive switching

in Cu/HfO_x/Pt devices, a 3D kinetic Monte Carlo simulation approach, Journal of Applied Physics 123 (15) (2018) 154501.

[3] S. Aldana, P. García-Fernández, R. Romero-Zaliz, F. Jiménez-Molinos, F. Gómez-Campos, J.B. Roldán, Analysis of conductive filament density in resistive RAMs, a 3D kinetic Monte Carlo approach, Journal of Vacuum Science & Technology B 36 (6) (2018) 62201.

[4] S. Aldana, P. García-Fernández, R. Romero-Zaliz, M.B. González, F. Jiménez-Molinos, F. Gómez-Campos, F. Campabadal, J.B. Roldán, Resistive switching in HfO2 based valence change memories, a comprehensive 3D kinetic Monte Carlo approach, Journal of Physics. D, Applied Physics 53 (22) (2020) 225106.

[5] M. Azzaz, A. Benoist, E. Vianello, D. Garbin, E. Jalaguier, C. Cagli, et al., Improvement of performances HfO2-based RRAM from elementary cell to 16kb demonstrator by introduction of thin layer of Al_2O_3, Solid-State Electronics 125 (2016) 1–7.

[6] W. Banerjee, Q. Liu, H. Lv, S. Long, M. Liu, Electronic imitation of behavioral and psychological synaptic activities using TiO_x/Al_2O_3-based memristor devices, Nanoscale 9 (2017) 14442–14450.

[7] G. Bersuker, D.C. Gilmer, D. Veksler, P. Kirsch, L. Vandelli, A. Padovani, L. Larcher, K. McKenna, A. Shluger, V. Iglesias, M. Porti, M. Nafría, Metal oxide resistive memory switching mechanism based on conductive filament properties, Journal of Applied Physics 110 (12) (2011) 124518.

[8] M. Bocquet, D. Deleruyelle, C. Muller, J.-M. Portal, Self-consistent physical modeling of set/reset operations in unipolar resistive-switching memories, Applied Physics Letters 98 (26) (2011) 263507.

[9] M. Bocquet, D. Deleruyelle, H. Aziza, C. Muller, J.-M. Portal, Compact modeling solutions for OxRAM memories, in: 2013 IEEE Faible Tension Faible Consommation, IEEE, 2013, pp. 1–4.

[10] M. Bocquet, H. Aziza, W. Zhao, Y. Zhang, S. Onkaraiah, C. Muller, et al., Compact modeling solutions for oxide-based resistive switching memories (OxRAM), Journal of Low Power Electronics & Applications 4 (1) (2014) 1–14.

[11] G.W. Burr, R.M. Shelby, C.D. Nolfo, J.W. Jang, R.S. Shenov, P. Narayanan, et al., Experimental demonstration and tolerancing of a large-scale neural network (165,000 synapses), using phase-change memory as the synaptic weight element, in: 2014 International Electron Devices Meeting, IEDM, IEEE, 2014, pp. 29.5.1–29.5.4.

[12] B. Butcher, G. Bersuker, D.C. Gilmer, L. Larcher, A. Padovani, L. Vandelli, et al., Connecting the physical and electrical properties of Hafnia-based RRAM, in: 2013 International Electron Devices Meeting, IEDM, IEEE, 2013, pp. 22.2.1–22.2.4.

[13] K.A. Campbell, K.T. Drake, E.H. Barney Smith, Pulse shape and timing dependence on the spike-timing dependent plasticity response of ion-conducting memristors as synapses, Frontiers in Bioengineering and Biotechnology 4 (97) (2016) 1–11.

[14] R. Carboni, D. Ielmini, Stochastic memory devices for security and computing, Advanced Electronic Materials 5 (9) (2019) 1900198.

[15] P.-Y. Chen, S. Yu, Compact modeling of RRAM devices and its applications in 1T1R and 1S1R array design, IEEE Transactions on Electron Devices 62 (12) (2015) 4022–4028.

[16] Y. Chen, B. Lin, I.-T. Wang, T.-H. Hou, J. Ye, J. Vrudhula, et al., Mitigating effects of non-ideal synaptic device characteristics for on-chip learning, in: 2015 ACM International Conference on Computer-Aided Design, ICCAD, IEEE, 2015, pp. 194–199.

[17] F.-C. Chiu, A review on conduction mechanisms in dielectric films, Advances in Materials Science and Engineering 2014 (2014) 1–18.

[18] E. Covi, S. Brivio, A. Serb, T. Prodromakis, M. Fanciulli, S. Spiga, HfO_2-based memristors for neuromorphic applications, in: Proceedings. IEEE International Symposium on Circuits and Systems (ISCAS), IEEE, 2016, pp. 393–396.

[19] S. Dirkmann, J. Kaiser, C. Wenger, T. Mussenbrock, Filament growth and resistive switching in hafnium oxide memristive devices, ACS Applied Materials & Interfaces 10 (17) (2018) 14857–14868.

[20] X. Fang, X. Yang, J. Wu, X. Yi, A compact SPICE model of unipolar memristive devices, IEEE Transactions on Nanotechnology 12 (5) (2013) 843–850.
[21] Z. Fang, X.P. Wang, J. Sohn, B. Bin Weng, Z.P. Zhang, Z.X. Chen, et al., The role of Ti capping layer in HfOx-based RRAM devices, IEEE Electron Device Letters 35 (9) (2014) 912–914.
[22] A. Fantini, D.J. Wouters, R. Degraeve, L. Goux, L. Pantisano, G. Kar, et al., Intrinsic switching behavior in HfO_2 RRAM by fast electrical measurements on novel 2R test structures, in: 2012 4th IEEE International Memory Workshop, IMW, IEEE, 2012, p. 6213646.
[23] J. Frascaroli, S. Brivio, E. Covi, S. Spiga, Evidence of soft bound behaviour in analogue memristive devices for neuromorphic computing, Scientific Reports 8 (2018) 7178.
[24] M.B. González, J.M. Rafí, O. Beldarrain, M. Zabala, F. Campabadal, Analysis of the switching variability in Ni/HfO_2-based RRAM devices, IEEE Transactions on Device and Materials Reliability 14 (2) (2014) 769–771.
[25] M.B. González, J. Martin-Martínez, M. Maestro, M.C. Acero, M. Nafría, F. Campabadal, Investigation of filamentary current fluctuations features in the high-resistance state of Ni/HfO_2-based RRAM, IEEE Transactions on Electron Devices 63 (8) (2016) 3116–3122.
[26] M.B. Gonzalez, M. Maestro-Izquierdo, F. Campabadal, S. Aldana, F. Jiménez-Molinos, J.B. Roldan, Impact of intrinsic series resistance on the reversible dielectric breakdown kinetics in HfO_2 memristors, in: 2020 International Reliability Physics Symposium, IRPS, IEEE, 2020, pp. 1–4.
[27] G. González-Cordero, J.B. Roldán, F. Jiménez-Molinos, J. Suñé, S. Long, M. Liu, A new compact model for bipolar RRAMs based on truncated cone conductive filaments, a Verilog-A approach, Semiconductor Science and Technology 31 (11) (2016) 1–13.
[28] G. González-Cordero, F. Jiménez-Molinos, J.B. Roldán, M.B. González, F. Campabadal, Transient SPICE simulation of Ni/HfO_2/Si-n^+ resistive memories, in: 2016 Conference on Design of Circuits and Integrated Systems, DCIS, IEEE, 2016, pp. 1–6.
[29] G. González-Cordero, J.B. Roldán, F. Jiménez-Molinos, Simulation of RRAM memory circuits, a Verilog-A compact modeling approach, in: 2016 Conference on Design of Circuits and Integrated Systems, DCIS, IEEE, 2016.
[30] G. González-Cordero, F. Jiménez-Molinos, J.B. Roldán, M.B. González, F. Campabadal, In-depth study of the physics behind resistive switching in TiN/Ti/HfO_2/W structures, Journal of Vacuum Science & Technology B 35 (1) (2017) 01A110.
[31] G. González-Cordero, M.B. González, H. García, F. Campabadal, S. Dueñas, H. Castán, F. Jiménez-Molinos, J.B. Roldán, A physically based model for resistive memories including a detailed temperature and variability description, Microelectronic Engineering 178 (2017) 26–29.
[32] G. González-Cordero, M.B. González, H. García, F. Jiménez-Molinos, F. Campabadal, S. Dueñas, H. Castán, J.B. Roldán, A physically based model to describe resistive switching in different RRAM technologies, in: 2017 Spanish Conference on Electron Devices, CDE, IEEE, 2017, pp. 1–4.
[33] G. González-Cordero, M.B. González, F. Jiménez-Molinos, F. Campabadal, J.B. Roldán, New method to analyze random telegraph signals in resistive random access memories, Journal of Vacuum Science & Technology B 37 (1) (2019) 012203.
[34] G. González-Cordero, M. Pedro, J. Martin-Martinez, M.B. González, F. Jiménez-Molinos, F. Campabadal, N. Nafría, J.B. Roldán, Analysis of resistive switching processes in TiN/Ti/HfO_2/W devices to mimic electronic synapses in neuromorphic circuits, Solid-State Electronics 157 (2019) 25–33.
[35] G. González-Cordero, M.B. González, F. Campabadal, F. Jiménez-Molinos, J.B. Roldán, A new technique to analyze RTN signals in resistive memories, Microelectronic Engineering 215 (2019) 110994.
[36] G. González-Cordero, M.B. González, A. Morell, F. Jiménez-Molinos, F. Campabadal, J.B. Roldán, Neural network based analysis of random telegraph noise in resistive random access memories, Semiconductor Science and Technology 35 (2020) 025021.

[37] B. Govoreanu, G.S. Kar, Y.-Y. Chen, V. Paraschiv, S. Kubicek, M. Jurczak, 10 × 10 nm^2 Hf/HfO_x crossbar resistive RAM with excellent performance, reliability and low-energy operation, in: 2011 International Electron Devices Meeting, IEDM, IEEE, 2011, pp. 31.6.1–31.6.4.
[38] X. Guan, S. Yu, H.S.P. Wong, On the switching parameter variation of metal-oxide RRAM—Part I: physical modeling and simulation methodology, IEEE Transactions on Electron Devices 59 (4) (2012) 1172–1182.
[39] X. Guan, S. Yu, H.S.P. Wong, A SPICE compact model of metal oxide resistive switching memory with variations, IEEE Electron Device Letters 33 (10) (2012) 1405–1407.
[40] X. Guan, S. Yu, H.S.P. Wong, On the variability of HfO_x RRAM: from numerical simulation to compact modeling, in: Technical Proceedings of the 2012 NSTI Nanotechnology Conference and Expo, NSTI-Nanotech 2012, Vol. 2, 2012, pp. 815–820.
[41] J. Guy, G. Molas, P. Blaise, M. Bernard, A. Roule, G. Le Carval, V. Delaye, A. Toffoli, G. Ghibaudo, F. Clermidy, B. De Salvo, L. Perniola, Investigation of forming, SET, and data retention of conductive-bridge random-access memory for stack optimization, IEEE Transactions on Electron Devices 62 (11) (2015) 3482–3489.
[42] A. Hardtdegen, C.L. Torre, F. Cüppers, S. Menzel, R. Waiser, S. Hoffmann-Eifert, Improved switching stability and the effect of an internal series resistor in HfO_2/TiO_x bilayer ReRAM cells, IEEE Transactions on Electron Devices 65 (8) (2018) 3229–3236.
[43] P. Huang, X.Y. Liu, B. Chen, H.T. Li, Y.J. Wang, Y.X. Deng, K.L. Wei, L. Zeng, B. Gao, G. Du, X. Zhang, J.F. Kang, A physics-based compact model of metal-oxide-based RRAM DC and AC operations, IEEE Transactions on Electron Devices 60 (12) (2013) 4090–4097.
[44] D. Ielmini, Filamentary-switching model in RRAM for time, energy and scaling projections, in: 2011 International Electron Devices Meeting, IEDM, IEEE, 2011, pp. 17.2.1–17.2.4.
[45] D. Ielmini, F. Nardi, S. Balatti, Evidence for voltage-driven set/reset processes in bipolar switching RRAM, IEEE Transactions on Electron Devices 59 (8) (2012) 2049–2056.
[46] D. Ielmini, Resistive switching memories based on metal oxides: mechanisms, reliability and scaling, Semiconductor Science and Technology 31 (6) (2016) 063002.
[47] D. Ielmini, V. Milo, Physics-based modeling approaches of resistive switching devices for memory and in-memory computing applications, Journal of Computational Electronics 16 (4) (2017) 1121–1143.
[48] D. Ielmini, Brain-inspired computing with resistive switching memory (RRAM): devices, synapses and neural networks, Microelectronic Engineering 190 (2018) 44–53.
[49] A.L. Jagath, C.H. Leong, T.N. Kumar, H.A.F. Almurib, Insight into physics-based RRAM models – review, The Journal of Engineering 7 (2019) 4644–4652.
[50] Z. Jiang, S. Yu, Y. Wu, J.H. Engel, X. Guan, H.-S.P. Wong, Verilog-A compact model for oxide-based resistive random access memory (RRAM), in: 2014 International Conference on Simulation of Semiconductor Processes and Devices, SISPAD, IEEE, 2014, pp. 41–44.
[51] Z. Jiang, Y. Wu, S. Yu, L. Yang, K. Song, Z. Karim, H.-S.P. Wong, A compact model for metal–oxide resistive random access memory with experiment verification, IEEE Transactions on Electron Devices 63 (5) (2016) 1884–1892.
[52] F. Jiménez-Molinos, M.A. Villena, J.B. Roldan, A.M. Roldan, A SPICE compact model for unipolar RRAM reset process analysis, IEEE Transactions on Electron Devices 62 (3) (2015) 955–962.
[53] S.H. Jo, T. Chang, I. Ebong, B.B. Bhadviya, P. Mazumder, W. Lu, Nanoscale memristor device as synapse in neuromorphic systems, Nano Letters 10 (2010) 1297–1301.
[54] S. Kim, S. Choi, W. Lu, Comprehensive physical model of dynamic resistive switching in an oxide memristor, ACS Nano 8 (3) (2014) 2369–2376.
[55] D. Kuzum, S. Yu, H-S.P. Wong, Synaptic electronics: materials, devices and applications, Nanotechnology 24 (2013) 382001.
[56] D. Kwon, K.M. Kim, J.H. Jang, J.M. Jeon, M.H. Lee, G.H. Kim, et al., Atomic structure of conducting nanofilaments in TiO_2 resistive switching memory, Nature Nanotechnology 5 (2) (2010) 148–153.

[57] M. Lanza, A review on resistive switching in high-k dielectrics: a nanoscale point of view using conductive atomic force microscope, Materials 7 (3) (2014) 2155–2182.
[58] M. Lanza, H.-S.P. Wong, E. Pop, D. Ielmini, D. Strukov, B.C. Regan, et al., Recommended methods to study resistive switching devices, Advanced Electronic Materials 5 (1) (2019) 1800143.
[59] L. Larcher, A. Padovani, L. Vandelli, A simulation framework for modeling charge transport and degradation in high-k stacks, Journal of Computational Electronics 12 (4) (2013) 658–665.
[60] H.Y. Lee, Y.S. Chen, P.S. Chen, P.Y. Gu, Y.Y. Hsu, S.M. Wang, et al., Evidence and solution of over-RESET problem for HfO_x based resistive memory with sub-ns switching speed and high endurance, in: 2010 IEEE International Electron Devices Meeting, IEDM, IEEE, 2010, pp. 19.7.1–19.7.4.
[61] H. Li, P. Huang, B. Gao, B. Chen, X. Liu, J. Kang, A SPICE model of resistive random access memory for large-scale memory array simulation, IEEE Electron Device Letters 35 (2) (2014) 211–213.
[62] H. Li, Z. Jiang, P. Huang, Y. Wu, H.-Y. Chen, B. Gao, et al., Variation-aware, reliability-emphasized design and optimization of RRAM using SPICE model, in: 2015 Design, Automation & Test in Europe Conference & Exhibition, DATE, IEEE, 2015, pp. 1425–1430.
[63] S. Long, X. Lian, C. Cagli, X. Cartoixà, R. Rurali, E. Miranda, et al., Quantum-size effects in hafnium-oxide resistive switching, Applied Physics Letters 102 (18) (2013) 183505.
[64] W. Lu, D.S. Jeong, M. Kozicki, R. Waser, Electrochemical metallization cells—blending nanoionics into nanoelectronics?, MRS Bulletin 37 (2) (2012) 124–130.
[65] M. Maestro-Izquierdo, M.B. Gonzalez, F. Campabadal, Mimicking the spike-timing dependent plasticity in HfO_2-based memristors at multiple time scales, Microelectronic Engineering 215 (2019) 111014.
[66] C. Márquez, N. Rodríguez, F. Gámiz, R. Ruiz, A. Ohata, Electrical characterization of random telegraph noise in fully-depleted silicon-on-insulator MOSFETs under extended temperature range and back-bias operation, Solid-State Electronics 117 (2016) 60–65.
[67] J. Martín-Martínez, J. Diaz, R. Rodriguez, M. Nafria, X. Aymerich, New weighted time lag method for the analysis of random telegraph signals, IEEE Electron Device Letters 35 (4) (2014) 479–481.
[68] J. Martín-Martínez, J. Díaz, R. Rodríguez, M. Nafria, X. Aymerich, E. Roca, et al., Characterization of random telegraph noise and its impact on reliability of SRAM sense amplifiers, in: 2014 European Workshop on CMOS Variability (VARI), IEEE, 2014, pp. 1–6.
[69] Y. Matveyev, R. Kirtaev, A. Fetisova, S. Zakharchenko, D. Negrov, A. Zenkevich, Crossbar nanoscale HfO_2-based electronics synapses, Nanoscale Research Letters 11 (147) (2016) 1–6.
[70] S. Menzel, B. Klopstra, C. Kügeler, U. Böttger, G. Staikov, R. Waser, A simulation model of resistive switching in electrochemical metallization memory cells (ECM), MRS Online Proceedings Library Archive 1160 (2009) 101–106.
[71] S. Menzel, P. Kaupmann, R. Waser, Understanding filamentary growth in electrochemical metallization memory cells using kinetic Monte Carlo simulations, Nanoscale 7 (2015) 12673.
[72] S. Menzel, U. Böttger, M. Wimmer, M. Salinga, Physics of the switching kinetics in resistive memories, Advanced Functional Materials 25 (2015) 6306–6325.
[73] E.A. Miranda, C. Walczyk, C. Wenger, T. Schroeder, Model for the resistive switching effect in HfO_2 MIM structures based on the transmission properties of narrow constrictions, IEEE Electron Device Letters 31 (6) (2010) 609–611.
[74] E. Miranda, D. Jiménez, J. Suñé, The quantum point-contact memristor, IEEE Electron Device Letters 33 (10) (2012) 1474–1476.
[75] E. Miranda, J. Muñoz-Gorriz, J. Suñé, K. Fröhlich, SPICE model for the current-voltage characteristic of resistive switching devices including the snapback effect, Microelectronic Engineering 215 (2019) 110998.

[76] T. Nagumo, K. Takeuchi, S. Yokogawa, K. Imai, Y. Hayashi, New analysis methods for comprehensive understanding of random telegraph noise, in: 2009 IEEE International Electron Devices Meeting, IEDM, IEEE, 2009, pp. 1–4.
[77] G. Niu, M.A. Schubert, S.U. Sharath, P. Zaumseil, S. Vogel, C. Wenger, et al., Electron holography on HfO_2/HfO_{2-x} bilayer structures with multilevel resistive switching properties, Nanotechnology 28 (21) (2017) 215702.
[78] N. Onofrio, D. Guzman, A. Strachan, Atomic origin of ultrafast resistance switching in nanoscale electrometallization cells, Nature Materials 14 (4) (2015) 440–446.
[79] A. Padovani, L. Larcher, O. Pirrotta, L. Vandelli, G. Bersuker, Microscopic modeling of HfO_x RRAM operations: from forming to switching, IEEE Transactions on Electron Devices 62 (6) (2015) 1998–2006.
[80] A. Padovani, J. Woo, H. Hwang, L. Larcher, Understanding and optimization of pulsed SET operation in HfO_x-based RRAM devices for neuromorphic computing applications, IEEE Electron Device Letters 39 (5) (2018) 672–675.
[81] F. Pan, S. Yin, V. Subramanian, A detailed study of the forming stage of an electrochemical resistive switching memory by KMC simulation, IEEE Electron Device Letters 32 (7) (2011) 949–951.
[82] F. Pan, S. Gao, C. Chen, C. Song, F. Zeng, Recent progress in resistive random access memories: materials, switching mechanisms and performance, Materials Science & Engineering R 83 (1) (2014) 1–59.
[83] D. Panda, P.P. Sahu, T.Y. Tseng, A collective study on modeling and simulation of resistive random access memory, Nanoscale Research Letters 13 (2018) 8.
[84] M. Pedro, J. Martín-Martínez, E. Miranda, R. Rodríguez, M. Nafria, M.B. Gonzalez, F. Campabadal, Device variability tolerance of a RRAM-based self-organizing neuromorphic system, in: 2018 IEEE International Reliability Physics Symposium, IRPS, IEEE, 2018, pp. P-CR.4.1–P-CR.4.5.
[85] E. Perez, A. Grossi, C. Zambelli, P. Olivo, C. Wenger, Impact of the incremental programming algorithm on the filament conduction in HfO_2-based RRAM arrays, IEEE Journal of the Electron Devices Society 5 (1) (2017) 64–68.
[86] E. Pérez, D. Maldonado, C. Acal, J.E. Ruiz-Castro, F.J. Alonso, A.M. Aguilera, F. Jiménez-Molinos, Ch. Wenger, J.B. Roldán, Analysis of the statistics of device-to-device and cycle-to-cycle variability in $TiN/Ti/Al{:}HfO_2/TiN$ RRAMs, Microelectronics Engineering 214 (2019) 104–109.
[87] S. Poblador, M.C. Acero, M.B. Gonzalez, F. Campabadal, Resistive switching with bipolar characteristics in $TiN/Ti/HfO_2/W$ devices, in: 2017 Spanish Conference on Electron Devices, CDE, IEEE, Barcelona, Spain, 2017, pp. 1–4.
[88] S. Poblador, M.B. Gonzalez, F. Campabal, Investigation of the multilevel capability of $TiN/Ti/HfO_2/W$ resistive switching devices by sweep and pulse programming, Microelectronic Engineering 187–188 (2018) 148–153.
[89] M. Prezioso, F. Merrikh Bayat, B. Hoskins, K. Likharev, D. Strukov, Self-adaptive spike-time-dependent plasticity of metal-oxide memristors, Scientific Reports 6 (2016) 21331.
[90] F.M. Puglisi, P. Pavan, A. Padovani, L. Larcher, G. Bersuker, RTS noise characterization of HfO_x RRAM in high resistive state, Solid-State Electronics 84 (2013) 160–166.
[91] F.M. Puglisi, P. Pavan, L. Vandelli, A. Padovani, M. Bertocchi, L. Larcher, A microscopic physical description of RTN current fluctuations in HfO_x RRAM, in: 2015 IEEE International Reliability Physics Symposium Proceedings, IRPS, IEEE, 2015, pp. 5B51–5B56.
[92] F.M. Puglisi, P. Pavan, L. Larcher, A. Padovani, Statistical analysis of random telegraph noise in HfO_2-based RRAM devices in LRS, Solid-State Electronics 113 (2015) 132–137.
[93] F.M. Puglisi, N. Zagni, L. Larcher, P. Pavan, Random telegraph noise in resistive random access memories: compact modeling and advanced circuit design, IEEE Transactions on Electron Devices 65 (7) (2018) 2964–2972.
[94] S. Realov, K.L. Shepard, Analysis of random telegraph noise in 45-nm CMOS using on-chip characterization system, IEEE Transactions on Electron Devices 60 (5) (2013) 1716–1722.

[95] J.B. Roldán, E. Miranda, G. González-Cordero, P. García-Fernández, R. Romero-Zaliz, P. González-Rodelas, A.M. Aguilera, M.B. González, F. Jiménez-Molinos, Multivariate analysis and extraction of parameters in resistive RAMs using the Quantum Point Contact model, Journal of Applied Physics 123 (2018) 014501.

[96] J.B. Roldán, F.J. Alonso, A.M. Aguilera, D. Maldonado, M. Lanza, Time series statistical analysis: a powerful tool to evaluate the variability of resistive switching memories, Journal of Applied Physics 125 (2019) 174504.

[97] K. Seo, I. Kim, S. Jung, M. Jo, S. Park, J. Park, H. Hwang, Analog memory and spike-timing-dependent plasticity characteristics of a nanoscale titanium oxide bilayer resistive switching device, Nanotechnology 22 (2011) 254023.

[98] M. Si, N.J. Conrad, S. Shin, J. Gu, J. Zhang, M.A. Alam, P.D. Ye, Low-frequency noise and random telegraph noise on near-ballistic III–V MOSFETs, IEEE Transactions on Electron Devices 62 (11) (2015) 3508–3515.

[99] S. Song, K.D. Miller, L.F. Abbott, Competitive Hebbian learning through spike-timing-dependent synaptic plasticity, Nature Neuroscience 3 (9) (2000) 919–926.

[100] S.H. Tan, P. Lin, H. Yeon, S. Choi, Y. Park, J. Kim, Perspective: uniform switching of artificial synapses for large-scale neuromorphic arrays, APL Materials 6 (2018) 120901.

[101] T. Tsuruoka, K. Terabe, T. Hasegawa, M. Aono, Forming and switching mechanisms of a cation-migration-based oxide resistive memory, Nanotechnology 21 (42) (2010) 425205.

[102] L. Vandelli, A. Padovani, G. Bersuker, D. Gilmer, P. Pavan, L. Larcher, Modeling of the forming operation in HfO_2-based resistive switching memories, in: 2011 IEEE International Memory Workshop, IMW, IEEE, 2011, pp. 1–4.

[103] D. Veksler, G. Bersuker, B. Chakrabarti, E. Vogel, S. Deora, K. Matthews, et al., Methodology for the statistical evaluation of the effect of random telegraph noise (RTN) on RRAM characteristics, in: 2012 International Electron Devices Meeting, IEDM, IEEE, 2012, pp. 9.6.1–9.6.4.

[104] M.A. Villena, F. Jiménez-Molinos, J.B. Roldán, J. Suñé, S. Long, X. Lian, F. Gámiz, M. Liu, An in-depth simulation study of thermal reset transitions in resistive switching memories, Journal of Applied Physics 114 (14) (2013) 144505.

[105] M.A. Villena, M.B. González, F. Jiménez-Molinos, F. Campabadal, J.B. Roldán, J. Suñé, E. Romera, E. Miranda, Simulation of thermal reset transitions in RRAMs including quantum effects, Journal of Applied Physics 115 (2014) 214504.

[106] M.A. Villena, J.B. Roldán, F. Jiménez-Molinos, E. Miranda, J. Suñé, M. Lanza, SIM^2 $RRAM$: a physical model for RRAM devices simulation, Journal of Computational Electronics 16 (4) (2017) 1095–1120.

[107] A.F. Voter, Introduction to the kinetic Monte Carlo method, in: Radiation Effects in Solids, Springer, Netherlands, 2007.

[108] I.-T. Wang, Y.-C. Lin, Y.-F. Wang, C.-W. Hsu, T.-H. Hou, 3D synaptic architecture with ultralow sub-10 f. energy per spike for neuromorphic computation, in: 2014 International Electron Devices Meeting, IEDM, IEEE, 2014, pp. 28.5.1–28.5.4.

[109] Z.-Q. Wang, S. Ambrogio, S. Balatti, D. Ielmini, A 2-transistor/1-resistor artificial synapse capable of communication and stochastic learning for neuromorphic systems, Frontiers in Neuroscience 8 (2015) 438.

[110] R. Waser, M. Aono, Nanoionics-based resistive switching memories, Nature Materials 6 (2007) 833–840.

[111] R. Waser, D. Ielmini, Resistive Switching from Fundamentals of Nanoionic Redox Processes to Memristive Device Applications, Wiley-VCH Verlag GmbH & Co., 2016.

[112] D.J. Wouters, A. Fantini, R. Degraeve, L. Goux, Y.Y. Chen, B. Govoreanu, et al., Analysis of complementary RRAM switching, IEEE Electron Device Letters 33 (8) (2012) 1186–1188.

[113] D.J. Wouters, S. Menzel, J.A.J. Rupp, T. Hennen, R. Waser, On the universality of the I-V switching characteristics in no-volatile and volatile resistive switching oxides, Faraday Discussions 231 (2019) 183–196.

[114] Y. Yang, W. Lu, Nanoscale resistive switching devices: mechanisms and modelling, Nanoscale 5 (2013) 10076–10092.

[115] S. Yu, Y. Wu, H.-S.P. Wong, Investigating the switching dynamics and multilevel capability of bipolar metal oxide resistive switching memory, Applied Physics Letters 98 (2011) 103514.

[116] S. Yu, X. Guan, H.-S.P. Wong, On the stochastic nature of resistive switching in metal oxide RRAM: physical modeling, Monte Carlo simulation, and experimental characterization, in: 2011 IEEE International Electron Devices Meeting, IEDM, IEEE, 2011, pp. 17.3.1–17.3.4.

[117] S. Yu, B. Gao, Z. Fang, H. Yu, J. Kang, H.-S.P. Wong, A neuromorphic visual system using RRAM synaptic devices with Sub-pJ energy and tolerance to variability: experimental characterization and large-scale modeling, in: 2012 IEEE International Electron Devices Meeting, IEDM, IEEE, 2012, pp. 10.4.1–10.4.4.

[118] S. Yu, Neuro-Inspired Computing Using Resistive Switching Devices, Springer, Switzerland, 2017.

[119] Y. Zha, Z. Wei, J. Li, Recent progress in RRAM technology: from compact models to applications, in: 2017 China Semiconductor Technology International Conference, CSTIC, IEEE, 2017, pp. 1–4.

[120] C. Zamarreño-Ramos, L.A. Camuñas-Mesa, J.A. Pérez-Carrasco, T. Masquelier, T. Serrano-Gotarredona, B. Linares-Barranco, On spike-timing-dependent-plasticity, memristive devices, and building a self-learning visual cortex, Frontiers in Neuroscience 5 (26) (2011) 1–22.

[121] L. Zao, H.-Y. Chen, S.-C. Wu, Z. Jiang, S. Yu, T.-H. Hou, H.-S.P. Wong, Y. Nishi, Multi-level control of conductive nano-filament evolution in HfO_2 ReRAM by pulse-train operations, Nanoscale 6 (2014) 5698–5702.

[122] M.A. Zidan, J.P. Strachan, W.D. Lu, The future of electronics based on memristive systems, Nature Electronics 1 (2018) 22–29.

Chapter 20

Analog circuit integration of backpropagation learning in memristive HTM architecture

Olga Krestinskaya[a] and Alex James[b]

[a]*King Abdullah University of Science and Technology, Thuwal, Saudi Arabia*, [b]*Indian Institute of Information Technology and Management, Kerala, India*

20.1 Introduction

HTM is an algorithm and an architecture inspired by the structure and functionality of the human neocortex [1]. HTM is more generalized than traditional artificial neural networks. There have been several software-based HTM applications and hardware implementations of HTM proposed and tested in recent years [2–4]. This chapter focuses in memristive HTM implementations integrated with backpropagation circuits on hardware with analog domain computation. The main advantages of analog memristive hardware implementation of HTM and analog learning are the small on-chip area and the lower power consumption. Such architectures are also suitable for edge device computing and near-sensor processing [5].

While the idea of HTM is not to use the backpropagation learning and to rely on simpler Hebbian learning, as is done in the human brain, the HTM concepts can be combined with backpropagation learning for different applications. For example, a variety of research work illustrates the application of sparse distributed representation of the input patters encoded by the HTM Spatial Pooler in a combination with traditional learning algorithms based on backpropagation learning for object classification and face recognition problems [6–8]. In this work, the HTM Spatial Pooler is used as a feature extractor for the input patters followed by traditional backpropagation learning and classification of input patterns.

As in the neural networks there are a lot of synapses which have certain weights, the hardware implementation of such systems is a challenging issue. With the development of emerging devices like memristors, the memristive crossbar-based analog dot product computation showed a potential to replace transistor-based memory with ability to move the processing into the memory. Memristive crossbars are able to store a large amount of information, which is

Mem-elements for Neuromorphic Circuits with Artificial Intelligence Applications
https://doi.org/10.1016/B978-0-12-821184-7.00029-3

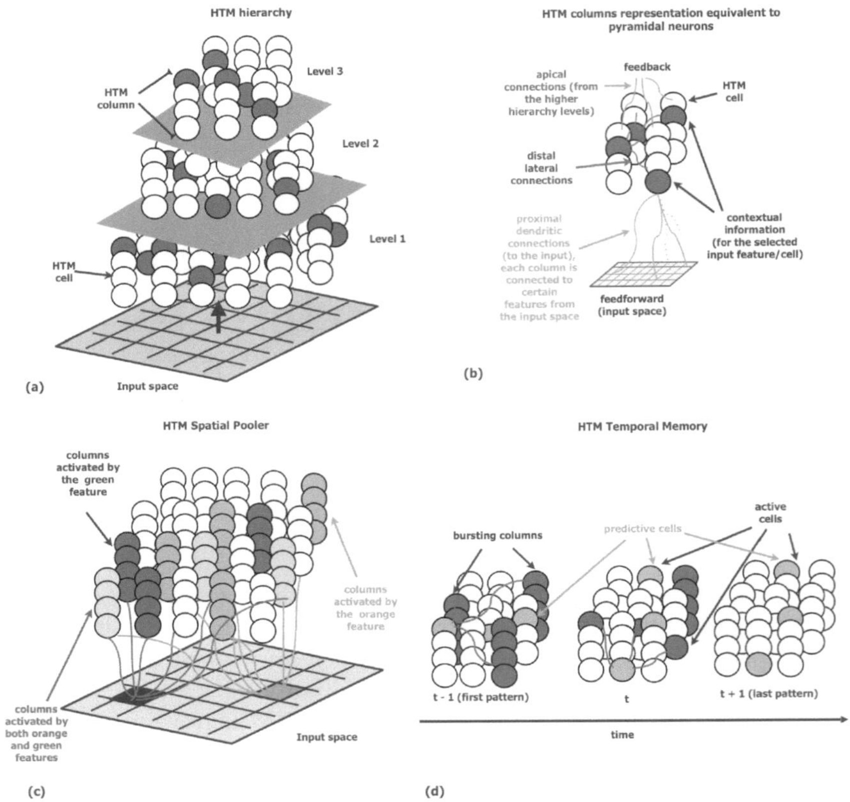

FIGURE 20.1 (a) Hierarchical structure of HTM. (b) HTM representation equivalent to the pyramidal neuron. (c) HTM SP. (d) HTM TM.

a promising solution for storage of the weights in neural networks and learning systems [9]. There have been several neural network implementations illustrating that memristive devices can be stable and can be utilized in neural network architectures for image processing and pattern recognition [10–12]. The HTM architecture can also be implemented on hardware using memristor-based circuits [13,14,8]. Such systems are covered in this chapter.

The chapter is organized in the following sections. Section 20.2 introduces the main principles of original HTM algorithm. Section 20.3 covers HTM architectures with memristive devices. Section 20.4 shows how analog backpropagation learning circuit is integrated into the HTM Spatial Pooler circuits used for the extraction of the input image features for pattern matching applications. Section 20.5 discusses the advantages of such architectures and covers the open problems. Section 20.6 concludes the chapter.

20.2 What is HTM?

HTM is a biologically plausible model of the neocortex that mimics its structure and functionality [15–17]. The human neocortex, often known as the storage of human intelligence, is divided into regions, which are connected together in a hierarchical manner. HTM can be considered as a more generalized neuronal network architecture, comparing to the traditional artificial neural networks which are based on the concept of data accumulation in the neuron and neuronal activation and are not structured as the networks of neurons in human brain. HTM is designed to process different sensory data in the same manner without differentiating between the sources of the data, whether it is visual data, audio signal or any other type of data fed into it. The main features of HTM are hierarchy, sparsity and modularity [18]. The hierarchical structure of HTM is illustrated in Fig. 20.1(a). HTM consists of several layers with columns. Each column consists of HTM cells. The lowest level receives the information from the input feature space. Also, HTM can have a topological structure, where the columns are divided into several regions. This is useful when the feature space is large, and the regions in the input space can be processed only by certain regions of HTM columns [19].

The connections between the columns and the layers in HTM architecture resemble the connections of the pyramidal neurons in the neocortex. The representation of the region of HTM columns and their connections, equivalent to the connections of the pyramidal neurons, are illustrated in Fig. 20.1(b). HTM columns have three different types of connections: proximal dendritic connections, apical connections and distal lateral connections. Proximal connections are the connections to the lower regions of hierarchy or sensory input space, which are different for each HTM columns. Apical connections represent the feedback from the higher levels of hierarchy. Meanwhile distal connections are the connections to the other columns at the same level of hierarchy.

HTM can be divided into two main parts: Spatial Pooling (SP) [20,21] and Temporal Memory (TM) [22]. The information that HTM processes is represented as Sparse Distributed Patterns (SDRs) [16]. SDRs are binary arrays where the information is sparse and can be represented as 1s or 0s, equivalent to ON and OFF states of HTM neurons [23,24]. HTM SP acts as a sensory encoder transferring the input data to sparse patterns containing the semantic information of this data and represented by the activity of the HTM neurons (columns). The sparsity concept refers to the percentage of the bits in the data representation that are ON. The main advantage of the sparse representations is the ability to store large amount of data in the arrays, where the sparse arrays have higher capacity than dense arrays.

The main idea of the HTM SP is to encode an input vector and convert it to SDR patterns, where each input pattern is represented as a set of active HTM columns. The HTM SP maintains the sparsity of the output and overlap properties of the input data, when the similar inputs produce similar outputs and vice versa. The HTM SP consists of initialization phase, overlap, inhibition and learning. The initialization phase is performed by only ones in the beginning. Certain HTM columns are selected randomly to receive the input data, where the permanence value of the connections (weight of the synapses) between the input space and the columns is set randomly. The permanence value of these synapses is compared to a certain threshold, and this determines whether the synapse is connected to the column or not. If the synapse is connected, it is identified as an active connected synapse. Fig. 20.1(c) shows that each cell (feature) in the input space can be connected to certain HTM columns, some columns are shared, while the others are not active at all. Each column has an overlap score, which is determined as a sum of all the features of the input space actively connected to this column. Based on the overlap scores the final set of actively connected columns is determined as a particular percentage of the total inhibition region. Each column also has a boosting factor, the multiplication factor that forces HTM columns to be more or less active. This is important for efficient representation of the input patterns and conversion to SDRs. Boosting forces the cellular activity to distribute over the columns. In the inhibition, the winning columns are selected and activated, while the others are inhibited. In the learning phase, the permanence values of the active columns are either increased, so the winning columns are learned (the connection is learned and becomes stronger) or decremented based on the input that is learned over time. Only active columns are updated during the learning phase, when the input space changes. Overlap, inhibition and learning phases are repeated each time new input space is fed to HTM.

The HTM TM (Sequence Memory in the later versions of cortical algorithms) is responsible for the continuous sequence learning of active cells in the columns over time and making the predictions of the next input. When HTM columns receive the input, it can predict the next input based on the learned sequences. The concept of the HTM TM is illustrated in Fig. 20.1 (d). The cells in HTM columns have distal connections to the cells in the other columns. These connections are learned and responsible for the prediction making. The HTM TM consists of two phases: the prediction of which columns will become active in the next time step, and the selection of the set of cells to be set to the predictive state. In the first time step or if the column has not formed any connections to the previously activated cells, the column is bursting. The cell in the column that will represent a new pattern is chosen as a cell with the least number of connections to the other cells. In the first phase, the cells that should be set to a predictive state are activated based on the context of the previous input. For the cells to become predictive, the distal connections to the other cells of the active columns are summed up across the synapses, and the summation result is compared with the threshold. In the second phase, this prediction is validated and the cells are switched to active state if predicted correctly.

HTM found its applications in various fields from simple pattern recognition tasks [25] to biomedical applications [26–28]. Recent work proposes to use HTM for real-time medical streams predictions [26], anomaly detection in ECG signal [28], time-series data analysis [29,30], document categorization [31], hydrological monitoring [32], licence plate recognition [25], face recognition [6–8] and pattern recognition tasks [13].

20.3 Memristive HTM architectures

There have been several attempts to implement HTM on various hardware platforms. FPGA implementation of HTM algorithm has been proposed recently to speed up the HTM computations [33]. There has been an attempt to integrate HTM cell to the Automata Processor [34] and map it to the SpiNNaker chip architecture [35]. Several digital [36,37], mixed signal [13] and analog designs [14,8] of HTM have been shown.

As the memristor-based neural networks and learning architectures gained the attention of the researches in the last decade [5,38], the memristor based HTM architecture is one of the promising solutions to reduce on-chip area and power consumption of HTM [39]. Memristive devices with small on-chip area, low power consumption and ability to be programmed enable fast and efficient processing in analog domain, especially for dot produce operation in the crossbar, reducing the requirements of having additional complex circuits for digital multiplication. Memristor based HTM architectures are proposed as one of the solutions to implement HTM on low power devices and edge device [39,5].

Recently, there have been several implementations of memristive HTM architectures [13,14,8]. In [13], the authors propose the implementation of the mixed-signal design of HTM architecture using memristive crossbars and spin devices. In this work, memristive crossbars are used to implement a hierarchy of the architecture and store the synaptic weights, and the system is successfully tested on handwritten digits recognition application using MNIST [40]. The development of CMOS-memristive circuits for the HTM SP implementation is illustrated in [14]. The system is tested on face recognition application with AR database [41]. The modified and improved HTM architecture with CMOS-memristive circuits and integrated Hebbian learning circuits, which was tested on face and speech recognition, is illustrated in [8]. Memristor-based HTM SP architecture with backpropagation learning for conventional pattern matching based face recognition is shown in [6]. The comparison of memristive HTM implementation with hybrid CNN-LSTM architecture is shown in [42].

20.4 Analog backpropagation circuit integration in memristive HTM

The integration of the analog learning circuit [43] to memristive HTM architecture is shown in [44,45]. Fig. 20.2 illustrates the integration of analog circuit blocks into two different memristive HTM architectures from [14] and [8]. In both of the cases, only the HTM SP is shown, which is used to extract the useful features and encode the images into binary SDRs for further processing

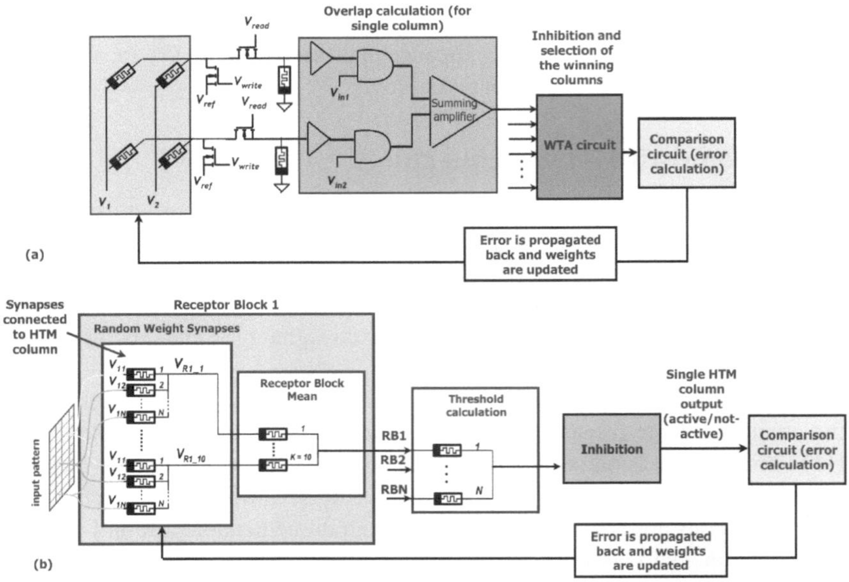

FIGURE 20.2 Integration of analog backpropagation circuit into (a) memristive HTM architecture and (b) modified memristive HTM architectures.

using conventional template-matching approach. The analog backpropagation circuit is integrated into the designs for adjusting the synaptic connections and make sure that obtained SDRs can be recognized using a conventional template-matching method. During the learning/training stages in such systems, the synapses are updated to perform the template-matching based on SDRs, where the ideal template is saved as an initial SDR pattern or an averaged image of several SDRs of the patterns from the same class [8]. During the inference/testing stage, the SDR of an input image pattern is compared to the stored templates of different classes using a pattern matcher circuit [8].

Fig. 20.2(a) shows a processing in a single HTM column in the HTM SP, where overlap is calculated based on the connected synapses in the crossbar and summed using a summing amplifier. After the overlap calculation and obtaining the summation score of all the connected synapses to the column, the winning columns are determined by a current-based Winner-Takes-All (WTA) circuit. The output of the WTA circuit is represented as a sparse pattern of 0-s and 1-s, so some of the outputs are inhibited and the others are active, which corresponds to the activation of the HTM columns and obtaining SDRs. The learning circuit consists of the comparison circuit performing the calculation of the error between the obtained SDR and the ideal template corresponding to a particular class of the input pattern. The synaptic connections represented by the memristors are updated based on the obtained error.

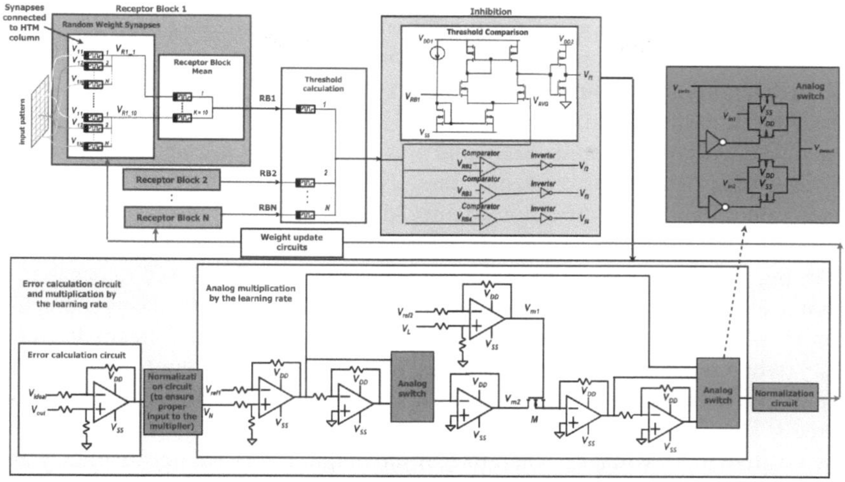

FIGURE 20.3 Analog backpropagation circuit integrated into the modified HTM architecture for pattern recognition applications, where HTM is used for the extraction of the meaningful features and recognition is performed using conventional template-matching method.

Comparing to Fig. 20.2(a) where the output sparse patterns are obtained using the WTA circuit (selecting maximum based on the output of the summing amplifier), the modified HTM circuit shown in Fig. 20.2(b) performs the selection of the output columns (SDRs) using mean calculation and comparing it to the threshold. This approach allows one to reduce on-chip area of the analog HTM circuit and power consumption keeping the recognition accuracy high [8], as the mean calculation requires only the memristive devices (see Threshold calculation in Fig. 20.2(b)). In the modified HTM circuit, the memristive synapses in the receptor block are set randomly to the R_{ON} or R_{OFF} state in the initialization state, and then averaged by the Receptor Mean Block to ensure completely random connections. The mean value serves as a threshold for inhibition calculation. The inhibition block either inhibits the columns, or outputs are set to a high value, comparing the outputs of the receptor blocks RB to the threshold value from the threshold calculation circuit. The output/activity of each HTM column is represented as an output of a single inhibition block. As in the previous example (Fig. 20.2(a)), the error between ideal and real SDRs is calculated for comparison of the patterns in the template-matching method. The memristors in receptor blocks are updated based on this error.

The detailed implementation of an analog backpropagation circuit for the HTM SP proposed in [6] is illustrated in Fig. 20.3. The example shows the integration of an analog backpropagation circuit into the modified HTM architecture. As SDRs are binary and the HTM SP circuit does not have several layers, the backpropagation circuit is simplified comparing to the one presented in [44] for the neural networks. The circuit involves the error calculation and multiplication by the learning rate to update the memristive devices in the receptor blocks. The output of the inhibition block representing an SDR of the input pattern is fed to the error calculation circuit, which consists of the difference amplifier calculating the difference between the output SDR V_{out} from the inhibition block and saved template V_{ideal}. The saved template can be binary, if it consists of a single SDR, and non-binary, if it is averaged from several SDRs of the patterns of the same class. Therefore, the difference amplifier is used rather than the comparator circuit. After the calculation of the error, it is multiplied by the learning rate using the analog multiplication circuit. As the learning rate can change, the analog multiplication circuit is used rather than a simple amplification circuit. The multiplication is performed by a single transistor M, where the output is proportional to the product of input voltages V_{m1} and V_{m2} (when transistor is in the linear region). To ensure the input voltage to be small, the normalization circuit scales down the output of the error calculation circuit. This is a two quadrant multiplier, and it performs multiplication only for the positive input voltages. Therefore, if the output errors are negative, they are converted to positive using an analog switch, and then converted back after the multiplication using the other analog switch. If the input error is positive, the analog switch does not convert it. The operating principle of the analog switch is based on the thresholding of the input voltage. If the input voltage V_{swin} drives the inverter to the high state, V_{swout} is set to V_{in1}, and if the inverter output is set to a low state, $V_{swout} = V_{in2}$. The difference amplifiers before the analog switch ensure the correct input to the analog switch shifting the input signal V_N by the reference voltage V_{ref1} considering the threshold of the inverters (which depends on the transistor geometry and the technology). The learning rate is represented by V_L and is also shifted before applying to the transistor M by V_{ref2}. The output of the multiplication circuit is applied to the normalization circuit to amplify the output after the initial normalization. The output of the second normalization circuit is used to update the memristive synapses in the receptor blocks.

The application of analog backpropagation circuit in the template-matching based learning with the HTM SP for feature extraction showed that the results achieved for face recognition application with AR database [41] can reach 90%, which is higher than the results illustrated in previous equivalent work on the HTM SP application for meaningful feature extraction [6].

20.5 Discussion and open problems

The ability of the HTM SP to convert input patterns to SDRs has been proven to be useful for feature extraction applications. The HTM SP concept has been successfully used for the extraction of useful features from the images converting them to sparse patterns followed by the conventional backpropagation learning and pattern matching based classification. Even though the hardware based systems do not implement the complete HTM system, the concept of sparsity of HTM neurons and SDRs can be useful for different applications.

The implementation of the HTM in analog hardware with memristive devices has such advantages, as low power consumption, small on-chip area and scalability. In order to build large learning systems for more complex applications, the memory devices and processing circuits should be scaled. Due to the small size, the memristor-based synapses can be easily scaled, comparing to CMOS-based implementations, which can be scaled only up to a certain extent. Also, due to the small size and reduced power consumption memristor-based analog HTM architectures can potentially be integrated into edge devices and be applied for near-sensor processing [46,47].

Besides a certain progress in the implementation of analog learning circuits and HTM in analog hardware, there are plenty of open problems in this area. The large-scale hierarchical on-chip implementation of HTM for more complex problems is one of the open challenges, as scaling down the architecture is useful for implementing it on low power edge devices, rather than power-hungry GPUs and CPUs. Even though there are several implementations of HTM in the analog domain, the complete hierarchical HTM architecture with Sequence Learning based on analog computation has not been shown yet. If the digital HTM architecture has been recently proposed [37], the implementation of large-scale HTM architecture in the analog domain with memristive device, which fully emulates the structure and functionality of HTM, has not been shown yet.

The memristor-based HTM architecture also has several limitations, such as limited number of levels in memristive devices, device variation, non-linear distribution of the resistive states and device aging, which should be solved in memristive HTM hardware. There have been several solutions proposed for these issues. The devices with more than 100 memristive states have been manufactured recently [11]. The variation-aware [48,49] and aging resilient [50] designs and architectures have been proposed. And the quantization methods for the neural networks to reduce the effect of non-linear distribution of the memristive states have been shown [51].

20.6 Conclusions

In this chapter, the introduction to the integration of analog backpropagation learning to the memristive HTM architecture has been discussed. In particular, the application of analog backpropagation learning circuit for the architectures with the HTM Spatial Pooler based feature extraction and conventional learning for pattern matching classification has been shown. Even though the backpropagation learning is not a part of the original idea of HTM, the HTM concepts can be useful in such applications. In the circuits covered in this chapter, the HTM SP is used as a feature extractor that outputs the SDRs of the input data, which allows one to perform conventional template-matching based classification of the faces better than processing of the original images. The backpropagation learning circuit in these architectures is used to learn the HTM synaptic weights to recognize these SDRs better. The advantages of such systems are small on-chip area and low power consumption, comparing to the tradition implementations of such complex algorithms on conventional GPUs and CPUs.

The system also can be scaled and is applicable for near-sensor processing and edge computing. The open problems include the implementation of a complete analog HTM architecture maintaining several levels of hierarchy, implementation of the Temporal Memory (Sequence learning) process, and alleviation of the effects of memristor non-idealities.

References

[1] J. Hawkins, Hierarchical temporal memory (htm), whitepaper, 2011.
[2] S. Ahmad, D. George, J.L. Edwards, W.C. Saphir, F. Astier, R. Marianetti, Hierarchical temporal memory (htm) system deployed as web service, US Patent 8,732,098 (May 20, 2014).
[3] J. Hawkins, S. Ahmad, D. George, F. Astier, R. Marianetti, Architecture of a hierarchical temporal memory based system, US Patent App. 11/351,437 (Aug. 16 2007).
[4] D. Rozado, F.B. Rodriguez, P. Varona, Optimizing hierarchical temporal memory for multivariable time series, in: International Conference on Artificial Neural Networks, Springer, 2010, pp. 506–518.
[5] O. Krestinskaya, A.P. James, L.O. Chua, Neuromemristive circuits for edge computing: a review, IEEE Transactions on Neural Networks and Learning Systems (2019).
[6] O. Krestinskaya, A.P. James, Analog htm: memristive spatial pooler learning with backpropagation, in: 2019 IEEE International Conference on Artificial Intelligence Circuits and Systems (AICAS), IEEE, 2019, pp. 262–266.
[7] T. Ibrayev, U. Myrzakhan, O. Krestinskaya, A. Irmanova, A.P. James, On-chip face recognition system design with memristive hierarchical temporal memory, Journal of Intelligent & Fuzzy Systems 34 (3) (2018) 1393–1402.
[8] O. Krestinskaya, T. Ibrayev, A.P. James, Hierarchical temporal memory features with memristor logic circuits for pattern recognition, IEEE Transactions on Computer-Aided Design of Integrated Circuits and Systems 37 (6) (2017) 1143–1156.
[9] O. Krestinskaya, I. Dolzhikova, A.P. James, Memristive hierarchical temporal memory, in: Deep Learning Classifiers with Memristive Networks, Springer, 2020, pp. 181–194.
[10] F. Cai, J.M. Correll, S.H. Lee, Y. Lim, V. Bothra, Z. Zhang, M.P. Flynn, W.D. Lu, A fully integrated reprogrammable memristor–CMOS system for efficient multiply–accumulate operations, Nature Electronics 2 (7) (2019) 290–299.
[11] C. Li, M. Hu, Y. Li, H. Jiang, N. Ge, E. Montgomery, J. Zhang, W. Song, N. Dávila, C.E. Graves, et al., Analogue signal and image processing with large memristor crossbars, Nature Electronics 1 (1) (2018) 52.
[12] C. Li, Z. Wang, M. Rao, D. Belkin, W. Song, H. Jiang, P. Yan, Y. Li, P. Lin, M. Hu, et al., Long short-term memory networks in memristor crossbar arrays, Nature Machine Intelligence 1 (1) (2019) 49–57.
[13] D. Fan, M. Sharad, A. Sengupta, K. Roy, Hierarchical temporal memory based on spin-neurons and resistive memory for energy-efficient brain-inspired computing, IEEE Transactions on Neural Networks and Learning Systems 27 (9) (2015) 1907–1919, https://doi.org/10.1109/TNNLS.2015.2462731.
[14] A.P. James, I. Fedorova, T. Ibrayev, D. Kudithipudi, Htm spatial pooler with memristor crossbar circuits for sparse biometric recognition, IEEE Transactions on Biomedical Circuits and Systems 11 (3) (2017) 640–651.
[15] J. Hawkins, S. Blakeslee, On Intelligence: How a New Understanding of the Brain Will Lead to the Creation of Truly Intelligent Machines, Macmillan, 2007.
[16] N. Inc, Biological and Machine Intelligence (BAMI), A living book that documents hierarchical temporal memory (htm), 2017.
[17] Y. Dauletkhanuly, O. Krestinskaya, A.P. James, Htm theory, in: Deep Learning Classifiers with Memristive Networks, Springer, 2020, pp. 169–180.

[18] A.M. Zyarah, Design and analysis of a reconfigurable hierarchical temporal memory architecture, Master's thesis, Rochester Institute of Technology, 1 Lomb Memorial Dr, Rochester, NY 14623, 2015.

[19] J. Hawkins, S. Ahmad, Why neurons have thousands of synapses, a theory of sequence memory in neocortex, Frontiers in Neural Circuits 10 (2016) 23.

[20] Y. Cui, S. Ahmad, J. Hawkins, The htm spatial pooler—a neocortical algorithm for online sparse distributed coding, Frontiers in Computational Neuroscience 11 (2017) 111.

[21] J. Mnatzaganian, E. Fokoué, D. Kudithipudi, A mathematical formalization of hierarchical temporal memory's spatial pooler, Frontiers in Robotics and AI 3 (2017) 81.

[22] J.C. Hawkins, D. George, C. Curry, F.E. Astier, A. Raj, R.G. Jaros, Hierarchical temporal memory system with higher-order temporal pooling capability, US Patent 8,407,166 (Mar. 26 2013).

[23] S. Ahmad, J. Hawkins, How do neurons operate on sparse distributed representations? A mathematical theory of sparsity, neurons and active dendrites, arXiv preprint, arXiv:1601.00720, 2016.

[24] S. Ahmad, J. Hawkins, Properties of sparse distributed representations and their application to hierarchical temporal memory, arXiv preprint, arXiv:1503.07469, 2015.

[25] P.P. Afeefa, P.P. Thulasidharan, Automatic license plate recognition (ALPR) using HTM cortical learning algorithm, in: 2017 International Conference on Intelligent Computing and Control (I2C2), 2017, pp. 1–4, https://doi.org/10.1109/I2C2.2017.8321947.

[26] N.O. El-Ganainy, I. Balasingham, P.S. Halvorsen, L. Arne Rosseland, On the performance of hierarchical temporal memory predictions of medical streams in real time, in: 2019 13th International Symposium on Medical Information and Communication Technology (ISMICT), 2019, pp. 1–6, https://doi.org/10.1109/ISMICT.2019.8743902.

[27] N. Mitri, W. Marrouche, M. Awad, R. Habib, Irregular breathing detection in cpap assisted patients using hierarchical temporal memory, in: 2017 IEEE Symposium Series on Computational Intelligence (SSCI), 2017, pp. 1–6, https://doi.org/10.1109/SSCI.2017.8285370.

[28] W. Midani, Z. Fki, M. BenAyed, Online anomaly detection in ecg signal using hierarchical temporal memory, in: 2019 Fifth International Conference on Advances in Biomedical Engineering (ICABME), 2019, pp. 1–4, https://doi.org/10.1109/ICABME47164.2019.8940307.

[29] A. Anandharaj, P.B. Sivakumar, Anomaly detection in time series data using hierarchical temporal memory model, in: 2019 3rd International conference on Electronics, Communication and Aerospace Technology (ICECA), 2019, pp. 1287–1292, https://doi.org/10.1109/ICECA.2019.8821966.

[30] J. Wu, W. Zeng, F. Yan, Hierarchical temporal memory method for time-series-based anomaly detection, Neurocomputing 273 (2018) 535–546.

[31] D. Shah, P. Ghate, M. Paranjape, A. Kumar, Application of hierarchical temporal memory theory for document categorization, in: 2017 IEEE SmartWorld, Ubiquitous Intelligence Computing, Advanced Trusted Computed, Scalable Computing Communications, Cloud Big Data Computing, Internet of People and Smart City Innovation (SmartWorld/SCALCOM/UIC/ATC/CBDCom/IOP/SCI), 2017, pp. 1–6, https://doi.org/10.1109/UIC-ATC.2017.8397473.

[32] H. Luo, S. Jia, W. Zhang, Hierarchical temporal memory based anomaly detection for hydrological monitoring of unmanned surface vehicle, in: 2019 IEEE 2nd International Conference on Information Communication and Signal Processing (ICICSP), 2019, pp. 420–424, https://doi.org/10.1109/ICICSP48821.2019.8958534.

[33] M. Kerner, K. Tammemäe, Hierarchical temporal memory implementation on fpga using lfsr based spatial pooler address space generator, in: 2017 IEEE 20th International Symposium on Design and Diagnostics of Electronic Circuits Systems (DDECS), 2017, pp. 92–95, https://doi.org/10.1109/DDECS.2017.7934577.

[34] M. Putic, A.J. Varshneya, M.R. Stan, Hierarchical temporal memory on the automata processor, IEEE MICRO 37 (1) (2017) 52–59, https://doi.org/10.1109/MM.2017.6.

[35] F. Walter, M. Sandner, F. Rcöhrbein, A. Knoll, Towards a neuromorphic implementation of hierarchical temporal memory on spinnaker, in: 2017 IEEE International Symposium on Circuits and Systems (ISCAS), 2017, pp. 1–4, https://doi.org/10.1109/ISCAS.2017.8050983.
[36] W. Li, P. Franzon, Hardware implementation of hierarchical temporal memory algorithm, in: 2016 29th IEEE International System-on-Chip Conference (SOCC), IEEE, 2016, pp. 133–138.
[37] A.M. Zyarah, D. Kudithipudi, Neuromorphic architecture for the hierarchical temporal memory, IEEE Transactions on Emerging Topics in Computational Intelligence 3 (1) (2019) 4–14, https://doi.org/10.1109/TETCI.2018.2850314.
[38] C. Sung, H. Hwang, I.K. Yoo, Perspective: a review on memristive hardware for neuromorphic computation, Journal of Applied Physics 124 (15) (2018) 151903.
[39] O. Krestinskaya, I. Dolzhikova, A.P. James, Hierarchical temporal memory using memristor networks: a survey, IEEE Transactions on Emerging Topics in Computational Intelligence 2 (5) (2018) 380–395.
[40] L. Deng, The mnist database of handwritten digit images for machine learning research [best of the web], IEEE Signal Processing Magazine 29 (6) (2012) 141–142.
[41] A. Martinez, R. Benavente, Ar face database, 2000.
[42] K. Smagulova, O. Krestinskaya, A.P. James, Who is the winner? Memristive-cmos hybrid modules: cnn-lstm versus htm, IEEE Transactions on Biomedical Circuits and Systems (2019).
[43] O. Krestinskaya, K.N. Salama, A.P. James, Analog backpropagation learning circuits for memristive crossbar neural networks, in: 2018 IEEE International Symposium on Circuits and Systems (ISCAS), IEEE, 2018, pp. 1–5.
[44] O. Krestinskaya, K.N. Salama, A.P. James, Learning in memristive neural network architectures using analog backpropagation circuits, IEEE Transactions on Circuits and Systems I, Regular Papers 66 (2) (2018) 719–732.
[45] O. Krestinskaya, Slam: Spatiotemporal learning with analog memristive circuits for htm, Master's thesis, Nazarbayev University School of Engineering, 2018.
[46] H. Wu, P. Yao, B. Gao, H. Qian, Multiplication on the edge, Nature Electronics 1 (1) (2018) 8–9.
[47] A.P. James, A hybrid memristor–cmos chip for AI, Nature Electronics 2 (7) (2019) 268–269.
[48] B. Liu, H. Li, Y. Chen, X. Li, Q. Wu, T. Huang, Vortex: variation-aware training for memristor x-bar, in: Proceedings of the 52nd Annual Design Automation Conference, 2015, pp. 1–6.
[49] H. Manem, G.S. Rose, X. He, W. Wang, Design considerations for variation tolerant multilevel cmos/nano memristor memory, in: Proceedings of the 20th Symposium on Great Lakes Symposium on VLSI, 2010, pp. 287–292.
[50] S.N. Mozaffari, K.P. Gnawali, S. Tragoudas, An aging resilient neural network architecture, in: Proceedings of the 14th IEEE/ACM International Symposium on Nanoscale Architectures, 2018, pp. 25–30.
[51] J. Lin, L. Xia, Z. Zhu, H. Sun, Y. Cai, H. Gao, M. Cheng, X. Chen, Y. Wang, H. Yang, Rescuing memristor-based computing with non-linear resistance levels, in: 2018 Design, Automation & Test in Europe Conference & Exhibition (DATE), IEEE, 2018, pp. 407–412.

Chapter 21

Multi-stable patterns coexisting in memristor synapse-coupled Hopfield neural network

Mo Chen, Cheng-jie Chen, Bo-cheng Bao, and Quan Xu
School of Microelectronics and Control Engineering, Changzhou University, Changzhou, China

21.1 Introduction

Hopfield neural network (HNN) is a well-known artificial neural network that has been analyzed in great mathematical detail [1,2]. It shows great potentials in the applications of life science and engineering, such as associating memory [3,4], medical imaging [5], information storage [6], cognitive study [7], and supervised learning [8]. Since dynamical analyses of neural networks are of significant importance for better understanding human activity and memory [9–11], several works have been carried out on improving the architectures of HNNs and revealing the complex dynamics residing in these HNNs. Therefore, some specific dynamical behaviors were uncovered by theoretical analyses, numerical simulations, and/or hardware experiments, such as self-excited [12–14] and hidden chaotic attractors [15], coexisting bistable or multi-stable patterns [16–19], and so on.

Nowadays, memristor is becoming an excellent building block for hardware implementations of neuromorphic circuits [10,20,21]. Numerous models for memristor-based single neuron and its network were proposed by taking memristor as a synaptic weight between the adjacent neurons [22–26] or characterizing the effects of electromagnetic induction and radiation [27–30], which leads to the emergence of more complex dynamical behaviors. For examples, Li et al. [22] studied a small HNN with a memristive synaptic weight, which exhibited rich complex dynamics of quasi-periodic orbits, chaos, and hyperchaos. Parastesh et al. [26] studied the network of hyperbolic-type memristor-based tri-neuron HNNs and uncovered the partial chimera state, chimera state, synchronization, imperfect synchronization, and oscillation death. By employing a threshold flux-controlled memristor to generate the electromagnetic induction current, Bao et al. [30] reported a 3-dimensional memristive Hindmarsh-Rose model with global hidden bursting oscillations. Obviously, memristor synapse

Mem-elements for Neuromorphic Circuits with Artificial Intelligence Applications
https://doi.org/10.1016/B978-0-12-821184-7.00030-X

with neural functions is increasingly becoming an essential component of neuromorphic circuits.

Moreover, focusing on investigations of multi-stable firing patterns in HNNs, various memristive HNN models were constructed and explored in recent years [31–34]. For example, Chen et al. [31,32] proposed a memristive bi-neuron HNN model under the action of electromagnetic induction current produced by the membrane potential difference of two adjacent neurons, and multi-stable firing patterns were disclosed under different memristor initial conditions. Lin et al. [33] reported hidden extreme multi-stability in 3D HNN affected by electromagnetic radiation. Leng et al. [34] emulated synaptic crosstalk effect of HNN using coupled hyperbolic memristors and discovered multi-stability, asymmetry attractors, and anti-monotonicity under various crosstalk strengths.

Inspired by the work reported in [31,32], a memristor-based tri-neuron HNN model was constructed through coupling an electromagnetic induction current, caused by the potential difference of two adjacent neurons, to link these two neurons bidirectionally [35], in which the initial-dependent tri-stability phenomena were revealed under different memristor coupling strengths. In this chapter, in-depth theoretical and numerical investigations for the parameter- and initial condition-related dynamics and their dynamical mechanisms are further performed. It reveals that the memristor coupling strengths and synaptic connection weighs all exert great influence on the dynamics of the proposed memristive HNN model.

The rest of the chapter is arranged as follows. Section 21.2 proposes a memristive tri-neuron HNN model and analyzes the stabilities of its equilibrium points. In Section 21.3, multiple analytical tools are utilized to simulate the bifurcation behaviors and coexisting patterns relying on the coupling strengths, synaptic weights, and initial conditions. PSIM circuit simulations manifest these distinctive dynamical behaviors in Section 21.4. Finally, the conclusions are summarized in Section 21.5.

21.2 Memristor synapse-coupled HNN with three neurons

In the Hopfield neural network (HNN), the potential difference between two adjacent neurons can induce a bidirectional stimulus current [32] described by an electromagnetic induction current I_M following through a non-ideal memristor synapse. The dimensionless mathematical model of the memristor synapse can be described as

$$\begin{cases} I_M = W(\varphi)V_M = k\varphi V_M \\ \dot{\varphi} = f(V_M, \varphi) = V_M - \varphi \end{cases} \tag{21.1}$$

in which φ denotes the inner state variable of the memristor synapse, $W(\varphi) = k\varphi$ is the memductance function with memristor coupling strength k, and V_M is the potential difference between the two adjacent neurons.

Introducing the non-ideal memristor synapse described in (21.1) into a tri-neuron HNN to couple two adjacent neurons in bi-directions [35], a 4-dimensional memristive HNN can be established. The corresponding mathematical model is formulated as

$$\begin{cases} \dot{x}_1 = -x_1 + w_{11}\tanh(x_1) + w_{21}\tanh(x_2) + w_{31}\tanh(x_3) + k\varphi(x_1 - x_2) \\ \dot{x}_2 = -x_2 + w_{12}\tanh(x_1) + w_{22}\tanh(x_2) + w_{32}\tanh(x_3) - k\varphi(x_1 - x_2) \\ \dot{x}_3 = -x_3 + w_{13}\tanh(x_1) + w_{23}\tanh(x_2) + w_{33}\tanh(x_3) \\ \dot{\varphi} = x_1 - x_2 - \varphi \end{cases} \tag{21.2}$$

The non-ideal memristor synapse induced electromagnetic induction current is bi-directionally applied to the first and second neurons. In (21.2), $\tanh(x_i)$ ($i = 1, 2, 3$) is the neuron activation function and its coefficient represents the synaptic connection weight. Referring to [35], the synaptic weight matrix is specified as

$$\mathbf{W} = \begin{pmatrix} w_{11} & w_{21} & w_{31} \\ w_{12} & w_{22} & w_{32} \\ w_{13} & w_{23} & w_{33} \end{pmatrix} = \begin{pmatrix} w_{11} & 2.8 & 0.5 \\ -1.5 & 1.2 & 0 \\ 0 & w_{23} & 0 \end{pmatrix} \tag{21.3}$$

To simplify quantitative analyses, only unilateral connections are considered for the third neuron. In later sections, the memristor coupling strength k, the synaptic self-connection weight w_{11}, the synaptic inter-connection weight w_{23}, and initial conditions $(x_1(0), x_2(0), x_3(0), \varphi(0))$ are taken as the main varying parameters. The other synaptic connection weights in (21.3) are kept unchanged. Note that the typical control parameters of the proposed HNN model (21.2) are $w_{11} = 1.5$, $w_{23} = -20$, and $k = 1$.

Set the left side of (21.2) to 0. The equilibrium point of the memristive HNN can be expressed as

$$P = (\xi_1, \xi_2, w_{23}\tanh\xi_2, \xi_1 - \xi_2) \tag{21.4}$$

in which ξ_1 and ξ_2 are solutions of the following two functions:

$$\begin{cases} -\xi_2 - 1.5\tanh(\xi_1) + 1.2\tanh(\xi_2) - k(\xi_1 - \xi_2)^2 = 0 \\ -\xi_1 - \xi_2 + (w_{11} - 1.5)\tanh(\xi_1) + 4\tanh(\xi_2) + 0.5\tanh[w_{23}\tanh(\xi_2)] = 0 \end{cases} \tag{21.5}$$

The Jacobin matrix at P is deduced as

$$\mathbf{J} = \begin{bmatrix} -1 + w_{11}\,\mathrm{sech}^2(\xi_1) + k\xi_\varphi & 2.8\,\mathrm{sech}^2(\xi_2) - k\xi_\varphi & 0.5\,\mathrm{sech}^2(\xi_3) & k\xi_\varphi \\ -1.5\,\mathrm{sech}^2(\xi_1) - k\xi_\varphi & -1 + 1.2\,\mathrm{sech}^2(\xi_2) + k\xi_\varphi & 0 & -k\xi_\varphi \\ 0 & -w_{23}\,\mathrm{sech}^2(\xi_2) & -1 & 0 \\ 1 & -1 & 0 & -1 \end{bmatrix} \tag{21.6}$$

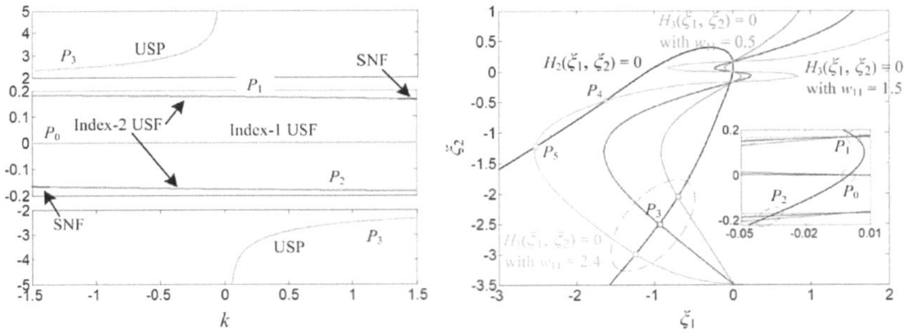

FIGURE 21.1 Equilibrium points and stabilities: (a) Equilibrium point traces and stabilities for $k = [-1.5, 1.5]$; (b) equilibrium points determined by inspecting the intersections of functions $H_2(\xi_1, \xi_2) = 0$ and $H_3(\xi_1, \xi_2) = 0$ with $w_{11} = 0.5/1.5/2.4$. (For interpretation of the colors in the figure, the reader is referred to the web version of this chapter.)

in which, $\xi_3 = w_{23}\tanh(\xi_2)$ and $\xi_\varphi = \xi_1 - \xi_2$. Based on (21.6), the eigenvalues at P are readily calculated and then the stability of P can be evaluated. The characteristics of the equilibrium P are changed with the variations of k, w_{11}, and w_{23}, which consequentially affect the exhibited dynamics of the proposed memristive HNN model (21.2).

When k is selected as the unique changing variable and the other two control parameters are fixed as $w_{11} = 1.5$ and $w_{23} = -20$, ξ_2 can be solved from the simplified function given as

$$H_1(k, \xi_2) = -\xi_2 - 1.5\tanh(\xi_1) + 1.2\tanh(\xi_2) - k(\xi_1 - \xi_2)^2 = 0 \tag{21.7}$$

in which ξ_1 is expressed using ξ_2 as

$$\xi_1 = -\xi_2 + 4\tanh(\xi_2) + 0.5\tanh[-20\tanh(\xi_2)] \tag{21.8}$$

When k increases from -1.5 to 1.5, the equilibrium points of the memristive HNN model are solved using 'ezplot' function of MATLAB® program and the traces of ξ_2 are depicted in Fig. 21.1(a). It can be seen that, for $k = 0$, the memristive HNN has three determined equilibriums, including one zero equilibrium point P_0 and two symmetric equilibrium points P_1 and P_2. Meanwhile, for $k \neq 0$, there exist one zero equilibrium point P_0 and three asymmetric equilibrium points P_1, P_2, and P_3. Stabilities of these equilibrium points are evaluated based on their eigenvalues and represented by different colors in Fig. 21.1(a). The red color indicates an index-2 unstable saddle-focus (USF) with a pair of conjugate complex roots with positive real parts and two negative real roots. The magenta color represents an index-1 USF with a positive real root, a pair of conjugate complex roots with negative real parts, and a negative real root. The green color stands for an unstable saddle point (USP) with one positive real root and three negative real roots. The blue color represents a stable node-focus

(SNF) with a pair of conjugate complex roots with negative real parts and two negative real roots. It is demonstrated that the memristive HNN possesses multiple unstable equilibrium points for a specified value of k, which will lead to several multi-stable states.

If w_{11} is selected as the varying parameter and the other varying parameters are fixed as $k = 1$ and $w_{23} = -20$, the equilibrium points of the memristive HNN are solved by detecting the intersections of the following two functions via the graphic method:

$$\begin{cases} H_2(\xi_1, \xi_2) = -\xi_2 - 1.5\tanh(\xi_1) + 1.2\tanh(\xi_2) - (\xi_1 - \xi_2)^2 = 0 \\ H_3(\xi_1, \xi_2) = -\xi_1 - \xi_2 + (w_{11} - 1.5)\tanh(\xi_1) + 4\tanh(\xi_2) \\ \qquad\qquad +0.5\tanh[-20\tanh(\xi_2)] = 0 \end{cases} \tag{21.9}$$

With reference to the trajectories of $H_2(\xi_1, \xi_2) = 0$ and $H_3(\xi_1, \xi_2) = 0$ depicted in Fig. 21.1(b), the memristive HNN model possesses one zero equilibrium point P_0 and three asymmetric equilibriums P_1, P_2, and P_3 for $w_{11} = 0.5$ and 1.5. More equilibrium points are observed when w_{11} is further increased. For example, one zero equilibrium point P_0 and five asymmetric equilibriums P_1, P_2, P_3, P_4, and P_5 are found for $w_{11} = 2.4$. Eigenvalues and stabilities of these equilibrium points are summarized in Table 21.1. Obviously, the synaptic connection weights also change the characteristics of the determined equilibrium points and further affect the dynamics of the memristive HNN model (21.2).

21.3 Bifurcation behaviors with multi-stability

In this section, the bifurcation behaviors relying on the memristor coupling strength, synaptic weights, and initial conditions are simulated using 2-dimensional and 1-dimensional bifurcation diagrams, Lyapunov exponent spectra, phase plane orbits, and time-domain waveforms.

21.3.1 Dynamics depended on the coupling strength k

Take the memristor coupling strength k and the second initial condition $x_2(0)$ as two main control parameters, the 2-dimensional bifurcation diagram within $k = [-1.5, 1.5]$ and $x_2(0) = [-10^{-5}, 10^{-5}]$ is simulated and presented in Fig. 21.2. The rest three initial conditions of the memristive HNN model (21.2) are all set to zero. The MATLAB ODE45 algorithm with time step 0.01 and time interval [2500, 3000] is used to depict the bifurcation diagrams in the whole chapter.

Glanced from the Fig. 21.2, the 2-dimensional bifurcation diagram takes on the odd symmetry pattern about the origin. The red and white regions, labeled as CH and Inf, represent chaotic and infinite states, respectively. The other colors represent periodic states with different periodicities. Note that P0 indicates fixed point and CH refers to attractors having more than 8 periodicities. The periodic

TABLE 21.1 Characteristics at the equilibrium points for $w_{11} = 0.5$, 1.5, and 2.4.

w_{11}	Equilibrium points	Eigenvalues	Stabilities
0.5	P_0: (0, 0, 0, 0)	1.3597, −1.3299±j2.4996, −1	Index-1 USF
	P_1: (0.0031, 0.1704, −3.3754, −0.1673)	−0.3964±j1.9998, −0.8921, −0.9839,	SNF
	P_2: (−0.0405, −0.1968, 3,8860, 0.1563)	0.0257±j1.9664, −0.9943, −1.0906,	Index-2 USF
	P_3: (−0.7084, −2.0521, 19.3506, 1.3437)	2.8829, −1, −1.2210, −1.5837	USP
1.5	P_0: (0, 0, 0, 0)	1.6062, −0.9531±j2.3986, −1	Index-1 USF
	P_1: (0.0027, 0.1693, −3.3540, −0.1666)	0.0987±j2.0025, −0.8767, −0.9877	Index-2 USF
	P_2: (−0.0369, −0.1814, 3.5887, 0.1445)	0.5145±j2.0051, −0.9923, −1.0882	Index-2 USF
	P_3: (−0.9448, −2.5018, 19.7332, 1.5770)	3.4659, −0.9464, −1, −1.6894	USP
2.4	P_0: (0, 0, 0, 0)	1.9306, −0.6653±j2.2375, −1	Index-1 USF
	P_1: (0.0030, 0.1683, −3.3346, −0.1653)	0.5467±j1.8940, −0.8676, −0.9897,	Index-2 USF
	P_2: (−0.0336, −0.1694, 3.3560, 0.1358)	0.9553±j1.9298, −0.9901, −1.0854,	Index-2 USF
	P_3 (−1.2444, −2.9975, 19.9006, 1.7531)	3.7945, −0.8549, −1, −1.7421	USP
	P_4 (−1.6321, −0.4771, 8.8784, −1.1550)	0.4221, −2.2140±j1.1128, −1	Index-1 USF
	P_5 (−2.5276, −1.2251, 16.8230, −1.3025)	−2.3846±j0.8837, −0.4243, −1	SNF

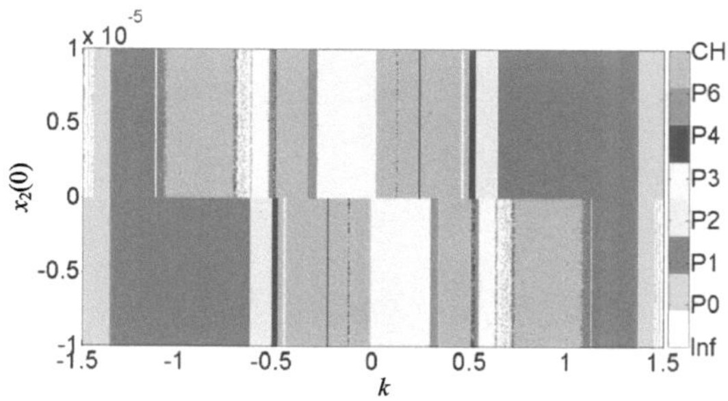

FIGURE 21.2 2-dimensional bifurcation diagrams in the $k - x_2(0)$ plane with fixed $x_1(0) = x_3(0) = \varphi(0) = 0$. (For interpretation of the colors in the figure, the reader is referred to the web version of this chapter.)

states of P5, P7, and P8 are ignored since these states are not found within this parameter region.

In the region of $x_2(0) = [0, 10^{-5}]$, with the decrease of k, the memristive HNN begins with the stable point behavior and then turns into periodic behaviors at $k = 1.37$. Thereafter, the dynamics evolves into the chaotic behaviors via period-doubling bifurcation route, and jumps to the periodic behaviors at $k = -1.1$, and then reverses back to the stable behaviors at $k = -1.37$, and finally turns into infinite state. Within the chaotic region, several periodic windows are uncovered. Particularly, for $x_2(0) = [-10^{-5}, 0]$, the exhibited bifurcation behaviors are odd symmetric with those reveal in the positive region of $x_2(0)$.

To further verify the symmetrical characteristic, the phase portraits of six pairs of symmetric attractors are plotted in the x_1–x_3 plane through tuning $x_2(0)$ and k, as shown in Fig. 21.3. Since the initial conditions are implicit parameters in the mathematical model of the memristive HNN, their dynamical effects are hard to be explained.

Inspired by the incremental integral transformation method [36,37], we integrate the four equations of the memristive HNN given in (21.2) from 0 to τ, and the four differential equations can be transformed into the integral forms as

$$\begin{cases} x_1 = \int_0^\tau [-x_1 + w_{11}\tanh(x_1) + w_{21}\tanh(x_2) + w_{31}\tanh(x_3) \\ \qquad + k\varphi(x_1 - x_2)]\,\mathrm{d}\zeta + x_1(0) \\ x_2 = \int_0^\tau [-x_2 + w_{12}\tanh(x_1) + w_{22}\tanh(x_2) + w_{32}\tanh(x_3) \\ \qquad - k\varphi(x_1 - x_2)]\,\mathrm{d}\zeta + x_2(0) \\ x_3 = \int_0^\tau [-x_3 + w_{13}\tanh(x_1) + w_{23}\tanh(x_2) + w_{33}\tanh(x_3)]\,\mathrm{d}\zeta + x_3(0) \\ \varphi = \int_0^\tau (x_1 - x_2 - \varphi)\,\mathrm{d}\zeta + \varphi(0) \end{cases} \tag{21.10}$$

In (21.10), $x_1(0)$, $x_2(0)$, $x_3(0)$, and $\varphi(0)$ are the initial conditions of four state variables. For the memristive HNN model (21.2), there exists the invariance property under the transformation $(x_1, x_2, x_3, \varphi, k, x_1(0), x_2(0), x_3(0), \varphi(0)) \leftrightarrow$

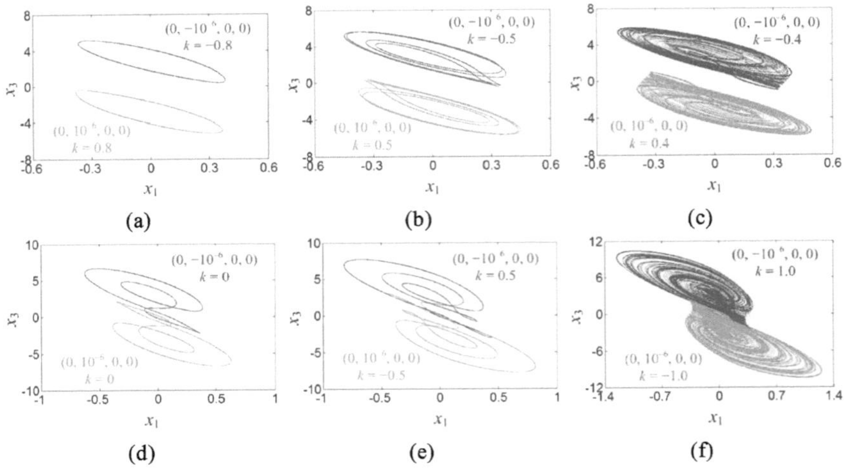

FIGURE 21.3 Symmetric attractors in the x_1–x_3 plane: (a) symmetric period-1 limit cycles, (b) symmetric period-4 limit cycles, (c) symmetric chaotic attractors, (d) symmetric period-3 limit cycles, (e) symmetric period-6 limit cycles, (f) symmetric chaotic attractors.

$(-x_1, -x_2, -x_3, -\varphi, -k, -x_1(0), -x_2(0), -x_3(0), -\varphi(0))$. In other words, if $(x_1, x_2, x_3, \varphi, k, x_1(0), x_2(0), x_3(0), \varphi(0))$ is a system solution, $(-x_1, -x_2, -x_3, -\varphi, -k, -x_1(0), -x_2(0), -x_3(0), -\varphi(0))$ is also a system solution. Consequently, when k, $x_2(0)$ are turned into $-k$, $-x_2(0)$ and other initial conditions are settled down to 0, the generated attractor and the original attractor are odd symmetric with the origin.

For the memristive systems with no extra nonlinear terms besides the memristor term, the original system can easily be explicitly expressed using the newly introduced state variables in the integral domain [36]. Thereafter, quantitative analyses for the dynamical effect of each initial condition can readily be performed. However, the time integrals of $\tanh(x_i)$ $(i = 1, 2, 3)$ and $\varphi(x_1 - x_2)$ in (21.10) are hard to be explicitly formulated. The dynamical mechanism of the odd symmetric bifurcation behaviors revealed in Figs. 21.2 and 21.3 is only briefly interpreted via the symmetric property of the implicit integration formulas given in (21.10).

21.3.2 Dynamics depended on the synaptic weights

21.3.2.1 2-dimensional bifurcation diagrams

The synaptic connection weighs also exert influences on the dynamical model [13,14,16]. In the proposed memristive HNN model (21.2), the synaptic self-connection weight w_{11} and inter-connection weight w_{23} are taken as two bifurcation parameters and the coupling strength is fixed as $k = 1$. When the initial conditions are assigned as $(0, 10^{-6}, 0, 0)$ and $(0, -10^{-6}, 0, 0)$, two sets of 2-dimensional bifurcation diagrams are numerically stimulated in Fig. 21.4(a) and

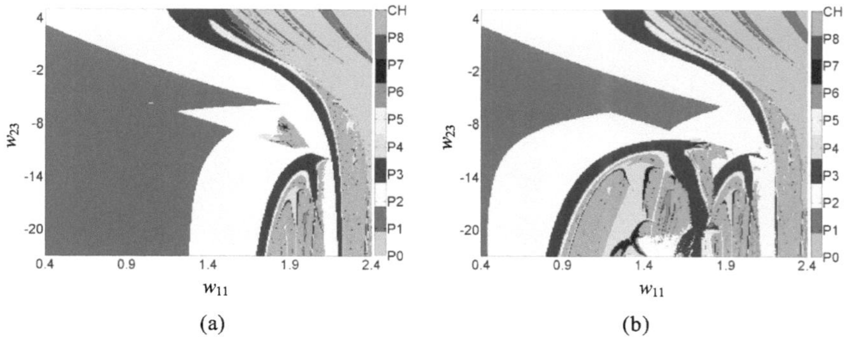

FIGURE 21.4 2-dimensional bifurcation diagrams with respect to w_{11} and w_{23} under two sets of initial conditions: (a) (0, 10^{-6}, 0, 0), (b) (0, -10^{-6}, 0, 0). (For interpretation of the colors in the figure, the reader is referred to the web version of this chapter.)

(b), respectively. The colors defined in Fig. 21.4 are almost the same as these used in Fig. 21.2 except that the colors for P3 and P4 states are interchanged to achieve better visual effect. In Fig. 21.4, the multi-period and chaotic states are mainly emerged in the region of $w_{11} = [1.7, 2.4]$ and $w_{23} = [-23, -2]$. And the dynamics of Fig. 21.4(b) is more complicated than that of Fig. 21.4(a), confirming that both synaptic connection weighs and initial conditions can affect the dynamics significantly.

21.3.2.2 Dynamics depended on the synaptic self-connection weight w_{11}

Take the synaptic self-connection weight w_{11} as a bifurcation parameter and fix the inter-connection weight and coupling strength as $w_{23} = -20$ and $k = 1$. On account of the 2-dimensional bifurcation diagrams, the coexisting bifurcation routes under two slightly difference initial conditions are illustrated using the 1-dimensional bifurcation diagrams and Lyapunov exponent spectra, as depicted in Figs. 21.5(a) and 21.5(b), respectively. MATLAB ODE45-based Wolf's method [38] with time step 1 and time end 20000 is employed to draw the Lyapunov exponent spectra. The forward and backward continuation method [39] could also be employed to plot the bifurcation diagrams. Using this method, the time-step and time interval may affect the revealed dynamics due to the randomness of the initial states for the next step simulation.

In Fig. 21.5(a), the purple and green bifurcation traces are triggered under initial conditions (0, 10^{-6}, 0, 0) and (0, -10^{-6}, 0, 0), respectively. In the top of Fig. 21.5(a), a long point attractor region with negative largest Lyapunov exponent is observed within $w_{11} = [0.40, 1.30]$. Thereafter, the dynamics turns into period-1 limit cycle, and then enters chaotic region via period-doubling bifurcation scenario. At $w_{11} = 2.10$, the dynamics jumps into period-1 limit cycle, and then returns back to chaotic region through another period-doubling bifurcation route, and finally settles down in the infinite state.

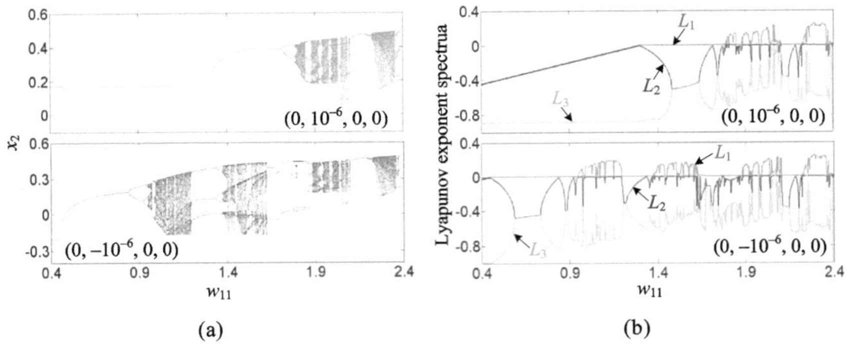

FIGURE 21.5 Dynamics with respect to w_{11} under two sets of initial conditions (0, 10^{-6}, 0, 0) and (0, -10^{-6}, 0, 0): (a) 1-dimensional bifurcation diagrams of the state variable x_2, (b) Lyapunov exponent spectra. (For interpretation of the colors in the figure, the reader is referred to the web version of this chapter.)

However, with the green trajectory depicted in the bottom of Fig. 21.5(a), different bifurcation scenarios are observed under the initial condition (0, -10^{-6}, 0, 0). In the region of $w_{11} = [0.40, 1.84]$, the dynamics starts with a short point attractor region, and then goes into chaotic region via period-doubling bifurcation scenario, and finally turns into periodic behaviors. Within the chaotic region, several periodic windows are emerged through tangential bifurcation and ended by period-doubling bifurcation scenario or chaos crisis. The Lyapunov exponent spectra in Fig. 21.5(b) match with the bifurcation scenarios revealed in Fig. 21.5(a).

Four representative coexisting attractor sets are plotted in Fig. 21.6 through tuning $x_2(0)$ and w_{11}. In Fig. 21.6(a), coexisting chaotic and point attractors are observed. In Figs. 21.6(b) and 21.6(d), coexisting chaotic and periodic attractors are revealed. In Fig. 21.6(c), coexisting periodic attractors with different topological structures and locations are uncovered. It is emphatically illustrated that the synaptic self-connection weight w_{11} has significant influence on the dynamic behaviors.

21.3.2.3 Dynamics depended on the synaptic inter-connection weight w_{23}

The synaptic inter-connection weighs are usually regard as a variable parameter that can induce complex dynamical behaviors [13,14,16]. When fixing $k = 1$ and $w_{11} = 1.5$, the coexisting bifurcation behaviors under two typical initial conditions are illustrated using 1-dimensional bifurcation diagrams and Lyapunov exponent spectra, as depicted in Figs. 21.7(a) and 21.7(b), respectively. In Fig. 21.7(a), the fuchsia and dark-blue bifurcation traces are triggered under the initial conditions (0, 10^{-6}, 0, 0) and (0, -10^{-6}, 0, 0), respectively.

Glance at the top of the Fig. 21.7(a), the system orbit begins with a period-1 limit cycle, and finally evolves into period-3 limit cycle via period-doubling

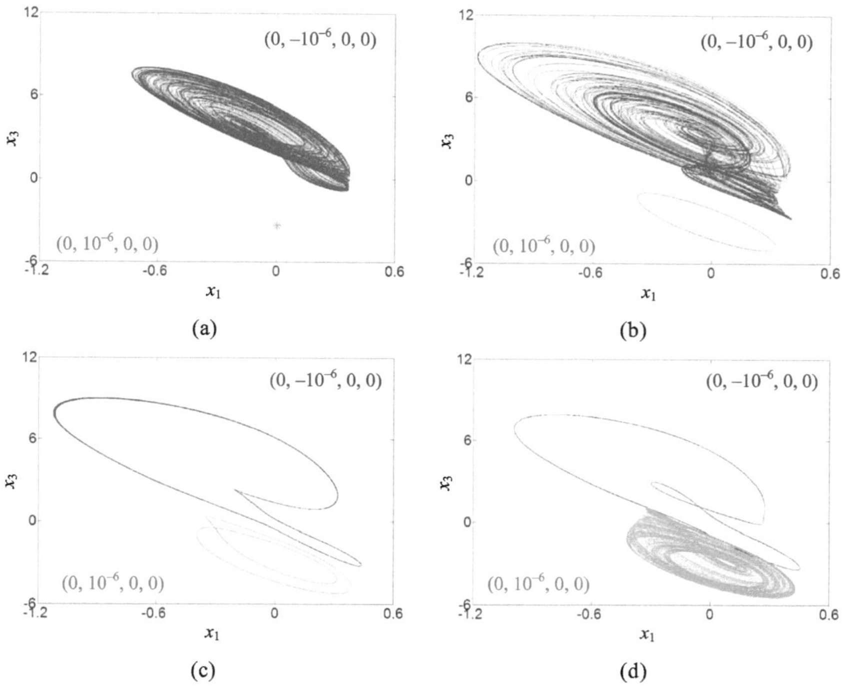

FIGURE 21.6 Representative coexisting attractors in the x_1–x_3 plane with fixed $k = 1$ and $w_{23} = -20$: (a) $w_{11} = 1$, (b) $w_{11} = 1.5$, (c) $w_{11} = 1.74$, (d) $w_{11} = 1.82$.

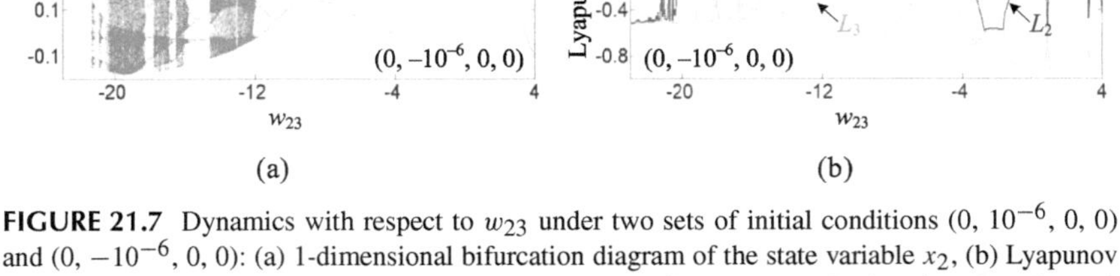

FIGURE 21.7 Dynamics with respect to w_{23} under two sets of initial conditions (0, 10^{-6}, 0, 0) and (0, -10^{-6}, 0, 0): (a) 1-dimensional bifurcation diagram of the state variable x_2, (b) Lyapunov exponent spectra. (For interpretation of the colors in the figure, the reader is referred to the web version of this chapter.)

bifurcation scenario and periodic bubbles. We only observe periodic and stable point attractors in this bifurcation region. These bifurcation behaviors are verified by Lyapunov exponent spectra in the top of Fig. 21.7(b) and 2-dimensional

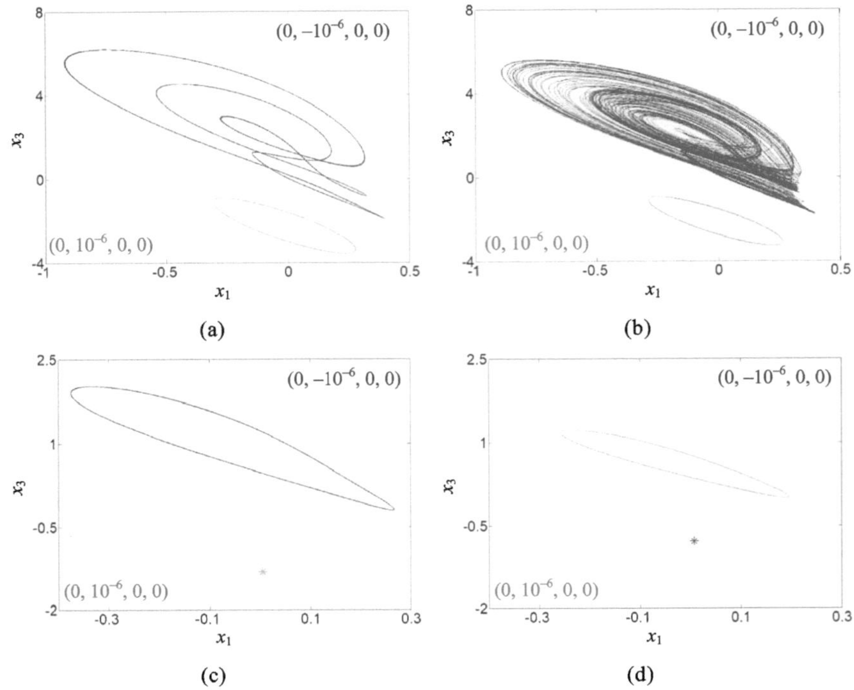

FIGURE 21.8 Representative coexisting attractors in the x_1–x_3 plane with fixed $k = 1$ and $w_{11} = 1.5$: (a) $w_{23} = -15$, (b) $w_{23} = -14$, (c) $w_{23} = -9$, (d) $w_{23} = -7$.

bifurcation diagrams in Fig. 21.4(a). In the bottom of Fig. 21.7(a), diversified bifurcation behaviors are observed in the region of $w_{23} = [-21.54, -8.40]$. With the decreasing of w_{23}, the bifurcation traces trigger a period-1 limit cycle first, and then turns into chaotic attractor suddenly via chaos crisis, and finally returns back to period-1 limit cycle via several tangential bifurcation and forward/reverse period doubling bifurcation routes. The Lyapunov exponent spectra in the bottom of Fig. 21.7(b) match well with the bifurcation diagrams.

Four representative coexisting attractor sets are plotted in Fig. 21.8 through tuning $x_2(0)$ and w_{23}. In detail, Fig. 21.8(a) displays coexisting multi-periodic and period-1 attractors. Fig. 21.8(b) presents coexisting chaotic and period-1 attractors. Figs. 21.8(c) and 21.8(d) give coexisting periodic and point attractors with different locations and shapes. As a result, the inter-connection synapse weights of the memristive HNN model (21.2) also have significant influences on the dynamic behaviors.

21.3.3 Dynamics depended on the initial conditions

The initial condition-relying dynamical behaviors of the memristive HNN are further discussed under different memristor coupling strengths. Four memristor

TABLE 21.2 Eight kinds of attractors and its colors in Fig. 21.9. (For interpretation of the colors in the table, the reader is referred to the web version of this chapter.)

Colors	Colors	Colors	Attractors
	Upper period-3		Lower chaos
	Lower period-3		Lower multi-period
	Divergent		Lower period-2
	Upper-scroll chaos		Upper period-2

coupling strengths, i.e. $k = 0$, 0.45, 0.6, and -0.6, are selected as representative examples with fixed $w_{11} = 1.5$ and $w_{23} = -20$. The attraction basins in the $x_1(0)$–$x_2(0)$ plane and phase plots of the corresponding coexisting attractors projected on the x_1–x_3 plane are numerically simulated and presented in Fig. 21.9 to demonstrate the variation of the initial condition-relying dynamical behaviors. The other two initial conditions $x_3(0)$ and $\varphi(0)$ are specified as 0. For better identification, the attracting regions and corresponding phase plots are illustrated using different colors as defined in Table 21.2. Note, for better visualization, the upper period-3 limit cycle is indicated using yellow color in the attraction basins, but its phase portraits are plotted using khaki color.

For $k = 0$, the electromagnetic effect of the potential difference between two adjacent neurons is neglected, and the memristive HNN model exhibits bistability consisting of upper and lower period-3 limit cycles, as depicted in Fig. 21.9(a). Moreover, since the transformed memristive HNN model (21.8) is invariant under the transformation $(x_1, x_2, x_3, \varphi, k = 0, x_1(0), x_2(0), x_3(0), \varphi(0)) \leftrightarrow (-x_1, -x_2, -x_3, -\varphi, k = 0, -x_1(0), -x_2(0), -x_3(0), -\varphi(0))$, the two attraction regions are odd symmetric with the origin.

When k is increased, the force of the electromagnetic induction current I_M gradually enhances, and thus asymmetrically distributed multiple disconnected attraction basins are uncovered in the memristive HNN. In Fig. 21.9(b), four attraction regions of lower multi-period limit cycle, two asymmetrical chaotic attractors, and divergent mode are observed under $k = 0.45$; while in Figs. 21.9(c) and 21.9(d), three attraction regions of lower/upper period-2 limit cycle, upper/lower period-3 limit cycle, and divergent mode are disclosed for $k = 0.6/-0.6$. In addition, due to the symmetry property revealed by (21.10), the attraction bastions for $k = 0.6$ and -0.6 are odd symmetric with the origin.

Besides, the attraction basins demonstrated in Fig. 21.9 are highly related to the equilibrium point distributions. With reference to Fig. 21.1(a), the memristive HNN model (21.2) possesses two symmetric index-2 saddle-foci for $k = 0$, which leads to two symmetric attraction regions, as shown in Fig. 21.9(a). Meanwhile, for nonzero values of k, the memristive HNN model (21.2) pos-

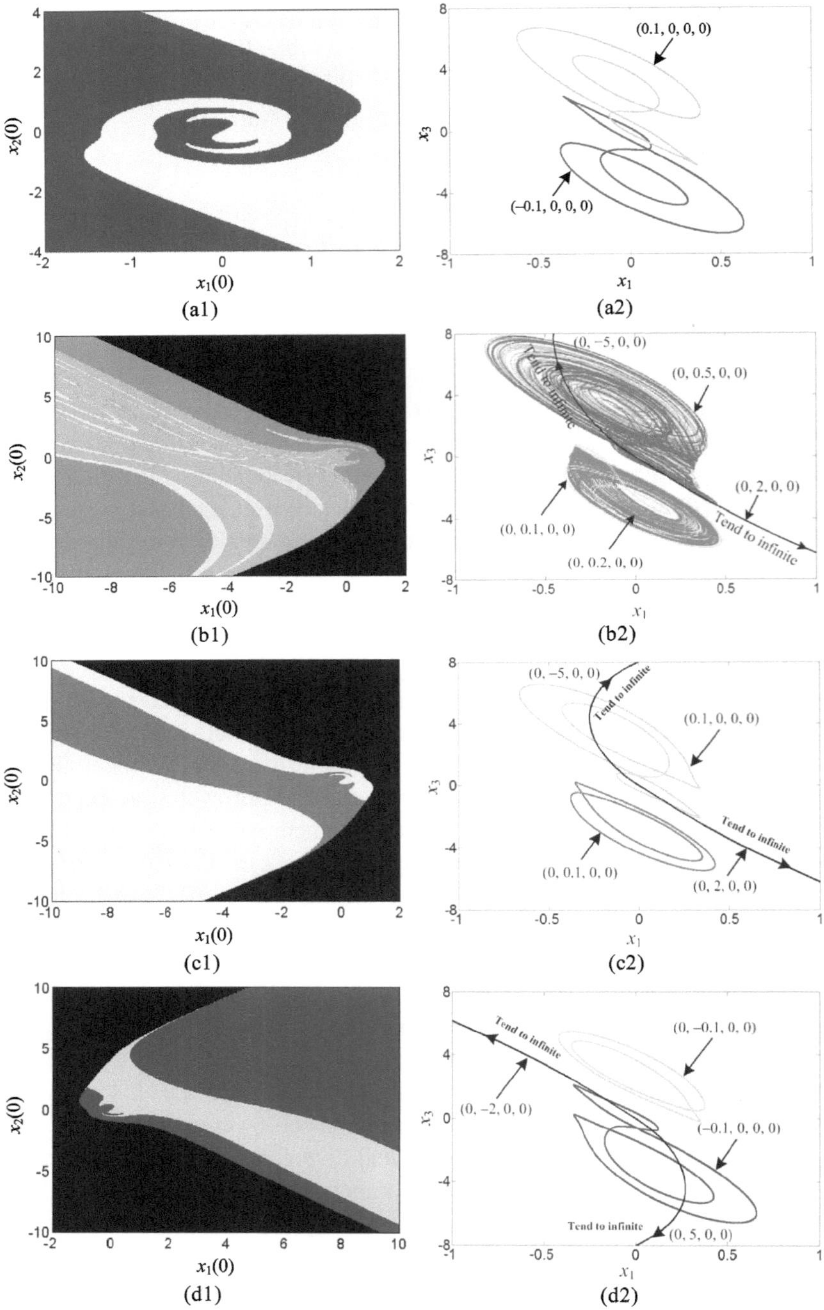

FIGURE 21.9 Initial condition-dependent dynamics with different values of the coupling strength k, in which (a1)–(d1) are attraction basins in the $x_1(0)$–$x_2(0)$ plane, and (a2)–(d2) are the corresponding coexisting attractors in the x_1–x_3 plane: (a) $k = 0$, (b) $k = 0.45$, (c) $k = 0.6$, (d) $k = -0.6$. (For interpretation of the colors in the figure, the reader is referred to the web version of this chapter.)

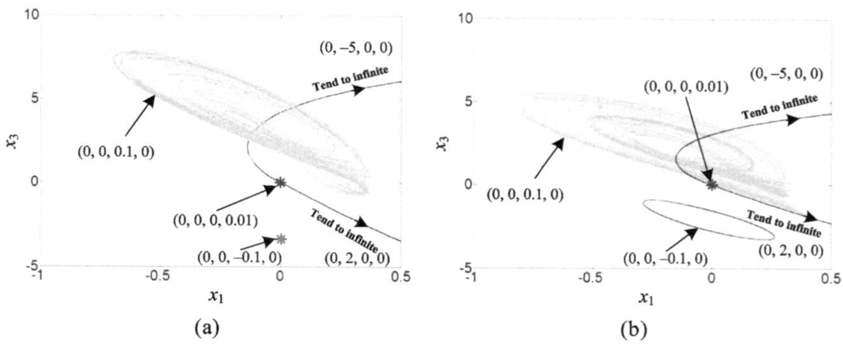

FIGURE 21.10 Initial condition-dependent dynamics with different values of the synaptic weights are coexisting in the x_1–x_3 plane: (a) $w_{11} = 1, w_{23} = -20$, (b) $w_{11} = 1.5, w_{23} = -14$.

sesses several asymmetrically distributed index-2 saddle-foci, which result in several asymmetric attraction regions, as plotted in Figs. 21.9(b) to 21.9(d).

When fixing the coupling synapse as $k = 1$ and changing the synaptic weights w_{11} and w_{23}, then multi-stable patterns in the memristor synapse-coupled Hopfield neural network (21.2) are elaborated in Fig. 21.10. In detail, Fig. 21.10(a) displays four coexisting states of chaotic spiral attractor, two fixed points with different locations, and infinite traces; while Fig. 21.10(b) illustrates four coexisting states of chaotic spiral attractor, fixed point, period-1 limit cycle, and infinite traces. As a result, multi-stable patterns of different combinations are found when varying the memristor coupling strength and synaptic connection weights.

21.3.4 Long-term transient chaotic behaviors

In the proposed memristive tri-neuron HNN, long-term transient chaotic behaviors [40–42] randomly appear under certain coupling strengths and initial conditions. When the synaptic inter-connection weight w_{23} is assigned as -20.4 under initial conditions $(0, -10^{-6}, 0, 0)$, the long-term transient chaotic behaviors are disclosed by the time-domain sequences of x_1 within $\tau = [0, 4000]$ and phase plane plot, as seen in Figs. 21.11(a) and (b). In Fig. 21.11(a), the orbit initiates a chaotic behavior and enters into a periodic behavior in the end. Its phase plane plot is given in Fig. 21.11(b), in which the upper chaotic attractor transforms into the lower periodic limit cycle with the evolution of time.

When the memristor coupling strength and initial conditions are specified as 0.6 and $(0, 0.5, 0, 0)$, respectively, and the other control parameters are kept unchanged, the time-domain sequence of x_1 within $\tau = [0, 2000]$ is plotted in Fig. 21.12(a). It is illustrated that the operation mode of the memristive HNN starts from chaos and finally settles down to period-3 limit cycle for $\tau > 1410$. Specially, to reveal the initial conditions that induce transient chaos, the attraction basin for $k = 0.6$ is redrawn within $\tau = [400, 800]$ and presented in

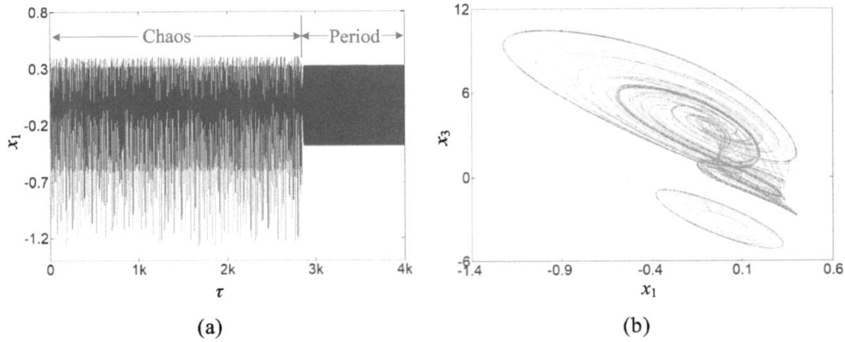

FIGURE 21.11 Long-term transient dynamics with $w_{23} = -20.4$: (a) time sequence of the state variable x under the initial conditions $(0, -10^{-6}, 0, 0)$, (b) the phase plane plot.

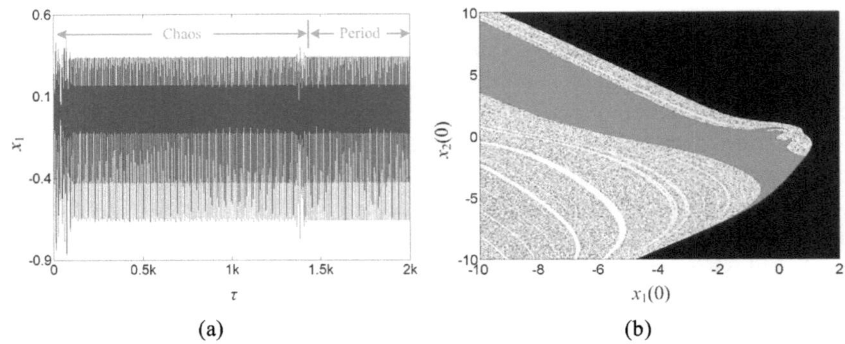

FIGURE 21.12 Transient chaos-induced extreme event at $k = 0.6$: (a) time sequence of the state variable x under the initial conditions (0, 0.5, 0, 0), (b) attraction basin in the $x_1(0)$–$x_2(0)$ plane with the time interval [400, 800], in which the red, green, yellow and black colors represent the transient chaos, upper period-3 limit cycle, lower period-2 limit cycle, and divergent modes, respectively. (For interpretation of the colors in the figure, the reader is referred to the web version of this chapter.)

Fig. 21.12(b), in which the transient chaotic mode is represented by red color. It can be seen that the lower period-2 limit cycle and divergent attraction regions are pure, while the transient chaotic mode and upper period-3 limit cycle attraction regions are mixed and riddled [43,44]. With the increase of simulation time, the initial conditions leading to transient chaos gradually decrease, and finally pure upper period-3 limit cycle attraction region is observed under the simulation time interval of [2500, 3500], as shown in Fig. 21.9(c). As a result, we can say that the extreme event may be induced by the transient chaos but not the property of the system itself.

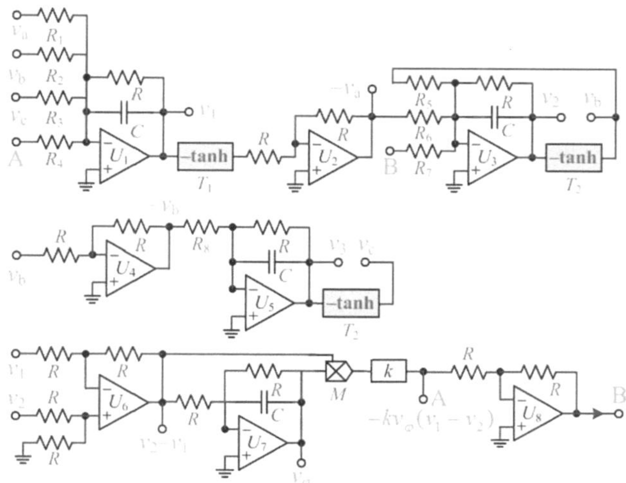

FIGURE 21.13 Realization circuit of the memristive HNN model.

21.4 Circuit synthesis and PSIM simulation

An equivalent realization circuit of the proposed memristive tri-neuron HNN model (21.2) is designed as given in Fig. 21.13, in which the top two circuits are the main circuits to realize the first three differential equations of (21.2) and the bottom one is the non-ideal memristor synapse emulator [32]. The circuit equations are formulated as

$$\begin{cases} RC\frac{dv_1}{dt} = -v_1 + \frac{R}{R_1}\tanh(v_1) + \frac{R}{R_2}\tanh(v_2) + \frac{R}{R_3}\tanh(v_3) + \frac{R}{R_4}kv_\varphi(v_1 - v_2) \\ RC\frac{dv_2}{dt} = -v_2 - \frac{R}{R_6}\tanh(v_1) + \frac{R}{R_5}\tanh(v_2) - \frac{R}{R_7}kv_\varphi(v_1 - v_2) \\ RC\frac{dv_3}{dt} = -v_3 - \frac{R}{R_8}\tanh(v_2) \\ RC\frac{dv_\varphi}{dt} = v_1 - v_2 - v_\varphi \end{cases} \tag{21.11}$$

In (21.11), v_1, v_2, v_3, and v_φ are capacitor voltages of the four integrators and $RC = 10\ \text{k}\Omega \times 100\ \text{nF} = 1$ ms is the time constant of these integrators. For typical synaptic connection weights defined in (21.3), the other circuit elements are determined as $R_1 = R/1.5 = 6.67\ \text{k}\Omega$, $R_2 = R/2.8 = 3.57\ \text{k}\Omega$, $R_3 = R/0.5 = 20\ \text{k}\Omega$, $R_4 = 10\ \text{k}\Omega$, $R_5 = R/1.2 = 8.33\ \text{k}\Omega$, $R_6 = R/1.5 = 6.67\ \text{k}\Omega$, $R_7 = 10\ \text{k}\Omega$, and $R_8 = R/20 = 0.5\ \text{k}\Omega$.

To better acquire the multi-stable patterns coexisting in the implementation circuit of the memristive HNN, PSIM (a circuit simulation software package) is utilized to perform circuit simulations. For simplifying, three two-port math models $-\tanh(\cdot)$ are directly used in PSIM simulations [31]. For the details for the implementation circuit of the $-\tanh(\cdot)$ function one may refer to [19,45]. When the specified memristor coupling strength k and initial conditions used in

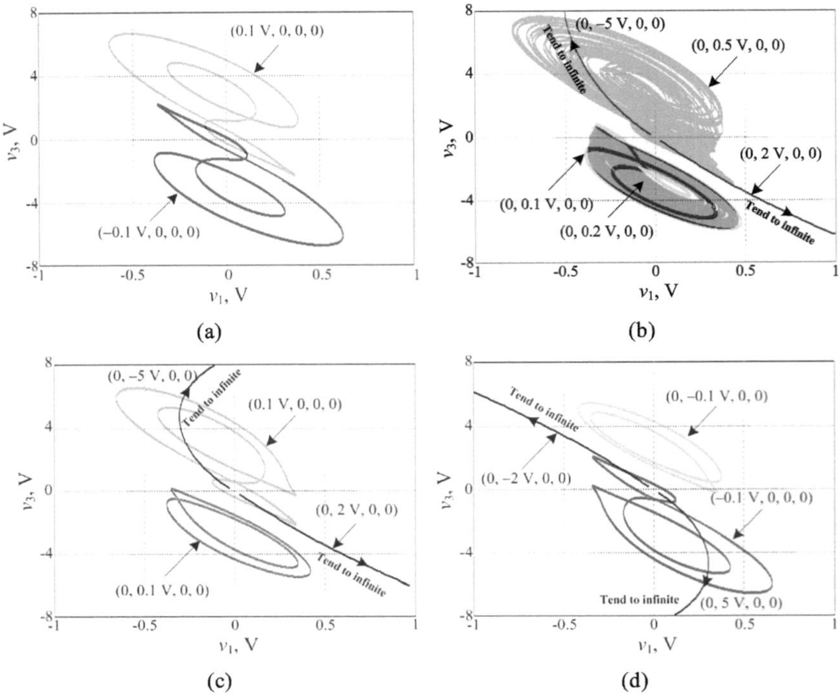

FIGURE 21.14 Simulated coexisting attractors in the v_1–v_3 plane with different values of the coupling strength k: (a) $k = 0$, (b) $k = 0.45$, (c) $k = 0.6$, (d) $k = -0.6$.

Fig. 21.9(a2) to 21.9(d2) are assigned, the screen shots of PSIM circuit simulations are displayed in Fig. 21.14, which confirm the numerical results given in Fig. 21.9.

Furthermore, when the specified synaptic connection weights and initial conditions used in Fig. 21.10(a) and 21.10(b) are assigned through tuning the weight-related resistances R_1 and R_8 and k is adjusted as 1, the screen shots of PSIM circuit simulations are displayed in Fig. 21.15, which confirm the numerical results given in Fig. 21.10 also.

21.5 Conclusion

This chapter presents a four-dimensional tri-neuron autonomous memristive HNN, which is constructed through bidirectionally coupling two neurons of the HNN with a non-ideal flux-controlled memristor synapse. In this way, the dynamical effects of the potential difference between the connected two neurons are modeled using the two-way electromagnetic induction current generated by the memristor synapse. When different memristor coupling strengths or synaptic connection weights are assigned, the proposed memristive HNN exhibits sev-

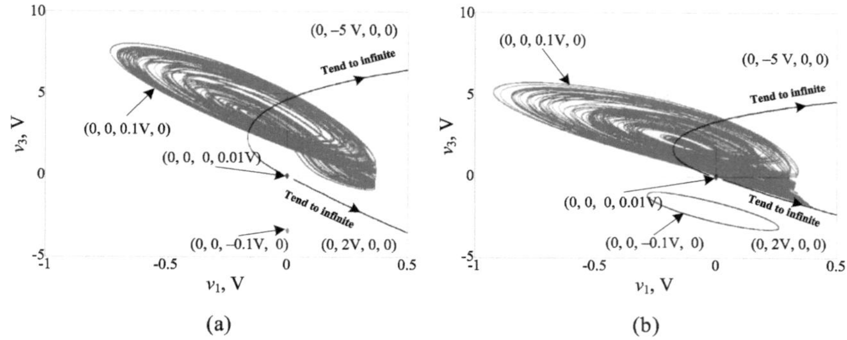

FIGURE 21.15 Simulated coexisting attractors in the v_1–v_3 plane with different values of the synaptic weights: (a) $R_1 = R/w_{11} = 10\ \text{k}\Omega$, $R_8 = R/|w_{23}| = 0.5\ \text{k}\Omega$, (b) $R_1 = R/w_{11} = 6.67\ \text{k}\Omega$, $R_8 = R/|w_{23}| = 0.71\ \text{k}\Omega$.

eral asymmetrically distributed equilibrium points, leading to the coexistence of asymmetric spiral chaotic and periodic patterns with different topological structures or different locations. Moreover, the long-term transient chaos behaviors are uncovered in the memristive HNN. Specially, parts of the simulated attraction regions are mixed and riddled under normal simulation interval, which can evolve into pure attraction regions through setting the simulation time interval as [2500, 3000]. The numerical results and circuit simulations well validate the coexisting multi-stable patterns caused by the electromagnetic effects of the potential difference between two neurons of the tri-neuron-based HNN.

Acknowledgment

This work was supported by the grants from the National Natural Science Foundations of China under 51777016, 61801054, and 61601062, and the Natural Science Foundation of Jiangsu Province, China, under Grant BK20191451.

References

[1] J.J. Hopfield, Neurons with graded response have collective computational properties like those of two-state neurons, Proc. Natl. Acad. Sci. 81 (10) (1984) 3088–3092.

[2] P.V.D. Driessche, X. Zou, Global attractivity in delayed Hopfield neural network models, SIAM J. Appl. Math. 58 (6) (1998) 1878–1890.

[3] H. Tang, Z. Feng, Y.H. Wang, P.C. Lai, C.Y. Wang, Z.Y. Ye, C.K. Wang, Z.Y. Shi, T.Y. Wang, Y. Chen, J. Gao, X.M. Jin, Experimental quantum stochastic walks simulating associative memory of Hopfield neural networks, Biophys. Rev. Appl. 11 (2) (2019) 024020.

[4] T. Deb, A.K. Ghosh, A. Mukherjee, Singular value decomposition applied to associative memory of Hopfield neural network, Mater. Today: Proceedings 5 (1) (2018) 2222–2228.

[5] K.S. Cheng, J.S. Lin, C.W. Mao, The application of competitive Hopfield neural network to medical image segmentation, IEEE Trans. Med. Imaging 15 (4) (1996) 560–567.

[6] J.G. Liu, H.C. Tseng, Information storage in Hopfield model with reduced complexity, Inf. Sci. 111 (1–4) (1998) 347–362.

[7] N. Rasheed, S.H.M. Amin, U. Sultana, A.R. Bhatti, Hopfield net spreading activation for grounding of abstract action words in cognitive robot, Biol. Inspir. Cognit. Archit. 21 (2017) 37–46.
[8] C.H. Wang, L. Xiong, J.R. Sun, W. Yao, Memristor-based neural networks with weight simultaneous perturbation training, Nonlinear Dyn. 95 (4) (2019) 2893–2906.
[9] H. Korn, P. Faure, Is there chaos in the brain? II. Experimental evidence and related models, C. R. Biol. 326 (9) (2003) 787–840.
[10] J. Ma, J. Tang, A review for dynamics in neuron and neuronal network, Nonlinear Dyn. 89 (3) (2017) 1569–1578.
[11] S. Panahi, Z. Aram, S. Jafari, J. Ma, J.C. Sprott, Modeling of epilepsy based on chaotic artificial neural network, Chaos Solitons Fractals 105 (2017) 150–156.
[12] Z.T. Njitacke, J. Kengne, Complex dynamics of a 4D Hopfield neural networks (HNNs) with a nonlinear synaptic weight: coexistence of multiple attractors and remerging Feigenbaum trees, AEU, Int. J. Electron. Commun. 93 (2018) 242–252.
[13] Q. Xu, Z. Song, H. Qian, M. Chen, P.Y. Wu, B.C. Bao, Numerical analyses and breadboard experiments of twin attractors in two-neuron-based non-autonomous Hopfield neural network, Eur. Phys. J. Spec. Top. 227 (7–9) (2018) 777–786.
[14] Z.T. Njitacke, S.D. Isaac, J. Kengne, A.N. Negou, G.D. Leutcho, Extremely rich dynamics from hyperchaotic Hopfield neural network: hysteretic dynamics, parallel bifurcation branches, coexistence of multiple stable states and its analog circuit implementation, Eur. Phys. J. Spec. Top. 229 (2020) 1133–1154.
[15] M.F. Danca, N. Kuznetsov, Hidden chaotic sets in a Hopfield neural system, Chaos Solitons Fractals 103 (2017) 144–150.
[16] Z.T. Njitacke, J. Kengne, T.F. Fozin, B.P. Leutcha, H.B. Fotsin, Dynamical analysis of a novel 4-neurons based Hopfield neural network: emergences of antimonotonicity and coexistence of multiple stable states, Int. J. Dyn. Control 7 (3) (2019) 823–841.
[17] K. Rajagopal, J.M. Munoz-Pacheco, V.T. Pham, D.V. Hoang, F.E. Alsaadi, F.E. Alsaadi, A Hopfield neural network with multiple attractors and its FPGA design, Eur. Phys. J. Spec. Top. 227 (7–9) (2018) 811–820.
[18] H.R. Lin, C.H. Wang, W. Yao, Y.M. Tan, Chaotic dynamics in a neural network with different types of external stimuli, Commun. Nonlinear Sci. Numer. Simul. 90 (2020) 105390.
[19] B.C. Bao, C.J. Chen, H. Bao, X. Zhang, Q. Xu, M. Chen, Dynamical effects of neuron activation gradient on Hopfield neural network: numerical analyses and hardware experiments, Int. J. Bifurc. Chaos 29 (2019) 1930010.
[20] M.P. Sah, H. Kim, L.O. Chua, Brains are made of memristors, IEEE Circuits Syst. Mag. 14 (2014) 12–36.
[21] W. Ma, M.A. Zidan, W.D. Lu, Neuromorphic computing with memristive devices, Sci. China Inf. Sci. 61 (2018) 060422.
[22] Q.D. Li, S. Tang, H.Z. Zeng, T.T. Zhou, On hyperchaos in a small memristive neural network, Nonlinear Dyn. 78 (2) (2014) 1087–1099.
[23] V.T. Pham, S. Jafari, S. Vaidyanathan, C. Volos, X. Wang, A novel memristive neural network with hidden attractors and its circuitry implementation, Sci. China, Technol. Sci. 59 (2016) 358–363.
[24] G. Zhang, C.N. Wang, F. Alzahrani, F.Q. Wu, X.L. An, Investigation of dynamical behaviors of neurons driven by memristive synapse, Chaos Solitons Fractals 108 (2018) 15–24.
[25] F. Xu, J.Q. Zhang, M. Jin, S.F. Huang, T.T. Fang, Chimera states and synchronization behavior in multilayer memristive neural networks, Nonlinear Dyn. 94 (2) (2018) 775–783.
[26] F. Parastesh, S. Jafari, H. Azarnoush, B. Hatef, H. Namazi, D. Dudkowski, Chimera in a network of memristor-based Hopfield neural network, Eur. Phys. J. Spec. Top. 228 (10) (2019) 2023–2033.
[27] X.Y. Hu, C.X. Liu, L. Liu, J.K. Ni, Y.P. Yao, Chaotic dynamics in a neural network under electromagnetic radiation, Nonlinear Dyn. 91 (3) (2018) 1541–1554.

[28] M.Y. Ge, Y. Jia, Y. Xu, L.J. Yang, Mode transition in electrical activities of neuron driven by high and low frequency stimulus in the presence of electromagnetic induction and radiation, Nonlinear Dyn. 91 (1) (2018) 515–523.
[29] L.L. Lu, Y. Jia, Y. Xu, M.Y. Ge, L.J. Yang, X. Zhan, Energy dependence on modes of electric activities of neuron driven by different external mixed signals under electromagnetic induction, Sci. China, Technol. Sci. 62 (3) (2019) 427–440.
[30] H. Bao, A.H. Hu, W.B. Liu, B.C. Bao, Hidden bursting firings and bifurcation mechanisms in memristive neuron model with threshold electromagnetic induction, IEEE Trans. Neural Netw. Learn. Syst. 31 (2) (2020) 502–511.
[31] C.J. Chen, J.Q. Chen, H. Bao, M. Chen, B.C. Bao, Coexisting multi-stable patterns in memristor synapse-coupled Hopfield neural network with two neurons, Nonlinear Dyn. 95 (4) (2019) 3385–3399.
[32] C.J. Chen, H. Bao, M. Chen, Q. Xu, B.C. Bao, Non-ideal memristor synapse-coupled bi-neuron Hopfield neural network: numerical simulations and breadboard experiments, AEÜ, Int. J. Electron. Commun. 111 (2019) 152894.
[33] H.R. Lin, C.H. Wang, Y.M. Tan, Hidden extreme multistability with hyperchaos and transient chaos in a Hopfield neural network affected by electromagnetic radiation, Nonlinear Dyn. 99 (3) (2020) 2369–2386.
[34] Y. Leng, D.S. Yu, Y.H. Hu, S.S. Yu, Z.B. Ye, Dynamic behaviors of hyperbolic-type memristor-based Hopfield neural network considering synaptic crosstalk, Chaos 30 (3) (2020) 033108.
[35] M. Chen, C.J. Chen, B.C. Bao, Q. Xu, Initial sensitive dynamics in memristor synapse-coupled Hopfield neural network, J. Electron. Inf. Technol. 42 (4) (2020) 870–877.
[36] M. Chen, M.X. Sun, H. Bao, Y.H. Hu, B.C. Bao, Flux-charge analysis of two-memristor-based Chua's circuit: dimensionality decreasing model for detecting extreme multistability, IEEE Trans. Ind. Electron. 67 (3) (2020) 2197–2206.
[37] M. Chen, Y. Feng, H. Bao, B.C. Bao, Y.J. Yu, H.G. Wu, Q. Xu, J. Wang, State variable mapping method for studying initial-dependent dynamics in memristive hyper-jerk system with line equilibrium, Chaos Solitons Fractals 115 (2018) 313–324.
[38] A. Wolf, J.B. Swift, H.L. Swinney, J.A. Vastano, Determining Lyapunov exponents from a time series, Phys. D 16 (3) (1985) 285–317.
[39] B.C. Bao, J.Y. Luo, H. Bao, C.J. Chen, H.G. Wu, Q. Xu, A simple nonautonomous hidden chaotic system with a switchable stable node-focus, Int. J. Bifurc. Chaos 29 (12) (2019) 1950168.
[40] C.C. Strelioff, A.W. Hübler, Medium-term prediction of chaos, Phys. Rev. Lett. 96 (4) (2006) 044101.
[41] M.F. Danca, Hidden transient chaotic attractors of Rabinovich–Fabrikant system, Nonlinear Dyn. 86 (2) (2016) 1263–1270.
[42] X.S. Yang, Q. Yuan, Chaos and transient chaos in simple Hopfield neural networks, Neurocomputing 69 (1–3) (2005) 232–241.
[43] A. Saha, U. Feudel, Riddled basins of attraction in systems exhibiting extreme events, Chaos 28 (3) (2018) 033610.
[44] G.Y. Wang, F. Yuan, G.R. Chen, Y. Zhang, Coexisting multiple attractors and riddled basins of a memristive system, Chaos 28 (1) (2018) 013125.
[45] S. Zhang, J.H. Zheng, X.P. Wang, Z.G. Zeng, S.B. He, Initial offset boosting coexisting attractors in memristive multi-double-scroll Hopfield neural network, Nonlinear Dyn. 102 (2) (2020) 2821–2841.

Chapter 22

Fuzzy memristive networks

Samaneh Soradi-Zeid[a] and Hadi Jahanshahi[b]
[a]*Faculty of Industry and Mining (Khash), University of Sistan and Baluchestan, Zahedan, Iran,*
[b]*Department of Mechanical Engineering, University of Manitoba, Winnipeg, Canada*

22.1 Introduction

Leon Chua in 1971 [21], found that there were only three basic circuit elements namely the resistor, capacitor, and inductor between relating voltage and current, charge and voltage and current and magnetic flux, respectively, however, there was no charge related to the primary circuit and magnetic flux. So, he firstly introduced the memristor portmanteau of memory and resistor as the fourth circuit element and showed that this memristor has different characteristics from the other three fundamental circuits. A few years later, Chua and Kang [25] generalized new concepts of memory to describe a wide range of memory systems. After the physical realization of the first memristive device or memristor [20], the field of simulation of the brain was addressed. This is important because there are many similarities between memristors and synapses in the brain [22,23]. The memristor is a passive device that can be modified by applying appropriate voltage or current to properties such as resistance (known as memristance) or conductance (known as memductance). So, by adjusting the memory of these devices, analog values such as synaptic weight can be stored on this device. Fortunately, unlike capacitor, the memristor can hold its memory for a long time without the voltage or current applied to it [24]. Because memristor-based circuits use memristors in a computer to save time in reloading data and could lead to human-like learning, they have recently become a major topic for research. The memristor, a thin semiconductor film between two metallic contacts with variable resistance that automatically changes as the voltage changes, can be used as an adjustable parameter in control systems [62].

Over the past decades, many researchers have been trying to build an intelligent system with computational power comparable to the human brain. These efforts are often divided into two general sections called Artificial Neural Networks (ANNs) and Fuzzy Logic. With an overview you will notice that most of this research is focused on software only and it is difficult to find a good example of hardware running a smart system. The importance of having practical, scalable hardware becomes clear when we look at the number of neurons in the human brain and the complexity of the connections between them. According

Mem-elements for Neuromorphic Circuits with Artificial Intelligence Applications
https://doi.org/10.1016/B978-0-12-821184-7.00031-1

to the nature of computation and memory in brain, now it is a well-accepted fact that this hardware should be in analog form since in this case it can work much faster than conventional digital circuits and will consume much less energy. However, theretofore, there was a big obstacle in front of reaching this goal. In fact, there is no simple inactive element that can be used to store and manipulate data such as a synaptic weight. Note that although analog values can be stored as voltage or charge in capacitors, the stored values are not easily readable and usable without modification.

In previous nonlinear analog circuits of neural networks, the weights of self-feedback connection and weights of connection are usually performed using traditional resistors. If the traditional resistors have been replaced by memristors, a new class of neural network models will emerge called models-memristor-based neural networks or briefly memristive neural networks (MNNs). Recently, many researchers have been fascinated by dynamics analysis of MNNs in the fields of applied sciences such as signal processing, pattern recognition, reconfigurable computing, programmable logic, brain–computer interfaces and control systems. Many significant developments on MNNs have been reported, for instance, image processing [18], feature extraction [19] and associative memory [17]. Since chaotic phenomena have appeared extensively in mathematics, physics, secure communication and engineering science, the dynamical behaviors of chaotic neural networks have also received much attention. Synchronization implies that two or more systems are sharing a common dynamical behavior in either chaotic or periodic form by coupling or external force. Synchronization of chaotic systems has been the focus of many researchers because chaotic synchronization has had successful results in a number of engineering fields such as image processing, information science, and secure communications. To ensure exponential synchronization results of MNNs, the authors created some new sufficient conditions using some of Lyapunov's analytical techniques and appropriate functions in [26,27]. Exponential delay synchronization for time-delayed MNNs was investigated by constructing new Lyapunov–Krasovskii novel functions in [28,29]. The aforementioned properties of Memristor have encouraged some researchers to develop new brain-like computing architectures and techniques where the primary focus was on hardware simplicity. To investigate uncertainties in human cognitive processes and better modeling of neural systems, Yang and Yang [16] applied fuzzy logic to the traditional neural network. Studies in this regard can be divided into two main categories. The first category is for work that attempts to use fuzzy inference methods using memristive hardware and the second category comprises studies focused on the implementation of hardware for neural networks and their learning methods. For example in the first state, the authors in [61] showed that relationships (describing the relationship between fuzzy concepts of input and output) using the Hebbian learning method could be effectively created on memstore boundary structures. In the second state we can refer for example Spike Timing-Dependent Plasticity using memristor crossbar structures (see Figs. 22.1 and 22.2) [48,49].

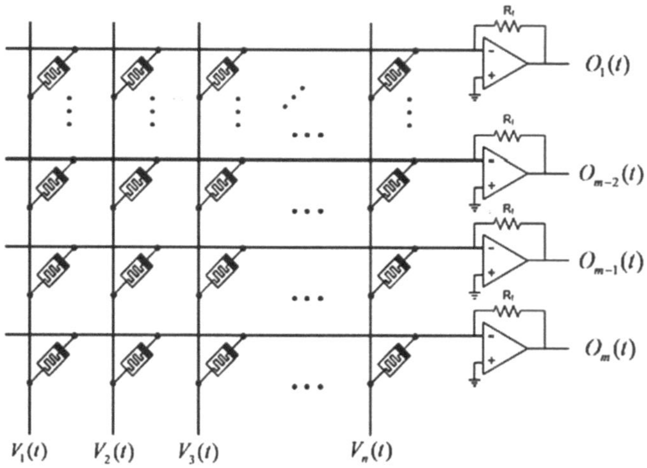

FIGURE 22.1 Memristor crossbar-based circuit.

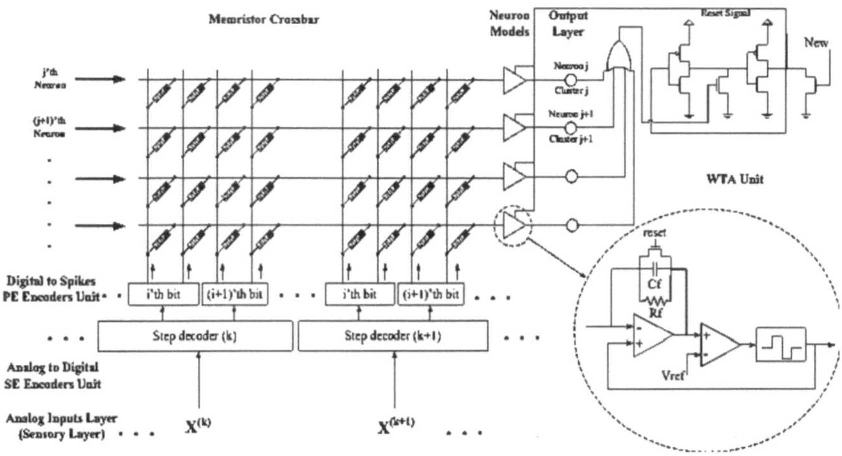

FIGURE 22.2 Memristor crossbar-based neuro-fuzzy clustering circuit.

In recent years, there has been no field in science and engineering that has remained untouched by fuzzy calculus. In fact, many researchers are currently paying attention to the fuzzy calculus concepts in chemistry, electromagnetic waves, quantitative finance, quantum evolution of complex systems, chaos and fractals, robotics, control systems, etc. [50–59]. Fuzzy calculus captures the history of the variable, or, in other words, shows memory, contrary to integer-amount calculus, which involves local operators. This characteristic makes them an important tool in the modeling of memory-intense and delay systems. Meanwhile, many systems are known to display fuzzy-value dynamics, such as viscoelastic systems, electrode–electrolyte polarization and complex adaptive systems in biology. The main reason is that fuzzy-value equations are naturally related to systems with memory which exists in most biological systems. Recently, the authors [60] have shown that using a fuzzy logic instead of digital logic, a superior soft-computing system with the capability of learning can be designed. They implemented the proposed analog system, which was very fast and fully compatible with the analog memory nature. In addition, the proposed fuzzy neural computing system had the advantage that in its hierarchical structure, memory units were absorbed into computing units, similar to the human brain.

Therefore, mathematical modeling has proven to be valuable in understanding the dynamics of systems and, in particular, the fuzzy-value systems have been used to model the biological systems such as neuronal communication. Having a good model is very interesting in computational neurosciences such as Parkinson's disease. Deep brain stimulations based on the existing models can be improved if we dispose of more realistic models based on the fuzzy systems. All the closed control strategies developed in the case of "deep brain stimulations" are based on ordinary differential equations and can be improved in the case of fuzzy models. Furthermore, delay plays an important role in the process of spreading infectious diseases; it can be used to simulate the incubation period of infectious diseases, the period of patients infected with disease, period of patients' immunity to disease, and so on. The basic fact reflected by the specific mathematical model with time delay is that the change of trajectory about time t not only depends on the t moment itself but also is affected by some certain conditions before, even the reflection of some certain factors before. This kind of circumstance is abundant in the objective world. In the last two decades, the elements of soft-computing such as neural networks and fuzzy systems have been widely applied for function approximation in theoretical and engineering applications. In most proposed algorithms for solving fuzzy programming problems, the fuzzy problem is converted to a crisp problem. From a point of view, converting a fuzzy problem to a crisp problem is not adequately soft for real-world applications. So, the researchers intend to solve the fuzzy programming problem directly, without changing it to a crisp program. In doing so, they need a direct approach for ordering fuzzy numbers.

Also, the theory of fractional-order integral and derivative was mainly developed in the last three decades of the nineteenth century. The denotation *fractional order* is actually a misnomer and actually encompasses all non-integer order numbers like fractions, irrational numbers, and complex numbers. In this respect the terminology *non-integer order* is more apt. In recent years, a number of scholars have been raised research interests to applied fractional-order calculus to the control field. With the improvement continuously of this concept, numerical methods and algorithms, and suggesting of various fractional-order analysis and control strategies, it is more impetuses in the application of fractional-order control theories and rapid development. In this chapter, solution techniques for fuzzy fractional equations are discussed, analyzed and applied to solve an initial boundary value problem involving fuzzy equation with fractional order [109].

The rest of this chapter is organized as follows. The notion of fuzzy and fractional calculus is briefly reviewed in Section 22.2. Application of the memristor for the implementation of fuzzy systems is discussed in Section 22.3. Section 22.4 introduces the delay memristive fuzzy systems and proposes the existing methods for its construction. Recent work on memristive fuzzy systems based on fractional calculus is presented in Section 22.5 and, finally, Section 22.6 concludes with a general overview of the area and future trends.

22.2 Requirement

Here, we briefly propose some definitions regarding fuzzy calculus and fractional operators permitting us to introduce memristive systems.

22.2.1 Fuzzy calculus concepts

The fuzzy set concept was introduced by Zadeh [1], which is an extended version of the concept of classical set. Fuzzy calculus is one of the main branches of fuzzy mathematics [2]. For the first time, Dubois and Prade introduced the preliminaries and concepts of fuzzy calculus in 1982 [3–5]. These concepts have been supplemented by other researchers over the years [6–15]. Here, some basic notions of fuzzy calculus are presented.

The formal mathematical representation of a fuzzy set is based on the fact that any classic subset can be characterized by a function as follows:

$$\chi_A = \begin{cases} 1 & \text{if} \quad x \in A \\ 0 & \text{if} \quad x \in A \end{cases} \tag{22.1}$$

for all $x \in U$ in which U a non-empty set and $A \subseteq U$. However, there are cases where an element is partially in a set which means we cannot always say that an element completely belongs to a given set or not. A simple extension of this concept is the so-called membership function μ_A of the fuzzy set A which models the idea that the statement x belongs to A is not necessarily true or false only. On the contrary, it may be a graded idea. Thus, we have

$$0 \leq \mu_A(x) \leq 1, \quad \text{for any } x \in U, \tag{22.2}$$

where $\mu_A(x)$ represents the degree of truth of the statement x belongs to A. Also, in general, a fuzzy number has a membership function which increases monotonically from 0 to 1 on the left-hand side, thereafter, there is a single top or a plateau at the level 1 and, finally, the membership function decreases monotonically to 0 on the right-hand side. The reader who wants greater details about fuzzy set and fuzzy number should consult [110,111].

Let us define $\mathbb{R}_F$ as the set of fuzzy subsets of the real axis (i.e. $u : \mathbb{R} \to [0, 1]$) for which the following conditions apply:

(i) $\forall u \in \mathbb{R}_F$, u is upper semi-continuous on $\mathbb{R}$.
(ii) $\forall u \in \mathbb{R}_F$, u is fuzzy convex.
(iii) $\forall u \in \mathbb{R}_F$, u is normal.
(iv) $cl\{x \in \mathbb{R}|u(x) > 0\}$ is compact, where cl denotes the closure of a subset.

Then $\mathbb{R}_F$ is called the fuzzy numbers' space. Apparently, $\mathbb{R} \subset \mathbb{R}_F$. Assuming $0 < t \leq 1$, we denote $[u]^r = \{x \in \mathbb{R}|u(x) \geq r\}$, and $[u]^0 = cl\{x \in \mathbb{R}|u(x) > 0$. From (i) to (iv), it can be concluded that, for each $r \in [0, 1]$, the r-level sets of $u \in \mathbb{R}_F$ are non-empty and closed intervals. Additionally, a triangular fuzzy number can be defined as a fuzzy set in $\mathbb{R}_F$ which is determined by $u = (a, b, c) \in \mathbb{R}^3$, where $a \leq b \leq c$, such that $\underline{u}(r) = a + (b - a)r$ and $\overline{u}(r) = c - (c - b)r$ are the endpoints of r-level ($r \in [0, 1]$) sets. Moreover, the generalized Hukuhara (gH) difference, as a fuzzy number ω, can be defined as

$$u \ominus_{gH} v = \omega \Leftrightarrow \begin{cases} (i) & u = v \bigoplus \omega, \\ (ii) & v = u \bigoplus (-1)\omega, \end{cases} \tag{22.3}$$

in which $u, v \in \mathbb{R}_F$ are two fuzzy numbers. It can be easily proven that conditions (i) and (ii) are both valid if and only if ω is a crisp number. Now, consider a fuzzy-valued function as:

$$f : [a, b] \subseteq \mathbb{R} \to \mathbb{R}_F \tag{22.4}$$

The r-level representation of the fuzzy-valued function f given by $f(x; r) = [\underline{f}(x; r), f(x; r)]$, $x \in [a, b]$, $r \in [0, 1]$. If

$$\forall \epsilon > 0,\ \exists \delta > 0;\ \forall x (0 < |x - x_0| < \delta \Rightarrow D(f(x), L) < \epsilon),\ x_0 \in [a, b] \tag{22.5}$$

Then $L \in \mathbb{R}_F$ is the limit of f in x_0 ($x_0 \in [a, b]$), it can be denoted as

$$\lim_{t \to x_0} f(x) = L. \tag{22.6}$$

Additionally, the function f in the interval $[a, b]$ can be considered as a continuous function if

$$\lim_{t \to x_0} f(x) = f(x_0), \tag{22.7}$$

such that the continuity is one-side at points a, b.

Now, we delineate some concepts of fuzzy logic that will be a foundation for researchers. This topic is of great interest in the resolution methods of the relational equations and of the systems based on fuzzy rules, such as Mamdani's controllers. The rule base fulfills the role of translate mathematically the information that form the knowledge base of the fuzzy system. The reader who wants greater details about fuzzy logic for process modeling should consult [112–114].

22.2.2 Fractional calculation concepts

Perhaps one of the most commonly used concepts in applied sciences is the derivative concept used to describe the rate of change of a particular function, and is more commonly used to construct mathematical equations that describe the behavior of real-world problems. However, given the complexities of real-world problems, this concept has been updated to mean fractional derivatives that are more suitable for modeling real-world problems than local derivatives. The most popular of definitions are as follows [89–92,99]:

(i) The fractional integral (or the Riemann–Liouville integral) with order $\alpha > 0$ of $f(t)$ is defined by

$$ {}_{t_0}I_t^{\alpha} f(t) = \frac{1}{\Gamma(\alpha)} \int_{t_0}^{t} (t-s)^{\alpha-1} f(s) ds, \tag{22.8} $$

where $\Gamma(.)$ is Euler's gamma function.

(ii) The left and right Grunwald–Letnikov derivatives with order $\alpha > 0$ for a given function $f(t)$ are defined, respectively, by

$$ {}_{t_0}^{GL}D_t^{\alpha} f(t) = \lim_{h\to 0} h^{-\alpha} \sum_{j=0}^{N} (-1)^j \begin{pmatrix} \alpha \\ j \end{pmatrix} f(t-jh),\ h = \frac{t-t_0}{N} \tag{22.9} $$

and

$$ {}_{t}^{GL}D_{t_f}^{\alpha} f(t) = \lim_{h\to 0} h^{-\alpha} \sum_{j=0}^{N} (-1)^j \begin{pmatrix} \alpha \\ j \end{pmatrix} f(t+jh),\ h = \frac{t_f-t}{N}. \tag{22.10} $$

(iii) For $n-1 \leq \alpha < n, n \in \mathbb{N}$, the left and right Riemann–Liouville derivatives with order $\alpha > 0$ of $f(t)$ are defined, respectively, by

$$ {}_{t_0}^{RL}D_t^{\alpha} f(t) = \frac{1}{\Gamma(n-\alpha)} \frac{d^n}{dt^n} \int_{t_0}^{t} (t-s)^{n-\alpha-1} f(s) ds, \tag{22.11} $$

and

$$ {}_{t}^{RL}D_{t_f}^{\alpha} f(t) = \frac{(-1)^n}{\Gamma(n-\alpha)} \frac{d^n}{dt^n} \int_{t}^{t_f} (s-t)^{n-\alpha-1} f(s) ds. \tag{22.12} $$

(iv) For $n-1 \leq \alpha < n, n \in \mathbb{N}$, the left and right Caputo derivatives with order $\alpha > 0$ of $f(t)$ are defined as

$$ {}_{t_0}^{C}D_t^{\alpha} f(t) = \frac{1}{\Gamma(n-\alpha)} \int_{t_0}^{t} (t-s)^{n-\alpha-1} f^{(n)}(s) ds \tag{22.13} $$

and

$$ {}_{t}^{C}D_{t_f}^{\alpha} f(t) = \frac{(-1)^n}{\Gamma(n-\alpha)} \int_{t}^{t_f} (s-t)^{n-\alpha-1} f^{(n)}(s)ds, \tag{22.14}$$

respectively.

Generally speaking, the above definitions of fractional derivatives are not equivalent. The differences and relations are discussed in detail in [93–95], and we just list some cases as follows:

$$ {}_{t_0}^{RL}D_{t}^{\alpha} f(t) = {}_{t_0}^{C}D_{t}^{\alpha} f(t) + \sum_{k=0}^{n-1} \frac{f^{(k)}(t_0)(t-t_0)^{k-\alpha}}{\Gamma(k+1-\alpha)}, \tag{22.15}$$

$$ {}_{t}^{RL}D_{t_f}^{\alpha} f(t) = {}_{t}^{C}D_{t_f}^{\alpha} f(t) + \sum_{k=0}^{n-1} \frac{f^{(k)}(t_f)(t_f-t)^{k-\alpha}}{\Gamma(k+1-\alpha)}, \tag{22.16}$$

where $f \in C^{n-1}[t_0, t]$ and $f^{(n)}$ is an integrable function on $[t_0, t]$. Furthermore, if $f \in C^n[t_0, t]$, then ${}_{t_0}^{GL}D_t^{\alpha} f(t) = {}_{t_0}^{RL}D_t^{\alpha} f(t)$ [96]. Also, if $f(t) \in AC^n([t_0, t])$ (absolutely continuous functions), $\alpha > 0$ and $n = \lceil \alpha \rceil + 1$, we have

$$ {}_{t_0}I_t^{\alpha}\, {}_{t_0}^{C}D_t^{\alpha} f(t) = f(t) - \sum_{j=0}^{n-1} \frac{f^{(j)}(t_0)}{j}(t-t_0)^j, \tag{22.17}$$

$$ {}_{t}I_{t_f}^{\alpha}\, {}_{t}^{C}D_{t_f}^{\alpha} f(t) = f(t) - \sum_{j=0}^{n-1} \frac{(-1)^j f^{(j)}(t_f)}{j}(t_f-t)^j. \tag{22.18}$$

If $\alpha > 0$ and $f(t)$ be a continuous function on $[t_0, t_f]$, then

$$ {}_{t_0}^{C}D_t^{\alpha}\, {}_{t_0}I_t^{\alpha} f(t) = f(t), \qquad {}_{t}^{C}D_{t_f}^{\alpha}\, {}_{t}I_{t_f}^{\alpha} f(t) = f(t). \tag{22.19}$$

22.3 Memristive fuzzy logic systems

In 1965, when Zadeh first proposed fuzzy sets [37], his vision was set on giving more control over decision making, with his Fuzzy Logic an immeasurable amount of decision making situations could be easily modeled whereas hard logic, using true or false values, could not. This opened a new era in decision making with fuzzy sets. Fuzzy systems transform the knowledge base into a mathematical formulation that has proven to be very efficient in many applications. Recently, Takagi–Sugeno fuzzy systems have been used successfully in several engineering fields, including near space vehicles, sewage treatment processes and nonlinear active suspension systems. In the last decade, significant progress has been made in the theoretical analysis of Takagi–Sugeno fuzzy

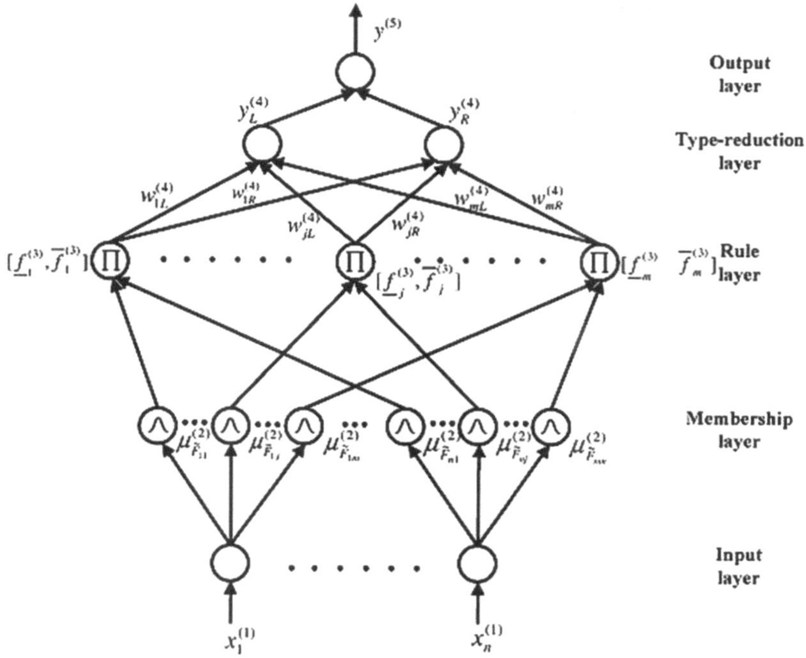

FIGURE 22.3 Structure of fuzzy neural networks.

systems, for example, stability [46], passivity [38], adaptive control [39], and sliding mode control [40]. Especially with the introduction of Takagi–Sugeno fuzzy logic in neural networks, many outstanding achievements in fuzzy neural networks (FNN), which have the capability of artificial neural networks to learn from processes, have been popularly addressed (see Fig. 22.3). For instance, Shi et al. [41] investigated H_∞ and passive filtering of discrete-time FNNs with Markovian jumps by means of theories of stochastic analysis, Lyapunov method, and matrix decomposition techniques. Finite-time cluster synchronization of complex networks with nonlinear coupling strengths and probabilistic coupling delays was discussed in [42] via rigorous analysis techniques. The first memristor which was proposed by Chua [115] is well-known as the fourth basic passive nonlinear circuit element (see Fig. 22.4). Wen et al. [43] studied master–slave exponential lag synchronization of fuzzy MNNs, where connection weights of slave neural networks were designed through adaptive strategies. Lately, global stabilization of fuzzy MNNs with mixed time-varying delays was studied in [44]. Meanwhile, Lagrange exponential stability of fuzzy MNNs with bounded and Lurie-type activation functions was concerned in [45]. Zhang et al. [67] investigated exponential synchronization of the MNNs based on periodically intermittent control and obtain the sufficient conditions of exponential synchronization by using the methods of non-smooth analysis, Lyapunov func-

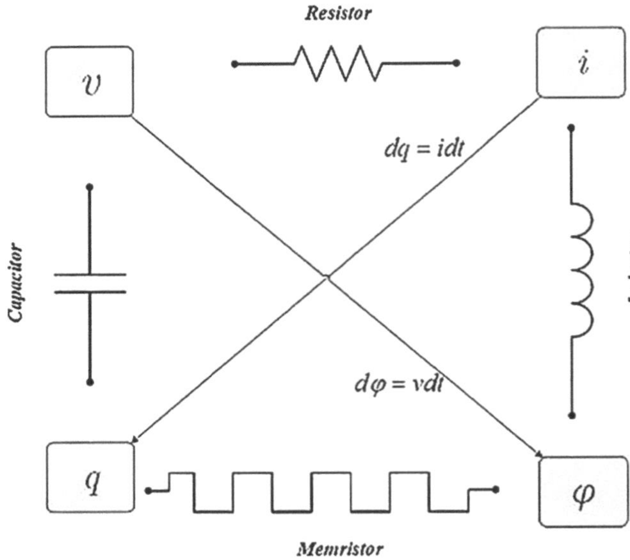

FIGURE 22.4 The position of memristor in four basic circuit elements.

tion, and inequalities. Shet et al. [69] investigated the suboptimal control for memristive system. They considered the state-dependent properties of the memristor and employed a parallel distributed compensation fuzzy model to linearize the complicated memristive chaotic system with only two subsystems. In [84] the authors studied the problem of exponential synchronization and stabilization of MNNs fuzzy model by adopting intermittent state feedback control. A novel neuro-fuzzy computing system is proposed in [86] and applied for design a simple memristor crossbar-based analog circuit to implement this neuro-fuzzy system where its learning is based on the creation of fuzzy relations by using a new implication method without utilizing any exact mathematical techniques. The problem of exponential stabilization and synchronization for fuzzy model of MNN is investigated by using periodically intermittent control in [104].

In general, fuzzy logic controllers are applicable to designs that are poorly understood mathematically and where experienced human operators are available to provide a qualitative "rule of thumb". Here, there has been a growing interest in systematic design methods for a class of nonlinear systems using fuzzy adaptive control programs. An adaptive fuzzy system is a fuzzy logic system that follows a training algorithm based on an adaptive controller made from a set of fuzzy IF-THEN rules and the parameters of the membership functions that specify linguistic terms in the IF-THEN rules are changed according to some adaptive rules to control the plant for tracking the reference trajectory. The fuzzy technique is also a powerful tool, especially, for chaos coordination in cases of disturbance [108]. Recently, Zhong et al. [88] introduced a memristor-based chaotic systems by replacing Chua's diode in the well-known Chua's chaotic system as follows:

$$\begin{cases} \dot{x}_1(t) = \alpha(x_2(t) - W(x_4(t)))x_1(t), \\ \dot{x}_2(t) = x_3(t) - x_1(t), \\ \dot{x}_3(t) = -\beta x_2(t) - \gamma x_3(t), \\ \dot{x}_4(t) = x_1(t). \end{cases} \tag{22.20}$$

They applied the Takagi–Sugeno fuzzy model for this memristor-based chaotic system and investigated its impulsive stabilization. In general, fuzzy control has multidimensional control input and requires all system status information. On the other hand, in practical engineering, it is not possible to get all the information about the status of the system. Also, multidimensional control not only increases the cost of control but also leads to the difficulty of entering the disorder. The effect of parameter mismatch on the impulsive synchronization for Takagi–Sugeno fuzzy model of the memristor-based chaotic system is investigated in [103] based on the linear decomposition and comparison system methods. The authors in [70] focus on the fuzzy synchronization based on the theory and Lyapunov stability theory for a new memristive chaotic system with disturbances as

$$\begin{cases} C_1 \dfrac{dV_{C_1}(t)}{dt} = i_{L_1}(t) - W(\varphi(t))V_{C_1}(t), \\ L_1 \dfrac{di_{L_1}(t)}{dt} = V_{C_2}(t) - V_{C_1}(t) + R_1 i_{L_1}(t), \\ C_2 \dfrac{dV_{C_2}(t)}{dt} = \dfrac{1}{R}(V_{C_3}(t) - V_{C_2}(t)) - i_{L_1}(t), \\ C_3 \dfrac{dV_{C_3}(t)}{dt} = \dfrac{1}{R}(V_{C_2}(t) - V_{C_3}(t)) - i_{L_2}(t), \\ L_2 \dfrac{di_{L_2}(t)}{dt} = V_{C_3}(t) - R_2 i_{L_2}, \\ \dfrac{d\varphi(t)}{dt} = V_{C_1}(t), \end{cases} \tag{22.21}$$

and built the fuzzy model for this memristive chaotic system. In the context of the Takagi–Sugeno fuzzy models, the authors in [72,83] addressed fuzzy modeling and impulsive control of the memristor based Chua chaotic system. An observer-based fuzzy control scheme based on the Takagi–Sugeno fuzzy model of the Chua systems is presented in [71] and depicted in Fig. 22.5. More recently, a new fuzzy model of the memristor-based Lorenz circuit, used to synchronize with the memorial-based Chua circuit, was investigated in [74], which mainly focused on the stability or synchronization of memristive chaotic circuit systems rather than state estimation. Also, the stability and stabilization of a class of Markovian chaotic systems with general uncertain transition rates have been studied in [75] via fuzzy sampled-data control. A powerful neuro fuzzy approach, fully implementable on the memristor-crossbar was described in [80].

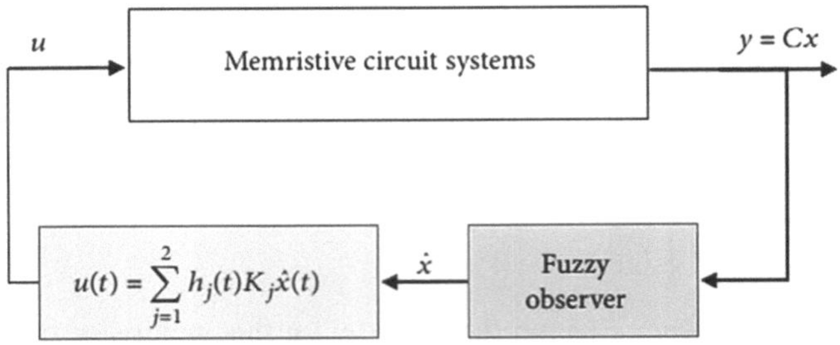

FIGURE 22.5 Observer-based fuzzy control structure diagram.

In [81] a novel neuro-fuzzy approach, where each neuron creates a flexible convex membership function for its specific cluster in the input space, based on a novel spike encoding scheme and a compatible learning algorithm, is proposed and applied on the hybrid memristor-crossbar platform. The memristor-based and Takagi–Sugeno fuzzy-based control protocols are adopted to solve the passivity and passification for a fuzzy memristor-based inertial neural networks in [87] on time scales. The problem of exponential lag adaptive synchronization control of MNNs was investigated via fuzzy method and applied in pseudo random number generator in [100]. In [105], Zhang et al. demonstrate a new strategy to tackle the memristor-based neuromorphic hardware stochasticity via introducing fuzziness into neural networks, that is, by fuzzifying the device parameters governing the learning processes based on the cycle-to-cycle and device-to-device variations.

22.4 Delay memristive fuzzy systems

Time delay which is a source of instability, chaos, and oscillation is often observed in linear or nonlinear systems such as communication circuits, power systems, electronic circuits, chemical processes, and bio systems. One of the major advantages of the memristor is that its resistance depends on the magnitude and polarity of the voltage and the duration of use of the voltage (see Fig. 22.6). Indeed, when the voltage is off, a memristor remembers its last value until the next turn on. Based on this feature, applications have expanded the potential of these systems by significantly increasing the speed of starting a computer, extending the battery life of mobile phones, enabling programmable analog circuits and implementing MNNs simulates the function of the human brain [76,77]. Due to the limited switching speed of the neuron amplifiers and the limited amount of signal delivery in the neural networks and electronic circuits, time delay should be used as these factors weaken these systems. On the other hand, given the efficient effect of fuzzy logic in image processing and correct pattern recognition issues, many authors have focused their attention on fuzzy neural networks with a variety of delays. The Lyapunov–Krasovici functions, M-matrix theory, and linear matrix inequality approach have been considered by some authors to

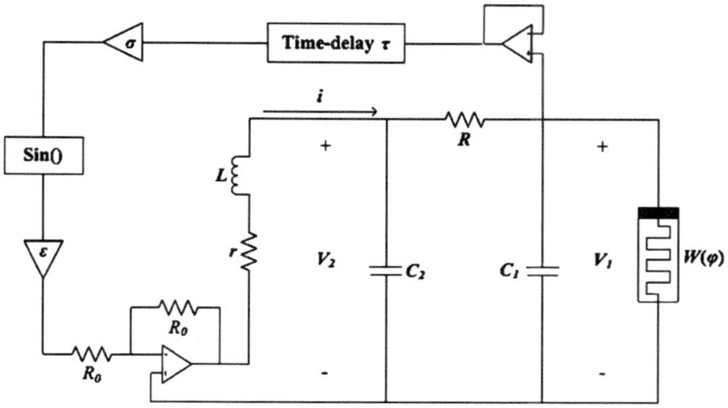

FIGURE 22.6 Time delay memristor.

investigate the global asymptotic / exponential stability of time-delayed fuzzy cellular neural networks [30,31]. Yang [35] has investigated the use of matrix theory, inequality analysis technique, and contraction mapping principle for the existence and stability of a periodic global solution for fuzzy Cohen–Grossberg two-way associative neural networks with different timing and different time delays. On the other hand, due to the wide applications of Cohen–Grossberg neural networks, for example in cellular, Hopfield and shunting inhibitory neural networks, stability and synchronization studies are essential and important for memristor-based fuzzy Cohen–Grossberg bidirectional associative memory neural networks with mixed delays and impulses. So, the exponential stability in the mean squares and the approximate exponential stability of the stochastic fuzzy delayed Cohen–Grossberg neural networks are studied by Zhu and Li [32] based on the use of Lyapunov–Krasovskii functional and functional analysis theory. The problems of estimating and synchronizing the non-brittle state for MNNs with different time delays have been investigated by the authors in [33,34] using the theory of differential components and Lyapunov–Krasovy stability theory. However, based on the information obtained, only the stability and synchronization of fuzzy neural networks or MNNs have been investigated by the authors in [36]. In [47], the authors attempted to design a fuzzy state feedback controller to investigate the exponential stability of a general class of fuzzy MNNs with hybrid unbounded time-varying delays. Wen et al. [66], based on a new parallel distributed compensation fuzzy model which is employed to linearize a complicated memristive system with only two subsystems, proposed an event-based approach to updating the control law to stabilize memristive systems and extending the results to systems with signal quantization and networked induced delays.

Since it is difficult to achieve satisfactory performance in time delay fuzzy systems, it is important to examine the stability of these systems, which are widely used in many physical and engineering phenomena in the real world such as Takagi–Sugeno fuzzy systems H_∞ control [63] and stability and stabilization criteria for fuzzy neural networks [64]. Fuzzy modeling and synchronization of different memristor-based perturbation circuits are presented in [65]. In [66], the authors considered a time delay memristor-based Chua circuit based on Takagi–Sugeno fuzzy controller as follows:

$$\begin{cases} \dot{v}_1(t) = \Big(-\dfrac{1}{RC_1} - \dfrac{W(\varphi(t))}{C_1}\Big)v_1(t) + \dfrac{1}{RC_1}v_2(t) \\ \dot{v}_2(t) = \dfrac{1}{RC_2}v_1(t) - \dfrac{1}{RC_2}v_2(t) + \dfrac{1}{C_2}i(t) \\ \dot{i}(t) = -\dfrac{1}{L}v_2(t) - \dfrac{r}{L}i(t) - \dfrac{\varepsilon}{L}sin(\sigma v_1(t-\tau)) \\ \dot{\varphi}(t) = v_1(t). \end{cases} \tag{22.22}$$

In addition, to synchronize two different circuits from the Chua delay controller and the secured communication programs based on the proposed chaotic synchronization structure, the fuzzy control vector is designed. In the context of the Takagi–Sugeno fuzzy model Cafagna and Grassi presented a novel fractional-order memristor-based chaotic system and carried out the theoretical analysis of the system dynamics [73]. In [78,79], a global exponential synchronization was investigated for a class of recurrent memristor-based neural networks with different time delays based on fuzzy theory and the Lyapunov method.

In [82], synchronization of fuzzy circuits with chaos delay based on Chua memristor models studied and the signal was applied to chaos secure communication using this synchronization scheme. Investigating the fuzzy model of MNNs under parameter perturbations provide a new perspective to analyze the complicated MNNs and propose the universality of robust adaptive lag synchronization for various memristive neural networks. The issue about robust global adaptive lag synchronization of uncertain MNNs with discrete delays via fuzzy model has been investigated by Liu et al. [85]. A general class of MNNs with hybrid time delays are concerned in [98] which are described by the following differential equations:

$$\dot{x}_i(t) = -a_i x_i(t) + \sum_{j=1}^{n} b_{ij}(x_i(t)) f_j(x_j(t)) + \sum_{j=1}^{n} c_{ij}(x_i(t)) g_j(x_j(t-\tau_j(t))) + \sum_{j=1}^{n} d_{ij}(x_i(t)) \int_{t-\sigma_j(t)}^{t} h_j(x_j(s))ds, \tag{22.23}$$

where the positive integer n is the amount of neurons in the neural network. In this chapter, through fuzzy blending, they investigated the stabilization for a class of Takagi–Sugeno fuzzy MNNs with mixed time delays that can be expressed by the following fuzzy differential equations:

$$
\begin{aligned}
\dot{x}_i(t) = \sum_{l=1}^{L} \mu_l(\theta(t))\Big\{ -a_i^{(l)} x_i(t) + \sum_{j=1}^{n} b_{ij}(x_i(t)) f_j(x_j(t)) \\
+ \sum_{j=1}^{n} c_{ij}(x_i(t)) g_j(x_j(t-\tau_j(t))) + \sum_{j=1}^{n} d_{ij}(x_i(t)) \int_{t-\sigma_j(t)}^{t} h_j(x_j(s))ds \Big\},
\end{aligned}
\tag{22.24}
$$

where the real value L is the number of fuzzy IF-THEN rules. Wang et al. [101] investigated the following problem of Takagi–Sugeno fuzzy MNN with discrete and distributed delays:

$$
\begin{aligned}
\dot{r}_i(t) = -d_i^{<q>} r_i(t) + \sum_{j=1}^{n} a_{ij}(r_i(t)) f_j(r_j(t)) \\
+ \sum_{j=1}^{n} b_{ij}(r_i(t)) g_j(r_j(t-\tau_j(t))) + \sum_{j=1}^{n} c_{ij}(r_i(t)) \int_{t-\sigma_j(t)}^{t} h_j(r_j(s))ds,
\end{aligned}
\tag{22.25}
$$

where $q \in \mathbb{Q}$ and $d_i^{<q>}$ is the self-feedback coefficient. Under a designed fuzzy feedback controller, several algebraic conditions have been derived to ensure the global synchronization of this fuzzy MNN by employing non-smooth analysis theory. The authors of [102] investigated the global exponential stability in Lagrange sense for Takagi–Sugeno fuzzy MNNs with time-varying delays on time scales based on inequality scaling techniques and matrix-norm strategies. Fixed-time synchronization issues with cellular neural networks for a time-varying delay memristive fuzzy bidirectional associative memory, described by the following differential equations, have been investigated in [107]:

$$
\begin{aligned}
\dot{x}_i(t) = -d_i^{(1)}(x_i(t))x_i(t) + \sum_{j=1}^{m} a_{ij}^{(1)}(x_i(t)) f_j(y_j(t)) \\
+ \sum_{j=1}^{m} b_{ij}^{(1)}(x_i(t-\sigma(t))) f_j(y_j(t-\tau(t))) + \sum_{j=1}^{m} c_{ij}^{(1)} w_j^{(1)} \\
+ \bigwedge_{j=1}^{m} \alpha_{ij}^{(1)} f_j(y_j(t-\tau(t))) \\
+ \bigvee_{j=1}^{m} \beta_{ij}^{(1)} f_j(y_j(t-\tau(t))) + \bigvee_{j=1}^{m} S_{ij}^{(1)} w_j^{(1)} + \bigwedge_{j=1}^{m} T_{ij}^{(1)} w_j^{(1)} + I_i^{(1)},
\end{aligned}
$$

$$
\begin{aligned}
\dot{y}_i(t) = & -d_j^{(2)}(y_j(t))y_j(t) + \sum_{i=1}^{n} a_{ji}^{(2)}(y_j(t))g_i(x_i(t)) \\
& + \sum_{i=1}^{n} b_{ji}^{(2)}(y_j(t-\tau(t)))g_i(x_i(t-\sigma(t))) + \sum_{i=1}^{n} c_{ji}^{(2)} w_i^{(2)} \\
& + \bigwedge_{i=1}^{n} \alpha_{ji}^{(2)} g_i(x_i(t-\sigma(t))) \\
& + \bigvee_{i=1}^{n} \beta_{ji}^{2)} g_i(x_i(t-\sigma(t))) + \bigvee_{i=1}^{n} S_{ji}^{(2)} w_i^{(2)} + \bigwedge_{i=1}^{n} T_{ji}^{(2)} w_i^{(2)} + I_j^{(2)}, \ t \geq 0,
\end{aligned}
\tag{22.26}
$$

where $x(t)$ and $y(t)$ denote the states of neurons in X-layer and Y-layer, respectively. Indeed, Li et al. [107], studied this problem with the help of the definition of fixed-time synchronization, differential inclusion theory and valuable maps.

22.5 Fractional memristive fuzzy systems

Fractional calculus, a combination of derivatives and integrals of arbitrary degrees, has found many applications in the fields of physics, applied mathematics and engineering. In addition, in the real world, many physical systems with fractional differential equations, i.e., combinations of integer and non-integer derivatives, are better introduced and characterized. Some systems have been found to be more accurate when using fractional derivatives. Recently, due to its potential applications in secure communication and control processing, the study of chaos synchronization in fractional-order dynamical systems and related phenomena is receiving growing attention. FNN with fractional derivative (FFNN), either combined with a memristor (MFNN) or in synchronous combination of a fractional derivative and a memristor (MFFCNN) is still in the early stages of development and may lead to more complex dynamic behaviors of neural networks. Unfortunately, so far we have not found the relevant research results on the MFFCNN.

Although fractional arithmetic theory has received much attention, it has not been widely studied in the industry or is not fully applicable in practice. Compared to the integer order chaotic memristive systems, fractional ones have more complex dynamic characteristics. The advantage of fractional-order systems over integer order systems is that fractional-order systems can produce infinite memory. Therefore, merging the memristors into a class of fractional-order neural networks is highly anticipated. There is not a general method to analyze the dynamic behavior of fractional-order memristive systems such as stability and synchronization. In [68], Liu et al. proposed an adaptive fuzzy control strategy to realize the projective synchronization of two fractional-order memristive systems. Liu et al. were concerned in [97] with the global stabilization for a class of fractional-order MNNs with time delay as follows:

$$\begin{aligned} {}_{t_0}^{C}D_t^{\alpha} x_i(t) = -x_i(t) + \sum_{j=1}^{n} a_{ij}(x_j(t))g_j(x_j(t)) \\ + \sum_{j=1}^{n} b_{ij}(x_j(t))f_j(x_j(t-\tau(t))) + u_i(t), \end{aligned} \quad (22.27)$$

where $0 < \alpha < 1$ and n is the number of neurons in the networks. They employed the state feedback control law and also the output feedback control law to stabilize this class of fractional-order memristive systems. Li et al. in [106] considered the following memristor-based fractional-order FNN work model:

$$\begin{cases} D^q x_i(t) = -\tilde{d}_i(x_i(t))x_i(t) + \sum_{j=1}^{n} \tilde{a}_{ij}(x_j(t))f_j(x_j(t)) \\ \quad + \sum_{j=1}^{n} b_{ij}v_j + \bigwedge_{j=1}^{n} \alpha_{ij}g_j(x_j(t-\tau)) + \bigwedge_{j=1}^{n} T_{ij}v_j + \bigvee_{j=1}^{n} S_{ij}v_j \\ \quad + \bigvee_{j=1}^{n} \beta_{ij}g_j(x_j(t-\tau)) + I_i, \quad t \in [0, T], \\ x_i(t) = \varphi_i(t), \quad t \in [-\tau, 0],\ i = 1, 2, \cdots, n. \end{cases} \quad (22.28)$$

where $0 < q < 1$ and n is the number of neurons. The primary objective of [106] is to analysis finite-time stability and MFFCNN synchronization problems of Filipov's meaning with the help of tuned maps, including differential components, Banach constant point theorem, Gronwall–Bellman inequality, and finite stability and harmony definitions.

22.6 General overview of the area and future trends

It can be noted that the number of publications for designing fuzzy controllers has been increasing each year and that this increasing trend will continue in the future because memristive systems have been used more frequently in the applications, and this will require designing more complex fuzzy systems, which in turn will need even better techniques. As the work done in this chapter progressed, many ideas came to mind, which could be addressed in future research, such as:

- Exponential delay synchronization for a class of mixed time delay MNNs can be investigated through fuzzy control methods.
- Inertial MNNs with time delay with discontinuous activation functions can be considered by fuzzy adaptive control strategy.
- Design the optimal update laws for the uncertain parameters of the master-slave systems to achieve desired results and make a more practical fuzzy

model of uncertain MNNs to extend the applications of MNNs to the fields of optimal computation, biological fuzzy systems, and fuzzy secure communications.
- Dynamical properties of fuzzy MNNs with hybrid unbounded discrete and distributed time delays can be addressed in future work.
- Design optimal updating fuzzy rules for slave weights for optimal results.
- Deal with the problem of synchronization of FMNNs with discrete and distributed time-varying delays.

References

[1] L.A. Zadeh, Fuzzy sets, Information and Control 8 (1965) 338–353.
[2] A. Armand, T. Allahviranloo, Z. Gouyandeh, Some fundamental results on fuzzy calculus, Iranian Journal of Fuzzy Systems 15 (2018) 27–46.
[3] D. Dubois, H. Prade, Towards fuzzy differential calculus part 3: differentiation, Fuzzy Sets and Systems 8 (1982) 225–233.
[4] D. Dubois, H. Prade, Towards fuzzy differential calculus part 2: integration on fuzzy intervals, Fuzzy Sets and Systems 8 (1982) 105–116.
[5] D. Dubois, H. Prade, Towards fuzzy differential calculus part 1: integration of fuzzy mappings, Fuzzy Sets and Systems 8 (1982) 1–17.
[6] L.T. Gomes, L.C. Barros, Fuzzy calculus via extension of the derivative and integral operators and fuzzy differential equations, IEEE, pp. 1–5.
[7] O. Kaleva, The calculus of fuzzy valued functions, Applied Mathematics Letters 3 (1990) 55–59.
[8] F. Terrier, Z. Chen, Fuzzy calculus applied to real time scheduling, IEEE, p. 1905–10.
[9] M. Ma, M. Friedman, A. Kandel, A new fuzzy arithmetic, Fuzzy Sets and Systems 108 (1999) 83–90.
[10] B.S. Mordukhovich, Y. Shao, Fuzzy calculus for coderivatives of multifunctions, Nonlinear Analysis: Theory, Methods & Applications 29 (1997) 605–626.
[11] L. Stefanini, L. Sorini, M.L. Guerra, Parametric representation of fuzzy numbers and application to fuzzy calculus, Fuzzy Sets and Systems 157 (2006) 2423–2455.
[12] R. Goetschel Jr, W. Voxman, Elementary fuzzy calculus, Fuzzy Sets and Systems 18 (1986) 31–43.
[13] Q. Lei, Z. Xu, Fundamental properties of intuitionistic fuzzy calculus, Knowledge-Based Systems 76 (2015) 1–16.
[14] L.C. De Barros, F. Santo Pedro, Fuzzy differential equations with interactive derivative, Fuzzy Sets and Systems 309 (2017) 64–80.
[15] T.G. Bhaskar, V. Lakshmikantham, Set differential equations and flow invariance, Applicable Analysis 82 (2003) 357–368.
[16] T. Yang, L.B. Yang, The global stability of fuzzy cellular neural network, IEEE Transactions on Circuits and Systems I, Fundamental Theory and Applications 43 (10) (1996) 880–883.
[17] Y.V. Pershin, M. Di Ventra, Experimental demonstration of associative memory with memristive neural networks, Neural Networks 23 (7) (2010) 881–886.
[18] S. Duan, X. Hu, Z. Dong, L. Wang, P. Mazumder, Memristor-based cellular nonlinear/neural network: design, analysis, and applications, IEEE Transactions on Neural Networks and Learning Systems 26 (6) (2014) 1202–1213.
[19] S. Choi, J.H. Shin, J. Lee, P. Sheridan, W.D. Lu, Experimental demonstration of feature extraction and dimensionality reduction using memristor networks, Nano Letters 17 (5) (2017) 3113–3118.
[20] D.B. Strukov, G.S. Snider, D.R. Stewart, R.S. Williams, The missing memristor found, Nature 453 (1 May 2008) 80–83.

[21] L.O. Chua, Memristor – the missing circuit element, IEEE Transactions on Circuit Theory CT-18 (5) (1971) 507–519.
[22] G. Snider, R. Amerson, D. Carter, H. Abdalla, M.S. Qureshi, J. Leveille, M. Versace, H. Ames, S. Patrick, B. Chandler, A. Gorchetchnikov, E. Mingolla, From synapses to circuitry: using memristive memory to explore the electronic brain, Computer 44 (2) (2011) 21–28.
[23] S.H. Jo, T. Chang, I. Ebong, B.B. Bhadviya, P. Mazumder, W. Lu, Nanoscale memristor device as synapse in neuromorphic systems, Nano Letters 10 (4) (March 2010) 1297–1301.
[24] T. Chang, S.-H. Jo, W. Lu, Short-term memory to long-term memory transition in a nanoscale memristor, ACS Nano 5 (9) (2011) 7669–7676.
[25] L.O. Chua, S.M. Kang, Memristive devices and systems, Proceedings of the IEEE 64 (2) (1976) 209–223.
[26] A. Abdurahman, H. Jiang, New results on exponential synchronization of memristor-based neural networks with discontinuous neuron activations, Neural Networks 84 (2016) 161–171.
[27] H. Bao, J.H. Park, J. Cao, Matrix measure strategies for exponential synchronization and anti-synchronization of memristor-based neural networks with time-varying delays, Applied Mathematics and Computation 270 (2015) 543–556.
[28] A. Abdurahman, H. Jiang, Z. Teng, Exponential lag synchronization for memristor-based neural networks with mixed time delays via hybrid switching control, Journal of the Franklin Institute 353 (13) (2016) 2859–2880.
[29] N. Li, J. Cao, Lag synchronization of memristor-based coupled neural networks via ω-measure, IEEE Transactions on Neural Networks and Learning Systems 27 (3) (2015) 686–697.
[30] P. Balasubramaniam, M. Kalpana, R. Rakkiyappan, Global asymptotic stability of BAM fuzzy cellular neural networks with time delay in the leakage term, discrete and unbounded distributed delays, Mathematical and Computer Modelling 53 (5–6) (2011) 839–853.
[31] Q. Zhang, Y. Shao, J. Liu, Analysis of stability for impulsive fuzzy Cohen–Grossberg BAM neural networks with delays, Mathematical Methods in the Applied Sciences 36 (7) (2013) 773–779.
[32] Q. Zhu, X. Li, Exponential and almost sure exponential stability of stochastic fuzzy delayed Cohen–Grossberg neural networks, Fuzzy Sets and Systems 203 (2012) 74–94.
[33] K. Mathiyalagan, R. Anbuvithya, R. Sakthivel, J.H. Park, P. Prakash, Non-fragile H_∞ synchronization of memristor-based neural networks using passivity theory, Neural Networks 74 (2016) 85–100.
[34] R. Sakthivel, R. Anbuvithya, K. Mathiyalagan, P. Prakash, Combined H_∞ and passivity state estimation of memristive neural networks with random gain fluctuations, Neurocomputing 168 (2015) 1111–1120.
[35] W. Yang, Periodic solution for fuzzy Cohen–Grossberg BAM neural networks with both time-varying and distributed delays and variable coefficients, Neural Processing Letters 40 (1) (2014) 51–73.
[36] W. Yang, W. Yu, J. Cao, F.E. Alsaadi, T. Hayat, Global exponential stability and lag synchronization for delayed memristive fuzzy Cohen–Grossberg BAM neural networks with impulses, Neural Networks 98 (2018) 122–153.
[37] H.T. Nguyen, M. Sugeno, Fuzzy Systems: Modeling and Control, vol. 2, Springer Science & Business Media, 2012, pp. 338–353.
[38] Q. Song, Z. Zhao, J. Yang, Passivity and passification for stochastic Takagi–Sugeno fuzzy systems with mixed time-varying delays, Neurocomputing 122 (2013) 330–337.
[39] T. Wang, Y. Zhang, J. Qiu, H. Gao, Adaptive fuzzy backstepping control for a class of nonlinear systems with sampled and delayed measurements, IEEE Transactions on Fuzzy Systems 23 (2) (2014) 302–312.
[40] H. Li, J. Wang, P. Shi, Output-feedback based sliding mode control for fuzzy systems with actuator saturation, IEEE Transactions on Fuzzy Systems 24 (6) (2015) 1282–1293.
[41] P. Shi, Y. Zhang, M. Chadli, R.K. Agarwal, Mixed H-infinity and passive filtering for discrete fuzzy neural networks with stochastic jumps and time delays, IEEE Transactions on Neural Networks and Learning Systems 27 (4) (2015) 903–909.

[42] X. Yang, D.W. Ho, J. Lu, Q. Song, Finite-time cluster synchronization of T-S fuzzy complex networks with discontinuous subsystems and random coupling delays, IEEE Transactions on Fuzzy Systems 23 (6) (2015) 2302–2316.
[43] S. Wen, Z. Zeng, T. Huang, Y. Zhang, Exponential adaptive lag synchronization of memristive neural networks via fuzzy method and applications in pseudorandom number generators, IEEE Transactions on Fuzzy Systems 22 (6) (2013) 1704–1713.
[44] Y. Sheng, H. Zhang, Z. Zeng, Stabilization of fuzzy memristive neural networks with mixed time delays, IEEE Transactions on Fuzzy Systems 26 (5) (2017) 2591–2606.
[45] Q. Xiao, Z. Zeng, Lagrange stability for T-S fuzzy memristive neural networks with time-varying delays on time scales, IEEE Transactions on Fuzzy Systems 26 (3) (2017) 1091–1103.
[46] L. Zhang, T. Yang, P. Shi, M. Liu, Stability and stabilization of a class of discrete-time fuzzy systems with semi-Markov stochastic uncertainties, IEEE Transactions on Systems, Man, and Cybernetics: Systems 46 (12) (2016) 1642–1653.
[47] Y. Sheng, F.L. Lewis, Z. Zeng, Exponential stabilization of fuzzy memristive neural networks with hybrid unbounded time-varying delays, IEEE Transactions on Neural Networks and Learning Systems 30 (3) (2018) 739–750.
[48] I.E.P. Afrakoti, S.B. Shouraki, F.M. Bayat, M. Gholami, Using a memristor crossbar structure to implement a novel adaptive real-time fuzzy modeling algorithm, Fuzzy Sets and Systems 307 (2017) 115–128.
[49] M. Bavandpour, S. Bagheri-Shouraki, H. Soleimani, A. Ahmadi, B. Linares-Barranco, Spiking neuro-fuzzy clustering system and its memristor crossbar based implementation, Microelectronics Journal 45 (11) (2014) 1450–1462.
[50] D.H. Rouvray, Fuzzy Logic in Chemistry, Academic Press, 1997.
[51] W. Liu, Y. Lou, Global exponential stability and existence of periodic solutions of fuzzy wave equations, Advances in Difference Equations 2020 (1) (2020) 1.
[52] S.X. Yu, Z. Xu, Generalized Intuitionistic Multiplicative Fuzzy Calculus Theory and Applications, Springer-Verlag, Singapore, 2020.
[53] F.R. Matenda, Constant proportion portfolio insurance strategies in fuzzy financial markets, Journal of Mathematics and Statistical Science 2016 (2016) 48–63.
[54] G. Alfonso, A.R.L. de Hierro, C. Roldán, A fuzzy regression model based on finite fuzzy numbers and its application to real-world financial data, Journal of Computational and Applied Mathematics 318 (2017) 47–58.
[55] S.N. Mayburov, Commutative fuzzy geometry and geometric quantum mechanics, JPhCS 1051 (1) (2018) 012022.
[56] S. Wang, S. Bekiros, A. Yousefpour, S. He, O. Castillo, H. Jahanshahi, Synchronization of fractional time-delayed financial system using a novel type-2 fuzzy active control method, Chaos, Solitons and Fractals 136 (2020) 109768.
[57] P. Muthukumar, P. Balasubramaniam, K. Ratnavelu, T-S fuzzy predictive control for fractional order dynamical systems and its applications, Nonlinear Dynamics 86 (2) (2016) 751–763.
[58] A. Boubellouta, A. Boulkroune, Synchronization of incommensurate fractional-order chaotic systems with input nonlinearities using a fuzzy variable-structure control, in: International Conference on Electrical Engineering and Control Applications, Springer, Cham, 2017, November, pp. 128–142.
[59] X. Cui, X. Zhao, Y. Guo, X. Li, P. Hou, Adaptive fuzzy terminal sliding mode synchronization of uncertain Newton–Leipnik chaotic system, in: 2020 6th International Conference on Control, Automation and Robotics (ICCAR), IEEE, April 2020, pp. 309–313.
[60] F. Merrikh-Bayat, S.B. Shouraki, A. Rohani, Memristor crossbar-based hardware implementation of the IDS method, IEEE Transactions on Fuzzy Systems 19 (6) (2011) 1083–1096.
[61] F. Merrikh-Bayat, S.B. Shouraki, F. Merrikh-Bayat, Memristive fuzzy edge detector, Journal of Real-Time Image Processing 9 (3) (2014) 479–489.

[62] F.Z. Wang, L. Li, L. Shi, H. Wu, L.O. Chua, ϕmemristor: real memristor found, Journal of Applied Physics 125 (5) (2019).
[63] P. Vadivel, R. Sakthivel, K. Mathiyalagan, P. Thangaraj, Robust stabilisation of non-linear uncertain Takagi–Sugeno fuzzy systems by H_∞ control, IET Control Theory & Applications 6 (16) (2012) 2556–2566.
[64] K. Mathiyalagan, R. Sakthivel, S.M. Anthoni, New stability and stabilization criteria for fuzzy neural networks with various activation functions, Physica Scripta 84 (1) (2011) 015007.
[65] S. Wen, Z. Zeng, T. Huang, Y. Chen, Fuzzy modeling and synchronization of different memristor-based chaotic circuits, Physics Letters A 377 (34–36) (2013) 2016–2021.
[66] S. Wen, Z. Zeng, T. Huang, Event-based control for memristive systems, Communications in Nonlinear Science and Numerical Simulation 19 (10) (2014) 3431–3443.
[67] G. Zhang, Y. Shen, Exponential synchronization of delayed memristor-based chaotic neural networks via periodically intermittent control, Neural Networks 55 (2014) 1–10.
[68] H. Liu, S.G. Li, Y.G. Sun, H.X. Wang, Adaptive fuzzy synchronization for uncertain fractional-order chaotic systems with unknown non-symmetrical control gain, Acta Physica Sinica 64 (7) (2015) 070503.
[69] S.H.E.N. Jun, W.U. Wei, X.Y. Lou, Suboptimal fuzzy control for memristive chaotic systems, DEStech Transactions on Engineering and Technology Research (2016) (icca).
[70] B. Wang, L.L. Chen, New results on fuzzy synchronization for a kind of disturbed memristive chaotic system, Complexity 2018 (2018).
[71] Q. Ye, X. Lou, Observer-based fuzzy control for memristive circuit systems, Complexity 2018 (2018).
[72] Z. Qi-Shui, Y. Yong-Bin, Y. Jue-Bang, Fuzzy modeling and impulsive control of a memristor-based chaotic system, Chinese Physics Letters 27 (2) (2010) 020501.
[73] D. Cafagna, G. Grassi, On the simplest fractional-order memristor-based chaotic system, Nonlinear Dynamics 70 (2) (2012) 1185–1197.
[74] M. Gao, W. Yang, Y. Xue, Z. Gao, F. Yuan, G. Dou, Y. Li, Multistability in a physical memristor-based modified Chua's circuit, Chaos: An Interdisciplinary Journal of Nonlinear Science 29 (4) (2019).
[75] R. Zhang, X. Liu, D. Zeng, S. Zhong, K. Shi, A novel approach to stability and stabilization of fuzzy sampled-data Markovian chaotic systems, Fuzzy Sets and Systems 344 (2018) 108–128.
[76] J.M. Munoz-Pacheco, Infinitely many hidden attractors in a new fractional-order chaotic system based on a fracmemristor, The European Physical Journal Special Topics 228 (10) (2019) 2185–2196.
[77] Z. Meng, Z. Xiang, Passivity analysis of memristor-based recurrent neural networks with mixed time-varying delays, Neurocomputing 165 (2015) 270–279.
[78] Z. Guo, J. Wang, Z. Yan, Global exponential synchronization of two memristor-based recurrent neural networks with time delays via static or dynamic coupling, IEEE Transactions on Systems, Man, and Cybernetics: Systems 45 (2) (2014) 235–249.
[79] Y. Cao, S. Wang, Z. Guo, T. Huang, S. Wen, Synchronization of memristive neural networks with leakage delay and parameters mismatch via event-triggered control, Neural Networks 119 (2019) 178–189.
[80] M. Cui, Y. Zhang, Memristive synaptic circuits for deep convolutional neural networks, in: IEEE International Symposium on Circuits and Systems, 2019, pp. 1–5.
[81] D. Przyczyna, M. Lis, K. Pilarczyk, K. Szaciłowski, Hardware Realization of the Pattern Recognition with an Artificial Neuromorphic Device Exhibiting a Short-Term Memory, Molecules 24 (15) (2019).
[82] T.C. Lin, F.Y. Huang, Z. Du, Y.C. Lin, Synchronization of fuzzy modeling chaotic time delay memristor-based Chua's circuits with application to secure communication, International Journal of Fuzzy Systems 17 (2) (2015) 206–214.

[83] J. Huang, C. Li, X. He, Stabilization of a memristor-based chaotic system by intermittent control and fuzzy processing, International Journal of Control, Automation, and Systems 11 (3) (2013) 643–647.
[84] S. Yang, C. Li, T. Huang, Exponential stabilization and synchronization for fuzzy model of memristive neural networks by periodically intermittent control, Neural Networks 75 (2016) 162–172.
[85] Y. Liu, C. Li, T. Huang, X. Wang, Robust adaptive lag synchronization of uncertain fuzzy memristive neural networks with time-varying delays, Neurocomputing 190 (2016) 188–196.
[86] F. Merrikh-Bayat, S.B. Shouraki, Memristive neuro-fuzzy system, IEEE Transactions on Cybernetics 43 (1) (2012) 269–285.
[87] Q. Xiao, T. Huang, Z. Zeng, Passivity and passification of fuzzy memristive inertial neural networks on time scales, IEEE Transactions on Fuzzy Systems 26 (6) (2018) 3342–3355.
[88] H. Kizmaz, U.E. Kocamaz, Y. Uyaroğlu, Control of memristor-based simplest chaotic circuit with one-state controllers, Chinese Physics Letters 28 (01) (2019).
[89] S.G. Samko, A.A. Kilbas, O.I. Marichev, Fractional Integrals and Derivatives: Theory and Applications, Gordon & Breach Science Publishers in Switzerland, Philadelphia, PA, USA, 1993.
[90] C.P. Li, W.H. Deng, Remarks on fractional derivatives, Applied Mathematics and Computation 187 (2007) 777–784.
[91] C.P. Li, X.H. Dao, P. Guo, Fractional derivatives in complex planes, Nonlinear Analysis TMA71 (2009) 1857–1869.
[92] C.P. Li, Z.G. Zhao, Numerical approximation of nonlinear fractional differential equations with subdiffusion and superdiffusion, Computers & Mathematics with Applications (2011), https://doi.org/10.1016/j.camwa.2011.02.045.
[93] G.S. Teodoro, J.T. Machado, E.C. De Oliveira, A review of definitions of fractional derivatives and other operators, Journal of Computational Physics 388 (2019) 195–208.
[94] C.P. Li, D.L. Qian, Y.Q. Chen, On Riemann–Liouville and Caputo derivatives, Discrete Dynamics in Nature and Society 2011 (2011) 562494.
[95] A.A. Kilbas, H.M. Srivastava, J.J. Trujillo, Theory and Applications of Fractional Differential Equations, Elsevier, Netherlands, 2006.
[96] S. Yuzbasi, Bessel polynomial solutions of linear differential, integral and integro-differential equations, Master Thesis, Graduate School of Natural and Applied Sciences, Mugla University, 2009.
[97] L. Liu, A. Wu, X. Song, Global $O(t^{-\alpha})$ stabilization of fractional-order memristive neural networks with time delays, SpringerPlus 5 (1) (2016) 1034.
[98] S.A. Karthick, R. Sakthivel, C. Wang, Y.K. Ma, Synchronization of coupled memristive neural networks with actuator saturation and switching topology, Neurocomputing 383 (2020) 138–150.
[99] S.S. Zeid, Approximation methods for solving fractional equations, Chaos, Solitons and Fractals 125 (2019) 171–193.
[100] S. Wang, Y. Cao, T. Huang, Y. Chen, S. Wen, Event-triggered distributed control for synchronization of multiple memristive neural networks under cyber-physical attacks, Information Sciences 518 (2020) 361–375.
[101] L. Wang, H. He, Z. Zeng, Global synchronization of fuzzy memristive neural networks with discrete and distributed delays, IEEE Transactions on Fuzzy Systems (2019).
[102] Y. Sheng, F.L. Lewis, Z. Zeng, T. Huang, Lagrange stability and finite-time stabilization of fuzzy memristive neural networks with hybrid time-varying delays, IEEE Transactions on Cybernetics (2019).
[103] S. Yang, C. Li, T. Huang, Impulsive synchronization for T-S fuzzy model of memristor-based chaotic systems with parameter mismatches, International Journal of Control, Automation, and Systems 14 (3) (2016) 854–864.
[104] Y. Wu, J. Zhu, W. Li, Intermittent discrete observation control for synchronization of stochastic neural networks, IEEE Transactions on Cybernetics 50 (2019) 2414–2424.

[105] T. Zhang, M. Yin, X. Lu, Y. Cai, Y. Yang, R. Huang, Tolerance of intrinsic device variation in fuzzy restricted Boltzmann machine network based on memristive nano-synapses, Nano Futures 1 (1) (2017) 015003.
[106] M. Zheng, L. Li, H. Peng, J. Xiao, Y. Yang, Y. Zhang, H. Zhao, Finite-time stability and synchronization of memristor-based fractional-order fuzzy cellular neural networks, Communications in Nonlinear Science and Numerical Simulation 59 (2018) 272–291.
[107] M. Zheng, L. Li, H. Peng, J. Xiao, Y. Yang, Y. Zhang, Fixed-time synchronization of memristive fuzzy BAM cellular neural networks with time-varying delays based on feedback controllers, IEEE Access 6 (2018) 12085–12102.
[108] K. Rajagopal, H. Jahanshahi, M. Varan, I. Bayır, V.T. Pham, S. Jafari, A. Karthikeyan, A hyperchaotic memristor oscillator with fuzzy based chaos control and LQR based chaos synchronization, AEU International Journal of Electronics and Communications 94 (2018) 55–68.
[109] H. Jahanshahi, A. Yousefpour, J.M. Munoz-Pacheco, I. Moroz, Z. Wei, O. Castillo, A new multi-stable fractional-order four-dimensional system with self-excited and hidden chaotic attractors: dynamic analysis and adaptive synchronization using a novel fuzzy adaptive sliding mode control method, Applied Soft Computing 87 (2020) 105943.
[110] R. Belohlavek, G.J. Klir (Eds.), Concepts and Fuzzy Logic, MIT Press, 2011.
[111] J. Kacprzyk, Studies in Fuzziness and Soft Computing 295 (2013).
[112] P. Sussner, E. Esmi, L.C. Barros, Controlling the width of the sum of interactive fuzzy numbers with applications to fuzzy initial value problems, in: 2016 IEEE International Conference on Fuzzy Systems (FUZZ-IEEE), IEEE, 2016, July, pp. 1453–1460.
[113] W. Pedrycz, F. Gomide, An Introduction to Fuzzy Sets: Analysis and Design, MIT Press, 1998.
[114] G.J. Klir, B. Yuan, Fuzzy sets and fuzzy logic: theory and applications, Possibility Theory versus Probab. Theory 32 (2) (1996) 207–208.
[115] L. Chua, Memristor – the missing circuit element, IEEE Transactions on Circuit Theory 18 (5) (1971) 507–519.

Chapter 23

Fuzzy integral sliding mode technique for synchronization of memristive neural networks

Amin Yousefpour[a], Hadi Jahanshahi[b], and Dongming Gan[c]
[a]*School of Mechanical Engineering, College of Engineering, University of Tehran, Tehran, Iran,* [b]*Department of Mechanical Engineering, University of Manitoba, Winnipeg, MB, Canada,* [c]*School of Engineering Technology, Purdue University, West Lafayette, IN, United States*

23.1 Introduction

Nowadays, memristor systems have attracted a lot of attention due to their potential applications in new technologies. The memristor, a contraction of memory and resistor, was firstly proposed by Chua [1], and in 2008, it was realized by HP laboratory. One of the most important features of the memristors is their ability to memorize the direction of electric charge flow in the past, which can act like remembering and forgetting processes in human brains [2,3]. Also, neural networks which can be easily constructed by nonlinear circuits have been extensively investigated in the literature because of their pivotal roles in various research areas [4–8].

Notably, in the twenty-first century, we can see an explosion of utilizing neural networks in various applications [9]. For example, different neural networks such as cellular neural networks, Hopfield neural networks, and Cohen–Grossberg neural networks have widely been used for associative memory, solving linear and nonlinear programming problems, computing technology, image processing, and so on [10–12]. Moreover, in the literature, it has been confirmed that memristive neural networks have more information capacity and computation power than ordinary neural networks. These properties would impressively increase the applications of neural networks for information processing and associative memory in the future [13,14]. So far, several fruitful and practical applications of these systems have been reported [15]. For instance, since this class of neural networks is a useful model to emulate the human brain, dynamical behaviors of memristive neural networks have attracted considerable attention [12,15].

Memristive neural networks possess a special class of governing equations with discontinuous right-hand side [16]. Hence, designing a controller for the

Mem-elements for Neuromorphic Circuits with Artificial Intelligence Applications
https://doi.org/10.1016/B978-0-12-821184-7.00032-3

synchronization of these systems has not been routine work. Up to now, various methods have been proposed to overcome this issue [17–19]. For instance, synchronization of memristive neural networks has been studied by using algebraic methods in Refs. [18,20]. Also, as a breakthrough in this field of study, the concept of Filippov regularization has been utilized for the synchronization of memristive neural networks [21,22]. By means of Filippov regularization [22], differential equations with discontinuous right-hand side can be transformed into a differential inclusion [23]. This way, several research studies have used Filippov regularization in order to stabilize [24,25] and synchronize these systems [26–29].

Among all nonlinear control techniques which have been proposed for nonlinear systems, sliding mode-based controller due to its advantageous properties such as simplicity in implementation, guaranteed stability, and robustness against parameter variations has received much attention [30–34]. Moreover, various techniques, such as adaptive sliding mode (ASMC) [35,36], terminal sliding mode (TSMC) [37,38], and integral sliding mode control (ISMC) [39,40] methods, have been proposed to improve the performance of the conventional sliding mode control [41]. By adding an integral term to the conventional sliding surface, the reaching phase of the siding surface will be eliminated [42]. Nevertheless, there are a few studies on the control of memristive neural networks using sliding mode methods [43].

The main problem of the sliding mode controller is the chattering phenomena due to the discontinuous function sgn (.) [44,45]. For instance, Chen et al. [46] have proposed a fixed-time controller for inertial memristor-based neural networks. Although their study proposed a proper controller in terms of convergence and settling time, it has a significant drawback in which discontinuous function in the control signal may result in chattering in the system. As it is evident, chattering could completely ruin the effective performance of such a delicate system. In order to reduce the chattering, some researchers have used a saturation function instead of sign(.) function [47,48]. However, this solution may result in a steady-state error. Actually, using this method is a tradeoff between tracking error and the control smoothness [49,50]. To better address this issue, a fuzzy logic system is combined with sliding mode control in the current study. In addition, by using fuzzy logic system, some control gains are tuned based on the distance of the system to the sliding surface and its derivative. Therefore, tracking errors and control efforts are reduced.

Motivated by the aforementioned discussions, firstly, the model dynamic of a memristive neural network is described in this chapter. Then a fuzzy integral sliding mode control (FISMC) is designed, and the stability of the closed-loop system is proven via the Lyapunov stability theorem. Finally, the performance of the proposed control scheme is compared with the conventional ISMC controller, and the synchronization results are presented.

23.2 Model description

In recent years, numerous research studies have been carried out on the simulation of memristors [51] and memristive neural networks [52]. For instance, on the basis of dynamic analysis of memristive neural networks, the complex brain functionalities can be studied [14]. Hence, various dynamical and statistical features of memristor-based neural networks have been considered, including stability [53] synchronization and control [54], passivity [55] and dissipativity [56].

The model of a memristive neural network can be represented by [57]

$$\dot{x}_i(t) = -cx_i(t)\sum_{j=1}^{n} a_{ij}\left(x_j(t)\right) f_j\left(x_j(t)\right) + I_i \tag{23.1}$$

where $i = 1, 2, \ldots, n$ and n denotes the number of units in a neural network, $x_i(t)$ corresponds to the state of the ith unit at time t. Besides, c_i is a positive constant, $f_j(x_j(t))$ represents a nonlinear function, and I_i stands for external input. In addition, a_{ij} is connection memristive weight, which is defined as follows:

$$a_{ij}\left(x_j(t)\right) = \begin{cases} \hat{a}_{ij} & \left|x_j(t)\right| > T_j \\ \tilde{a}_{ij} & \left|x_j(t)\right| < T_j \end{cases} \tag{23.2}$$

where $\hat{a}_{ij}$ and $\tilde{a}_{ij}$ are constant parameters and $a_{ij}\left(\pm T_j\right) = \hat{a}_{ij}$ or $\tilde{a}_{ij}$. Also, T_j denotes a switching jump.

Remark 1. Since $a_{ij}(x_j(t))$ has a discontinuous value, the classical definitions for differential equations cannot be applied to system (23.1). To overcome this problem, the concept due to Filippov is used. According to Filippov [22], a differential equation with a discontinuity has the same solution as a certain differential inclusion.

Based on the theories of set-valued maps and differential inclusions [58], Eq. (23.1) can be rewritten as

$$\dot{x}_i(t) \in -cx_i(t)\sum_{j=1}^{n} co\left(a_{ij}\left(x_j(t)\right)\right) f_j\left(x_j(t)\right) + I_i \tag{23.3}$$

where co indicates the convex closure of $a_{ij}\left(x_j(t)\right)$ and can be defined as follows:

$$co\left(a_{ij}\left(x_j(t)\right)\right) = \begin{cases} \hat{a}_{ij} & \left|x_j(t)\right| > T_j \\ co\left(\hat{a}_{ij}, \tilde{a}_{ij}\right) & \left|x_j(t)\right| = T_j \\ \tilde{a}_{ij} & \left|x_j(t)\right| < T_j \end{cases} \tag{23.4}$$

Or there exist $\gamma_{ij}(x_j(t)) \in co[a_{ij}(x_j(t))]$ such that

$$\dot{x}_i(t) = -cx_i(t)\sum_{j=1}^{n}\gamma_{ij}\left(x_j(t)\right) f_j\left(x_j(t)\right) + I_i \tag{23.5}$$

As a result, based on Ref. [22], if $x(t) = [x_1, x_2, \ldots, x_n]$ is absolutely continuous on any compact interval of $[0, +\infty)$ and satisfies the differential inclusions (23.3) or (23.5), then $x(t)$ is a solution of system (23.1). Therefore, to design a controller we can consider a continuous model (23.5) instead of the model (23.1), but the problem is that some parameters and functions such as γ_{ij} are completely unknown in this model. Thus, designing a robust control scheme is crucial for such a system.

23.3 Controller design

As it was mentioned, up to now, researchers have developed many types of memristors [59]. Due to the considerable applications of memristive neural networks in a wide variety of fields, their control and synchronization have received a lot of attention. There are complex behaviors even in a simple memristive network; hence, it is of crucial necessity to design a reliable controller to synchronize this system [60,61]. To this end, in this section, a new FISM is designed for nonlinear memristive neural networks, and the stability of the closed-loop system is proven.

23.3.1 ISMC

Consider the following systems in the presence of uncertainties and disturbance:

$$\dot{x}(t) = f(x) + \Delta f(x) + (g(x) + \Delta g(x))\,u + d_0(t) \tag{23.6}$$

where $x = [x_1, x_2, ..., x_n]^T$ and $d_0 = [d_{0\_1}, d_{0\_2}, ..., d_{0\_nt}]^T$ stand for state vector and external disturbance, respectively. Additionally, $f(x)$ and $g(x)$ are nonlinear functions representing system dynamics. Also, Δf and Δg represent uncertainties. Vector $u = [u_1, u_2, ..., u_n]^T$ is the control input. Considering the external disturbance (d_0) and the uncertain functions as a single disturbance term $d(t)$, the equation of system (23.6) can be written as

$$\dot{x}(t) = f(x) + g(x)u + d(t) \tag{23.7}$$

$$d(t) = \Delta f(x) + \Delta g(x)u + d_0(t)) \tag{23.8}$$

Also, consider the master system:

$$\dot{y}(t) = h(y) \tag{23.9}$$

where $y = [y_1, y_2, ..., y_n]^T$ represents the state vector of the master system, and $h(x)$ is a nonlinear function. The vector of the tracking error is given by

$$e(t) = [e_1(t), e_2(t), \ldots, e_n(t)]^T = x(t) - y(t) \tag{23.10}$$

To design the integral sliding mode tracking controller, the integral sliding surface is defined as follows [62]:

$$s(t) = e(t) + \alpha \int_0^t e^{p/q}(\tau) d\tau \tag{23.11}$$

in which q and p are odd integers and $q > p > 0$. When $s(t) = 0$, Eq. (23.11) results in

$$e(t) = -\alpha \int_0^t e^{\frac{p}{q}}(\tau) \, d\tau \tag{23.12}$$

By differentiating each side of Eq. (23.12), Eq. (23.13) can be derived as follows:

$$\dot{e}(t) = -\alpha e^{p/q} \tag{23.13}$$

Integrating Eq. (23.13) results in

$$\alpha \int_{T_i}^{T_f} d\tau = -\int_{e(T_i)}^{e(T_f)} \frac{1}{e^{p/q}} de \tag{23.14}$$

Hence, the convergence time of the error dynamic is given by

$$T_f = T_i + \frac{e(T_i)^{1-p/q}}{\alpha^{1-p/q}(1-\frac{p}{q})} = \frac{e(T_i)^{1-p/q}}{\alpha^{1-p/q}(1-\frac{p}{q})} \tag{23.15}$$

taking the first derivative of $s(t)$ and considering trajectories of system (23.7) result in

$$\dot{s}(t) = \dot{e} + \alpha e^{\frac{p}{q}}(t) = f(x) + g(x)v + d(t) - h(y) + \alpha e^{\frac{p}{q}}(t) \tag{23.16}$$

The control signal for uncertain system (23.6) is as follows:

$$u = g^{-1}(x)\left(h(y) - f(x) - K \,\text{sign}(s) - \delta s - \alpha e^{\frac{p}{q}}(t)\right) \tag{23.17}$$

In the proposed controller $K = dig([k_1, k_2, \ldots, K_n])$ and $\delta = dig([\delta_1, \delta_2, \ldots, \delta_n])$ in which k_i and δ_i $(i = 1, 2, \ldots, n)$ are positive parameters. Also, $k_i > d_i$ where d_i is the bound of the disturbances and uncertainties.

Theorem 1. *The proposed control law* (23.17) *synchronizes the master and slave system even when there are bounded disturbances and uncertainties.*

Proof. Let a Lyapunov function candidate be

$$V_0 = \frac{1}{2} s^T s \tag{23.18}$$

Taking the time derivative of the Lyapunov function, and considering the model of system (23.7) as well as control law (23.17), we have

$$\begin{aligned}
\dot{V}_0 &= s^T \dot{s} = s^T \left(f(x) + g(x)u - h(y) + d(t) + \alpha e^{\frac{p}{q}}(t) \right) \\
&= s^T \left(f(x) + \left(h(y) - f(x) - K \operatorname{sign}(s) - \delta s - \alpha e^{\frac{p}{q}}(t) \right) \right. \\
&\quad \left. - h(y) + d(t) + \alpha e^{\frac{p}{q}}(t) \right) \\
&= s^T \left(-K \operatorname{sign}(s) - \delta s + d(t) \right)
\end{aligned} \tag{23.19}$$

considering $k_i > d_i$ yields

$$\dot{V}_0 \leq -\delta s^T s \tag{23.20}$$

Thus, the developed controller fulfills the Lyapunov condition. Accordingly, it can be confirmed that all states of the system asymptotically converge to the desired value. □

23.3.2 FISMC

One of the most crucial issues which should be taken to account in designing sliding mode control is the chattering phenomena [63]. In this study, to reduce the chattering produced by the discontinuous function sgn(.) a fuzzy controller has been designed. Also, the proposed fuzzy engine will increase the speed of convergence of the system. Indeed, when the fuzzy inference engine adaptively tunes the gains of the controller, the performance of the controller will be improved [4,64–66]. The proposed fuzzy engine maps the input variables s and $\dot{s}$ to the output variables F_s and ρ. Here F_s will be used instead of the discontinues function sign(.), and ρ is a design parameter of the controller. Ultimately, the proposed FISMC is as follows:

$$u = g^{-1}(x) \left(h(y) - f(x) - K F_s - (\delta + \rho) s - \alpha e^{\frac{p}{q}}(t) \right) \tag{23.21}$$

in which F_s and ρ will be obtained by the fuzzy logic engine, and the rest of the parameters must be designed to satisfy the stability condition. Based on the stability conditions of the designed ISMC, the negativeness or positiveness of F_s is matched with sign(s). In the current study, the fuzzy system is modeled by Mamdani's minimum operator, the conjunction operator is Min, and the Max operator is utilized for the aggregation of the rules. Also, the Gaussian membership functions which are used for the fuzzy controller are depicted in Figs. 23.1 and 23.2.

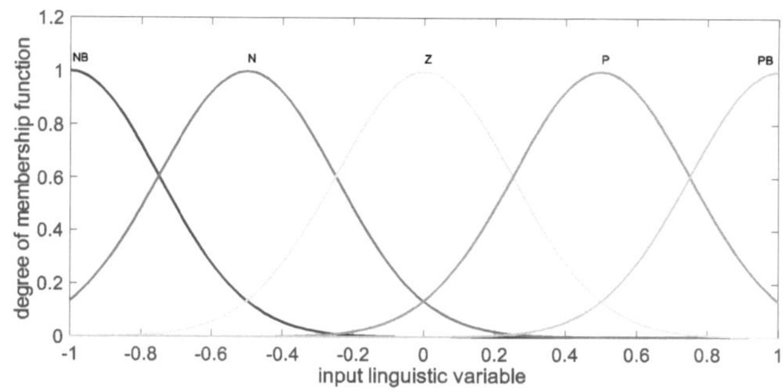

FIGURE 23.1 Membership functions of input variables s and $\dot{s}$.

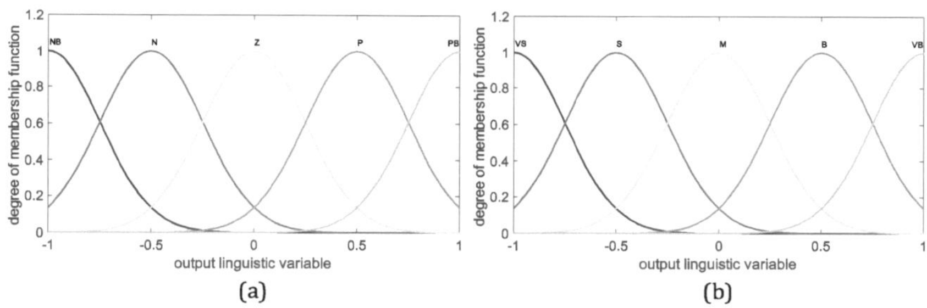

FIGURE 23.2 Membership functions of (a) output linguistic variables F_s. (b) Output linguistic variables ρ.

Five fuzzy partitions are considered for variables F_s, s, and $\dot{s}$ and the following symbols have been used: PB (positive big), P (positive), Z (zero), N (negative), and NB (negative big). The fuzzy sets for F_s, s, and $\dot{s}$ are normalized in the interval $[-1, 1]$. Moreover, for the positive variable ρ, the membership functions have been considered as VB (very big), B (big), M (medium), S (small), and VS (very small); also, its fuzzy set is normalized in the interval $[-1, 1]$. The fuzzy rules which are implemented in this study are presented in Tables 23.1 and 23.2.

Fig. 23.3 demonstrates the procedure of FISMC. To enhance the effectiveness of the control scheme in the presence of uncertainties, as is shown in this figure, the fuzzy engine adaptively feeds ISMC based on the value of s and $\dot{s}$. In the rest of this chapter, the proposed control scheme will be applied to memristive neural networks.

23.4 Numerical results

The governing equation of the uncertain memristive neural networks with control input and in the presence of external disturbances is given by

$$\dot{x}_i(t) = -cx_i(t) \sum_{j=1}^{n} a_{ij}\left(x_j(t)\right) f_j\left(x_j(t)\right) + I_i + d_i + u_i \tag{23.22}$$

TABLE 23.1 Fuzzy rule base for ρ.

		$\dot{s}$				
	ρ	NB	N	Z	P	PB
	NB	VB	VB	B	VB	VB
s	N	B	B	S	B	B
	Z	M	S	VS	S	M
	P	B	B	S	B	B
	PB	VB	VB	B	VB	VB

TABLE 23.2 Fuzzy rule base for F_S.

		$\dot{s}$				
	F_S	NB	N	Z	P	PB
	NB	NB	NB	N	NB	NB
s	N	NB	N	N	N	NB
	Z	N	N	Z	P	P
	P	PB	P	P	P	PB
	PB	PB	PB	P	PB	PB

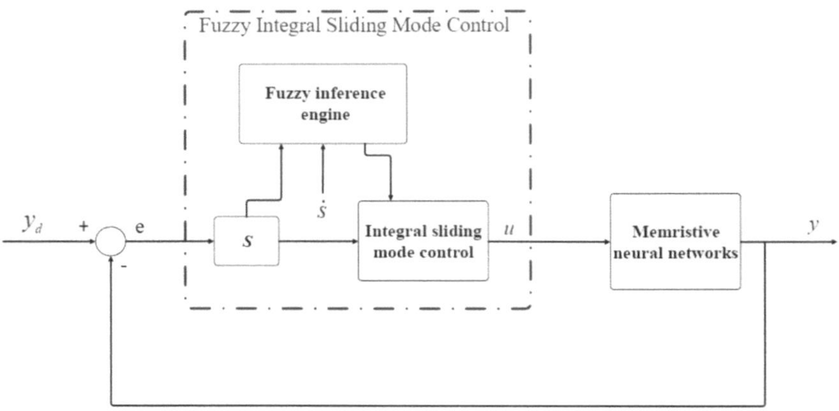

FIGURE 23.3 Block diagram of FISMC.

23.4.1 Memristive neural network without controller

For all numerical simulations, the parameters and functions of the memristive neural networks are considered as $f_j\left(x_j(t)\right) = \tanh\left(x_j(t)\right)$, $n = 3$, $c_1 = c_2 = c_3 = -1$, $a_{12} = 1$, $a_{13} = -9$, $a_{21} = -9$, $a_{23} = 1$, $a_{31} = 1$, and $a_{32} = -9$. Moreover, the discontinuous parameters are

$$a_{11} = \begin{cases} 2 & |x_1| < 1 \\ -2 & |x_1| > 1, \end{cases}$$

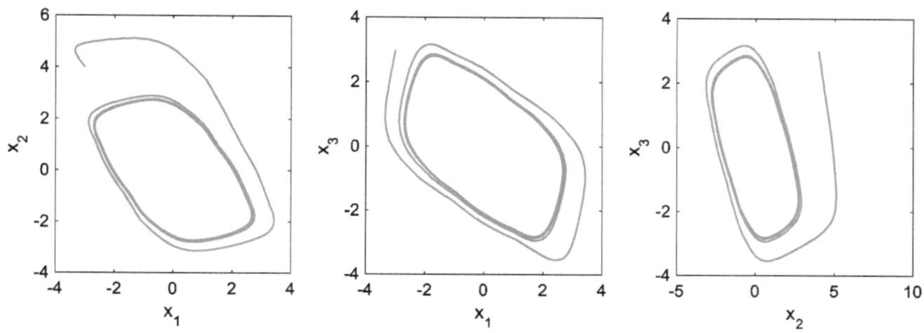

FIGURE 23.4 2D phase portraits of memristive neural network (23.1) with $(x_1(0), x_2(0), x_3(0)) = (-3, 4, 3)$.

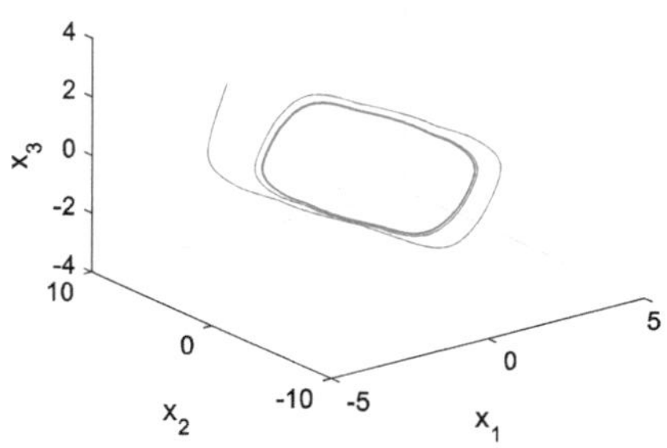

FIGURE 23.5 3D phase portraits of memristive neural network (23.1) with $(x_1(0), x_2(0), x_3(0)) = (-3, 4, 3)$.

$$a_{22} = \begin{cases} 1.71 & |x_2| < 1 \\ -1.71 & |x_2| > 1, \end{cases}$$

$$a_{33} = \begin{cases} 1.1 & |x_3| < 1 \\ -1.1 & |x_3| > 1. \end{cases}$$

Figs. 23.4 and 23.5, respectively, demonstrate the 2D and 3D phase portraits of the nonlinear memristive neural network without control input.

23.4.2 Stabilization and comparison of the proposed method with the ISMC

In this part, the memristive neural network is controlled, and the results of the designed controller are compared with ISMC. In this regard, the design parameters of the proposed controller are chosen as $\alpha = [1, 1, 1]$, $K = [10, 10, 10]$, $\delta = [1, 1, 1]$, $p = [1, 1, 1]$, and $q = [3, 3, 3]$.

By considering the initial conditions as $[2, -1, 0.1]$, the time history of the system is depicted in Fig. 23.6. Based on this figure, it could be easily con-

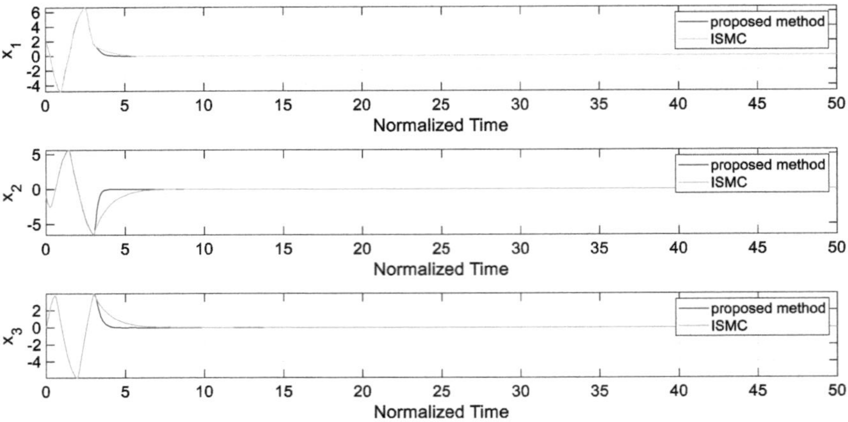

FIGURE 23.6 The states of the memristive neural networks based on the proposed control scheme and ISMC (the control signals are turned on at $T_{start} = 3$).

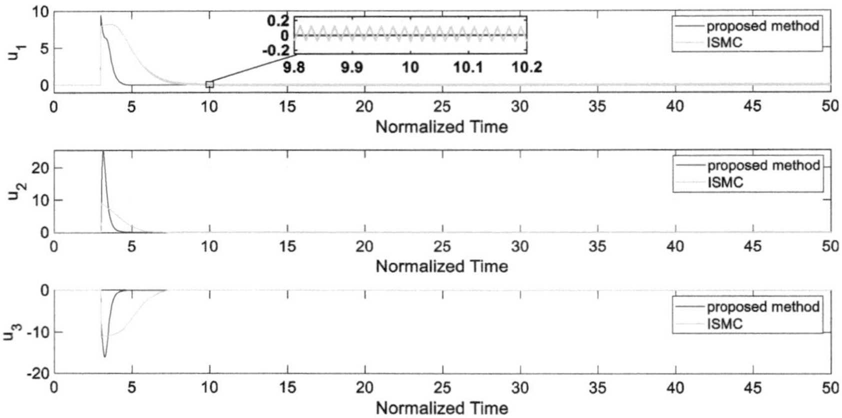

FIGURE 23.7 The control input of the memristive neural networks based on the proposed control scheme and ISMC (the control signals are turned on at $T_{start} = 3$).

firmed that the proposed controller is faster than the ISMC. Fig. 23.7 shows the time history of the control inputs. Using the proposed FISMC, the control input convergence to zero significantly sooner in comparison with ISMC. Also, as is shown in this figure, the proposed control scheme avoids chattering. On the other hand, in the ISMC, there exists a lot of chattering due to the sign(.) function. In the real application, high-frequency chattering will negatively affect actuators. Hence it can be concluded that the fuzzy engine improves the performance of the controller in terms of convergence speed and quality of the control inputs.

Tables 23.3–23.5 list the results of the proposed controller and ISMC. Based on the values of the rise time (T_r) and settling time (T_r), which are listed in

TABLE 23.3 The settling time (T_s) and the rise time (T_r).

	$T_r(x)$	$T_s(x)$	$T_r(y)$	$T_s(y)$	$T_r(z)$	$T_s(z)$
Proposed method	0.7127	4.2587	0.3853	3.6966	0.6327	4.1536
ISMC	1.7874	5.5411	2.0719	6.3566	2.1950	6.8998

TABLE 23.4 The norms of the regulation errors.

	$\|e_x\|_2$	$\|e_x\|_\infty$	$\|e_y\|_2$	$\|e_y\|_\infty$	$\|e_z\|_2$	$\|e_z\|_\infty$
Proposed method	9.5977	1.6521	26.2918	6.5239	20.3174	3.9757
ISMC	11.7125	1.6521	45.9315	6.5119	28.3788	3.9557

TABLE 23.5 The norms of the control signals.

	$\|u_x\|_2$	$\|u_x\|_\infty$	$\|u_y\|_2$	$\|u_y\|_\infty$	$\|u_z\|_2$	$\|u_z\|_\infty$
Proposed method	53.8321	9.4453	127.1587	25.5123	97.8079	16.0042
ISMC	111.2282	8.3061	83.1211	9.0871	146.7752	11.0197

Table 23.3, using the proposed controller, the responses of the system reach zero in a shorter time in comparison with the SMC. Table 23.4 confirms the norms of the regulation errors have been considerably decreased using the fuzzy controller. Additionally, Table 23.5 shows that the control inputs values are less in the proposed controller scheme. Actually, in the ISMC, chattering phenomena cause vibrations in the control input, which results in large control efforts. Therefore, based on the numerical simulation, it can be confirmed that the fuzzy engine considerably enhances the performance of the control scheme.

23.4.3 Synchronization of uncertain memristive neural networks

Here, we consider the uncertain memristive neural networks (23.23) as the slave system, and the master system is considered as follows:

$$\dot{y}_i(t) = -cy_i(t)\sum_{j=1}^{n} a_{ij}\left(y_j(t)\right) f_j\left(y_j(t)\right) + I_i \tag{23.23}$$

The parameters are identical for the slave and the master systems. However, the initial condition for the slave system is $[2, -1, 0]$, and the initial condition of the master system is $[1, 5, -5]$. For simulation results, the control inputs are turned on at $T_{start} = 10$. The time history of synchronization and the synchronization errors are illustrated in Figs. 23.8 and 23.9, respectively. As can be seen in these figures, the proposed FISMC synchronizes the slave and master system in a short period of time. Moreover, Fig. 23.10 shows the control signals which have appropriate amplitudes. To recapitulate briefly, the numerical simulations evidently confirm that the designed controller successfully pushes the states of the slave systems to the desired trajectories.

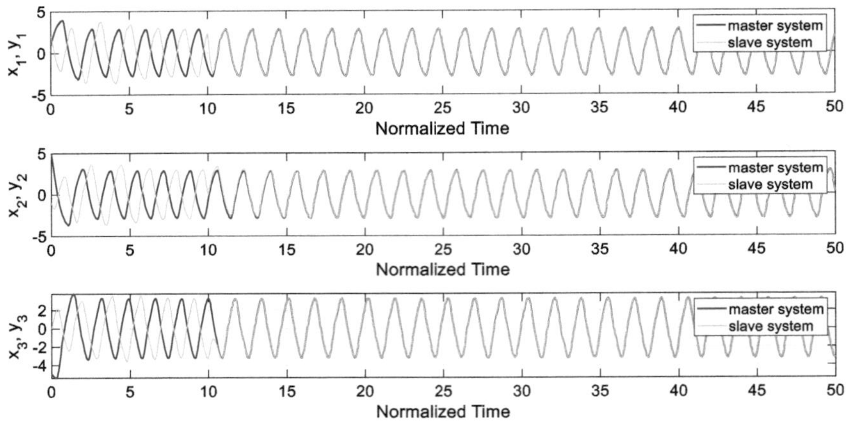

FIGURE 23.8 Synchronization results of memristive neural networks using the FISMC (the control signals are turned on at $T_{start} = 10$).

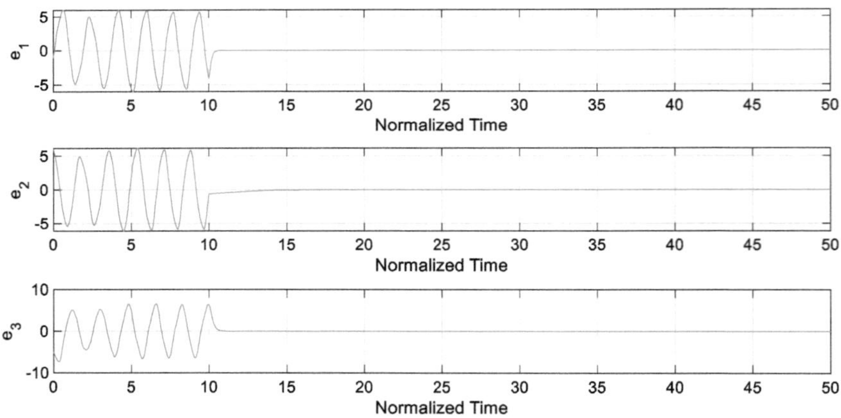

FIGURE 23.9 Time history of the synchronization error using FISMC (the control signals are turned on at $T_{start} = 10$).

23.5 Conclusions

In this chapter, the synchronization of memristive neural networks using FISMC was studied. To address the chattering phenomenon, which is one of the most challenging problems in the sliding mode approaches, fuzzy logic concepts were combined with conventional ISMC. Moreover, by tuning some parameters of the controller using the fuzzy interval, the performance of the controller remarkably was improved, and the speed of the convergence was enhanced. Also, by implementing Lyapunov stability theory, the stability of the closed-loop system was proven. Finally, numerical simulations were performed to suppress and

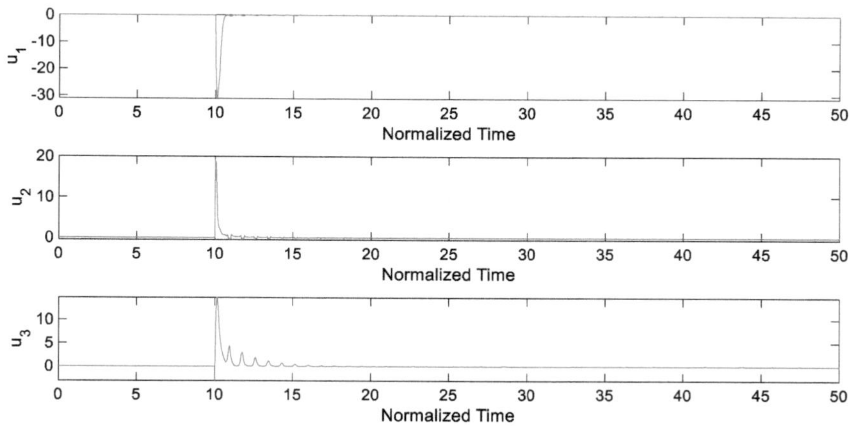

FIGURE 23.10 The control input for memristive neural networks using FISMC (the control signals are turned on at $T_{start} = 10$).

synchronize the system, and the superiority of the proposed method over the conventional ISMC method was demonstrated.

References

[1] L. Chua, Memristor – the missing circuit element, IEEE Transactions on Circuit Theory 18 (1971) 507–519.

[2] S. Yang, Z. Guo, J. Wang, Robust synchronization of multiple memristive neural networks with uncertain parameters via nonlinear coupling, IEEE Transactions on Systems, Man and Cybernetics: Systems 45 (2015) 1077–1086.

[3] S. Vaidyanathan, C. Volos, Advances in Memristors, Memristive Devices and Systems, Springer, 2017.

[4] H. Jahanshahi, M. Shahriari-Kahkeshi, R. Alcaraz, X. Wang, V.P. Singh, V-T. Pham, Entropy analysis and neural network-based adaptive control of a non-equilibrium four-dimensional chaotic system with hidden attractors, Entropy 21 (2019) 156.

[5] L. Skrzypek, Y. You, Dynamics and synchronization of boundary coupled FitzHugh–Nagumo neural networks, arXiv preprint, arXiv:2005.05394, 2020.

[6] Y. Wang, J. Lu, X. Li, J. Liang, Synchronization of coupled neural networks under mixed impulsive effects: a novel delay inequality approach, Neural Networks (2020).

[7] T. Zhang, F. Deng, Adaptive finite-time synchronization of stochastic mixed time-varying delayed memristor-based neural networks, Neurocomputing (2020).

[8] C. Zhou, C. Wang, Y. Sun, W. Yao, Weighted sum synchronization of memristive coupled neural networks, Neurocomputing (2020).

[9] M. Egmont-Petersen, D. de Ridder, H. Handels, Image processing with neural networks—a review, Pattern Recognition 35 (2002) 2279–2301.

[10] H. Ye, A.N. Michel, K. Wang, Qualitative analysis of Cohen–Grossberg neural networks with multiple delays, Physical Review E 51 (1995) 2611.

[11] M. Fitzsimmons, Kunze H. Combining, Hopfield neural networks, with applications to grid-based mathematics puzzles, Neural Networks 118 (2019) 81–89.

[12] M. Itoh, L. Chua, Memristor cellular automata and memristor discrete-time cellular neural networks, in: Handbook of Memristor Networks, Springer, 2019, pp. 1289–1361.

[13] A. Wu, Z. Zeng, Exponential passivity of memristive neural networks with time delays, Neural Networks 49 (2014) 11–18.

[14] Y.V. Pershin, M. Di Ventra, Experimental demonstration of associative memory with memristive neural networks, Neural Networks 23 (2010) 881–886.

[15] A. Thomas, Memristor-based neural networks, Journal of Physics D, Applied Physics 46 (2013) 093001.

[16] X. Liu, J. Cao, Local synchronization of one-to-one coupled neural networks with discontinuous activations, Cognitive Neurodynamics 5 (2011) 13–20.

[17] K. Rajagopal, H. Jahanshahi, M. Varan, I. Bayır, V-T. Pham, S. Jafari, et al., A hyperchaotic memristor oscillator with fuzzy based chaos control and LQR based chaos synchronization, AEU International Journal of Electronics and Communications 94 (2018) 55–68.

[18] X. Yang, J. Cao, W. Yu, Exponential synchronization of memristive Cohen–Grossberg neural networks with mixed delays, Cognitive Neurodynamics 8 (2014) 239–249.

[19] X. Liu, W. Yu, Quasi-synchronization of delayed coupled networks with non-identical discontinuous nodes, Springer, 2012, pp. 274–284.

[20] Z. Guo, J. Wang, Z. Yan, Global exponential synchronization of two memristor-based recurrent neural networks with time delays via static or dynamic coupling, IEEE Transactions on Systems, Man and Cybernetics: Systems 45 (2014) 235–249.

[21] Z. Cai, L. Huang, D. Wang, L. Zhang, Periodic synchronization in delayed memristive neural networks based on Filippov systems, Journal of the Franklin Institute 352 (2015) 4638–4663.

[22] A.F. Filippov, Differential Equations with Discontinuous Righthand Sides: Control Systems, Springer Science & Business Media, 2013.

[23] P. Balasubramaniam, R. Chandran, S.J.S. Theesar, Synchronization of chaotic nonlinear continuous neural networks with time-varying delay, Cognitive Neurodynamics 5 (2011) 361–371.

[24] K. Mathiyalagan, R. Anbuvithya, R. Sakthivel, J.H. Park, P. Prakash, Reliable stabilization for memristor-based recurrent neural networks with time-varying delays, Neurocomputing 153 (2015) 140–147.

[25] G. Zhang, Y. Shen, J. Sun, Global exponential stability of a class of memristor-based recurrent neural networks with time-varying delays, Neurocomputing 97 (2012) 149–154.

[26] Z. Guo, S. Yang, J. Wang, Global exponential synchronization of multiple memristive neural networks with time delay via nonlinear coupling, IEEE Transactions on Neural Networks and Learning Systems 26 (2014) 1300–1311.

[27] S. Vaidyanathan, A memristor-based hyperchaotic system with hidden attractor and its sliding mode control, in: Applications of Sliding Mode Control in Science and Engineering, Springer, 2017, pp. 343–369.

[28] S. Yang, C. Li, T. Huang, Exponential stabilization and synchronization for fuzzy model of memristive neural networks by periodically intermittent control, Neural Networks 75 (2016) 162–172.

[29] H. Wu, R. Li, X. Zhang, R. Yao, Adaptive finite-time complete periodic synchronization of memristive neural networks with time delays, Neural Processing Letters 42 (2015) 563–583.

[30] H. Jahanshahi, Smooth control of HIV/AIDS infection using a robust adaptive scheme with decoupled sliding mode supervision, The European Physical Journal Special Topics 227 (2018) 707–718.

[31] H. Jahanshahi, K. Rajagopal, A. Akgul, N.N. Sari, H. Namazi, S. Jafari, Complete analysis and engineering applications of a megastable nonlinear oscillator, International Journal of Non-Linear Mechanics 107 (2018) 126–136.

[32] N.N. Sari, H. Jahanshahi, M. Fakoor, Adaptive fuzzy PID control strategy for spacecraft attitude control, International Journal of Fuzzy Systems 21 (2019) 769–781.

[33] A. Yousefpour, H. Jahanshahi, J.M. Munoz-Pacheco, S. Bekiros, Z. Wei, A fractional-order hyper-chaotic economic system with transient chaos, Chaos, Solitons and Fractals 130 (2020) 109400.

[34] H. Jahanshahi, N.N. Sari, V-T. Pham, F.E. Alsaadi, T. Hayat, Optimal adaptive higher order controllers subject to sliding modes for a carrier system, International Journal of Advanced Robotic Systems (2018) 15, https://doi.org/10.1177/1729881418782097.
[35] S. Wang, A. Yousefpour, A. Yusuf, H. Jahanshahi, R. Alcaraz, S. He, et al., Synchronization of a non-equilibrium four-dimensional chaotic system using a disturbance-observer-based adaptive terminal sliding mode control method, Entropy 22 (2020) 271.
[36] H. Jahanshahi, A. Yousefpour, J.M. Munoz-Pacheco, S. Kacar, V-T. Pham, F.E. Alsaadi, A new fractional-order hyperchaotic memristor oscillator: dynamic analysis, robust adaptive synchronization, and its application to voice encryption, Applied Mathematics and Computation 383 (2020) 125310.
[37] A. Vahidi-Moghaddam, A. Rajaei, R. Vatankhah, M.R. Hairi-Yazdi, Terminal sliding mode control with non-symmetric input saturation for vibration suppression of electrostatically actuated nanobeams in the presence of Casimir force, Applied Mathematical Modelling 60 (2018) 416–434.
[38] A. Yousefpour, A. Haji Hosseinloo, M. Reza Hairi Yazdi, A. Bahrami, Disturbance observer-based terminal sliding mode control for effective performance of a nonlinear vibration energy harvester, Journal of Intelligent Material Systems and Structures (2020), https://doi.org/10.1177/1045389X20922903.
[39] A. Rajaei, A. Vahidi-Moghaddam, A. Chizfahm, M. Sharifi, Control of malaria outbreak using a non-linear robust strategy with adaptive gains, IET Control Theory & Applications 13 (2019) 2308–2317.
[40] A. Rajaei, A. Vahidi-Moghaddam, M. Ayati, M. Baghani, Integral sliding mode control for nonlinear damped model of arch microbeams, Microsystem Technologies 25 (2019) 57–68.
[41] A. Yousefpour, H. Jahanshahi, Fast disturbance-observer-based robust integral terminal sliding mode control of a hyperchaotic memristor oscillator, The European Physical Journal Special Topics 228 (2019) 2247–2268.
[42] H. Chen, S. He, A.D.P. Azucena, A. Yousefpour, H. Jahanshahi, M.A. López, et al., A multistable chaotic jerk system with coexisting and hidden attractors: dynamical and complexity analysis, FPGA-based realization, and chaos stabilization using a robust controller, Symmetry 12 (2020) 569.
[43] S. Song, X. Song, I.T. Balsera, Mixed H∞ and passive projective synchronization for fractional-order memristor-based neural networks with time delays via adaptive sliding mode control, Modern Physics Letters B 31 (2017) 1750160.
[44] H. Jahanshahi, A. Yousefpour, Z. Wei, R. Alcaraz, S. Bekiros, A financial hyperchaotic system with coexisting attractors: dynamic investigation, entropy analysis, control and synchronization, Chaos, Solitons and Fractals 126 (2019) 66–77.
[45] A. Yousefpour, A. Vahidi-Moghaddam, A. Rajaei, M. Ayati, Stabilization of nonlinear vibrations of carbon nanotubes using observer-based terminal sliding mode control, Transactions of the Institute of Measurement and Control (2019), https://doi.org/10.1177/0142331219881547.
[46] C. Chen, L. Li, H. Peng, Y. Yang, Fixed-time synchronization of inertial memristor-based neural networks with discrete delay, Neural Networks 109 (2019) 81–89.
[47] M. Chen, W.H. Chen, Sliding mode control for a class of uncertain nonlinear system based on disturbance observer, International Journal of Adaptive Control and Signal Processing 24 (2010) 51–64.
[48] G. Bartolini, A. Ferrara, E. Usai, Chattering avoidance by second-order sliding mode control, IEEE Transactions on Automatic Control 43 (1998) 241–246.
[49] J.-J.E. Slotine, W. Li, Applied Nonlinear Control, Prentice Hall, Englewood Cliffs, NJ, 1991.
[50] G. Bartolini, A. Ferrara, E. Usai, V.I. Utkin, On multi-input chattering-free second-order sliding mode control, IEEE Transactions on Automatic Control 45 (2000) 1711–1717.
[51] L. Wang, H. Li, S. Duan, T. Huang, H. Wang, Pavlov associative memory in a memristive neural network and its circuit implementation, Neurocomputing 171 (2016) 23–29.
[52] Q. Duan, Z. Jing, X. Zou, Y. Wang, K. Yang, T. Zhang, et al., Spiking neurons with spatiotemporal dynamics and gain modulation for monolithically integrated memristive neural networks, Nature Communications 11 (2020) 1–13.

[53] A. Wu, Z. Zeng, Lagrange stability of memristive neural networks with discrete and distributed delays, IEEE Transactions on Neural Networks and Learning Systems 25 (2013) 690–703.
[54] X. Yang, D.WC. Ho, Synchronization of delayed memristive neural networks: robust analysis approach, IEEE Transactions on Cybernetics 46 (2015) 3377–3387.
[55] J. Xiao, S. Zhong, Y. Li, New passivity criteria for memristive uncertain neural networks with leakage and time-varying delays, ISA Transactions 59 (2015) 133–148.
[56] J. Xiao, S. Zhong, Y. Li, Relaxed dissipativity criteria for memristive neural networks with leakage and time-varying delays, Neurocomputing 171 (2016) 708–718.
[57] A. James, Memristor and Memristive Neural Networks, BoD – Books on Demand, 2018.
[58] J. Henderson, A. Ouahab, Fractional functional differential inclusions with finite delay, Nonlinear Analysis: Theory, Methods & Applications 70 (2009) 2091–2105.
[59] J. Kengne, Z. Njitacke Tabekoueng, V. Kamdoum Tamba, Nguomkam Negou A. Periodicity, Chaos, and multiple attractors in a memristor-based Shinriki's circuit, Chaos: An Interdisciplinary Journal of Nonlinear Science 25 (2015) 103126.
[60] Y. Cao, S. Wang, Z. Guo, T. Huang, S. Wen, Synchronization of memristive neural networks with leakage delay and parameters mismatch via event-triggered control, Neural Networks 119 (2019) 178–189.
[61] B. Sun, S. Wen, S. Wang, T. Huang, Y. Chen, P. Li, Quantized synchronization of memristive neural networks with time-varying delays via super-twisting algorithm, Neurocomputing 380 (2020) 133–140.
[62] C-S. Chiu, Derivative and integral terminal sliding mode control for a class of MIMO nonlinear systems, Automatica 48 (2012) 316–326.
[63] H. Jahanshahi, A. Yousefpour, J.M. Munoz-Pacheco, I. Moroz, Z. Wei, O. Castillo, A new multi-stable fractional-order four-dimensional system with self-excited and hidden chaotic attractors: dynamic analysis and adaptive synchronization using a novel fuzzy adaptive sliding mode control method, Applied Soft Computing 87 (2020) 105943.
[64] A. Kosari, H. Jahanshahi, S.A. Razavi, An optimal fuzzy PID control approach for docking maneuver of two spacecraft: orientational motion, Engineering Science and Technology, an International Journal 20 (2017) 293–309.
[65] A. Kosari, H. Jahanshahi, S.A. Razavi, Optimal FPID control approach for a docking maneuver of two spacecraft: translational motion, Journal of Aerospace Engineering 30 (2017) 04017011.
[66] M.J. Mahmoodabadi, H. Jahanshahi, Multi-objective optimized fuzzy-PID controllers for fourth order nonlinear systems, Engineering Science and Technology, an International Journal 19 (2016) 1084–1098.

Chapter 24

Robust adaptive control of fractional-order memristive neural networks

Amin Yousefpour[a], Hadi Jahanshahi[b], Stelios Bekiros[c,d], and Jesus Manuel Muñoz-Pacheco[e]

[a]*School of Mechanical Engineering, College of Engineering, University of Tehran, Tehran, Iran,* [b]*Department of Mechanical Engineering, University of Manitoba, Winnipeg, MB, Canada,* [c]*European University Institute, Department of Economics, Florence, Italy,* [d]*Rimini Centre for Economic Analysis (RCEA), Wilfrid Laurier University, Waterloo, ON, Canada,* [e]*Faculty of Electronics Sciences, Meritorious Autonomous University of Puebla, Puebla, Mexico*

24.1 Introduction

Memristor is a nonlinear memory element which due to its property of material inertia may be a sturdy candidate for replacement of Chua's nonlinear diode [1–3]. During the past several years, memristor systems have been applied in many application domains, such as nonvolatile static random-access memory, and hardware implementation of artificial neural networks [4,5].

Fractional calculus is a mathematical idea that allows non-integer order integration and differentiation. Despite a long history of fractional calculus, its applications are a relatively new subject of interest [6–8]. The ability to describe the hereditary characteristics of the system, as well as their memory, are the most important advantages of fractional calculus over integer calculus. Hence, fractional calculus has been the subject of many research studies [9–12].

Over the past few decades, neural networks have been attracted considerable attention due to their wide applications in various fields such as associate memories, signal processing automatic control, combinational optimization, and so forth [13–20]. Also, memristor-based neural networks as a special kind of neural networks have been an interesting research subject. The memristor-based neural networks are a switching nonlinear system which sometimes results in undesirable dynamical behaviors such as instability, transient chaos, and oscillators [21,22].

Due to the wide applications of the memristive neural networks in various fields, the synchronization of these systems has attracted significant interests [23–26]. Thus, many control schemes such as delayed feedback control [27],

https://doi.org/10.1016/B978-0-12-821184-7.00033-5

fuzzy control [28,29], sliding mode technique [30], adaptive control [31] have been applied to control memristors. The significant issue that should be taken into account in the control and synchronization of complex systems is the existence of uncertainties. Hence, recently, several studies have been carried out on the application of adaptive control to nonlinear systems [32,33].

In the control of memristive neural networks, it is usually the case that some parameters of the system, are not precisely known [34]. Moreover, some features of systems may change over time due to environmental conditions. Hence, it could be beneficial to use an adaptive controller for memristive neural networks [35,36]. Also, most of the systems in real applications are in the presence of unexpected disturbances; therefore, robust techniques should be employed for these systems [37–45].

In this chapter, an adaptive controller is designed to synchronize two fractional-order memristive neural networks. Adaptive controllers are effective methods for the control and synchronization of real systems. It is supposed that the parameters of the slave system are completely unknown and an adaption mechanism is designed to adjust the parameters of the controller. Thereafter, the stability of the closed-loop system through the Lyapunov stability theorem and Barbalat's Lemma has been proven. Hence, the proposed control scheme is robust against external disturbances and uncertainties. Finally, numerical simulations are presented.

24.2 Model of the system and preliminary concepts

Since the advent of fractional calculus, several definitions have been proposed for fractional derivatives. In this chapter, Caputo definition is used for all of the calculations.

Definition 1 ([46]). The fractional integral of function $f(t)$ is given by

$$ {}_0I_t^p f(t) = \frac{1}{\Gamma(p)} \int_0^t (t-\tau)^{p-1} f(\tau) d\tau \tag{24.1}$$

where $\Gamma(\cdot)$ indicates the Gamma function.

Definition 2 ([46]). The Caputo fractional derivative of function $f(t)$ is defined as

$$ {}_0^C D_t^p f(t) = \begin{cases} \frac{1}{\Gamma(k-p)} \int_0^t \frac{f^{(k)}(\tau)}{(t-\tau)^{p+1-k}} d\tau & k-1 < p < k \\ \frac{d^k}{dt^k} f(t) & p = k \end{cases} \tag{24.2}$$

Consider the model of fractional-order memristor-based neural networks [22]

$$ D^\alpha x_i(t) = -c_i x_i(t) + \sum_{j=1}^{n} a_{ij}(x_j(t)) f_j(x_j(t)) + I_i \tag{24.3}$$

where $i = 1, 2, ..., n$ and n denotes the number of units in a neural network, $x_i(t)$ corresponds to the state of the ith unit at time t. In addition, c_i is a positive constant and $f_j(x_j(t))$ represents the nonlinear function of the system, I_i stands for the external input. a_{ij} are connection memristive weights which are defined as follows:

$$a_{i,j}(x_j(t)) = \begin{cases} \hat{a}_{ij} & |x_j(t)| > T_j \\ \tilde{a}_{ij} & |x_j(t)| < T_j \end{cases} \tag{24.4}$$

where $a_{i,j}(\pm T_j) = \hat{a}_{ij}$ or $\tilde{a}_{ij}$; also, weights $\hat{a}_{ij}$ and $\tilde{a}_{ij}$ are constant parameters and T_j is switching jump.

Remark 1. Since $a_{ij}(x_j(t))$ are discontinuous, the classical definition for differential equations in system (24.3) cannot apply. To overcome this problem, the concept due to Filippov is used. According to Filippov [47], a differential equation with a discontinuity has the same solution as a certain differential inclusion.

Based on the theories of set-valued maps and differential inclusions [48], from (24.3), it follows that

$$D^{\alpha} x_i(t) \in -c_i x_i(t) + \sum_{j=1}^{n} co\left(a_{ij}(x_j(t))\right) f_j(x_j(t)) + I_i \tag{24.5}$$

where co indicates the convex closure of a set and it is

$$co\left(a_{i,j}(x_j(t))\right) = \begin{cases} \hat{a}_{ij} & |x_j(t)| > T_j \\ co\{\hat{a}_{ij}, \tilde{a}_{ij}\} & |x_j(t)| = T_j \\ \tilde{a}_{ij} & |x_j(t)| < T_j \end{cases} \tag{24.6}$$

or there exist $\gamma_{ij}(x_j(t)) \in co[a_{ij}(x_j(t))]$ such that

$$D^{\alpha} x_i(t) = -c_i x_i(t) + \sum_{j=1}^{n} \gamma_{ij}(x_j(t)) f_j(x_j(t)) + I_i \tag{24.7}$$

Definition 3 ([47]). Function $x(t)$ is a solution of system (24.3). If $x(t)$ is absolutely continuous on any compact interval of $[0, +\infty)$ and fulfills the differential inclusions (24.5) or (24.7).

The following assumption is considered to ensure the existence and uniqueness of solutions of system (24.3). Function f_i are Lipschitz-continuous on R.

Hence, for all $u, v \in R$, we have $|f_i(u) - f_i(v)| \leq L_i |u - v|$ where $L_i > 0$ and $i = 1, 2, ..., n$ (for more details Refs. [22,49] may be consulted).

24.3 Controller design

According to the previous section, for designing a controller, instead of system (24.3) we could consider its equal system, which is described in Eq. (24.7). In this section, an adaptive controller for the synchronization of two fractional-order memristor-based neural networks is proposed. It is supposed that the function of the system is unknown but locally Lipschitz. Theorem 1 presents the Lyapunov stability theorem for fractional-order systems.

Theorem 1 ([50]). *Let a commensurate fractional-order system be as follows:*

$$ {}_0^C D_t^\alpha x(t) = f(t, x) \tag{24.8} $$

where $x = 0$ is an equilibrium point of system (24.8). Consider a positive definite function $V(t, x(t))$ as a Lyapunov function candidate, which satisfies the following condition:

$$ {}_0^C D_t^\alpha V(t, x(t)) \le 0 \tag{24.9} $$

Then, as a result, system (24.8) will be asymptotically stable.

Lemma 1 ([50]). *Consider $x(t) \in \Re$ be a continuously differentiable function. Then one can obtain*

$$ \frac{1}{2} {}_{t_0}^C D_t^\alpha x^2(t) \le x(t) {}_{t_0}^C D_t^\alpha x(t), \forall \alpha \in (0, 1) \tag{24.10} $$

Barbalat's lemma for fractional-order systems is given now.

Theorem 2 ([51]). *Suppose $\phi : \Re \to \Re$ be a uniformly continuous function on the interval $[t_0, \infty)$. If there exist positive constants p and M such that ${}_0 I_t^\alpha |\phi|^p \le M$, then*

$$ \lim_{t\to\infty} \phi(t) = 0 \tag{24.11} $$

The master system is considered as follows:

$$ {}_0^C D_t^\alpha x_i = f_i(t, \mathbf{X}) \tag{24.12} $$

Also, the slave system is given by

$$ {}_0^C D_t^\alpha y_i = f_i(t, \mathbf{Y}) + u_i(t) \tag{24.13} $$

$\mathbf{X} = [x_1 \ \cdots \ x_n]^T$ and $\mathbf{Y} = [y_1 \ \cdots \ y_n]^T$ are the states vector of master and slave systems. The term $f_i(\cdot)$ represents the nonlinear functions of the systems which are defined in the previous section. In the proposed controller, it is assumed that $f_i(\cdot)$ is unknown but locally Lipschitz. This condition is illustrated by the following inequality:

$$ |f_i(t, \mathbf{X}) - f_i(t, \mathbf{Y})| \le m_i, \quad \max_{1\le k\le n} |x_k - y_k| = m_i \, \|e\|_\infty \tag{24.14} $$

Moreover, $u_i(t) \in \Re$ is the control signal. By defining the synchronization error $e_i = x_i - y_i$ the error dynamics is described in the following form:

$$ {}_{0}^{C}D_t^{\alpha} e_i = f_i(t, \mathbf{X}) - f_i(t, \mathbf{Y}) - u_i(t) \tag{24.15} $$

Then the proposed adaptive control is

$$ u_i = k_i e_i \tag{24.16} $$

$$ {}_{0}^{C}D_t^{\alpha} k_i = \gamma_i e_i^2 \tag{24.17} $$

where k_i and γ_i are adaptation coefficients.

Theorem 3. *If control law (24.16) and adaptation laws (24.17) are applied to system (24.13), the synchronization of two fractional-order memristor-based neural networks will be fully achieved, and the slave system mimics the behavior of the master system.*

Proof. Suppose we have the following Lyapunov function:

$$ V = \frac{1}{2} \sum_{i=1}^{n} \left(e_i^2 + \frac{1}{\gamma_i} (M - k_i)^2 \right) \tag{24.18} $$

where M is a positive constant and is selected as $M > \sum_{i=1}^{n} m_i$. Considering Lemma 1, the time derivative of the Lyapunov function (24.18) yields

$$ \begin{aligned} {}_{0}^{C}D_t^{\alpha} V &= \frac{1}{2} \sum_{i=1}^{n} \left({}_{0}^{C}D_t^{\alpha} e_i^2 + \frac{1}{\gamma_i} {}_{0}^{C}D_t^{\alpha} (M - k_i)^2 \right) \\ &\le \sum_{i=1}^{n} \left(e_i {}_{0}^{C}D_t^{\alpha} e_i - \frac{1}{\gamma_i} (M - k_i) {}_{0}^{C}D_t^{\alpha} k_i \right) \end{aligned} \tag{24.19} $$

Substituting Eqs. (24.15)–(24.18) into Eq. (24.19) results in

$$ \begin{aligned} {}_{0}^{C}D_t^{\alpha} V &= \frac{1}{2} \sum_{i=1}^{n} \left({}_{0}^{C}D_t^{\alpha} e_i^2 + \frac{1}{\gamma_i} {}_{0}^{C}D_t^{\alpha} (M - k_i)^2 \right) \\ &\le \sum_{i=1}^{n} \left(e_i \left(f_i(t, \mathbf{X}) - f_i(t, \mathbf{Y}) - k_i e_i \right) - \frac{1}{\gamma_i} (M - k_i) \gamma_i e_i^2 \right) \\ &= \sum_{i=1}^{n} \left(e_i \left(f_i(t, \mathbf{X}) - f_i(t, \mathbf{Y}) \right) - M e_i^2 \right) \end{aligned} \tag{24.20} $$

Considering condition (24.14), we have

$$
\begin{aligned}
{}_0^C D_t^\alpha V &\le \sum_{i=1}^{n}\left(m_i \left\|e\right\|_\infty^2\right) - M\sum_{i=1}^{n} e_i^2 \le \sum_{i=1}^{n}\left(m_i \left\|e\right\|_2^2\right) - M\sum_{i=1}^{n} e_i^2 \\
&= \left(\sum_{i=1}^{n} m_i\right)\left\|e\right\|_2^2 - M\left\|e\right\|_2^2 = \left(\sum_{i=1}^{n} m_i - M\right)\left\|e\right\|_2^2
\end{aligned} \tag{24.21}
$$

Since parameter M is designed such that $M > \sum_{i=1}^{n} m_i$, by defining $\eta = \left(M - \sum_{i=1}^{n} m_i\right) > 0$, inequality (24.21) is written as follows:

$$
{}_0^C D_t^\alpha V \le -\eta \left\|e\right\|_2^2 \tag{24.22}
$$

Integrating both sides of Eq. (24.22) results in

$$
\begin{aligned}
&{}_0I_t^\alpha\, {}_0^C D_t^\alpha V = V(t) - V(0) \le -{}_0I_t^\alpha\left(\eta\left\|e\right\|_2^2\right) = -\eta\, {}_0I_t^\alpha\left(\left\|e\right\|_2^2\right) \Rightarrow \\
&V(t) + \eta\, {}_0I_t^\alpha\left(\left\|e\right\|_2^2\right) \le V(0)
\end{aligned} \tag{24.23}
$$

Based on Eq. (24.23) we have the following inequality:

$$
\eta\, {}_0I_t^\alpha\left(\left\|e\right\|_2^2\right) \le V(0) \Rightarrow {}_0I_t^\alpha\left(\left\|e\right\|_2^2\right) \le \frac{V(0)}{\eta} \tag{24.24}
$$

As is well known $\left\|e\right\|_2^2 = \sum_{i=1}^{n} e_i^2$; therefore, the following inequality can be obtained:

$$
{}_0I_t^\alpha\left(\sum_{i=1}^{n} e_i^2\right) \le \frac{V(0)}{\eta} \Rightarrow {}_0I_t^\alpha\left(e_i^2\right) \le \frac{V(0)}{\eta}\ (1 \le i \le n) \tag{24.25}
$$

Hence, according to inequality (24.25) the synchronization error become zero as time approaches infinity. This way, the control objective is completely achieved. □

24.4 Simulation results

Firstly, in Figs. 24.1 and 24.2, the response of the master system with different fractional-order derivative is illustrated. Then, to demonstrate the performance of the proposed method, numerical simulations of synchronization and control of the system are presented. Finally, in Section 24.4.2, the proposed controller is compared with PI control.

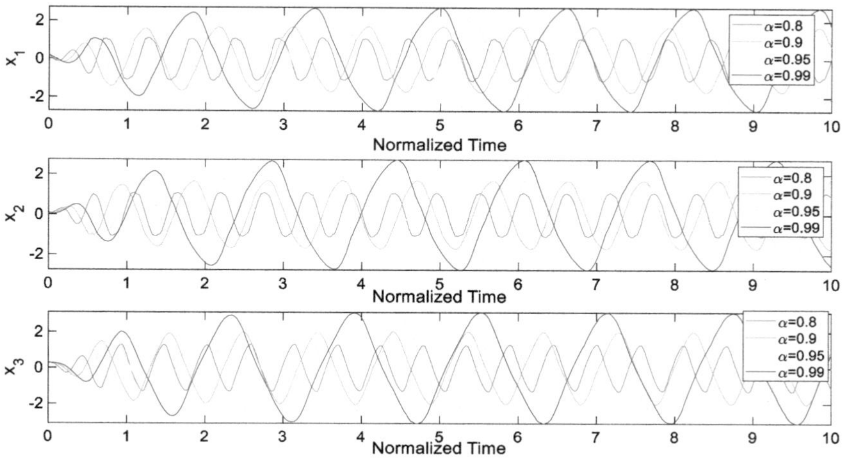

FIGURE 24.1 Time history of fractional-order memristive neural networks with different fractional-order derivative. (For interpretation of the colors in the figure, the reader is referred to the web version of this chapter.)

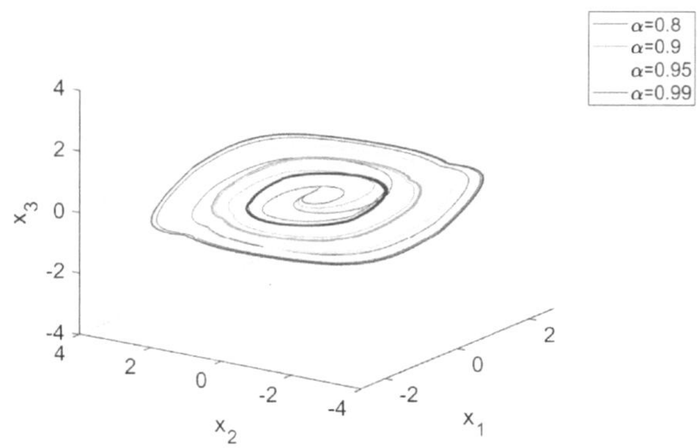

FIGURE 24.2 3D plot of fractional-order memristive neural networks with different fractional-order derivative. (For interpretation of the colors in the figure, the reader is referred to the web version of this chapter.)

24.4.1 Synchronization of fractional-order memristive neural networks

Both the slave and master systems are fractional-order memristor-based neural networks (24.3) with different initial conditions. The parameters of the slave system are considered to be unknown. For numerical simulation, the parameters of the system are set to $n = 3$, $\alpha = 0.98$, $c_1 = c_2 = c_3 = -1$, $f_j(x_j(t)) =$

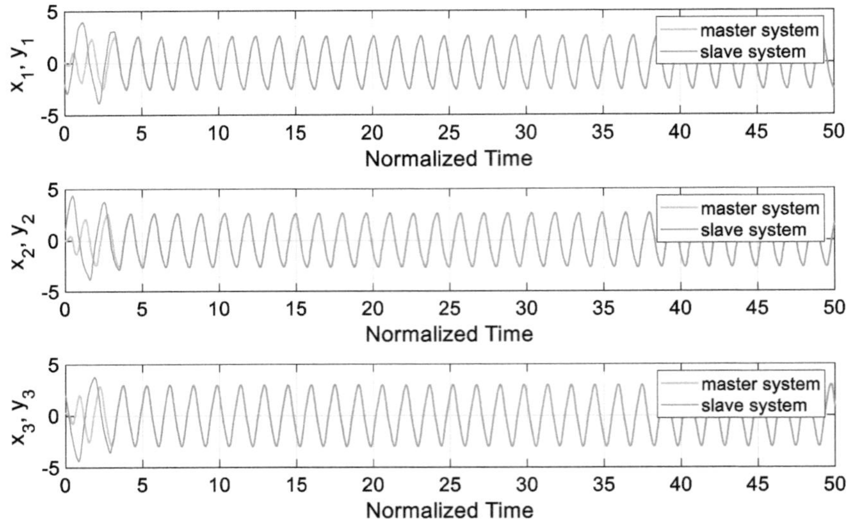

FIGURE 24.3 Time history of the master and slave systems (the control input is turned on at $t = 3$). (For interpretation of the colors in the figure, the reader is referred to the web version of this chapter.)

$\tanh(x_j(t))$. In addition,

$$a_{11} = \begin{cases} 2 & |x_1| < 1 \\ -2 & |x_1| > 1, \end{cases}$$

$$a_{22} = \begin{cases} 1.71 & |x_2| < 1 \\ -1.71 & |x_2| > 1, \end{cases}$$

$$a_{33} = \begin{cases} 1.1 & |x_3| < 1 \\ -1.1 & |x_3| > 1, \end{cases}$$

$a_{12} = 1$, $a_{13} = -9$, $a_{21} = -9$, $a_{23} = 1$, $a_{31} = 1$, $a_{32} = -9$, and the adaptation coefficients are considered as $\gamma = [0.1 \quad 0.1 \quad 1]^T$. Also, the initial conditions are set to $\mathbf{X}(0) = [0.2 \quad 0.1 \quad 0.3]^{\mathrm{T}}$, $\mathbf{Y}(0) = [-2 \quad 1 \quad 2]^T$. The parameters of the controller are chosen as $\gamma_1 = 0.3$, $\gamma_2 = 0.08$, and $\gamma_3 = 5$.

Figs. 24.1 and 24.2, respectively, illustrate time history and 3D plot of the master system with different fractional-order derivative. According to these figures, the value of fractional-order plays a pivotal role in the memristive neural networks' behavior. Hence, using the fractional model, we could investigate the behavior of the system more accurately.

Figs. 24.3–24.5 illustrate the synchronization results of the memristive neural networks when the fractional-order derivative is $\alpha = 0.98$. To be more specific, Fig. 24.3 depicts the time history of the master and slave systems. The synchronization error is plotted in Fig. 24.4. Based on these figures, it is evident

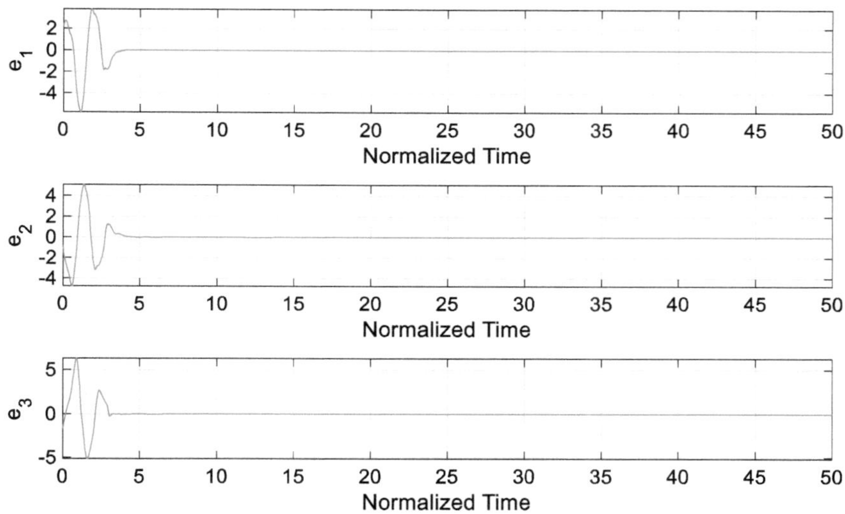

FIGURE 24.4 Time history of the synchronization error (the control input is turned on at $t = 3$).

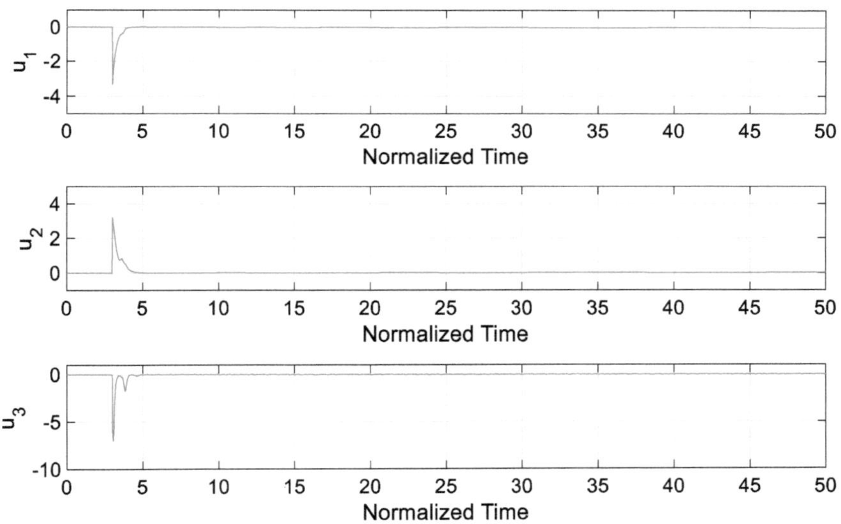

FIGURE 24.5 Time history of the control inputs (the control input is turned on at $t = 3$).

that after less than 2 time units, the state of the slave system mimics those of the master system, and the synchronization purpose is achieved, and the synchronization error reaches zero. Fig. 24.5 demonstrates the time history of the control input. As is shown in this figure, since after the synchronization occurs, the control inputs converge to zero. To recapitulate briefly, the slave system precisely mimics the master system, and the closed-loop system is stabilized.

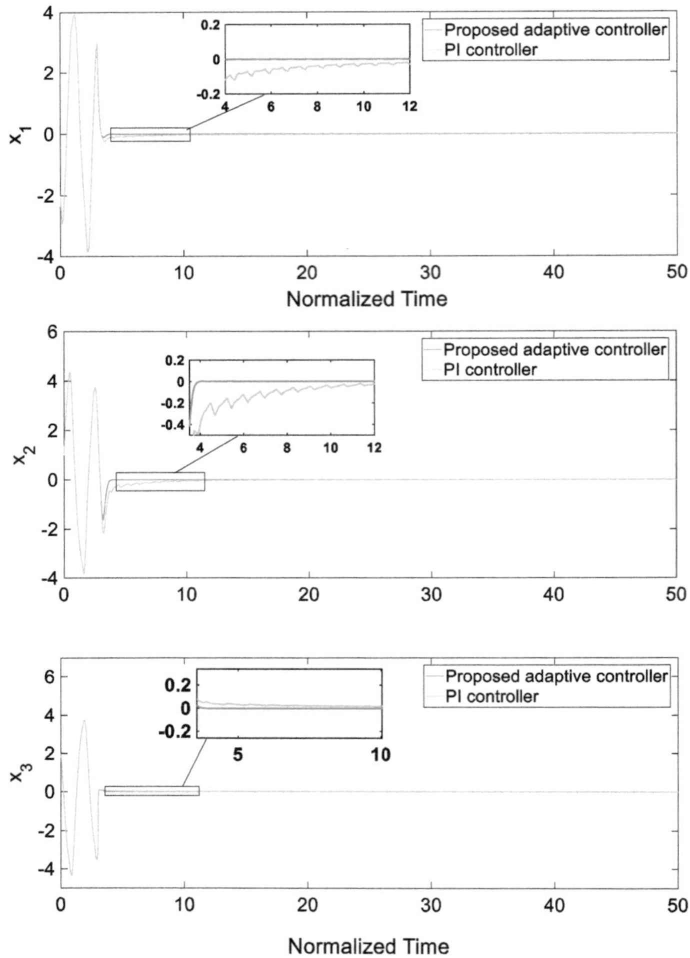

FIGURE 24.6 The states of the fractional-order memristive neural networks based on the proposed adaptive controller and PI controller (the control inputs are applied to the system at $t = 3$). (For interpretation of the colors in the figure, the reader is referred to the web version of this chapter.)

24.4.2 Comparison of the proposed adaptive controller method with PI control

The performance of the proposed adaptive controller is compared with PI controller. The parameters of the proposed controller are equal to the previous simulations. PI controller is considered as follows:

$$u_i = \mathrm{K}_{\mathrm{p}_i} e_i + K_{Ii} \int e_i dt, \quad i = 1, 2, 3 \tag{24.26}$$

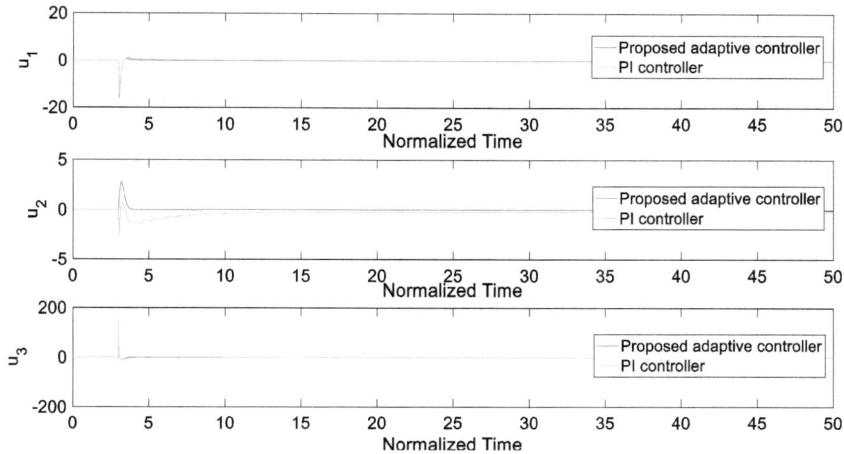

FIGURE 24.7 The control inputs based on the proposed adaptive controller and PI controller (the control inputs are applied to the system at $t = 3$). (For interpretation of the colors in the figure, the reader is referred to the web version of this chapter.)

TABLE 24.1 The values of the settling time (T_s) and the rise time (T_r).

	$T_r(x_1)$	$T_s(x_1)$	$T_r(x_2)$	$T_s(x_2)$	$T_r(x_3)$	$T_s(x_3)$
Proposed adaptive controller	0.1666	3.6665	0.0810	3.6763	0.1577	3.0330
PI controller	0.1660	5.2184	0.0812	8.5331	0.1570	3.3671

TABLE 24.2 The norms of errors and control inputs.

	$\|e_1\|_2$	$\|e_2\|_2$	$\|e_3\|_2$	$\|u_1\|_2$	$\|u_2\|_2$	$\|u_3\|_2$
Proposed adaptive controller	44.1526	47.2094	45.4358	41.8473	14.3619	164.1748
PI controller	44.2515	48.3599	45.4504	39.9725	26.2999	171.488

in which $\mathrm{K_{p_1}} = 5$, $\mathrm{K_{p_2}} = 2$, $\mathrm{K_{p_3}} = 100$, $\mathrm{K_{I_1}} = 0.05$, $\mathrm{K}_{I_2} = 2$, $\mathrm{K_{I_3}} = 1$, these parameters have been chosen in a way that the value of control inputs are approximately the same in both control methods. Figs. 24.6 and 24.7 show the time histories of the system as well as control inputs by applying both controllers. As is evident, the proposed controller shows better performance in terms of settling time and error of the system.

To better investigated the performance of the proposed controller, the results of both control schemes are summarized in Tables 24.1 and 24.2. The Euclidian norm is represented by the symbol $\|\,\|_2$.

Table 24.1 demonstrates the values of the settling time and the rise time. As one can observe, by applying the proposed controller, the response of the system is achieved to the desired value in a shorter time in comparison with the PI controller. Table 24.2 presents the norms of errors and control inputs. As is shown, by applying the proposed controller, the errors of the system are less. Also, based on these results, the control inputs are almost equal in both the control scheme. Therefore, according to Figs. 24.6 and 24.7, as well as Tables 24.1 and 24.2, it can be corroborated that the proposed adaptive control technique shows effective performance for fractional-order memristive neural networks. Another important advantage of the proposed controller in comparison with conventional methods such as PI controller is its guaranteed stability. The PI controller sometime may ruin the stability of the system. On the other hand, the proposed controller guarantees the stability of the closed-loop system, which is a significant concern in real-world applications.

24.5 Conclusion

In this chapter, a factional-order model of a memristor-based neural was studied. Firstly, some basic concepts of fractional calculus and the model of memristive neural networks were presented. Thereafter, using the Filippov concept, a robust adaptive controller was designed for the synchronization of memristive neural networks. Based on the Lyapunov stability theorem and Barbalat's lemma, the stability of the proposed control technique was proven. Lastly, the simulation results of the memristor-based neural networks with different fractional-order derivative, as well as synchronization results, were illustrated. For the synchronization of the systems, it was considered that both master and the dynamic of the slave system are completely unknown. Numerical results clearly show the proposed robust adaptive control successfully synchronizes the fractional-order memristive neural networks. Moreover, the performance of the proposed controller was corroborated by comparing it with PI controller. Also, the significant superiority of the proposed method over conventional methods such as PI controller is its guaranteed stability. This way, the developed control technique can be applied for the synchronization of a broad range of uncertain fractional-order systems.

References

[1] S. Vaidyanathan, C. Volos, Advances in Memristors, Memristive Devices and Systems, Springer, 2017.

[2] R. Rakkiyappan, A. Chandrasekar, J. Cao, Passivity and passification of memristor-based recurrent neural networks with additive time-varying delays, IEEE Transactions on Neural Networks and Learning Systems 26 (2014) 2043–2057.

[3] M. Jiang, S. Wang, J. Mei, Y. Shen, Finite-time synchronization control of a class of memristor-based recurrent neural networks, Neural Networks 63 (2015) 133–140.

[4] A. Yousefpour, H. Jahanshahi, Fast disturbance-observer-based robust integral terminal sliding mode control of a hyperchaotic memristor oscillator, The European Physical Journal Special Topics 228 (2019) 2247–2268.

[5] Z. Wang, S. Ding, Z. Huang, H. Zhang, Exponential stability and stabilization of delayed memristive neural networks based on quadratic convex combination method, IEEE Transactions on Neural Networks and Learning Systems 27 (2015) 2337–2350.

[6] K. Rajagopal, H. Jahanshahi, S. Jafari, R. Weldegiorgis, A. Karthikeyan, P. Duraisamy, Co-existing attractors in a fractional order hydro turbine governing system and fuzzy PID based chaos control, Asian Journal of Control (2020).

[7] S. Wang, S. He, A. Yousefpour, H. Jahanshahi, R. Repnik, M. Perc, Chaos and complexity in a fractional-order financial system with time delays, Chaos, Solitons and Fractals 131 (2020) 109521.

[8] H. Jahanshahi, A. Yousefpour, J.M. Munoz-Pacheco, I. Moroz, Z. Wei, O. Castillo, A new multi-stable fractional-order four-dimensional system with self-excited and hidden chaotic attractors: dynamic analysis and adaptive synchronization using a novel fuzzy adaptive sliding mode control method, Applied Soft Computing 87 (2020) 105943.

[9] S. Wang, S. Bekiros, A. Yousefpour, S. He, O. Castillo, H. Jahanshahi, Synchronization of fractional time-delayed financial system using a novel type-2 fuzzy active control method, Chaos, Solitons and Fractals 136 (2020) 109768.

[10] H. Jahanshahi, A. Yousefpour, J.M. Munoz-Pacheco, S. Kacar, V-T. Pham, F.E. Alsaadi, A new fractional-order hyperchaotic memristor oscillator: dynamic analysis, robust adaptive synchronization, and its application to voice encryption, Applied Mathematics and Computation 383 (2020) 125310.

[11] S. Soradi-Zeid, H. Jahanshahi, A. Yousefpour, S. Bekiros, King algorithm: a novel optimization approach based on variable-order fractional calculus with application in chaotic financial systems, Chaos, Solitons and Fractals 132 (2020) 109569.

[12] A. Yousefpour, H. Jahanshahi, J.M. Munoz-Pacheco, S. Bekiros, Z. Wei, A fractional-order hyper-chaotic economic system with transient chaos, Chaos, Solitons and Fractals 130 (2020) 109400.

[13] H. Jahanshahi, M. Shahriari-Kahkeshi, R. Alcaraz, X. Wang, V.P. Singh, V-T. Pham, Entropy analysis and neural network-based adaptive control of a non-equilibrium four-dimensional chaotic system with hidden attractors, Entropy 21 (2019) 156.

[14] J. Cao, G. Chen, P. Li, Global synchronization in an array of delayed neural networks with hybrid coupling, IEEE Transactions on Systems, Man and Cybernetics Part B Cybernetics 38 (2008) 488–498.

[15] J. Cao, Y. Wan, Matrix measure strategies for stability and synchronization of inertial BAM neural network with time delays, Neural Networks 53 (2014) 165–172.

[16] J. Lu, D.W.C. Ho, J. Cao, J. Kurths, Exponential synchronization of linearly coupled neural networks with impulsive disturbances, IEEE Transactions on Neural Networks 22 (2011) 329–336.

[17] M-J. Park, O-M. Kwon, J.H. Park, S-M. Lee, E-J. Cha, Synchronization criteria for coupled stochastic neural networks with time-varying delays and leakage delay, Journal of the Franklin Institute 349 (2012) 1699–1720.

[18] Q. Song, Synchronization analysis of coupled connected neural networks with mixed time delays, Neurocomputing 72 (2009) 3907–3914.

[19] Z. Wang, Y. Wang, Y. Liu, Global synchronization for discrete-time stochastic complex networks with randomly occurred nonlinearities and mixed time delays, IEEE Transactions on Neural Networks 21 (2009) 11–25.

[20] X. Yang, Q. Zhu, C. Huang, Lag stochastic synchronization of chaotic mixed time-delayed neural networks with uncertain parameters or perturbations, Neurocomputing 74 (2011) 1617–1625.

[21] S. Yang, Z. Guo, J. Wang, Robust synchronization of multiple memristive neural networks with uncertain parameters via nonlinear coupling, IEEE Transactions on Systems, Man and Cybernetics: Systems 45 (2015) 1077–1086.

[22] H-B. Bao, J-D. Cao, Projective synchronization of fractional-order memristor-based neural networks, Neural Networks 63 (2015) 1–9.

[23] J. Chen, B. Chen, Z. Zeng, Global asymptotic stability and adaptive ultimate Mittag–Leffler synchronization for a fractional-order complex-valued memristive neural networks with delays, IEEE Transactions on Systems, Man and Cybernetics: Systems 49 (2018) 2519–2535.

[24] L. Zhang, Y. Yang, F. Wang, Projective synchronization of fractional-order memristive neural networks with switching jumps mismatch, Physica A: Statistical Mechanics and its Applications 471 (2017) 402–415.

[25] J. Xiao, S. Zhong, Y. Li, Xu F. Finite-time, Mittag–Leffler synchronization of fractional-order memristive BAM neural networks with time delays, Neurocomputing 219 (2017) 431–439.

[26] X. Huang, J. Fan, J. Jia, Z Wang, Y. Li, Quasi-synchronisation of fractional-order memristor-based neural networks with parameter mismatches, IET Control Theory & Applications 11 (2017) 2317–2327.

[27] L. Wang, Y. Shen, Y. Sheng, Finite-time robust stabilization of uncertain delayed neural networks with discontinuous activations via delayed feedback control, Neural Networks 76 (2016) 46–54.

[28] K. Rajagopal, H. Jahanshahi, M. Varan, I. Bayır, V-T. Pham, S. Jafari, et al., A hyperchaotic memristor oscillator with fuzzy based chaos control and LQR based chaos synchronization, AEU International Journal of Electronics and Communications 94 (2018) 55–68.

[29] S. Wen, Z. Zeng, T. Huang, Y. Zhang, Exponential adaptive lag synchronization of memristive neural networks via fuzzy method and applications in pseudorandom number generators, IEEE Transactions on Fuzzy Systems 22 (2013) 1704–1713.

[30] S. Vaidyanathan, A memristor-based hyperchaotic system with hidden attractor and its sliding mode control, in: Applications of Sliding Mode Control in Science and Engineering, Springer, 2017, pp. 343–369.

[31] Z. Cai, L. Huang, Finite-time stabilization of delayed memristive neural networks: discontinuous state-feedback and adaptive control approach, IEEE Transactions on Neural Networks and Learning Systems 29 (2017) 856–868.

[32] Z. Tu, N. Ding, L. Li, Y. Feng, L. Zou, W. Zhang, Adaptive synchronization of memristive neural networks with time-varying delays and reaction–diffusion term, Applied Mathematics and Computation 311 (2017) 118–128.

[33] X. Han, H. Wu, B. Fang, Adaptive exponential synchronization of memristive neural networks with mixed time-varying delays, Neurocomputing 201 (2016) 40–50.

[34] X. Li, J-a. Fang, H. Li, Exponential stabilisation of stochastic memristive neural networks under intermittent adaptive control, IET Control Theory & Applications 11 (2017) 2432–2439.

[35] D.A. Prousalis, C.K. Volos, I.N. Stouboulos, I.M. Kyprianidis, Hyperchaotic memristive system with hidden attractors and its adaptive control scheme, Nonlinear Dynamics 90 (2017) 1681–1694.

[36] D.A. Prousalis, C.K. Volos, I.N. Stouboulos, I.M. Kyprianidis, Extreme multi-stability in hyperjerk memristive system with hidden attractors and its adaptive synchronisation scheme, International Journal of Simulation and Process Modelling 13 (2018) 433–445.

[37] H. Jahanshahi, N.N. Sari, V-T. Pham, F.E. Alsaadi, T. Hayat, Optimal adaptive higher order controllers subject to sliding modes for a carrier system, International Journal of Advanced Robotic Systems 15 (2018), https://doi.org/10.1177/1729881418782097.

[38] A. Yousefpour, A. Vahidi-Moghaddam, A. Rajaei, M. Ayati, Stabilization of nonlinear vibrations of carbon nanotubes using observer-based terminal sliding mode control, Transactions of the Institute of Measurement and Control 42 (2020) 1047–1058.

[39] H. Jahanshahi, A. Yousefpour, Z. Wei, R. Alcaraz, S. Bekiros, A financial hyperchaotic system with coexisting attractors: dynamic investigation, entropy analysis, control and synchronization, Chaos, Solitons and Fractals 126 (2019) 66–77.

[40] H. Jahanshahi, K. Rajagopal, A. Akgul, N.N. Sari, H. Namazi, S. Jafari, Complete analysis and engineering applications of a megastable nonlinear oscillator, International Journal of Non-Linear Mechanics 107 (2018) 126–136.

[41] S. Wang, A. Yousefpour, A. Yusuf, H. Jahanshahi, R. Alcaraz, S. He, et al., Synchronization of a non-equilibrium four-dimensional chaotic system using a disturbance-observer-based adaptive terminal sliding mode control method, Entropy 22 (2020) 271.

[42] H. Chen, S. He, A.D.P. Azucena, A. Yousefpour, H. Jahanshahi, M.A. López, et al., A multistable chaotic jerk system with coexisting and hidden attractors: dynamical and complexity

analysis, FPGA-based realization, and chaos stabilization using a robust controller, Symmetry 12 (2020) 569.

[43] A. Kosari, H. Jahanshahi, S.A. Razavi, An optimal fuzzy PID control approach for docking maneuver of two spacecraft: orientational motion, Engineering Science and Technology, an International Journal 20 (2017) 293–309.

[44] A. Kosari, H. Jahanshahi, S.A. Razavi, Optimal FPID control approach for a docking maneuver of two spacecraft: translational motion, Journal of Aerospace Engineering 30 (2017) 04017011.

[45] H. Jahanshahi, Smooth control of HIV/AIDS infection using a robust adaptive scheme with decoupled sliding mode supervision, The European Physical Journal Special Topics 227 (2018) 707–718.

[46] I. Podlubny, Fractional Differential Equations. An Introduction to Fractional Derivatives, Fractional Differential Equations, Some Methods of Their Solution and Some of Their Applications, Academic Press, San Diego – New York – London, 1999.

[47] A.F. Filippov, Differential Equations with Discontinuous Righthand Sides: Control Systems, Springer Science & Business Media, 2013.

[48] J. Henderson, A. Ouahab, Fractional functional differential inclusions with finite delay, Nonlinear Analysis: Theory, Methods & Applications 70 (2009) 2091–2105.

[49] J. Chen, Z. Zeng, P. Jiang, Global Mittag–Leffler stability and synchronization of memristor-based fractional-order neural networks, Neural Networks 51 (2014) 1–8.

[50] Y. Li, Y. Chen, I. Podlubny, Stability of fractional-order nonlinear dynamic systems: Lyapunov direct method and generalized Mittag–Leffler stability, Computers & Mathematics with Applications 59 (2010) 1810–1821.

[51] R. Zhang, Y. Liu, A new Barbalat's lemma and Lyapunov stability theorem for fractional order systems, in: 2017 29th Chinese Control and Decision Conference (CCDC), IEEE, 2017, pp. 3676–3681.

Chapter 25

Learning memristive spiking neurons and beyond

Olga Krestinskaya[a] and Alex James[b]

[a]*King Abdullah University of Science and Technology, Thuwal, Saudi Arabia,* [b]*Indian Institute of Information Technology and Management, Kerala, India*

25.1 Introduction

The main advantage of biologically inspired spike-based neuromorphic architectures such as Spiking Neural Networks (SNNs) [1,2] is the functional emulation of neuron firing and spatio-temporal event-based information processing mechanism of human brain. Hardware-based implementation of SNN introduces the parallel processing of the data and allows one to mimic human brain architecture and neural processing of information. In addition, hardware implementation of SNN ensures low power consumption [3,4] and allows one to increase the information processing density, comparing to conventional pulse-based Artificial Neural Networks (ANNs). The SNN implementation in the digital domain achieved high performance results in recent years, and several SNN based processing chips have been developed [5,6]. For example, IBM TrueNorth SNN chip [5] incorporates 1 million digital neurons and 256 million synapses; and Intel Loihi SNN chip has 128 neuromorphic cored fabricated using 14 nm FinFET transistors [6]. Even though the digital domain SNN implementation achieved impressive results, the scalability and online-training of such systems is still an open problem, both from the algorithmic and from the hardware complexity point of view.

As SNN attempts to mimic human brain processing, the high-speed real-time parallel processing of large amount of data is a critical design issue. In digital SNN systems, SNN processing speed is limited by the performance of analog-to-time or analog-to-digital converters, analog-to-digital converters and encoders responsible for input data encoding process with transformation of analog signal to sets of spikes. Most of the implemented systems illustrate that this transformation is performed without the consideration of biological spike shape. The realistic biological spikes have a certain shape that is still of analog nature. The low power processing circuits of the signals in analog domain is an open research opportunity.

Mem-elements for Neuromorphic Circuits with Artificial Intelligence Applications
https://doi.org/10.1016/B978-0-12-821184-7.00034-7

In recent years, one the promising hardware efficient solution for spike-based neuromorphic architecture has been the use of memristive devices, which ensure a small on-chip area, low power dissipation and scalability of neuromorphic architectures. In such neuromorphic architectures, online on-chip learning is a critical requirement. Most of the memristive neuromorphic architectures with on-chip learning are based on pulse domain processing, while the hardware implementation of spike domain learning architectures for memristive neural network is still a challenging problem. Furthermore, the implementation of learning algorithms in analog domain is a natural solution to spike-based architectures that, however, poses several design complexities to overcome. This work provides an explicit overview of spike domain processing and learning architectures in memristive neuromorphic architectures. There are ongoing possibilities to incorporate spike based learning for different neuromorphic architectures, such as Deep Learning Neural Networks (DNNs) [7], Long Short Term Memory (LSTM) [8,9] and Hierarchical Temporal Memory (HTM) [10,11]. The memristor based approaches for SNN systems are faced with several challenges [12] for its practical realization, and are yet not considered fully ready for reliable industrial applications.

25.2 Spike domain data processing and learning

The inspiration for spike-based processing comes from the spike based functional models of the human brain. Fig. 25.1 illustrates that the design of a neuromorphic architecture includes the following: (1) modeling and implementation of neurons, which include stochastic neurons and integrate and fire neurons, (2) efficient learning algorithms, (3) network topology and overall structure of SNN chip, and (4) the way of information transmission between the neurons. The dynamic behavior of the neurons, spike-based transmission of the data, various conversion schemes and network topology are important aspects to consider in the design of neuromorphic spike-based architectures.

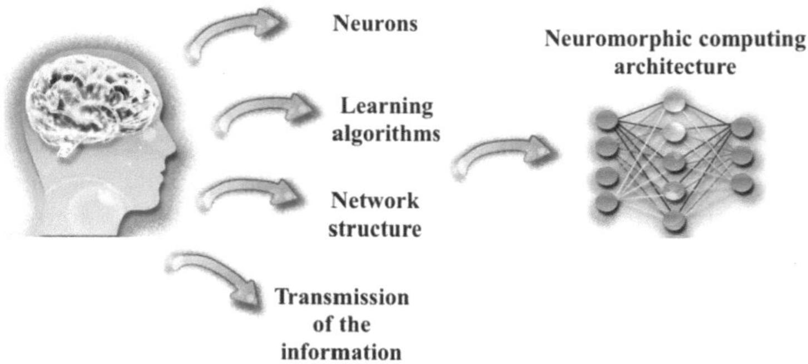

FIGURE 25.1 Main directions in brain-inspired architectures.

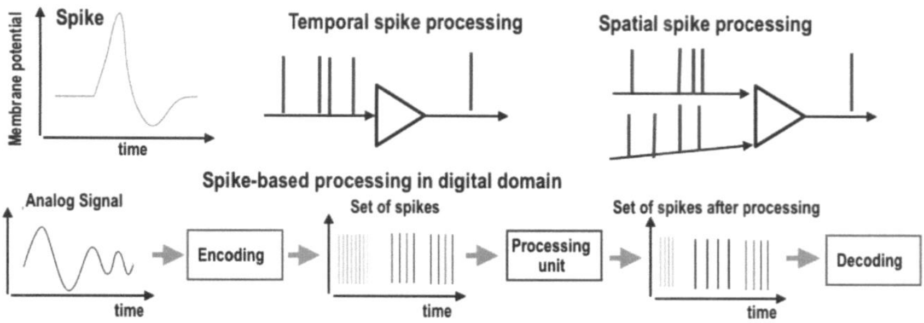

FIGURE 25.2 Spikes and spike-based processing.

The dynamic behavior of spiking neuron is a function of membrane potential and external input (Fig. 25.2). However, in most of the existing digital and software-based implementations of large-scale spike-based neuromorphic systems, the biological nature and spike shape is not entirely considered, as biological shape of a spike is complex and contains several sets of variable parameters. It is difficult to incorporate accurately the biological shape of the spike into a large-scale architecture for real data processing without reducing the processing speed. Furthermore, the selection of hyperparameters is a critical issue and the processing of the neuronal behavior becomes a computationally complex problem. Therefore, there are several existing hardware-based solutions, which are designed to move from software-based sequential processing of the data to parallel computation in hardware [5,6].

The learning in spike-domain processing systems and SNNs can be divided into two main algorithms: rate coding and temporal coding of the information (Fig. 25.2). In a rate encoding approach, the information processing is carried out by the counting of the rate of spikes and temporal averaging of data inputs. In temporal coding, if the stimulus is stronger the spike happens earlier. Temporal coding considers the exact time when neurons fire and a spike occurs. The advantages of temporal coding are the higher processing speed, count of each spike and the absence of average calculation process [13].

In comparison to ANN, the development of learning rules and algorithms for SNN is an ongoing issue [14–16], while the hardware implementation is still an open research problem. Majority of the spike domain processing systems are conceptually based on thresholding logic, where the membrane potential exceeds the threshold, the spike occurs [7]. Therefore, the recent SNN implementations were based on unsupervised learning algorithms and the update process is based on the spike count. In most of the systems, the unsupervised STDP and Hebbian learning rule are used, which implies that the increase in the connection between the neurons (Long Term Potentiation), where pre-synaptic spike causes post-synaptic spike, and vice versa (Long Term Depression) [17].

The application of backpropagation learning in SNN is complicated, as neuron firing and spikes are asynchronous. As there is no continuous data flow, gradient descent algorithm is considered to be not efficient. Also, in the systems based on perfect digital spikes, it is practically impossible to compute differential of the signal. In most of the proposed systems, the application of backpropagation rule is approximated and the backpropagation algorithm is modified or simplified [7,18–20]. Several gradient descent modifications for spike-based processing in SNN were proposed in [21] and [22]. There were attempts to combine STDP rule with backpropagation learning [16] and investigation of stochastic approach for SNN in training [23]. The research work [14] showed that the gradient descent training can be successfully performed for temporal coding SNN, comparing to rate-coding approach.

25.3 Learning in memristive neuromorphic architectures

There are only a limited type of neural architectures that exclusively uses the concept of spike domain learning. The most popular implementations relate with that of spiking neural networks, while there are also attempts to use LSTM and HTM architectures in spike domains.

25.3.1 Spiking neural networks

There are various memristive neural network architectures proposed in recent years, however, the spike domain processing was investigated mostly in simplistic DNN architectures. One of the early CMOS-memristive SNN chips for event-based processing is shown in [24], where the problems, such as wiring density of memristors, limitations in the range of resistive levels, component mismatch and the lack of accurate memristor model that can fully emulate the behavior of memristive device are discussed. The most often used learning rule for SNN is STDP. The implementation of feed-forward memristive neural network architecture with STDP without any additional learning mechanism is shown in [25].

The recent research work [26] proposed to use hardware friendly modified STDP rule to perform the weight update in memristive crossbar robust to the variations and noise. The implementation of memristive crossbar-based SNN is shown in Fig. 25.3 and consists of excitatory and inhibitory represented as G_{off} and G_{on} of the memristor, respectively. The learning in this system is based on modified STDP learning rule based on the sign of memristive synapse. Even though the memristive synapses and neurons are implemented on-chip, the learning algorithm requires additional control circuit. The control circuitry and processing unit for the memristor weight update in such a network is an open problem. The memristor crossbar-based SNN architecture with online STDP learning is proposed in [27]. The architecture is illustrated in Fig. 25.4 and is based on the improved implementation of Integrate-and-Fire Neuron design. The other STDP spiking learning is investigated in [28] for digital SNN design and memristor-based spiking neurons.

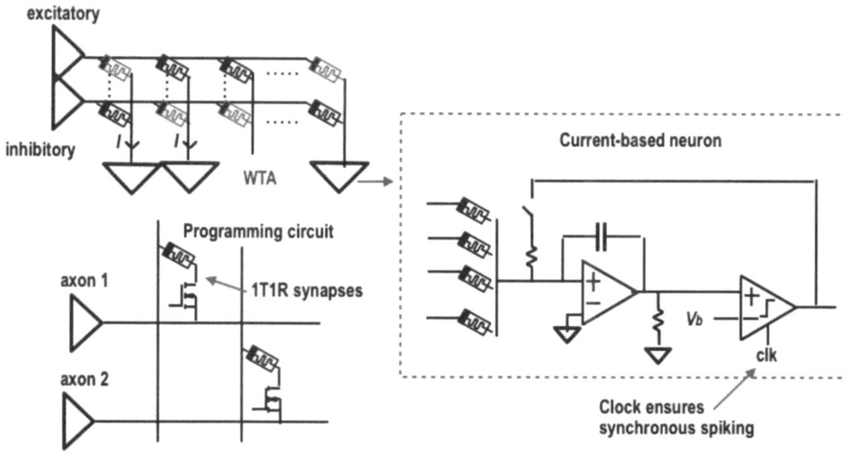

FIGURE 25.3 Hardware friendly modified STDP rule [26].

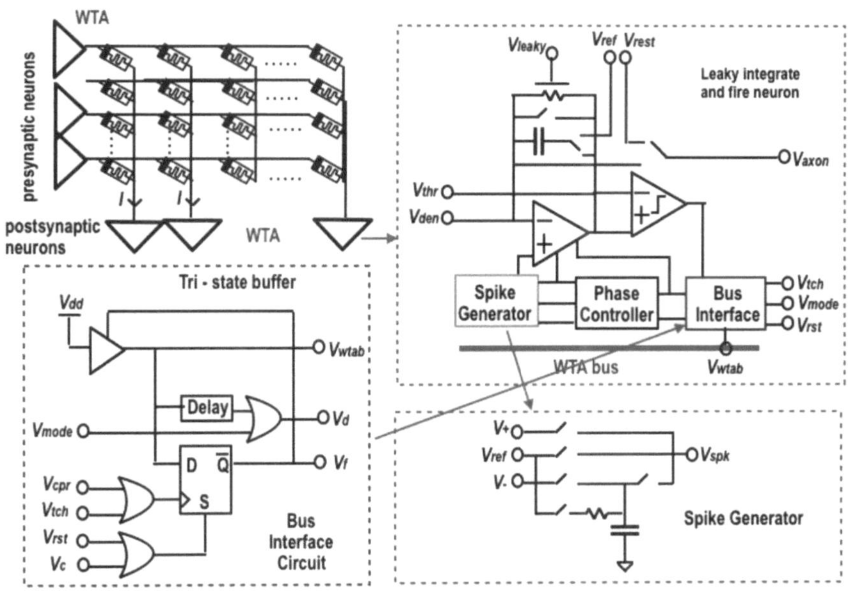

FIGURE 25.4 SNN architecture with online STDP learning [27].

In the recent years, the investigation of stochasticity, non-linearity and non-ideal behavior in memristive systems became popular, as the memristive technology is not mature and memristive behavior incorporates natural variabilities and stochasticity. One of such systems is illustrated in [29] and presents the application of stochastic memristive neuron for spike-based processing in WTA Neural network shown in Fig. 25.5. The stochastically firing WTA neurons are

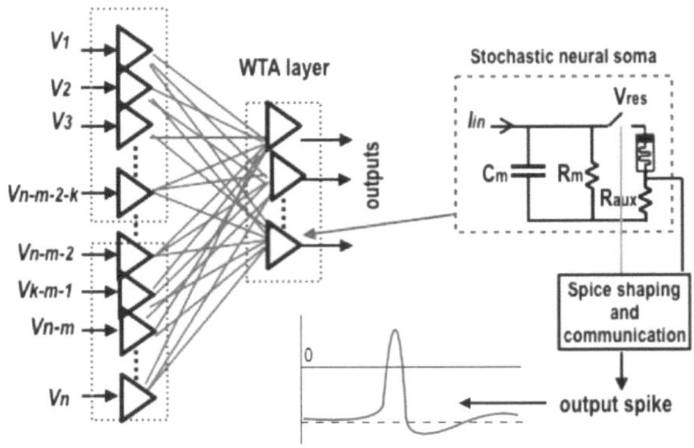

FIGURE 25.5 Application of stochastic memristive neuron for spike-based processing in WTA Neural network [29].

updated using STDP learning rule. The other ReRAM-based stochastic neuron design for SNN is shown in [30]; however, the neuron circuit is complex and scalability of SNN with such neurons is a critical issue.

There are several implementations of memristive SNN considering back-propagation learning approach. The hardware implementation of SNN error backpropagation algorithm and supervised learning is shown in [31]. The circuit level implementation is shown in Fig. 25.6. The implementation of the architecture is based on three-terminal ferroelectric memristor synapses. The spiking backpropagation method is used to update the weights, however, the circuit level design of error calculation technique is an open problem. The simulation of deep stochastic SNN illustrated in [32] exploits stochastic gradient descent approach for network training. The architecture is based on fully connected memristive crossbar and stochastic memristor neuron, however, the weight update circuits are also not shown.

25.3.2 Complex SNN typologies: long-short term memory and hierarchical temporal memory

The implementation of complex SNN topologies can be computationally complex and hardware implementation of such systems is challenging as it relies on spike-based processing between the layers and in feedback paths. The implementation of complex SNN topologies and spike-based processing in such topologies is still an open problem. One of such topologies are recurrent neural networks with feedback connections, like LSTM, and HTM architectures with hierarchical modular structure.

LSTM and HTM are more complex biologically inspired architectures, which aim to mimic the structure and information processing in a human brain. HTM aims to mimic human neocortex structure and information processing, while LSTM incorporates the gated information processing and ability of human

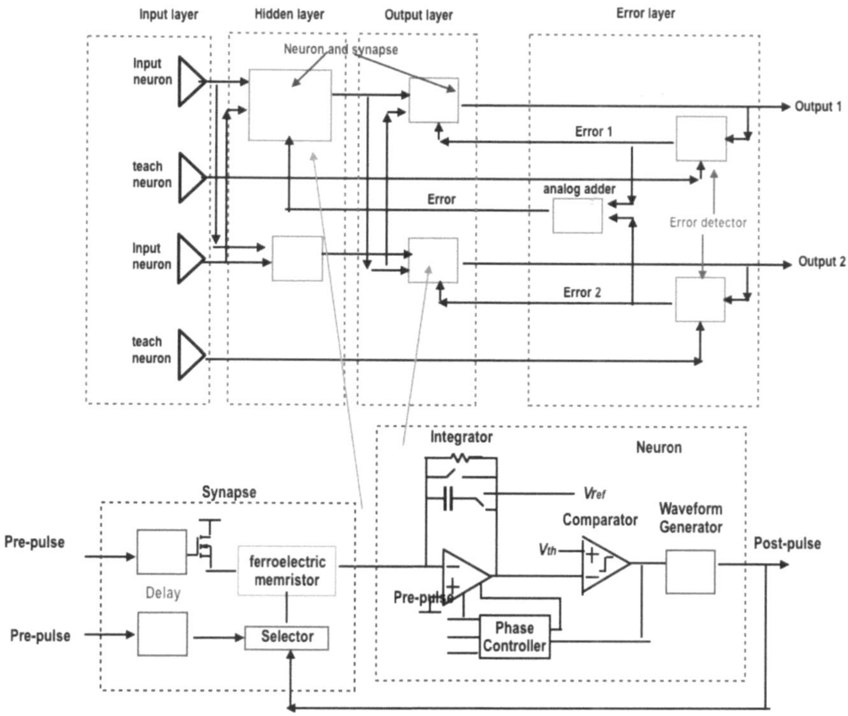

FIGURE 25.6 Hardware implementation of SNN error backpropagation algorithm and supervised learning [31].

brain to forget certain information. HTM is a hierarchical modular architecture, where the information processing is sparse [33]. LSTM is a gated modification of recurrent neural network. The information that is kept and forgotten is determined by gates of the architecture in time [34]. There exist CMOS-memristive pulse-based hardware implementations of HTM and LSTM. However, the spike-based learning in such architectures has not been fully investigated yet, mainly that can do online on-chip learning. It has been shown that the algorithmic implementation of spike-based LSTM algorithm can be efficient [35]. In essence, it is indicative that there exist several opportunities to develop spike-based memristive hardware implementations for more complex topologies, like HTM and LSTM, that can have several practical implications in the development of predication systems.

25.3.3 Recent advances in memristor-based SNN architectures

Recently, the memristor-based SNN architecture has been shown in [36]. This architecture is a fully integrated SNN with single layer of analog integrate and fire neurons implemented with analog memristors and fabricated in 130 nm

CMOS technology. In addition, several SNN designs have been recently shown in [37,38]. ANN to SNN conversion based on diffusive memristors is shown in [39]. Tuning of memristor conductance variation for SNNs is shown in [40]. Learning in SNN and software-hardware co-design are discussed in [41]. The integration of SNN and ANN into hybrid network for energy efficiency is discussed in [42].

25.4 Open problems and research directions

The open problems and ongoing research in SNN and architectures based on spiking neurons include the following:

- Efficient learning algorithms for spiking networks.
- Reduction of dynamic power consumption in SNNs.
- Implementation of more complex topologies in spiking architectures.
- Hardware implementation of efficient conversion schemes to ensure the low power consumption of peripheral circuits.
- Investigation of the challenges on asynchronous design.
- Hybrid architectures combining the best functionalities of ANN and SNN.
- Solving issues of device non-idealities, and introducing robust architectures.
- Efficient ways of integration of SNN chips to the architectures.
- Integration of the networks into dynamic vision sensors (DVS) and image sensor for edge processing.

The implementation of general purpose analog and mixed-signal memristor-based SNN neuromorphic learning chip is an open problem. The efficient and on-chip learning algorithms for SNN are important. As most of the existing memristive architectures with spike-based learning are based on STDP and simple Hebbian learning. It is important to investigate the possibility to implement the other recently proposed algorithms, like modified spiking backpropagation, on memristor-based hardware. Even though the recently proposed backpropagation algorithms are considered to be less effective than the STDP rule due to the asynchronous spiking and inability to differentiate digital spikes, the analog or mixed signal implementation of such systems using memristive circuits can solve the problem of scalability of the existing modified backpropagation algorithms for spike-based processing. In addition, the circuit level implementation of the backpropagation algorithm may allow one to transfer the processing directly to the CMOS-memristive chip without relying on additional complex digital or software-based approaches. As traditional backpropagation methods are not applicable for SNN, the modifications of backpropagation methods have been explored recently to ensure efficient learning in spike-based neural network designs [43,44,2,45].

The hardware implementation of complex spike-based network topologies, like LSTM and HTM, has not been proposed yet. The incorporation of spike-based processing to the architectures that imitate hierarchical modular structure and data processing in the human brain opens up a new direction in the research of neuromorphic architectures. The low power consumption of spike domain processing and ability of memristive HTM and LSTM to imitate the structure and information flow in human brain have a potential to make these architectures more effective than conventional Spiking ANN and DNN architectures.

The implementation of spike-based processing in memristive neuromorphic architectures may require efficient voltage-to-time conversion techniques. The hardware implementation of both rate-based learning and temporal learning approaches required additional circuits for spike generation or analog-to-spike converters, especially if the system is based on biological spikes. The possibility to incorporate the analog nature of biological spikes to train spike-based neuromorphic architectures can also the studied. The feasibility of application and hardware implementation of both rate-based learning and temporal learning in memristive neuromorphic architectures should also be investigated. To develop the spike-based processing architectures, the improvement of the scalability of spiking neurons should be considered, and optimization of data pre-processing techniques for analog and mixed-signal spike-based CMOS-memristive neuromorphic systems for supervised and unsupervised learning should be developed. Moreover, the issues of asynchronous design should be taken into account.

25.5 Conclusion

This chapter provides an overview of existing spike-domain processing memristive architectures and highlights the scientific challenges and opportunities in this emerging field. While a large majority of existing research is based on memristive SNN implementations there is also a growing interest in the development of more complex spiking LSTM and HTM architectures that have been demonstrated as biologically plausible models. Also, the hybrid network-based combination of ANN and SNN has a potential to improve energy efficiency when developed as spike-based architectures.

The other important research aspect to consider in spiking memristive neuromorphic architectures is the memristor variability issue, which can be correlated with the synaptic unreliability of neurons. In the conventional pulse-based ANN, the memristor variabilities can degrade the system performance, whereas the variability in stochastic spike-based neural networks could increase the accuracy of the system. The memristor variability issues, changes in R_{on} and R_{off} values of memristors, probability of switching, and the effect of the fabrication issues on the performance of the memristive devices are open research topics.

References

[1] A. Taherkhani, A. Belatreche, Y. Li, G. Cosma, L.P. Maguire, T.M. McGinnity, A review of learning in biologically plausible spiking neural networks, Neural Networks 122 (2020) 253–272.

[2] A. Tavanaei, M. Ghodrati, S.R. Kheradpisheh, T. Masquelier, A. Maida, Deep learning in spiking neural networks, Neural Networks 111 (2019) 47–63.

[3] P. Wijesinghe, A. Ankit, A. Sengupta, K. Roy, An all-memristor deep spiking neural computing system: a step toward realizing the low-power stochastic brain, IEEE Transactions on Emerging Topics in Computational Intelligence 2 (5) (2018) 345–358.
[4] N. Zheng, P. Mazumder, A low-power hardware architecture for on-line supervised learning in multi-layer spiking neural networks, in: 2018 IEEE International Symposium on Circuits and Systems (ISCAS), IEEE, 2018, pp. 1–5.
[5] F. Akopyan, J. Sawada, A. Cassidy, R. Alvarez-Icaza, J. Arthur, P. Merolla, N. Imam, Y. Nakamura, P. Datta, G.-J. Nam, et al., TrueNorth: design and tool flow of a 65 mW 1 million neuron programmable neurosynaptic chip, IEEE Transactions on Computer-Aided Design of Integrated Circuits and Systems 34 (10) (2015) 1537–1557.
[6] M. Davies, N. Srinivasa, T.-H. Lin, G. Chinya, Y. Cao, S.H. Choday, G. Dimou, P. Joshi, N. Imam, S. Jain, et al., Loihi: a neuromorphic manycore processor with on-chip learning, IEEE MICRO 38 (1) (2018) 82–99.
[7] A. Tavanaei, M. Ghodrati, S.R. Kheradpisheh, T. Masquelier, A.S. Maida, Deep learning in spiking neural networks, arXiv preprint, arXiv:1804.08150, 2018.
[8] K. Smagulova, O. Krestinskaya, A.P. James, A memristor-based long short term memory circuit, Analog Integrated Circuits and Signal Processing 95 (3) (2018) 467–472.
[9] S.P. Gibson, Y. Liu, Long short-term memory using a spiking neural network, US Patent App. 14/527,679 (Feb. 4 2016).
[10] O. Krestinskaya, T. Ibrayev, A.P. James, Hierarchical temporal memory features with memristor logic circuits for pattern recognition, IEEE Transactions on Computer-Aided Design of Integrated Circuits and Systems 37 (6) (2018) 1143–1156.
[11] F. Walter, M. Sandner, F. Rcöhrbein, A. Knoll, Towards a neuromorphic implementation of hierarchical temporal memory on spinnaker, in: 2017 IEEE International Symposium on Circuits and Systems (ISCAS), IEEE, 2017, pp. 1–4.
[12] A. Mehonic, A. Sebastian, B. Rajendran, O. Simeone, E. Vasilaki, A.J. Kenyon, Memristors – from in-memory computing, deep learning acceleration, spiking neural networks, to the future of neuromorphic and bio-inspired computing, arXiv preprint, arXiv:2004.14942, 2020.
[13] T.U. Krause, P.D.R. Würtz, Rate coding and temporal coding in a neural network, Ph.D. thesis, M.Sc. thesis, Electrical Engineering, Univ. of Bochum, Germany, 2014.
[14] H. Mostafa, Supervised learning based on temporal coding in spiking neural networks, IEEE Transactions on Neural Networks and Learning Systems (2017).
[15] O. Booij, H. tat Nguyen, A gradient descent rule for spiking neurons emitting multiple spikes, Information Processing Letters 95 (6) (2005) 552–558.
[16] A. Tavanaei, A.S. Maida, Bp-stdp: approximating backpropagation using spike timing dependent plasticity, arXiv preprint, arXiv:1711.04214, 2017.
[17] N. Caporale, Y. Dan, Spike timing-dependent plasticity: a Hebbian learning rule, Annual Review of Neuroscience 31 (2008) 25–46.
[18] T.P. Lillicrap, D. Cownden, D.B. Tweed, C.J. Akerman, Random synaptic feedback weights support error backpropagation for deep learning, Nature Communications 7 (2016) 13276.
[19] F. Zenke, S. Ganguli, Superspike: supervised learning in multilayer spiking neural networks, Neural Computation 30 (6) (2018) 1514–1541.
[20] S.M. Bohte, J.N. Kok, H. La Poutre, Error-backpropagation in temporally encoded networks of spiking neurons, Neurocomputing 48 (1–4) (2002) 17–37.
[21] D. Huh, T.J. Sejnowski, Gradient descent for spiking neural networks, arXiv preprint, arXiv:1706.04698, 2017.
[22] J.H. Lee, T. Delbruck, M. Pfeiffer, Training deep spiking neural networks using backpropagation, Frontiers in Neuroscience 10 (2016) 508.
[23] B. Gardner, I. Sporea, A. Grüning, Learning spatiotemporally encoded pattern transformations in structured spiking neural networks, Neural Computation 27 (12) (2015) 2548–2586.
[24] T. Serrano-Gotarredona, T. Prodromakis, B. Linares-Barranco, A proposal for hybrid memristor-cmos spiking neuromorphic learning systems, IEEE Circuits and Systems Magazine 13 (2) (2013) 74–88.

[25] J.A. Pérez-Carrasco, C. Zamarreno-Ramos, T. Serrano-Gotarredona, B. Linares-Barranco, On neuromorphic spiking architectures for asynchronous stdp memristive systems, in: Proceedings of 2010 IEEE International Symposium on Circuits and Systems (ISCAS), IEEE, 2010, pp. 1659–1662.

[26] N. Zheng, P. Mazumder, Learning in memristor crossbar-based spiking neural networks through modulation of weight-dependent spike-timing-dependent plasticity, IEEE Transactions on Nanotechnology 17 (3) (2018) 520–532.

[27] X. Wu, V. Saxena, K. Zhu, A cmos spiking neuron for dense memristor-synapse connectivity for brain-inspired computing, arXiv preprint, arXiv:1506.01069, 2015.

[28] A.M. Zyarah, N. Soures, D. Kudithipudi, On-device learning in memristor spiking neural networks, in: 2018 IEEE International Symposium on Circuits and Systems (ISCAS), IEEE, 2018, pp. 1–5.

[29] M. Al-Shedivat, R. Naous, G. Cauwenberghs, K.N. Salama, Memristors empower spiking neurons with stochasticity, IEEE Journal on Emerging and Selected Topics in Circuits and Systems 5 (2) (2015) 242–253.

[30] J. Lin, J.-S. Yuan, Analysis and simulation of capacitor-less reram-based stochastic neurons for the in-memory spiking neural network, IEEE Transactions on Biomedical Circuits and Systems 99 (2018) 1–14.

[31] Y. Nishitani, Y. Kaneko, M. Ueda, Supervised learning using spike-timing-dependent plasticity of memristive synapses, IEEE Transactions on Neural Networks and Learning Systems 26 (12) (2015) 2999–3008.

[32] P. Wijesinghe, A. Ankit, A. Sengupta, K. Roy, An all-memristor deep spiking neural network: a step towards realizing the low power, stochastic brain, arXiv preprint, arXiv:1712.01472, 2017.

[33] J. Hawkins, S. Ahmad, D. Dubinsky, Hierarchical temporal memory including htm cortical learning algorithms, Technical report, Numenta, Inc., Palo Alto, 2010, http://www.numenta.com/htmoverview/education/HTM_CorticalLearningAlgorithms.pdf.

[34] S. Hochreiter, J. Schmidhuber, Long short-term memory, Neural Computation 9 (8) (1997) 1735–1780.

[35] G. Bellec, D. Salaj, A. Subramoney, R. Legenstein, W. Maass, Long short-term memory and learning-to-learn in networks of spiking neurons, arXiv preprint, arXiv:1803.09574, 2018.

[36] A. Regev, A. Bricalli, G. Piccolboni, A. Valentian, T. Mesquida, G. Molas, J.-F. Nodin, Fully-integrated spiking neural network using SiO_x-based RRAM as synaptic device, in: 2020 2nd IEEE International Conference on Artificial Intelligence Circuits and Systems (AICAS), IEEE, 2020, pp. 145–148.

[37] Z. Zhao, L. Qu, L. Wang, Q. Deng, N. Li, Z. Kang, S. Guo, W. Xu, A memristor-based spiking neural network with high scalability and learning efficiency, IEEE Transactions on Circuits and Systems II, Express Briefs 67 (5) (2020) 931–935.

[38] J. Lin, J.-S. Yuan, A scalable and reconfigurable in-memory architecture for ternary deep spiking neural network with reram based neurons, Neurocomputing 375 (2020) 102–112.

[39] R. Midya, Z. Wang, S. Asapu, S. Joshi, Y. Li, Y. Zhuo, W. Song, H. Jiang, N. Upadhay, M. Rao, et al., Artificial neural network (ann) to spiking neural network (snn) converters based on diffusive memristors, Advanced Electronic Materials 5 (9) (2019) 1900060.

[40] B. Chen, H. Yang, F. Zhuge, Y. Li, T.-C. Chang, Y.-H. He, W. Yang, N. Xu, X.-s. Miao, Optimal tuning of memristor conductance variation in spiking neural networks for online unsupervised learning, IEEE Transactions on Electron Devices 66 (6) (2019) 2844–2849.

[41] K. Roy, A. Jaiswal, P. Panda, Towards spike-based machine intelligence with neuromorphic computing, Nature 575 (7784) (2019) 607–617.

[42] C. Lee, A. Kosta, A.Z. Zhu, K. Chaney, K. Daniilidis, K. Roy, Spike-flownet: event-based optical flow estimation with energy-efficient hybrid neural networks, arXiv preprint, arXiv:2003.06696, 2020.

[43] C. Lee, S.S. Sarwar, K. Roy, Enabling spike-based backpropagation in state-of-the-art deep neural network architectures, arXiv preprint, arXiv:1903.06379, 2019.

[44] D. Roy, P. Panda, K. Roy, Synthesizing images from spatio-temporal representations using spike-based backpropagation, Frontiers in Neuroscience 13 (2019) 621.
[45] A. Tavanaei, A. Maida, Bp-stdp: approximating backpropagation using spike timing dependent plasticity, Neurocomputing 330 (2019) 39–47.

Index

C

D

W

X